# MEDICINAL BIOCHEMISTRY

# MEDICINAL BIOCHEMISTRY

*By*

**Dr. G.P. Garg**

**DISCOVERY PUBLISHING HOUSE PVT. LTD.**

**NEW DELHI-110 002**

*Published by:*
**Tilak Wasan**
**DISCOVERY PUBLISHING HOUSE PVT. LTD.**
4383/4B, Ansari Road, Darya Ganj
New Delhi-110 002 (India)
*Phone* : +91-11-23279245, 43596064-65
*Fax* : +91-11-23253475
*E-mail* : discoverypublishinghouse@gmail.com
namitwasan9@gmail.com
sales@discoverypublishinggroup.com
*web* : www.discoverypublishinggroup.com

***Reprinted:* 2018**
***First Edition:* 2010**

**ISBN: 978-81-8356-644-5**

**Medicinal Biochemistry**



*Printed at:*
Infinity Imaging Systems
Delhi

# Preface

Biochemistry today has made spectacular progress in unraveling the mysteries of animate nature. This progress has allowed us to gain deeper insight into the principles of vital activity and has to a very significant extent stimulated the development of applied disciplines, especially medicine. The present title *"Medicinal Biochemistry"* is intended for those who wish to understand living organisms, especially man. Biochemistry is essential for this purpose, but it would be almost impossible for a student to survey on his own the massive body of existing knowledge, constantly augmented by a remarkable torrent of brilliat discoveries. The purpose of the book, then is to organize our knowledge into something that can be comprehended in a relatively short time and still convey a reasonable complete picture of the chemical structure and function of man. Readability without sacrifice of coverage has been a prime goal. An important device in gaining that goal is to keep attention constantly focused on function, with repeated use of rationalization to show that the chemical facts are not isolated, but part of a whole.

Biochemistry has two major goals as a fundamental science, it treats the vital functions from the standpoint of physical chemistry, and as an applied discipline, it points out practical applications for the wealth of scientific knowledge it has acquired. The dual purpose has been, as far as possible, take into account in the writing of this book. We have also tried to summarize our pedagogical experience in teaching biochemistry to students specializing in medicine and pharmacy. The material of this book has been organized according to the principle of functionally in order to trace the close relationship between the functions of the living organism and the structures of its constituents molecules as well as the chemical and physico-chemical process in which they are involved.

The aim of this book is to present a core of biochemical knowledge that is desirable for undergraduate and postgraduate students and also those involved in the field of medical, microbiology, biotechnology and pharmaceutical. Every attempt has been made to keep abrest of the advances in the subject and at the same time to include the fundamentals.

To make the work more comprehensive and informative, the author has consulted many authoritative books, research journals, abstracts, monographs etc. He is grateful to all those great scholars whose work are cited or substantially reproduced.

There can be no claim to originality except in the manner of treatment and much of the information has been obtained from the books and scientific journals available in the different libraries.

The author expresses his thanks to his friends and colleagues whose continue inspirations have initiated him to bring out this book.

The author expresses his gratitude to Mr. Wasan and Staff of M/s Discovery Publishing House Pvt. Ltd., for their whole hearted co-operation in the publication of this book.

In the mean time, the author will remain sincerely responsible for any shortcomings of the book and be grateful to the readers for their suggestions and constructive criticism for the continuous betterment of the book. He takes this opportunity to appeal to the readers to send their suggestions straightway to his publisher.

**Author**

# CONTENTS

Splenocytes—Antibody incubation—Microbead incubation—Flow Cytometry Staining—Stain cells—Incubate Stain—Wash—Secondary Antibody—Fix—Stimulation and Cytokine Response of Cells After Polyclonal Stimulation—Preparation of the Culture Plate—(Antigen-Specific T Cells, Using Attenuated, L. Monocytogenes as a Model Organism—Inoculation—Cell Preparation—Culture—Antigen Presenting Cells (APCs)—Whole Splenocytes—Loaded B Cells—P815 Cell line as APCs—Intracellular Staining for Cytokines after Brefeldin A Treatment—Time Course of Response—Adoptive Transfer—Adoptive Transfer Utilizing Congenic Strains of Mice—Fluorescent-labeled MHC Class I Pentamers and MHC Class II Tetramers—Defensive System—Principles of Evolution—Antigen-Binding Units—Hypervariable Loops—Specific Molecules—X-Ray Analyses—Antigens and Antibodies with Numerous Interactions—Gene Rearrangements—Antibody Diversity—Combinatorial Association and Somatic Mutation—Oligomerization of Antibodies—Different Classes of Antibodies—Major-Histocompatibility-Complex Proteins—Peptides Presented by MHC Proteins—Antibody-like Proteins—Cytotoxic T Cells—Class II MHC Proteins—Foreign Peptides on Antigen-Presenting Cells—Diverse MHC Proteins—Human Immunodeficiency Viruses—Responses Against Self-Antigens—Positive and Negative Selection in the Thymus—Autoimmune Diseases—The Immune System and Cancer—Maple syrup urine Disease and Jamaican Vomiting Sickness—Sulfur Compounds of Garlic, Onions, Skunks, ETC.

CHAPTER

# 1 Dietary Antioxidants

## Introduction

This chapter discusses evidence linking antioxidant nutrients with the inhibition of low-density lipoprotein (LDL) oxidation. The possibility that increasing antioxidant intake could protect against coronary heart disease (CHD) is reviewed. The chapter also provides information regarding what ideal antioxidant intakes should be.

Hypercholesterolemia is universally accepted as a major risk factor for athero-sclerosis. However, at any given concentration of plasma cholesterol, there is still great variability in the occurrence of cardiovascular events. One of the major breakthroughs in atherogenesis research has been the realization that oxidative modification of LDL may be a critically important step in the development of the atherosclerotic plaque. The formation of foam cells from monocyte-derived macrophages in early atherosclerotic lesions is not caused by native LDL but only following modification of LDL by various chemical reactions such as oxidation.

The cholesterol-laden foam cell is a characteristic feature of the atherosclerotic lesion. The rapid uptake of oxidatively modified LDL occurs through scavenger receptors, which are not down regulated by cholesterol accumulation. Recognition of LDL by the scavenger receptor depends on alteration of key lysine residues of apolipoprotein B, which can be brought about by aldehydes produced during the spontaneous decomposition of lipid hydroperoxides.

Some reports have suggested the presence of oxidatively modified LDL in plasma, but most oxidation is believed to occur in the arterial wall. There, LDL may be in a microenvironment where the antioxidants, which normally prevent lipid peroxidation, can become depleted. All the cells of the vessel wall—endothelial cells, smooth muscle cells, macrophages, and lymphocytes—can modify LDL in vitro. LDL oxidation is believed to be caused by highly reactive free radicals but the nature and source of these have yet to be fully defined. Several mechanisms are likely to be involved, including transition metal ion-mediated generation of hydroxyl radicals, production of reactive oxygen species by enzymes such as myeloperoxidase and lipoxygenase, and direct modification by reactive nitrogen species.

Oxidized LDL may also be atherogenic by mechanisms other than its rapid uptake into macrophages the scavenger receptor. Oxidized forms of LDL are chemotactic for circulating macrophages and smooth muscle cells and facilitate monocyte adhesion to the endothelium and entry into the subendothelial space. Oxidized LDL is also cytotoxic toward arterial endothelial cells and inhibits the release of nitric oxide and the resulting endothelium-dependent vasodilation. There is therefore a potential role for oxidized LDL in altering vasomotor responses, perhaps contributing to vasospasm in diseased vessels. In addition, oxidized LDL is immunogenic; autoantibodies against various epitopes of oxidized LDL have been found in human serum, and immunoglobulin (IgG) specific for epitopes of oxidized LDL can be found in lesions. Oxidized LDL may be able to induce arterial wall cells to produce chemotactic factors, adhesion molecules, cytokine, and growth factors that have an important role in the development of the plaque.

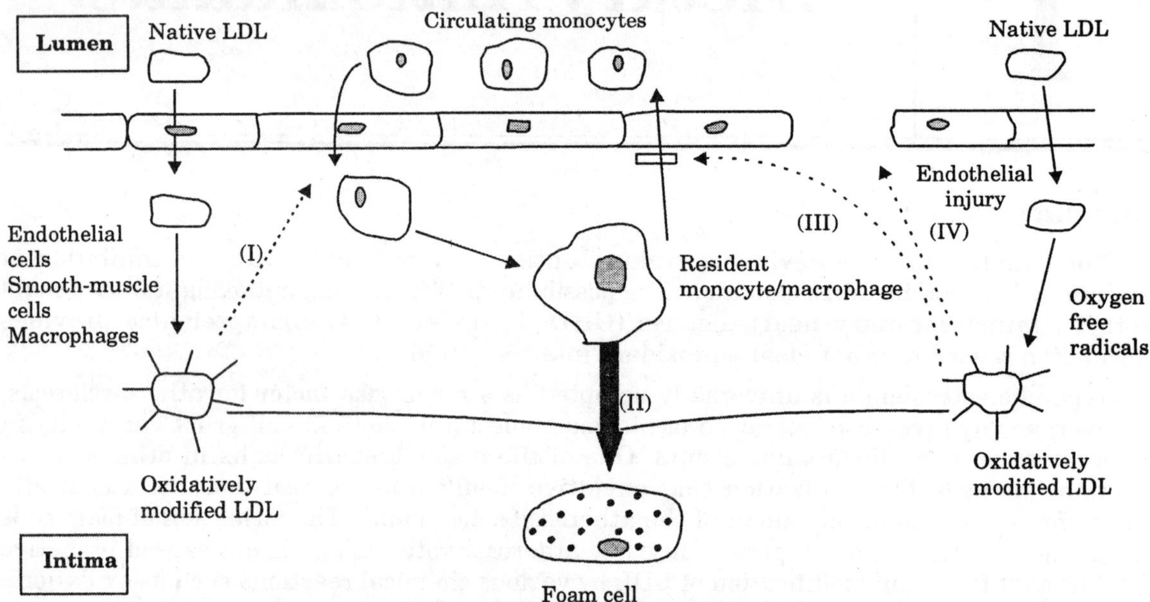


**Fig. 1.1. Mechanisms by which oxidation of LDL may contribute to atherogenesis. Oxidized LDL is chemotactic for circulating monocytes (I), which are phenotypically modified and become macrophages. Oxidized LDL is recognized by the scavenger receptor on the macrophage and becomes internalized rapidly. As more lipid is ingested by the macrophage, a foam cell is formed (II). This eventually bursts and a fatty streak, the first their ability to leave the intima (III). Oxidized LDL is cytotoxic to endothelial cells, leading directly phase of an atherosclerotic lesion, results. Oxidized LDL inhibits the motility of resident macrophages and therefore to endothelial cell damage (IV). Oxidation can occur via the effects of reactive oxygen species or due to the oxidation of the cell's own lipids.**

Evidence for LDL oxidation in vivo is now well established. In immunocyto-chemical studies, antibodies against oxidized LDL stain atherosclerotic lesions but not normal arterial tissue. LDL extracted from animal and human lesions has been shown to be oxidized and is rapidly taken up by macrophage scavenger receptor. In young myocardial infarction (MI) survivors, an association has been demonstrated between increased susceptibility of LDL to oxidation and the degree of coronary atherosclerosis, while the presence of ceroid, a product of lipid peroxidation, has been shown in advanced atherosclerotic plaques. A recent study, however, has suggested that atherosclerotic plaques contain very little oxidized LDL compared to the amounts of activated complement and enzymatically altered LDL an observation that requires further study.

## DIETARY ANTIOXIDANTS AND LDL OXIDATION

The role of dietary factors in protecting against the change from native to oxidized LDL has received considerable attention. The antioxidant vitamins are derived from fresh fruits and vegetables, and from vegetable oil and polyunsaturated margarine to which vitamin E is usually added as an antioxidant; they cannot be synthesized from simple precursors. Thus, dietary intake, absorption, metabolism, and storage determine concentrations of vitamins in plasma and body tissues. An overview of epidemiological research suggests that individuals with the highest intake of antioxidant vitamins, whether through diet or supplements, tend to experience 20 – 40% lower risks of CHD than those with the lowest intake or blood levels. Vitamin E is the major lipid-soluble antioxidant present in LDL, preventing the formation of lipid hydroperoxides from polyunsaturated fatty acids. Vitamin C can scavenge free radicals in the aqueous phase and may also regenerate vitamin E. β-Carotene, a vitamin A precursor, does not have a confirmed antioxidant mechanism although it and other carotenoids are contained within LDL. There is some evidence that carotenoids may protect LDL against oxidation more efficiently at low $pO_2$ levels, which could have relevance to the levels of protection provided in the arterial wall in vivo.

When LDL is exposed to oxidative stress in vitro, lipid peroxidation can only proceed after the sequential loss of its antioxidants in the order ubiquinol-10, α-tocopherol, γ-tocopherol, lycopene, and β-carotene. Accordingly, LDL supplemented with vitamin E in vitro is much harder to oxidize in vitro. Similarly, the α-tocopherol content of the LDL has been shown to be the most important determinant of susceptibility to oxidation in an in vitro model.

In addition to antioxidant content of LDL, several other factors can influence the susceptibility to oxidation. Polyunsaturated fatty acids appear to be the most vulnerable moiety following the application of oxidative stress. The fatty acid composition of the diet is, therefore, an important factor determining the susceptibility of LDL to oxidation, with monounsaturated fatty acids protecting LDL against oxidation. Small dense LDL particles are also easier to oxidize than more buoyant particles. The binding of LDL particles to proteoglycans, glycation, and the presence of preformed lipid peroxides are also important factors.

## INTERVENTION STUDIES IN ANIMALS

Studies of antioxidant supplementation in laboratory animals have provided evidence of the importance of oxidized LDL in vivo. Many such studies have been performed, and they generally provide support for the antioxidant hypothesis. For example, in a study in atherosclerotic rabbits, uptake of LDL into foam cells was approx 4 times greater for oxidized LDL than for native LDL. Adding of vitamin E to the system resulted in a 30 – 55% decrease in accumulated radio-labeled LDL in plaques and foam cells. The effects of a diet high in saturated fat with and without fruits and vegetables and antioxidant vitamins (C, E, and β-carotene) on oxidative stress and development of atherosclerosis has also been evaluated in rabbits. Blood lipid peroxide levels decreased significantly in rabbits supplemented with fruits and vegetables or antioxidant vitamins. In contrast, blood lipid peroxide levels increased significantly in the unsupplemented groups and precipitated coronary thrombosis.

## INTERVENTION STUDIES IN HUMANS WITH BIOCHEMICAL ENDPOINTS

A number of studies have evaluated the effects of vitamin E on copper-catalyzed LDL oxidation in healthy volunteers. In one study, men supplemented with 268, 537, or 805 mg vitamin E per day for 8 wk showed a decreased susceptibility of LDL to oxidation. There was no significant effect of daily supplementation with 40 or 1.34 mg. In a study of the effects of low-dose vitamin E supplementation (100 mg/d for 1 wk, then 200 mg/d for 3 wk), there was a significant increase in lag time before the onset of LDL oxidation and a significant decrease in the propagation rate. Princen and associates have evaluated the minimal supplementary dose of vitamin E necessary to protect LDL against oxidation in vitro in healthy young adults. Resistance of LDL to oxidation increased in a dose-dependent manner with resistance time differing significantly from baseline even after ingestion of only 17 mg/d of vitamin E. However, the progression of lipid peroxidation in LDL was only reduced after intake of 268 and 536 mg/d. In a study of 200 mg α-tocopherol supplementation over a 2-mo period in smoking men, lag time was increased in LDL + VLDL after both copper induction and hemin/hydrogen peroxide induction. Plasma α-tocopherol, VLDL + LDL α-tocopherol, and LDL total antioxidant capacity were all increased in the intervention group. Vitamin E supplementation appears to promote a clear reduction in the susceptibility of LDL to oxidation. The case for (3-carotene and vitamin C, however, is less clear cut.

The effect of high-dose vitamin C supplementation (1000 mg/d) on LDL oxidation was evaluated in a study of 19 smokers. The vitamin C-supplemented group had a significant reduction in the susceptibility of LDL to oxidation after 4 wk. Two months of vitamin C supplementation with freshly squeezed orange juice (estimated 500 mg/d) also produced a significant increase in lag time in 36 healthy males consuming a diet high in saturated fatty acids. Jialal and associates found that supplementation with β-carotene (1-2 μmol) inhibited the oxidative modification of LDL in healthy subjects. By contrast, Lin and colleagues reported that a natural-food diet containing 0.93 (μmol β-carotene per day was insufficient to alter either plasma or LDL (3-carotene or carbonyl groups in LDL. A higher dose of 6.2 μmol β-carotene per day did increase both plasma and LDL β-carotene and reduce LDL carbonyl production.

When the effectiveness of β-carotene, vitamin C, and vitamin E supplements were compared by Reaven and associates, susceptibility of LDL to oxidation did not change during β-carotene supplementation (60 mg/d) but decreased 30-40% with the addition of a vitamin E supplement (1600 mg/d). Addition of vitamin C supplementation (2 g/d) did not further reduce the susceptibility to oxidation. In another study, a combined supplement—β-carotene (30 mg/d), vitamin C (1 g/d), and vitamin E (530 mg/d)—was given to men for 3 mo. It produced a twofold prolongation of the lag phase of LDL oxidation and a 40% reduction in the oxidation rate. The effects of combined antioxidant supplementation were not significantly different from the effects of vitamin E supplementation alone. Finally, we have recently shown that a combination of low-dose antioxidant vitamins (150 mg ascorbic acid, 67 mg α-tocopherol, 9 mg β-carotene daily) over a period of 8 wk significantly prolonged the lag time to oxidation.

The effects of antioxidant supplementation on LDL oxidation may depend on smoking status. In a group of smokers and nonsmokers, resistance of LDL to oxidation increased significantly and the rate of LDL oxidation decreased significantly after vitamin E supplementation (671 mg/d for 7 d). There was some indication of a small increase in the resistance of LDL to oxidation in smokers after supplementation with β-carotene (20 mg/d for 2 wk, then 40 mg/d for 12 wk).

In summary, therefore, experimental evidence suggests that antioxidant vitamins can reduce the susceptibility of LDL to oxidation in vitro. Vitamin E would appear to be the most effective antioxidant; both β-carotene and vitamin C have produced extensions in lag time to oxidation only in a minority of studies, although it remains possible that they may have a beneficial effect in individuals with poor baseline status.

## EPIDEMIOLOGICAL STUDIES LINKING ANTIOXIDANTS AND LDL OXIDATION

### Vitamin E

Two large longitudinal studies in the United States examined the association between vitamin E intake and risk of CHD. In a group of 39, 910 male health professionals, those who took vitamin E supplements in doses of at least 100IU per day for over 2 yr had a 37% lower relative risk of CHD compared to men who did not take vitamin E supplements, after adjustment for age, coronary risk factors, and intake of vitamin C and β-carotene. In the Nurses' Health Study of 87, 245 female nurses, women who took vitamin E supplements for more than 2 yr had a 41 % lower relative risk of major coronary disease. This effect persisted after adjustment for age, smoking, obesity, exercise, blood pressure, plasma cholesterol, and use of postmenopausal estrogen replacement, aspirin, vitamin C, and β-carotene.

It must be noted that this effect was limited to vitamin E supplement use. High vitamin E intakes from dietary sources were not associated with a significant decrease in risk, although even the highest dietary vitamin E intakes were far lower than intakes among supplement users.

Support from case-control studies based on biological samples is sparse, although Gey and coworkers found that plasma levels of vitamin E in men aged 40 – 49 yr correlated strongly and inversely with the age-specific mortality from CHD in 16 European regions. A population case-control study also evaluated the relation between undiagnosed angina pectoris and plasma antioxidant levels in men aged 35 – 54 yr. Plasma levels of vitamin C, vitamin E, and carotene manifested a significant inverse correlation with undiagnosed angina. The inverse association between vitamin E levels and angina remained signifi-cant after adjustment for smoking habits, age, blood pressure, relative weight, and blood lipid levels. However, these results have been offset by several negative studies. There was no association between plasma vitamin E and prevalence of CHD in a cross-sectional survey of 1132 Finnish men. Similarly, most nested case-control studies found no relationship between plasma vitamin E levels and subsequent coronary mortality or risk of MI. The reason for these disparate results is unknown, but may include changes in diet following disease diagnosis, poor classification of controls, and lack of variation in plasma levels within populations not using supplements.

### Vitamin C

The evidence linking the water-soluble vitamin C with cardiovascular disease is less strong than that for vitamin E. In the Physicians' Follow-Up Study, a high intake of vitamin C was not associated with a lower risk of CHD in men, whereas in women from the Nurses Health Survey, an initial effect was attenuated after adjustment for multivitamin use. Only one prospective study, which involved 11,348 adults, demonstrated an inverse relationship between vitamin C intake and cardiovascular mortality. This effect was due largely to the use of vitamin C in supplements and may have been a reflection of other antioxidant vitamins in multivitamin preparations. A link between intake and carotid artery wall thickness has also been suggested. Plasma levels were not correlated with coronary mortality rates among four European populations or with prevalent coronary disease in Finland. In the Basle Prospective Study, low levels of vitamin C alone did not increase the risk of CHD, although the risk of disease at low levels of both vitamin C and β-carotene was greater than that for β-carotene alone. However, a prospective population study of 1605 healthy men aged 42 – 60 in Finland has recently shown that men who had vitamin C deficiency had a relative risk of MI of 2.5 after adjusting for the main risk factors for MI.

## β-Carotene

There is some indication that increased dietary intake of β-carotene is associated with reduced risk of CHD although again the evidence is less convincing than that for vitamin E. In the prospective Nurses Health Survey, consumption of vitamin A and β-carotene in food and supplements weakly predicted the incidence of CHD; Gaziano and Hennekens reported a 22% risk reduction for women in the highest quintile of β-carotene compared with those in the lowest. No adjustment was made for the potentially confounding effect of other antioxidant vitamins in multivitamin preparations. However, a small prospective study on 1271 elderly people also demonstrated an inverse relationship between β-carotene intake in fruit, and vegetables and subsequent cardiovascular death. Similar findings have been shown for serum carotenoid level and CHD risk, and carotenoid intake and carotid artery plaque thickness. Several studies indicate that dietary and circulating levels of β-carotene affect smokers more than nonsmokers. In the Health Professionals Follow-Up Study, high β-carotene intake was associated with reduced CHD risk in current smokers and exsmokers (70% risk reduction) but not never-smokers, after adjustment for cardiovascular risk factors and vitamin C and E intake. It was suggested that a high dietary intake of β-carotene is especially important in smokers who have both an increased demand for antioxidants (to combat smoking-induced free radicals) and a correspondingly lower circulating level for a given dietary intake. However, (3-carotene supplements in smokers may have harmful effects as discussed in the next section.

## Intervention Studies in Humans with Clinical Endpoints

Observational studies, of course, only infer a cause-and-effect relationship. Even in the most well-designed observational studies, the amount of uncontrolled and uncontrollable confounding could easily be as large as the small to moderate reductions in risk that are most plausible. Individuals who select higher intakes of antioxidants may also adopt other dietary or nondietary lifestyle characteristics that account for the apparent benefits of antioxidants seen in observational studies. Prospective studies with a clinical endpoint are required, and only randomized trials of sufficient sample size, dose, and duration of treatment and follow-up can provide reliable data. The limitations of intervention trials are that they can often only be interpreted for the particular study population, and for the antioxidant dose provided during the trial. Numerous intervention studies designed to test the hypothesis that increased antioxidant intake will protect against atherosclerosis are currently in progress. In general, these studies are using supplements of antioxidant vitamins or other antioxidants, and results so far have not been encouraging.

The Alpha-Tocopherol, Beta-Carotene Cancer Prevention Trial (ATBC), conducted among 29, 133 male heavy smokers in Finland, found no reduction in CHD morbidity or mortality during 5-8 yr of treatment with vitamin E (50 mg/d) or (3-carotene (20 mg/d). Those assigned vitamin E had no significant decrease in CHD deaths, but a 50% excess of deaths from cerebral hemorrhage, whereas those assigned to β-carotene experienced an 11% increase in CHD deaths. In a further analysis, a subgroup of the original subjects with a previous MI were considered. The endpoint of this substudy was the first major coronary event after randomization. The proportion of major coronary events was not decreased with either α-tocopherol or β-carotene supplements. In fact, β-carotene conferred an excess of fatal CHD (75% increase in risk). There was a beneficial effect of vitamin E on nonfatal MI with a risk reduction of 38%.

The CARET (Beta-Carotene and Retinol Efficacy Trial), designed to test the effects of a combined supplement of 30 mg (3-carotene and 25,000 IU retinol daily among 18, 314 cigarette smokers and individuals with occupational asbestos exposure, was ended early when researchers detected an elevated risk of death from lung cancer in those receiving β-carotene and, again, no beneficial effect on cardiovascular disease (C VD) was found. For C VD mortality, there was a nonsignificant 26% increase in the treated group ($p = 0.06$).

The Physicians Health Study followed more than 22,000 US male doctors treated with 50 mg β-carotene or placebo every other day for an average of 1.2 yr. The trial appears to have been conducted meticulously and its results would seem to seriously question any beneficial effect with such supplementation on CVD in well-nourished populations. There were no significant effects on indi-vidual outcomes, or on a combined endpoint of nonfatal MI, nonfatal stroke, and cardiovascular death, for which the relative risk was 1.0. There was also no evidence of harm (or benefit) among the 11% of participants who were current smokers at baseline.

Greenberg and associates studied the effect of β-carotene supplementation (50 mg/d) in 1720 male and female subjects for a median period of 4.3 yr with a median follow-up of 8.2 yr. Subjects whose plasma levels of β-carotene were in the highest quartile at the beginning of the study had the lowest risk of death from all causes compared with those in the lowest quartile. Supplementation, however, had no effect on either all-cause or cardiovascular mortality.

Thus for β-carotene supplementation, it would appear that there are no overall benefits among individuals with a good nutritional status who are at low or average risk of developing CHD. The situation may be different, however, for those with a previous history of such disease.

## Intervention Studies in Humans—Secondary Prevention

Hodis and coworkers have shown a reduction in CHD progression (as measured angiographically) in men given 100 IU vitamin E daily, although no benefit was found for vitamin C. Singh and colleagues found that a combi-nation of vitamins A, C, E, and β-carotene administered within a few hours after acute MI and continued for 28 d led to significantly fewer cardiac events and a lower prevalence of angina pectoris in the supplemented group. The Cambridge Heart Antioxidant Study (CHAOS), a trial of vitamin E supplementation on 2002 patients with angiographic evidence of coronary disease, was carried out with a mean treatment duration of 1.4 yr. This short-term supplementation with α-toco-pherol (268 or 537 mg/d) reduced CHD morbidity in patients (77% decreased risk of subsequent nonfatal MI). No benefit was found, however, in terms of cardiovascular mortality, with a nonsignificant excess among vitamin E-allocated participants. The recently published GISSI-P trial investigated the independent and combined effects of n-3 PUFA and vitamin E on morbidity and mortality after MI. Although n-3 PUFA reduced the primary combined endpoint of death, nonfatal MI, and stroke, vitamin E had no benefit. When each supplement was compared with no treatment in a four-way analysis, however, there was a nonsignificant reduction of 11 % in events by vitamin E (the study was designed so that only a reduction of 20% in events would be statistically significant).

## Summary and Evaluation of Human Intervention Studies

Thus, for vitamin E in Western populations, the only available trial data in primary prevention are from the ATBC trial which shows a negative effect. In secondary prevention, the accumulating trial data for vitamin E are more encouraging, particularly in terms of nonfatal endpoints, but far from conclusive in demonstrating net benefits. Each of these studies is open to specific criticism—the CHAOS study was too small to measure mortality, hemodynamic data were limited to only 706 of the 2002 patients, the amount of vitamin E was changed during the trial, and the patients were only followed up for 17 mo on average. Subjects in the ATBC study received a relatively low dose (50 mg) of vitamin E; there are doubts about the length of treatment; and it is uncertain whether the potential vitamin E-mediated protective effect could have any impact in such high-risk subjects (middle-aged heavy smokers). This last uncertainty also applies to the CARET study where those treated were already at elevated risk for lung cancer due to exposure to asbestos or cigarette smoking; therefore, preclinical cancerous change may already have been present.

How should we interpret the discordance between data from cohort studies and the results so far available from clinical trials? In general, it may be that the duration of clinical trials is too short to show a benefit, and that antioxidant intake over many years is required to prevent atherosclerosis. Thought needs to be given to trial design; with dose, duration of treatment and follow-up period, initial antioxidant levels and dietary intake, and extent and distribution of existing atherosclerosis being taken into consideration. Animal models have nearly always tested the effects of antioxidants on the early atherosclerotic lesions. Whether or not antioxidants have inhibitory effects on the later stages remains to be seen. In addition, the complex mixture of antioxidant micronutrients found in a diet high in fruit and vegetables may be more effective than large doses of one or a small number of antioxidant vitamins. It could be that several of these compounds work together but have no effect individually, or that other dietary components (*e.g.*, trace elements) may be effectors of carotenoid action.

The significant results linking antioxidant intake with CHD risk observed in cohort studies may be due to confounding by other lifestyle behaviours. Slattery and associates examined dietary antioxidants and plasma lipids in the Coronary Artery Risk Development in Young Adults (CARDIA) study and found that a higher intake of antioxidants was associated with other lifestyle factors such as physical activity and not smoking. Plasma concentrations of antioxidants are linked with social class, being higher in more affluent groups. Although these variables can be individually controlled for in analyses, it may be that a complex lifelong behaviour pattern needs to be studied before conclusions regarding antioxi-dants and CHD can be made. For example, passive smoking has recently been shown to have an atherogenic effect on LDL, yet exposure to smoke is a difficult lifestyle variable to control for in cohort analyses and is rarely measured.

### Further Clinical Trials of Antioxidant Supplementation

There are several clinical trials of antioxidant supplementation underway at present that are designed to clarify the effects of antioxidant supplementation on CVD. In each of these trials, doses of vitamin E greater than 200 mg/d are being used, which should be sufficient to increase serum levels at least two- to three-fold. The Women's Health Study is a primary prevention trial investigating the effects of vitamin E, β-carotene, and aspirin on CVD and cancer in 40,000 women aged 50 yr and over. In France, the Supplementation vitamins, minerals and antioxidant (SU.VI. MAX) Trial is testing a combination of antioxidant vitamins including vitamin E, vitamin C, and β-carotene in 15,000 healthy men and women. The Heart Protection Study in Oxford is investigating the effects of vitamin E, vitamin C, and β-carotene in 18,000 subjects with above average risk of MI, and a secondary prevention trial using the same three vitamins in 8000 women has been established in the United States (Women's Antioxidant Cardiovascular Disease Trial, WACDR). The Heart Outcomes Protection Study is also assessing vitamin E in 9000 persons with previous MI, stroke, or peripheral vascular disease and diabetic patients.

## WHOLE FOODS AND LDL OXIDATION

Although intervention trials are important in evaluating possible beneficial effects of antioxidants against development or progression of CHD, they have limitations and should be considered as only one component in the totality of available research evidence. It may be that a lifetime of intake is required to show a protective effect, or that a mixture of natural antioxidants found in fruits and vegetables provide the necessary protective mixture. A number of studies of people who eat a diet rich in fruit and vegetables, and therefore rich in antioxidant nutrients, have tried to test the hypothesis that fruits and vegetables lower the risk of CHD. In general, observational studies of vegetarians and those with diets rich in fruits and vegetables support the hypothesis that such a diet might lower the risk of CHD. Vegetarians generally have high intakes of cereals, nuts, vegetable oils, vegetables, and fruit. However, vegetarians differ from the rest of the population in a number of important ways: they tend to smoke less, have a lower BMI and alcohol intake, and come predominantly from higher social classes, all of which are known to confer a health advantage.

Few studies have looked at the effects of whole foods on biochemical end-points. A recent study asked subjects with normal lipid concentrations who ate three or fewer servings of fruit and vegetables daily to consume eight servings per day. Plasma concentrations of vitamin C, retinol, α-tocopherol, α- and β-carotene, lipids, and lipoproteins were assessed before and after an 8-wk intervention period. The plasma vitamin C, α-carotene, and β-carotene concentrations increased, whereas concentrations of retinol, α-tocopherol, lipids and lipoproteins remained unchanged, despite some increase in dietary vitamin E and a small reduction in saturated fat intake. An interesting addition to the results would have been the inclusion of data on the susceptibility of LDL to oxidation. The authors concluded that more specific dietary advice to modify fat intake may be necessary to reduce the risk of CVD.

By contrast, Singh and colleagues found over a 12-wk period that fruit and vegetable administration to subjects at high risk of CHD lowered total and LDL cholesterol and triglyceride levels, and increased HDL-cholesterol. Another study by Wise and coworkers using dehydrated fruit and vegetable extracts over a period of 28 d in 15 healthy adults aged 18 – 53 yr produced increases of 50 – 2000-fold in plasma carotenoid and tocopherol levels. During the same intervention period, plasma lipid peroxides decreased fourfold, with much of this lowering taking place during the first week.

Other antioxidant micronutrients may be important in increasing the resistance of LDL to oxidation. Flavonoids are plant-derived compounds that inhibit in vitro copper-catalyzed LDL oxidation (100 – 103). The inhibition of oxidation of human LDL by consumption of red wine or tea has been attributed to the presence of antioxidants such as flavonoids and other polyphenols in red wine and catechin in tea. Isoflavonoids such as genistein also reportedly increase LDL resistance to oxidation, although this has not been confirmed, whereas aged garlic extract has recently been shown to have antioxidant properties. Ubiquinol-10 is another effective lipid-soluble antioxidant that inhibits LDL oxidation due to aqueous or lipid-phase peroxyl radicals. The effect of antioxidant supplementation may be sex-specific with 17 β-estradiol at physiological levels increasing the resistance of LDL to oxidation in some studies. Further studies, especially in humans, are required to validate the role of these antioxidants in inhibiting LDL oxidation.

## OTHER ANTI-ATHEROGENIC EFFECTS OF ANTIOXIDANTS

The difficulty in linking inhibition of LDL oxidation with inhibition of atherosclerosis may stem from the dynamic nature of CHD, CHD involves not only the development of an atherosclerotic plaque, but also plaque rupture, vasoconstriction, and local thrombosis, resulting in partial or total arterial obstruction. Some antioxidants may limit the clinical expression of atherosclerosis by stabilizing the plaque rather than by affecting its size. Scavenger receptor activity has been shown to be downregulated in macrophages after incubation with α-tocopherol. Several other mechanisms can contribute to the vascular effects of α-tocopherol: it has been shown to reduce monocyte adhesion and transmigration into the intima, it can inhibit the proliferation of smooth muscle cells, and it prevents the cytotoxic effects of oxidized LDL by reducing endothelial cell damage.

Oxidized LDL also impairs the release of nitric oxide from normal arteries, which usually prevents inappropriate adhesion of leukocytes and platelets, and vasospasm. Thus, the presence of oxidized LDL might contribute to the platelet adhesion and vasospasm that are involved in the pathogenesis of acute coronary syndromes. LDL derived from patients treated with Probucol (a synthetic antioxidant) and oxidized in vitro does not impair the action of nitric oxide to the same degree as LDL derived from normal subjects. Antioxidant protection of LDL by Probucol and improved endothelial function has also been found in patients treated for hypercholesterolemia.

These alternative vascular effects of antioxidants and decreased LDL oxidation may explain the positive findings for antioxidants in the prevention of secondary CHD or the clinical expression of established CHD, rather than in primary prevention or the initial development of atherosclerotic lesions.

## LEVELS OF ANTIOXIDANT INTAKE FROM FOOD SOURCES

In the absence of conclusive evidence linking antioxidant intake with a reduced risk of CHD, dietary recommendations are difficult. Vitamin C and β-carotene are available from fruits and vegetables, and vitamin E from vegetable oils. Results from southern European countries consuming the classical Mediterranean diet show that high plasma levels of these antioxidants can be achieved. This diet is characterized by a preference for fresh products and frequent consumption of fruits, vegetables, legumes, and oils with a high vitamin E content. In contrast, major parts of populations in the United States or in northern parts of Europe do not consume optimal amounts of antioxidant nutrients. The availability of lower-priced convenience foods in the United States acts against the consumption of freshly prepared foods. Thus, only 40% of Americans con-sume five servings of fruit and vegetables daily, as recommended by the US national food guide, the Food Guide Pyramid. Only a quarter consumed fruits or vegetables rich in vitamin C or the carotenoids, and on any given day 54% ate no fruit at all. A recent reanalysis of the Second National Health and Nutrition Examination Survey (NHANES II) data showed that vitamin supplements are the major contributors of the principal antioxidant micronutrients in the US diet (28% of vitamin C and 46% of vitamin E).

Estimates of the requirement for micronutrients are based on the minimum quantity necessary to prevent a deficiency. The translation of minimum requirement to a dietary recommendation for population groups necessitates that allowance be made for a number of variables, including periods of low intake, increased utilization, individual variability, and bioavailability. In the United States, the Recommended Daily Allowances (RDAs) incorporate "margins of safety" intended to be sufficiently generous to encompass the variability in the minimum requirement among people, and bioavailability from different food sources.

To date, recommendations on micronutrient intake have therefore been primarily intended to prevent clinically overt deficiencies such as scurvy. If, however, observational and experimental evidence continues to show that the prevention of slow multistage processes such as CVD and cancer might require a higher intake of some essential antioxidants, then a recommended intake should be devised referring to amounts considered sufficient for the avoidance of these disease states. Gey suggests that the present recommendations will require either an upgrading or an additional term, *e.g.,* a recommended optimum intake (ROI) that will vary with gender and age, with special requirements for smokers, pregnancy, and the elderly. The ROI could be defined as sufficient (culture-and/or regionspecific) intake to achieve blood levels associated with the observed minimum relative risk of disease. For example, Carr and Frei have suggested a doubling of the current RDA for vitamin C for optimum reduction in chronic disease risk. A ROI would simply quantify specific dietary constituents of conceivably crucial importance within the still desirable "five servings of fruit and vegetables daily."

## CONCLUSION

There is strong evidence to support a link between antioxidant intake and a protective effect on the susceptibility of LDL to oxidation, but the relevance of this to CHD, and the possibility of other important bioactive micronutrients and phytochemicals in vegetables and fruit, require further research.

CHAPTER

2

# Sterilisation of Disinfection

Sterilisation is the action of eliminating microorganisms from a medium. Sterility is the absence of any detectable and viable microbes in a culture medium or in the gas phase. Sterilisation is a process that destroys all living organisms, spores and viruses in a pressurised vessel at high temperature. In the food and dairy industries, sterilisation is commonly used to preserve food products. At the laboratory scale, huge steel vessels with live stream at 105 kPa (15 psig) are commonly used for 20-30 minutes. This is a closed system known as an autoclave; therefore it is batch sterilisation. The temperature is raised to 121 °C when air is initially flashed out and all live streams are replaced. Wet steam is usually used for effective autoclaving. The high temperature and long duration may kill all living microorganisms, spores and viruses.

When media is prepared, to eliminate bacterial and fungal contaminants it must be heat treated at high pressure. Even at high temperatures the fungal spores may survive if only heat is used. Therefore, media is autoclaved at 121 °C and 105 kPa. In fact, an autoclave is used like a pressure cooker, which is convenient and economic autoclaving equipment. Overheatmg the prepared media may have negative impact, causing the media pH to be unstable.

Acid pH is very sensitive to overheating. Overheating in media containing sugars cuses the carbohydrate to be caramelised. This media may have reverse impact, reducing bacteriological performance. Gelatinous media or any nutrient agar at acidic pH is hydrolysed. This is due to the acidic condition of catalytic activities or excess amounts of protons breaking down the solid media and extra sugars forming. This may cause substrate inhibition on growth of organisms on solid slant media.

## BATCH STERILISATION

Batch sterilisation uses steam to eliminate living organisms. Heat losses, heating and cooling are major steps, and it is a time-consuming process. It requires air to be evacuated replaced with steam. At first the chamber is flashed with pressurised steam. With this technique we should get rid of the existing air. Steam sterilisation is performed in a jacketed vessel by supplying steam and maintaining the set pressure and temperature at a constant level for a fixed duration. Batch sterilisation wastes energy and overcooks the medium. There is no conservation of energy; therefore the process may not be economical for implementation at a large scale. Batch sterilisation is common at bench and laboratory scales. Batch media sterilisation is done in an autoclave.

Basically, it is a huge steam cooker. Steam enters the jacket of the surrounding chamber. When the pressure from the steam has been built up in the jacket, a venting valve for the outlet of chamber air is closed and the inlet valve allows the steam to enter the chamber. The pressure of the chamber is increased to 105 kPa. (15 $P_{sig}$) At this point the sterilisation time begins to count down. Usually 15-30 minutes are used, depending on the type and volume of media. For instance, sterilisation of one litre of media at 121 °C requires 20 minutes. Larger volumes may require longer retention times for sterilisation. There are recommended holding times of 20, 10 and 3 minutes for 121, 126 and 134 °C, respectively. The high pressure in a closed container allows the temperature to exceed 121 °C.

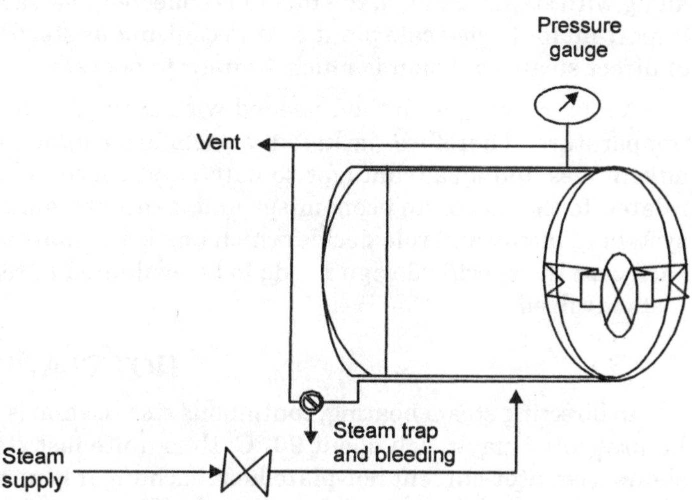

**Fig. 2.1. Pressure steam sterilizer (autoclave).**

## CONTINUOUS STERILISATION

One of the methods for continuous sterilisation of medium for fermentation is the direct use of live steam by injection of steam into the medium. The heat exchanger is eliminated. The medium stays in a loop for a predetermined holding time until the entire medium is sterile. The problem with directly injecting steam is dilution of media since it is initially cold. However, it has better heat economy, since it comes from substituting heat exchangers for direct steam injection. To utilise all the supplied heat, preheaters or heat economisers are used. Instead of having a cold water stream to cool the sterile media, the lower temperature, unsterile media stream takes the primary heat from the warm stream, cooling the sterile media.

Continuous sterilisation has a holding coil for detention long enough to kill all the microorganisms. The medium from a make-up vessel flowing through the exchanger is held in the coil, and then goes through the heat exchanger, heating more unsterile medium while become cool itself, as it is collected in a sterile fermenter. Presents the sequence of piping media goes through,

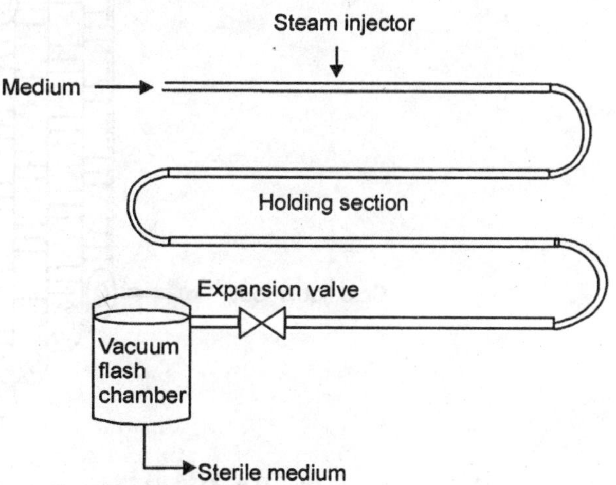

**Fig. 2.2. Continuous sterilization.**

along with steam. Finally, the media is collected in a vacuum flash drum. Heat economisers are important for large-scale plant unit in continuous sterilisation. For a small-scale bioreactor use of direct steam injection is much simpler to operate.

A heat exchanger is then needed with cooling water to bring the medium back to ambient temperature. Therefore, jacketed vessels are commonly used with cooling water. There are advantages and disadvantages to batch and continuous sterilisation. The energy savings are related to the use of an economiser and direct stream or indirect sterilisation. In a largescale system economy and role decide which one is the most suitable process to be implemented. For each case the specific design needs to be evaluated in terms of the thermal efficiency and fixed costs involved.

## HOT PLATES

In directing steam heating, continuous sterilisation is preferable. The medium passed through La preheater may reach about 90 °C; then quite fast sterilisation may take at 140 °C. Fig. 2.3 shows a counter-current hot-plate heat exchanger is normally used for continuous sterilisation. Hot-plate heat exchangers are extremely efficient and easily maintained if any fouling or scale deposition takes place. Normally fouling is serious problem, occuring by deposition of proteins on the hot surface of the exchanger, which causes reduction in the overall heat transfer coefficient. Plate exchangers are easily separated and cleaned. Special care must be taken when the media contains any agglomerated particles. Correlation is required to calculate the suitable residence time for sterilisation.

The choice of a suitable sterilisation process is based on economics and cost of the heat exchangers to reduce energy consumption. In a suitable exchanger, the larger the heat transfer coefficient, the greater the energy recovery.

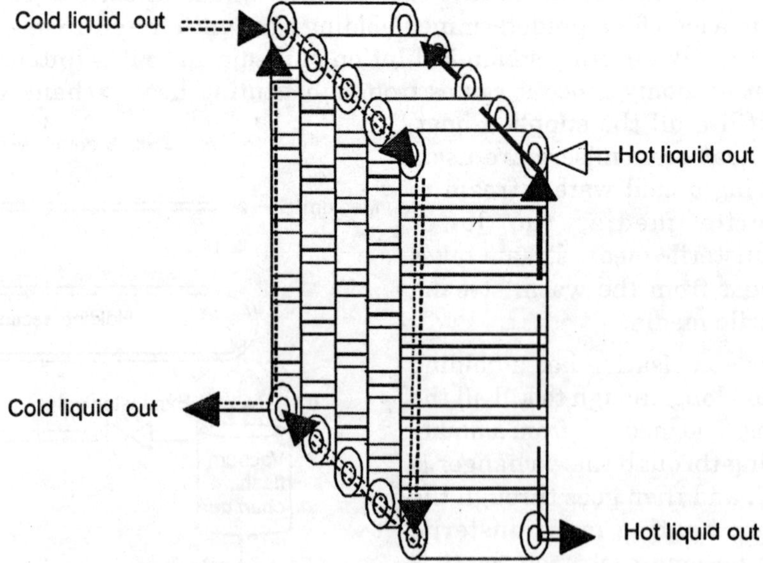


**Fig. 2.3. Counter current hot plate heat exchanger.**

## HIGH TEMPERATURE STERILISATION

Fast sterilisation is performed at high temperature. High-temperature sterilisation requires short holding times. This technique is used in the fast preparation of nutrient media for industrial bioprocesses and in pasteurising milk. A short retention time may favour media with heat-sensitive proteins and cause less damage to the biochemical composition of the media than more prolonged times at lower temperatures. The ability of high temperatures to perform rapid sterilisation is related to activation energies. That is affected by how fast bacteria are killed in an elevated temperature.

On the other hand, long sterilisation at high temperatures may destroy the protein and biochemical composition of the media. Short-duration sterilisation at high-temperatures are more lethal to organisms and less chemically damaging than are longer sterilisation processes at lower temperatures. Sterilisation at high temperature is recommended for 3 minutes at 134 °C. It is preferable to 20 minutes at 115 °C in conventional operation.

## STERILISED MEDIA FOR MICROBIOLOGY

Sterile media are used for pure culture. The media used to culture microorganisms depend on the living conditions of the microorganisms. Media compositions must be identified based on the needs of the organisms for carbon, nitrogen and phosphorus. Nutrients such as protein and sugars are added at wide ranges of pH. The acidic, neutral and alkaline conditions are defined based on differentiation of microbes in their biochemical reactions. Colour indicators are used to observe any transition of pH. Buffer is used in the media to maintain constant pH. Growth factors and growth stimulants are also used to accelerate microbial growth and to maintain exponential growth.

The most common media in the liquid phase is nutrient broth; in the solid phase, nutrient agar is used to propagate microorganisms. When microbiological media is prepared, it has to be sterilised before any inoculation of organisms. Sterilisation is used to eliminate any microbial contamination that may originate from air, glassware or hands. Microbes propagate quite quickely: within a few hours there will be thousands of organisms reproduced in the media. To control the growth of unwanted organisms, the media has to be sterilised quickly before any microbes start to utilise the nutrients.

The sterilisation process makes the media absolutely free of contaminants and guarantees that the medium will stay sterile, so that is used only for the desired organisms. The kinetics of culture media sterilisation describe the rate of destruction of microorganisms by steam using a first-order chemical reaction rate model. As the population of microorganisms ($N$) decreases with time, the rate is defined by the following equation:

$$-\frac{dN}{dt} = k_d N \quad ...(2.1)$$

where $N_0$ is the number of viable organisms present in the culture media, $t$ is the retention time or sterilisation time, and $k_d$ is the reaction rate constant as it is known for a specific death rate. Using separation of variables and integrating with initial condition, the following useful expression is obtained:

$$N(t) = N_0 e^{-k_d t} \quad ...(2.2)$$

where $N_0$ is the number of viable organisms present in the media at the starting point before sterilisation. Now take the natural log; it is reduced to a linear model:

$$\ln \frac{N(t)}{N_0} = -k_d t \qquad ...(2.3)$$

The graphical presentation of the equation shows a straight line with a negative slope for $k_d$. As the death rate constant follows Arrhenius' law, the death rate constant is temperature dependent. The value of $k_d$ is about 0.02 min$^{-1}$ at 100 °C, the death rate constant increases by 10-fold at 110 °C and 100-fold at 120 °C.[2]

$$k_d = k_0 e^{-E/RT}$$

where $k_0$ is the death rate constant at a reference temperature also known as Arrhenius' constant, $R$ is the gas constant, $T$ is the absolute temperature, $E$ is the activation energy, which is 60-70 kcal per mole for microorganisms, 100-150 kcal per mole for spores, and 30-40 kcal per mole for media with vitamins and protein solution.[2]

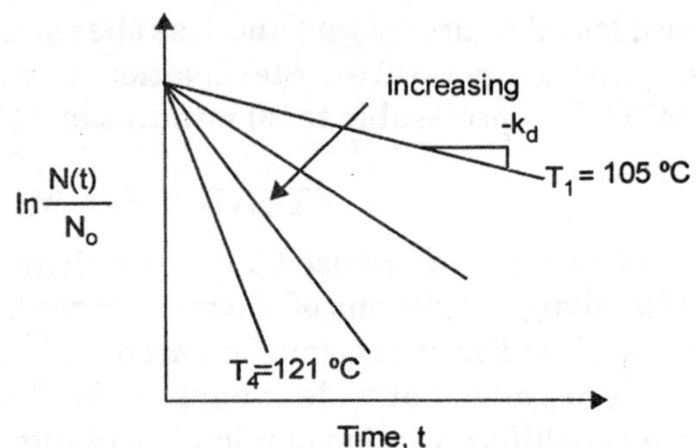


**Fig. 2.4. The loss of cell viability at various temperatures.**

The linear model for the loss of cell viability at various temperatures. As the temperature increases from 105 to 121 °C, the value for the slope of the line increases. This means that the number of viable cells at a fixed time of sterilisation will drastically decrease as the temperature increases by 16 °C.

## Sterilisation of Media for Stoke Cultures

To clear slant tubes, culture agar solution is prepared. To have clear agar solution, it has to boil. Agar is a polysaccharide. It is not soluble in cold water, so heating helps to dissolve it in water. Once the solution is dispensed in the tube, it goes for sterilisation. Agar is in the solid state at room temperature, about 35 °C.

## Sterilisation of Bacterial Media

Often, sterilisation of culture media is done in an autoclave at temperatures between 121 and 134 °C. It is good to know the damage caused to the medium by heating it. Heating the culture media, which contains peptides, sugars, vitamins, minerals and metals, results in nutrient destruction, either thermal degradation or reaction between the components of the media. Protein in the media is denatured by high temperature, or sugars are easily caramelised. During the heat treatment, toxic compounds are formed, which may retard or inhibit the microbial growth. For minimal damage to the ingredients of the media, it is important to optimise and minimise the heating and holding time of the sterilisation process, respectively.

## Sterilise Petri Dishes

There are several ways to handle Petri dishes. The standard dishes are 15-20 mm deep and 12-15 cm in diameter. They are normally available in glass and also transparent plastic. Petri dishes are individually wrapped. The media is separately sterilised while we add the sterilised media to the Petri dishes in front of a flame. It is recommended to use 20-25 ml of medium in each Petri dish. These are all called culture dishes. The media contains about 1.5-2% agar. After the media cools off it solidifies. The Petri dishes with the nutrient agar should be sterilized and stored in a refrigerator. It is common to sterilise the media and Petri dishes separately. When the medium is cooled to about 55 °C, in front of a flame or in a laminar flow chamber, lift the lid of the dish enough to pour about 25 ml of the medium to the desired depth and lower the lid in place.

It is best to gently move the Petri dish in way that spreads a thin layer of agar uniformly without any air bubbles. Distribution of media in the Petri dishes should be done in front of a flame. Most plastic Petri dishes are made of polystyrene and are not autoclaveable. Plastic Petri dishes are easily deformed during sterilisation at high temperature. Some plastic dishes can be autoclaved, but they are more expensive. Please follow the instructions given by the manufacturer or obtain information from catalogues. It is recommended to autoclave glass dishes and medium separately. If your dishes are autoclaveable, you may dispense the agar into the dish and then autoclave it. In this case it is best to cool the dishes in the autoclave or in the pressure cooker to reduce the amount of water that will condense on the cover of the dish.

## DRY HEAT STERILISATION

Dry heat sterilisation is used for equipment that can withstand high temperature and dry heat but cannot withstand wet or steam autoclave. This method is often used for glassware as it dries and sterilises in one operation. The pipets must be wrapped in dustproof aluminum foil or placed in metal pipette cans. The can lids are removed during heating and replaced after sterilisation, that is before any dust can get in the can. Disposable items are not recommended for dry heat sterilisation. This method may only be good for permanent reusable glass pipettes. Normal laboratory glassware must first be washed and cleaned. It has to be rinsed with deionised water.

The clean glassware is sterilised in an oven set at 200 °C for 1–4 hours. It is suitable to cover glassware with aluminum foil to maintain aseptic conditions after removing the glassware from the oven. If aluminum foil is not available, special heat-resistant wrap paper can be used. The sterile glassware must be protected from the air, which has micro-flora, or any contaminants. Avoid the use of any plastic caps and papers. Detach any labelling tape or other flammable materials, as they are fire hazards.

## STERILISATION WITH FILTRATION

Certain media components are susceptible to heat; it is going to be denatured if it is heated. Therefore they must be added to the media after autoclaving. To do so, it is necessary to carry out filtration, the components using a 0.22 μm pore size filter that is appropriate to the solvent used. Filters are normally available from Whatman, or Fisher Scientific. It is recommended you consult filter experts or suppliers about the solvent and the special application you are intending to implement for the desired filters for the process. Normally filtration of media is used instead of sterilised media with an autoclave. To filter sterilised media, filter the water using a 0.45 μm pore size filter, before using a 0.22 μm pore size filter for final sterilisation.

## MICROWAVE STERILISATION

Rapid sterilisation of media is achieved by using microwave ovens. Most plant tissue culture media can be sterilised using a microwave, although it may not be suitable with a medium containing complex additives like oatmeal.

## ELECTRON BEAM STERILISATION

Electron beam sterilisation is a high-voltage potential established between a cathode and an anode in an evacuated tube. The cathode emits electrons, as a cathodic ray or electron beam. A high intensity of electrons is produced. These electrons are accelerated to extremely high velocities. These accelerated electron intensities have great potential as a bacteriocide. Most electron beams operate in a vacuum. As a result the unwanted organisms in the media vanish and the media is sterilized. The electron accelerator equipment producing the high-voltage beam also has various applications in the medical field and surgical supplies. Electron beam sterilisation is a successful technology used for sterilisation of disposable medical appliances and devices with a wide range of densities. The electron beam inactivates microorganisms that cause destruction of biomolecules, and results in death of microbes by an indirect chemical reaction. Irradiation by gamma rays is a similar mechanism. Advanced electronics precisely control the use of electron beams in the sterilisation of medical devices. Electron beam sterilisation results in less material degradation than with gamma irradiation. Medical products are sterilised in the original shipping containers, saving lots of time and maintaining integrity of the original package.

## CHEMICAL STERILISATION

Chemical agents are used to sterilise heat-sensitive equipment. Chemical solutions are used as a suitable method for sterilising long pipettes and glassware. Normally at pilot scale in the absence of life, steam and chemical agents are often used. Application of an oxidising agent such as 10% chlorox for 20 minutes or longer proves the system operates without any contamination. Excess amounts of chemical agent have to be removed; otherwise organisms are able to grow in a toxic environment. The bleach (sodium hypochlorite) then needs to be removed by rinsing with sterile water or rubbing with ethyl alcohol, 70% ethanol-water (pH = 2) or an alcohol solution of 70% *iso*-propanol without recontaminating the glassware or tools. Ethanol is commonly used for cleaning surfaces. Use of bleach on metal devices is not suitable as it corrodes metals rapidly.

Ethylene oxide in the gas phase is commonly used. This is a special chemical effectively used for column bioreactors. It is a volatile compound and strong oxidizing agent. It boils at ambient temperature, therefore the solution of ethylene oxide (liquid phase) must be stored in a refrigerator (4°C). An excellent oxidizing agent such as a 3% sodium hypochlorite is used for chemical sterilization of equipment.

### Tyndallization

John Tyndall, a famous British scientist, noticed that a period of about 24 hours at ordinary room temperatures is usually sufficient to enable dormant and heat-resistant spores to germinate and grow into the vegetative state in which they are sensitive to heat. He devised a process of sterilization based on these observations. It consists in steaming for a few minutes at 100° C, on three or four successive occasions, separated by 24-hour intervals at a temperature favorable to spore germination. The intervals permit the dormant, resistant spores to become active, vulnerable vegetative cells, readily killed by 100° C. This process renders an infusion sterile, whereas one single continuous boiling for one hour may not, since many spores remain in their dormant and resistant state during this time. This intermittent process is sometimes put to practical use in the home canning of foods. A disadvantage of the method is the time required to achieve complete sterilization; an advantage is that it requires no special apparatus. In some fluids, such as water, spores may not grow out promptly. Also, if the material is freely exposed to air, anaerobic spores may not germinate and may survive the process. If not freely exposed to air, aerobic spores will not grow out freely.

**Table 2.1. Boiling point at various altitiudes**

| *Location* | *Altitude** | *Boiling Point*** |
|---|---|---|
| New York City | 0 | 100 |
| Chicago, Ill | 589 | 98.9 |
| Denver, Col | 5,280 | 94.3 |
| Fort Laramie, Wyo | 7,390 | 92.2 |
| Tahoe, Nev | 10,000 | 89.1 |
| *Feet above sea level. | | |
| Degrees C. | | |

* Feet above sea level.

** Degrees C.

## Compressed Steam: Autoclaving

Anyone familiar with the operation of a home pressure cooker is familiar with an *autoclave,* because the cooker is a simple form of autoclave. In the autoclave, be it a small and simple home pressure cooker or an apparatus large enough to fill a room and fitted with various gauges, pipes, valves, clocks and wheels, the object of both is alike: to heat the articles to be sterilized by means of steam under considerable pressure.

Steam under pressure is hotter than boiling water or free-flowing steam such as is used in tyndallization. The higher the steam pressure, the higher the temperature will be.

It must be remembered that it is the compressed steam (moisture, hydration) that sterilizes and not compressed air (dry and usually not as hot as steam).

For example, pure steam at 15 pounds pressure has a temperature of 121° C. If the steam is mixed with an equal amount of air, at the same pressure the temperature is only 110° C, while if the mixture is two thirds air the temperature is only 109° C.

Steam *hydrates* and thus promotes coagulation; air does neither. Steam, being water vapor, also produces *hydrolysis* at autoclave temperatures. Dry air cannot do this at any pressure or temperature. In autoclaving, therefore, as in using a pressure cooker, a valve is left open for the escape of all air before the steam pressure is allowed to rise.

The actual amount of water present as steam in an autoclave is small, and articles soon dry off after removal, especially if removed from the autoclave while hot.

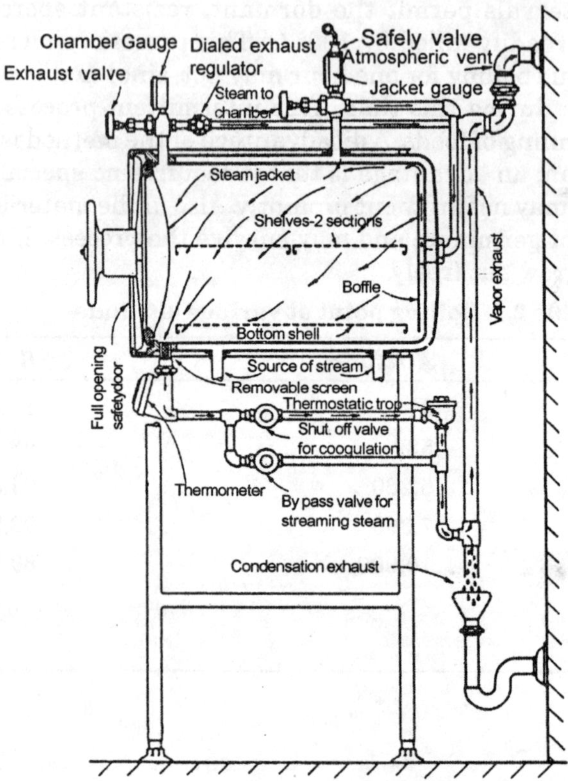


**Fig. 2.5. Diagrammatic illustration of steam jacketed autoclave. Steam enters the *jacket,* a double-walled shell, at two places just beneath the cylinder. It passes out the top through a pipe to which are attached: a wheel valve admitting the steam to the inner *chamber,* a *safety valve,* a gauge showing *pressure in the jacket.* The steam enters the inner chamber at the right of the diagram, filling the upper portion. Its pressure registers on the *chamber gauge.* It may be allowed to escape rapidly by the *exhaust valve.* If this is closed, steam pushes the cooler air in the lower portion out at the bottom (left), where the *thermometer* registers proper temperature only when the air is gone and is followed by the hot steam. The escaping steam may be allowed to flow out without building up any pressure if the *by-pass valve* is fully opened. If the bypass valve is closed and the *shut-off valve* is opened, steam passes through the *thermostatic trap* where the heat shuts off all but a pinhole opening. This causes pressure to build up in the chamber, yet prevents stagnation by permitting a constant minute flow of steam through the apparatus.**

The *thermometer* on the autoclave is the important guide to the process of autoclaving, not the pressure gauge. However, the latter, as well as a steam-escape or safety valve, is essential to safety.

The common practice in autoclaving fluids or freely exposed surfaces such as those of dishes and instruments is to apply 115° to 125° C (10 to 20 pounds' pressure) for 20 minutes. The pressure must be allowed to subside *slowly* after the heating is over or *superheated* fluids in open (cotton-plugged) vessels will boil over. Any large, solid masses must be heated a longer time to allow for heat penetration. Packages must be spaced so as to allow free circulation of steam. Substances such as mineral oil or petrolatum, sand, or any dry objects in tight jars, or substances that are impervious to moisture, cannot be satisfactorily sterilized in the autoclave. The temperature may rise as high as 125° C, but in the absence of moisture this is ineffective. Such materials are more effectively sterilized by dry heat in an oven.

Acid materials, such as canned tomatoes, acid fruits, pickles or sauerkraut (pH more acid than 4.5), require much shorter periods or lower temperatures for sterilization than low acid foods (pH 6.5 to 7) such as milk, corn and meats. This is because of the synergistic action between pH and temperature.

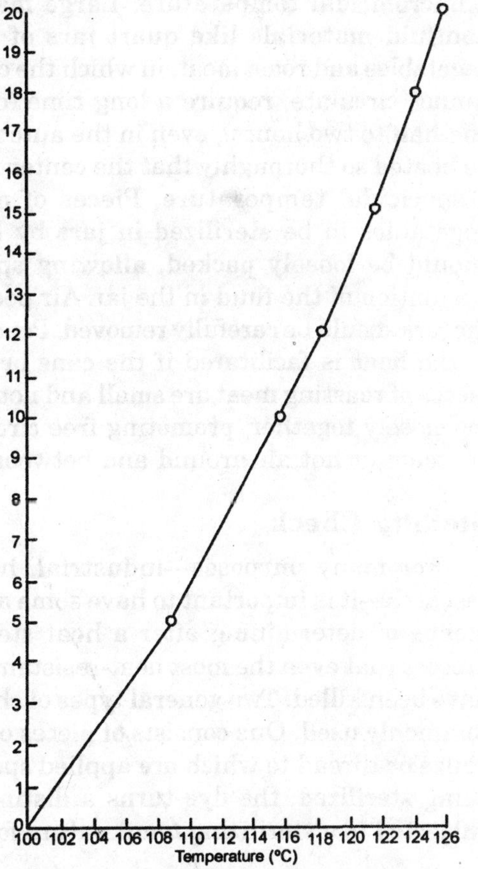


**Fig. 2.6. Relation between temperature and pressure of pure steam.**

## DRY HEAT

Dry heat is used in oven sterilization. It is necessary to bear in mind that significant coagulation does not occur when moisture is not present.

Articles in ovens are very dry and, therefore, in order to be freed of live spores, must reach a very high temperature (165° to 170° C; 329° to 338° F). It is customary to apply 165° C for a period of two hours. This accomplishes no coagulation but, what is more effective, slight charring.

A home oven can easily be used for sterilization. A "moderate" temperature (330° F) is satisfactory, and the heating should be allowed to proceed for two hours *after* reaching that temperature. Paper wrappings should be slightly browned but not brittle; muslin or string should be faintly yellow due to the heat.

Only dry articles not injured by baking (glassware, bandages, instruments, mineral oils, petrolatum, talcum powder and the like) may be thus sterilized. Solutions containing water, alcohol or other volatile substances will, of course, boil away and be ruined.

## TIME AND CONTROLS

### Thorough Heating Necessary

In any process of disinfection or sterilization by heat it is absolutely essential that the object be heated through and that the center of the object be held at a killing temperature long enough to destroy the bacteria.

Thus, in home canning, a quart jar of spinach may be held in free steam (100° C) for five to ten minutes and when grasped with the hand will feel very hot, yet the center may be well below a microbicidal temperature. Large masses of nonfluid materials like quart jars of canned vegetables and roast meat, in which the contents cannot circulate, require a long time (one and one-half to two hours), even in the autoclave, to be heated so thoroughly that the center reaches a sporicidal temperature. Pieces of meat or vegetables to be sterilized in jars by heating should be loosely packed, allowing space for circulation of the fluid in the jar. Air pockets in the jars should be carefully removed. Penetration of the heat is facilitated if the cans or jars or pieces of roasting meat are small and not packed too closely together, promoting free circulation of steam or hot air around and between them.

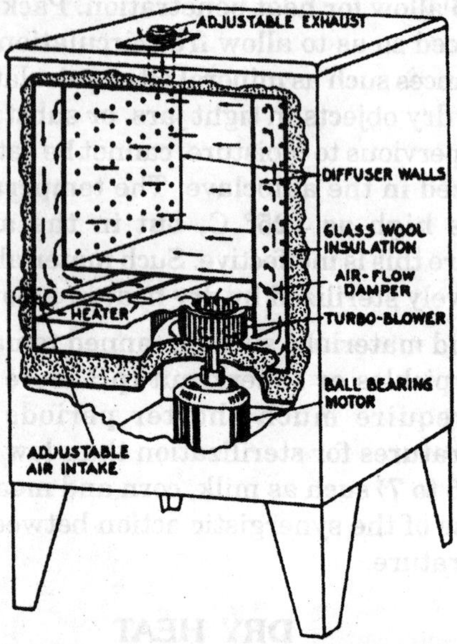

**Fig. 2.7. A type of sterilizing oven for microbiology. The motor at the bottom of the oven forces streams of hot dry air through the chamber of the oven, as indicated by the arrows.**

## Sterility Check

For many purposes—industrial, hospital, research—it is important to have some absolute means of determining after a heat sterilizing process that even the most heat-resistant spores have been killed. Two general types of check are commonly used. One consists of pieces of paper, gauze or thread to which are applied spots of some thermolabile dye. Inserted in the materials being sterilized, the dye turns a distinctive color if the contents of the sterilizer have been subjected to a sterilizing temperature for a sufficient time.

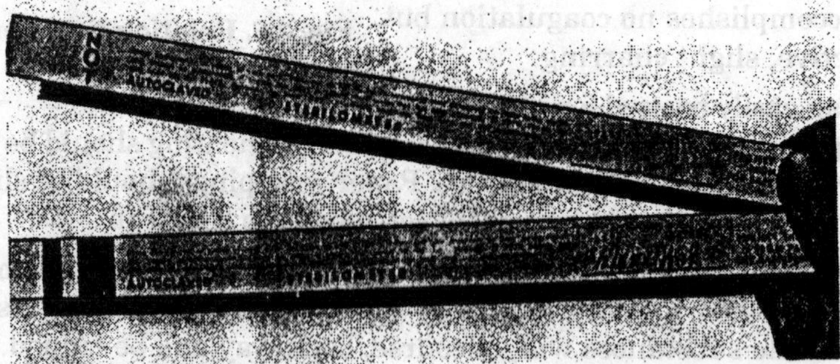

**Fig. 2.8. One type of device for indicating proper autoclaving. Before heating the word "NOT" appears clearly as on the upper indicator. After partial heating the leftmost black bar appears as seen on the lower indicator. Not until autoclaving is complete both as to time and temperature is the word "NOT" obliterated by a second dark bar on the lower indicator.**

A more direct check (though more time-consuming) consists of strips of filter paper having on them the very heat-resistant spores of *Bacillus subtilis.* This is a harmless aerobic bacterium. The spore-impregnated strips are inserted (enclosed in sterile envelopes) inside materials being sterilized. Afterward the strips are removed to tubes of culture medium and incubated. If growth occurs in the tubes within seven days then it is evident that the sterilization process in which they were involved was inadequate to kill the spores. In a modification of this process, sealed ampules (commercially available) containing culture medium with dormant spores of harmless *Bacillus stearothermophilus* already in it are inserted in the materials being sterilized. After sterilization the complete ampules are incubated at 55° C (since *B. stearothermophilus* is a strict thermophile). If no growth occurs in seven days it is proof of effective sterilization, since spores of *B. stearothermophilus* are among the most heat-resistant of known living things.

## SOME USEFUL DISINFECTANTS

There are hundreds of disinfectants in the market and it would obviously be impossible to discuss even a small part of them here. A few commonly used representatives of several classes will illustrate general principles and usages. For convenience we may group the disinfectants as follows: halogens and halogen compounds, compounds of heavy metals, phenol and its derivatives, alcohols, detergent disinfectants, microbicidal gases and radiations.

## HALOGENS AND HALOGEN COMPOUNDS

Iodine and chlorine are the most widely used of this group. Chlorine gas is used to disinfect filtered water at all municipal water-purification plants and many sewage-disposal plants. It is usually handled in tanks or tubes like oxygen. It is applied to drinking water in a final concentration of about 1.0 part per million.

A more convenient form of chlorine for the individual and household user is calcium or sodium hypochlorite. Solutions (5 per cent) of sodium hypochlorite are purchasable in all grocery stores under various names (*e.g.,* Clorox). They have a multitude of household and sanitary uses. They depend for their effectiveness probably on their liberation of free chlorine. Directions for use of these solutions are on each bottle. Four points need to be remembered in their use: the chlorine tends to evaporate from the solution; they give a bad odor to the hands; chlorine is poisonous; and it is readily inactivated by all organic matter.

### Chloride of Lime

Chloride of lime (0.5 to 5 per cent aqueous solution) is excellent for similar purposes and is inexpensive. The ordinary chloride of lime of commerce is unstable and soon loses most of its free chlorine. Many organic chlorine compounds which liberate their chlorine more slowly are very effective. *Azochloramid* is one of these; *dichloramine toluol* is another. The odor of chlorine may be objectionable.

In addition to the above substances, there are many new proprietary chlorine compounds on the market based on principles given above.

### Lodine

The most actively antimicrobial of the halogens is iodine. However, it is not commonly used for the same large-scale purposes as chlorine because of its physical properties and its cost. Formerly used widely as a local antiseptic in a 7 per cent tincture (alcoholic solution), this strength is far too caustic and poisonous for local application and does more harm than good. The 2 per cent tincture is much better and is the most generally useful for small cuts and abrasions. If the alcohol causes too much pain, the tincture may be diluted with two to four volumes of water.

An aqueous solution containing approximately 1.85 per cent of iodine and 2.2 per cent of potassium iodide is excellent but has a high surface tension.

## Lodophors

Nonirritating, nonstaining, virtually odorless and very effective solutions of organic compounds of iodine in which the iodine is loosely combined with a surfactant organic *wdophor* (Gr. *phor* = bearer or carrier) are available commercially. The iodophor releases the iodine slowly and also lowers the surface tension of the solution. Some idea of the amount of iodine remaining in such solutions is readily gained by comparing the diminishing brown color with that of the original full-strength solution. This is not possible with colorless chlorine solutions. Examples of such iodine products are Wescodyne, Iodide and Betadine. In such forms iodine may be used for a variety of purposes, from clinical uses to sanitization of dishes, with aqueous solutions containing 200 parts per million. Under certain conditions of concentration and temperature such solutions may be sporicidal. They appear to kill bacteria, molds and some viruses quickly. Iodine is a much more effective germicide than chlorine.

Iodine, like chlorine, combines indiscriminately with organic matter; its solutions are unstable and, like chlorine, it is poisonous.

# COMPOUNDS OF HEAVY METALS

The first scientifically designed, specifically acting drug was based on the heavy metal arsenic. Arsenic was compounded with an organic group to diminish its poisonous properties. Prior to 1907, Paul Ehrlich and his coworkers had synthesized and tried many such compounds in the treatment of spirochetal diseases, notably syphilis, without success. The 606th compound ("SOS" for "606" or Salvarsan or arsphenamine), discovered in 1907, was very effective and, with various modifications such as neoarsphenamine, was used for years in the treatment of syphilis and other spirochetoses. It fell into disuse after 1943 when it was found by Mahoney et al., in the U. S. Public Health Service, that penicillin is much more active against the same organisms and almost nontoxic.

Other metals in common use as microbicides today are iron and zinc, previously discussed, and mercury, silver and copper. All act mainly by coagulative action except in very low concentrations (*e.g.*, 1: 10,000 and lower) in which they appear to have specific molecular toxic effects (see *oligodynamic action* and *ionic effects*).

Mercury compounds were first used in the fifteenth century specifically to treat syphilis. More recently zinc in a chelated form called *Zinc Omadine* has been shown to have potent microbicidal properties, as well as the sodium salt of *Omadine*.

## Bichloride of Mercury

Long used as a general disinfectant and as an antiseptic, bichloride of mercury (mercuric chloride) is now much less used than formerly. For general disinfection it may be prepared in dilutions of 1: 1000 to 1:5000. However, it is very toxic to tissues, very poisonous internally and corrosive to materials, especially metals (hence its older name of *corrosive sublimate*). In many situations it fails because it forms a thick coating of coagulum on the outer surface particles of mucus, tissues and feces, and thus protects the microorganisms within. As described previously, in dilutions of 1:10,000 or more its action is more bacteriostatic than coagulative. This action may be reversed by sulfur compounds unless coagulation has occurred.

## Organic Molecules

Attempts have been made, following Ehrlich's tead with arsenic, to decrease the toxic, corrosive and irritating qualities of mercuric disinfectants by incorporating mercury in complex organic molecules. This has yielded a number of products that are less toxic and irritating than bichloride of mercury, among them *Mercumchrome, Merthiolate* and *Metaphen.* In the presence of organic matter they are more effective than bichloride of mercury. Mercurochrome is often used for local, superficial application for disinfecting cuts, wounds and skin. Some of these organic mercury compounds are used as preservatives in biological materials. Other useful organic mercurials are *phenylmercuric nitrate,* very effective and of sufficiently low toxicity for external use, and *ammoniated mercury,* long used as a 10 per cent ointment for external wounds and fungal infections of the skin. These mercurials all mainly act microbistatically, and their action, like that of bichloride of mercury, is reversed by thioglycollate and hydrosulfuric acid. All mercury compounds are poisonous internally to some degree, and mercury is absorbed through the skin.

## Silver Nitrate

Silver nitrate (1 per cent aqueous solution) is used principally for application to infants' eyes at birth to prevent infection. Silver is also used as a local disinfectant in the form of organic colloidal preparations such as Argyrol or Protargol, in which form it is nonirritating and may be used in adult eyes at 20 per cent concentration. It is not for use in infants' eyes in place of silver nitrate. The use of silver nitrate (1 per cent aqueous) or an equivalent, approved antibacterial in the eyes of infants at birth is required by law in most states as a safeguard against blindness of the newborn *(ophthalmia neonatorum).* However, antibiotic treatment may be satisfactory in some situations.

## Copper Sulfate

Copper sulfate is used chiefly to control the growth of algae in open water reservoirs and as a fungicide in garden sprays *(Bordeaux mixture).* Organic compounds of iron are also used as fungicides in garden sprays.

# PHENOL AND ITS DERIVATIVES

Phenol ($C_6H_5OH$) (carbolic acid) in the pure state is not commonly used as a disinfectant because it is expensive and because there are more than a dozen derivatives that are not only less costly but more effective. All act similarly to phenol (*i.e.,* mainly by coagulation, partly by little understood lytic and toxic effects). Unlike the halogens, which are indiscriminately effective against a wide range of microorganisms, susceptibility to phenolic disinfectants varies greatly. Some organisms use phenol as a food.

## Surface (Residual) Disinfection

Phenol and, more especially, many of its derivatives are surface tension reducers and are commonly used in solutions containing additional surfactants to improve wetting properties.

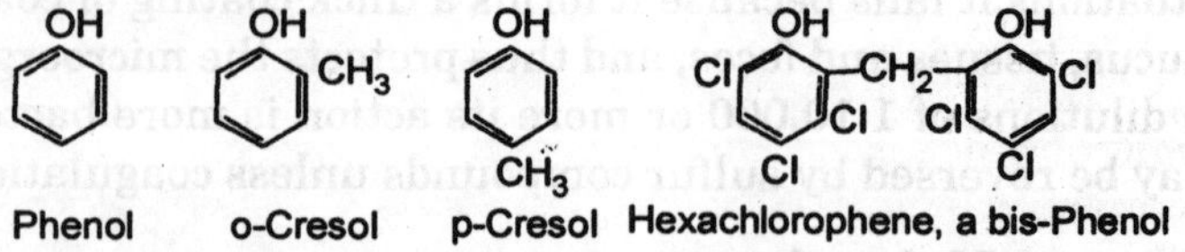

Since they are surfactants themselves, the phenolic disinfectants tend to be adsorbed in thin, and more or less durable, films on inert surfaces to which they are applied, resulting in what is termed *surface,* or *residual,* disinfection. As compared with the volatile halogens or alcohols their action is thus prolonged, often for many hours. This is an especial value of certain preparations of phenol derivatives for surgical and hospital hand washes.



**Fig. 2.9. Structures of phenol, representative cresols and one commonly used bisphenol (hexa-chlorophene).**

## Types of Phenolic Disinfectants

Phenolic disinfectants are of many types, common among which are: the *cresols,* the molecules of which are like phenol but have methyl groups attached; and the *bis-phenols,* the molecules of which consist of two phenolic groups joined directly or through some other radicle.

## Crude Cresols

Crude cresols form colloidal (milky) suspensions in water and are therefore especially effective as disinfectants, because each colloidal droplet consists of *concentrated* cresol. This is true of many colloidal disinfectants. There are many excellent cresol preparations on the market. Some have a clinging odor. Most of them contain a surface tension reducer in addition to the disinfectant. They are commonly used for disinfecting floors, furniture, barns and stables in from 1 to 5 per cent concentrations. Some may also be used in temporary contact with human tissues in 0.5 to 1.0 per cent strengths. Good examples are Lysol and saponated cresol solution, *U.S.P.* The preparation of such solutions is an important industry.

## Bis-Phenols

Like the cresols, bis-phenols have surfactant properties. In order to increase their solubility in water they are commonly used in solutions with other surfactants or detergents. In general they are less toxic than the cresols. Among well-known bis-phenols is one combined with chlorine (one of the class of *halogenated bis-phenols),* called Hexachlorophene (G 11). Because it is a surfactant it provides good residual surface disinfection of the skin, and it is not irritating if used properly. It is incorporated especially in surgical soaps and hand washes (Gamophen, pHisohex and Hex-O-San).

## Orthophenylphenol

A similar compound orthophenylphenol, is used in mixtures such as *O-syl. Chlorothymol* and *Chlorohexidine* are other halogenated bis-phenols which are similar in general properties. All are broadly germicidal, and each has distinctive properties that adapt it to particular situations. Some can be used effectively as irrigations in surgical wounds in 1:1000 to 1:10,000 concentrations.

The activity of all phenolics is markedly depressed in the presence of extraneous organic matter, and they are highly microbistatic.

## ALCOHOLS

There are many substances that chemically are alcohols: many of these are effective microbicides. Like various organic disinfectants, their microbicidal activity is directly related to increased length of their carbon chain and molecular weight; beyond their critical point (different for each type): microbicidal activity decreases. The two alcohols most commonly used for disinfection or sanitization are ethyl ("grain") alcohol and isopropyl ("rubbing") alcohol. They are mild disinfectants and are nontoxic on external application. Methyl or "wood" alcohol and its vapor are extremely poisonous and should never be used in any bodily contact whatever.

Ethyl alcohol is a surface tension reducer, a lipid solvent and a coagulating agent, properties that should contribute to microbicidal potency. However, it is also a potent dehydrating agent, a property that interferes greatly with its coagulating power and probably also with its other antimicrobial properties. Because of its dehydrating power it is best used as a 70 per cent aqueous solution. Stronger concentrations (95 or 100 per cent "absolute") are much less effective. As a coagulating agent it tends to combine nonspecifically with extraneous organic matter and can produce a thick, protective coating of coagulum around microorganisms. As a lipid solvent it is useful as a *cleansing* agent ("disinfectant"?) for skin prior to hypodermic injections. It adversely affects viruses and cells with envelope lipids. As a surface tension reducer it is a potent adjuvant of other disinfectants such as phenolics and quaternaries. Similar considerations apply to isopropyl alcohol.

## DETERGENT DISINFECTANTS

As just mentioned, not all detergents are disinfectants. However, several compounds combine the properties of detergent, surface tension reducer and disinfectant.

### Soaps

Among these versatile compounds are the common soaps; potassium or sodium esters of higher fatty acids (stearic or glyceric). They are actively lethal for certain species of microorganisms: *Treponema pallidum,* the syphilis spirochete; *Diplococcus pneumoniae,* a cause of pneumonia; some streptococci; and a few other organisms. Unfortunately, with these few exceptions, soaps are relatively ineffective as general disinfectants. However, soaps are potent surface tension reducers and consequently are good emulsifying and foaming or "sudsing" agents. Foaming results from the fact that, since the surface tension is greatly weakened, the surface can be extended almost indefinitely as foam or as droplets in an emulsion; compare with mercury. When soapy water is applied to skin or instruments, microorganisms are carried into the emulsified grease droplets and removed by rinsing. Detergency is the chief virtue of neutral soaps. Soluble soaps are precipitated by acids, ions of heavy metals, and by $Ca^{++}$. Hand soap is often used as a carrier for phenolic disinfectants, as previously discussed.

### Synthetic Detergents

As generally understood, this term comprises a group of potent surface tension reducers that are not typical soaps. They are synthesized commercially and much used for household and commercial cleansing and laundry purposes. They are available in "large, economy" packages in all grocery stores. Most of them are long-chain, alkyl-benzyl-sulfonates. Few are markedly microbicidal. They are *anionic* compounds (see following paragraphs).

Another group of synthetic compounds, classed as detergents because they are strong surface tension reducers, are much more important as disinfectants and sanitizing agents. The first was discovered in 1936 by Domagk (Nobel Prize winner), who described the properties of dodecyl-dimethyl-benzyl-ammonium chloride, now familiar as *Roccal* or *Zephiran*. This is a quaternary alkyl ammonium salt, and is representative of a number of similar compounds that have since been developed commercially as disinfectants or detergents. As a group they are commonly called *quaternaries* or "quats." Basically, they are ammonium halides (for example, ammonium chloride) in which the hydrogen atoms have been replaced by various alkyl groups.

$$\left[ R-\underset{R}{\overset{R}{N}}-R \right]^{+} Cl^{-} \text{ (}^{-}Br\text{, etc.)}$$

A. General formula of alkyl ammonium halides

$$\left[ CH_3-(CH_2)_{11}-\underset{H_2}{C}-\overset{CH_3}{\underset{CH_3}{N^+}}-CH_2-C_6H_5 \right] Cl^-$$

B. Alkyl-dimethyl-benzyl-ammonium chloride; a cationic, microhicidal detergent

$$\left[ CH_3-(CH_2)_{11}-O-\overset{O}{\underset{O}{S}}-O^- \right] {}^{+}Na$$

C. Sodium lauryl sulfate; an anionic mirobicidal detergent.

**Fig. 2.10. Forms of microbicidal detergents.**

## Chemical Structure and Activity

In general, the more effective of these detergent-disinfectant compounds are those in which at least one of the alkyl groups contains a chain of 12 to 16 carbon atoms, whereas the less effective are those of lower or higher molecular weight. This represents a common observation previously mentioned; *i.e.*, the germicidal efficacy of many organic compounds is greatly affected by chemical structure, especially length of carbon chains.

## Cationics, Anionics and Nonionics

There are three types of microbicidal detergents: (1) those in which the surfactant portion is the *cation* (*e.g.*, quaternary alkyl ammonium halides those in which the surfactant portion is the *anion* (*e.g.*, complex soaps like sodium lauryl sulfate; and (3) those that do not ionize. These compounds are classified as *cationic, anionic* and *nonionic* detergents, respectively. The cationic compounds appear to be the most generally effective as disinfectants and are what is generally meant by the term "quaternary." Soaps are anionic detergents. The nonionic compounds are relatively weak disinfectants. They are sometimes used with anionic compounds in household detergents as adjuvants.

## Mechanism of Action

Since detergents like the quaternaries and sodium lauryl sulfate are generally alkyl compounds of chlorine or sodium, they are bipolar, *i.e.*, they exhibit both fat-soluble (lipophilic) and water-soluble (hydrophilic) sides. When dissolved in aqueous culture media, the longer alkyl chains tend to be oriented toward, and to dissolve in, the cell lipids. The negatively charged residue (chlorine or sodium ion) is oriented away from the cell and is drawn toward the surrounding aqueous fluid. The surface tension reducer molecules are systematically arranged side by side at the cell-aqueous medium interface, as closely placed as possible.

**Since the reducer molecules at any surface have a lower mutual attraction than the water molecules in which they are dissolved, the surface tension of their solution is lowered. Lipids upon which detergents in aqueous solution become oriented tend to become divided into smaller and smaller particles, each coated with the surface tension reducer (*i.e.*, to be *emulsified);* detergents thus characteristically "cut" (remove) grease. The lipophilic property of quaternaries and alkyl sulfates is thought to disrupt the lipids, and therefore the whole structures, of bacterial cell walls and membranes, especially of those rich in lipids like many gram-negative bacteria. Disruption of the cell walls and membranes would permit outward passage of vital intracellular components and passage inward of the surface tension reducer. Electron micrographs show the destruction of bacterial cell walls and inner structures by antimicrobial substances which reduce surface tension. Probably these detergents also act in part by occluding the bacterial surfaces by being adsorbed there. In use, soaps and other anionic compounds should not be mixed with cationic compounds, because the oppositely charged molecular complexes neutralize each other and the agents are precipitated or inactivated. For the same reasons hard, acid and iron-rich waters tend to inactivate them.**

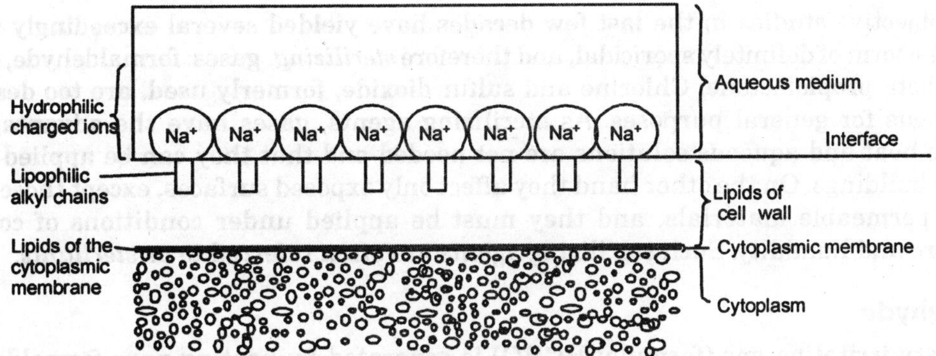


**Fig. 2.11. Orientation of polar groups of an *anionic,* surface-active detergent, in this case sodium lauryl sulfate [$Na^+(C_{12}H_{25}OSO_{,1})^-$], in relation to the surrounding aqueous medium and a cell having lipids in its cell wall or cytoplasmic membrane. The hydrophilic group (here represented by $Na^+$) orients at the fluid-cell interface; the hydrophobic group (here represented by the lauryl sulfate group⁻) is oriented in the lipids of the cell wall or membrane.**

**At the air-water interface the hydrophilic groups orient themselves within the surface water while the hydrophobic alkyl groups are at the air-water interface, weakening the intermolecular attraction between water molecules at the surface; *i.e., lowering the surface tension* of the water. Accumulation of molecules of a surface-active, polarized agent at any interface between solids and fluids or between immiscible fluids is called *adsorption.***

**A very similar relationship would be that between a cell and a *cationic* detergent of the class of quaternary ammonium compounds; *e.g.,* cetyl pyridinium chloride (Ceepryn):**

$$\left[\bigcirc\!\!N-C_{16}H_{33}\right]^+Cl^-.$$

**Unlike many other disinfectants, quaternaries retain much of their activity in the presence of considerable organic matter. The quaternaries tend to be microbistatic or slowly lethal rather than to kill instantly. When the effectiveness of quaternaries as disinfectants is being measured,**

their marked microbistatic action must be offset by the use of special inactivators to allow the microorganisms to grow in the subculture. *Tinctures* of quaternaries appear to be very effective indeed against tubercle bacilli, and some are effective against molds, but they are of limited value against certain viruses.

Since they are emulsifying agents for lipids quaternaries should be effective against viruses that have lipid-rich components, *e.g.*, envelopes, and there are data in support of this view. Opinions differ concerning other viruses. Pi-cornaviruses are resistant. Quaternaries are not sporicidal. They are much used for disinfection of skin and for purposes of sanitization in restaurants and hospitals.

## MICROBICIDAL GASES

Attempts to cure and prevent disease with vapors and smokes date from the earliest human records, when "aromatics," smudges, perfumes and the like were endowed by popular supposition with almost miraculous, and generally quite factitious, powers to drive away evil spirits and "noxious humors."

More objective studies in the last few decades have yielded several exceedingly valuable results in the form of definitely sporicidal, and therefore *sterilizing,* gases: formaldehyde, ethylene oxide and beta-propiolactone. Chlorine and sulfur dioxide, formerly used, are too destructive and poisonous for general purposes. As sterilizing agents, gases have the advantages that destructive heat and aqueous solutions are not needed and that they can be applied in large volumes in buildings. On the other hand they affect only exposed surfaces, except those made of porous or permeable materials, and they must be applied under conditions of controlled temperature and humidity. Such sterilizing agents are often referred to as *sterilants.*

### Formaldehyde

This very irritating gas (formula: HCHO) is generated by heating para-formaldehyde or concentrated solutions of formaldehyde. Unless very carefully applied at raised temperatures, formaldehyde tends to polymerize as paraformaldehyde or *paraldehyde* in a thin white film over all surfaces, and hence has definite limitations to places not containing valuable furniture or equipment. Its action is to some extent microbistatic and reversible with sulfite ions. In some solutions it is definitely sporicidal and therefore sterilizing but, like bichloride of mercury, it tends to form protective coatings of coagulum around microorganisms if extraneous organic matter is present. For best results a relative humidity of about 70 per cent and a temperature of about 22° C, are requisite. Formaldehyde vapor has low penetrability into such materials as blankets. It is much less used now than formerly.

### Ethylene Oxide

This substance (formula: $CH_2 \cdot O \cdot CH_2$) is liquid at temperatures below 10.8° C (51.4° F), its boiling point. It is relatively inexpensive and is handled in metal tubes or bottles like other compressed gases. In the pure state it is very toxic, irritating and explosive. For use it is mixed with other substances: ethylene oxide, 10 per cent and carbon dioxide, 90 per cent (sold as Carboxide); ethylene oxide, 20 per cent and carbon dioxide, 80 per cent (sold as Oxyfume); ethylene oxide, 11 per cent and halogenated petroleum, 89 per cent (sold as Cryoxcide). Each preparation has special advantages in certain situations.

The desired mixture is generally applied in autoclaves or other closed containers where humidity, temperature and pressure may be controlled. The concentration of the gas is generally expressed in milligrams of pure ethylene oxide per liter of space. A typical use situation is 500 mg. of gas per liter at about 58° C (130° F) and a relative humidity of 40 per cent for four hours. If one factor is varied the others must be varied accordingly. For example, if the concentration is doubled, the time may be halved. At a relative humidity of 30 per cent the action of the gas is about 10 times as fast as at 95 per cent relative humidity.

In using ethylene oxide in an autoclave, a typical procedure involves seven steps, all of which can be carried out by fully automatic equipment now available for the purpose:

(1) Withdraw almost all air with a vacuum pump.

(2) Admit a controlled volume of steam or water vapor.

(3) Admit ethylene oxide in required concentration.

(4) Adjust the temperature as indicated and hold constant.

(5) After the required time turn off heat.

(6) Take out all gas with vacuum pump.

(7) Admit sterilized air to normal pressure.

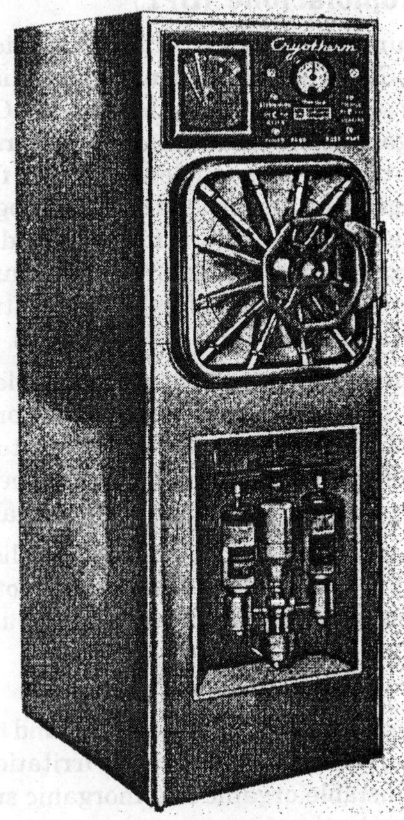


**Fig. 2.12. Autoclave arranged for sterilization by sporicidal ethylene oxide gas (Cryoxcide). Below the steel door are seen the containers of gas and a water inlet so arranged that the vapors can be mixed and humidified (much as gasoline, air and water vapor are mixed in a modern automobile carburetor) before admission of the sterilizing vapor mixture to the chamber of the autoclave. Above the door are the dials controlling sterilizing temperature and time, vacuum pump, gas and humidity inlet and a maximum and minimum recording thermometer. Operation is almost completely automatic.**

Ethylene oxide is a very effective general microbicide and kills bacterial endospores. It is therefore a sterilizing agent. It acts by alkylation or introduction of an organic group in place of a hydrogen atom, in various compounds in protoplasm. It does not injure even delicate organic materials that are injured by heat and aqueous solutions. It penetrates well, does not condense on objects and soon dissipates from all materials.

It can also be used in the liquid state in 1 per cent concentration to sterilize unstable organic fluids, such as bacteriological culture media, by mixing fluid ethylene oxide with them at 10° C. After a few hours the ethylene oxide is driven off by a short incubation at 37° C.

## Beta-Propiolactone (BPL)

In pure form at 20° C this is a liquid with a sweetish but very irritating odour [formula: $(CH_2)_2 \cdot CO$]. Unlike ethylene oxide, it is not explosive. It may be stored at 4° C, but is more stable in sealed glass containers at –25° C. Its aqueous solutions are very unstable. It kills most microorganisms, including bacterial spores, and is therefore a sterilizing agent. It probably acts by alkylation, as does ethylene oxide. In fluid form it has been used to sterilize vaccines, tissues used for grafting and other delicate biological materials that could not be sterilized in any other way. Like ethylene oxide, it must be used in closed chambers where temperature and humidity can be accurately controlled. A relative humidity of 70 to 80 per cent and a temperature of about 25° C are optimal, with a final concentration of beta-propiolactone of about 2 to 4 mg. per liter of air and a sterilizing time of 2 to 3 hours.

When properly applied, beta-propiolactone is much more active than either formaldehyde (HCHO) or ethylene oxide. It is noncorrosive, and does not condense on surfaces as does formaldehyde. However, beta-propiolactone lacks the penetrating power of ethylene oxide and is more like formaldehyde in this respect. It also requires higher relative humidities. It is generally recommended to replace formaldehyde for use in buildings.

Since they are so effective, gaseous disinfectants such as ethylene oxide, formaldehyde, and beta-propiolactone have tremendous potential value in combating massive and widespread dissemination of infectious agents such as might be employed in biological warfare.

## Toxic and Destructive Properties

Experience shows that prolonged and repeated exposure to any chemically active disinfectant, or any sterilant, may cause tissue irritation or destruction, and may damage articles containing certain unstable organic and inorganic substances such as magnesium (instrument housings) and some plastics. Users must always be alert to this possibility.

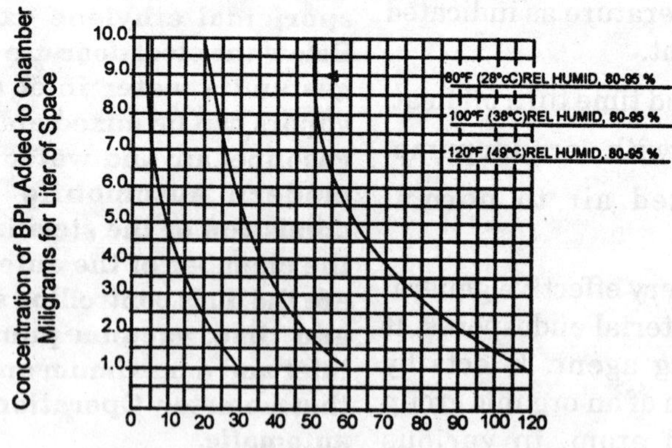


**Fig. 2.13. Relation between concentration of beta-propiolactone and time required to kill spores of *Bacillus globigii* on filter paper strips. Curves are shown for three different temperatures at a relative humidity of 80 to 95 per cent.**

They must also be alert to toxic properties, especially dangerous potentialities of newly developed disinfectants. For example, with continued study of beta-propiolactone, data have appeared suggesting that it may be carcinogenic as well as irritating. The toxic or carcinogenic properties of a disinfectant do not necessarily militate against its use; otherwise, such substances as chlorine, formaldehyde, bichloride of mercury and phenol, and radiations such as ultraviolet, x-rays and gamma rays would be banned. However, proper precautions must be taken in handling such agents to prevent ingestion or undue contact.

## RADIATIONS

### Ultraviolet Light

As described, ultraviolet light is actively microbicidal when applied properly. It has very slight powers of penetration. Exposure must be direct, intense and sufficiently prolonged. Eyes must be protected, as they can be severely damaged. It is sometimes used in hospitals, especially in operating rooms, reportedly with good results, to kill microorganisms on instrument tables, in the air and on floors and walls. Ultraviolet light is used on a large scale to prevent growth of molds in meatpacking houses, bakeries and other commercial and industrial establishments.

Another application is in the treatment of biological fluids, such as blood plasma and vaccines, to kill contaminating viruses. The rays are directed on fluid that is made to flow in a micro-thin film that ultraviolet rays can penetrate. The wave length commonly used in disinfectant applications of ultraviolet light is 2537 A, because it is most practical. Shorter waves, such as 1849 A, are much more lethal but are difficult to handle and apply. Ultraviolet kills spores as readily as vegetative cells.

### Cathode Rays

A recent development in radiation sterilization, cathode rays (electron beams) are expensive to apply, but are being developed on a commercial scale. They can be made to penetrate thin metal, paper or plastic sheets and have been used successfully to sterilize sealed packages for surgical use, packaged meats, vegetables and other foods.

## INHIBITION OF MICROORGANISMS (MICROBISTASIS)

As previously indicated, microbistasis denotes a condition of microorganisms in which, although they are alive (or *viable),* they do not multiply. Their enzymic or other vital functions are in a static condition.

In one sense the dormant ascospores of yeasts, conidia or molds, cysts of some protozoa and the endospores of bacteria represent microbistasis as a part of their normal life cycles. However, microbistasis is usually thought of as affecting vegetative cells and as being produced artificially. Microbistasis is exceedingly important in all aspects of microbiology. There is even published discussion of preserving important(?) people in "deep-freeze" ("Homostasis"?).

### Physical Methods

As mentioned elsewhere, there are several physical methods of inducing microbistasis. Most of them are fundamentally methods of depriving the cell of liquid water: desiccation; immobilizing intracellular water by changing it into ice; drawing most of the water out of the cell by immersion in a fluid of high osmotic pressure; and combinations of these.

In addition, mild irradiations with ultraviolet light and x-rays are often microbistatic instead of lethal. Their microbistatic action is not clearly understood but it is not related to dehydration. Apparently, oxidations are among the most important reactions that occur.

### Reversibility

The action of microbistatic agents is *characteristically reversible;* that is, the organisms can be reactivated if the inhibiting action has not been so prolonged that the cells have gradually died off. We have seen in foregoing chapters how the microbistatic effect of irradiations may be reversed by visible light and certain chemicals, and how bichloride of mercury poisoning may be "cured" by glutathione. Tubercle bacilli dried in a vacuum were found fully viable and virulent after 17 years; diphtheria bacilli and scarlet fever streptococci, after 25 years. Microbial spores (natural microbistasis) will survive for decades. Viruses are commonly held unaltered for months or years frozen in carbon dioxide ice at –76° C. Cattle breeders and others preserve sperm by freezing. In all cases the totally inactive cells can be reactivated.

## MICROBISTASIS AND DISINFECTION

The distinction between chemical microbistasis and disinfection is interesting but entirely arbitrary; a matter of semantics. The differentiation depends on what one means by killing "quickly" (*i.e.,* disinfection) and what one may mean by killing "slowly" (*i.e.,* microbistasis). Even under the most effective conditions of microbistasis, some of the weaker cells die fairly soon; presumably the others are not actually immortal and die eventually, even though only after decades or centuries. As applied to bacteria, one could say that death within a time less than the maximal for one generation (*i.e.,* one cell-fission) under optimal growth conditions might be regarded as more than mere prevention of multiplication and, therefore, true killing or disinfectant action. If death were delayed beyond one generation-time, one might say that multiplication had been delayed or that the organisms had been in a static condition for at least a short time. This last situation could represent true bacteriostasis. There are many obvious difficulties in making such a distinction. The term bacteriostasis (or microbistasis) is, however, widely used and usually means "long" delayed death (*i.e.,* the cells die only after some hours, days or decades), rather than in, say, 10 to 30 minutes.

## CHEMICAL MICROBISTASIS

We may note three general types of chemical microbistatic agents: certain aniline dyes; sulfonamide and similarly acting drugs; and antibiotics.

### Aniline Dyes

We have already referred to selectively bacteriostatic culture media containing dyes. A representative dye used in such media is crystal violet, commonly used in the Gram stain. Since it is a basic dye, it has a marked affinity for acidic constituents of the bacterial cell, notably nucleic acids. With these the dye forms unstable salts that (in the absence of iodine) soon dissociate, and are either disposed of by the cell or can be removed by washing the cell with a harmless solvent such as water. Cells stained in this way with crystal violet, especially gram-positive cells, are in a static condition. They will not grow in media containing such dyes. They can be quickly reactivated by washing.

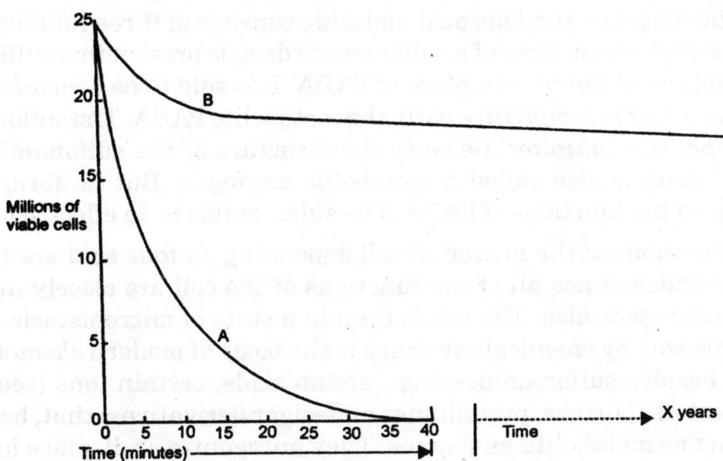


**Fig. 2.14. Comparison of disinfectant action with microbistasis. The microorganism concerned has a generation time of 40 minutes at 37° C in broth at pH 7.2. Curve *A* shows complete sterilization within one generation time resulting from action of an applied microbicidal agent such as heat or disinfectant. Curve *B* shows initial rapid diminution of numbers caused merely by the action of the microbistatic agent such as freeze-drying. Bacteriostatic action is characterized by very slow decline in numbers.**

On the other hand, gram-negative cells in contact with crystal violet are not stained and generally grow well on media containing such dyes as crystal violet, basic fuchsin or methylene blue-eosin. Such dye-containing media are therefore commonly used for *selective* bacteriostasis; especially to isolate gram-negative bacteria from materials contaminated with gram-positive bacteria.

## Sulfonamide Drugs

The sulfonamides are excellent examples of bacteriostatic agents. They were discovered by Domagk (Nobel Prize winner) about 1936. Sulfonamide drugs are derived from the same basic material (coal tar) as aniline dyes. All are derivatives of sulfanilamide like basic dyes they are effective mainly (with a few notable exceptions: *Nesseria, Shigella, Proteus, Klebsiella* and *Escherichia)* against gram-positive bacteria.

## Sulfonamide Drugs and Metabolite Antagonism

The true sulfonamide drugs are 'those that readily release the sulfanilamide residue in the living cell. Sulfanilamide so closely resembles paraaminobenzoic acid (PABA) that it can displace PABA in the absolutely essential vitamin, folic acid. This important vitamin is necessary to a co-enzyme (CoA) involved in synthesis of amino acids and hence of proteins. Such absolutely essential specific substances, necessary to the completion of a particular functional part of the cell, but not used as sources of energy or bulk, are sometimes called *essential metabolites* or *micronutrients.* In the present discussion PABA is considered to be an essential metabolite. When we compare the diagrams of the molecules of sulfanilamide and PABA, the resemblance between the major parts of the two is made clear.

As shown in the diagram, the folic acid molecule consists of three functional groups: pterin, PABA and glutamic acid. If any form of a sulfonamide drug is present in a sufficient amount, part of it goes into the folic acid complex in place of PABA. It is said to be a *metabolite antagonist.* In this case it *antagonizes,* or is *competitive* with, the metabolite PABA. The sulfonamide is therefore often called a *competitive inhibitor.* Because the structure of the sulfonamide is analogous to that of PABA, the drug is also called a *metabolite analogue.* But no form of sulfonamide is capable of carrying on the functions of PABA. The sulfonamide is, in effect, a molecular imposter.

The enzymic functions of the microbial cell depending on folic acid are therefore promptly stopped by sulfonamides. Since all of the functions of the cell are closely interdependent, the whole cell mechanism stops also. The cell is then in a state of microbistasis. The production of microbistasis in this way by chemicals or drugs is the basis of modern *chemotherapy*. There are many substances besides sulfonamides (*e.g.,* amino acids, certain ions [see ion antagonism], organic acids and esters, purines, pyrimidines and sugar derivatives) that, because of similarity of structure, can act as metabolite analogues. They antagonize or displace ions or molecules of vital cell structures such as proteins, enzymes and coenzymes or nucleic acids. In the nucleic acids nucleotide analogues can produce mutations.

### Reversal of Metabolite Antagonism

It is important to note that, like most other microbistatic reactions, the enzyme blockage produced by sulfonamide is not fatal and may be reversed in several ways—for example, by adding enough PABA to displace and exclude the sulfonamide; by supplying new folic acid ready-made in the medium; by supplying, in the culture medium, the product that would have been formed by the blocked enzyme (or any subsequent enzyme in the chain); or by supplying the finished product of the entire enzymic series, in this case certain amino acids.

## DRUG RESISTANCE

The cell itself may be capable of bypassing or overcoming the effects of metabolite antagonists. For example, let us suppose that in a parallel series of enzyme reactions enzyme D is blocked by some microbistatic drug. However, let us assume that some of the substance usually made by the blocked enzyme D is produced as a side reaction or intermediate stage in the function of another enzyme, H. This product becomes available to E and thus the blockage at D is bypassed. Such a cell is more or less *drug-resistant,* depending on how much of the product needed by E is available from H.

Such enzymic rearrangements and bypassings *(alternate metabolic pathways)* are quite common. From the clinical standpoint, sulfonamide resistance of an infecting organism means that the patient must receive some other drug.

### Resistant Mutants

It has already been shown that among any large number of microbial cells in a given pure culture, mutant forms may be found. Among the large numbers of bacterial cells of a given species infecting a patient there are, therefore, likely to be mutants with enzymic arrangements that will be unaffected by a microbistatic drug such as sulfonamide. In the presence of the drug these mutants will thrive, while all the other microbial cells are killed off. We thus have bred up a sulfonamide-resistant strain of the infecting organism which is dangerous to the patient and to those who contract the resistant organism from him.

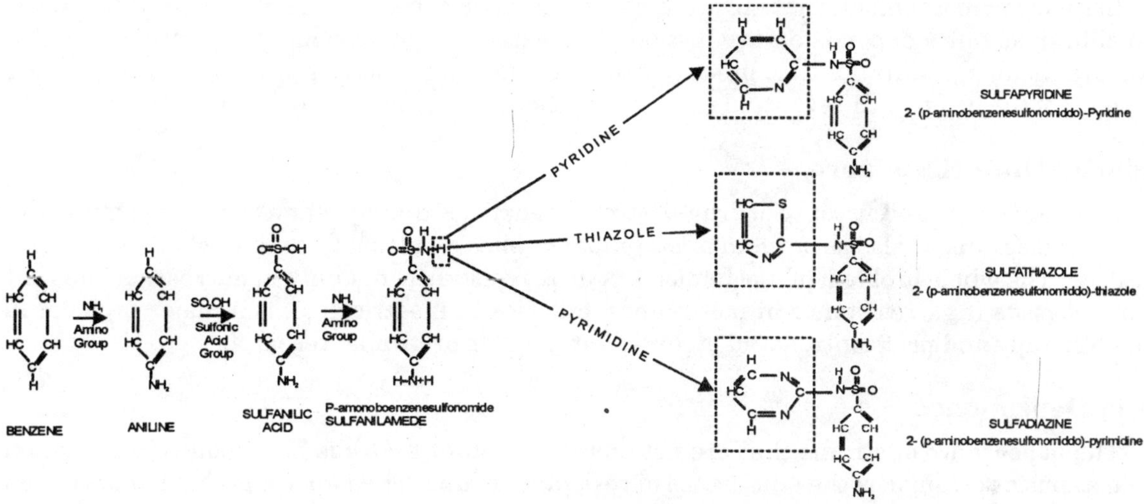


**Fig. 2.15. How the sulfonamide drugs are derived from benzene. First, an amino group is added, yielding aniline (a base from which dyes are synthesized). The addition of a sulfonic acid group to aniline yields sulfanilic acid. The addition of another amino group yields sulfanilamide. The addition to this, in place of one hydrogen atom, of various other chemical groups yields other sulfanilamide drugs, three of which are shown in the diagram.**

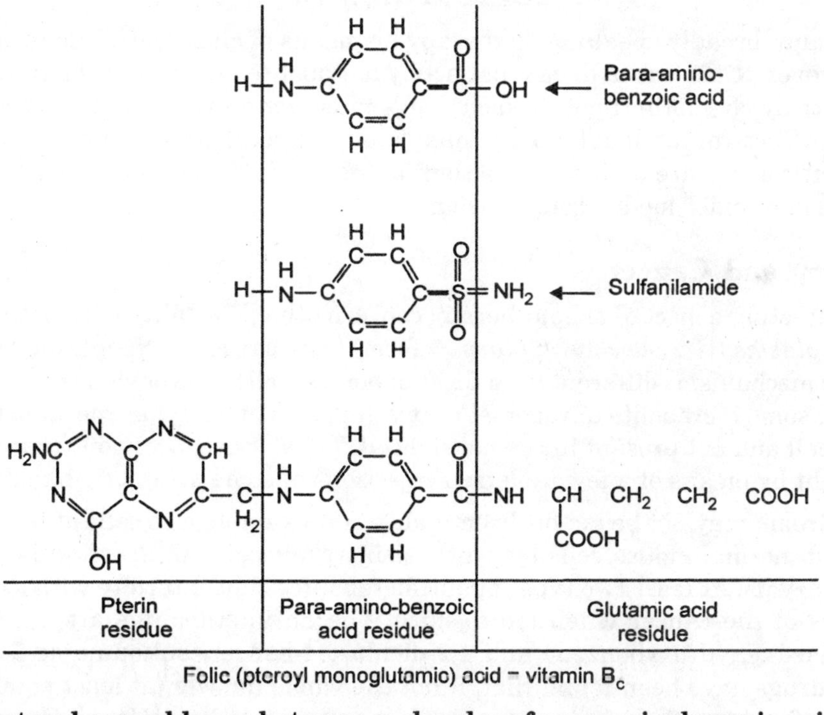

Para-amino-benzoic acid

Sulfanilamide

| Pterin residue | Para-amino-benzoic acid residue | Glutamic acid residue |
|---|---|---|

Folic (pteroyl monoglutamic) acid – vitamin $B_c$

**Fig. 2.16. Structural resemblance between molecules of paraaminobenzoic acid (PABA) and sulfanilamide (above), and the antagonism of PABA by sulfanilamide in the folic acid molecule (below).**

In some microorganisms that are ordinarily susceptible to penicillin, notably *Staphylococcus,* penicillin-resistance depends on production of an extracellular enzyme *(ptenicillinase)* which destroys the drug. Mutants that produce such enzymes are able to grow unhampered in the presence of the drug, to the eventual exclusion of the nonenzyme-producing sensitive mutants.

### Induced Drug Resistance

In addition to the occurrence of drug-destroying enzymes that appear as a result of mutation, such enzymes appear also as *induced* or *adaptive enzymes* resulting from contact of the cell with the drug. The whole problem of mechanisms of drug resistance, not only in microorganisms but in insect pests (*e.g.,* resistance of mosquitoes and flies to insecticides), is a major problem in chemotherapy and pest control, and an excellent field for profitable research.

### Drug Dependence

The appearance of strains that are not merely resistant to drugs but wholly *dependent* on them seems as strange as the emergence of resistant strains. The two are probably not related except that they are two manifestations of the phenomenon of mutation. Drug-dependent mutants obviously could not survive and develop significantly except in an environment containing the drug on which they are dependent. It is only because we have tried to cultivate microorganisms in contact with antimicrobial drugs that these curious mutant forms have been revealed.

## CHEMOTHERAPY

Chemotherapy, broadly speaking, is therapy by means of chemicals. The term is generally restricted, however, to the use of drugs that act by metabolite antagonism or enzyme blockage. In order to be truly chemotherapeutic such drugs must act on the parasite in concentrations that have no effect (or at least no serious "side" effects) on the host. Some valuable chemotherapeutic drugs are toxic to the patient as well as to the parasite and so must be used with great care and under medical supervision.

### Chemotherapy and Cancer

A very interesting aspect of chemotherapy concerns the possibility of selectively poisoning the cells of *neoplasms* (Gr. *neo* = new; *plasm* = form) (*e.g.,* cancers). Neoplasms are thought to have metabolic mechanisms different from those of normal cells. The obvious possibility presents itself of finding some metabolite antagonist that will poison an enzymic mechanism peculiar to the neoplasm cell and not present in the normal cell. Thus, it is conceivable that we could cure cancer overnight by means of a few tablets or injections of some antibiotic-like drug.

This rosy dream may not be too far from reality. For example, the cells of certain cancers of mice differ from normal mouse cells by synthesizing cyanocobalamin (vitamin $B_{12}$), a cobalt-containing coenzyme. At least two types of antimetabolites that interfere with synthesis of $B_{12}$ produced cures of the cancer when fed to mice. The antimetabolites are 1,2-dimethyl-4(P-carboxyphenylazo)5-hydroxybenzene and 1,2-dichloro-4-benzenesulfonamide-5-nitrobenzene. Several other drugs have been found that, if not the whole answer, at least point definitely to the possibilities for chemotherapy in research on cancer metabolism. Honor, Fame and Fortune are to be won in this field!

CHAPTER

# 3 Biochemistry of Parasites

The youngest basic-science diseipline applied to the study of parasite and parasitic diseases is molecular biology. The ability to clone parasite genes and express them in *Escherichia coli* or yeast has provided the first glimmer of hope for the development of vaccines against some of the parasites. Before the availability of recombinant DNA technology, there was no conceivable way in which sufficient antigenic material could be produced from any of the medically important parasites for use in vaccines. In only a few years the very rapid application of the new methods led to cloning of plasmodial genes, large-scale antigen production, and phase I and phase II clinical trials of an antisporozoite vaccine. Although recombinant vaccines against other plasmodial stages and other parasites have not yet progressed this far, their development is underway. From experience with the sporozoite vaccine, it is clear that more than cloning and expressing genes of interest may be required to obtain a successful vaccine. Yet the availability of large amounts of antigenic material and the primary sequence data that can be derived by application of molecular techniques have revolutionized the field.

Perhaps the only alternatives to this approach in vaccine development, given the paucity of antigenic material available directly from the parasite, are antiidiotype vaccines (such vaccines employ antibodies [antiidiotypic antibodies] that mimic the structural conformation of an epitope.) An additional practical application of molecular technology has been in the area of rapid diagnosis. With specific cDN A probes, it is now possible not only to determine the presence of leishmania in skin biopsies, but also to speciate them. The effectiveness of this diagnostic method is facilitated by the recognition that the extrachromosomal DNA of the kinetoplast contains so-called minicircles, which are highly divergent within the genus, allowing for species discrimination by kinetoplast DNA (kDNA) hybridization techniques. This achievement is an important practical development because conventional diagnostic methods, which require parasite identification by microscopy or isolation of parasites in culture of biopsy material, are cumbersome and often unreliable. Plasmodia also can be detected in blood in this manner.

Although the current level of sensitivity of the molecular method of diagnosis is no greater than that achievable by microscopic examination of blood smears, modifications (that might include amplification by the polymerase chain reaction method) may increase sensitivity. The application of molecular techniques to malaria diagnosis presumably will have greater application in epidemiological surveillance efforts than in clinical practice in areas where malaria is endemic.

Application of molecular biology approaches to the pharmacology of certain parasitic diseases has spurred great excitement and optimism for the idea of developing new antimicrobials by rational design rather than by the traditional and inefficient approach of drug screening. The strategy of this approach is to identify critical enzymes of the parasite that are distinctive from that of the host. Isolating the gene encoding the enzyme, expressing it in *E. coli*, and gleaning the fine structure of the protein by X-ray crystalography—especially its substrate-binding domain—provide the data base.

With this information, drugs might then be designed that can bind avidly to this domain and competitively block access by the substrate. Insights into the genetic basis of drug resistance also have been revealing. For example, resistance to folate antimetabolites in *Leishmania* species has been found to result from extraordinary amplification of the gene that encodes the enzymes that are the targets of such antimicrobial agents. Molecular pharmacology, as this new discipline is called, could hold great promise for winning the race of medicine against parasites such as *Plasmodia,* which can develop multidrug resistance at a rate that exceeds our ability, by conventional means, to find effective alternative agents. One intriguing strategy is to prepare derivatized oligonucleotides that aneal with specific parasite mRNA, prevent translation, and thereby incapacitate parasite metabolism. Although practical application of the new techniques has understandably received the greatest attention, emphasis on basic aspects of the molecular biology of certain parasites has uncovered some unexpected findings of broad biological and evolutionary significance.

For example, comparisons of primary ribosomal RNA sequences suggest that the plasmodia causing human malaria are more closely related, in evolutionary terms, to rodent and avian parasites than to parasites causing malaria in other primates. Similar ribosomal RNA analysis in Kinetoplastida (such as African trypanosomes and *Leishmania)* suggests that these eukaryotes may be among the most distantly divergent, in evolutionary terms, perhaps having branched even before plants. Another striking finding has been the unique manner in which these organisms transcribe genetic information: by a noncontiguous construction of mRNA. In contrast to the conventional cis-splicing of most eukaryotes, these protozoa can carry out trans-splicing, a process whereby mRNA transcribed from genome on remote sites on a chromosome—or even on different chromosomes—forms a hybrid mRNA that then is translated. Many interesting biological questions about specific parasites have yet to be subjected to molecular scrutiny. One of the most obvious is gene regulation in the transformation of the parasite from one stage to another. Efforts in this direction have been initiated.

For example, expression of the gene for α- and β-tubulin has been found to be expressed as leishmania transform from the intracellular (aflagellate) amastigote stage to the extracellular (flagellate) promastigote stage. Tubulin is the major protein in promastigotes, perhaps not surprisingly so in view of the presence of an extensive subpellicular microtubular system that includes the flagellum. What triggers gene tubulin gene expression now awaits definition. Along similar lines, an important challenge to the biology of malaria is to determine the molecular basis of gametocytogenesis (the emergence from plasmodial merozoites of gametocytes, the sexual stages) Since gametocytes are responsible for initiating transmission in the mosquito, knowledge that could lead to strategies for preventing their genesis in the—human host could have distinct practical value. A molecular understanding of the manner whereby specific environmental factors trigger interstage transformation in schistosomes is but one example of a worthy goal in molecular helminthology. Indeed one particularly exciting aspect of parasite molecular biology is that contemplation of parasite cell biology can engender numerous interesting questions that are now, or soon will be, amenable to molecular biological investigation.

Furthermore, the parasitic relationship may provide a fertile ground for gaining insights into fundamental evolutionary aspects of gene structure and regulation. As the molecular biology of parasites matures, it is likely that many important discoveries will be made as a result of asking the right biological questions—with the assistance of some serendipity. Inasmuch as none of the disciplines in modern parasite biology stands alone, new knowledge in the cellular biology of parasites and the immunology of parasitic diseases will no doubt influence perspectives and research directions in molecular biology studies. By the same token, these other areas can be expected to be influenced by progress in basic molecular parasitology.

## BIOCHEMISTRY OF PLASMODIUM

Protozoan parasites of the genus *Plasmodium* are the causative agent of malaria, a disease that today affects more than 200 million people in the tropics. Of the four human malarias, those due to *Plasmodium falciparum* and *Plasmodium vivax* are the most common. This chapter will focus on the role of molecular biology in the efforts to understand and eradicate malaria and discuss the molecular biology relating to diagnosis, taxonomy, genetics, gene analysis, and vaccine production.

### Evolution

Members of the genus *Plasmodium* have historically been grouped according to the hosts that they infect, then subdivided according to morphological and biological characteristics. Thus, for example, the species are classified as primate, rodent, avian, and lizard malarias. This classification scheme implies that the parasites have evolved with their hosts and that there is a greater relatedness among parasites in related hosts than among those greatly separated in evolution. A major exception to this scheme is the human malarial organism P. *falciparum;* this parasite species appears to be more closely related to parasites that cause avian and rodent malarias than to those that cause primate malaria. This exception was suggested on the basis of the similarities in the structure and composition of the genomes of *Plasmodia* causing avian and rodent malaria with those of *P. falciparum.* More recently these relationships have been confirmed by comparing the primary sequence of ribosomal RNAs from the different species of malarial parasite. Ribosomal RNA sequence comparisons constitute a generally accepted measure of evolutionary relationship and thereby add weight to these conclusions.

### Diagnosis of Malaria

In the standard, time-consuming diagnosis of malaria, a skilled technician does a microscopic examination of blood smears. In many areas diagnosis is omitted altogether; people with malaria-like symptoms are simply treated with antimalarial drugs empirically. The detection of malarial infections in patients by the development and use of a DNA hybridization assay could be a valuable tool in the control of the disease. The success of these assays depends on finding DNA probes that hybridize to the DNA of a particular malarial parasite in a sample of blood and not to the DNA of the host or of other microbes that might be in the same sample. In comparison to molecular diagnosis, microscopy by a skilled technician can detect a concentration of about 20 parasites/1 microliter (µl) of blood. In addition, he or she can determine by microscopy the species of the parasite and degree of anemia. Refining of a molecular diagnosis for malaria to obtain this information will be difficult but clearly possible.

The degree of sensitivity that has already been reported for molecular probes approaches 50 parasites per microliter, and increased sensitivity in assays of this type can be expected in the future. Species-specific probes certainly will be available soon; it is clear that potential diagnostic probes include ones that detect the highly repetitive sequences of *P. folciparum* not found in the other species. It is even possible that similar, although less sensitive, molecular assays may be developed for detecting such other characteristics of a parasite as drug resistance. Perhaps the strength of molecular diagnosis is not in performing better than a skilled microscopist but in allowing a large number of assays to be done with great consistency. A perfected molecular form of diagnosis should allow relatively untrained personnel to do a large number of assays every day with great accuracy.

## Plasmodium Genes

This chapter focuses on the three *Plasmodium* genes that have received the greatest attention and discusses reasons for interest in a particular gene as well as the technology involved in its study. An overlying theme in this discussion is how gene structure relates to rapid changes in immunodominant portions of a protein.

## The Circumsporozoite Protein

Interest in the circumsporozoite protein (CSP) of malarial parasites is related mainly to the need for a malarial vaccine. The important sporozoite antigen, being studied for its ability to elicit a protective response, is a single protein, the circumsporozoite (CS) protein. It covers the entire surface of the sporozoite. The reaction of a fluorescein-tagged anti-CS protein antibody with the sporozoite surface. The dominant epitope recognized by antibodies to the CS protein and the one involved in the protective response is included in a single immunodominant region of the protein.

The fact that a single antigen could elicit a protective response suggested that a molecular biological approach could provide a source of material for a vaccine. The important gene would be cloned and the product produced by microbes in the laboratory.

## Cloning of the Circumsporozoite Gene

Cloning of the CS protein gene presented a particular problem to molecular biologists because of the lack of availability of the sporozoite stage of the parasite. The first CS protein gene was cloned from the simian malaria *P. knowlesi*. To do this, a large number of sporozoites were collected from the salivary gland of infected mosquitoes. *P. knowlesi* was chosen because in the laboratory it produces approximately 10 times more sporozoites in an infected mosquito than any of the human malarial species. From these sporozoites, mRNA was isolated and used to make a cDNA library in *E. coli* that contained copies of sporozoite-specific messages. The cDNA containing a fragment of the CS protein gene was identified with a monoclonal antibody to the CS protein. This gene segment was then used to identify the entire CS protein gene. Unfortunately, the P. *knowlesi* gene was of little use in identifying CS genes of other species of malaria, because there were no apparent sequence similarities among genes carried by different species of parasites.

**Fig. 3.1. Immunofluorescent microscopy of a sporozoite with the use of anticircumsporozoite antibody. The distribution of the fluorescinated antibody reveals that the circumsporozoite antigen is distributed over the surface of the parasite. Monoclonal antibodies to the circumsporozoite protein identified this surface component as the immunodominant epitope and facilitated cloning of the gene that encodes the protein.**

It therefore appeared that the most direct approach to isolating genes from different species was to start from the beginning and collect sporozoites from each of the parasites of interest. Asexual parasites were grown in culture and their DNA was used to prepare an expression library containing intact genes, including those from all stages of the parasite's life cycle. This work was made possible by an earlier discovery indicating that the mung bean nuclease cuts *Plasmodium* DNA in areas flanking genes, thereby reducing the genome to fragments containing intact genes. Bacterial clones containing the CS protein genes were detected by anti-CS protein antibody and standard screening techniques. The clones detected with this approach contained the complete CS protein gene of *P. falciparum.*

## Sequence Evolution of the Circumsporozoite Protein

The cloning and DNA sequencing of the *P. falciparum* and *P. vivax* CS protein gene allowed comparison of these sequences with that of the *P. knowmlesi* gene. Although there were only two small regions of DNA sequence homology among the three, there did appear to be a common structural relationship. Evolutionary conservation of structure at some level was expected among analogous proteins that are presumably responsible for the same functions. For example, concentrations of hydrophobic amino acids at the amino and carboxy terminus of the protein probably serve as signal and anchor regions for the CS proteins. Regions of the three CS proteins that contain a concentration of charged amino acids can also be aligned and probably result from structural restraints on the molecule. The two regions of sequence homology between the *P. falciparum* gene and the other two CS genes are small and flank the repeated or immunodominant portion of the molecule. It seems possible that the strong conservation of sequence in these two areas reflected their involvement in some essential parasite function such as reception for liver invasion.

The sequence homology between the *P. vivax* and *P. knowlesi* genes is much more extensive and extends over the molecule, except for the central repeating area. In general, at least small regions of conserved sequence in each of the CSP genes flank a species-specific repeating epitope. In schematic of the CS gene shows the position of the highly conserved regions (regions I and II) surrounding the species-specific repeats. This repeated area in the center of the molecule is the immunodominant area of the protein and the subject of the most intensive investigation regarding vaccines. Some typical DNA sequencing data from the analysis of a CS protein gene are shown in Fig. 3.3. The data shown in the figure have been read from the arrow on the left in are ascending order to the arrow on the right. The translation of raw data to the gene sequence and then to the protein sequence of the CS repeating epitope. Notice that although the primary DNA sequence varies somewhat from repeat to repeat, only one nucleotide change in the five segments results in an amino acid change.

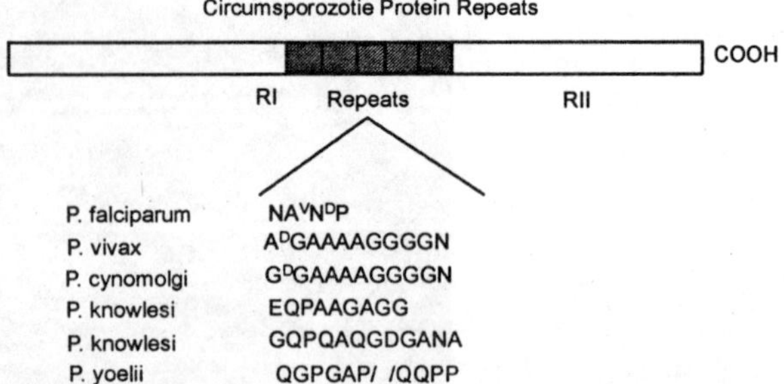

**Fig. 3.2. The immunodominant repetitive epitopes in the circumsporozoite protein of several different species of *Plasmodia*. In contrast to the species specificity of the repeating units, the regions (RI and RII) that flank them are conserved in the genus.**

The other changes are in silent third positions, which do not result in the amino acid change. The extent and rate of variability in the repeat region are important questions in vaccine research. Changes in the immunodominant target would nullify the effectiveness of a vaccine. Perhaps much can be learned from the work done on *P. falciparum* S-antigen, discussed later, which has a repeating immunodominant region similar to that found in the CS protein. Clearly the repeat area of S-antigen varies so rapidly that there are antigenically distinct populations of *P. falciparum.* The CS proteins do not appear to vary as rapidly as the S-antigens, but there are indications of a similar type of drift. The analysis of CS-protein genes from malarial parasites that infect monkeys has shown intraspecies variation of the immunodominant region of the protein. Although to date no variation has been seen in isolates of *P. vivax* or *P. falciparum,* this work raises at least two questions. Are there reasons to believe that the *P. falciparum* CS protein gene will be more stable than the CS protein of corresponding monkey malarias? What effect would small subpopulations containing variant CS protein epitopes have on the effectiveness of a vaccine?

## The S-Antigen

The S-antigen is a soluble, heat stable protein that is released from the malarial infected erythrocyte as it ruptures. S-antigens circulate in the blood of infected individuals, and the serum of some immune individuals recognize these molecules. Although the expression of a single type of S-antigen is a stable genetic characteristic of a parasite, there are probably a large number of immunologically distinct types. Apparently these immunologically distinct types of S-antigen have a worldwide distribution, since antibody to a mixture of types can be found in a single immune individual and a given type of S-antigen can be found in endemic areas throughout the world. If, as evidence suggests, the frequency of a single type of S-antigen in a particular area varies greatly with time, the dominant form in a single isolated area may vary from yea

The S-antigen also may be a target for protective immunity. Studies of patients in The Gambia. West Africa, show a high frequency of different S-antigen serotypes occurring in sequential natural infection, which could indicate that immunity to parasites expressing one S-antigen protect the individual from reinfection by parasites carrying that particular type. Antibodies to S-antigen may then play a role in the gradual development of immunity to malaria. This type of immunity would not extend to protection from the parasites expressing the S-antigen types that the host immune system has not seen before. It may also be that the frequency of individual S-antigens oscillates in a particular area in response to the development of immunity to specific S-antigens by members of the host population. The general approach that resulted in the isolation of the S-antigen gene was to isolate any gene that produced a product recognized by antibodies of people who are immune to malaria.

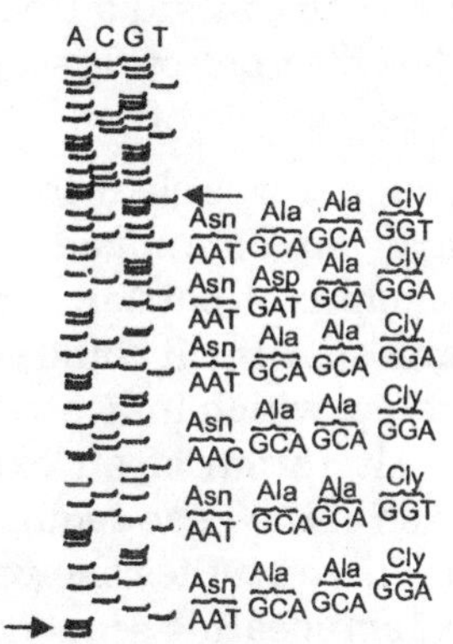


**Fig. 3.3. Representative DNA sequence analysis of a circumsporozoite protein gene. The sequence has been read in ascending order from the arrow on the left to the arrow on the right. Translation of the codons into the amino acid sequence is shown. This type of analysis permits derivation of the primary peptide structure from knowledge of the gene structure.**

Individual genes were then characterized and identified by assessing the properties of the protein product of the gene. The S-antigen gene was identified in this manner because the protein produced in the parasite had previously been extensively characterized. The series of events leading to the cloning of the S-antigen gene proceeded in the following manner. A cDNA library was made with mRNA from a mixture of the asexual stages of *P. falciparum.* The library vector, bacteriophage lambda-gt 11, allowed the expression of *P. falciparum* cDNAs as peptides after infection in *E. coli.* Bacterial colonies expressing the *P. falciparum* proteins encoded on the lambda vector were detected with serum from humans that were immune to *P. falciparum* infection. This procedure allowed these researchers to isolate a number of bacterial colonies that contained clones of individual *P. falciparum* genes whose proteins were being recognized by the host immune system. Immunologically specific serum recognizing only a single *P. falciparum* product could then be produced by injecting the bacterially produced product into mice. Having monospecific serum allowed researchers to localize the position of the unknown *P. falciparum* protein on the malarial parasite by an immunofluorescence assay and to characterize its time of expression. It was clear from the immunofluorescence study that one clone from the library produced a product that was synthesized in schizonts, that seemed to be released upon rupture of the infected erythrocyte, and that was isolate-specific.

All these features indicated that the clone carried an S-antigen gene. They then used their monospecific serum to identify *P. falciparum* proteins on Western blots. They founi that it recognized single proteins of variable size in several cloned lines of parasites. These parasite lines produced S-antigens that were antigenically similar but physically distinguishable by size. By correlating specificity and size they were able to match the serum's specificity to the S-antigen protein. This series of experiments selected for clones carrying and expressing the most immunogenic portion of the S-antigen molecule. When the cDNA clone of S-antigen was analyzed it was shown to contain a DNA sequence that encoded a sequence of 11 amino acids repeated 23 times. The only nucleotide sequence variation that occurred in the repeating element was located in the third position of codons and did not alter the encoded repeating peptide unit.

Although it was clear from this study that there were other areas of the S-antigen molecule that were not composed of this repeat, it was suggested that the repeats were the immunodominant portion of the protein. This work also showed that the nucleotide sequence was isolate-specific. It could be found only in strains of *P. falciparum* expressing one particular antigen type, which indicated that the difference between S-antigen genes was not simply the result of alteration of an existing sequence by point mutation insertions or deletions. The comparison of the gene sequences of antigenically distinct S-antigens of *P. falciparum* has been detailed. These studies showed that the repeating element of S-antigen was bounded on either side by a nonrepeating sequence. Although the repeating elements of the two genes were different, the nonrepeating sequences were very similar. This pattern is similar to that found in comparisons of CSP genes from different species of *Plasmodium*. Hence, S-antigen genes are variable between strains and are therefore varying faster than the species-specific CSP genes.

The mode of change, however, appears to be similar. Variation in the S-antigen form does not appear to be occurring by rearrangement of sequences, in that the repeated sequence of one gene cannot be found in the genome of another line of parasite expressing a different S-antigen. Although the precise mechanism that generates diverse forms of S-antigen is not understood, it seems to involve simple sequence alterations as one repeating unit of a gene is spread to other units. For example, any changes to the common theme or base pattern of a repeating element are generally reproduced exactly in several other units in the string of repeats. Since it is unlikely to alter each of several units identically by chance, a mechanism of spreading of single sequence alterations has been proposed. Sequences surrounding the coding portions of the genes were homologous; which indicates that the genes are allelic and that polymorphism exists, because the repeated areas can change more rapidly than their surroundings. The study of the S-antigen genes has revealed several important tacts. (1) The structure of the immunodominant epitope has been elucidated. This information has allowed researchers to compare the S-antigen gene structure with other DNA structures that are constructed similarly. (2) Variations in types of S-antigen are not linked to rearrangement and expression of preexisting sequences, as is the case with mammalian immunoglobulin genes or the trypanosome variant surface antigen.

A single parasite therefore does not have an infinite repertoire of changes with regard to the S-antigen. (3) The definition of differences between antigenically distinct types indicates that different S-antigens consist of alleles of a single gene rather than a family of different genes. (4) The mechanism that creates diversity has been partially revealed. We know that the repeating immunodominant epitope is changing faster than the rest of the molecule. The mechanism apparently disseminates local changes in one repeat unit to neighboring units. The mechanism and rate of change of the S-antigen may give us a valuable lesson by providing an extreme example of the way in which other repetitive antigens can vary.

It is unlikely that the accumulation of point mutations in the repeating element of S-antigen occurs any more rapidly than it does in any other repeating element. The question is, Why is the result of this spreading of the initial base change through the S-antigen repeats apparently more frequent? In a population of parasites a great variety of amino acid sequences may be acceptable in S-antigen, whereas sequence variation in the circumsporozoite gene, for example, may be selected against. It also may be that the rate of change is inversely proportional to the amount of host immune pressure. In any case, the different repetitive antigens of *Plasmodium* probably change by similar mechanisms.

## Histidine-Rich Proteins

Proteins of unknown function in several species of malarial parasites are characterized by an astonishingly high content of histidine. *Plasmodium lophurae* (an avian parasite) synthesizes one such protein (HRP), which has a histidine content of 73 percent. It may constitute as much as 50 percent of the total protein in asexual forms of the parasite. The protein is found in cytoplasmic granules. One of the things about the HRP that has particularly interested researchers is the report that this antigen protects ducklings against infection with P. *lophurae.* Another parasite, P. *falciparum,* has at least three histidine-rich proteins. One of these proteins (PfHRP-I) is associated with the formation or function of the knoblike protrusions on *P. falciparum* - infected erythrocytes. An example of localization of PfHRP-I on these knoblike protrusions by immunoelectron microscopy is shown in Fig. 7.4.

In panel A, an antibody specific for the PfHRP-I protein has been labeled with an opaque tag, allowed to react with an intact parasite, and the antibody reactivity localized at the knob is visualized by electron microscopy. Panel B shows a similar experiment using normal mouse sera as antibody; the resulting microscopy shows no localization of antibody over knobs as expected. Panel C repeats the experimental protocol used in panel A, except that the parasite used does not have knobs and hence there is no reactivity.

The association between knobs and PfHRP-I is of interest because the knobs mediate cytoadherence of infected erythrocytes to endothelium. Another histidine-rich protein from P. *falciparum* is designated PfHRP-II. It is synthesized throughout the asexual cycle and is released as a soluble protein into culture supernatants from intact cells. There is at least one other histidine-rich protein from P. *falciparum* that remains uncharacterized. The three histidine-rich protein genes of *P. falciparum* have been isolated. An interesting comparison can be made between the gene sequences of PfHRP-II and PfHRP-III. The sequence of the intron and sequence beyond the carboxy terminus of the protein are the same in both genes. Both genes also contain a similar hexapeptide repeat. The hexapeptide repeats vary between genes in only one of six positions.

Presumably, one set of repeats arose from the other. Divergence of one region of hexapeptide repeats from the other seems to occur in a uniform way: each repeat unit of the series is altered in the same way. Another set of repeats that are unrelated to the hexapeptide repeats occurs in PfHRP-III. The implication is that both genes were derived from a common precursor by a gene duplication and then diverged independently and assumed different functions. This process, at the very least, establishes the event of gene duplication in *Plasmodium.* It also outlines a mechanism whereby two proteins can share a common immunodominant epitope.

## Merozoite Surface Antigens

The merozoite stage of *Plasmodium* invades host erythrocytes to start a new round of asexual replication. The process may involve several independent steps, including (1) initial attachment of the merozoite to the erythrocyte surface; (2) reorientation of the attachment, so that the apical end of the merozoite contacts the surface of the erythrocyte; (3) release of a substance into the erythrocyte from an apical organelle of the merozoite (at this point the shape of the erythrocyte is altered and a vacuole forms opposite the apical end of the merozoite); and (4) entry of the merozoite into the vacuole. Antibody against epitopes of important antigens involved in any of these steps could be a target for protective immunity. The genes of three *P. falciparum* antigens that are potentially important for vaccine development and are associated with the merozoite stage of development have been cloned. All three can serve as immunogens and give a protective response that blocks merozoite invasion when administered with Freund's complete adjuvant. This does not mean that they would necessarily be useful candidates for a vaccine. It is not known in any of these cases if the protective response is purely antibody mediated, which would be the ideal situation in terms of a vaccine.

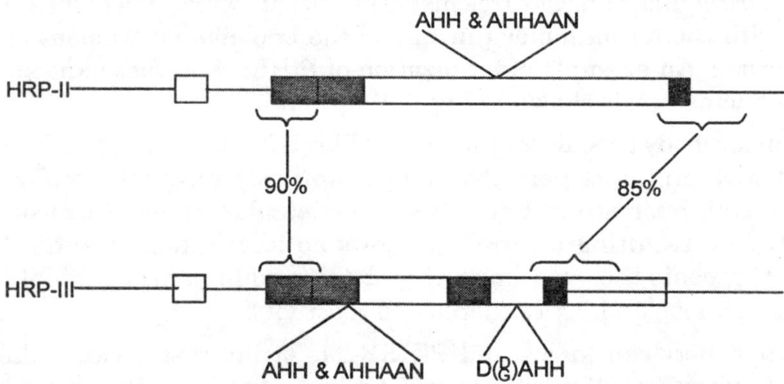


**Fig. 3.4. A comparison of two histidine-rich protein (HRP) genes from *P. falciparum*. It is believed that the two genes were derived from a common precursor by gene duplication and then diverged independently, leading to the differences in the sequence of one region of the hexapeptide repeats shown.**

Furthermore, not all antibody responses would be useful in a vaccine. Ideally, a neutralizing antibody that can block invasion at low titers is needed. Glycophorin A appears to be a merozoite receptor molecule on the surface of the erythrocyte. A glycophorin A binding protein (GBP) of *P. falciparum* has been identified which is probably involved in the initial binding of the merozoite to the red-cell surface. This protein is on the surface of the merozoite, binds glycophorin, and appears to recognize the surface of the erythrocyte. Antibodies to this protein block invasion of erythrocytes; however, the involvement of this particular protein in reception is unproved. The gene for the GBP has been cloned and the DNA sequenced. The protein contains a 50-amino-acid repeating segment. Preliminary evidence suggests that the structure of the repeat does not vary among isolates of *P. falciparum*. A merozoite-associated protein has been identified that may be localized to one of the apical organelles involved in erythrocyte invasion.

Antibody against this protein stained the surface of ring-infected erythrocytes, but the stain disappeared as the parasite matured. Again, antibodies against the protein block invasion. A protein of molecular weight ($M_r$) 195,000 that is synthesized in schizonts is the precursor to three major surface proteins of the merozoite. The function of the proteins is unknown. Preparations of the precursor molecule can induce protective immunity against blood-stage parasites. One of the three proteins ($M_r$ 83,000) is shed when the merozoite invades the red cell.

This molecule contains a repeating DNA sequence that may represent an isolate-specific epitope. Two other proteins remain in associated with the ring after invasion. The particular antigenic sites that induce protection have not been defined. Whether or not isolate-specific sites are involved in the protective response also has not been clarified.

## Asexual Erythrocyte Parasites

The proteins that are produced by the malarial parasite but reside on the surface of the infected erythrocyte may be an important target of protective immunity. For example, parasite-encoded proteins may be localized in knoblike protrusions on the surface of infected cells that bind to host endothelial cells (a process called cytoadherence). Cytoadherence prevents erythrocytes containing mature parasites from circulating in the peripheral blood and thereby protects them from filtration by the spleen. Analysis of genes whose products are directly involved in this binding might provide a good vaccine target. Antibodies to one of these proteins might block sequestration of parasites and allow them to be cleared from circulation by the spleen.

## Sexual-Stage Parasites

Other possible vaccine targets are proteins associated with the sexual cycle of the parasite. It has been shown that the transmission of malaria through the mosquito can be blocked by antisera directed against malarial gametes and zygotes. The specific malarial proteins involved in eliciting such a response have been identified. A vaccine involving these protein products would not prevent the disease directly but would prevent its spread. The hope is that the reduction of transmission of the disease would lead to its eradication.

## Genetics of Malaria

The production of an appropriate antigen is one step in producing an effective vaccine. Presently four different sources of antigen are being investigated: chemical synthesis, microbial production, vaccinia vaccine, and avirulent *Salmonella* vaccine. Many of the advantages and disadvantages of each approach relate directly to the biology of the vaccine target. The specific example used here is the production of antigen for a circumsporozoite-derived vaccine, and the relationship of approach to CS protein biology. The specific considerations involved in the production of an antigen will of course vary with different vaccine targets. Chemical synthesis of the repeating epitope of the CS protein appears to be the most direct approach to producing a vaccine antigen. One major objection to this approach, based on work in other systems, is that short, chemically synthesized peptides often do not display an antigenic site in the manner that a complete folded protein would. Initial experiments with chemically synthesized peptides from the repeating epitope of the CS protein indicated that as few as two repeats containing seven amino acids weakly mimicked the antibody-binding site of the complete CS protein.

Three repeats or twelve amino acids appear to constitute a complete epitope. This suggested that overall configuration of the CS protein was not necessary to display the appropriate antigenic site and that short, chemically synthesized molecules might be useful in a vaccine. This will not necessarily be the case with other antigens. Microbial production of antigen based on a recombinant DNA approach has been used by two groups. The P. *knowlesi* CS protein has been produced in yeast cells. A protein that is approximately the size of the precursor CS protein was detected.

These researchers report production of 1 microgram (μg) of stable CS protein per 20 milligrams (mg) of total protein. The bacterial fermentation approach is also being investigated.

The complete CS protein could not be synthesized in bacteria as it was in the yeast system. Clearly products were rapidly broken down in vivo. But when a smaller segment of the immunodominant region was expressed in the bacterial vector, the products were stable and the yields were nearly 5 percent of total protein, or several thousand times better than the yeast system. These researchers went on to show that antisera raised to their product block hepatoma cell invasion by sporozoites in vitro. Work directed toward a vaccinia-derived sporozoite vaccine is also proceeding. The gene coding for the CS protein of *P. knowlesi* has been inserted into the *Vaccinia* genome. Cells infected with these recombinant viruses produced a protein that reacted with monoclonal antibodies against the CS protein. Also, rabbits infected with virus produced antibodies against sporozoites.

To date, studies of successful vaccinations have not been reported. Some success in producing recombinant vaccinia vaccines for other diseases makes this system look like a promising approach for a malarial vaccine. Another potential "live vaccine" involves the use of nonvirulent, nonreverting strains of *Salmonella* to deliver antigens. These strains of bacteria could be genetically engineered with any antigen or group of antigens. They grow in the laboratory but cannot grow in the host, because they depend on media supplements that do not exist in host tissue. They apparently exist long enough to elicit an immunogenic response in the host but do not grow and divide.

Vaccines potentially could be delivered orally, and there has already been a great deal of testing of the system in the veterinary field. A live bacterial vaccine, however, is another level of increased complexity over the vaccinia system. The fear of loss of control over a "live vaccine" therefore increases. Although antigen for a sporozoite vaccine can be produced in a number of ways, host immune responses to varied products will not be identical. Thus, a few considerations with regard to antigen production for a sporozoite vaccine relate to the epidemiology of malaria. The first consideration is the dynamics of infection. The sporozoites that are injected into the blood reach the liver within minutes, where they may be protected from the host immune response. If only a single parasite escapes the immune response, the infected individual will contract malaria.

Subsequent stages of the parasite's life cycle would not be affected by antibodies directed against sporozoites. Antibody tilers must then be held at a very high level to be effective. The immune system must be more than primed to respond: it must be continually producing specific antibodies. The hope for effective long-term protection is that vaccination will produce a lasting response and that there will be natural boosting of this response by the bite of infected mosquitoes. Clearly, the possibility exists that the vaccine response to constructs consisting of less than the entire CS protein may not be boosted by sporozoites contacted in the field. Even the structure of various CS protein may be sufficiently different so that sporozoites from one strain of *P. falciparum* will not boost the response to a CS protein from another strain of *P. falciparum*. In terms of antigen production, these facts favor the live vaccines that use complete gene sequences. The host responses to live vaccines also tend to be more enduring.

Another factor that relates to a sporozoite vaccine is that a purely antibody-mediated response can be protective. This does not exclude the possibility that cell-mediated immunity may be much more effective in protection.

Many of the standard arguments against the clean, inexpensive peptide vaccines therefore may not apply in this situation. For example, if only an antibody response is necessary, then the peptide vaccines become attractive even though they would not elicit certain types of T-cell responses. A final consideration relates to the fact that the group at highest risk is children. Even a short lived vaccine that had to be continually boosted would be of great value in saving lives in this age range. There appears to be no simple definition of the ideal vaccine. Both the live and peptide vaccines have advantages, and a sense of urgency may dictate the parallel development of both.

## Drug Resistance

Little work in molecular biology has been done on the origin of drug resistance in malaria. There are two likely candidates for work in this area, pyrimethamine resistance and chloroquine resistance. Pyrimethamine is a chemotherapeutic agent that is used alone in the prevention of malaria and in combination with sulfonamides in the treatment of malaria. This drug inhibits the dihydrofolate reductase (DHFR) of the blood schizonts of avian, rodent, and primate malarias at concentrations that have little effect on the corresponding host enzyme. Some drug-resistant parasites can tolerate levels of the drug at 1000 times this concentration. Although DHFR is a classic example of an amplifiable gene, restriction analysis of DNA from pyrimethamine-susceptible or resistant lines of parasites showed no obvious gene amplification.

Such analysis indicates that resistance is likely to be the result not of an increase in gene copy number but of altered properties of the protein conferring both a different affinity for the drug and an increased stability. Chloroquine is another commonly used drug in the prevention and treatment of malaria. Most resistant forms that have been studied carry only a two or threefold level of resistance to the drug, although some greatly increased amounts of resistance have been reported lately. The mechanism of resistance is unknown. Drug resistance seems to be acquired gradually in response to drug pressure. It is assumed that increases in resistance are the result of multiple mutations.

The origins of chloroquine resistance may be much more difficult to find than those of pyrimethamine resistance but are of great practical importance. Contributions of molecular biology to the discovery of new drugs may well be significant in the future.

Analysis of proteins such as tubulin may help us focus on central differences between comparable host and parasite enzymes. Potentially, defining these differences will help researchers identify drugs that will selectively kill parasites. Another possibility for the creation of new drugs rests with the use of derivatized oligonucleotides as drugs that will selectively anneal with parasite messenger RNAs and prevent their translation. Neither of these approaches has yielded a new drug to date, but both may have future potential.

## Genetics of Malaria

The techniques commonly used to study the genetics of free-living organisms are difficult to apply to malarial parasites, which have a complex cycle involving two hosts; therefore, the amount of classical genetics done in malarial parasites is limited. The number of classical genetic markers is small, and no markers have been reported to be linked. Selectable genetic markers, which would make the recovery of recombinant clones easier, are mainly limited to chloroquine and pyrimethamine resistance. Each generic cross is an arduous task. Parents with readily definable markers must be cloned and then mated in the mosquito. Progeny of each mating must then be cloned, propagated and analyzed. Even in the face of these difficulties, several important facts have been revealed by classical genetics in malaria.

Several markers have been shown to segregate in a normal Mendelian fashion. These studies have also indicated that the blood-stage parasites are haploid and that the diploid stage of the parasite resides in the mosquito after fusion of the gametes to form a zygote. The use of pulsed-field gradient gel electrophoresis will help resolve some of the genetic questions in the field of malaria. This technique allows researchers to electro-phoretically resolve the chromosomes of *P. falciparum* and analyze them individually. Two advances have come from this work to date. A large number of genetically linked markers have been established, and it has been shown that the chromosome sizes of different geographical isolates can vary. The latter may allow the firm demarcation of various taxonomic categories, such as an isolate, strain, and subspecies. Analysis of the parents and progeny of mated parasites will also yield significant information about the genetics of *P. falciparum*.

## Coda

*Plasmodium* is not an especially good model system for studying molecular biology. The complete life cycle of the parasite is difficult to maintain, classical genetic techniques are difficult to apply to the system, and no procedure for transformation has been developed. The research effort has been directed toward understanding and controlling malaria rather than understanding molecular mechanisms in general. The link between understanding molecular mechanisms and understanding the biology of the parasite should, however, not be underestimated. Of considerable importance in this regard is understanding more fully the way in which this parasite evades the host immune system, the amount of natural genetic diversity of the parasites and how it arose, and the evolutionary relationship among the *Plasmodium* species. Most of the genes that have been studied encode antigens that are on the surface of the parasite or are actually released from the parasite.

The unique features of these *Plasmodium* genes seem to be the ubiquity of the short repeating epitopes and the frequent sharing of cross-reacting epitopes among different antigens. Clarifying the reasons for these unusual features of *Plasmodium* genes may enhance our understanding of the way that these parasites evade host immunity and may contribute to the production of an effective vaccine.

## Biochemistry of Trypanosomes

Through the application of recombinant DNA technology, we have recently gained some insight into the inner workings of the trypanosomes. Few, if any, other biological systems have revealed such a remarkable collection of extraordinary phenomena in such a short period of time. This is a testimony not only to the power of the techniques being used but also to the unusual position these organisms occupy on the evolutionary tree. This chapter will present an overview of the areas where some or the more interesting findings have been made. As there are many recent reviews on antigenic variation in the African trypanosomes, this topic will receive proportionately less attention. Instead, areas that are just as unusual and interesting but that have not received the same coverage will be emphasized. The chapter is divided into six parts, each representing a functional area of study: genome and genetics, transcription, translation, antigenic variation, mitochondria and respiration, and glycolysis. Most of the work described has been obtained from studies of *Trypanosoma brucei,* but some of the results extend to other species in the order Kinetoplastida.

## Genome and Genetics

Like most other eukaryotes, all trypanosomes have DNA within two organelles, the nucleus and mitochondrion. The mitochonrial DNA is highly unusual and will be discussed in detail in the section entitled Mitochondria and Respiration. The genome of *T. brucei* is essentially diploid, with a total complexity of about $4.3 \times 10^7$ base pairs (bp). Unfortunately, the chromosomes of trypanosomes do not visibly condense during mitosis (possibly owing to an apparent lack of histone Hl); their analysis therefore has awaited the arrival of technologies that do not depend on cytology. Such procedures, termed pulse-field gradient gel electrophoresis, or orthogonal field alternating gel electrophoresis, have recently been developed. These techniques are similar to ordinary agarose gel electrophoresis but are capable of resolving much larger DNA pieces.

The method exploits the fact that the time taken for large DNA to reorient in an electrical field that has been rotated through 90 degrees depends on the length of the molecule (although the analogy is biophysically inaccurate, this is akin to the molecules having to repeatedly turn a tight corner). Thus, by switching the orientation of the electrical field from north-south to east-west in repeated pulses DNA of up to several million base pairs can be resolved.

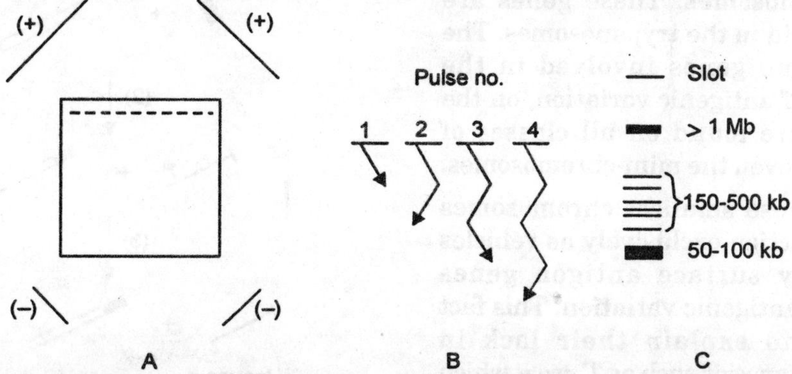


**Fig. 3.5. Chromosome configuration. Part A shows the principle behind orthogonal field alternating gel electrophoresis (OFAGE). A horizontal agarose gel is placed between four electrodes, as shown. Each pair of electrodes (+ and -) is connected to a voltage source in alternating sequence. Under pulse times of a few seconds to many minutes, the DNA slowly migrates into the gel. Ultimately, conventional bands of discretely sized DNA are produced without enormous distortion of a given lane because of the vector sum of the voltage fields, as shown in part B. Part *C* shows a schematic of trypanosome chromosomes analyzed in this way.**

With the use of these techniques, the chromosomes of trypanosomes have been found to fall into four general size classes. The largest is well over $10^6$ bp (perhaps even over $10^7$ bp) and comprises about 60 percent of the genome. The second set of about 3-6 chromosomes is $1 - 2 \times 10^6$ bp. Most, if not all, of these two size classes are apparently present at two copies per nucleus. The third class of chromosomes is about $2 - 7 \times 10^5$ bp. In T. *brucei,* these number about six and are apparently present at only one copy per nucleus. The mini-chromosomes form the fourth and smallest class, with an average size of about $0.5 - 1.5 \times 10^5$ bp. This class, which is unique to the African trypanosomes, is comprised of the most members, totaling about 100 chromosomes per nucleus (or about 15 percent of the total nuclear DNA).

The concept of ploidy is somewhat misleading when applied to members of the mini-chromosome class: they are highly redundant and repetitive, carrying related surface antigen genes, as discussed later. The chromatin of trypanosomes is apparently conventionally organized in nucleosomes comprised of a histone octamer and about 200 bp of DNA. As yet, however, no H1 histone or equivalent cross-linker has been identified. The chromosome ends (referred to as telomeres) also appear to be conventional, possessing over the last few hundred base pairs a repeated hexanucleotide with occasional nicks.

The sequence of this hexanucleotide is similar but not identical to that seen in other protozoa. Preceding this is a subtelomeric sequence that is apparently conserved within the species. Only a few genes have been mapped to a given chromosome or even class of chromosomes. However, even at this early stage, a trend is clearly emerging, so that "housekeeping genes," or genes involved in routine cellular processes common to most eukaryotic cells, are located on the largest class of chromosomes. These genes are therefore diploid in the trypanosomes. The surface-antigen genes involved in the phenomenon of antigenic variation, on the other hand, are found on all classes of chromosomes, even the mini-chromosomes.

Indeed, these smallest chromosomes appears to function exclusively as vehicles for the many surface antigen genes necessary for antigenic variation. This fact also serves to explain their lack in Kinetoplastida species such as *T. cruzi* which do not undergo extensive antigenic variation. Little is known about the sexual cycle of *T. brucei*, but it is now clear that such a cycle does occur during transmission by the insect vector, the tsetse fly. On the basis of findings to date, it appears that the entire sexual cycle, from meiosis and gametogenesis to zygote formation and subsequent asexual expansion of the population, occurs within the fly. The details of the ploidy of the different stages in this process is still a matter of debate. During meiosis, it may be that the mini-

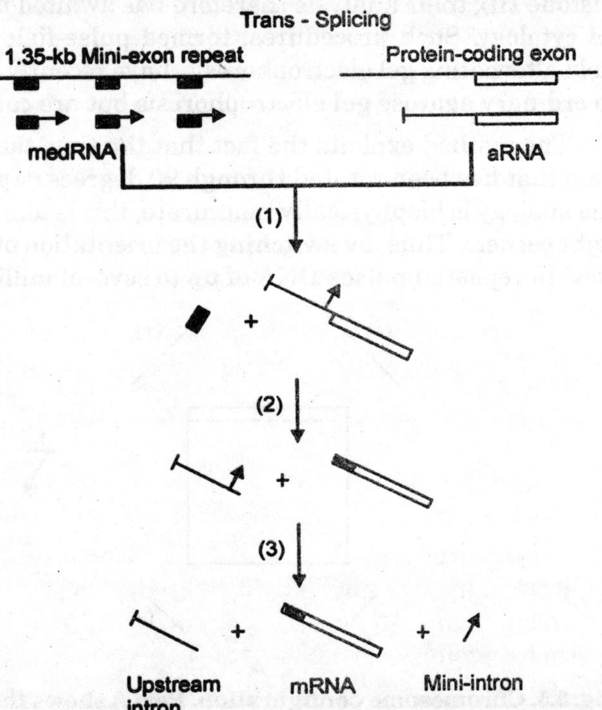


**Fig. 3.6. Model of trans-splicing. The mini-exon repeats direct the synthesis of mini-exon -derived RNA (medRNA); the protein-coding exon directs the transcription of the acceptor RNA (aRNA). The first step in trans-splicing involves formation of abranched species (comprising the intron portion of medRNA (minRNA) and aRNA and free miniexon. In the second step, the two exons are joined so that the two introns are released as a discrete, branched species. The third step comprises the debranching of the branched species into the two component introns. This reaction yields three final products: the upstream intron of aRNA, minRNA, and the mature mRNA.**

chromosomes are not evenly segregated (owing to their haploid state in an otherwise diploid nucleus).

However, this would not impair the viability of a daughter cell so long as it received sufficient genes to enable it to undergo several cycles of antigenic variation in the mammalian host and thereby survive long enough to be picked up by another fly. Subsequent genetic exchange during the sexual cycle in the fly would enable it to collect a new set of surface antigen genes when the pool is reasserted. There is no definitive evidence for a sexual cycle in the other Kinetoplastida so far studied.

## Transcription

One of the most unusual features of the trypanosomes is their mode of transcribing nuclear genes. This process involves the production of a chimeric mRNA by transcription of two segments of DNA that are not physically contiguous. The first indications of this unusual process came from sequence analysis of the 5′-end of the mRNAs coding for surface antigens. This showed that roughly the 5′-most 35 nucleotides of the mRNAs encoding different antigens were identical and not contiguously encoded with the protein-coding portion of these genes.

Synthetic oligonucleotide probes were constructed and used to determine where the mini-exon coding for this spliced leader was located. The results indicated that the mini-exon is part of a 1.35-kilobase (kb) unit (the mini-exon repeat), which is tandemly repeated about 200 times in the genome and is unlinked to the protein-coding exon for these surface antigen genes. In some cases, the mini-exon repeats are on different chromosomes from the surface antigen genes. It is now known that the mini-exon sequence is present at the 5′-end of most, probably all, mRNAs in all stages of all Kinetoplastida. Although the mini-exon repeat may vary in size and overall composition, the mini-exon sequence is highly conserved. Its exact size in *T. brucei* has recently been shown to be 39 nucleotides plus the usual eukaryotic 7-methyl guanosine in 5′-5′ tri-phosphate linkage.

The original estimates of 3 5 nucleotides were incorrect as a result of the highly modified state of the first four nucleotides, which probably affected the assays used. Given their abundance and presumed functioning in the generation of so many mRNAs, it was not surprising to find that each mini-exon repeat is independently transcribed to give a short, discrete RNA of about 140 nucleotides (in *T. brucei,* shorter in other Kinetoplastida species), the first 39 of which correspond to the mini-exon sequence. Based on these observations, two basic schemes for the synthesis of chimeric mRNAs were originally proposed. The first involved the use of the med (mini-exon-derived) RNA as a primer in a transcriptional reinitiation event upstream of the protein-coding exon followed by cis-splicing of the resulting molecule. The second model supposed that some sort of bimolecular mechanism operated to combine the mini-exon portion of medRNA with the exon portion of an independently generated transcript from protein-coding genes. Some sort of splicing must be involved because only 39 nucleotides of the medRNA are ultimately found on the mature mRNA.

This splicing apparently uses much of the same basic machinery used by higher eukaryotes. This fact is apparent from looting at the sequences located at the exon - intron boundary in the medRNA and intron-exon boundary in the coding gene. These conform to the absolute consensus sequences found in nuclear-encoded genes of all other eukaryotes. Moreover, it has recently been found that the Kinetoplastida have several of the mRNAs involved in cis-splicing in higher eukaryotes. Recent evidence strongly suggests that the second model, trans-splicing, is correct.

This model supposes that the medRNA and an acceptor RNA are spliced by a pathway similar to the more conventional cis-splicing, except that the two RNAs are never found as a single, covalently linked precursor. Evidence for this model came from the ability of a debranching activity of HeLa cells to release the tail portion of medRNA (the mini-intron sequence) from higher-molecular-weight RNA. It is now clear that the analogy to cis-splicing is accurate because the linkage between the miniintron and the acceptor intron is a 2′-5′-phosphodiester bond and the trypanosomes have a debranching enzyme with the same activity as the HeLa enzyme. The unusual process of trans-splicing in Kinetoplastida is critically important to our understanding of the way trypanosome genes are regulated.

Trans-splicing may have equally important significance, at least to the molecular biologist, in telling us something of the evolution of RNA splicing, an exciting and contentious subject. This suggestion stems from the fact that trypanosomes are also unusual in having no known conventional (cis) introns in the many genes now studied; that is, no protein-coding exons are interrupted by introns. This total lack is without precedent in eukaryotes and begs the question of whether trypanosomes have somehow removed their introns. An alternative suggestion is that they never had them. If true, this idea argues that trans-splicing predates cis-splicing and that the cis-machinery adapted that of the trans-process for its own purposes. What is the function of trans-splicing and what is the role of the mini-exon sequence itself? One way in which trans-splicing is apparently used to advantage is in the generation of multiple, capped mRNAs from a single, polycistronic precursor transcript. This process would explain the high frequency of trypanosome genes found tandemly arranged on the chromosome, sometimes with the same gene(s) repeated several times.

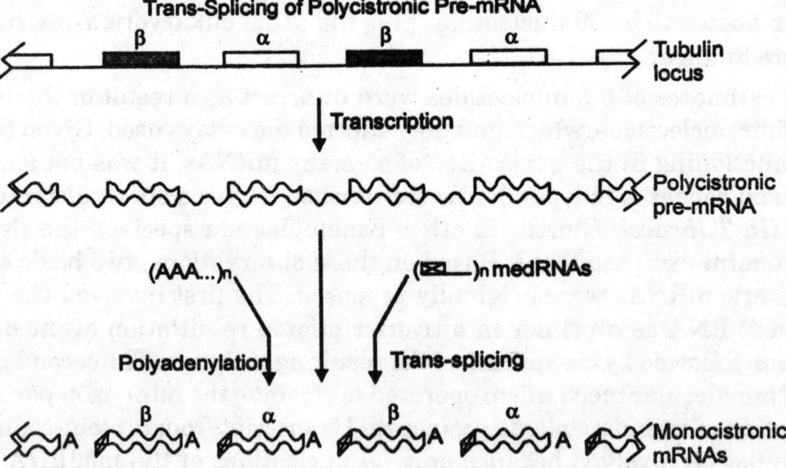


**Fig. 3.7. Model of polycistronic precursor transcripts. The model proposes that a polycistronic RNA is transcribed from a locus composed of multiple coding regions (in this case, the tubulin locus of *T. brucei* encoding both α– and β-exons). Multiple monocistronic mRNAs are generated from this transcript by trans-splicing to yield a capped (medRNA derived) mini-exon sequence at the 5′-end, and by cleavage and polyadenylation to yield the polyadenylated 3′-end. This processing could occur post- or co-transcription-ally, and the order of trans-splicing versus polyadenylation is not known.**

The model proposes that transcription begins at a promoter upstream of the cluster and that through a combination of trans-splicing and polyadenylation, mature, capped mRNAs are generated, each coding for a single polypeptide. By this pathway, each mRNA would acquire a cap, which is believed to be a prerequisite for stability and translatability of eukaryotic mRNAs. (Note that without trans-splieing, acquisition of a cap could not occur by any known pathway, because caps can be added only to transcription initiation sites, probably because of their unique 5′-triphosphate end) In addition to the tandem reiteration of many genes, the evidence for this pathway is fourfold.

1. Heat shock causes the accumulation of polycistronic transcripts.
2. Pulse-chase experiments suggest a precursor - product relationship between polycistronic transcripts and mRNAs.
3. Intergenic regions are transcribed in nuclear run-on assays.
4. Kinetics of UV inactivation indicate some genes are 50 kb downstream of their promoter, with multiple coding regions intervening.

The question of the function of the mini-exon sequence itself is intangible at present because we know of no exception to the rule that all trypanosome mRNAs include this sequence. That such a sequence will prove essential in the translation of these mRNAs seems very likely but tells us little about their function. It would simply confirm that the translational machinery and mini-exon sequence have co-evolved and are now functionally interdependent, just as the bacterial ribosome requires its sequence-specific binding site. However, in the absence of in vitro translation or DNA transformation, it has not been possible to test this idea.

## Translation

As in all other eukaryotes, translation in Kinetoplastida is the responsibility of the ribosome, a complex organelle composed of many proteins and RNAs (rRNA). Most of the rRNA in trypanosomes is encoded by a tandemly repeated segment of DNA (about 22 kb per repeat in *T. brucei*. In most eukaryotes the rDNA is transcribed as a single long precursor that is subsequently processed by several cleavages into the 18S, 5.8S, and 28S rRNAs. The Kinetoplastida adhere to this basic pattern, except that the small subunit rRNA (18S-like) is in fact the largest of any known, at about 21S, and the large subunit rRNA (28S-like) is further processed into two medium and several small RNAs.

The machinery involved in this processing is not vyet identified, but all the cleavage sites have been precisely located. Interestingly, there is no apparent commonality in either primary or predicted secondary structure between all sites. The signal for cleavage therefore remains cryptic. Cleavage of the 28S rRNA into two halves is known to occur in a few other eukaryotic systems (*e.g.*, *Drosophila* and *Plasmodium berghei)* but in no other system does this result in the generation of several small, stable RNA fragments that remain associated with the mature ribosome. Eukaryotic ribosomes generally contain another small RNA with a sedimentation coefficient of about 5S (120 nucleotides). As in other eukaryotes, the trypanosome 5S RNA is encoded by its own tandemly repeated DNA rather than as part of the rDNA repeat described above. Ribosomal RNA is among the most conserved RNA and as such is frequently used to deduce evolutionary relatedness between organism.

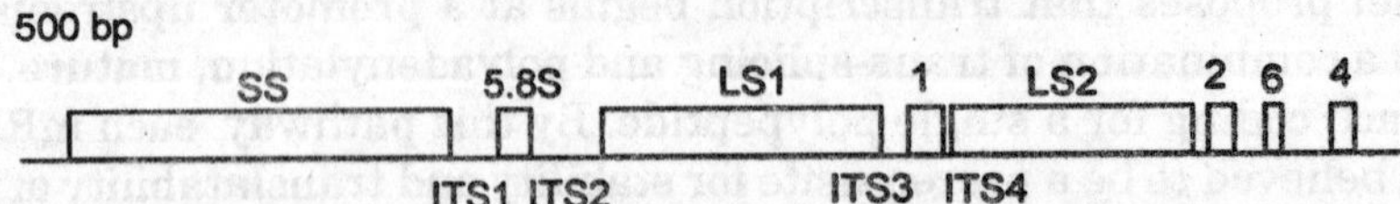


**Fig. 3.8. Organization of ribosomal DNA. A portion of a single rDNA repeat of *T. brucei* is shown. The regions represented in stable RNAs are boxed and identified as small subunit RNA (SS), large subunit RNA (LSI and LS2), 5.8S RNA, and the small ribosomal RNAs (1, 2, 4, and 6). The intervening transcribed spaers (ITS), which are processed out and rapidly degraded, are also shown. The final component of the rRNA, 5S RNA, is encoded by its own DNA repeat.**

The complete sequence of 5S rRN for two kinetoplastid species has been obtained. This shows that in addition to being the largest small subunit rRNA yet characterized, the order Kinetoplastida is one of the most distantly diverged (relative to the main eukaryotic line), apparently branching off before even the split into plants, fungi, and animals. As we progress further in our understanding of the basic molecular biology of the Kinetoplastida, information on their position on the phylogenetic tree will be increasingly important in enabling us to place new findings in an evolutionary context.

## Antigenic Variation

This portion of the chapter deals with the extraordinary way in which the African trypanosomes successfully elude the immune system of their mammalian hosts. Antigenic variation describes the ability of the African trypanosomes to alter the antigenic nature of their surface coat. On an individual trypanosome, this coat is generally composed of a single type of abundant glycoprotein molecule known as the variant surface glycoprotein, or VSG. Each parasite contains many hundreds of genes coding for different VSGs. They effect antigenic variation by activating these genes one at a time, so that during the course of an infection initiated by a single parasite, the host is faced with a succession of antigenically distinct populations. No known stimulus triggers this switch; it appears to be spontaneous, occurring approximately once in every $10^6$ divisions.

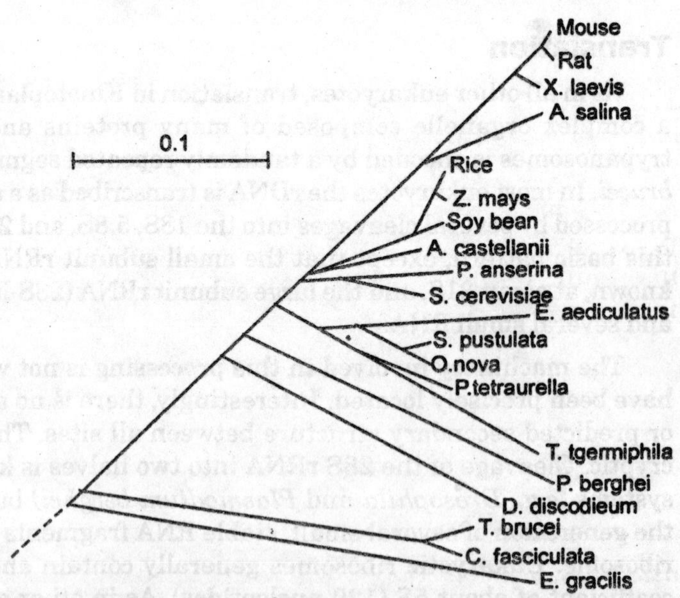


**Fig. 3.9. Placement of trypanosomes on the evolutionary tree. Using the small subunit rRN A sequence as a measure of sequence variation, investigators have devised this phylogenetic tree, showing the placement of the *T. brucei* relative to other organisms, including plasmodia.**

Antigenic variation is sustained through the appearance of new antigenic 'types before the preceding wave has been completely eliminated. It is slow enough, however, to prevent the full repertoire from being revealed to the host too quickly. This prolongment is further ensured by expressing the VSG genes in a nonrandom way, so that for a given strain, there is about a 40 percent chance of one antigenic type being followed by a second, defined type, with the remainder of the switched types largely limited to a small subset of the repertoire. The molecular basis of this phenomenon is now partially known. First, the expression of a gene is somehow controlled at the level of transcription, with only one VSG gene typically transcribed at any one time. In order for a VSG gene to be active, it must reside in a limited number of expression sites, which are invariably found very near the ends of the trypanosome chromosomes (the telomeres, discussed earlier).

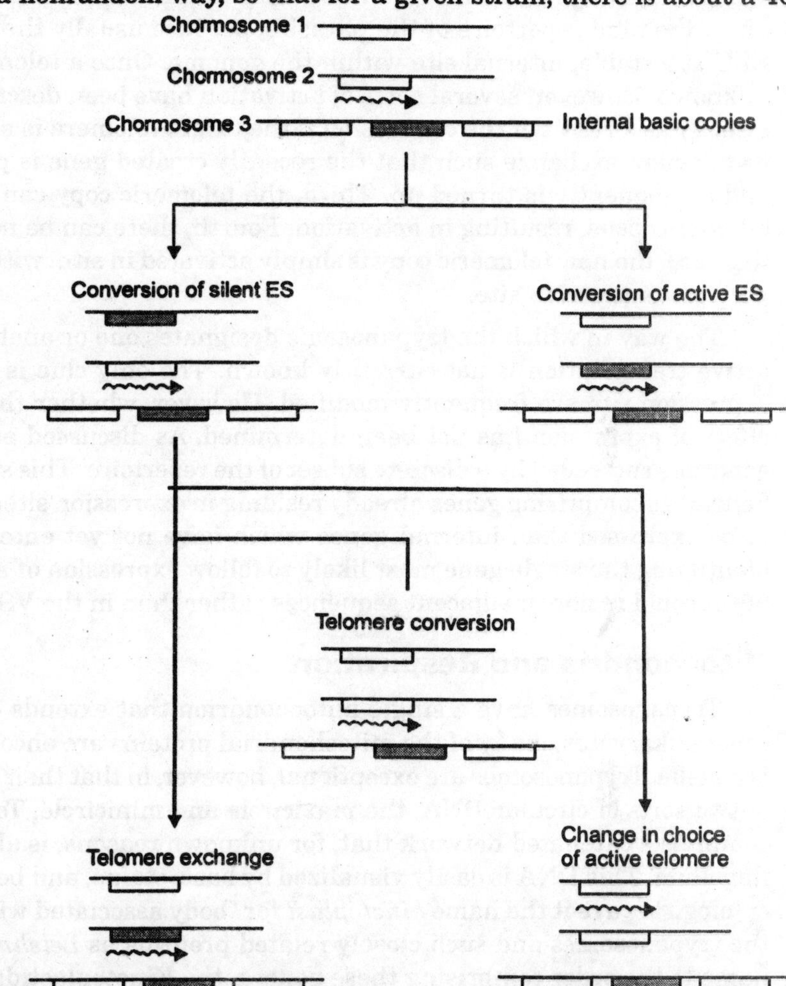


**Fig. 3.10. Models for molecular genetics of antigenic variation. Activation of VSG genes requires that they first be duplicated and transposed into a telomeric expression site. In this drawing, several VSG genes (represented by the boxes) are shown on three chromosomes. The upper drawing represents a parasite expressing the "white" gene in expression site 1 (ES1), as indicated by the arrow. Activation can proceed by a one-step process if the basic copy of the "black" gene is copied into the active ES1, as shown by the right-hand pathway. Alternatively, the black gene could convert a silent ES (2), as shown in the left-hand pathway. This would be phenotypically silent with continued expression of the white gene. Following this (by any amount of time), the black gene could be activated by telomere exchange (such that the ES2 becomes activated), telomere conversion (such that a third copy of the black gene is generated, now in the active ES1) or a change in the choice of active telomere with no rearrangements.**

A VSG gene enters an expression site by gene conversion a process whereby a gene in one locus (here the expression site) is replaced by a copy of a gene normally found in a different locus. The repertoire of VSG genes in trypanosomes is generally maintained as a collection of basic copy, or BC, genes, which are located at chromosome-internal positions. During gene conversion a duplicate of the basic-copy sequence replaces a gene in an expression site, the new copy being designated an expression linked copy, or ELC. The displaced ELC is permanently lost in this process, but this loss does not generally affect the total repertoire of the parasite, because usually there is a BC corresponding to this ELC at a stable, internal site within the genome. Once a telomere, the process of activation is unknown. However, several sorts of activation have been described. First, activation can occur directly as a result of the conversion if the target telomere is already active. Second, there can be telomere exchange such that the recently created gene is placed on an active chromosome and consequently is turned on. Third, the telomeric copy can in turn convert another, active telomeric locus, resulting in activation. Fourth, there can be no detectable rearrangements; in this case, the new telomeric copy is simply activated in situ, with corresponding switch-off of the previous expression site.

The way in which the trypanosome designates one or another of these expression sites for active transcription is not currently known. The only clue is that the nucleotides of a silent expression site are frequently modified. However, whether this modification is a cause or an effect of expression has not been determined. As discussed earlier, a given antigenic type is generally succeeded by a discrete subset of the repertoire. This subset seems likely to be partially defined as comprising genes already residing in expression sites. They are, it seems, more likely to be expressed than internal genes which have not yet entered an expression site. The tag identifying the single gene most likely to follow expression of any given gene is not yet known, but it could reside in adjacent sequences rather than in the VSG-coding portion itself.

## Mitochondria and Respiration

Trypanosomes have a single mitochondrion that extends over the length of the cell. Like other eukaryotes, many of the mitochondrial proteins are encoded by circular DNA within this organelle. Trypanosomes are exceptional, however, in that their mitochondrial DNA is comprised of two sorts of circular DNA, the maxicircle and minicircle. Together, these two types of DNA comprise a catenated network that, for unknown reasons, is always found near the base of the flagellum. This DNA is easily visualized by basic stains, and before its composition was known, cytologists gave it the name *kinetoplast* for "body associated with motion." As it is found only in the trypanosomes and such closely related protozoa as *Leishmania* and *Crithidia,* it gives its name to the order comprising these genera, the Kinetoplastidal.

It also gives its name to the mitochondrial DNA of trypanosomes, which is usually referred to as kinetoplast DNA or kPNA. Despite the unusual structure of its DNA, the fact remains that the mitochondrion must perform its essential function of generating ATP through oxidative respiration. In the trypanosomes, this process is highly regulated. When in the bloodstream of the mammalian host, there is an ample supply of readily converted energy in the form of glucose. The glycolysis necessary to convert the energy in glucose to ATP is handled by an unusual organelle, the glycosome, which will be discussed in greater detail later. Within the actively dividing, long-slender forms of the bloodstream, the mitochondrion is largely quiescent, with no detectable cytochromes or mitochondrial ATP generation.

Morphologically, the mitochondrion in this stage is a long, narrow tube with few cristae; biochemically, most of the enzymes and cytochromes necessary for respiration are absent. In the insect vector, the situation is very different, with proline and other amino acids being the major carbon source. In this stage of its life cycle, respiration is fully switched on and the mitochondrion is expanded with many cristae. Transcription of the mitochondrial genome is generally increased in this stage and there is a concomitant increase in the amount of the enzymes and cytochromes necessary for oxidative phosphorylation. The sorts of kDNA are very different in both structure and function.

The mini-circles are covalently closed circles of 0.7-2.5 kb, depending on the biological species. They have no known coding function and no substantial open-reading frame, but they may be transcribed to give one or more short transcripts. There are about 5000-10,000 minicircles in each kinetoplast network, but these are generally not homogeneous even within a single network. Instead, they are known to be highly variable in sequence, with only short, conserved regions likely to be involved in their replication. Notable exceptions to this variability are *Trypanosoma equiperdum* and *Crithidia fasciculata,* which have nearly homogeneous populations of minicircles. The other component of kDNA is the maxi-circle, which ranges in size from 20 kb to 38 kb. In other systems this component would be termed the mitochondrial genome. It contains genes for the ribosomal RNAs of the mitochondria (although no such ribosomes have yet been detected), several mitochondrial enzymes or subunits thereof, and some evolutionarily conserved open-reading frames of unknown coding function.

As with other mitochondrial genomes, the maxi-circle uses UGA to code for tryptophan (as opposed to using it as a stop codon in the "universal" code. Comparisons with the mitocondrial genomes of other organisms reveal several properties unique to the Kinetoplastida. First, the putative ribosomal RNAs encoded by the maxi-circle are substantially smaller than those found in any other organism, prokaryote or eukaryote (about 600 and 1150 nucleotides for the 9S and 12S RNAs, respectively). Second, no tRNA genes have yet been identified on any kinetoplastid maxicircle. The third and most unusual property of the maxi-circle is that several of its genes are "incomplete" yet functional.

The paradox describes the fact that transcripts from several of the genes are nonfunctional unless subjected to a remarkable process of uridine addition. This process of RNA editing, which has been observed in three disparate Kinetoplastida species, *T. brucei, Leishmania tarentolae,* and C. *fasciculata,* involves the addition of several (over 300 in some cases) uridine residues at precise positions within a transcript both in and out of the protein-coding region. Without such additions, the coding region is not a continuous open-reading frame and may even lack the start codon, AUG (*i.e.,* the corresponding gene encodes only AG at this position, so a U insertion is necessary for functionality). The mechanism of U addition, or even whether it is co- or posttranscriptional, is not yet known. Interestingly, the phenomenon is apparently developmentally regulated in *T. brucei,* so that in the bloodstream forms where respiration is quiescent, the transcripts lack the added Us and are thus nonfunctional. In the insect forms, on the other hand, the transcripts possess the Us and are fully functional, encoding essential mitochondrial enzymes. Transcription of the maxi-circle is not well understood.

No promoters have been definitively identified, and the primary transcription units have not been mapped. The ribosomal RNAs appear to be derived from longer primary transcripts (as is the case in mammalian mitochondria) based on pulse/chase experiments. The situation for the unedited precursors of the protein-encoding mRNAs is not yet known. The function of the maxi-circle is clear: it encodes several, but not all, of the mitochondrial enzymes, the remainder of these enzymes presumably being encoded in the nucleus. The function of the mini-circle is more enigmatic. Given its variability and lack of open-reading frames, it may be sewing as purely structural DNA, perhaps as a scaffolding on which to array the maxi-circle. The structure of such a scaffold could be a stack of parallel, elongated ovals (perhaps supercoiled), like baguettes in a baker's shop. This pattern is consistent with electron micrographs, which show a striated layer with a width equal to the predicted length of one or two mini-circles.

At least one of the "sharp turns" could be due to a periodicity in the sequence, wherein a run of adenosines is known to be present at about every 10 nucleotides. Such a nonrandom interval (approximately equivalent to the number of residues in one helix turn) can cause the DNA to bend a few degrees at each adenosine run; the cumulative effect is a substantial turn in a single plane. Replication of mini-circles first requires that they be freed from the complex by a topoisomerase and then replicated by a Cairns-type (theta) mechanism. The resulting two progeny molecules are reintegrated into the complex, where the nicks that remained after replication are sealed. Maxi-circles, on the other hand, replicate while still part of the network by a rolling-circle mechanism.

The linear extension of the rolling circle is cleaved at a discrete site into monomeric units which are not part of the network. The manner by which the linear units are circularized and reattached to the network is not yet known. The short-stumpy bloodstream forms, which are thought to be the forms infectious to the fly, also have increased levels of maxi-circle transcripts, consistent with the notion that they are preadapting to life in the fly.

## Glycosomes

Trypanosomes rely on glycolysis for energy production when in the glucose-rich bloodstream of their mammalian hosts. This process enables them to aerobically metabolize glucose, producing two molecules of the three-carbon pyruvate. This process requires the consumption of four ATP molecules while generating six, resulting in a net gain of only two ATPs per glucose molecule metabolized. Because this process is the major source of ATP to these very motile parasites, the trypanosomes have evolved an unusually efficient system of glycolysis, which metabolizes glucose about 50 times faster than the system in mammals. This efficiency is attained through the enzymes involved being not only highly abundant but also concentrated into discrete, membrane-limited organelles called the glycosomes. These organelles, which are analogous to the micro-bodies found in other organisms, are about 03 microns ($\mu$m) across and number 200-300 per cell. They are bounded by a single membrane and constitute about 4 percent of the total volume of the trypanosome and 9 percent of the total cell protein.

The nine enzymes involved in glycolysis are the major components of the glycosome. In addition, however, the enzyme machinery involved in other metabolic processes, such as synthesis of ether lipids, beta-oxidation of fatty acids, and biosynthesis of pyrimidines, is also present, although in much lower amounts. One clear advantage to the compartmentalization of glycolysis is that as the intermediates in the glycolytic pathway are generated, they cannot diffuse too far before being acted on by the next enzyme in the pathway. The intriguing possibility of channeling of the intermediates from one enzyme directly to the next (without diffusion) has proved not to be the case. The genes coding for the different glycolytic enzymes in trypanosomes are nuclear-encoded (there is no detectable nucleic acid in the glycosome).

In general, the genes for the different enzymes are not physically linked. Interestingly, however, like many other nuclear genes in the trypanosomes, some glycolytic enzymes are encoded by two or more tandemly reiterated genes. In at least one case (phosphoglycerate kinase), the individual genes comprising a group are not absolutely identical, with their respective products being localized to the glycosome and cytoplasm, respectively. They are also present in different amounts in the various life stages. The glycosome enzymes have exceptionally high pis (9-10) compared with their homologues in the trypanosome cytoplasm and other organisms. For each enzyme, the extra, positively charged residues are found in two topological clusters spaced 40-50 angstroms (Å) apart. It has been suggested that these clusters are the tags that cause these enzymes to be transported into the glycosomes. This intriguing possibility is yet to be tested experimentally.

## Coda

Clearly the application of molecular biology to the kinetoplastid organisms has revealed many unusual and unprecedented phenomena. But where do we go from here, apart from filling in the many missing steps in these unusual processes? Determining where trypanosomes fit into the larger picture of evolutionary biology is going to be just as exciting an area for future research. We know they are exceptional in so many processes, but how isolated are they as a group? The trypanosomes promise much in what they can tell us about the evolution of many complex processes whose origins are unclear from studies of more recently evolved systems. All this is in addition to the potential of applying a detailed knowledge of such fundamentally different pathways to effecting new means of controlling this critically important class of parasites. Exciting times await us.

## BIOCHEMISTRY OF LEISHMANIA

The *Leishmania* organisms are single-celled members of the order Kinetoplastida. The leishmania has two distinct stages in its life cycle, the promastigote stage, in which the parasite is extracellular, flagellated, and motile, and the amastigote stage, in which the parasite is intracellular, nonmotile, and nonflagellated. The promastigote form is found in the insect vector, whereas the amastigote resides in macrophages of the mammalian host. The parasite multiplies in both stages of its life cycle. Human leishmaniasis is caused by at least 14 different species and subspecies of the genus *Leishmania*. Epidemiological studies over the last 25 years have demonstrated that in general leishmaniasis is a zoonotic disease; the parasite is transmitted to humans from a reservoir mammalian host by a phlebotomine sandfly vector during a blood meal. Only a subset of the numerous species of the genus *Leishmania* found to infect mammals also infects and causes disease in humans, presumably because of a combination of factors, including the intrinsic susceptibility of humans and the feeding habits of the sandfly vector. Enormous effort has been devoted to the isolation and characterization leishmanias that infect humans and to the identification of the principal mammalian reservoirs and species of phlebotomine vector.

The result of many such studies has been the correlation of particular clinical manifestations with certain species or subspecies of the parasite. *Leishmania* species identification is based on a variety of ecological, biological, biochemical, and immunological criteria. Previously, each cultured isolate of the parasite was analyzed by one or more of these criteria and grouped into species and subspecies. Controversies remain as to whether organisms isolated in distant geographic locations but sharing certain common properties belong to the same or distinct subspecies of the genus *Leishmania.* Whether these organisms represent strains of the same subspecies or distinct subspecies cannot be resolved because there is no single generally accepted method for species identification in the genus *Leishmania.* There is no defined sexual stage, so that traditional criteria for species identification cannot be applied. This problem complicates comparison of the disease epidemiology in distinct geographic locations and represents a potential limitation on the transfer of control measures from one geographic location to another. Recently there has been an interest in molecular mechanisms that might be unique to the parasite and therefore useful either in understanding the parasite itself or certain facets or the host-parasite interaction.

The obvious long-term practical goal of such work is the identification of potential new targets for chemotherapeutk attack of the parasite. This area should receive focus because although vaccine work is in progress, both the diversity of *Leishmania* sp. and the potential problems of antigenic change will significantly hamper those developments. Several questions are of fundamental interest to molecular biologists with regard to the parasite and its life cycle. These include organization, kinetoplast DNA structure and function, developmental control of gene expression, and novel biochemical and molecular processes. Recent work has defined the chromosomal organization of the parasite, but there is no information on the exchange of genetic information in these organisms and the establishment of genetic systems for analysis. This area is just beginning to develop. No sexual stage has been defined for these parasites, and controversy remains concerning even the most basic questions, such as whether the organism is haploid or diploid in either promastigote or amastigote stages.

The kinetoplast DNA of the parasite has stimulated particular interest; this fascinating set of DNA molecules is unique to the order Kinetoplastida, and the most extensive molecular characterization of any DNA molecule in the leishmania parasite has been of the kDNA maxi-circle and mini-circle. Control of gene expression in the various stages of the parasite's life cycle has been a focus of research since the parasite exists in vastly different environments within the insect vector and the mammalian host. One major interest for molecular biologists is to determine which adaptive mechanisms the parasite has developed to survive in two distinct environments: the sandfly gut and the host macrophage. Novel biochemical and molecular pathways that are specific for the parasite have attracted attention because their study could lead to new strategies for chemotherapeutic attack.

## Genetic Organization and Function

The genome size of *Leishmania donovani* has been determined by Cot analysis and is estimated to be $5 \times 10^7$ base pairs (bp) per haploid genome. Recent work using pulse-field gel electrophoresis has demonstrated that the *Leishmania* species have a minimum of 16 chromosomes or chromosome-sized DNA molecules. Cot analysis of the ribosomal RNA demonstrates that there are approximately 150 copies of ribosomal RN A per haploid genome. The organization of the genomic DNA is in the very early stages of analysis. Separation of large chromosome-sized DNA molecules by the technique of alternating-field gel electrophoresis has allowed the identification of 16 to 23 chromosome-sized DNA molecules in the *Leishmania* species. In general, the chromosomes range in size from 150 kilobases (kb) to greater than 1000 kb. Before this work, it had not been possible to visualize the parasite's chromosomes because, unlike mammalian chromosomes, they are not visible as condensed chromosomes at mitosis.

Several laboratories have analyzed chromosomes from different leishmania species and have discovered that there is significant size polymorphism of certain chromosomes between species. This result is not unexpected; different yeast species also show chromosome size polymorphism. In addition, Southern blot analysis demonstrates species-specific variation in the size of specific chromosomes. Several structural genes have now been localized onto specific chromosomes-sized DNA molecules. In this way, a molecular karyotype of several *Leishmania* organisms has been derived. At least five chromosomes have been defined by the location of alpha and beta tubulin genes, rRNA genes, and the gene for the 5′-spliced leader sequence. The mapping of alpha tubulin genes and ribosomal genes demonstrated that each mapped to a distinct large chromosome in a variety of isolates. Both the beta tubulin gene and the 5′-spliced-leader sequence show evidence of size variation and variation in the number of chromosomes with homologous sequences. This pattern indicates a certain plasticity in the arrangement of genes on chromosomes and may have implications with regard to the evolution of the various strains and species.

## Structure and Function of Kinetoplast DNA

In addition to the nuclear DNA, leishmanias have an unusual mitochondrial DNA, called the kinetoplast DNA, that comprises 10 percent to 15 percent of the total DNA of the cell. It is composed of two types of circular DNA molecules. The first, referred to as the maxi-circle (20,000 bp long), is the mitochondrial DNA of the parasite; the second type, the mini-circle (800-1200 bp), has no known coding function. The maxi-circle DNA is present in approximately ten copies per kinetoplast; the mini-circle DNA is present in 10,000-50,000 copies. Replication of both the kDNA maxi-circle and mini-circle has been extensively studied.

Elegant work has demonstrated that the maxi-circle replicates by the rolling-mechanism, whereas the mini-circle replicates as individual circles, using an RNA primer at the site of the putative origin of mini-circle replication. Several mitochondrial proteins have been identified by cross-hybridization with genes from yeast and mammalian cells. In addition, DNA sequence analysis of maxi-circle DNA has identified several open reading frames on the maxi-circle DNA. Transcripts of the maxi-circle genes have been identified in both the promastigote and amastigote stages. Developmental regulation of maxi-circle gene expression is less pronounced than in a related kinetoplastid organism, *Trypanosome brucei,* where several maxi-circle transcripts are developmentally regulated in their expression.

## Kinetoplast DNA Mini-Circle

One of the most puzzling DNA molecules in nature is the kinetoplast DNA mini-circle. It has been studied extensively in both *Leishmania* and *Trypanosome* species, (both African and South American). It represents approximately 10 percent of the total cell DNA and is repeated 10,000 to 100,000 times per kinetoplast. Its function is unknown; no protein product or transcript has been demonstrated, although many mini-circle sequences contain a small open-reading frame of 30 - 50 amino acids. A unique 13-nucleotide sequence is contained in all mini-circles sequenced to date and is proposed to be the origin of DNA replication.

All mini-circles have a unique structural feature: a DNA sequence that contains a natural bend. This feature was first described because of anomalous migration of the mini-circle DNA in agarose and acrylamide gels. The DNA sequence itself is not conserved, but the structural

**feature of the bend is conserved. The function of this bend in the DNA is unknown, but one proposal is that it serves as a protein binding site important for the function or structure of the kDNA. DNA mini-circles appear to undergo relatively rapid DNA sequence divergence, presumably because of the noncoding nature of the DNA. This observation has led to a useful application of molecular biology to the differentiation of *Leishmania* species and subspecies and subsequently to the direct diagnosis of leishmaniasis in patients.**

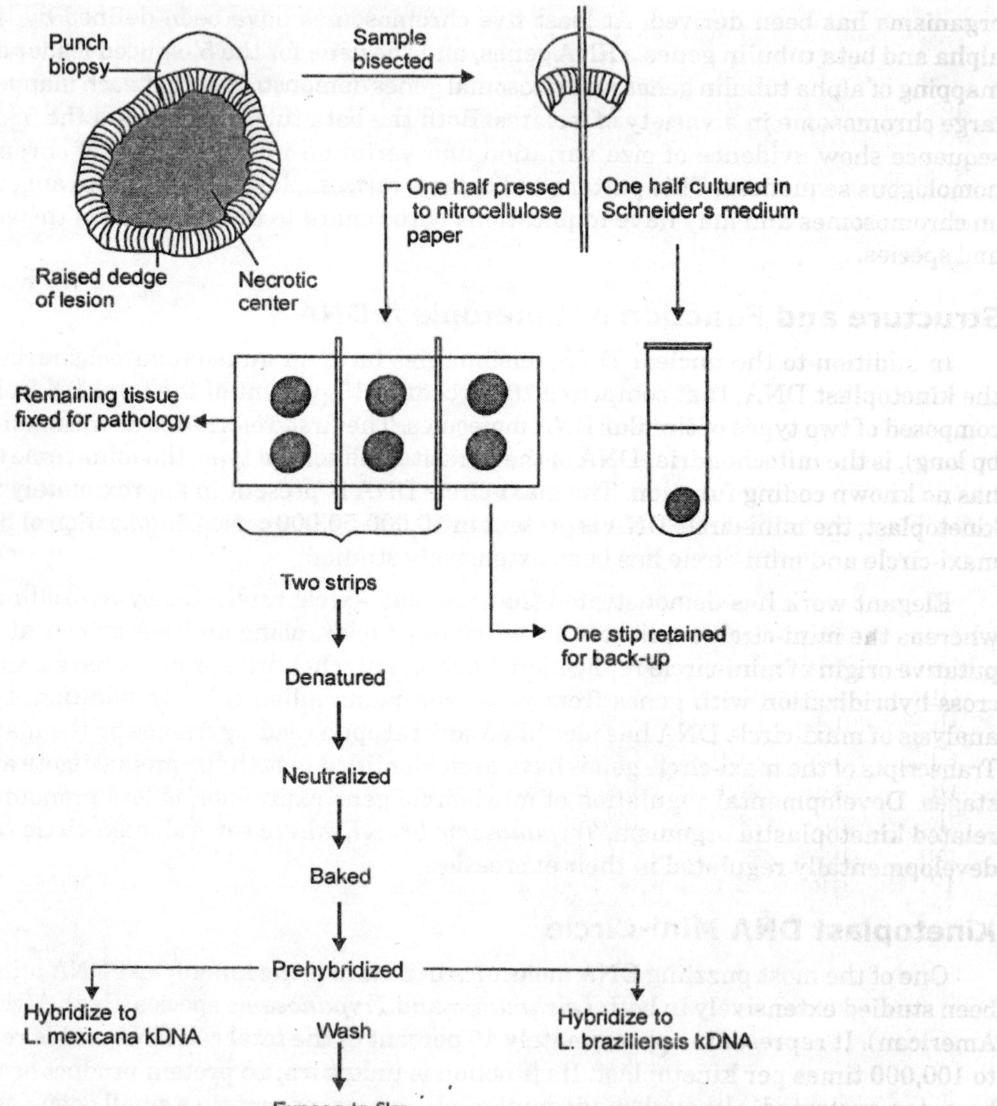


**Fig. 3.11. A schematic representation of the method used for the DNA probe diagnosis of cutaneous leishmaniasis. A biopsy of the cutaneous lesion is taken and is divided in half; the tissue is then spotted directly onto a DNA-binding membrane. This membrane is then treated to denature the DNA and is hybridized with the specific probe for either *Leishmania mexicana* or *Leishmania braziliensis*. Specific hybridization is detected by x-ray film.**

In the New World *Leishmania* organisms, the kDNA mini-circle isolated from L. *mexicana* does not share any sequence homology with those isolated from L. *braziliensis*. This difference in DNA sequence has provided the basis for a DNA probe that can distinguish these two *Leishmania* species directly when material obtained from a lesion is applied to nitrocellulose. This methodology has now been used to diagnose over 300 patients with cutaneous leishmaniasis offering both a more rapid and specific diagnosis than is currently available with any other methodology. Work to develop subspecies-specific diagnostic probes and molecular analysis of kDNA have shown species-related restriction site heterogeneity within the kDNA and several groups have cloned species- or isolate-specific kDNA fragments from visceral and Old World cutaneous strains of *Leishmania*. This work demonstrates that within the kDNA network, fragments of DNA exist with different taxonomic specificities. These cloned subspecies-specific DNA probes are important for the development of future diagnostic probes but are also interesting with regard to the kDNA sequence divergence. Perhaps one of the fascinating molecular questions in kDNA analysis is the mechanism of DNA sequence divergence within the kDNA mini-circle.

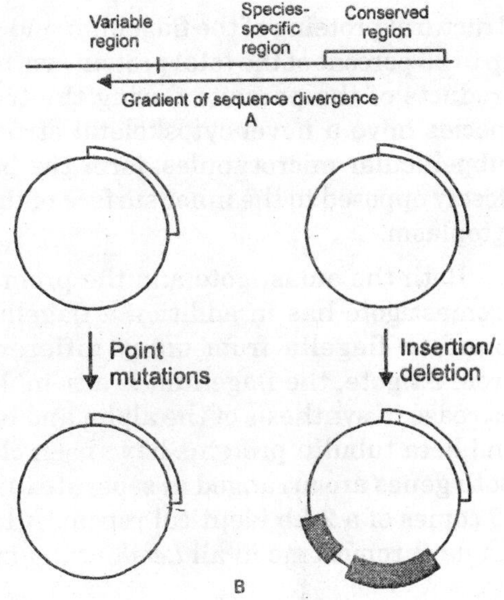


**Fig. 3.12. Models of kDNA sequence divernce.**

Within a single mini-circle, there are regions of conserved DNA sequence and regions of highly divergent DNA sequence. The 10,000 mini-circles within a single *Leishmania* organism are heterogeneous and fall into a number of different sequence classes. Within a given sequence class, the mini-circle sequences are identical. Sequence analysis of different classes of mini-circles in *Leishmania tarentole* and *T. brucei* has revealed that mini-circles of different classes share 150 bp of homology but are widely divergent throughout the rest of their length. In, addition, in *L. mexicana amazonensis,* isolates. The maintenance of homogeneous sequences with different taxonomic specificities are present in a defined order along the length of a single cloned mini-circle. There appears to be a gradient of sequence divergence on the length of the mini-circle. There is a 230-bp region which is conserved across species boundaries, an adjacent 170-bp region that is found only in the *L. mexicana* complex and a 400 bp region which is highly divergent and found only in a single or small number of *L. mexicana amazonenesis* isolates.

The maintenance of homogeneous sequence classes in the face of rapid sequence variation presumably requires a mechanism whereby mutations in a single mini-circle are transmitted to other mini-circles. In addition, some mechanism must exist for the generation of rapid sequence divergence in one portion of the mini-circle molecule with maintenance of relatively conserved regions within the same molecule.

## Developmentally Regulated Gene Expression

To study the developmental regulation of gene expression, it was important to choose the major proteins that are regulated during the transition from amastigote to promastigote. The major biosynthetic change is the increase in the synthesis of alpha and beta tubulin, the major structural proteins of the flagellum and subpellicular microtubules. These proteins can represent up to 25 percent of the total protein synthesized; they therefore represent the major transcriptional products of the parasite during the transition from amastigote to promastigote. *Leishmania* species have a novel cytoskeletal structure: organized arrays of microtubules, referred to as subpellicular microtubules, form the basis for the parasite structure. These microtubules are closely opposed to the inner surface of the plasma membrane and completely outline the parasite cytoplasm.

Both the amastigote and the promastigote stages contain subpellicular microtubules; the promastigote has in addition a flagellum with the standard 9 + 2 microtubule arrangement found in flagella from many different organisms. During transition from amastigote to promastigote, the flagellum is assembled, and this process presumably accounts for the large increase in synthesis of the alpha and beta tubulin proteins. The genes encoding both the alpha and beta tubulin proteins have been cloned and analyzed in *L. enriettii,* a rodent leishmania. Both genes are arranged in separate tandem arrays. The alpha tubulin gene is composed of 10 - 15 copies of a 2-kb identical repeat. All the copies of the alpha tubulin gene are contained on a single chromosome in all *Leishmania* species that have been analyzed.

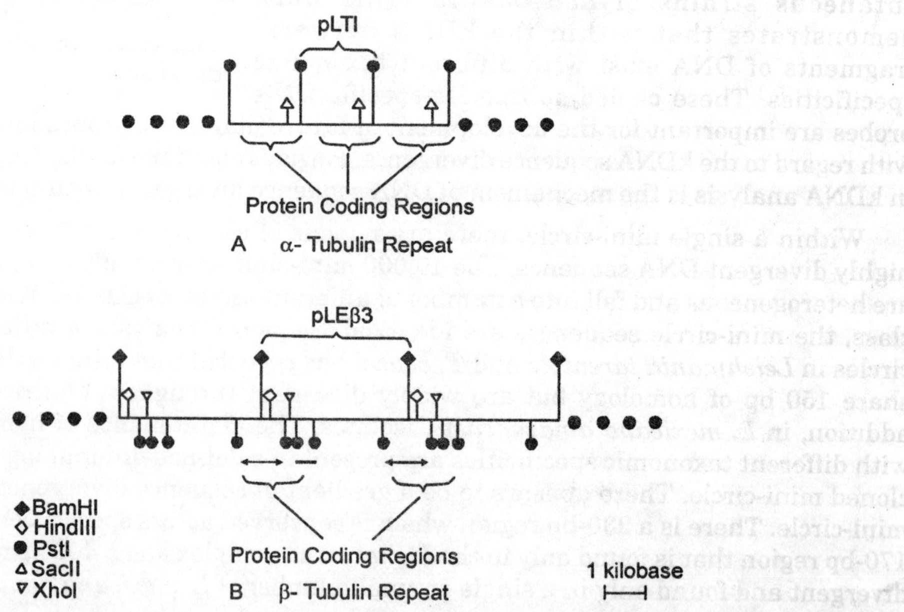


**Fig. 3.13. Models for the arrangement of α- and β-tubulin genes in the genome of *L. enriettii*. A. The structure of tandemly repeated α-tubulin genes is indicated, along with selected restriction sites. The 2-kb *PstI* fragment that forms the insert for the α-tubulin genomic clone pLT1 is marked in brackets, and the α-tubulin protein-coding region is delineated by the bar. The direction of transcription is from left to right. B. The structure of the tandemly repeated β-tubulin genes is displayed, along with selected restriction sites. The 4-kb BamHI fragment that constitutes the insert of the *b*-tubulin genomic clone pLEB3 is indicated by the brackets, and the protein coding region is marked by the narrow bars. The direction of transcription is from right to left.**

In contrast, the beta tubulin gene, which is approximately 4 kb long, is present both as a tandem array of 10-15 genes on a single chromosome and in separate copies on at least two other chromosomes. The exact function of the beta tubulin genes not contained in the repeat is unknown. They may be nonfunctional pseudogenes or they may be genes required for special functions, such as mitosis. Why the alpha tubulin gene complex does not have these genes outside the repeat is a curiosity.

The expression of alpha and beta tubulin in *L. enriettii* is controlled at the level of mRNA accumulation. Both of the tubulin mRNAs are present in 8-10-fold greater quantity in the promastigote than the amastigote. In another species, *L. mexicana,* translational control also may play a role in the regulation of tubulin protein expression. Approximate sizes of the alpha and beta tubulin mRNAs were determined by fractionating total RN A on denaturing gels and sequentially hybridizing the same Northern blot with the cloned probes for alpha and beta tubulin.

Both mRNAs appear to be slightly greater than 2 kb long, with the beta tubulin slightly larger that the alpha tubulin mRNA. It is interesting to note that although the beta tubulin gene is almost twice as large as the allpha tubulin gene, the final mRNA molecules are very similar in size. The sizes of the tubutin RNAs isolated from promastigotes are identical to those isolated from amastigotes. Detailed transcriptional analysis of the tubulin mRNAs has revealed that each mRNA is modified at its 5′-end to include the identical 35-nucleotide sequence. This 35-nucleotide sequence is not encoded upstream from the gene, but is spliced onto the structural gene RNA by a novel mechanism, described later, which results in the formation of a hybrid mRNA composed of transcripts from two distinct genes on separate chromosomes. DNA sequence and SI mapping analysis have led to the identification of putative gene control or RNA processing sequences.

The 35-nucleotide sequence is added onto the 5′-end of the structural gene mRNA at the position of a consensus splice site in both the alpha and beta tubulin upstream regions. There is a single splice site in the alpha tubulin and two putative splice sites in the beta tubulin. The most striking feature of the upstream region of both the alpha and beta tubulin genes is a tandem repeat of the dinucleotide AC just upstream from the RNA splicing site. The AC runs occur at different distances from the splice site in the two genes, but their position relative to the actual transcription initiation sites is unknown. Purine-pyrimidine runs of the type observed here may have the potential to form zDNA, but the role of DNA structure in gene expression in these parasites remains to be determined. In the alpha tubulin sequence, there is consensus polyadenylation site at the position of the 3′-end of the mRNA.

## Other Protein Genes

Analysis of structural protein genes in *Leishmania* organisms has been limited mainly by the workers in the field. mRNA and DNA sequences homologous to the heat-shock proteins (HSP 70) have been identified both in African trypanosomes and *Leishmania* species. As described next, the genes encoding dihydrofolate reductase/thymidylate synthetase bifunctional enzyme have been isolated from drug-resistant parasites.

# NOVEL BIOCHEMICAL AND MOLECULAR MECHANISMS

## RNA Processing

The *Leishmania* species have an unusual mechanism of mRNA processing. Each mRNA is a hybrid RNA molecule containing RNA sequences derived from two distinct parts of the genome, in fact from two different chromosomes. Each mRNA contains both a unique sequence, which encodes a structural protein, and a common sequence, which is an identical 35-nucleotide cap at its 5′-end. In the *L. enriettii,* this cap, or spliced leader sequence (SLS), is encoded by a 438-bp gene arranged in a tandem array of approximately 200 copies. These spliced-leader genes are transcribed to give a 85-nucleotide precursor that has the 35-nucleotide cap sequence at its 5′-end. The genes for the SLS are located on a separate and distinct chromosome from that of the alpha and beta tubulins.

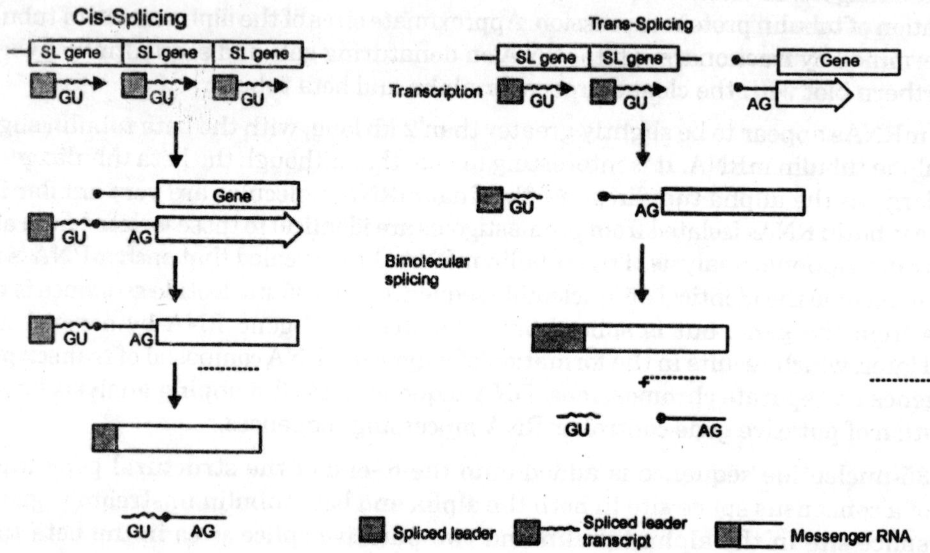


**Fig. 3.14. Models of RNA processing after discontinuous transcription. In cis-splicing, the spliced-leader transcript could prime transcription of a gene, resulting in a precursor with an intron. Unimolecular splicing produces a single RNA by-product containing genomic sequence upstream to the AG consensus splice site and the remainder of the spliced-leader transcript. Alternatively, in cis-splicing, a ligation reaction after two unlinked transcription events could result in a unimolecular splicing event. In trans-splicing, independent transcription events followed by bimolecular splicing produce the two RNA by-products shown and mature mRNA. One RNA by-product contains a spliced-leader genomic sequence, and a separate by-product contains a transcribed genomic sequence from the gene processed to mRNA. Primed transcription after removal of the 3′-end of the spliced-leader transcript would also result in a RNA processing product comprising the free 3′-end of the spliced-leader transcript. GU represents the nucleotides of the consensus eukaryotic splice junction found in the spliced-leader transcript immediately 3′ to the spliced leader. AG represents the nucleotides of the consensus eukaryotic splice junction found immediately 5′ to the splice junction in all trypanosomatid mRNA genes examined to date.**

An analogous gene was originally identified in *Trypanosoma brucei,* and work in that system has demonstrated that this 5 LS is added onto all mRNAs in the cell. The exact function of this SLS is not known. It may be a novel mechanism for the capping of mRNAs, or it may be important as a signal in the translation of proteins. Determination of function must await a transfection system that will allow site-specific mutagenesis of the gene. One of the major areas of current research is the elucidation of the mechanism of this novel hybrid mRNA formation. DNA sequence and primer extension analysis have demonstrated the presence of consensus splice sites both in the SLS transcript and in the upstream regions of the alpha and beta tubulin genes. There are two basic models for the addition of the SLS to mRNA precursors, the cis-splicing model and the trans-splicing model. In the cis-splicing model, the SLS is covalently attached to the primary transcript of the tubulin gene, either as a primer during transcription or posttranscriptionally by ligation, thereby creating a hybrid intron.

Subsequently unimolecular splicing occurs similar to that found in eukaryotic systems. This model predicts that there will be two products of splicing: a mature hybrid mRNA, with the SLS attached to the tubulin structural RNA, and the hybrid intron, consisting of the downstream region of the SLS transcript and the upstream region of the tubulin primary transcript. Evidence is accumulating that the splicing is more likely to be through a trans-splicing mechanism. In this model, the SLS gene and the tubulin structural gene are independently transcribed and splicing occurs between two separate transcripts. No covalent hybrid intron structure is formed. This model predicts that there will be three products of splicing: the hybrid mRNA, the downstream region of the SLS transcript, and the upstream region of the tubulin gene.

The downstream region of the SLS gene has been identified in *L. enriettii,* and recent work in *T. brucei* and *L. enriettii* has demonstrated that incubation with extracts known to debranch intron lariat structures results in the accumulation of the downstream region of the SLS transcript. What remains to be demonstrated is the presence of free upstream transcript from the tubulin or any structural gene. Future experiments in this area include the identification of splicing complexes and the isolation of enzymes responsible for. this novel RNA processing mechanism.

## Drug Resistance

Major areas of interest both for genetic analysis and for the more practical consideration of drug therapy are the mechanisms of drug resistance in the *Leishmania* organisms. Work in several laboratories has demonstrated that gene amplification plays a major role in the development of drug resistance. At least one mechanism of resistance is the overproduction of the target gene product in the resistant parasite, a pattern that is similar to the mechanism found in many drug-resistant mammalian cells and that may be an important in vivo step in the development of drug resistance. *Leishmania major* parasites have been selected for their resistance to the folic acid analogue methotrexate; in separate experiments, they have been selected for resistance to an inhibitor of thymidylate synthetase, 10-propargyl-5,8-dideazafolate (CB3717). These resistant parasites were selected after stepwise increases in drug concentration. *L. major* has a bifunctional enzyme that contains both DHFR and TS activity, and resistant parasites show an increased activity of the bifunctional enzyme. In mutants selected with methotrexate, this increased enzyme activity is accompanied by the amplification of two regions of the DNA: one of 30 kb, the R region, and the other of 80 kb, the H region.

In mutants selected with CB3717, the TS inhibitor, only the R-region DNA is amplified. These amplified DNA regions exist as circular extrachro-mosomal "factors" demonstrated by migration in chromosome gels and by visualization in the electron microscope. Initially, these circles were unstable in the absence of selective pressure; a decrease in the amount of amplified DNA was accompanied by a desrease in the bifunctional enzyme activity. After prolonged selective periods, stable mutants have been isolated. Analaysis of both stable and unstable resistant mutants by orthogonal field alternating gel electrophoriesis (OFAGE) demonstrates extrachromosomal circles homologous to the R and H regions. Thus, these circles must contain sequences for replication and stable inheritance. The structural gene for the bifunctional enzyme is located on the 30-kb amplified circle, and its protein sequence is derived from the cDNA sequence and genomic sequences.

It is interesting that the DHFR portion of the protein, located at the amino terminus has little primary sequence homology with mammalian or bacterial enzymes, whereas the TS or carboxyl portion of the enzyme has regions of significant homology with thymidylate synthetase isolated from other organisms. The DHFR portion of the protein does share secondary structure homology and known substrate binding sites with other DHFR proteins. Mutant parasites resistant to tunicamycin, an inhibitor of N-acetylglucosminyl-transferase in the dolichol pathyway of protein glycosylation, have been isolated in a similar stepwise manner. These resistant parasites have an increased level of the target enzyme; there is also evidence from a resistance-specific banding pattern of *Hind*III-digested, ethidium bromide-stained DNA that resistant parasites have amplified DNA.

The gene product of this amplified DNA has yet to be determined, however, it is interesting to note that these resistant mutants have increased virulence. Gene amplification of enzyme structural genes appears to be at least one important mechanism in the development of drug resistant in Leishmania species. An important question that remains to be addressed is the molecular mechanism of this circular amplification and its role in the development of drug resistance in the field.

## Coda

The major development necessary to continue molecular genetic analysis of leishmanias is the development of a transfection system to allow the introduction and analysis of modified genes in these parasites. Initial attempts at this work in a related Kinetoplastida species, *Crithidia facisulata*, involved the introduction of a drug-resistant gene (G418) carried on a plasmid. Resistant parasites were isolated, but rearrangement of the plasmid NDA in these parasites made the interpretation of these results ambiguous. Several areas of research need to be pursued in order to develop a successful transfection system. First, some mechanism of introducing DNA into the cells is necessary.

Work in other eudaryotic systems suggests that electroporation, microinjection, or direct DNA addition should be tried. Second, a selectable marker, either drug resistance or an auxotrophic marker, is necessary. Finally, in order to achieve stable transformation, either an independent origin of replication or an integrating plasmid will be required. Highly repeated genes such as the SLS or tubulin genes may be good candidates for integration and the kDNA molecule or the amplified stable circles may provide the function of the origin of replication. One this transfection system is developed, of the structure and function relationships of the various putative controlling elements in gene expression can be directly assayed. This system also may help to define important regions of receptors or other virulence factors and could be the breakthrough that will take molecular work with this parasite from description to functional explanation.

## BIOCHEMISTRY SCHISTOSOMES AND FILARIAE

Filariae and schistosomes rank among damaging helminth parasites infective to humans. Approximately 300 million people harbor schistosomes, 800,000 of whom die each year; filariae affect 300,000 individuals, and although they account for little mortality, they lead to blinding and disfigurement of significant numbers of people. Moreover, infections by these parasites tend to follow a chronic course that immeasurably saps the vitality and compromies the productivity of millions of children and adults. The study of the molecular biology of parasitic helminthes is a new discipline that arose from the opportunity to apply technologies offered by molecular genetics to questions raised by workers investigating the immunology, biochemistry, physiology, and biology of the organisms. Although it is clear that the use of recombinant DNA techniques in the study of schistosomes and filariae will primarily serve as an amplification system for providing quantities of proteins whose availability is otherwise limited, preliminary indications suggest that application of the new technologies also may aid in unraveling basic biological questions.

## Vaccine Development

The goal of conferring sterilizing, or even partial, immunity to humans against schistosomes and filariae is obviously a desirable objective, and the search for safe, inexpensive vaccines that can impart this protection is a high priority. Such vaccines would complement current efforts to control these diseases with chemotherapy, vector control, and education. Although scientific progress to date is encouraging, scientific and economic considerations make if currently unclear whether such vaccines can be developed or if a large-scale vaccination program could be implemented. A major motivation to develop vaccines is the realization that in the absence of protective immunity, individuals treated with antihelminthic drugs tend to become reinfected. The life span of schistosomes and filariae within a vertebrate host is extraordinarily long; schistosomes have been documented to remain viable and fecund for up to 35 years. They stimulate a host immune response of variable intensity, and how they survive host attack has intrigued many scientists. While investigators for decades have attempted to identify antigens having immunoprohylactic potential, recent molecular and immunological methodologies have provided several powerful approaches to the problem.

Different laboratories have tackled the issue in different ways, with some degree of success. Convinced that target antigens should be routinely exposed to the immune system, some workers have proceeded to isolate and genetically express components of the parasites' surface or excretory/secretory antigens, and are testing their prophylactic efficacy in murine models. Others have produced and characterized monoclonal antibodies that were chosen for their ability either to kill worms in vitro or to confer resistance to infection after passive transfer into immunologically naïve recipients. Antibodies identified in this manner have then been used as reagents for isolating the protective antigen itself, usually from recombinant DNA sources. Others have concentrated on the identification of antigens recognized by antibody from hosts naturally resistant to infection but not recognized by antibodies of susceptible hosts.

The most thorough approach has involved fractionation of parasite extracts, assessment of the protective value of each fraction, and purification of active components. None of the candidate antigens characterized to date has proved capable of engendering sterile immunity, although this failure may reflect suboptimal stimulation of appropriate T-cell populations or of B cell of the right isotype rather than identification of an unsuitable antigen.

## Schistosomes

The characterization of protective antigens with monoclonal or monospecific polyclonal antibodies that are cytotoxic to schistosomes has yielded a perplexing catalog of molecules. For example, preliminary data indicate that enzymes of the parasite's glycolytic pathway may represent important vaccine candidates. Adult schistosomes and filariae are anaerobes that use glucose as their sole source of energy. Their consumption of the sugar is prodigious: schistosomes consume the equivalent of 20 percent of their dry weight per hour. The importance of glycolysis to worm survival has been known for many years, and trivalent antimonials (once used as antischistosomal agents) appear to kill the organisms by interfering with the action of phosphofructokinase. Independent research groups have characterized three apparently protective antigens and identified them as aldolase, glyceraldehyde-3 phosphate dehydrogenase, and triose phosphate isomerase. Curiously, the enzymes catalyze sequential reactions that form the core of the glycolytic pathway.

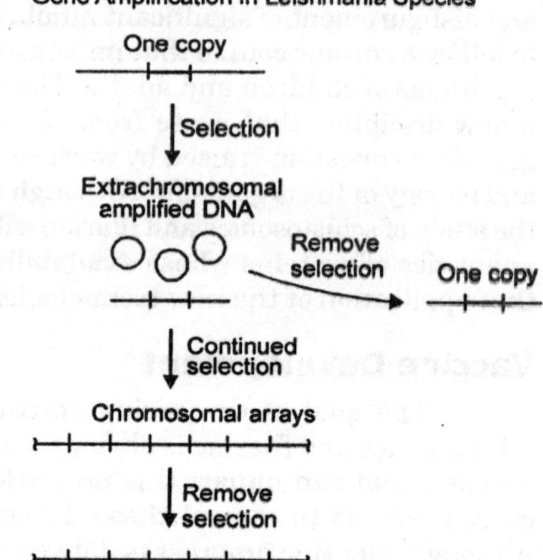


**Fig. 3.15. Summary of the molecular structure of amplified DNA in unstable and stable *Leishmania* lines resistant to methotrexate.**

It has recently been demonstrated that aldolase also is a potential vaccine target of *Plasmodium falciparum.* Should these proteins indeed impart protection when used as vaccines, their interplay with the immune system would raise several questions. For example, since these enzymes are not rate-limiting to the pathway (whereas membrane transport and the three kinases are), it is surprising and unclear why immune responses to them have antihclminthic consequences. Furthermore, the fact that the enzymes are normally found in the cytoplasm of eukaryotic cells raises the perplexing question of how the immune system can launch an attack on them in intact organisms. Finally, the amino acid sequences of the three proteins are very similar to host enzymes, which raises the concern that vaccines based on these proteins might elicit autoimmune responses. Another empirically determined class of schistosome vaccine candidates consists of glutathione metabolizing enzymes. One of these proteins was selected on the basis of the observation that mice resistant to *Schistosoma japonicum* infection mount a response to an antigenic enzyme, whereas mice that are susceptible to infection do not.

The antigen has been cloned and expressed in bacteria, and has been found to be a 26-kilodalton (kD) isoenzyme of glutathione S-transferase. Unlike the case for the glycolytic enzymes, the protein demonstrates only a modest degree of resemblance to the host homologue. Mice infected with *S. japonicum* produce antibodies that cross-react with the enzyme from *Schistosoma mansoni,* and sequencing of the S. *mansoni* homologue has revealed that they are nearly identical in structure differ significantly from a 20-kD isoenzyme isolated from *S. mansoni*. The similarity between the enzymes of the two parasite species is much greater than that noted for their heat-shock protein, HSP-70 (which ranks among the phylogenetically most conserved proteins), and is reflected in the sequence homology of 5′ untranslated portions of their RNAs, which begin with an unusual string of polyT.

```
Aldolase

Sm ald  RHAADESTATMGKRLQQIGVENNEENRRLYRQLLFSADHKLAENISGVILFEETLHQ
RM ald  IL*******GSIA****S**T**T*****F*****LTA*DRVNPC*G*****H***Y*
HL ald  IL******VG**N***R*K***T*****QF**I***V*SSINQS*G*****H***Y*

Sm ald  KSDDGKTLPTLLAERNIIPGIKVDKGVVPLAGTDNETTTQGLDDLASRCAEYWRLGC
RM ald  *A***RPF*QVIKSKGGVV*************NG******G*SE***Q*KKD*A
HL ald  KDSQGKLFRNILKEKGIVV***L*Q*G******NK***I****G*SE***Q*KKD*V

Sm ald  RFAKWRCVLKISSHTPSYLAMLEMLMYL
RM ald  D***********GE****A**IM*NANV*
HL ald  D*G***A**R*ADQC**S**IQ*NANA*

Glyceraldehyde-3-Phosphate Dehydrogenase
        1                                                          60
SCH     GSRAKVGINGFRGIGRLVLRAAFLKNTVDVVSVNDPFIDLEYMVYMIKRDSTHGTFPGEV
HUM     MGKV***VN**GR*****T****NSGK**I*AI*******N*****FQY*****KFH*T*

                                                                   120
SCH     STENGKLKVNGKLISVHCERDPANIPWDKDGAEYVVESTGVFTTIDKAQAHIKNNRAKKV
HUM     KA*****VI**NP*TIFQ****SK*K*GDA***************ME**G**LQGG-**RV

                                                                   180
SCH     IISAPSADAPMFVVGVNENSYEKSMSVVSNASCTTNCLAPLAKVIHDKFEIVEGLMTTVH
HUM     **************M***HEK*DN*LKII*************************N*G**********
```

**Fig. 3.16. Inferred amino acid sequence of the amino termini of two putative schistosome vaccine target antigens. In the top sequence, the primary sequence of a clone identified as encoding the schistosome aldolase (sch ald) is compared to that of clones encoding rat muscle aldolase (RM ald) and human, liver aldolase (HL ald). In the bottom panel, the sequence of the schistosome glyceraldehyde-3-phosphate dehydrogenase is compared to an isoenzyme from human beings. Asterisks represent regions of identity.**

A second isoenzyme of the glutathione S-transferase of *S. mansoni* also has prophylactic potential. The protein has a size of 28 kD, shows little sequence similarity to the 26-kD isoenzyme, and induces production of antibodies that do not cross-react with the 26-kD homologue or with homologues of the host. The enzymes are normally found in the cytoplasm or microsomal fraction of metazoan cells; however, there are indications that they may also be found on the surface of the parasite. The functional role of the schistosome isoenzymes is unknown, but their presence in the tegument suggests that they may play a role in protecting the organisms from oxygen-mediated damage. Protective responses stimulated by the antigen appear to involve the production of specific IgGZa antibodies that direct eosinophil-mediated cytotoxicity. Another unexpected schistosome vaccine candidate is paramyosin.

The peptide was identified by fractionation of parasite extracts and testing the degree of protection conferred by each fraction in a mouse model. A cDNA encoding part of the schistosome paramyosin has been used to isolate a filarial homologue, and this homologue has been shown to confer partial resistence to filariae in immunized animals. Paramyosin is found in invertebrate catch muscles, where, in association with myosin, it forms complexes that permit strong, protracted muscle contraction, as seen in the adductor muscles of clams. Immunocytochemical studies indicate that the protein is also found in nonfibrillar form within elongate membrane-bound structures within the worm's tegument, although it is apparently not exposed on the surface. Like schistosome myosin, sequenced portions of the parasite's paramyosin indicate that it consists of a series of seven amino acid repeats (a,b,c,d,e,f,g), in which residues a and d are generally hydrophobic, and which are part of a 28-amino-acid unit.

**Table 3.2**

Inferred amino the amino acid sequences of the amino termini of the 26-kilodalton (kD) glutathione S-transferase of *S-Japonicum* (Sj26), the rat yb isoenzyme-(RYb), the rat yb the rat yc (Ryc), and a 28-kD isoenzyme from S.mansoni (Sm 28) compared to the 26(kD) protens Mansoni (Sm 26). Asterisks represent regions of idecutety

| | |
|---|---|
| Sm26 | MAPKLGYWKIKGLVQPTELLLEYL |
| Sj26 | TKL * I*********************** |
| RYB | * PMI ****NVR**TH*I******T |
| SYC | * PGG*PVLHYFDGRGRMEP I RWL* |
| Sm28 | * AGEHIKVIFDGRGRAES I RMT |
| Sm26 | (#68) DVKLTQSMA I LRY I ADKHNMLGGCPKERAEISMLEGA |
| Sj26 | (#68) ***********I******************************* |
| RYb | (#67) SR * I ***N**M**L*R*HLC*ETEE***IRADIV*NQ |
| RYc | (#62) GM **V*TR***N***T*YALY*KDM****L*D*YSEC |
| Sm28 | (#65) VKWMLE*L **A**M*K*H**M*ETDE*YYSVEK*I*Q |

Acidic and basic periods within the 28 amino acid units are almost out of phase with each other, so that each segment can be subdivided into zones of alternating charge. In general, it appears that the molecule, like myosin, evolved from a 28-amino-acid peptide whose gene underwent a series of duplications and that the register between the molecule and myosin is mediated by ionic interactions. The amino acid sequence of paramyosin predicts that the molecule forms a classic a helix, where an antiparallel second strand can make periodic contacts at regularly spaced intervals where amino acids having a nonpolar side chain are positioned. Amino acids facing the aqueous environment of paramyosin dimers tend to have hydrophilic side chains. Once the complete amino acid sequences of schistosome myosin and paramyosin are determined, it will become possible to construct and test models on how the two proteins interact.

Unlike the case of glutathione S-transferases, paramyosin appears to stimulate T lymphocytes to produce interferon gamma, which activates macrophages to kill larval parasites. Notably, it seems that antibodies play no role in this cytotoxic reaction. The protein also may elicit an inflammatory reaction in sensitized hosts, which could wall off invading larvae. Another schistosome antigen — myosin — has been cloned and is recognized by antibodies generated by animals that are poorly resistant to infection by sera from susceptible animals. Antibodies against the antigen rise dramatically in patients after chemotherapeutic cure. The protein is unique in being encoded by a single-copy gene (all organisms studied to date have their myosins encoded by a multigene family).

Preliminary indications are that, unlike paramyosin, schistosome myosin is not a promising vaccine candidate. Taken as a whole, these findings add a new dimension to our understanding of schistosome antigens susceptible to cytotoxic arms of the immune response. Particularly puzzling is the mechanism by which host antigen processing cells manage to recognize cytoplasmic components of living schistosomes and subsequently translate this information into an immune response capable of homing in on molecules enclosed by a plasma membrane. Also surprising is the fact that nearly all potentially protective antigens sequenced to date share significant blocks of homology with host homologues, yet they apparently do not elicit an autoimmune response.

Whether this is attributable to special modes of presentation of the antigens by the parasites or to an inherent lack of immunogenicity of evolutionary conserved domains is unknown, but the question has obvious relevance to vaccine development. The observation that schistosome antigens contain regions whose structure is highly conserved suggests that these may have important functional purposes.

## Filariae

As is the case with schistosomes, experimental evidence suggests that it is possible to develop protective immunity to filariae. Several workers have reported that immunization with irradiation-attenuated $L_3$ larvae can lead to a significant depression of adult worm burden and microfilaremia. Monoclonal antibodies against 70-kD and 75-kD stage-specific surface antigens of *Brugia malayi* have been developed that upon passive transfer lower the microfilaremia of infected animals. Immunization of gerbils with a soluble extract of *B. malayi* induces a 50 percent reduction in adult worm burden and a 80 percent to 100 percent decrease in parasitemia on challenge.

The development of resistance correlates with the appearance of IgG antibodies directed at proteins of 25 kD, 42 kD, 60 kD, and 112 kD. Immunization of animals with purified 25 kD and 60 kD proteins leads to their ability to lower the number of circulating microfilariae. A cDNA encoding the 60-kD antigen has recently been cloned and sequenced. The predicted size of the cDNA's translation product is 63 kD, and the antigen does not have any consensus glycosylation sequences or any hydrophobic domains suggestive of membrane-anchoring sites. Interestingly, the mRNA for the protein begins with a 5′ untranslated leader sequence that is identical to the trans-spliced leader of some actin genes of *Caenorhabditis elegans* (a free-living nematode). The 22-nucleotide spliced leader is present in several locations in the parasite's genome, including (as in *C. elegans)* within the 5s rRNA gene repeat.

The spliced leaders are located at the 5′-end of a 109-base pair (bp) transcriptional unit of nonpolyadenylated RNA. The role of the spliced leaders, and why some genes acquire this sequence (as opposed to the trans-spliced leaders in trypanosomes, which is a universal feature of the mRNAs of the organisms) is unclear and awaits the development of appropriate assay systems which can evaluate their significance.

## Chemotherapy

The drugs of choice currently used in the treatment of schistosomiasis and filariasis are praziquantel and ivermectin, respectively. Praziquantel is effective in a single oral dose and appears to paralyze the organisms by perturbing their tegument and by affecting calcium transport; the immune system probably plays a role in killing and clearing organisms debilitated by the drug. Praziquantel is also effective against other trematode and cestode parasites but not against nematodes whose cuticle blocks the action of the drug. Verapamil, a slow calcium blocker, also disrupts the schistosome tegument and may prove a less expensive alternative to praziquantel. The semisynthetic antibiotic mixture, ivermectin, similarly affects calcium fluxes at neuromuscular junctions, presumably by increasing the secretion of gamma amino butyrate (GABA) and enhancing the binding of the neurotransmitterf to its receptor.

Although not a perfect drug, ivermectin is a welcome alternative to diethylcarbamazine, in that it kills microfilariae more slowly, thereby decreasing the possibility of inducing potentially life-threatening Mazzoti-like reactions. The latter are thought to be allergic responses triggered by a sudden appearance of high levels of antigen released by dying microfilariae. The perceived need to develop new antihelminthic drugs stems from three problems. First, researchers suspect that the parasites will, in time, develop resistance to currently available drugs. Second, current therapeutic regimens are too expensive to apply on a large scale, and reinfection is a constant threat. Finally, in the case of filariasis, available drugs have serious side effects and are not completely effective.

It has been recently noted that veterinary use of ivermectin may introduce new ecological problems because the compound is not biodegradable and is toxic to a wide variety of invertebrates. Combining methods in molecular biology with emerging ones in computer modeling can in theory strengthen and facilitate innovative approaches to the rational development of therapeutic drugs. Antihelminthic drugs currently in use are all products of empirical research, an inefficient method considering that on the average only 1 in 10,000 tested compounds shows any activity.

The concept behind rational drug design is to take advantage of biochemical idiosyncrasies that distinguish parasite from host metabolism. This might be accomplished in at least two ways. First, methods in molecular genetics can be used to clone and express active parasite enzymes in more manipulate hosts (yeast and bacteria). If enzymatically active, the enzyme can then be studied by conventional biochemical techniques. If the gene can genetically complement a yeast strain deficient in the enzyme, it is conceivable that the protein's properties can also be studied in vivo. In the other approach, the primary sequence of the protein is determined, and if the three-dimensional structure of a homologue from another organism is known, predictions may be made about the conformation of the parasite peptide.

After determination of the probable active site is made, the molecules can be modified by in vitro mutagenesis in order to test the validity of the predicted model. Several laboratories are currently developing algorithms to predict the structure of chemicals that might interfere with the function of active sites of known morphology and chemistry. One group, for example, has developed a model that correctly predicted that methotrexate should inhibit the activity of the dihydrofolate reductase-thymidylate synthetase of leishmania. One of the better examples of rational identification of drug targets has centered on how purine nucleoside pools are formed in *S. mansoni*. This organism is incapable of synthesizing these compounds de novo and must rely on salvage pathways involving acquisition of precursors from the host. Purines and their respective nucleosides gain entry into the parasites by carrier-mediated mechanisms involving at least five transporters.

The surface of the worm contains nucleotidases that may be coupled to nucleoside transport. Culture media preincubated with adult schistosomes (which regularly slough off their outer surface) are capable of converting adenosine to adenine, inosine, and hypoxanthine, suggesting the existence of a salvage pathway consistent with the scheme and noteworthy in three ways: (1) the organisms lack a nucleotide kinase, so that the initial substrate for nucleic acid synthesis is always a free base; (2) adenosine deaminase activity is developmentally regulated, being absent in larval worms, which therefore cannot interconvert adenosine and guanosine nucleosides; and (3) a single enzyme, hypoxanthine guanidyl phosphoribosyl traosferase (HGPRTase), catalyzes the ribosylation of guanidine and hypoxanthine.

The proposed pathway indicates that the availability of guanosine nucleotides is strictly dependent on the activity of the parasite's HGPRTase. Studies on the schistosome enzyme

HGPRTase have shown that it is structurally different from mammalian homologues, which makes it a plausible drug target. A cDNA encoding a schistosome HGPRTase has recently been cloned and sequenced. The message for the protein is approximately 1.5 kb in size and hybridizes to a cDNA encoding a mammalian counterpart under conditions of moderate stringency. The predicted size of the protein is 26 kD, indicating that the active 105,000-dal-ton (D) protein is a tetramer.

Alignment encoding the predicted aminojacidseduence of the cDNA for the schistosome enzyme with that of humans and *Plasmodium falciparum* indicates that the enzyme is more closely related to that of humans (47.9 percent versus 36.9 percent). The areas of greatest identity between the human and schistosome enzyme correspond to the putative purine and phosphoribosyl binding sites. Interestingly, despite considerable differences in amino acid sequence the predicted α-β-α structure of the two enzymes is very similar and demonstrates considerable differences from that of malarial parasites. The 13 amino acid sequence stretch corresponding to the putative binding sites of the human and schistosome HGPRTase differs by one conservative and one nonconservative substitution, suggesting that development of inhibitors acting at this site will prove difficult. Several critical proteins of filariae that differ significantly in structure from those of mammals have been identified. These include their respiratory electron pathway complex III, tubulins, chitin, proteinases, amino acyl tRNA synthetases, and glutathione-folic acid metabolizing enzymes.

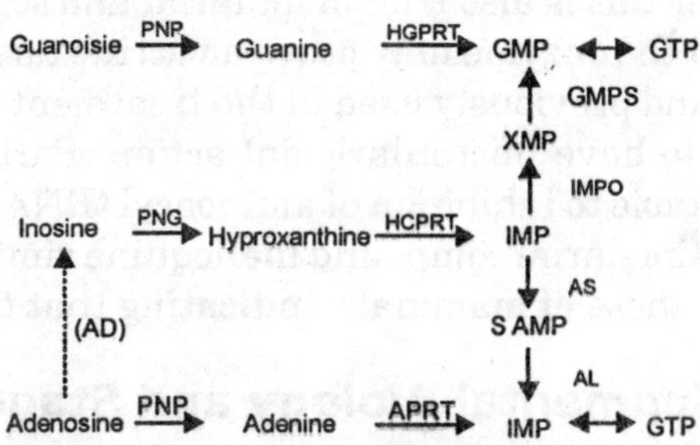


**Fig. 3.17. Purine salvage pathways of schistosomes. Abbreviations: PNP, purine nucleoside phosphorylase; HGPRT, hypoxanthine-guanine phosphoribosyl transferase; GMPS, guanosine monophosphate transferase; IMPD, inosine monophosphate dehydrogenase; AS, adenylosuccinyl AMP synthetase; AL, adenylosuccinyl AMP lyase; AD, adenine deaminase. Solid lines represent general pathways; dashed lines indicate stage-specific ones.**

Parasites

Putative Phosphoribosylphosphate Binding Site
Schistosome KDKNVLVVEDI IDTGKTITKL
Mouse TG**** I ***********MQT*
Human TG**** I ***********MQT*

**Fig. 3.18. Amino acid sequences of the putative purine and phosphoribosyl-phosphate-binding sites of the schistosome-XGPRTase are compared with those of the enzyme from mice and humans. Asterisks represent points of identity; periods denote the presence of amino acids that it do not appear to participate in binding.**

There are indications that the filarial aldolase exerts a rate-limiting action on glycolysis since its activity is lower than that of phosphofructokinase, an enzyme that can catalyze 10 times more substrate than that present at steady-state levels. The filarial aldolase thus appears to be a nonequilibrium regulatory enzyme, and may prove a good chemotherapeutic target; even slight inhibition of such enzymes can greatly reduce the metabolic flow through the pathway.

Whether this is also true in malarial and schistosome parasites, where preliminary indications are that the enzyme may prove a vaccine target, is unclear. Derivatives of amoscanate, a noxious compound previously used in the treatment of hookworm and schistosome infections, have been shown to have microfilaricidal action. Part of the activity of these compounds appears to be attributable to inhibition of aminoacyl tRNA synthetases specific for arginine, lysine, and serine. The antimalarial compound mefloquine similarly inhibits filarial aminoacyl tRNA transferases, but not those of mammals, indicating that these enzymes represent a potential drug target.

## Developmental Biology and Stage-Specific Gene Expression

Filaria and schistosome worms display a range of intricate developmental plans involving permutations of (1) life in aquatic, vertebrate, and invertebrate habitats; (2) shifts in temperature, ambient gas phases, and available nutrients; and (3) changes in the identity of hostile environmental pressures. The mechanisms whereby the organisms find it possible to transform from one stage to another are poorly understood, and how these are regulated at the level of gene expression is completely unknown. Several laboratories have recently begun to address the issue, primarily by analyzing stage- and tissue-specific transcripts. The parasites under discussion are genetically programmed to sequentially and specifically survive within widely different habitats. During these transitions, they display both non-stage-specific (glycolysis, DNA replication, transcription and translation, and gluconeogenesis) and stage-specific (respiratory pathways used by larval forms but not adults) metabolic strategies. While the inference of differential gene expression is easy to understand in the case of stage-specific gene expression, many workers have suggested that common metabolic pathways may be mediated by isoenzymes subject to stage-specific appearance.

Studies in comparative enzymology have revealed that protein-catalyzed reactions are subject to ionic, pH, and temperature defined modulation. Temperature is particularly important, as it influences rates of reactions, rates of protein denaturation, affinity toward substrate, and pH dependence. These considerations may critically affect the function of near-equilibrium, regulatory, or membrane-bound enzymes (changes in temperature affect membrane fluidity. How schistosomes and filariae have solved the problem of different stages adapting to different environments (for example, see Chapter 6) is unknown. The solution might involve the expression of separate genomic segments, the synthesis of enzymes that have broad physical optima, or synthesis of developmentally regulated factors that, depending on need, modulate enzyme function.

Schistosome glyceraldehyde 3-phos-phate dehydrogenase is encoded by a single-copy gene, indicating that the enzyme expressed during development in the snail host derives from the same genetic locus as that expressed during the parasitism of mammalian hosts. Some preliminary observations that deal with the developmental biology of the organisms are discussed next.

## Developmentally Regulated Proteins of Unknown Function

Stage-specific gene expression by filariae and schistosomes requires a broad range of intrinsic genetic versatility, allowing the organisms to alternate from a poikilothermic to a homeothermic environment. Although investigations are still limited by technology (*e.g.*, development of systems for genetic analysis, DNA transformation, and in vitro culture systems), examination of the subject has in principle been made more accessible by recent advances in molecular biology. In one study, an expression cDNA library was screened with serum from schisto-some-infected patients. Of 21 cDNA clones expressing parasite antigens, three proved to be developmentally regulated. Northern analyses of adult worm, egg, and cercarial poly $A^+$ RNA revealed that two of the three stage-specific cDNAs hybridized to mRNAs which change in length during development.

While this observation in itself is not particularly unusual, the observation that the two RNAs appear to be transcribed from the same (single copy) gene is significant. How, and to what extent, schistosomes manage to produce multiple transcripts from a single gene is unknown; however, preliminary results indicate that the size variations noted above are due to the use of alternate start sites and alternative RNA processing.

## Proteinases

One of the more surprising features of the schistosome proteins that stimulate a mammalian immune response is that most of these are synthesized by invertebrate as well as vertebrate hosts. A notable exception to this trend is the organism's proteinases, which appear to play specialized roles at different phases of development. Helminth proteinases are interesting for several reasons: in addition to playing a role in parasite nutrition, they may be essential for invasion of mammalian hosts, they may play a role in stimulating inflammatory reactions, they may be important for differentiation, and they may participate in the production of egg yolk.

It has been hypothesized that proteases secreted by the parasite or anchored onto its membranes may serve to evade the immune response by hydrolyzing bound immunoglobulin so that the Fc portion, necessary for recognition by effector cells, is removed. As many of these proteases are immunogenic, they may prove to be valuable diagnostic reagents and vaccine candidates. Two hydrolases, referred to as hemoglobinases, responsible for the digestion of hemoglobin by schistosomes have recently been cloned and expressed in *Escherichia coli*. One of the enzymes is a thiol proteinase that displays enzymatic similarities to vertebrate cathepsin $B_2$ and is synthesized by cells lining the parasite gut. Products of its activity are hematin, which is regurgitated, and a large number of oligopeptides, which result from digestion of hemoglobin.

The fate of the oligopeptides is unknown but it has been hypothesized that they are shuttled into the next anatomical compartment — the vitellaria, in the case of female worms, which consume the bulk of the hemoglobin — where they serve as egg yolk for the soon-to-develop miracidium. The enzyme is not essential to the parasites: they can survive for several months in protein-free, chemically defined media. On the other hand, worms maintained in culture quickly lose the ability to deposit eggs, and female schistosomula grown in media containing red blood cells reach sexual maturation, whereas those grown in the absence of the cells do not.

Moreover, unisexual female worms, or female worms grown in nonpermissive hosts, contain little pigment in their guts and do not achieve sexual maturity. The inferred ammino acid sequence of one of the cloned cDNAs encoding the enzyme is not homologous to any known protease. This is somewhat surprising as cystinyl proteases from slime molds to humans share recognizable blocks of homology, especially around regions forming the active site. The other shows clear homology to vertebrate cathepsins. Northern analysis indicates that female worms make more of the enzymes than do male worms and that only trace levels are produced during the intramolluscan stage of the life cycle.

The enzymes are predictable immunogenic in natural infections, induce a hypersensitivity response, and are one of the more promising substrates for development of an immunodiagnostic assay system. Another stage- and cell-specific schistosome proteinase has been cloned from a sporocyst cDNA library. The cDNA encodes the cercarial elastase which facilitates the schistosome breach of the human skin barrier, and is arguably the best-characterized macromolecule of a parasitic helminth.

It is a calcium-dependent serine protease that is exquisitely designed to fragment the major components of skin. In addition to elastase, the enzyme is capable of digesting keratin, types IV and VIII collagen, proteoglycan, fibronectin, and laminin. Molecular cloning of the cercarial elastase has provided new information about how the enzyme is regulated, how it functions, and where it is synthesized. Nucleotide-sequence analyses of the cDNA predicts that its mRNA untranslated leader, which is decidedly AT-rich, is unusually long (> 220 bp). The translation product of the longest open-reading frame indicates that the protein possesses an amino terminal extension of 27 amino acids not present in the mature enzyme. Because the extra amino acids are predominantly hydrophobic, they probably represent a peptide leader that maintains the enzyme in an inactive (zymogen) form. The length of the leader is also consistent with its potential functioning as a signal peptide that directs export of the elastase into secretion granules.

Notably, activation of the enzyme is predicted to occur at a site recognized by the enzyme itself, suggesting that once a few zymogen molecules are activated by autocatalysis, a massive burst of autocatalytic activation might occur. The preproenzyme consists of 244 amino acids, which, aside from regions flanking the active site histidine, aspartate, and serine, and regions corresponding to probable protein-folding areas, shares little homology with that of other serine proteases. The mature protein starts with the sequence IsoArgSerGly, which differs radically from the consensus ValValGlyGly, characteristic of serine proteases from invertebrate and vertebrate sources. Together these data suggest that the immune response stimulated by the enzyme should not cross-react with host tissues. In situ nucleic acid hybridization studies indicate that the cercarial protease mRNA is transcribed by the postacetabular gland cells (also known as the escape glands or penetration glands) during early stages of cercarial development, and studies involving monoclonal antibodies jndicate that the inactive form of the enzyme is stored within these cells in secretory vesicles.

Southern analyses indicate that the protein is encoded by one or a small number of genes that, because of the presence of intervening sequences, extend over several kilobase. Understanding the biosynthesis and activation of the schistosome penetration — elastase suggests that the potentiation of schistosome penetration of the host skin and its subsequent access to the circulatory system follows this course of events. At the 32-cell stage of cercarial development (when the acetabular cells, which are the largest cells in the organism, first become microscopically obvious) or shortly thereafter, the cercarial elastase gene is transcriptionally activated in acetabuler cells and an mRNA is expressed whose translational product is inactive and contains signals that direct it toward a subset of actabular secretion granules.

Elastase mRNA synthesis ceases sometime before full maturation of the cercaria. Upon contact with skin lipids and stimulation by heat, the cercaria secretes the storage granules and somehow a small number of preproelastase molecules become activated, perhaps by specific cleavage by another unidentified parasite protease. The remaining share of elastase zymogen would then be activated by autocatalytic processing. The observation that the active enzyme can be isolated from cercariae artificially exposed to linoleic acid suggests that host factors are not

**important in its activation. Preliminary data indicate that interference with the activity of cercarial elastase may be of practical value.**

```
                                  -1
SE       MSNRWFVVVVTLFTYCLTFERVSTWL
REI       MLRFLVFDASLV*YGHSTQDFPETNAR
REII    MIRTLLLSALVGALSCGYPTYEVQHDVS

        1                  20                  40                  60
SE      IRSGEPVQHPAEFPF-----IAFLTTERTMCTGSLVSTRAVLTAGHCVCSPLPVIRVSFL
REI     VVG*AEARR-NSW*SQISLQYLSGGSWYHT*G*TIRRNW*M**A***S---------
REI     VVG*QEASP-NSW*WQVSLQYLSSGKWHHT*G****ANNW****A**IS--------

                           80                  100                 120
SE      TLRNGAQQGIHHQPSGVKVGYMPSCMSARQRRPIAQTLS--GFDIAIVMLAQMVNIQS
REI     SQMTF*VVVGD*NLSQNDGTEQ*VSVQKIMVHPTWNSNNVA-A*Y***LLR***S*T**NN
REII    NS*TYRVLLGR*SLSTSESGSLAVQVSKLVVHEKWN**K**NN*N***L*K**SP*A*T*

                           140                 160                 180
SE      GIRVISLPQPSDIPPPGTGVFIVGTGRDDDND---RDPSRLNGGILKKGRATIMECRHAT
REI     YVQLAV****EGT*:LANNNPCY*T*Q**TRT*GQLSQTLQQAYLPSVDYSICSSSSYWGS*
REII    K*QTAC**PAGT*LPNNYPCYVT*W**LQT*GAT*DVLQQGRLLVVDYATCSSASWWGSS

                           200                 220                 240
SE      NGNPICVKAGQNFGQLPAPGDSGGPLLPSLQGPVLGVVSHGVTLPNLPDIIVEYASVARM
REI     VKTTTM*C**GDGVRSGCQ*******HCL---VNGQYSV****SFVSSMGCNVSKKPTVF
REII    VKTTNM*C**GDGVTSSCN*******NCQ--ASNGQWQV**I*FFGSTLGCNYPRKPSVF

SE      VDFVRSNI
REI     TRVSAYISWMNNVIAYT
REII    TRVSNYIDWINSVIAKN
```

**Fig. 3.19. The predicted schistosome elastase (SE) amino acid sequence is compared with that of rat pancreatic elastases I and II. Abbreviations: REI, rat pancreatic elastase. I; REII, rat pancreatic elastase II. Dashes indicate gaps and asterisks regions of identity. Components of the catalytic triad (histidine 46, asparate 106, and serine 202) are shown in bold lettering, and the numbering system is arbitrary. The leader sequences are shown on the top lines.**

**Monoclonal antibodies that both inhibit the enzyme activity and are cytotoxic to schistosomula have been developed, suggesting that the enzyme may represent a realistic vaccine target. Studies involving the use of chemicals that inhibit the enzyme indicate that the enzyme is essential for successful parasite invasion; parasites whose elastases are irreversibly inactivated do not reach the vasculature and instead die in the skin. Interestingly, the interstitial fluids of vertebrate skin contain proteins that are broad-spectrum inactivators of serine proteases. These antiproteases, which exist in multiple forms, are synthesized in the liver, from which they are subsequently secreted into the plasma.**

**The reactive center of the inhibitors lies near their carboxy terminus and acts as bait for foreign proteases that covalently bind to it and thereby inactivated. The finding that these proteins appear to have evolved in an accelerated Darwinian manner has led to the suggestion that they have been selected for their ability to recognize tissue-invasive parasites such as *Pseudomonas* organisms, schistosomes, and filariae. Whether helminth proteases secreted during tissue migration also have evolved in an accelerated manner is unknown, but it is noteworthy that the major protease of *Schistosomatium douthitti,* another digenetic trematode, is completely different from that of *mansoni.* Schistosomes possess two other proteases that appear to be stage-specific and that are thought to have a specialized function.**

**One is a surface-bound or secreted enzyme that degrades immunoglobulins so that the Fab fragment remains bound to the worm's tegument while the Fc fragment, which is the portion**

recognized by effector cells, is released into the circulation. The bound immunoglobulin therefore presents a hostlike barrier to the immune system. Some of the peptides generated by degradation of immunoglobulins appear to be immunosuppresive and may therefore further facilitate survival in the vertebrate host. The other enzyme is a protease synthesized within schistosome eggs. Although the function of this protease is unknown, some workers believe that it may play a role in assuring that eggs laid in the circulatory system manage to work their way into the lumen of the intestine. Less is known about the proteases of filariae.

Microfilariae of *Onchocerca volvulus* contain a collagenase that may facilitate their movement through interstitial spaces. The enzyme is immunogenic, shares epitopes with bacterial collagenases, and may act as a virulence factor by degrading host elastin and collagen fibers, thereby-leading to loss of elasticity of the skin and the "hanging groin" syndrome. The enzyme also may expose new epitopes of host collagen molecules to the immune system, and therefore lead to autoimmune manifestations. *O. cervicalis* contains a metaloprotease of 60 kD and a serine protease of 30 kD which display collagenase activity. Extracts of *Dirofilaria immitis* microfilariae contain a 22-kD and a 76-kD protease. Proteolytic activity has also been noted in third-stage larvae of *Brugia pahangi* and B. *malayi*.

The role of the latter in skin invasion has not been determined. In general, the observation that tissue-invading helminths possess and secrete collagenolytic or elastinolytic enzymes is consistent with their migratory behavior. These classes of enzymes are notable for having a broad spectrum of substrate specificities; they therefore can degrade a large number of different proteins. The relative importance of these enzymes in the dissolution of tissue barriers, however, has proved difficult to assess. On an equally conjectural plane, it is possible that these enzymes play an additional role in preventing activation of host blood clotting mechanisms by interfering with proteolytic events in the activation of coagulation cascade.

## Eggshell Proteins

The schistosome eggshell is a highly resilient, somewhat pliable biopolymer that appears to be composed of proteins cross-linked with each other by quinone bridges. The complex is highly resistant to proteases and contains pores that presumably allow the influx of nutrients required by developing miracidia and that permit the secretion of macromolecules, which lead to granuloma formation. Acid hydrolyzates of purified eggshells indicate that they are rich in glycine, aspartate, histidine, and lysine but contain unexpectedly low levels of tyrosine. The predicted amino acid sequence of a cDNA encoding a probable component of the schistosome eggshell shows areas of considerable similarity to keratin and silkworm chorion proteins.

The protein is unusual in that it contains regular repeats of GlyGlyGlyTyr, GlyGlyGlyCys, GlyGlyAsp, GlyGlyAsn, and GlyGlyLys, which indicates that it tends to form β-pleated sheets. The protein is tyrosine-rich, a feature consistent with the probability that the eggshells are cross-linked by quinone tanning, a process involving the hydroxylation of tyrosine, the oxidation of hydroxyl residues on the aromatic ring, and covalent binding of the latter to free amino groups of other amino acids. Production of the eggshell mRNA appears to be mediated by a small gene family that is under transcriptional control and that, unlike silkworm chorion genes, is not amplified during expression. The predicted amino acid sequence of another cDNA encoding a probable eggshell protein of 48 kD also demonstrates significant identity with chorion proteins. It consists of a series of pentapeptide repeats having a consensus sequence of GlyTyrAspLysTyr.

The repeats would be predicted to lead toward a structure consisting of alternating beta turns and beta sheets, and the carboxy terminal portion of the peptide contains a histidine-rich region likely to form an alpha helix. The protein is remarkably rich in tyrosine, a feature that indicates that it is not an abundant component of the eggshell, as hydrolysis of the complex yields little of this amino acid. Treatment of the protein with phenol oxidase results in the formation of covalently linked polymers.

## Epidemiology

Several laboratories have applied techniques of molecular biology toward the development of probes that might be useful in identifying infected mosquito, blackfly, or mollusc vectors or in assessing the occurrence, species, and distribution of larvae within in large areas. The Onchocerchiasis Control Program, for example, has, over the last 13 years, relied on vector-control measures to manage onchocerchiasis. It has assessed its success by assaying the prevalence of infective larvae in the blackfly population over a huge area of Africa. In recognition of the fact that light infections are difficult to detect and that onchocerchiasis has a relatively long prepatent period, the program has not relied on determination of the incidence of infection in the human population.

## Repetitive Sequences

The major portion of the genome of eukaryotic cells consists of nontranscribed repetitive sequences whose function is poorly understood. Some workers theorize that these sequences play a structural role, somehow stabilizing the genome and perhaps even regulating gene expression, while others feel that these sequences provide a plasticity that may accelerate evolutionary processes. Repetitive sequences include duplicated pseudogenes, transposons, and satellite DNA. They can be clustered or dispersed throughout unique sequence DNA, and since they lack constraints imposed on functional genes, they are evolutionarily unstable. In most respects, the composition of helminth genomes has proven similar to that of other invertebrates. Their genome size is on the order of $10^8$ bp, and their GC content ranges from 43 percent to 47 percent, compared with roughly 38 percent to 47 percent in mammals and 30 percent to 45 percent in invertebrates (gross exceptions exist).

Modified bases have not been found in parasitic helminths, suggesting that transcriptional regulation through methylation may not be of general importance. Although the genome size of parasitic helminths tends to be roughly twice that of free-living relatives (which posess some of the smallest genomes among the metazoa), the fraction consisting of repetitive DNA tends to be lower. Thermal denaturation studies indicate that the increase in genome size of these parasites is due to the introduction of low-copy-number sequences. Several workers have taken advantage of the inherent variability of repetitive sequences to develop nucleic acid probes that allow the identification of filariae in infected flies or mosquitoes, in blood samples, or in skin biopsies. Since these sequences exist in large copy number, these probes have the potential of being sensitive.

Although repetitive sequences have a high rate of drift, they may show considerable homology within a genus, thus limiting their usefulness. In one study, detailed later, this problem was circumvented by constructing oligonucleotide probes that spanned areas of divergence between different species, thus allowing species-specific identification of the parasites. In the case of

*Onchocerca* species, an important question that might be clarified by the use of repetitive DNA probes is whether forest and savannah strains of the parasite have different biological characteristics. The ability to discriminate between the two forms of the parasite also may assist in determining whether outbreaks of infection are due to invasion of forest- or savannah-dwelling flies.

Preliminary indications are that the geographically separated strains may lead to different clinical manifestations and that they might be separable on the basis of differences in isoenzyme patterns. Early studies indicate that probes hybridizing to repetitive DNA may help to speciate members of the genus *Onchocerca* and to discriminate between forest and savannah types. The probes are sensitive enough to identify a single infective larva. In the case of *Brugia* spp., it is felt that development of species-specific probes may further our understanding of the parasite host range. *B. malayi* is thought to infect humans, nonhuman primates, and several other vertebrates, whereas *B. timori* is thought to infect only humans. *B. pahangi* is known to infect domestic animals and is a suspected human pathogen.

A genomic clone representing a repetitive sequence of *B. malayi* has been isolated. Twenty-five percent of the parasite's genome consists of repetitive DNA, and the isolated repeat, which consists of a tandemly arrayed 322-bp sequence repeated 30,000 times per haploid genome, accounts for approximately half of this DNA. Under stringent conditions, the probe can easily distinguish *Brugia* from other genera of filariae, but it differs from the major repeat of *B. pahangi* by only 11 percent. As differences between the two repeats are clustered in pockets, workers have designed oligonucleotides that can easily discriminate between the two species.

## Ribosomal Genes and their RNAs

Eukaryotic ribosomal RNA genes are typically part of multigene families whose members are tandemly linked (mRNA) in the genome. In general the arrangement follows a plan wherein the gene encoding the 18S RNA subunit lies 5' to that encoding the 5,8S mRNA subunit, the latter being placed 5' to that encoding the 28s mRNA snbunit. The three subunits are transcribed by RNA polymerase I as a polycistronic unit whose spacer regions are subsequently removed by RNA splicing. Processed ribosomal RNAs are highly conserved in nucleotide sequence, a feature that has assisted in clarification of the taxonomic relationships between helminth parasites.

The intergenic regions, on other hand, vary in length and have evolved more freely. Since these spacer regions represent a homogeneous family of repeated sequences, they may present-species-specific differences that can be exploited toward the development of highly discriminating probes. Polymorphisms in RNA genes also have been exploited as a means of determining the species, sex, and strain of schistosomes within their molluscan host. There exist 17 known species of schistosomes, which are categorized into three groups on the basis of egg morphology. In addition, there appear to exist several strains of the parasite, a phenomenon that may account for differences in infectivity, virulence, isoenzyme patterns, and drug resistance among schistosomes.

Although these biological characteristics are unlikely to be related to polymorphisms in the structure of rRNA genes, their fixation within discrete subpopulations may nonetheless allow strain identification and thereby provide a marker that can be correlated later with differences in pathogenicity, longevity, and overall biological fitness. Coding sequences of helminth rRNA genes also have been used to determine phylogenetic distances between the organisms.

Sequence analyses of the 5′-end of the 28s rRNA of various nematodes and schistosomes indicate that the differences in size and sequence of these molecules occur within a few hotspots that have retained the same position relative to the core of all mRNAs examined to date. Preliminary data indicate that nematodes radiated a long time ago: the rRNA sequence of *Nematospiroides dubius* is as similar to that of *O. gibsoni* and *B. pahangi* as it is to that of *S. mansoni.*

## Immunodiagnosis

Definitive diagnosis of schistosomiasis rests on the demonstration of parasite ova in stool or urine specimens. Similarly, detection of filariasis depends on the visual detection of microfilariae in blood or skin biopsy. Available methods to detect larval parasites are time consuming and insensitive, especially when applied in large epidemiological surveys — problems that have stimulated the development of immunologically based assay systems. Antigens that have rendered assays reproducible, sensitive, and specific have been identified: four of these have been cloned; two are proteolytic enzymes that are discussed elsewhere in this chapter.

Peliminary studies on the antibody response to schistosome peptides indicate that the response is erratic, demonstrates little correlation with clinical status, and persists for a long period after chemotherapeutic cure, thus limiting the information garnered by assay of specific antibodies. Nevertheless, the search for immunodiagnostic schistosome and filarial antigens has provided significant information relevant to understanding helminth biology. Strangely enough, it appears that immunodominant antigens of schistosomes include a family of macromolecules called heat-shock proteins (HSPs). HSPs are phylogenetically conserved proteins expressed universally by organisms in response to environmental stress. The function of HSPs is poorly understood, although they appear to protect cells from damage and at the same time rid them of denatured macromolecules.

Constitutively expressed HSP cognates are thought to play a role in mediating the assembly or disassembly of higher macromolecular structures; in surveying the cytoplasm for denatured, improperly modified, or foreign proteins; and in mediating the insertion of proteins into, or their translocation across, cell membranes. HSP cognates are subject to developmental regulation and are thought by some to be critical factors of differentiation. One of the major antigens of schistosome eggs is a 40-kD protein referred to as p40. The protein is also found in adult worms and elicits a strong immunological response in over 90 percent of infected humans. The antigen is actually a family of four nearly identical proteins that are differentially expressed during the schistosome life cycle.

One form of the peptide is produced by female worms, two in eggs, and four in adult worms. The sequence of p40, based on the nucleotide sequence of a cDNA, reveals that it is homologous to *Drosophila* and dogfish α-crystallins and a family of eukaryotic small heat-shock proteins. The primary sequence of the schistosome protein indicates that it is a descendant of a protein that was ancestral to the functionally different α-crystallins and small HSPs. The amino-terminal portion of the protein consists of a long stretch homologous to the α-crystallins, a central portion homologous to the carboxy terminus of small HSPs, an amino-proximal region homologous to α-crystallins, and a carboxy-terminal tail similar to that of some small HSPs. This sequence suggests that the protein is the result of an early intragenic duplication.

```
Schist P40     G E D G K V H F K V R F D A Q G F A P Q D I N V T S S E N - R V T V H A K K E - T T T D - - G
Sechist P40    G S G * * R L H V E V A V D P V Y K * E * L F * N V D S * - * * V V S G R H H K Q K S * Q H *
α-cryst 27     V R S D R D K * V I F L * V K H * S * E * L T * K V Q * D - F * E I H G * H N - E R Q * D H -
bHSP 22        I R H T A D R W R * S L * V N H * * * D E L T * K T K D G - V R E I T G * H * - E R Q * E H -
DHSP = 27      A T V N * D G Y * L T L * V K D * S - - E L K * K V L D E S V * L * E * * S * - Q Q E A E Q *
Schist p40 R   P A V * * D G * Q * C M * V S Q F K * N E L T * K V V D * - T * V * E G * H * - E R E * G H -

Schist p40     R - - - - - K C S R E F C R M V Q L P K S I D D S Q L K C R M T D D G V L M L E A P V K V - - -
Schist p40     * - - - - - S S S F A * * S Q S * A I * E T V * P L S V S A Q V - V G N T * V * * * * L E K - - -
α cryst A      - - - - - G Y I * * * * H * R * R * * S N V * Q * A * S * S L S A * * M * T F S G * K I P S G V
bhHSP 27       - - - - - G Y I * * * * T * K * T * * P G V * * P T * V S S S L S P E * T * T V * * * M P K L A T
dHSP = 22      - - - - - G Y S * * * H * L G R * V * * D G Y E A D K V S S S L S * * * * * T I S V * - N P P G V
dHSP = 27      - - - - - G M I Q * H * V * K * T * * * G F * P N E V V S T V S S * * * * T * K * * - P P P S K

Schist p 40    - D Q N Q S L T L N E S G Q V A V R P K S D N Q I K A V P A S Q A L V A K G V H G L S Y V D D
Schist p 40    - - - - - - - - - - - - - Q H A I T H
α cryst        D A G H S E R A I P V * R E E K P - - S S A P S S
bHSP = 27      Q - S * - E I * I P V T F E S R A Q L G G R S C K I R
d HSP = 22     O E T L K E R E V T I E Q T G E P A K * * A E E P * D K T * * *
dHSP = 27      E Q A K S E R I V Q I Q Q T G - P A H L S V K A P A P E A G D G K A E N G S G E K M E T S K
```

**Fig. 3.20. Regions of homology between schistosome egg antigen p40, bovine α-crystallin A2, *Drosophila* small heat-shock proteins (dHSP) 22 and 27, and human heat-shock protein 27 (hHSP 27). Sequences are aligned. The egg antigen is represented as follows: p40, ammo acids 136-262; p40, amino acids 263-354.**

The function of the protein is unknown, although its similarity to the α-crystallins and small HSPs suggests that it probably forms large aggregates that somehow stabilize cellular structures. Background studies on the human immune response to *S. mamoni* indicate that although the response is erratic, virtually all infected individuals produce sera that react to the two schistosome proteins, one having a relative molecular weight (Mr) of 70 kD and the other an Mr of 38 kD. Antibodies to these schistosome proteins are not detectable in uninfected individuals or in persons infected with other parasitic helminths (species of *Brugia, Onchocerca, Chlonorchis, Heterobilharzia, Ascaris, Trichuris, Ancylostoma,* or Echiococcus), indicating that they may prove to be of diagnostic value.

The 70-kD schistosome antigen has been identified as an HSP-70 homologue; epitope mapping has localized the immunogenic portion(s) of the molecule at the carboxy terminus of the protein, the least conserved domain of HSP-70s in general. Similar properties are displayed by the *S. japonicum* HSP-70 cognate. Oddly, however, the 70-kD schistosome immunogens do not cross-react with each other. As immunization of rabbits with chicken HSP-70 leads to a response that cross-reacts with the homologue of fruit flies and humans, these findings indicate that schistosomes present the protein to the immune system in a manner that avoids an autoimmune response. Along similar lines, it has recently been noted that *P. falciparum* produces two different HSP-70s that are immunodominant and non-cross-reactive. These findings are of interest because they suggest that the parasite species-specific nature of the protective immune response generated by immunization with live-attenuated vaccines is not attributable to fundamentally different processes or targets, but rather to differences in the structure of homologous parasite antigens.

It is possible, however, that HSP-70s, molecules whose three-dimensional configuration is unknown, selectively present their hypervariable carboxy termini to the immune system, a property lost when animals are immunized with the denatured protein. The function of the schistosome HSP-70 is unknown. Curiously, HSP-70 also has been found to be a dominant immunogen of filariae, trypanosomes, plasmodia, and mycobacteria. In other eukaryotes, HSP-

70 is considered to be a molecular chaperone that escorts and influences the conformation of proteins destined to become part of, or which are going to be removed from, higher molecular assemblies. HSP-70 also plays a role in the translocation of proteins across cellular membranes. This process is thought to involve hydrophobic interactions between HSP-70 and the protein to be translocated, and it appears to be associated with conformational changes mediated by the binding and hydrolysis of ATP by HSP. HSP-70 has been implicated as playing a central role in the recognition of denatured proteins and in collaboration with ubiquitin, tagging these with signals recognized by lysozomal proteases.

Interestingly, synthesis of HSP-70 increases during senescence of fruit flies, perhaps as a result of increases in the production of improperly formed or glycosylated proteins. In light of the finding that the schistosome life expectancy is unusually long for an invertebrate, it would be interesting to see whether the same holds true for them. The schistosome HSP-70 is found in the tegument of adult worms and in the hatching fluid, surface cells, and nervous system of miracidia. Synthesis of the protein is constitutive in adult worms and sporocysts, although apparently it does not appear to be made by cercariae. The latter nonetheless maintain a previously synthesized pool of the protein. Transforming schistosomula synthesize massive amounts of HSP during a period in which production of nearly every other protein is repressed. The cue for HSP gene transcription in transforming larvae is not heat, as cercariae subjected to heat shock do not synthesize the uprotein, whereas cercariae placed in mammalian culture media do. Synthesis of other proteins slowly resumes 16-hour posttransformation, eventually plateauing to steady state levels.

Whatever the function of HSP-70, its mRNA is one of the more abundant species in adult worms, accounting for 0.1 percent to 1.0 percent of the total message. The observation that the protein is synthesized primarily by tegumental cells suggests that it may play a role in maintaining the integrity of this anatomical compartment. The observation that despite their similarities, the HSP-70s of S. *mansoni* and S. *japonicum* stimulate production of non-cross-reacting antibodies during the course of natural infection can be taken as a good omen for vaccine development.

```
     445
Sm   ERALTKDNNLLGKFELSGIPPAPRGTPQIEVTFDIDANGILNVSAVDKGTGKQNKITI
Sj   ** TM***************C********V**********************A***E*****
Hs   *** M**********R**************V********************T*TKDS***A*****

Sm   TNDKGRLSKEEIERMVADADKYKAEDEKQRDRVA SKNSLESYVYTMKQQVEGE-LKEK
Sj   ***************D**INE******S******KN*IC************S***S****DEM*D*
Hs   **************QE*E*********V**E********A*****AFN**AFN****SA***EG**G*

Sm   IPESDH QVIISKCEDTISWLDVHQSAEKHETESKREELEKVCAP IITKVYQ-A----
Sj   *S****RKNR*L*****E**R*M*MN*L****E*F*E*KS********M*****AMNR-*----
Hs   *S*A*KKK* LD**QEV******ANTL****D*F*H***K****Q**N****SCL**G*----

Sm   -----GGMPGGMHEAS--GAGGSG-----------------------KGPTIEEVD
Sj   -GGVPS*******P--------****G*------------------*********
Hs   -----**-****-FG*Q-GPK****---------------------S*********
```

**Fig. 3.21. Comparison of the predicted carboxy-terminal amino acid sequences of HSP 70 of *Schistosoma japonicum* (Sj), and *Homo sapiens* (*Hs*) to that of *Schistosoma mansoni* (Sm). Asterisks denote points of identity with the *S. mansoni* sequence, and dashes represent gaps.**

It seems that in many experimental systems, immunity to the parasites is often specific to the species of sensitizing organism. The response to HSP-70 suggests that this may be due more to interspecific differences in the structure of parasite antigens than to different effector mechanisms or target antigens, and that once a protective antigen is found for one parasite, it may be possible to clone equally active homologues from species that are currently difficult or impossible to maintain in the laboratory. A case in point is the identification of paramyosin as a vaccine target in schistosomes, and subsequent demonstration of the prophylactic potential against filariasis of the nematode homologue.

## Coda

Although studies on the molecular biology of schistosomes and filariae are at a stage of infancy, they have significantly contributed to the understanding of both the biology of the organisms and their interaction with their hosts. Many of the parasite antigens have now been assigned a biological function, a significant advance over their previous indexing in the literature on the basis of molecular weight. At the same time, the field has posed a number of new questions to researchers interested in biological and immunological aspects of parasite relationships with their hosts. The answers to these questions are likely to refine efforts to develop vaccines and immunodiagnostic systems. In conjunction with DNA probe technologies, such as those using the polymerase chain reaction, these studies offer the potential to develop highly sensitive probes that could detect and determine the species (and perhaps even strain) of parasites obtained in small biological samples. At a more basic level, the observation that parasitic nematodes possess both cis- and trans-splicing activities makes them a convenient system for studying the mechanics of eukaryotic RNA processing.

CHAPTER

4

# Antibiotics

The term antibiotic was introduced by S. A. Waksman (Nobel Prize winner) in 1945, as the name of a class of substances of biological origin that are inimical to microorganisms. Strictly speaking, antibiotic means "against life." Since antibiotics, as commonly used, are inimical only to microorganisms they might logically be called "antimicrobiotics." Although substances of biological origin with antimicrobial properties were known before 1900, antibiotics as now defined first came into world-wide prominence about 1940. There followed a widespread search for new antibiotics. It was found that antibiotic substances are produced by a wide variety of living organisms, ranging from man to microbes.

The most important and widely used antibiotics at present are derived from microorganisms—mainly from various bacteria of the genera *Bacillus* and *Streptomyces* and certain common species of mold (*e.g., Penicillium*). Since the earliest studies with pure cultures, microbiologists had known that when certain airborne, saprophytic microorganisms contaminated their cultures they suppressed the growth of the desired species. The phenomenon was so commonplace and attention was so fixed on other problems that the antagonistic action of such contaminants was pushed aside as merely an inevitable nuisance. The true significance of the "nuisance" was largely overlooked until 1929, when Alexander Fleming (afterwards Sir Alexander Fleming, Nobel Prize winner with Florey and Chain) realized it, acted upon the basis of his idea, and discovered, named and described penicillin.

## PENICILLIN

The antagonistic airborne saprophytic organism that first attracted Fleming's attention was a colony of the common mold, *Penicillium notatum.* When it contaminated one of his cultures in a Petri plate, Fleming experimented with its antimicrobial action. He passed broth cultures of the mold through filters, removed the mold filaments and was thus able to study the activity of the soluble growth products alone as they occurred in the clear broth. He found that the clarified, sterile broth contained a highly potent antimicrobial principle, the activity of which was readily demonstrated in contact with sensitive microorganisms, especially gram-positive bacteria. He called this principle *penicillin.* In a much purified and refined form it is still one of the forms of penicillin used therapeutically today. Fleming realized the practical possibility of his discovery and made use of it in his laboratory to eliminate gram-positive contaminants from cultures but was not, at that time, in a position to develop it more fully.

For some years the value of such antibiotic phenomena remained relatively unknown. Interest was reawakened when, realizing that many nonsporeforming pathogenic organisms in the soil are rapidly destroyed, presumably by antimicrobial substances produced by saprophytic microorganisms, Waksman suggested that the search for antibiotic-producing microorganisms be carried to the soil. In 1939 Dubos found, in bogs, organisms *(Bacillus brevis)* that produced two valuable antibacterial substances *(gramicidin* and *tyrocidin)*. These antibiotics, while of enormous value therapeutically for surface application and of great interest scientifically, were too toxic for internal use. Dubos's discoveries stimulated new interest in Fleming's observations on *Penicillium notatum;* work on penicillin was begun in England on a large scale by Florey, Chain, Abraham and others (the Oxford Group) in 1940, just after the outbreak of World War II. Because of the exigencies of the war, work on mass production methods was transferred largely to the United States.

**Fig. 4.1. Sir Alexander Fleming, penicillin discoverer and Nobel Prize winner, points with his inoculating needle to a giant colony of *Penicillium notatum*, the organism that produces penicillin, on agar in a Petri dish.**

Since that time penicillin has played a role in the greatest war in human history; in the most far-reaching piece of cooperative research ever organized prior to atomic research; in a great industrial development; in the most complete control over a variety of diseases ever achieved by man in a short time; and, because of its impact on venereal diseases, in a tremendous new social and moral trend, the possibilities of which are still only partly realized. The military importance of penicillin in World War II, then just beginning, can hardly be estimated. The development of American facilities for mass production of penicillin, not then available in Europe, was soon of great benefit to all concerned.

## Production

Production of penicillin and other antibiotics is now a billion-dollar industry in the United States alone. In 1943 an ounce of penicillin cost around twenty thousand dollars. Now it costs about three dollars because of mass production methods.

Production of penicillin involves cultivation of *Penicillium notatum* under conditions most favorable to growth and penicillin production (*i.e.*, at about 24° C and at a pH between 7 and 8). As the mold is strictly aerobic, exposure to air is essential.

Commercially, a process using submerged growth in closed but vigorously aerated tanks holding thousands of gallons of medium most widely used today.

In all methods, suitable medium is first inoculated with suspensions of the conidia of *Penicillium notatum* or the closely related *P. chrysogenum,* which also produces penicillin. The entire procedure is a *pure-culture* process and necessitates costly aseptic technique at all stages.

During the 7 to 14 days of incubation in the large vat of aerated fluid medium the mold excretes at least three waste substances that are of importance: the yellow pigment *chrysogenin,* which must be removed by adsorption with charcoal or similar means; *penicillin;* and *notatin,* or *penicillin B.* The last occurs especially if the acidity of the medium is too great. It is removed during the final purification process. Other substances are also produced, some of them related to penicillin. In general, if a microorganism produces one antibiotic it produces others of the same nature, some valuable, others of little significance.

## Medium

Different formulas are doubtless in use by various manufacturers, but basically they are similar and are mainly modifications of the following:

| | |
|---|---|
| Corn-steep liquor | 40.0 ml. |
| Lactose monohydrate | 27.5 gm. |
| $NaNO_3$ | 3.0 gm. |
| $MgSO_4\,H_2O$ | 0.25 gm. |
| KH2$PO_4$ | 0.50 gm. |
| $ZnSO_4\,H_2O$ | 0.044 gm. |
| $MnSO_4\,H_2O$ | 0.020 gm. |
| Glucose monohydrate | 3.0 gm. |
| Water ad | 1000.0 ml. |

After about seven days, growth is complete, pH rises to 8.0 or above, and penicillin production ceases.

When no more penicillin is being formed, the masses of mold growth are separated from the culture fluid by centrifugalization and filtration. The complex process of extracting the penicillin from the clear fluid then begins. The method involves various extractions with organic solvents and recrystallizations. These are chemical engineering problems and do not concern us at present.

## Properties and Uses

The material extracted from the culture medium is in reality a mixture of six chemically related penicillins called X, G, two forms of F, dihydro F and K. The most important is penicillin G (benzyl penicillin), which is usually meant when the term penicillin is used. Penicillin is readily soluble in water and alcohol, but the latter inactivates it. It is quickly destroyed by acid, so that in the unmodified form it is ineffective if taken orally. In the crystalline form it is quite stable, even for several days at 100° C or for months refrigerated in the dark; however, in aqueous solution it is very unstable and quickly decomposes.

Penicillin acts mainly by inhibiting the enzymes concerned in the incorporation of muramic acid into the mucocomplex of the bacterial cell wall. Whether this action is of the nature of an antimetabolite is not known at present. Since synthesis of cell wall is carried on only by young, actively multiplying cells, the antibacterial action is manifested only during active growth. Because septum formation appears particularly vulnerable to penicillin, rod-shaped bacteria tend to form aseptate filamentous forms like mycelia in the presence of nonlethal concentrations of penicillin.

Among the most valuable properties of penicillin is its relatively low toxicity for man and animals, even in large doses, coupled with the fact that its antibacterial potency is such that a concentration of 0.000001 gm. per ml. will exert marked bactericidal effects. It may be administered subcutaneously, intravenously or locally. A particular disadvantage of penicillin is that it is a potent allergen in some persons and can cause fatal allergic reactions.

## Chemistry of Penicillins

The naturally occurring penicillins are compounds of the strong, monobasic acid, *penicillanic acid.* The penicillins usually occur as salts of sodium and potassium. The general chemical structure assigned. They are unstable to heat and to strong acids and alkalis. Laboratory synthesis of penicillin was announced shortly after the formula was known. The substance artificially synthesized was penicillin G.

## Biosynthesis of Penicillins

It has been found possible to induce *Penicillium* cultures to synthesize new forms of penicillin. The method consists of "feeding," to cultures of *Penicillium notatium,* synthetic substitutes for the normal "R's" (*i.e.,* the distinctive side chains) of the penicillin molecule. These substitutes, "prefabricated" molecules, are referred to as *precursors* of the new "R." In the presence of such abnormal precursors, the mold unwittingly conjugates them "as is" with penicillanic acid, thus forming new and more valuable penicillins that have been synthesized partly by man, partly by the mold. An important product is *phenoxymethyl penicillin (penicillin V),* produced by feeding the precursor phenoxyacetic acid to the mold. Penicillin V is fairly stable to gastric acidity and is less quickly excreted from the blood than other penicillins. It is widely used for oral administration.

This same method (*i.e.,* providing abnormal precursors) has been extended to other industrial processes in which a desired product is synthesized partly by man and partly by a living organism. The field of *biosynthesis* is clearly wide open for the ingenious, well-trained and energetic student.

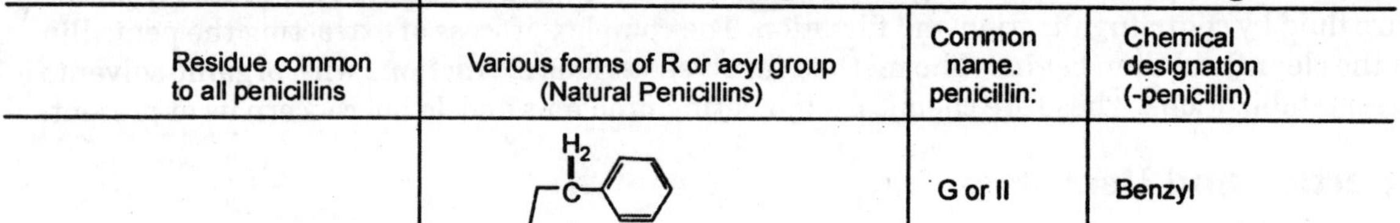

| Residue common to all penicillins | Various forms of R or acyl group (Natural Penicillins) | Common name. penicillin: | Chemical designation (-penicillin) |
|---|---|---|---|
| 6-amino-penicillanic acid residue: $H_3C$, $H_3C$, S, H H H, N, C=O, COOH · C(H), N–C=O | $-\overset{H_2}{C}-C_6H_5$ | G or II | Benzyl |
| | $-\overset{H_2}{C}-C_6H_4-OH$ | X or III | Para-hydroxybenzyl |
| | $-\overset{H_2}{C}-O-C_6H_5$ | V | Phenoxymethyl* |
| | $-\overset{H_2}{C}-\overset{H}{C}=\overset{H}{C}-\overset{H_2}{C}-C\equiv H_3$ | F or I | 2-pentenyl |
| | $-\overset{H_2}{C}-\overset{H_2}{C}-\overset{H}{C}=\overset{H}{C}-C\equiv H_3$ | F | 3-pentenyl |
| | $-\overset{H_2}{C}-\overset{H_2}{C}-\overset{H_2}{C}-\overset{H_2}{C}-\overset{H_2}{C}-\overset{H_2}{C}-C\equiv H_3$ | K or IV | n-heptyl |

**Fig. 4.2. Various forms of penicillin. They differ only in the form of *R* in the general formula. The different forms of *R* are shown in the second column.**

## Penicillinases

Several hydrolytic enzymes can attack and destroy penicillins. One, an inducible enzyme, is commonly called *penicillinase.* It is a type of penicillinacylase. This type of penicillinase is produced by a number of dangerous pathogenic bacteria, notably *Staphylococcus* and several species of gram-negative intestinal bacteria. By virtue of this enzyme these and other bacteria are penicillin-resistant and so defeat therapeutic uses of natural penicillins in many cases. These bacterial penicillinases attack the *β*-lactam bond of penicillins G, X and, to a lesser extent, V, and are often referred to as *β*-lactamases.

Another kind of penicillinase is produced by many common genera of fungi, including *Penicillium, Aspergillus* and *Mucor.* The penicillinases produced by these organisms differ from bacterial β-lactamase in that they attack the peptide bond between the 6-amino-penicillanic acid residue and the "R's". These enzymes are sometimes called *penicillin amidases.* They attack penicillins V and K most rapidly, G more slowly.

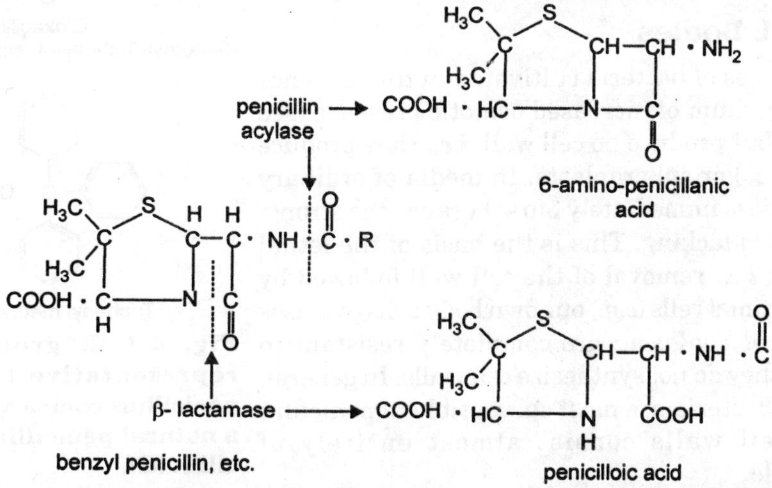

**Fig. 4.3. Action of two major types of penicillinase: penicillinacylase (penicillin amidase) and β-lactamase (commonly called penicillinase). The dotted lines and arrows show which bonds are broken.**

## Semisynthetic Penicillins

When it was discovered about 1958, that 6-amino-penicillanic acid, the "nucleus" of all penicillins, could be produced in commercial quantities, it became possible to attach various artificially synthesized acyl groups to 6-amino-penicillanic acid on a large scale. Hundreds of such semi-synthetic penicillins have been made and tried. Some are extremely valuable, *e.g.,* phenethycillin, which is quickly absorbed into the blood and is as effective as penicillin G; methicillin, cloxacillin and nafcillin, which are not readily destroyed by gastric acidity and which are resistant to *β*-lactamase; and ampicillin, which is more effective against a range of bacteria, both gram-positive and gram-negative, than other penicillins, and is effective in oral administration. The acyl groups of some of these comparison with natural penicillins.

## The Cephalosporins

These are a group of antibiotics, the first described type of which, cephalosporin C, is produced by the mold, *Cephalosporium acremonium.* Although a relatively ineffective antibiotic, cephalosporin C is of great interest because it has a formula very similar to that of penicillin. Like penicillin, cephalosporin C lends itself to production of a large number of synthetic derivatives. Knowledge of these processes is based on the work of Woodward (Nobel Prize winner) and many other skillful chemists and biologists. The central residue of the cephalosporins is indicated. Various acyl groups may be attached to this residue, with the production of much more valuable cephalosporins than cephalosporin C; *e.g., cephalothin.*

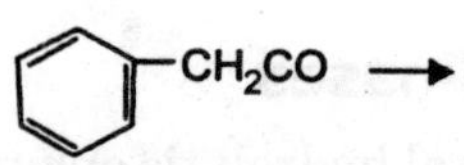

$C_6H_5-CH_2CO \rightarrow$

Benzyl penicilin (G)

$OCH_3$ / $CO \rightarrow$ / $OCH_3$

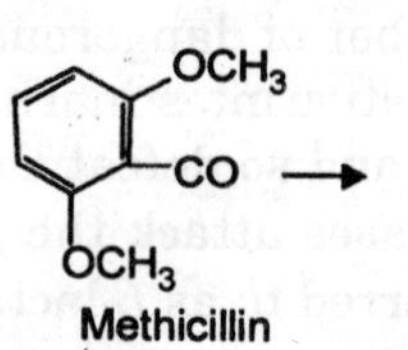

Methicillin
(Dimethoxyphenyl penicilline)

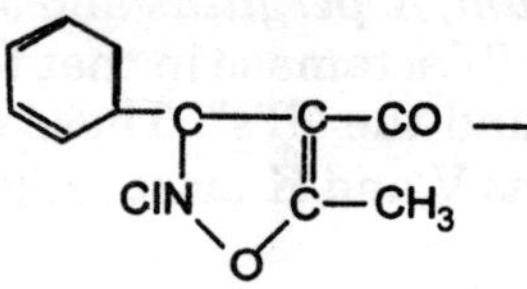

$C-C-CO \rightarrow$ / ClN / $C-CH_3$ / O

Cloxacillin
(5-methyl-3-phenyl-4-isoxyzolyl penicillin)

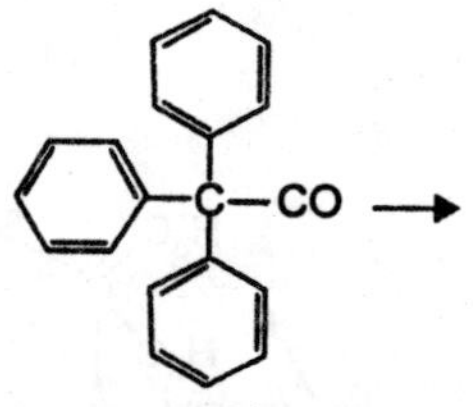

$C-CO \rightarrow$

Triphenyl methyl penicillin

**Fig. 4.4. R groups of some representative bio-synthetic penicillins compared with that of a natural penicillin, benzylpenicillin (G).**

## Penicillin and L Bodies

Numerous species of bacteria cultivated in the presence of penicillin in a medium of increased osmotic pressure (see Protoplasts) grow but produce no cell wall, *i.e.,* they produce protoplasts (L forms) or spheroplasts. In media of ordinary osmotic pressure these immediately burst because the strong, retaining cell wall is lacking. This is the basis of the lethal effect of penicillin; *i.e.,* removal of the cell wall followed by osmotic rupture. Animal cells (*e.g.,* our own body cells), viruses and the PPLO or *Mycoplasma* are completely resistant to penicillin because they do not synthesize cell walls. In general, the gram-positive bacteria are most susceptible to penicillin because their cell walls consist almost entirely of mucopolysaccharide.

## Adaptive Penicillinase

Penicillinase ($\beta$-lactamase) production may appear in some bacteria as a result of induced enzyme formation. For example, strains of some kinds of bacteria are ordinarily sensitive to penicillin because they do not produce penicillinase. However, contact with nonlethal concentrations of penicillin appears to induce penicillinase production. They are then penicillin resistant.

## Various Strains of Penicillin Notatum

Fleming's strain of *P. notatum* for a time was thought to be unique. It was later found that some "wild" strains of this organism were better producers of penicillin and that other closely related organisms, especially *P. chrysogenum,* also produce penicillin. These findings led to a very extensive search for better strains of *P. notatum* and for penicillin-producing strains of other organisms. Many thousands of cultures were tested, and it was found that the penicillin-producing property is widely distributed in the *P. notatum-chrysogenum* group and that some strains are better than others, especially for the various methods of production (*i.e.*, some were good in tanks, not so good on surfaces, and vice versa). Irradiation with ultraviolet light has furnished many thousands of mutant strains for testing, among them some very potent penicillin-producing mutants.

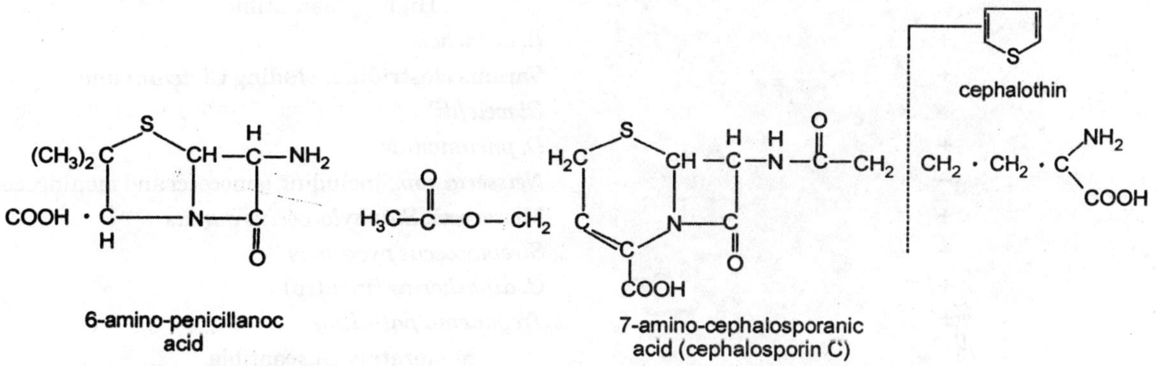

cephalothin

6-amino-penicillanoc acid

7-amino-cephalosporanic acid (cephalosporin C)

**Fig. 4.5. Comparison of formulas of 6-amino-penicillanic acid, 7-amino-cephalosporanic acid, and cephalothin as a derivative of cephalosporin.**

The same principle of hunting for new and more effective mutants for a variety of purposes can be applied in most fields of industrial microbiology, using yeasts, molds and bacteria.

## Development of Resistance Drug-Fastness and Dependence

An undesirable aspect of the continued clinical use of antibiotics (as well as of sulfa drugs) is the emergence of mutants having *increased resistance* to the drugs (*drug-fast* strains) or being wholly *dependent* on the drugs. Drug-fastness and drug dependence have been described elsewhere.

## Laboratory Uses of Penicillin

As mentioned, the value of penicillin as a selective bacteriostatic agent in laboratory bacteriology had been immediately recognized by Fleming and was adopted by others. It is a convenient aid to the isolation of penicillin-resistant organisms (such as the gram-negative whooping cough and influenza bacilli) from throat cultures, since it suppresses the growth of unwanted penicillin-sensitive (mainly gram-positive) organisms. It is also used for suppressing bacterial growth in animal tissue cultures used in studies of viruses, since viruses and animal cells are wholly resistant to most antibiotics.

## Range of Activity

The clinical uses of penicillin are beyond the scope of this discussion, but some of the common infectious organisms against which natural penicillins are more or less effective are listed.

Note that most of the susceptible species are gram-positive. Notable exceptions are *Treponema pallidum,* cause of syphilis, and gram-negative *Neisseria gonorrhoeae,* cause of gonorrhea; both very susceptible.

Natural penicillin is generally ineffective in infections caused by gram-negative bacilli, viruses, rickettsias or PPLO, and in infections with protozoa, yeasts and molds. As we have noted, several biosynthesized penicillins have much broader ranges of antimicrobial activity.

**Table 4.1. Penicillin Sensitivity of Some Common Pathogens**

| ***Gram Stain*** | ***Organisms**** |
|---|---|
| | Highly Susceptible |
| + | *B. anthracis* |
| + | Various clostridia including Cl. *tetani* and *Cl welchii* |
| + | *D. pneumoniae* |
| – | *Neisseria spp.*, including gonococci and meningococci |
| + | Micrococci, *Staphylococcius aureus* |
| + | *Streptococcus pyogenets* |
| + | *C. diphtheriae* (in vitro) |
| ± | *Treponema pallidum* |
| | Moderately Susceptible |
| – | *Borrelia* of relapsing fever |
| + | Various alpha (viridans) type hemolytic streptococci |
| – | *Leptospira spp.* |
| | Not Significantly Susceptible |
| – | Enterobacteriaceae |
| – | *Pseudomonas* species |
| – | Brucellaceae |

• Except nonconforming mutants.

## STANDARDIZATION OF ANTIBIOTICS

Standards of potency for all antibiotics are established by international agreement as to what shall constitute an *International Standard Sample* and an *International Unit.* Generally, the International Standard Sample of an antibiotic consists of a certain highly purified and tested lot of that antibiotic. For example, the 1960 International Standard Sample of the antibiotic tetracycline (see farther on) was a 500-gm. lot of highly purified tetracycline chloride furnished by an American manufacturer. The International Unit of tetracycline was defined as the antibiotic activity contained in 1.01 $\mu$g. of that particular standard sample; 1.01 $\mu$g. was equivalent in activity to 1 $\mu$g. of pure, crystalline tetracycline chloride.

## ASSAY AND SENSITIVITY TESTING

Several types of procedure are in use for assaying the potency of antibiotic preparations for therapeutic purposes. This is commonly done in manufacturers' control laboratories under the supervision of the U. S. Food and Drug Administration. These methods, conversely modified, are also used for measuring sensitivity *(sensitivity testing)* of "unknown" organisms to antibiotics. This is commonly done in hospital laboratories and is essential to rational chemotherapy, although not always feasible. There are sources of error in all of these methods, and so they must be used and interpreted by experts.

## ASSAY METHODS

### Serial Dilution in Broth

Serial dilutions of the antibiotic being assayed are made in uniform amounts of standard broth in culture tubes. These are then inoculated with a uniform number of cells of the test organism, which is selected on the basis of its known degree of sensitivity. After incubation, turbidity (or its absence) is noted and may be measured by means of a photoelectric turbidimeter. The turbidities (amounts of growth) are compared with a dilution series made in the same way but with an antibiotic reference standard of carefully measured potency.

The method is time consuming and sometimes requires preliminary sterilization of the sample, which destroys some of its potency. However, the dilution method is an accurate procedure, and results can be stated in exact terms of units of antibiotic.

### Sensitivity Testing by Broth Method

The serial dilution procedure—reversed—is often used to measure the sensitivity to various antibiotics of an organism freshly isolated from an infected patient. Graded dilutions of an antibiotic of known strength are placed in series of tubes. Each tube is inoculated with a drop of the organism whose sensitivity is being tested. If growth occurs in all tubes, the organism is resistant; if no growth occurs, the organism is sensitive. Growth in some of the tubes permits a measure of the sensitivity of the organism.

### Serial Dilution (Streak Assay) in Agar

The streak assay method is rapid, convenient and much used for preliminary assays of crude material in the search for new antibiotics.

Graded dilutions of the substance to be tested (for example, a plant extract thought to have antibiotic properties) are placed in a series of Petri dishes. Into each of these is poured about 10 ml. of melted and cooled (45° C) agar, which is thoroughly mixed with the antibiotic dilution by tilting the plates back and forth. After the agar has hardened, the plates are marked into several sectors, each of which is streaked with a different test organism. After incubation, the highest dilution of the antibiotic that inhibits each organism is recorded.

This method—reversed—also is often used for sensitivity testing by using *known* strengths of known antibiotics and *unknown* organisms.

### The Cylinder-Plate Method

A series of steel, glass or porcelain cylinders about 10 by 6 mm. in dimensions with perfectly square-cut ends (like a drainage tile) are heated and then sealed to the upper surface of already inoculated agar in a Petri plate by placing them, end up, on the solid agar. Dilutions of the single antibiotic to be tested are placed in the cylinders. The plate is then incubated. When they diffuse into the agar, the antibiotic dilutions reveal their potency by the width of the zone of inhibition of growth in the agar around each cylinder. The potency of different antibiotics cannot be compared on the basis of the width of their zones of inhibition, however, because a very potent antibiotic may not diffuse readily through the agar and will therefore produce relatively narrow zones, while a weak antibiotic may diffuse very well and produce deceptively wide zones. In such measurements the test organism, agar density, time, temperature and pH must be very exactly standardized. This is true of all the methods described.

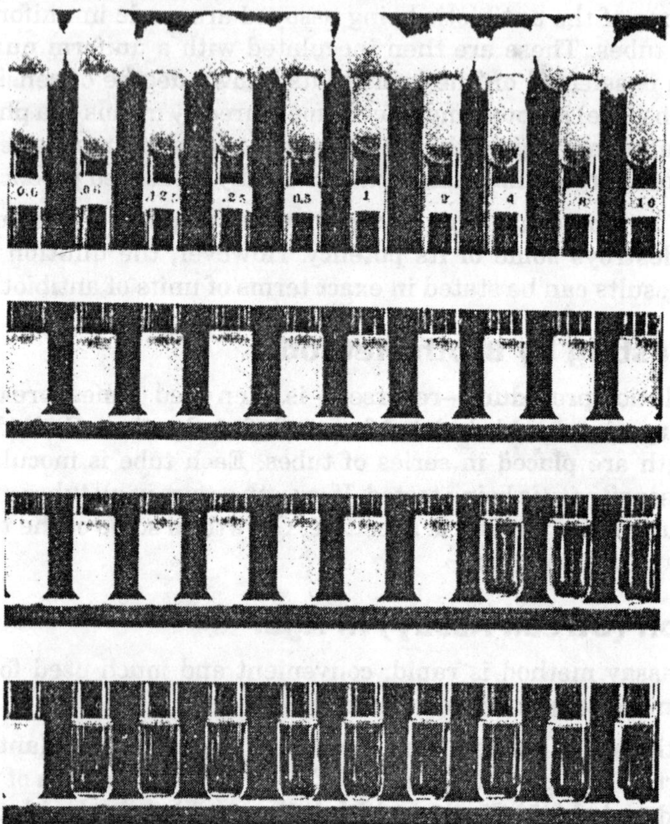

**Fig. 4.6. Testing sensitivity of a bacterium to antibiotics or other chemotherapeutic agents by the "tube dilution" method. The top row of tubes of broth culture contain agent "A" in increasing amounts as shown by the figures on the tubes. The second row contains agent "B" in the same amounts, while the third row contains agent "C" in the same dilutions. All tubes were inoculated with organism "X" from the same culture at the same time and incubated together. Luxuriant growth, as shown by white turbidity, as occurred even in the highest concentration of agent "A," showing that organism "X" (perhaps from a patient very ill with this infection) is not in the least affected by this drug. The drug of choice will be agent "C" which prevents growth of organism "X" even in the smallest concentration. Agent "B" is slightly effective. The tube at the left of each row is a control tube containing no inhibitory agent.**

## The Disk or Tablet Method

This is possibly a little less exact than the procedures just described but is reasonably accurate and is quick, convenient and economical. It is therefore a preferred and widely used method for many purposes not requiring high degrees of precision. The principle is the same in the cylinder-plate method, but the cylinders are replaced by disks or tablets.

Nutrient agar of appropriate composition is heavily inoculated with the desired organism all over the surface of the solidified agar or mixed with the agar while still fluid before pouring the plate. If an antibiotic solution of unknown potency is being assayed (*e.g.*, manufacturer's sample, blood, urine) the organism used is a stock strain of known sensitivity to standard doses of that antibiotic.

Measured strengths of the antibiotic solution to be tested are applied to the inoculated agar in disks of sterile filter paper or in small tablets, which are placed on the surface of the agar plate before incubating. Several devices are available for application of the disks. The antibiotic diffuses outward from each disk or tablet and inhibits growth in the agar around it.

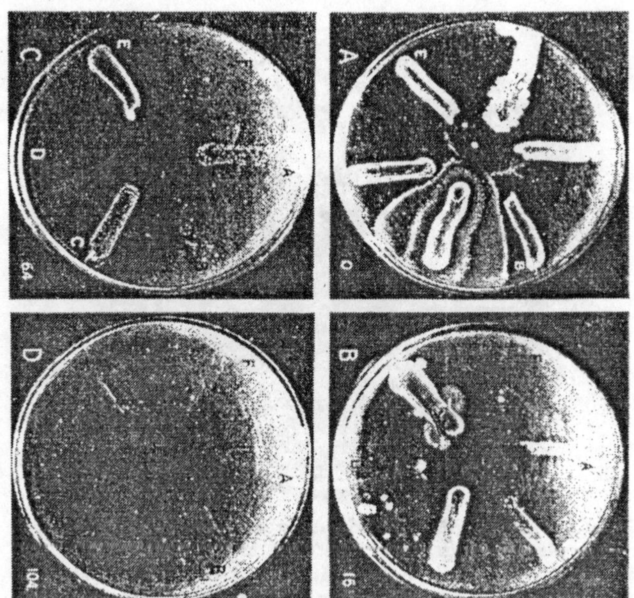

**Fig. 4.7. Testing sensitivity of several species of bacteria to three different concentrations of streptomycin by the agar-dilution method. Plate A contains no antibiotic and is inoculated with *(A) Escherichia coli, (B) Salmonella typhi,* (C) *Proteus* sp., *(D) Klebsiella pneumoniae, (E) Pseudomonas aeruginosa, (F) Mycobacterium tuberculosis.* In Plate B, containing 1.6 mg. of streptomycin per ml. of agar, organisms *D* and *F* have been virtually eliminated. Larger amounts of the antibiotic, as shown in Plates C and D, successively eliminate all but a few *resistant* colonies of the other organisms. (What is the significance of these resistant colonies?) Note the great sensitivity of this strain of *M. tuberculosis* to streptomycin.**

## Growth-Indicator Methods

In these methods some early-perceptible indicator of growth is looked for rather than the often-delayed growth itself. In a representative procedure, sterile blood agar is poured into a Petri dish to a depth of about 3 mm. This is the base layer.

Similar fluid agar is heavily inoculated with the organism being tested. After this is thoroughly mixed, it is poured over the base layer to an equal depth (seed layer). When solid, antibiotic disks are placed on the surface, and the plate is placed in the incubator. The plate may be examined after two hours and then every hour thereafter. Diffuse growth of the organism, whether actual colonies are visible or not, produces a marked discoloration or hemolysis of the blood, except where inhibited under and around disks containing effective antibiotic concentrations. Here the blood remains unchanged.

Modifications of this method employ indicators of growth other than discolorations of hemoglobin; *i.e.,* pH indicators such as phenol red, and oxidation-reduction indicators such as resazurin. In all of these methods the change in color of whatever indicator is used reflects growth of the organism hours before growth itself is apparent.

# ANTIBIOTICS FROM STREPTOMYCES

*Streptomyces* a genus of bacteria that has some of the physiological and gross morphological characteristics of molds like *Penicillium notatum. Streptomyces* species are, however, minute in size as compared with true molds and are entirely procaryotic in cellular structure.

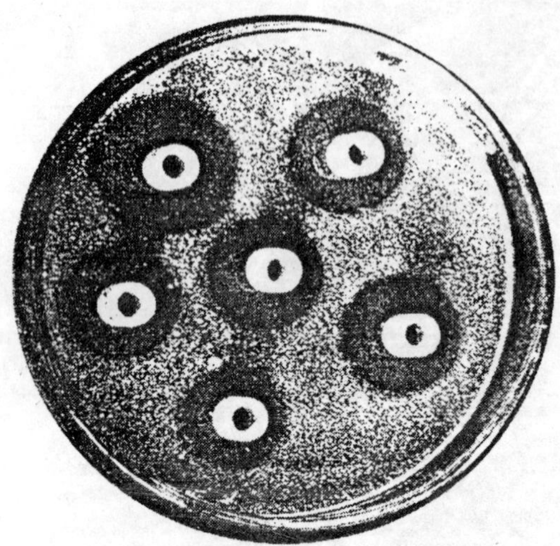

**Fig. 4.8. Cylinder plate standardization of penicillin. The culture plate contains a very carefully standardized medium, uniformly and heavily seeded with staphylococci or spores of *B. subtilis*. In each cylinder was placed a measured quantity of penicillin-containing extract. Following incubation under standard conditions, growth of bacterial colonies gave the medium the pebbled, gray appearance seen everywhere except around the cylinders. Here, growth has been inhibited by penicillin. Measurement of these zones of inhibition permits standardization of penicillin in Oxford units by comparison with zones produced under identical conditions of test by standard solutions of known unit strength. The sensitivity of an unknown organism to any antibiotic is tested in exactly the same way; inoculation of the plate is made with the "unknown" in place of the staphylococci or *B. subtilis* routinely used. Known amounts of various antibiotics may be placed in the cylinders.**

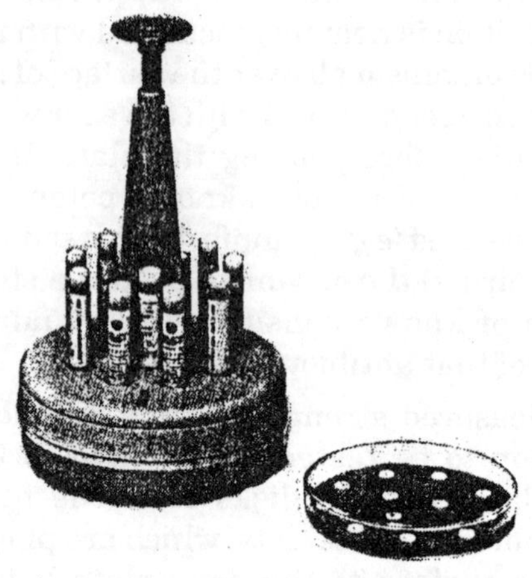

**Fig. 4.9. Antibiotic solutions are released automatically onto an inoculated plate when plunger of Sensi-Disc dispenser is pushed.**

A number of species of *Streptomyces* produce antibiotics of medical and commercial value. More are constantly being found. Some of the antibiotics produced by *Streptomyces* are said to be active against chlamydias, others against certain viral neoplasms and some against certain protozoa. A description of *streptomycin* will exemplify all the *Streptomyces*-derived antibiotics, although each has its special properties and values.

## Streptomycin

The discoverv of streptomycin dates from 1944, and followed naturally in a long series of researches by Waksman and his collaborators, Schatz and Bugie into the numbers and kinds of antagonistic microorganisms in the soil. The source organism *Streptomyces griseus,* produces moldlike conidia and aerial mycelia suggestive of some of the filamentous Ascomycetes *(e.g., Aspiergillus)* but is one species of the bacterial order of Actinomycetales.

The principles underlying the production, purification, assay and standardization of streptomycin are analogous to those of penicillin, differing mainly in respect to cultural details and technical procedures appropriate to the organism and antibiotic involved.

Aeration of the culture is essential for rapid streptomycin production, which is at its maximum in two to three days. Various strains of *Streptomyces griseus* are more or less active producers of the drug.

The molecular formula of streptomycin is $C_{21}H_{39}O_{12}N_7$. It is structurally much more complex than penicillin. Numerous salts and derivatives have been prepared, some of them of great therapeutic value. Dihydrostreptomycin, in which two hydrogen atoms are added is of especial importance. Both streptomycin and dihydrostreptomycin are much more stable than penicillin, both in dry form and in solution. In the refrigerator the solutions retain potency for months; at 37° C for about two weeks. Boiling for 10 minutes produces about 50 per cent loss of potency. Both drugs, *in vitro,* are bactericidal in commonly used concentrations.

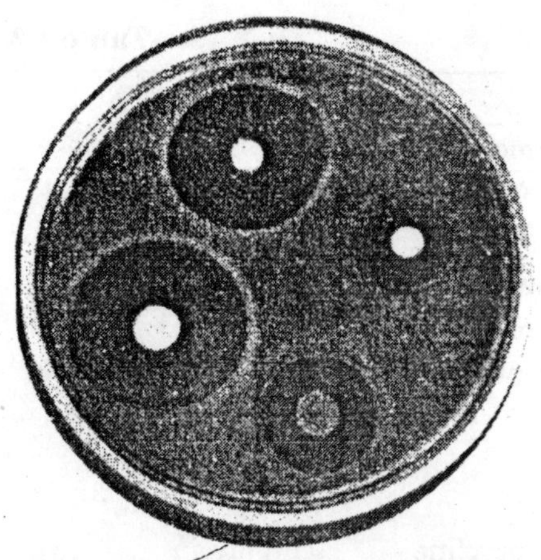

**Fig. 4.10. Testing sensitivity of a bacterium to antibiotics or other chemotherapeutic agents by the "disk method." The entire surface of agar medium in a Petri plate is inoculated with the organism to be tested. Paper disks of uniform thickness containing graded amounts of the agent to be tested (or the same amount of different agents if a comparison is desired) are then placed on the surface of the agar. The agent diffuses into the agar and prevents growth of the bacterium in a zone around the disk. The width of the zone indicates, roughly, the sensitivity of the organism to the agent or agents being tested, though the *presence* or *absence* of a zone is of greater significance. (About 1/2 actual size.)**

## Scope of Action

The therapeutic range of streptomycin is in some respects broader than that of penicillin; it is active against a variety of gram-negative and several gram-positive organisms, some of which are listed in Table 4.3. It has been used in tuberculosis with striking results and in many infections with species of gram-negative rods that are not susceptible to penicillin or sulfonamides. Streptomycin is not so rapidly destroyed in or excreted from the body as penicillin. Some toxic effects (vertigo, deafness) have been described. It must be used only under medical supervision.

## Mode of Action

Although the method is not entirely clear, streptomycin undoubtedly acts by interfering with enzyme action, notably enzymes concerned in early stages of the synthesis of proteins. Streptomycin has other, little-understood actions. Dihydrostreptomycin acts similarly. Both streptomycin and dihydrostreptomycin are much less active anaerobically than aerobically; *i.e.,* their action may therefore be related to aerobic oxidative metabolism that is dependent on the Krebs cycle. Like penicillin both are mainly bactericidal rather than bacteriostatic in their action.

No streptomycin-destroying enzyme has been reported, but resistant organisms appear and, unfortunately, often occur in patients during treatment. Strains of tubercle bacilli wholly dependent on streptomycin have also been found in treated patients.

**Table 4.2. Partial list of antibiotics***

| *Antibiotics* | | *Characteristics*** | *Source* |
|---|---|---|---|
| *Common Name* | *Trade Name* | | |
| Penicillin*** | | Gram-positive bacteria; *Treponema; Neisseria* | *Penicillum notatum* |
| Fumagillin | | *Entamoeba histolytica* | *Aspergillus fumigatus* |
| Paromomycin | Humatin | *E. histolytica* | *Streptomyces rimosus* |
| Streptomycin | | Myco. Tuberculosis; Gram-negative bacteria | *Streptomyces griseus* |
| | | Like streptomycin | Streptomycin; also some Species of *Streptomyces* |
| Tetracycline | Achromycin | Broad-spectrum | Chlortetracycline |
| Oxytetracyline | Terramycin | Broad-spectrum | *Streptomyces rimosus* |
| Chlortetracycline | Aureomycin | Broad-spectrum | *Streptomyces venezuelae* |
| Erythromycin | Ilotycin, Erythrocin | Braod-spectrum (not Enterobacteriaceae) | *Streptomyces erythreus* |
| Carbomycin | Magnamycin | Like erythromycin | *Streptomyces halstedii* |
| Oleandomycin | Matromycin | Broad-spectrum | *Streptomyces antibioticus* |
| Neomycin B*** | Flavomycin | Mycobacteria | *Streptomyces fradiae* |
| Viomycin | Viochin | Like penicillin | Str. Floridae; *Str. funiceus* |
| Oligomycin | | Fungi of plants | *Str. Diastatochromogenes* |
| Amphotericin B | Fungizone | *Candida* sp. | *Streptomyces nodosus* |
| Kanamycin | Kantrex | Broad-spectrum | *Streptomyces kanamyceticus* |
| Nystatin | | Pathogenic fungi | *Streptomyces noursei* |
| Cycloheximide | Actidione | Saprophytic fungi | *Streptomyces griseus* |
| Griseofulvin | Grifulvin | Pathogenic fungi | *Streptomyces griseus* |
| Bacitracin | | Like penicillin | *Bacillus subtilis* |
| Polymyxin B*** | | Gram-positive bacteria | *Bacillus polymyxa* |
| Pyocyanin**** | | Miscellaneous | *Pseudomonas aeruginosa* |

* Several not listed here are valuable commercially, agriculturally and horticulturally.

** Not necessarily the only activity.

*** Several of these antibiotics are in reality mixtures consisting of related compounds such as the penicillins, polymyxin A, B, C, D, carbomycin A and B and so on.

**** Not used medicinally. One of the first known antibiotic substances.

## Chloramphenicol

The organism producing chloramphenicol is much like other species of *Streptomyces*. It was found in 1947 by Paul R. Burkholder in soil collected near Caracas, Venezuela, and is called *S. venezuelae*. Like most other natural antibiotics, the drug is a by-product of growth. It may be produced in a manner analogous to the production of penicillin and streptomycin, but is now produced mainly synthetically.

**Table 4.3. Some Bacterial Species Sensitive to Streptomycin**

| Gram-Positive |
|---|
| *Mycobacterium tuberculosis* |
| *Staphylococcus aureus* |
| *Diplococus pneumoniae* |
| *Streptococcus species* |
| *Bacillus subtilis* and related species |
| **Gram-Negative** |
| *Salmonella typhi* |
| *Pasteurella tularensis* |
| *Klebsiella species* |
| *Brucella abortus* |
| *Proteus vulgaris* |
| *Salmonella paratyphi B* |
| *Bordetella pertussis* |
| *Haemophilus influenzae* |
| *Pseudomonas aeruginosa* |
| *Escherichia species* |

It affects many species of both gram-positive and gram-negative bacteria in high dilutions and may be given either orally or intravenously. Unlike either penicillin or streptomycin, it is also effective against the chlamydias and rickettsias. It is the first antibiotic discovered to have definite rickettsicidal properties. It has proved of inestimable value in the treatment of many infectious diseases, including especially typhoid fever. Because of its wide range of activity, it is classed as a *broad-spectrum* antibiotic.

## Mode of Action

As in the case of streptomycin, although chloramphenicol does not with certainty compete with any known metabolite, it appears to block several important enzymes, possibly by *noncompetitive combination* (*i.e.,* at a site other than the functional center in the co-enzyme). Whatever its mechanism or mechanisms, the most obvious effect is to stop synthesis of proteins. There is also interference with oxidations at higher concentrations. Chloramphenicol does not stop the synthesis of nucleic acids.

The therapeutic excellence of chloramphenicol led to intensive chemical studies, as in the case of penicillin. The formula was soon learned and the drug synthesized *(Chloromycetin)* in 1949 in the laboratories of Parke, Davis and Company. It is the only antibiotic of major clinical importance that can be made synthetically on a profitable basis. It must be used only under medical supervision.

## The Tetracyclines

This group includes three antibiotics: tetracycline *(Achromycin),* chlortetracycline *(Aureomycin)* and oxytetracycline *(Terramycin).* Their similar molecular structures. Chlortetracycline differs from tetracycline in having a chlorine atom attached to carbon atom

No. 7 in place of hydrogen. It is produced by *Streptomyces aureofaciens.* Oxy-tetracycline has a hydroxyl group attached to carbon atom No. 5 in place of hydrogen (encircled). It is produced by *S. rimosus.* Tetracycline is an artificial derivative of chlortetracycline. All have similar antibiotic properties.

Unlike penicillin and streptomycin, tetra-cyclines are bacteriostatic in clinical concentrations. They are effective against many gram-negative and gram-positive species of bacteria, chlamydias and some rickettsias. For this reason, they are also classed as broad-spectrum antibiotics and have similar ranges of therapeutic activity. All are effective orally or hypoderm-ically. However, in any given patient, any one of these drugs may at times show surprising irregularities and give unexpectedly brilliant results—or fail. They are not identical.

## Mode of Action

These antibiotics disrupt ribosomes and thus interfere with protein synthesis, especially in actively growing cells. They may also interfere with functioning of the Krebs cycle. Other noxious actions in the bacterial cell have been demonstrated but not yet explained.

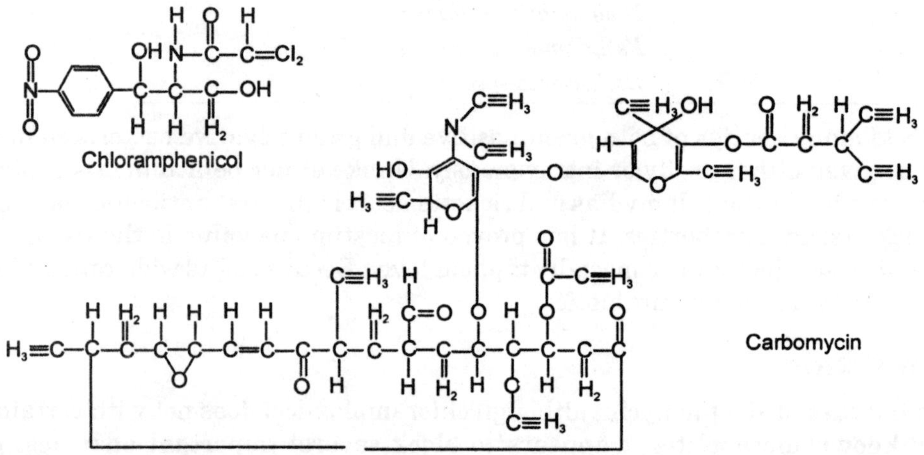

Chloramphenicol

Carbomycin

**Fig. 4.11. Structural formula of carbomycin, a typical macrolide. Compare with formulas of penicillin, cyclines and (above) chloramphenicol.**

Also, since they are active chelating agents, they may deprive the cell of essential metal ions.

## Erythromycin

This antibiotic, one of a group including erythromycins A and B, is representative of several valuable and widely used antibiotics, including *carbomycin* and *oleandomycin.* All are substances of a basic nature derived from *Streptomyces* species and having relatively high molecular weights. Because of their large, complex molecules, the group name *macrolides* has been given to them. They act principally against gram-positive bacteria, but also attack several species of gram-negative bacteria as well as other types of pathogens. They are therefore generally included with the broad-spectrum antibiotics. Their chief effect is to inhibit synthesis of proteins, possibly by blocking or destroying one or more types of RNA.

## ANTIBIOTICS FROM BACILLUS

Several useful antibiotics have been derived from gram-positive, aerobic, sporeform-ing bacteria closely related to, or identical with, *Bacillus subtilis*. The first to be described, *tyrothricin or gramicidin,* from *B. brevis,* was discovered by Dubos, as previously mentioned. Another is *bacitracin,* discovered by F. L. Meleney.

### Bacitracin

Bacitracin was first described in 1945. The source organism is a particular strain of *Bacillus subtilis,* a common and usually harmless organism widely distributed in dust. It was found contaminating a wound in a patient named Tracy, hence bacitraein. Bacitracin is a yellowish powder readily soluble in water. Like penicillin it is active clinically mainly against infections due to gram-positive bacteria. However, it also affects the gram-negative meningococcus *(Neisseria meningitidis;* cause of meningitis), gonococcus (cause of gonorrhea) and *Haemo-philus influenzae,* as well as *Treponema pallidum* (cause of syphilis). Unlike penicillin, streptomycin, chloramphenicol and the tetracyclines, bacitracin is quite toxic if given internally and is therefore used only for topical applications.

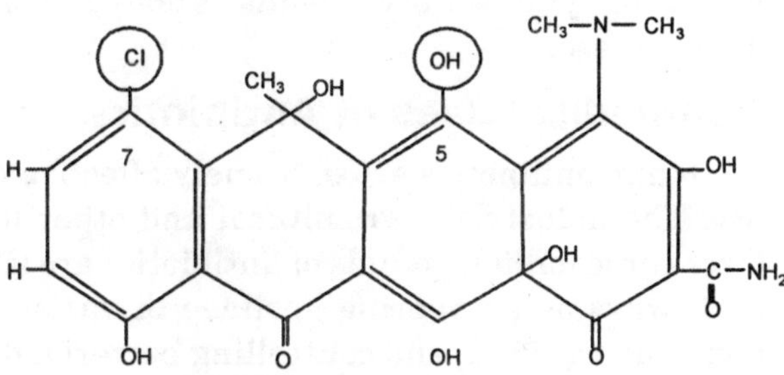


**Fig. 4.12. Composite formula of the tetracycline group of antibiotics. In tetracycline each of the numbered carbon atoms (5 and 7) has an attached hydrogen atom. In chlortetracycline, chlorine (circled) replaces the hydrogen of the number 7 carbon atom; in oxytetracycline an OH group (circled) replaces the hydrogen of the number 5 carbon atom.**

### Other Polypeptide Antibiotics

Bacitracin is representative of several *polypeptidc* antibiotics with similar properties, derived from species of *Bacillus:* the polymyxins A, B, C, D, E *(B. polymyxa);* subtilin and bacitracin *(B. subtilis);* the gramicidins and tyrocidin *(B. brevis);* bio-cerin *(B. cereus);* and circulin *(B. circulans).* Commercial bacitracin, and probably these related polypeptides, are all mixtures of related substances.

These are curious chemically in that they contain rare types of peptide linkages and amino acids of the D-series which have not been found elsewhere in nature except in bacterial cell walls. They are alike in being surface tension reducers, or surfactants.

### Mode of Action

Antibiotics such as bacitracin resemble other surfactants, such as the quaternaries and soap, in quickly destroying susceptible bacteria by attacking the cell membrane, possibly by acting as an emulsifying agent on the lipids in the membranes. They affect many gram-negative species, cell walls of which are rich in lipids. They act so quickly as to suggest that, unlikc most other antibiotics, interference with enzymes is not involved. Unlike penicillin, these antibiotics leave the cell wall intact.

Other polypeptide antibiotics (*e.g.,* streptogramin, vernamycin and ostreogricin) produced by *Streptomyces* sp., act by inhibiting protein synthesis but not synthesis of nucleic acids or cell walls.

## Lysostaphin

This antibiotic, discovered in 1964, is remarkably effective against coagulase-positive (pathogenic) *Staphylococcus aureus,* which is often resistant to other antibiotics. Instead of inhibiting synthesis of cell walls as does penicillin, lysostaphin destroys cell walls, regardless of penicillinase.

## Nonmedical Uses of Antibiotics

Many antibiotics are extremely effective as antimicrobials but are too toxic for medicinal use. The industrial, agricultural and other uses of such antibiotics have grown enormously. Over three million pounds of antibiotics are used for such purposes annually. They are used in such ways as preventing spoilage of various products; supplementing stock feeds, greatly increasing growth; and controlling bacterial diseases of plants.

CHAPTER

5

# Neuropathy

Peripheral neuropathies are the most common neurological syndromes associated with HIV infection. The spectrum of neuropathic syndromes is broad, and neuropathies may be encountered during any stage of the infection. Some disorders of the peripheral nerves in patients with HIV infection are presumed to be caused by pathological factors resulting from the virus itself, whereas others may result from responses of a competent immune system to viral antigens, opportunistic processes unleashed when immune deficiency has occurred, or as adverse effects of therapies for AIDS-related diseases.

To provide a context for understanding the toxic neuropathies, these other HIV-related neuropathies are briefly described in this first section. Inflammatory demyelinating polyneuropathies (IDP) are immune-mediated reactions directed at myelin peptides, which often present in association with a viral infection. They may occur acutely (Guillain-Barre syndrome or acute inflammatory demyelinating polyneuropathy [AIDP]) or follow a chronic fluctuating or progressive course (chronic inflammatory demyelinating polyneuropathy [CIDP].

Because they result from competent immune responses directed at antigens that are either derived from an invading pathogen or exposed by damage induced by the resulting inflammatory response, these syndromes tend to occur earlier in the course of HIV infection, even at the time of seroconversion. IDP is notable for a predominant picture of motor weakness and reflex loss, which can progress rapidly, compromising respiratory muscles and cranial nerve function as well as limb strength. Autonomic features may be present at onset, including tachycardia, blood pressure elevation, diaphoresis, and sphincter disturbances. In HIV infection, the clinical picture is indistinguishable from other settings, except for the presence of a moderate lymphocytic pleocytosis accompanying the elevation of protein levels in cerebrospinal fluid, in contrast to the usual increased protein in the absence of cells (albumino-cytological dissociation).

Electrophysiological studies are indicative of a demyelinating process, demonstrating severe slowing and blockage of nerve conduction velocities (NCV) and dispersion of compound action potentials. Variable degrees of axonal involvement may be seen and may have prognostic significance when present. Fortunately, the response to conventional therapy with plasmapheresis or pooled human immune globulin transfusions is usually successful. Mononeuntis multiplex is a multifocal process affecting multiple nerves. Inflammatory cells may be found in the epineurial and endoneurial perivascular spaces, however, vascular destruction is not seen.

**Table 5.1. Neuropathies Associated With HIV Infection.**

| |
|---|
| IDPs |
| Acute IDP (Guillain-Barre syndrome) |
| Chronic IDP |
| Monomeuritis multiplex |
| Inflammatory—vasculitic |
| Opportunistic (cytomegalovirus) |
| Polyradiculitis |
| Cytomegalovirus, Varicella-zoster, lymphomatous meningitis, syphilis |
| Dermatomal zoster radiculitis—may affect multiple dermatomes |
| DSP |
| Toxic neuropathies |
| Dideoxynucleosides (clinically DSP) |
| Acute neuromuscular weakness syndrome associated with hyperlactatemia and d4T |
| Other toxic neuropathies causes by treatment agents |
| Hydroxyurea, vincristine, INH, dapsone, thalidomide, etc |
| Diffuse infiltrative lymphocytosis syndrome |

The distribution of clinical features reflects the specific cranial and/or peripheral nerves involved. This type of neuropathy is characteristically asymmetric; affecting both motor and sensory functions, and may occur during the period of symptomatic HIV disease. It may occur because of vasculitis as part of a multiorgan systemic syndrome. In patients with AIDS, multiple mononeuropathy has been found in association with cytomegalovirus infection. An acute polyradiculopathy syndrome may be encountered in patients with AIDS, most commonly resulting from cytomegalovirus infection, although cases caused by varicella-zoster virus, lymphoma, and syphilis have also been reported.

Typical features include an ascending motor weakness with reflex loss and sphincter disturbances, usually urinary retention. Painful paresthesias in the perineum and lower limbs are common, and sensory levels or Babinski signs may indicate involvement of the spinal cord in addition to the nerve roots. Diagnosis is usually made by cerebrospinal fluid analysis, which commonly reveals a polymorphonuclear pleocytosis and evidence of specific viral DNA when tested with polymerase chain reaction amplification techniques. Although the clinical picture superficially resembles an IDP, electrophysiological testing demonstrates features of an axonal process, including diminished action potential amplitudes and increased recruitment patterns. Treatment with specific antiviral agents may be successful when initiated early in the course of the opportunistic infection.

Additional HIV-associated neuropathies include varicella-zoster virus ganglio-neuritis, which presents with pain in a dermatomal distribution, similar to that affecting patients without HIV. In some patients, multiple segments will be affected, and occasionally the syndrome can be seen in the absence of the characteristic vesicular rash. Among the rarer neuropathic entities is the diffuse infiltrative lymphocytosis syndrome, which is a painful axonal sensory-motor neuropathy that may be either symmetric, or multifocal and asymmetric. Nerve biopsies reveal an angiocentric infiltration of CD8 lymphocytes in the epineurium. Peripheral blood CDδ lymphocytosis and infiltration of salivary glands are characteristic.

Dorsal root ganglionopathies producing sensory ataxia, acute motor-sensory axonal neuropathies, and neuropathies related to nutritional deficits are also reported in HIV-infected patients. The most common of the HIV-associated neuropathies, however, is the distal sensory polyneuropathy (DSP), which is seen most often in patients with HIV infection that is more advanced, and causes neurological symptoms in up to one-third of such individuals. Typicalclinical symptoms include numbness beginning in the soles of the feet followed by burning or shooting pains, pin and needle-like paresthesias, and variable aching and cramping pains. Symptoms occur symmetrically, gradually ascend into the legs, and may result in sufficient loss of sensation to cause impairment of balance (sensory ataxia).

Eventually, numbness and paresthesias involve the distal upper extremities as well. Ankle reflexes are typically reduced or lost early, and reduction of other reflexes may follow. Characteristically, motor weakness is not a prominent complaint, although examination may disclose modest weakness in the small muscles of the feet and hands, especially the toe extensors. Pathologically, DSP is a dying-back axonopathy, affecting both small and large nerve fibers. Because conventional NCV and EMG studies are relatively insensitive to pathology in the smallest nerve fibers, they may be unrevealing in early stages of symptomatic neuropathy, despite severe pain and evidence of distal sensory loss. Alternatively, some individuals without neurological complaints or findings may be found to have evidence of neuropathy using neurophysiological testing.

Quantitative testing of sensory thresholds (QST), a computer assisted determination of perception dependent on patient response to repeated stimulus presentation at various intensities, may be more sensitive to small fiber involvement. Such testing may reveal abnormalities in individuals without clinical findings or complaints, sometimes foreshadowing the evolution of symptomatic DSP. In a study using QST, the pain was associated with diminished thresholds to mechanical noxious stimulation and increased responses to supramaximal mechanical stimuli. In contrast, although thermal detection thresholds were also increased in subjects with neuropathy, the levels were similar in subjects with HIV-associated neuropathy, regardless of whether the neuropathy was painful or not.

This selective relationship of pain to mechanical receptors was interpreted as indicative of a peripheral pain mechanism selectively involving a class of nociceptors responsive to pressur. Assessment of intraepidermal nerve fiber density in skin biopsy specimens has recently been used in clinical trials as a means of quantifying loss of small nerve fibers in patients with polyneuropathies. The extent of epidermal small fiber loss was shown to correlate with measures of pain intensity in a recent clinical therapeutic trial of nerve growth factor for HIV associated polyneuropathy. The pathological process causing DSP is not completely understood at present.

Direct HIV infection of nerves is not seen, although occasional macrophages containing HIV genetic material are found in biopsies of affected nerves and some biopsies also show lymphocyte infiltration in the perineurium. A current theory is that the neuropathy results from local secretion of proinflammatory cytokines by activated macrophages, resulting in cytokinemediated neurotoxicity. Recent studies have suggested that the prevalence of DSP is related to both CD4 count and plasma viral load, although epidemiological studies comparing periods before and after the common use of highly active antiretroviral therapy (HAART) have shown conflicting effects on the prevalence of DSP, possibly in relation to an increasing frequency of adverse effects of components of that therapy.

## TOXIC NEUROPATHIES

Early clinical studies of dideoxynucleoside reverse transcriptase inhibitors were complicated by a painful neuropathy, which proved to be a dose limiting toxicity. It has been estimated that approx 10% of patients receiving zalcitabine (ddC) or stavudine (d4T) and 1 to 2% of those receiving didanosine (ddl) will have to discontinue therapy because of a painful neuropathy. Affected individuals developed burning or shooting pains, or pins and needles paresthesias in the soles of the feet and toes associated with numbness and sometimes with aching and cramping pains in the feet and calves. The pain and paresthesias worsened over time, gradually ascending up the legs.

On detailed neurological examination, impairments of distal sensation to pain, temperature, light touch and vibration, variable impairment of balance as a result of the sensory loss, and allodynia (hypersensitivity to normally nonpainful stimuli), were accompanied by loss or reduction of ankle reflexes, and variable weakness of the small muscles in the feet. Symptoms could precede abnormal examination findings, and were remarkable for their resemblance to the HIV-associated DSP.

The initial symptoms might be aching or neuralgic pain or paresthesias, depending on the dosage of the neurotoxic agent. When the dideoxynucleoside treatment was discontinued, the majority of individuals experienced gradual improvement in the neuropathic symptoms, however, this improvement was often preceded by a period of up to 3 months following cessation of the medication during which symptoms persisted or might even worsen, a phenomenon referred to as coasting. EMG and NCV could be normal, depending on the severity of the neuropathy, but demonstrated features of an axonal neuropathy when abnormalities were present.

QST was more sensitive than nerve conduction velocities in one study evaluating both methods. Clinical trials have identified a number of factors associated with increased risk for the development of painful neuropathies in association with dideoxynucleoside therapy, in addition to the parameters of drug dose and administration. These include lower CD4 lymphocyte levels in some but not all studies, previous history of neuropathy from other cause, excessive use of alcohol, exposure to potentially neurotoxic medications such as isoniazid or vincristine, neuromuscular or neuropathic symptoms at onset of therapy, diabetes mellitus, or weight loss before entry.

Most of these factors are either causes of neuropathy in their own right, or are associated with increased risk of HIV DSP. Not all individuals who develop paresthesias or pain go on to develop clinical neuropathy. In a retrospective study of subjects at one center participating in a large clinical trial, 28% of individuals receiving either ddl or ddC experienced transient numbness, paresthesias, or pain, but did not develop signs of toxic neuropathy. Criteria have been suggested for attribution of a compatible clinical syndrome to a toxic neuropathy. These criteria emphasize the occurrence of symptoms after exposure to the agent and their improvement after its discontinuation.

A subacute progression is also considered suggestive of a toxic neuropathy, because HIV-associated DSP is typically slowly progressive. Involvement of the upper extremities is more common in DSP than in toxic neuropathy, but has been associated with nudeoside exposure as well. The descriptions of toxic DSP were derived from experience in early clinical trials of dideoxynucleoside agents, and reports of clinical anecdotes. It is immediately evident from these descriptions that HIV-associated DSP and painful toxic neuropathies commonly associated with exposure to dideoxynucleoside reverse transcriptase inhibitors are remarkably similar in their clinical symptoms and examination findings. This leads to the obvious question of whether evolution of such symptoms during treatment with one of these agents represents unmasking or aggravation of an occult DSP, or represents a primary effect of the potentially neurotoxic agent itself. In many of these early trials, subjects either did not undergo detailed screening for prevalent polyneuropathy at entry or were entered with symptoms suggestive of preexisting neuropathy. In the next section, we review this data for each of the relevant nucleoside analogs.

**Table 5.2. Reported Risk Factors for Toxic Neuropathy With Dideoxynucleoside Therapy**

| |
|---|
| Dose and regimen of dideoxynucleoside |
| Combination therapy with multiple dideoxynucleosides or with hydroxyurea |
| Lower CD4 levels |
| Hyperlactatemia |
| History of peripheral neuropathy before dideoxynucleoside exposure |
| Exposure to other potentially neurotoxic drugs |
| Heavy alcohol use |
| Diabetes mellitus |
| Diminished creatinine clearance |
| Vitamin $B_{12}$ deficiency |
| Weight loss during dideoxynucleoside therapy |

**Table 5.3. Criteria for Diagnosis of Dideoxynucleoside-Associated Toxic Neuropathy**

| |
|---|
| Symmetrical pain and/or paresthesias in distal lower extremities—may ascend over time Diminished distal sensation to vibration, pain, and/or light touch on neurological examination—may have distal contact hypersensitivity (allodynia) |
| Normal motor tone and normal strength, except for mild weakness in small distal muscles[a] |
| Absent or diminished ankle reflexes |
| Onset of symptoms associated with exposure to dideoxynucleoside agents and improvement associated with their discontinuation or dose reduction, allowing for a period of persistence or intensification within the first 3 mo after the change in regimen |
| Elevated serum lactic acid levels[b] |

[a] The recently described acute neuromuscular syndrome associated with hyperlactatemia and d4T therapy is characterized by acutely or subacutely progressive motor weakness and areflexia. The clinical pattern superficially resembles Guillain Barre syndrome rather than the more commonly seen distal sensory polyneuropathy (*see* text).

[b] Elevated serum lactic acid levels have recently been found to discriminate distal sensory polyneuropathy caused by dideoxynucleosides from that caused by HIV.

## DIDEOXYNUCLEOSIDE THERAPY

### Zalcitabine

Early phase I trials encountered painftil neuropathy as a major dose-limiting toxicity in ddC therapy. Yarchoan and colleagues administered doses of ddC ranging from 0.03 mg/kg to 0.25 mg/kg, every 8 h, to 20 subjects with AIDS or AIDS-related complex (ARC) and CD4 lymphocyte levels fewer than 350 cells/mm$^3$. A reversible painful neuropathy syndrome occurred in 10 of the participants, beginning between 6 and 14 wk after onset of therapy. This syndrome was characterized by aching pain in the feet followed by burning paresthesias 2 to 3 wk later, with impairment of light touch, temperature, vibration, and position senses, and diminished ankle reflexes on neurological examination. Three subjects also developed mild weakness of the toe and foot extensor muscles.

After discontinuation of ddC, worsening of neuropathic symptoms continued for up to 5 wk before improvement. Four weeks after discontinuing continuous ddC, one patient with toxic neuropathy who was treated with an alternating regimen of 200 mg zidovudine (ZDV) every 4 h for 1 wk and 0.03 mg/kg ddC every 4 h for the next week, experienced improvement in neuropathic symptoms that continued during the subsequent 5 wk.

Electromyography (EMG) and nerve conduction velocity (NCV) studies were obtained in seven of the ten subjects who developed neuropathy at baseline, or within the first 2 wk of therapy. Five of these patients had normal test results and two patients showed a preexisting neuropathy, one axonal and slowly progressive (HIV-associated DSP?), the other a polyneuropathy attributed to diabetes mellitus.

Ten additional studies were performed after the onset of neuropathic symptoms, which began concurrently with treatment in nine subjects, and 4 wk after discontinuation of ddC in the 10th patient. NCVs revealed markedly decreased or absent sural nerve action potentials in the lower extremities, and, in addition, diminished sensory action potentials from ulnar nerves in the upper extremities, with preserved distal latencies, of seven subjects in whom these tests were performed. EMGs demonstrated abnormal spontaneous activity, increased polyphasic potentials, and abnormal interference patterns. These features are consistant with an axonal neuropathy. Merigan and colleagues conducted a dose-finding study in 61 subjects with AIDS or ARC, excluding individuals with "significant neurological disease."

All subjects administered 0.06 or 0.03 mg/kg ddC every 4 h developed painful neuropathy. The time-to-onset was dose related, with the higher-dose group developing neuropathic symptoms in a mean of 7.4 wk, and the 0.03-mg/kg group developing neuropathic symptoms in a mean of 8.2 wk. Individuals receiving 0.01 mg/kg ddC every 4 h also developed painful neuropathy, which occurred in 80% of these subjects in a mean of 11.8 wk. Symptoms escalated with onset over weeks and involved the distal upper extremities in half of the cases. QST revealed abnormalities of vibratory perception coincident with the earliest svmptoms.

Even the group receiving the lowest dose of 0.005 mg/kg ddC every 4 h had a small incidence of neuropathy, which occurred in 17% of the patients, evolving more insidiously during a mean of 17.5 wk. Improvement on discontinuation of ddC was dose related, with the two lower dose groups experiencing complete recovery within 6 mo. Improvement was slower in the higher dose arms (requiring up to 1 yr), and incomplete in several subjects receiving the highest dose who continued to experience pain and sensory loss after 1 yr. Fifty-two of these subjects receiving ddC every 4 h at 0.06 mg/kg, 0.03 mg/kg, 0.01 mg/kg, or 0.005

mg/kg, were evaluated for toxic neuropathy using serial clinical and electrophysiological monitoring combined with symptom questionnaires.

Criteria for neuropathy were specified as at least two symptoms of moderate intensity on the questionnaire, or one symptom of severe or moderate intensity plus an objective abnormality on either the neuro-logical examination or the electrophysiological testing. In the two higher-dose arms, all subjects developed neuropathy with a mean onset of 7.7 wk. QST for vibration became abnormal before the appearance of clinical symptoms, with a mean onset of 7.3 wk. In the intermediate 0.01 mg/kg-dose arm, all subjects developed similar but less intensely painful symptoms, evolving more gradually with a mean onset at 9.3 wk.

Vibratory QSTs became abnormal in 70% of this group, coincident with the clinical symptoms. NCVs were obtained in only 20% of the total cohort. They revealed abnormalities in all subjects from the higher-dose arms after the development of clinical dysfunction. NCVs were mildly abnormal in 40% of the intermediate dose cohort. In the low dose group, one subject developed mild but transient neuropathic symptoms 26 wk after onset of therapy. Another developed asymptomatic abnormalities on vibratory QST and a loss of ankle reflexes, however, no further change occurred despite continued therapy with ddC. One might reasonably question whether this subject developed HIV-associated DSP rather than a toxic neuropathy.

In contrast, however, when ddC was discontinued in symptomatic participants, improvement occurred in 83% of patients, with a mean time-to-onset of improvement of 19 wk, 15 wk, and 11 wk for the high dose, intermediate dose, and low-dose groups, respectively, consistent with a toxic etiology. In an open-label study conducted in individuals taking ZDV for at least 48 wk, ddC administered in a dosage of 0.75 mg every 8 h was associated with symptoms of painful neuropathy in 25% of 59 subjects. The pain was characterized as moderate or severe in 17% of patients, and 10% of patients withdrew from the study because of the neuropathic symptoms.

Subjects entering the trial had CD4 counts of fewer than 200 cells/mm$^3$ and no known peripheral neuropathy, although the screening methodology is not elaborated in the report. A regression analysis found that neuropathic symptoms occurred more frequently in subjects with CD4 levels of fewer than 50 cells/mm$^3$ and in subjects with diminished creatinine clearances of less than 110 mL/min. Another open-label trial cornpared courses of 200 mg ZDV plus ddC at either 0.01 mg/kg or 0.03 mg/kg every 4 h alternating either weekly or monthly, 0.03 mg/kg ddC at administered on alternate weeks, and 200 mg ZDV every 4 h administered either on alternate weeks or continuously, in a cohort with a median CD4 level of 142 cells/mm$^3$ (mean, 164 ± 139 cells/mm$^3$) and no known previous neuropathy.

Neurological screening methods included a base line evaluation for neuropathy, a neuropathic symptom questionnaire administered every 2 wk, a focused examination for neuropathy every 4 wk, and a vibratory QST every 8 wk. The frequency of painful neuropathy in subjects on ddC was related to the dose. Among individuals in the three 0.03 mg/kg-dose arms combined, neuropathy occurred in 34% of patients compared with 15% of patients in the two 0.01 mg/kg-dose arms. Of patients on ZDV alone taken either continuously or on alternate weeks, 17% of patients developed neuropathy, which was comparable to the results for those taking lower-dose ddC on a weekly (10%) or monthly (21%) alternating regimen, and those taking higher dose ddC on the monthly alternating regimen (14%).

In subjects alternating ddC and ZDV weekly, 46% developed neuropathy, compared with 14% of those alternating ddC and ZDV monthly. In those who developed neuropathy, 68%

experienced the onset of symptoms within 24 wk of starting therapy. This cohort had relatively low CD4 levels at entry and it is not noted whether the time-to-onset of neurological symptoms differed between those on ddC and those taking only ZDV, or whether any increase in CD4 level in response to therapy had an independent relationship with the occurrence of neuropathy. A large community-based, open-label clinical trial compared 500 mg/d ddl with 2.25 mg/d ddC in subjects with up to 300 CD4 cells/mm$^3$ who had failed therapy with ZDV.

Peripheral neuropathy was an exclusion criterion for entry, and occurred during the study at a rate of 22.1 per 100 patient years with ddl compared with 45.1 per 100 patient years for ddC, a statistically significant twofold increased risk in the ddC arm. A subsequent study evaluated the frequency of toxic neuropathy in a clinical trial using ddC dosed at 2.25 mg/d (~0.03 mg/kg) alone, or in combination with ZDV, in subjects with CD4 counts of up to 200 cells/mm$^3$ or up to 300 cells/mm$^3$ with concurrent systemic symptomas of HIV. At one site, 76 sub-jects entered into this multicenter clinical trial were evaluated with serial neuro-logical examinations, neuropathy symptom questionnaires, and vibratory QST.

Individuals who reported moderate or severe symptoms of peripheral neuropathy on the initial questionnaire, or who had moderate deficits on examination (undefined) at screening were excluded. Operational criteria for toxic neuropathy included at least one distal sensory symptom and at least one objective sign on examination, persisting for at least 72 h, and appearing while the subject was receiving ddC. During a follow-up period of up to 73 wk, the incidence of neuropathy was 34% in the two groups receiving ddC compared with 4.3% in the group receiving ZDV alone, an increased risk of 7.9 times for the ddC groups. The mean time-to-onset of neuropathy was 16 wk, but varied widely, from 1 to 51 wk.

In the entire multicenter cohort, the rate of moderate or worse neuropathy was 22% in the ddC group and 23% in the combination arm, compared with 13% for the ZDV group alone. In addition, in the multicenter cohort, for those entering with CD4 lymphocyte counts greater than 150 cells/mm$^3$, moderate or worse neuropathy was significantly associated with ddC therapy alone (16%) or in combination (21%), compared with ZDV alone (9%). However, there was no significant association with toxic neuropathy for ddC therapy, alone or combined, compared with ZDV alone, for patients with CD4 cell levels of fewer than 150 cells/mm$^3$ at entry. Although there seemed to be a trend favoring a relationship to ddC, this result could also reflect a greater risk of HIV associated DSP in patients with reduced CD4 levels.

Interestingly, the rate for neuropathies characterized as severe in the parent study was similar in all groups, at 4% for ddC alone and 6% for combination therapy or ZDV alone. In the smaller single center cohort, a case-control study comparing baseline characteristics found only diabetes mellitus to be a significant risk factor for neuropathy, whereas weight loss during the study showed a trend suggesting a relationship but did not reach statistical significance. No effect of CD4 level was demonstrated in this smaller sample in contrast to the total cohort. Development of moderate or worse neuropathic symptoms was managed by dose interruption until improvement occurred, followed by resumption of treatment at half of the original dose.

Blum and colleagues observed subiective symptomatic improvement within an average of 10 wk in 67% of their subjects after dose reduction, but no reversal of abnormal findings on neurological examination. The persistent neurological abnormalities could represent either residual damage from a toxic neuropathy or emergence of HIV-associated DSP during the study.

## Didanosine

Some early dose-escalation studies with ddI were limited by a painful neuropathy similar to that seen with ddC. Lambert and colleagues encountered symptoms of neuropathy in almost one-third of a cohort of 37 subjects with CD4 levels of fewer than 400 cells/mm$^3$, who received doses of ddI greater than 12 mg/kg/d, but in only 1 of 15 subjects who received smaller doses. Subjects were not screened for preexisting neuropathy at entry. The investigators also noted an increased risk of neuropathy with total cumulative dose greater than 2 g/kg. The time-to-onset of symptoms was dose related and ranged from 55 to 201 d after initiation of therapy. Withdrawal of ddI resulted in resolution or improvement of pain and paresthesias during 4 to 8 wk in seven of eight affected individuals, three of whom tolerated rechallenge with ddI at 10 mg/kg/d for 4 to 12 wk without recurrence.

A subsequent follow-up, of these subjects and several additional individuals taking 250 to 750 mg/day of ddI, for a mean of 34 wk (2-72 wk) revealed neuropathy in two additional subjects, both taking ddI at 750 mg/d. One developed symptoms at 30 wk. The other, who entered with what was described as mild peripheral neuropathy symptoms (HIV-associated DSP?), developed intensified neuropathic discomfort at 18 wk. After an 8-wk suspension of ddI, however, the patient was able to continue treatment at a reduced dose of 500 mg/d. In contrast, a large multicenter trial comparing ddI with continued ZDV in individuals previously taking ZDV for at least 16 wk and having either AIDS, AIDS related complex (ARC), and fewer than 300 CD4 cells/mm$^3$ or fewer than 200 CD4 cells/mm$^3$without systemic symptoms, found no difference in the rates if of symptomatic neuropathy between the groups.

Neuropathy at entry was not noted to be an exclusion criterion, and it appears that only symptomatic neuropathic complaints were evaluated. Perhaps of some importance, CD4 counts rose in the two ddI-treatment arms but not in the continued-ZDV arm. In another dose-escalation study, a cohort of 58 subjects with AIDS or ARC and a median CD4 count of 47 cells/mm$^3$ (range, 4-267 cells/mm$^3$) was followed for a median of 8.9 mo (with 28 subjects followed for longer than 1 yr).

Yarchoan and coworkers encountered painful neuropathy in 15 subjects. Twelve of these individuals received doses greater than 9.6 mg/kg/d. Preexisting neuropathy was not evaluated at entry. A subsequent report from this group described a 5-yr follow-up of 72 subjects, including individuals from this study and others from phase I studies who were subsequently switched to doses of 200 or 250 mg ddI twice daily. The authors reported 11 cases of painful neuropathy during this period, affecting approx 15% of the cohort. Of these, four patients continued treatment without dose change, two were successfully rechallenged at the same dose after temporary suspension of therapy, two tolerated rechallenge at a lower dose, and three discontinued ddI and switched to ZDV.

No evaluations for persistent neurological abnormalities in those who experienced pain were described. In an open label, dose escalation study of ddI administered once daily to 34 individuals with CD4 cell levels fewer than 400 cells/mm$^3$ (mean, 125 ± 110 cells/mm$^3$ at entry), followed for an average of 15.8 wk (2-56 wk; median, 12 wk), severe painful neuropathy was reported in only one of six individuals receiving the highest dose, of 30.4 mg/kg ddI. After suspension of ddI, partial resolution was noted during the next 8 wk. An interesting aspect of this study is the observation of 11 subjects entering with grade 1 neuropathy (paresthesias not requiring treatment), 8 had resolution of their symptoms during the trial, whereas the other 3 were unchanged.

During the trial, CD4 levels rose to greater than 200 cells/mm$^3$ in the majority of the subjects, suggesting that neurological improvement of the baseline neuropathy (HIV-associated DSP?) might have occurred because of a therapeutic response induced by the ddl regimen. Evaluation of subjects from a single site participating in another multicenter ddl dose escalation study identified toxic neuropathy, confirmed by a neurologist, in 10 (23%) of 44 subjects who complained of pain or numbness or were noted to have abnormalities on study examinations. Neuropathic symptoms improved or resolved after suspension of ddl. Toxic neuropathy was defined as the occurrence of distal sensory symptoms in the lower extremities in with reduced or absent ankle reflexes and distal sensory impairments on examination, appearing in relation to ddl therapy and improving on its withdrawal.

One subject maintained at half of the original dose had resolution of the neuropathic symptoms during 6 mo follow up, whereas four others tolerated rechal lenge after suspension of therapy either without recurrence or with mild paresthesias. In these 10 subjects, the mean CD4 level at entry was 114 cells/mm$^3$ (0-212 cells/mm$^3$) and improved by an average of 49 cells/mm$^3$ (range, 0-170 cells/mm$^3$) in seven patients, whereas it worsened by a mean of 38 cells/mm$^3$ (range, 11-75 cells/mm$^3$) in three other patients. The Alpha trial, a large, double-blind European study, used two dose regimens of ddI, 10.5 mg/kg/d and 3.3 mg/kg/d.

Peripheral neuropathy at entry was not an exclusion criterion, except in French centers. Only neuropathy leading to study discontinuation was reported, with rates of 7.8% in the higher dose group and 5.9% in the lower-dose arm. An open label, community based study comparing ddl with ddC in 467 subjects with CD4 levels of up to 300 cells/mm$^3$ reported the development of neuropathy in 21.6% ofsubjects taking ddl at 500 mg/d.

## Stavudine

Painful neuropathy was a dose-limiting side effect in early studies of d4T, as with the other dideoxynucleosides. In a series of 41 subjects with CD4 cell levels of fewer than 400 cells/mm$^3$, 55% of patients developed symptoms of painful neuropathy. The time-to-onset ranged from 5 to 40 wk. Painful neuropathy was related to both dose and dosing frequency. Each increase of 1 mg in the dose unit was associated with a twofold increased risk of neuropathy. Dosing twice daily was associated with a fourfold increased risk of neuropathy compared with administering the same total dose as a thrice daily regimen. It appears from the report that no screening for neuropathy at baseline was incorporated into the entry criteria.

Of 20 subjects developing painful neuropathies, 4 were noted to have used alcohol excessively, 1 was taking INH, and 1 was taking amphotericin. Five additional subjects reported myalgias at baseline. Some subjects tolerated rechallenge at reduced dose. The maximally tolerated dose was determined to be 2 mg/kg/d. Skowron reviewed early trials of d4T and found painful neuropathy to be the principal dose-limiting side effect.

In a combined analysis, the rate of neuropathy at doses of 0.5 and 1.0 mg/kg/d was 21 per 100 person-years compared with 66 per 100 person-years for doses of 2 mg/kg/d. In addition to the factors noted in the Browne study existing neuropathy at entry was also found to be a risk factor for subsequent neuropathic adverse effects. In a phase II study of subjects with CD4 levels of up to 250 cells/mm$^3$, painful neuropathy was dose- and treatment duration-dependent and occurred up to 48 wk after initiation at 2.0 mg/kg/d. The rate of painful neuropathy was 17 per 100 person years at a dose of 0.5 mg/kg/d *h* compared with 41 per 100 person-years at a dose of 2 mg/kg/d.

Resolution of pain and tolerance to rechallenge with a lower dose of d4T was noted in 55% of subjects who were followed for a median of 32 wk. A large, double blind, multicenter trial compared 40 mg d4T twice daily with 200 mg ZDV thrice daily in a cohort of 882 subjects with CD4 levels from 50 to 500 cells/mm$^3$, and previous exposure to ZDV for at least 6 mo. Individuals with a history of peripheral neuropathy were excluded. The rate of painful neuropathy requiring dose modification in the d4T group was 12%, compared with a rate of 4% in patients taking ZDV.

Neuropathy was more common in those with AIDS at baseline, regardless of treatment assignment, with rates of 25% in the d4T group and 12% in the ZDV group, compared with 10% and 3%, respectively, for those without AIDS at entry. Of those receiving d4T who developed neuropathy, 63% resolved symptomatically within a median of 17 d after suspension of the medication (range, 4-57 d) and tolerated resumption of therapy at reduced dose for a median of 23 wk (range, 1-120 wk). Differences in neuropathy rates for those with AIDS, in addition to toxic neuropathy associated with the antiretroviral therapy may affect these results. A different pattern of neurotoxicity has recently been described in association with d4T therapy, which superficially resembles the acute inflammatory polyneuropathy, or Guillain-Barr'e syndrome, described in the first section of this chapter. This new neuromuscular syndrome has been attributed to hyperlactatemia related to dideoxynucleoside reverse transcriptase inhibitor therapy, particularly with d4T.

Individuals with this syndrome develop profound progressive motor weakness during several weeks, which may be associated with ataxia and numbness. The cranial nerves may be involved, resulting in facial weakness, and involvement of the respiratory muscles may lead to ventilatory failure. In contrast to typical Guillain-Barr'e syndrome, in which the primary pathology is demyelinating, electrophysiological studies in this syndrome reveal features of an acute axonal neuropathy. In some instances, neurological symptoms were preceded by symptoms of lactic acidosis syndrome, including malaise, nausea, emesis, and abdominal cramping. Elevated plasma lactate levels have been documented in some of the relatively few case reports thus far reported.

Recovery in those individuals who do improve may be prolonged up to 1 yr or longer. Cases have been described in pediatric patients. Recently, a patient presenting with the Miller-Fisher variant of Guillain-Barre syndrome consisting of the clinical triad of ophthalmoplegia, ataxia, and areflexia evolving in association with hyperlactatemia and after discontinuation of anti retroviral therapy including d4T was reported. As this chapter was being finalized, a report appeared describing a retrospective collection of 69 cases, including some from previous reports, meeting criteria of new acute or subacute neuromuscular limb weakness associated with lactic acidosis.

Cases were classified for diagnostic certainty based on the presence of confirmatory diagnostic studies and the exclusion of confounding causes of neuromuscular weakness. In 88% of the cohort, the dideoxynucleoside used was d4T, with a median duration of 10.5 mo of therapy (range, 1-71 mo). No association between serum lactate level and duration of d4T use was found. Of interest is the finding that cases were almost equally likely to occur in women as in men. Approximately one-third of the cases were associated with symptoms of lactic acidosis syndrome. An important finding was that the neuromuscular weakness syndrome occurred despite previous discontinuation of the antiretroviral therapy in 36% of the cases (median, 14 d). Respiratory failure requiring ventilator support occurred in 23% of the cohort.

Mortality was notable at 16% of the patients and was marginally associated with levels of serum lactate greater than 6 mmol/L, although levels greater than 10.08 mmol/L showed a strong trend to associate with death. Most cases evaluated more extensively were sensorimotor polyneuropathies, although both axonal and demyelinating pathologies were seen in the limited nerve biopsies obtained. Four of 15 muscle biopsies revealed evidence of mitochondria! dysfunction, and 3 revealed inflammatory infiltrates. The full spectrum of this syndrome, its underlying pathophysiology, and its relationship to nucleoside therapy are only beginning to be understood as this is written. Similarly, although isolated case reports have described improvement in association with various treatments no consistently effective therapy has yet been identified.

## TOXIC NEUROPATHY WITH COMBINED AGENTS

Several reports have evaluated cohorts or described neuropathy in individuals receiving combinations of dideoxynucleoside agents, or combination dideoxynucleoside agents with hydroxyurea (HU). An early case report concerned an individual participating in a clinical trial alternating ddC and ZDV monthly. Neurological findings at the onset of therapy were not described. After 16 wk of treatment, the patient developed paresthesias, followed by burning pain 4 wk later, at which time the ddC was discontinued. Two weeks later, he reported the neurological symptoms to be the same. The following week, another physician initiated ddl treatment, and 1 wk after that, the neurological symptoms worsened with ascending sensory impairments in the legs and increased pain. The ddI was stopped, and the symptoms resolved 1 mo later.

The authors suggested that ddl could potentiate ddC neurotoxicity. The combination of ddI and d4T was evaluated in a cohort of 52 treatmentnaive subjects with a mean baseline CD4 count of 330 cells/mm$^3$. During 24 wk, nine subjects developed neuropathic symptoms (17%). Two of these withdrew and the other seven continued therapy. In a 24 wk extension completed by 30 subjects, 1 additional subject withdrew because of painful neuropathy and 2 other subjects developed minor paresthesias. The authors noted that toxic neuropathy was the most common adverse event in the trial, but that only three subjects discontinued therapy because of it, and concluded that the combination was safe and well tolerated. In a dose finding, double blind study comparing the effect of combinations of ddI and d4T on plasma viral load, there were 13 adverse events in 86 subjects followed for a median of 349 d (1-728 d).

Patients with peripheral neuropathy and previous treatment with neurotoxic drugs were excluded, and most subjects were treatment naive, with CD4 çell counts greater than 200 cells/mm$^3$ (mean, 343 cells/mm$^3$). It seems, from the report, that determination of neurotoxicity was based solely on subject complaints and only two subjects were reported to develop neuropathyas a serious adverse event (2.3%). The addition of HU to dideoxynucleosides has been shown to increase neurotoxicity.

A cohort of 144 subjects with CD4 levels of 200 to 500 cells/mm$^3$ were randomized to receive ddl and d4T with or without HU. Seventy two subjects were initially randomized to the three-drug arm and 12 additional subjects switched to that arm after 12 wk. After 2 yr, only 25% of subjects remained in the trial, most withdrawing because of lack of efficacy or adverse events. Neuropathy occurred in 35% of the group taking all three drugs, compared with 15% of those taking only the ddl and d4T,a statistically significant difference. In a retrospective database review of 1116 patients treated with various regimens containing ddl, d4T, and HU in one large center, the risk of neuropathy was calculated for the agents alone and in combination.

Inclusion criteria for neuropathy required sensory symptoms in the lower extremities or all four extremities, beginning, recurring after resolution, or (if preexisting) worsening after the onset of therapy with the agents. The mean CD4 count was 270 cells/mm$^3$. The risk of developing neuropathy was related to lower CD4 count, history of non-drug-asso-ciated neuropathy, and, weakly, with age greater than 40 yr. Adjusting for these factors yielded a relative risk compared with ddl alone of 1.39 for d4T, 2.35 for ddl plus HU, 3.5 for ddI plus d4T, and 7.8 for ddI plus d4T plus HU. The crude incidence rate for ddl alone was 6.8 per 100 person-years of treatment. Another retrospective review found a frequency of peripheral neuropathy 27% of a small group of subjects administered ddI plus d4T in conjunction with HU, compared with 10% in a group administered ddl plus d4T.

Although the trend did not reach statistical significance, this report is consistent with the findings in the two studies described above. A single case report describes a patient with neuropathic toxicity attributed to lamivudine (3TC). This individual experienced onset of paresthesias while taking ddC plus ZDV. His symptoms improved after discontinuing the ddC, however, distal sensory impairment and mild distal weakness persisted, with electrophysiological findings consistent with a polyneuropathy.

Eighteen months later, he started taking 3TC, and, within 3 wk, he experienced recurrent painful dysesthesias. These symptoms improved within 4 wk after the 3TC was discontinued, but recurred on rechallenge, consistent with a toxic exposure aggravating the preexistent neuropathy.

## PATHOPHYSIOLOGY OF TOXIC NEUROPATHY

The mechanism of nucleoside induced neurotoxicity is not completely understood at the present time. Current hypotheses emphasize toxic effects of nucleoside reverse transcriptase inhibitors on mitochondria, and tissue specific vulnerabilities based on energy requirements. Nucleoside analogs can replace nucleotide bases and can be incorporated into newly synthesized DNA. In contrast to other forms of DNA synthesized from nuclear chromosomal genetic material, mitochondrial DNA (mtDNA) is inherited from the cytoplasm of the oocyte and replicates by random segregation rather than recombination.

Replication of mtDNA requires nuclear coded DNA polymerase-γ. In mtDNA, there are few introns and limited self-repair mechanisms, thus, mutations are more likely to affect coding sequences altering protein production, and dysfunctional mutations are more likely to persist with subsequent replication. This results in mixtures of mitochondria containing wild-type and mutant DNA in the same cells in varying proportion, referred to as heteroplasmy. The impact of heteroplasmy on a particular tissue may be a threshold effect, dependent on the severity of the resulting defect in oxidative phosphorylation in the abnormal mitochondria and the relative metabolic demands of the tissue for energy production.

The hypothesis, as articulated by Lewis and Dalakas, suggests that manifestations of nucleoside toxicity reflect:

1. Subcellular availability and abundance of nucleoside analog in the target tissue.
2. The ability of the cellular thymidine kinases to use the nucleoside analog as a

competitive alternative substrate for monophosphorylation and, subsequently, triphosphorylation.

3. The ability of the triphosphate nucleoside analog to inhibit DNA polymerase-γ either by acting as an alternative substrate for incorporation into mtDNA via competitive inhibition resulting in aberrant chain extension, or by noncompetitive incorporation and premature chain termination.

4. The relative metabolic requirements of the tissue for oxidative phosphorylation.

It has been suggested that depletion of the mtDNA induced by nucleoside analogs results in the depletion of enzymes encoded by the mtDNA, leading to impaired oxidative phosphorylation. An analogy has been drawn to multiple symmetric lipomatosis, a hereditary condition in which mtDNA mutations and deletions impairing oxidative phosphorylation complex IV have been identified. This condition results in hypertriglyceridemia, insulin resistance, and fat deposition abnormalities similar to those associated with HAART, and is associated with peripheral neuropathy in almost all cases. In vitro studies have demonstrated the relative potency of nucleoside agents to inhibit mtDNA in a T-lymphoblastoid cell line, CEM, to be ddC greater than d4T greater than ZDV greater than ddI.

The concentration of nucleoside analog required to inhibit mtDNA replication was less than that required to inhibit cell growth at 4 d for ddC, ddI, and d4T, but not for ZDV. In a neuronal cell model, using PC 12 cells derived from rat pheochromocytoma and stimulated to differentiate into neurons by incubation with nerve growth factor, ddC and ddI inhibited production of mtDNA and neurite generation at pharmacological concentrations, whereas ZDV and 3TC did not. In contrast, d4T inhibited neurite generation at concentrations that did not affect mtDNA synthesis, despite the ability of d4T to impede activity of DNA polymerase-γ, and despite the presence of d4T triphosphate in the mitochondria, suggesting a different mechanism of neurotoxicity.

The authors suggested that a repair mechanism, perhaps the 3′5′ exonuclease present in DNA polymerase-γ, might offset toxicity of d4T to a degree. Absence of phosphorylated 3TC in the isolated mitochondria was consistent with the hypothesis that the inability to transport phosphorylated 3TC accounted for its lack of toxic effect on mtDNA. The lack of neurotoxicity of 3TC may also lie in its ability to act as a substrate for the integral 3′5′ exonuclease activity of DNA polymerase-γ, making incorporation less likely. A consequence of mtDNA polymerase-γ dysfunction is a reduced capacity of the mitochondria to synthesize uridine nucleotides, which are integral to the synthesis of oligosaccharides on membrane glycoproteins and glycolipids.

Cytidine nucleotides, derived from uridine nucleotides, are involved in phospholipid synthesis. In PC 12 cultures, the addition of uridine and pyruvate can prevent ddC-induced reduction of mtDNA replication. Nucleoside analog neurotoxicity has been demonstrated in animal models. Axonal degeneration and reduction in fiber size, myelin degeneration manifested by myelin splitting and edema, and abnormal mitochondria in Schwann cells were seen in rabbits treated witn ddC. It was not possible to determine from the study whether the axonal changes were a primary toxic effect or were secondary to the myelin damage.

In rats treated with ddI, sections of sciatic nerve taken at various durations after onset of exposure demonstrated myelin toxicity manifested by edema and myelin splitting at 15 wk. Concentrations of ddI used resulted in peak plasma levels five to six times those in humans administered therapeutic doses. A recent study analyzed morphological changes in sural nerve biopsies taken from patients with ddC-induced toxic neuropathy, in comparison with biopsies from patients with HIV-associated distal sensory neuropathy, who either never received any antiretroviral therapy or received only ZDV, and from individuals with non-HIV-associated axonal neuropathies.

Dalakas and colleagues found abnormalities in mitochondria in axons and Schwann cells of subjects receiving ddC more frequently than in controls. No unique features were noted on light microscopy; however, using electron microscopy, abnormal mitochondria with interrupted inner morphology, enlarged size, osmophilic deposits between the inner and outer limiting membranes, and diminished cristae were noted in all three groups, with different degrees of severity. In the ddC-associated neuropathies, 55% of axonal mitochondria were abnormal, compared, with 28% in HIV-associated DSP and 9% of the other axonal neuropathies.

In Schwann cells, 47% of the mitochondria were abnormal in the ddC-associated neuropathies, compared with 5% in the HIV-associated DSP and 14% in the other neuropathic controls. Competitive polymerase chain reaction amplification was used to quantify mtDNA in the nerves, and demonstrated significant depletion of mtDNA in the ddC-treated group compared with the neuropathic controls. Results were considered to be supportive of the ability of ddC to induce a toxic mitochondrial neuropathy by depletion of neural mtDNA, consistent with the selective inhibition of DNA polymerase-γ demonstrated injieural cell lines. The potential role of acetyl-carnitine deficiency as a result of mitochondrial damage in toxic neuropathy was suggested by Famularo and colleagues, after a study in which they demonstrated diminished serum levels of acetylcarnitine but not total carmtine, in 12 subjects with symptoms of painful neuropathy while taking ddI, ddC, or d4T in comparison with 10 subjects takingjidl and 11 subjects taking ZDV, who had no symptoms of neuropathy.

Acetyl-L-carnitine is a transport molecule for longchain fatty acids into the mitochondria, where they undergo β-oxidation, generating adenosine triphosphate, and donate the acetyl group to acetyl coenzyme for subsequent participation in Krebs cycle reactions, generating additional adenosine triphosphate. The authors hypothesized that acetylation of carnitine is important to normal function of peripheral nerve, facilitating binding of nerve growth factor and enhancing neural repair, and that lower levels in serum reflected defects in carnitine metabolic pathways caused by mitochondrial injury. In contrast to Famularo's findings, in a sample of 232 subjects drawn from a clinical treatment trial of nerve growth factor for HIV-associated neuropathy, Simpson and colleagues found an increased proportion of subjects with abnormally low levels of total and free carnitine among the study subjects compared with normal HIV-negative controls (no HIV controls without neuropathy were assessed), but found no association between the levels of total or free carnitine, or acetylcarnitine levels and the severity of the peripheral neuropathy as measured by clinical symptoms, QST, or nerve fiber density in skin biopsy specimens.

No relationship of carnitine levels to current or recent use of dideoxynucleoside agents, as compared with subjects with no use or remote use (6 mo or longer before entry), was found. There was no difference in any of the neuropathy measures studied between subjects with decreased levels compared with subjects having normal levels of carnitine. A possible explanation for the discrepancies between these studies could lie in differences between tissue and plasma levels of carnitine compounds. A recent report describes results of open label acetyl-carnitine treatment in 21 subjects with nucleoside reverse transcriptase inhibitor associated toxic neuropathies.

Subjects were treated with 1500 mg acetyl-L-carnitine twice daily for a median of 14 mo (range, 5-33 mo) and followed both clinically and by serial skin biopsies taken from the distal leg. At baseline, biopsies in patients with DSP showed reduced total innervation of epidermis, dermis, and sweat glands in comparison with five healthy controls. Selective staining for small sensory (type C and Aδ) fibers revealed an 80% reduction in epidermis and a 25% reduction in the dermis. After 6 mo of therapy with acetyl-L-carnitine, the total epidermal and dermal nerve fiber innervations reached or exceeded normal levels, whereas the fibers around sweat glands increased by 75%. Selective small fiberjnnervation reached 40% of control in the epidermis and 175% in the dermis.

Further increases were seen on subsequent biopsies in those continuing therapy for longer periods. Clinical neuropathic symptoms improved in 76% of the subjects and did not change in the others. No significant changes in CD4 or CDS lymphocyte counts or HIV plasma viral load occurred during the study period and, thus, did not affect results. In a brief trial, 10 of 16 patients with painful DSP receivıng 500 to 1000 mg daily of acetyl-carnitine intramuscularly or intravenously reported improvement in pain symptoms after 3 wk. Five subjects reported no change in pain and one reported worsened pain. No large controlled clinical trial of acetylcarnitine therapy in HIV-associated or nucleoside-associated neuropathies has been published, to date.

## MANAGEMENT OF NEUROPATHY

Complaints of pain or paresthesias in the lower extremities of a patient taking ddI, ddC, or d4T suggests the possibility of a toxic neuropathy and warrants detailed evaluation. Compared with HIV-associated DSP, toxic neuropathies appear more abruptly and progress more quickly, thus, an individual who develops such progressive symptoms within several months after the introduction of a potentially neurotoxic antinucleoside agent is more likely to have a toxic neuropathy. The diagnosis, however, can be confirmed only if symptoms resolve on discontinuation of the agent. Because some subjects in clinical trials experience only transient paresthesias minor symptoms may not necessarily require immediate change in therapy.

Furthermore, in some instances, it may be preferable to continue the agents despite minor neuropathic symptoms because of a robust antiretroviral response. A detailed neurological examination to assess the extent of sensory and reflex impairment, and a detailed neurological history to pursue other causes of neuropathy is an important measure, which should be obtained at the onset of symptoms and is also a reasonable screening measure before institution of dideoxynucleoside therapy. Other potential treatable causes for neuropathy, such as vitamin $B_{12}$ deficiency, diabetes mellitus, thyroid disease, syphilis, Lyme disease, or exposure to other neurotoxic substances or medications should be sought.

Electrophysiological studies may be helpful in centers with the capability to do both QST and conventional EMG-NCV studies, however, routine EMG-NCV studies may be unrevealing untilafter clinical signs appear. Serial monitoring of the neurological symptoms and examination is indicated in any individual experiencing incident neuropathic symptoms or abnormal findings on examination while taking dideoxynucleosides. In the event the symptoms progress or are severe enough to limit function, and if other antiretroviral treatment options are available, consideration should be given to suspending therapy with the nucleoside analog to allow the neuropathy to improve.

The patient should be counseled about the possibility of symptoms worsening in the first months after discontinuation of the presumed offending agent (coasting). If substantial improvement occurs, rechallenge with the agent at a lower dose is reasonable after 2 to 3 mo if no alternative therapy is available. Neuropathic pain may be intermittent and tolerable for some patients and require only nonsteroidal antiinflammatory agents for relief. For many individuals, the pain is both continuous and severe enough to limit function. The World Health Organization has published a three-step analgesic ladder to provide a systematic and rational approach to pain management.

Although developed to treat cancer-related pain, the same approach is useful for addressing neuropathic pain associated with HIV and dideoxynucleoside therapy. Initially, adjuvants such as tricyclic agents (amitriptyline and others) or anticonvulsants (gabapentin, lamotrigine, carbamazepine, and others) are used in combination with nonopioid analgesics. In the second step, a second adjuvant with a differing mechanism of action is added. Many patients with moderate neuropathic pain can be controlled at this level; however, for those who prove refractory to these measures, step three adds opioid analgesics, which should be used in long-acting formulations (fentanyl, oxycodone, and others). A recent study comparing the analgesic efficacy of lamotrigine to placebo for HIV-associated neuropathic pain found self-reported benefit in subjects taking dideoxynucleoside agents, but not in those who were not using neurotoxic treatment regimens.

A randomized trial of amitriptyline and mexiletine failed to show benefit for either drug compared with placebo in subjects with painful HI V-associated neuropathies and varying exposure to dideoxynucleosides. Individuals with sensory ataxia caused by loss of postural sensation may require support for safe ambulation. Such patients are at increased risk of falling in the dark, when changing posture or direction, or when descending stairs. In contrast to the patient complaining of distal paresthesias, urgent evaluation is required for the individual who complains of acute motor weakness. These types of neuropathies can progress rapidly to paralysis and, potentially, to respiratory failure.

Plasma lactate levels should be drawn, and dideoxynucleoside agents discontinued, particularly if the neurological symptoms are associated with gastrointestinal symptoms. EMG-NCV studies may help to differentiate IDPs from axonal or mixed neuropathies that are more characteristic of toxic syndromes or opportunistic disease. In a patient with progressive weakness, hospitalization and serial monitoring of neurological and respiratory function is indicated until stability or sufficient recovery occurs to allow independent function. Cerebrospinal fluid samples should be obtained to exclude opportunistic infections or lymphoma.

Ventilatory support should be initiated for subjects with deteriorating respiratory status when vital capacity falls to 1 L, and prophylaxis for venous thrombosis should be initiated. Physical and occupational therapy to maintain range of motion and prevent contracture formation should be started early in individuals anticipated to have prolonged weakness. No specific therapy for the acute neuromuscular syndrome associated with nucleoside analogs and lactic acidosis has been shown to be effective. In one review of reported cases of pathological syndromes attributed to nucleoside-associated lactic acidosis, some individuals who improved were treated with cofactors such as thiamine, riboflavin, L-carnitine, prostaglandin E, or coenzyme Q. When neuropathic weakness improved, it did so very slowly, taking up to 1 yr after discontinuation of the nucleoside analog agent.

## IMPLICATIONS FOR FUTURE CLINICAL TRIALS

Current data on the toxicity of dideoxynucleoside analogs in humans has been drawn largely from subject complaints during clinical treatment trials and from case reports. No large carefully controlled prospective neurological trial focused specifically on nucleoside-associated neuropathy in HIV has been performed, to date. In many clinical treatment trials, previous neuropathy was not an exclusion factor at entry. In others, detailed neurological examination to determine the prevalence of asymptomatic neuropathy at entry, and to confirm the origin of symptomatic complaints during the study, were not incorporated into the protocols. As a result, errors in attribution of symptoms or in the recognition of occult neuropathy may have occurred. Simpson and colleagues reanalyzed data from a large clinical trial of ZDV, ddl, ddC, and placebo in four combinations.

Using criteria for DSP as described above, they reevaluated diagnoses made at the treatment sites. In 63% of reported cases of neuropathy, the site investigators were unable to determine the cause of the symptoms. On review, Simpson and colleagues were able to classify approximately half of these as DSP, and, therefore, potentially treatment related. Of subjects classified as having treatment related neuropathy, 27% did not have typical features of DSP as reflected in the criteria used by the reviewers, and were, therefore, considered unlikely to have medication-related toxic neuropathy. These findings suggest that data from studies not incorporating detailed baseline and serial neurological assessments may both underestimate or overestimate true treatment-related neurotoxicity.

Furthermore, in addition to modifying the frequency of DSP, successful HAART might also modify the time-to-onset and rate of progression of neuropathies induced by nucleoside analog therapy, confounding the temporal characteristics currently used to discriminate HIV-associated DSP from toxic neuropathy. The question of whether painful neuropathies appearing in association with nucleoside therapy represent exacerbation of a previously occult HIV-associated distal sensory neuropathy, or whether they arise *de nnvn* as a result of nucleoside toxicity is of significant clinical importance.

Neuropathy in advanced HIV infection seems to be common, and not infrequently asymptomatic. Marra and colleagues examined 226 subjects drawn from a large AIDS clinical trials unit and found DSP in 21%. Of the subjects with DSP, 71% had signs of neuropathy on examination but no symptoms. Individuals with DSP were more likely to have received dideoxynucleosides, they had more advanced HIV infection by Centers for Disease Control classification, and they had a trend toward lower CD4 counts. Schiffito and associates determined the frequency of DSP in a cohort selected for cognitive symptoms and CD4 counts of fewer than 300 cells/mm$^3$ or fewer than 200 cells/mm$^3$.

They found symptomatic DSP in 35% of subjects, and asymptomatic DSP in an additional 20% of subjects. In this cohort evaluations, 72% of subjects without neuropathy at entry developed incident DSP during a period of 30 mo, with a 1 yr incident of 52%. Among those with no neuropathy or asymptomatic neuropathy at entry, the 1-yr incidence of symptomatic DSP was 36%. Only 23% of the subjects were taking dideoxynucleosides and, interestingly, neither these subjects nor those with asymptomatic neuropathy were more likely to develop symptomatic DSP during the study. In a series of 251 HFV- infected subjects undergoing clinical and neurophysiological assessments, Tagliati and coworkers found evidence of DSP in 38% of the entire cohort, and in 48% of subjects with AIDS.

Of subjects with clinical features of DSP, 19% had normal electrodiagnostic studies, whereas, of those without clinical findings, 28% had abnormalities on the neurophysiological assessments suggestive of DSP. The limitations of previous data and the frequency of asymptomatic polyneuropathy demonstrated in these studies underscore the need for any future clinical trial using potentially neurotoxic therapies, including nucleoside analogs, to incorporate a sufficiently detailed neurological screening at baseline and during the observation period to accurately assess and attribute incident neuropathy, and to determine risk factors for the emergence of incident neuropathy. In some recent clinical trials incorporating dideoxynucleoside treatment regimens, fewer subjects have been noted to develop symptomatic painful neuropathy.

Additionally, some individuals seem to experience improvement of neuropathic symptoms in association with dideoxynucleoside therapy, further confounding the determination of frequency and etiology of painful neuropathies in patients with advanced HIV infection who are treated with potentially neurotoxic therapies. Factors contributing to the emergence of treatment-related neuropathic toxicity need to be elucidated. Do host factors, such as genetic susceptibilities, predispose to the development of toxic neuropathy? Are there particular interactions among the multiplicity of drugs taken by HIV-infected patients and the multiorgan pathologies, particularly those affecting the liver or kidneys, which combine to increase the toxicity of a standard therapeutic dose of a dideoxynucleoside in a given individual? Are the beneficial effects of successful therapy on DSP partially masking the neurotoxicity of dideoxynucleoside therapy? The risk of acquiring HIV-associated DSP s appears to be related to both increased plasma HIV viral load and decreased CD4 count.

Successful treatment with current HAART may have the capacity to improve peripheral nerve function, as demonstrated in a study using serial QST assessments, and as reported in an earlier study of ddl, described above. Thus, HIV-associated DSP may become less prevalent with successful retroviral suppression and control of other, as yet unidentified factors, offsetting an increased frequency of neuropathy induced by dideoxynucleoside components of HAART regimens. This offsetting effect was suggested in a retrospective study of a German cohort, in which the prevalence of DSP in a 2-yr period before the introduction of HAART was 42.5%, with a suspected rate of toxic neuropathy of 20.4%. In the following 2-yr period, during which subjects were placed on HAART in large numbers, the prevalence of DSP was reduced to 34.4%; however, the proportion of neuropathies attributed to nucleoside analog toxicity increased to 31.2%. Recently, a small prospective study assessed serum lactate levels in patients with DSP, comparing those taking d4T to those without exposure to dideoxynucleosides, as well as to HIV-infected individuals without neuropathy taking d4T and non-HIV-infected controls.

An elevated serum lactate level was found to be 90% sensitive and 90% specific in discriminating d4T-related toxic DSP from HIV-associated DSP. If results of this study are confirmed, the presence of a surrogate marker should contribute to the understanding of the pathogenesis of toxic DSP, and perhaps clarify the potential contributions of both the dideoxynucleoside- and HIV-related pathologies to the natural history of DSP. Future controlled clinicaltrials which include longitudinal assessments both of symptomatic and asymptomatic neuropathy from onset of antiretroviral therapy, and epidemiological studies of the natural history of HIV-associated neuropathies in relation to HAART therapy to identify factors associated with their occurrence, may help to clarify which subjects are at risk of severe toxic neuropathies and provide prognostic indicators to facilitate modification of therapeutic regimens.

Perhaps serum lactate levels will prove useful in this regard. It may be possible, with better understanding of the pathophysiology of these syndromes, to identify adjunctive therapeutic agents which protect against the emergence of toxic neuropathies, as has recently been suggested for acetylcarnitine. In addition to detailed serial clinical assessments, better measurement tools may be required, including electrophysiological tests that are more sensitive, such as QST assessments, and pathological measurements, such as quantification of intraepidermal and, perhaps, dermal nerve fiber density obtained by skin biopsy. Treatments relieving neuropathic pain may not result in improvement of clinical, electrophysiological, or pathological measures in neuropathy, as was seen in the recent clinical trial of nerve growth factor.

The natural history of such painful neuropathies relieved, but not resolved, after withdrawal of dideoxynucleosides, awaits elucidation in future studies. Will these prove to be aborted toxic neuropathies with incomplete resolution, or incident HIV-associated DSP exacerbated by a superimposed agent with some neurotoxicity, perhaps synergistic with primary neuropathological processes? Thus, most importantly, studies to improve our understanding of the mechanisms of toxicity in both HIV-associated DSP and nucleoside-associated toxic neuropathies are essential to provide the basis for future therapies that specifically target the pathophysiological mechanisms of these syndromes.

CHAPTER

# 6 Vaccines

Vaccines against infectious diseases are needed to increase global immunization compliance and to decrease the costs of delivery, while expanding participation by vaccine manufacturers. If implemented well, vaccines can provide great economic benefits, with cost-to-benefit ratios of up to 1:10. The technological revolution in vaccine development over the last half of the 20$^{th}$ century has centered on the use of mammalian cell cultures. This technology, now used throughout the developed and developing world, enables the production of more uniform and potent vaccines than is possible with other methods. However, subunit vaccines have penetrated the developing world to a much lesser degree than the developed world. This is for two reasons: IP challenges and technology entry barriers, whose costs increase along with the sophistication of molecular biology techniques. Thus, the development and implementation of new vaccines often involves substantial economic and logistic barriers that are difficult to overcome in many poor countries. Transgenic plant-derived proteins represent a promising strategy for vaccine development that combines innovations in medical science and plant biology for the creation of affordable vaccines.

A growing number of laboratories are investing in the development of plant-derived protein pharmaceuticals, expanding on the seminal works that first put forward this idea. The subject of plant-derived vaccines has been reviewed several times in recent years (e.g. and a comprehensive list of the plant-derived vaccine antigens that have been reported is provided in Table 6.1 and 6.2. While much research is still needed to optimize the production of vaccines in plants and to validate them in large-scale clinical trials, the results to date are very promising and suggest that the technology justifies commercial development. The oral delivery of transgenic plant tissue expressing vaccine antigens, typically mucosally targeted subunits, can promote specific mucosal secretory IgA (slgA) and serum IgG antibody responses via the gut lymphoid system. Although the protective efficacy of a plant-derived vaccine has yet to be determined in humans, some challenge studies in animals have shown promising results.

## VACCINE PRODUCTION IN PLANTS

Three major research strategies have emerged for the production of subunit vaccines in plants. The first, the direct transformation of plants, introduces genes encoding the antigenic proteins of human or animal pathogens stably into the plant genome. In this case, the genes can be inserted into either the nuclear or chloroplast genomes. The resulting plants accumulate

**Table 6.1. Vaccine Antigens Expressed in Plant Virus Epitope Display Systems**

| *Antigen* | *Expression system* | *Immunogenicity* | *Protective challenge* |
|---|---|---|---|
| Canine parvovirus (CPV) VP2 epitope | Epitope display on cowpea mosaic virus in cowpea leaf | Immunogenic in mice when delivered parenter ally or nasally. Dogs developed 3L17-and VP2-specific IgG; sera neutralized CPV *in vitro.* | Protective against lethal challenge in dogs immunized parenterally. |
| Canine parvovirus VP2 epitope | Epitope display on plum pox potyvirus in tobacco leaf | Mice developed CPV-specific antibodies that showed neutralizing activity. Immunogenic in mice and rabbits when delivered parenterally | — |
| Foot and mouth disease virus VP1 epitope | Cowpea mosaic virus in cowpea leaf | No immunogenicity assays performed | — |
| Hepatitis C virus (HCV) HVR[1] epitope of E2 envelope protein | Epitope display on cucumber mosaic virus in tobacco leaf | Cross reactive with a wide range of human anti HVRI antibodies. | — |
| Human immuno-deficinecy virus (HIV) type 1 ELKDWA epitope | Potato virus X in tobacco leaf | Sera from normal and hu-PBL-SCID mice showed anti-HIV-1 neutralizing acitivty. immunogenic in mice when delivered parenter-ally or nassaly. | — |
| Human rhinovirus type 14 VP1 epitope | Epitope display on Cowpea mosaic virus in cowpea leaf | Immunogenic in rabbits when delivered parenterally. | — |
| Mink enteritis virus VP2 epitope | Epitope display on cowpea mosaic virus in cowpea leaf | Immunogenic in mink when delivered parenterally. | Protective against challenge with virulent MEV. |
| *Pseudomonoas aeruginosa* memberane protein F | Epitope display on cowpea mosaic virus in cowpea leaf | Elicited specific antibodies. Immunogenic in mice when delivered parenterally. | Mice protected when challenged with model chronic pulmonary infection with *P. aeruginosa.* |
| *Pseudomonas aeruginosa* membrane protein F | Epitope display on tobacco mosaic virus in tobacco leaf | Elicited specific antibodies against 7 immuno-type strains. Immunogenic in mice when delivered parenterally. | Mice protected when challenged with model chronic pulmonary infection with *P. aeruginosa.* |

**Table 6.1. Continued**

| *Antigen* | *Expression system* | *Immunogenicity* | *Protective challenge* |
|---|---|---|---|
| Rabbit hemorrhagic disease virus VP60 epitope | Epitope display on plum pox potyvirus in tobacco leaf | Rabbits developed specific antibodies that showed neutralizing activity. Immunogenic when delivered to mice parenterally; immunogenic in rabbits when delivered parenterally | Rabbits immunized parenterally survived lethal challenge. |
| Rabies virus glycoprotein (G) and nucleo-protein (N) | Alfala mosaic virus and tobacco mosaic virus in tobacco and spinach leaf | Elicited specific virus-neutralizing antibodies in mice. Immunogenic in mice when delivered orally and parenterally; immunogenic in humans when delivered orally | Moderately protective against lethal challenge infection in mice. |
| Respiratory syncytial virus (RSV) G and F proteins | Alfalfa mosaic virus in tobacco leaf | High levels of serum antibodies specific for RSV-G. Immunogenic in mice when delivered parenterally | Protective against challenge with RSV Long strain. |
| *Staphylocaccus aureus* D2 epitope of fibronectin-binding protein (FnBP) | Cowpea mosaic virus in cowpea leaf; Potato virus X in tobacco leaf | Elicited FnBP-specific IgA and IgG. Immunogenic in mice and rats when delivered orally, nasally, or parenterally. | |

the subunit antigens, which can then either be purified from plant tissue for parenteral or mucosal delivery, or the plant tissue can be processed for oral delivery with the antigen as a component of the plant matrix. In the second research strategy, plant viruses have been utilized as pliable genetic platforms for protein expression.

Three formats have been developed, and the one that has undergone the most extensive evaluation is the display of epitopes on the surface of the virus as fusions with the viral coat protein. This epitope-display system usually involves subsequent purification of the viral particle for parenteral delivery. Another viral strategy uses cloned viral genomes to introduce antigenic protein-coding sequences driven by sub-viral promoters, and results in antigen expression as a byproduct of the viral replication cycle. A final variant is the integration of entire viral genomes into plant chromosomes by transformation, with the subsequent transcription of viral RNA as the initiating event in viral replication. A coding sequence for the antigen is included in this viral genome and is thus co-expressed. The potential advantage of viral expression systems compared to stable plant transformation is that viral replication can greatly amplify the template for protein synthesis resulting in high-level protein accumulation.

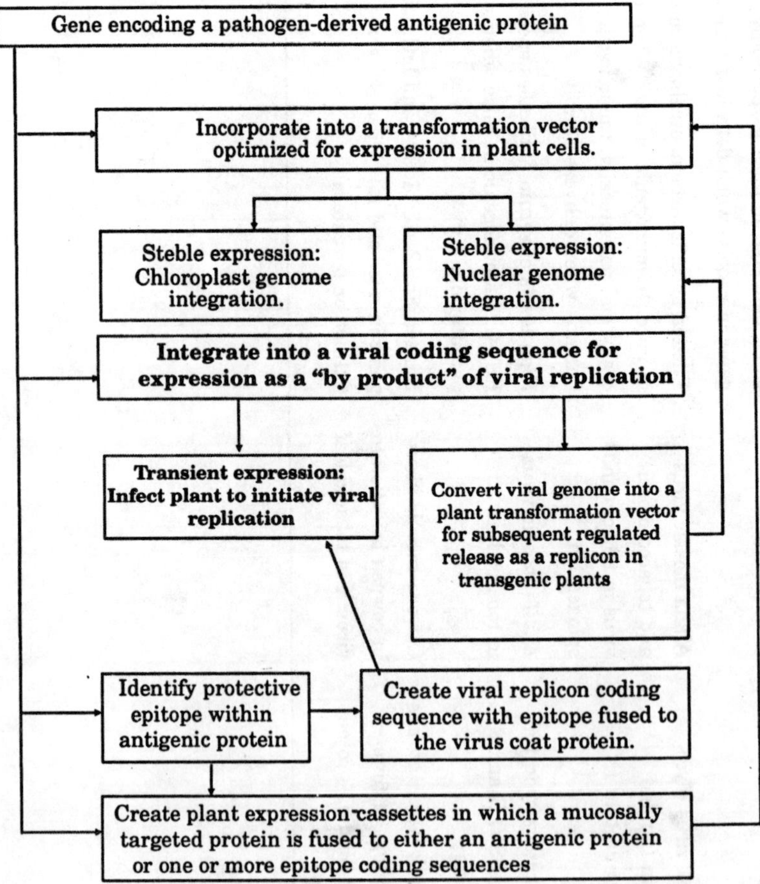


**Fig. 6.1. Plant-derived vaccines: research strategies.**

A third strategy for the production of plant-based vaccines is the design of mucosally targeted fusion proteins. We define this as a separate research strategy to emphasize the goal of creating new mucosal vaccines, even though such fusion proteins can be expressed in transgenic plants or using viral vectors as discussed above. We highlight mucosally targeted proteins because this captures the emerging research emphasis on protein engineering to obtain more effective vaccines, and the potentially unique role of plants for the production and accumulation of novel vaccine products. See Table 6.3 for details of fusion proteins expressed in transgenics.

# SYNTHESIS

Current vaccine technology frequently requires the purification of subunit proteins from mammalian cell cultures or tissues, yeast, fertilized eggs or bacterial fermentation systems to produce the immunogen. Typically, the product requires refrigeration during transport to its final point of use, adding significant cost to the vaccination program. One solution may be to use dried plant extracts containing subunit vaccines, since this provides a product that is stable at ambient temperatures, i.e. equivalent in storage and transport characteristics to dehydrated food products. In addition, the production of antigens in plants may improve product safety by removing animal cell-related contaminants and pathogens.

## Advantages of Plants

Plant-derived pharmaceuticals (PDPs) are proteins or organic compounds produced in plants via recombinant DNA technology, which are used to improve human or animal health. Subunit vaccines represent one category of PDPs that have been validated in a variety of studies, including human clinical trials. Current efforts in product formulation utilize food-processing technology to convert transgenic plant material into dried samples that can be delivered in unit doses with assured product uniformity and quality. Plant-derived vaccines appear to offer several product advantages, including oral delivery, heat stability, lower manufacturing costs for the active ingredient and suitability of the manufacturing technology for use in developing countries.

## Mucosal Immune Responses

Orally delivered, non-replicating subunit vaccines have not yet achieved commercial success using any means of manufacture. Possible hurdles facing the use of orally delivered immunogenic proteins include the likelihood that some proteins will be degraded after ingestion and that some immunogens may not be recognized efficiently at mucosal immune effector sites in the gut. As a result, higher concentrations of immunogen may be required in oral versus parenteral delivery. Although this is a potential limitation, the use of plants as a protein biomanufacturing system offers advantages in that the cost of obtaining the end product is comparatively low.

In addition, empirical evidence suggests that encapsulation of the immunogenic protein within the plant cell matrix during administration provides some protection from degradation by gastric enzymes and acids. The best candidates oral subunit vaccines for production in transgenic plants are the primary antigens of infectious pathogens that aggregate in forms that are recognized at mucosal sites where an immune response is triggered. These include viral surface proteins that co-assemble to form virus like particles (VLPs), and bacterial toxins that naturally aggregate to form mucosally targeted multimeric complexes. In addition, in several laboratories there are also ongoing efforts to produce a variety of fusion proteins that target immunoresponsive mucosal sites.

## Antigens Produced in Plants

Tables 6.1 and 6.2 shows the range of different plant and vector systems that have been used for the expression of antigens. The immunogenicity of PDPs was demonstrated in a number of studies, including those using hepatitis B surface antigen ($HB_SAg$), the S protein of transmissible gastroenteritis coronavirus (TGEV) and human immunodeficiency virus 1 (HIV-1) epitopes. Furthermore, successful challenge trials resulted after immunization with *Pseudemonas aeruginosa* epitopes the FP1 epitope of foot and mouth disease virus (FMDV) and *Escherichia coli* labile enterotoxin B subunit (LTB). Additional antigens expressed in plants include the respiratory syncytial virus (RSV) G and F proteins, the rota-virus VP6 protein and VP7 protein the measles virus hemagglutinin protein and an epitope from the major surface antigen of *Plasmodium falciparum* (PfMSPl). Although the immunogenicity of the plant-derived RSV G and F proteins and the measles virus hemagglutinin protein were demonstrated in animal trials, the immunogenicity of the VP6 antigen and the PfMSPl epitope were not tested.

## Targeting Antigens to Specific Tissues

Tissue-specific targeting of antigens can greatly facilitate the harvesting of proteins. Streatfield *et al.* expressed antigens specifically in maize kernels, Sandhu *et al.* In tomato fruit, and Wright *et al.* In tobacco seeds. Targeting the major glycoprotein (gB) of human cytomegalovirus (hCMV) to transgenic tobacco seeds was previously described and further investigations showed that recombinant gB was almost exclusively deposited in protein storage vesicles in mature tobacco seeds. Sojikul *et al.* and Huang & Mason reported the targeting of a fusion molecule to the endoplasmic reticulum.

## Expression Systems

The achievement of significant expression levels for recombinant proteins produced in plants often depends on the tailoring and optimization of the expression system. Unfortunately, except for a few rules of thumb, insights gained while studying one protein are not always readily transferable to others, and low levels of expression are found in some of these systems. A few studies have attempted to address the low recombinant protein expression levels that may sometimes occur in transgenic plants. The issues of how, where and when to express a transgene in order to achieve maximal accumulation of a functional protein remain priorities in this field. It is especially important in the case of plant-derived vaccines (where the goal is to achieve sufficient dosage with minimal processing of antigenic proteins in edible plant tissue) to ensure that sufficient quantities survive passage through the stomach to elicit a clinically relevant immune response. As discussed above, vaccine antigens can be produced in plants using two different systems: stable genetic transformation and transient expression.

Stable transformation produces a genetic line that can be propagated either by vegetative (stem) cuttings or by seeds, and was the method used in published edible vaccine clinical trials to date. Transient expression involves the use of a recombinant plant virus that carries the vaccine gene and, by systemic infection, causes the plant to express the antigen. The main advantage of transient expression with a plant virus system is that virus replication amplifies the gene copy number, typically resulting in a much higher level of expression than with stable transformation. However, the plant virus systems might in some cases suffer from instability and loss of foreign genes larger than 1 kb.

In addition, the need to inoculate each plant individually makes large-scale production rather labour intensive. Recombinant DNA is integrated into the nuclear or chloroplast genome through stable transformation. *Agrobacterium tumefaciens,* a plant pathogen, is typically used for nuclear transformation, and strains have been engineered with deletions of the virulence genes that cause tumor growth in plants. *A. tumefaciens* efficiently transports DNA into cells and then promotes nuclear chromosomal integration at random sites. Integrated nuclear transgenes are expressed and inherited in typical Mendelian fashion. Some agronomically important plant species (e. g. soybean and most cereal grains) are recalcitrant to *Agrobacterium* transformation, and a biolistic method (microprojectile bombardment) is frequently used for these plants.

DNA coated on micron-sized gold particles is propelled into plant cells using compressed helium gas and becomes incorporated into chromosomal DNA. Foreign DNA that integrates into the nuclear genome yields stable transformants that show Mendelian inheritance of the transgenes. The biolistic method, as traditionally carried out, usually results in higher-copy-number plants compared to those generated by *Agrobacterium,* which can enhance expression. However, excessive copy numbers or very high-level expression of nuclear genes can cause gene silencing, resulting in low protein accumulation. Thus it is important to select transgenic lines that carry only between one and three copies of transgene. The biolistic method is also used to introduce transgenes into the chloroplast genome.

Although such transplastomic plants have been produced for only a few species, the potential for chloroplast transformation in other crop plants is strong. The high chloroplast genome copy number in plant cells enhances recombinant protein expression, as shown for the cholera toxin B (CTB) subunit [52]. DeCosa *et al.* [53] expressed an insecticidal protein in the chloroplasts of mature tobacco leaves at levels up to 45% of total protein. Stable transgene expression was recently described in the chloroplast-derived chromoplasts of tomato fruit. A further advantage is that, as the chloroplast genome is inherited strictly maternally in most plants, there is minimal danger of unintended spread of transgenes by dissemination of pollen. Although chloroplast transformation is not yet a routine procedure, at least 25 different foreign genes have been expressed in transplastomic plants.

Chloroplast transformation is not suitable for some eukaryotic proteins that require post-translational modifications like glcosylation. Gene silencing, however, has not been observed in this system. Seed-specific production of LTB by Streatfield *et al.* resulted in expression levels of up to 1.8% total soluble protein (TSP), and two separate maize breeding programs have increased antigen production by fivefold and tenfold. The investigations of Chikwamba *et al.* regarding the expression of LTB in maize are among the first to include the use of particle bombardment transformation for the production of plant-derived vaccines.

## Mucosally-targeted Fusion Proteins

Most soluble proteins are rapidly hydrolyzed in the intestinal tract. Subunit vaccines are subject to the same proteolytic environment, and as a result some candidates may be ineffective at inducing the mucosal immune system. This is due to degradation prior to reaching immune effector sites in the gut-associated lymphoid tissues (GALT). In order to increase subunit vaccine mucosal immunogenicity, a number of investigators have used protein engineering to create fusion molecules consisting of the antigen (or epitopes) linked to a mucosally active carrier. Two studies have involved the fusion of an anitigen to β-glucuronidase (GUS). Gil *et al.* fused GUS to the 2L21 protective epitope from canine parvovirus, and Dus Santos *et al.* fused GUS to the protective epitope from FMDV. Transformants in both cases were selected on the basis of GUS activity, and both proved to be immunogenic.

Mice immunized orally or parenterally with the GUS-FMDV epitope fusion werecompletely protected against challenge with the native virus. Cholera toxin subunits have also been used to design targeted, plant-derived fusion vaccines. Using a tobacco mosaic virus (TMV) expression system, Nemchinov *et al.* Generated a *Vibrio choleras* CTB fusion to an epitope (HVR1) from the hepatitis C virus (HCV). Tobacco plants inoculated with the recombinant TMV produced the HVR1 epitope fused to a functionally active, pentameric CTB, which reacted with HVRl-specific monoclonal antibodies and sera from individuals infected with virus from four of the major genotypes of HCV. Nasal immunization of mice with a crude plant extract containing the recombinant CTB-HVR1 elicited both anti-CTB serum antibodies and anti-HVRl serum, antibodies that specifically bound to HCV-like particles. Sojikul *et al.* reported the expression of a fusion protein designed to improve the expression of the hepatitis B surface antigen (HBsAg). This was achieved by mimicking the process of HBsAg targeting to the endoplasmic reticulum of human liver cells during HBV infection.

The gene encoded a recombinant HBsAg modified to contain an N-terminal signal peptide from soybean vegetative storage protein A (VSPaS). This signal peptide directed the HBsAg-VSPaS fusion protein to the endoplasmic reticulum and resulted in greater fusion protein stability, enhanced protein accumulation and formation of more VLPs. Moreover, the HBsAg-VSPaS fusion stimulated higher levels of serum IgG when administered parenterally in mice compared to native HBsAg.

## Multivalent and Multicomponent Vaccines

Among vaccine products, those that stimulate several facets of the immune system such as the induction of strong humoral and mucosal responses as well as effective cellular immune responses – are highly desirable. Combination vaccines targeting multiple pathogens in one formulation are likewise preferred. Therefore, developing both multivalent and multicomponent plant-based vaccines provides both efficacious and cost-effective immunization strategies. This can be accomplished through the development of recombinant plants carrying transgenes that encode the antigens of several pathogens, either by direct simultaneous transformation, sexual crosses of individually transformed lines, or the blending of separately transformed plant tissues. One approach for the production of multicomponent vaccines was taken by Yu and Langridge and Arakawa *et al.* They fused peptides containing important protective epitopes derived from two enteric pathogens - enterotoxic *E. coli* (ETEC), which causes bacterial traveler's diarrhea, and rotavirus, which causes acute viral gastroenteritis – to the A2 and B subunits of cholera toxin, respectively.

**Table 6.2. Vaccine Antigens Expressed in Plant Systems: Subunit Antigen Expression in Transgenics.**

| *Antigen* | *Expression system* | *Immunogenicity* | *Protective challenge* |
|---|---|---|---|
| Bacillus anthracis protective antigen | Tobacco leaf | No immunogenicity assays | — |
| Bovine herpes virus type 1 (BHV1) glycopro tein D (gDc) | Tobacco leaf | Sustained and specific humoral and cellular im mune response in mice and cows when delivered once parenterally. | Cows challenged with infectious dose of BHV1 exhibited milder symptoms over a shorter duration. |
| Enteropathogenic *E. coli* pilus subunit A (BfpA) | Tobacco leaf | Immunogenic when delivered orally in mice. Anti-BfpA in fecal matter. | — |
| Enterotoxigenic E. coli B subunits of the heat labile toxin (LTB) | Maize seed | Elicited neutralizing antibodies. Immunogenic when administered orally. Serum and secretory immune responses in humans. | Partially protective in mouse gut fluid assay. |
| Enterotoxigenic *E. coli* B subunits of the heat labile toxin (LTB) | Potato tuber | Receptor-binding activity. Immunogenic in mice and humans when administered orally. | Partially protective in mouse gut fluid assay. |
| Enterotoxigenic *E. coli* B subunits of the heat labile toxin (LTB) | Tomato fruit and leaf | Offspring were passively immunized through oral immunization of parents. | — |
| *E. coli* 0157:H7 Intimin epitope | Tobacco callus | Mucosally immunogenic in mice when primed parenterally and boosted orally | Mive immunized parenterally and orally boosted showed reduced duration of colonization upon challenge. |
| Foot and mouth disease virus (FMDV)VP1 epitope | Arabidopsis leaf | Serum antibodies reacted strongly with intact FMDV particles. Immunogenic in mice when administered parenterally. | Protective aginst challenge with virulent FVDM. |
| Foot and mouth disease virus VP1 epitope | Potato | FMDV-VP1-specific antibody response. Immunogenic. | Protective against challenge. |
| Foot and mouth disease virus VP1 epitope | Alfalfa leaf | Mice developed specific antibody response to synthetic peptide, VP1 epitope. and intact FMDV particle. Immunogenic in mice when adminis tered parenterally or orally. | Mice protected against challenge with FMDV virus |

| *Antigen* | *Expression system* | *Immunogenicity* | *Protective challenge* |
|---|---|---|---|
| Hepatitis B surface antigen (HBsAg) | Cherry tomatillo leaf and fruit | Immunogenic in mice as an oral booster after parenteral priming. | — |
| Hepatitis B surface antigen (HBsAg) | Lettuce leaf | Immunogenic in humans when administered orally; 2 of 3 human subjects developed very modest serum antibody response titers in primary immunization. | — |
| Hepatitis B surface antigen (HBsAg) | Lupin callus | Immunogenic in mice when administered orally | — |
| Hepatitis B surface antigen (HBsAg) and middle protein | Potato tuber | Immunogenic in mice when administered orally. | — |
| Hepatitis B surface antigen (HBsAg) | Soybean cell culture | No immunogenicity assays performed. | — |
| Hepatitis B surface antigen (HBsAg) | Tobacco leaf | Mice vaccinated parenterally produced all IgG subclasses and IgM. | — |
| Hepatitis E virus (HEV) ORF2 epitope | Tomato leaf and fruit | No immunogenicity assays performed. | — |
| Human cytomegalo virus glycoprotein B | Tobacco seed | gB-specific antigenic activity. | — |
| Human immuno-Deficiency virus (HIV) type 1 p24 protein | Tomato bushy stunt virus in tobacco leaf | No immunogenicity assays performed. | — |
| Human immuno-deficiency virus (HIV) type 1 gp 120 (v3 loop) protein | Alfalfa mosaic virus in tobacco leaf | Elicited specific virus-neutralizing antibodies in mice when delivered parenterally. | — |
| Human immuno-deficiency virus (HIV) type 1 gp 120 (v3 loop) protein | Tomato bushy stunt virus in tobacco leaf | HIV epitope detected by V3-specific mono-clonal antibodies and human sera from HIV positive patients. Immunogenic in mice when delivered parenterally. | |
| Human papilloma virus type 11 major capsid protein L1 | Tobacco leaf, potato tuber | Reacted well with assembled capsids and isolated L1 capsomers Immunogenic in mice when administered parenterally with sub-sequent oral boosting. | — |

**Table 6.2. continued...**

| *Antigen* | *Expression system* | *Immunogenicity* | *Protective challenge* |
|---|---|---|---|
| Human papilloma virus (HPV) type 16 onco-protein E7 | Potato X in tobacco leaf (complete reading frame) | Immunogenic in mice when administered with surrogate (C3 cells). | Mice protected moderately when challenged parenterally. |
| Human papilloma virus (HPV) type 16 major capsid protein L1 | Tabacco leaf | Nutralizing monoclonal antibodies bounds plant-produced particles. | |
| Human papilloma virus (HPV) type 16 major capsid protein L1 | Tobacco and potato | Weak, transient anti L1 antibody response in 3 of 24 mice when administered orally. Immunogenic when administered orally after parenteral boost. | — |
| Infectious bronchitis virus (IBV) S1 glycoprotein | Potato tuber | Antibodies neutralized IBV from mice and chickens. Immunogenic in chickens when administered orally. nasally, nasally, and parenterally. | Protective after 3 immunizations in chickens when challenged with virulent IBV. |
| Measles virus hemag-glutinin (H) protein | Tobacco leaf | Mice produced serum anti-H antibodies that Neutralized measles virus in vitro. Immunogenic in mice when delivered orally or parenterally. | — |
| Measles virus hemag-glutinin (H) protein | Carrot leaf and root | IgG1 and IgG2 cross reacted strongly with measles virus and neutralized virus *in vivo*. Immunogenic in mice when delivered parenterally. | — |
| Measles virus hemag-glutinin (H) protein | Carrot leaf and root | H-specific antibody isotypes and subclasses in mouse sera; mouse sera neutralized wt virus. Immunogenic in mice when administered parenterally. | — |
| Murine hepatitis virus glycoprotein S 5B19 epitope | Tobacco mosaic virus in tobacco leaf | Mice developed serum lgG and IgA specific for 5B19. Immunogenic in mice when delivered nasally or parenterally. | Protective in mice immunized parenterally or orally when challenged with MHV strain JHM. |
| Norwalk virus (NV) capsid protein | Potato tuber; tobacco leaf | Mice and humans developed serum lgG and secretory lgA specific for recombinant NV. Immunogenic in mice and humans when administered orally. | |
| *Plasmodium falciparum* (PfMSP) | Tobacco leaf | Antigenic characteristics identical to *E. coli* expressed protein. Immunogenic characteristics identical to *E. coli* expressed protein. | |
| Porcine epidemic diarrhea virus (PEDV) spike protein | Tobacco leaf | Systemic and mucosal immune reponse in mice immunized orally. | — |

| *Antigen* | *Expression system* | *Immunogenicity* | *Protective challenge* |
|---|---|---|---|
| Rabbit hemorrhagic disease virus VP60 epitope | Potato leaf | VP60 specific antibodies elicited in rabbits. Immunogenic in rabbits when administered parenterally. | Rabbits protected when challenged with virulent RDHV. |
| Rabies virus glycoprotein (G) | Tomoto leaf and fruit | Intact protein forms. | |
| Respiratory syncytial virus fusion protein | Tomato fruit | Mice developed serum and mucosal RSV-F specific antibodies. Immunogenic in mice when delivered orally. | — |
| Rinderpest virus hemagglutinin(H) protein | Pigeon pea leaf | — | — — |
| Rinderpest virus hemagglutinin (H) protein | Tobacco leaf | High titers of neutralizing antibodies detected in immunized mice | |
| Rotavirus VP6 protein | Potato tuber | Anti-GAR antibodies detected. Immunogenic in mice when delivered parenterally. | — |
| Rotavirus VP6 protein | Tomato callus | No immunogenicity assays. | |
| Rotavirus VP7 protein | Potato tuber | Serum IgG and IgA specific for VP7; IgA had VP7-neutralizing activity. Immunoginic in mice when delivered orally. | — |
| Tetanus toxin fragment C | Tobacco leaves | Systemic IgG and local IgA responses. Immunogenic in mice when administered nasally. | — |
| Transmissible gastroenteritis coronavirus N-terminal domain of the spike glycoprotein (S) | Arabidopsis leaf | gS elicited virus-neutralizing antibodies. Immunogenic when administered parenterally. | - |
| Trunsmissble gastroenteritis coronavirus N-terminal domain of the spike glycoprotein(S) | Maize seed | Orally immunized piglets had a memory immune response leading to rapid accumulation of virus-neutralizing antibodies on challenge. | Reduced symptoms in swine immunized orally and challenged with virulent TGEV. |

**Table 6.2. continued...**

| *Antigen* | *Expression system* | *Immunogenicity* | *Protective challenge* |
|---|---|---|---|
| Transmissible gastro-enteritis coronavirus N-terminal domain of the spike glyco-protein (S) | Potato tuber | Mice developed serum antibodies specific for gS. Immunogenic in mice when delivered parenterally or orally. | - |
| Transmissible gastro-enteritis coronavirus N-terminal domain of the spike glyco-protein (S) | Tobacco leaf | TGEV-specific antibodies (in pigs) showed low levels of virus-neutralizing activity. Immunogenic in pigs when administered parenterally. | — |
| *V. cholerae* B subunits of the cholera toxin (CTB) | Potato | Receptor-binding activity. Immunogenic in | Mice challenged with CT showed up to 60% reduction in diarrheal fluid accumulation in small intestines |
| *V. cholerae* B subunits of the cholera toxin (CTB) | Tovacco leaf chloroplast | No immunogenicity assays performed. | — |
| B subunits of the cholera toxin (CTB) | Tomato loaf and fruit | No immunogenicity assays performed. | — |
| Canine parvovirus (CPV) VP2 epitope fused to GUS | Arabidopsis leaf | Specific antibodies detected and immunigenic in mice when delivered parenterally. | — |
| Enterotcxigenic *E. coli* epitope and totavirus epitope fused to CTB | Potato tuber | Mice developed detectable levels of serum and intestinal antibodies. Immunogenic in mice against ETEC, rotavirus, andV. cholerac when delivered orally. | Symptoms reduced in passively immunized mouse neonates following rotavirus challenge. |
| *E. coli* B subunit of the heat labile enterotoxin (LTB) fused toan immunicontraceptive epitope | Tomato fruit and leaves | No immunigenicity assays performed. | — |

**Table 6.2. continued...**

| *Antigen* | *Expression system* | *Immunogenicity* | *Protective challenge* |
|---|---|---|---|
| Hepatitis B surface antigen (HBsAg)fused to soybean vegetative protein (VSPaS) | Tobacco leaf | VSPaS-HBsAg generatee higher levels of serum lgG than native HBsAg in mice. | — |
| Hepatitis C virus (HCV) HVR1 epitope fused to *v. choleare* CTB | Tobacco leaf | Reacted with HRV1-specific monoclonal antibodies and sera from individuals infected with HCV. HVR1 epitope formed fused to CTB was immunogenic in mice when administered nasally. | — |
| Mannuheimia haemolytica A1 leukotoxin 50 fused to green fluourscent protein | White clover leaf | Rabbits produced antibodies that recoginized and neutralized leukotoxin. Immunogenic in rabbits after drying when delivered parenterally. | — |
| Ricin B fusee to green fluorescent protein (GFP) | Tobacco leaf and hairy root culture | GfP-specific lgG present in serum; lg A in serum and fecal matter. | — |
| Rotavirus enterotoxin protein NSP4 fused to CTB | Potato | Fusion proteins assembled into cholera holotoxin-like structures that retained enterocyte-binding affinity. Orally immunized mice generated levels of serum and mucosal antibodies specific for the native antigen. Induced TH1 immune response. | — |

The two recombinant cholera toxin subunit fusions were expressed from a single bidirectional promoter, ensuring coordinated expression for the two gene fusions and potentially facilitating the assembly of the chimeric holotoxin. In this approach, cholera toxin provides a scaffold for presentation of the protective epitopes, acts as a mucosal targeting molecule without toxic effects due to use of the nontoxic A2 and B subunits, and is itself a vaccine candidate. The recombinant protein represents a trivalent vaccine that can elicit significant mucosal and humoral responses against *Vibrio cholerae,* ETEC, and rotavirus. Mice, orally immunized with potatoes expressing these recombinant antigens, developed immune memory B cells as well as helper T cell type 1 (Thl) responses, which are indicators of successful immunization. Further, pups of immunized dams were protected from challenge with rotavirus, with a significantly lower morbidity rate as compared to controls. These results provide robust evidence for a vaccine strategy employing chimeric microbiological proteins expressed in plants and for multicomponent vaccines.

## Stability and Processing

Lee *et al.* carried out preliminary stability studies on dehydrated plant tissues. Clover plants expressing the *Mannheimia haemolytica* A1 leukotoxin 50 fusion protein were harvested and allowed to dry at room temperature and ambient humidity for 1– 4 days. After 3 days, the clover tissue retained approximately 20% of its initial fresh weight, but no significant degradation of the fusion protein was observed. Hence, the fusion protein did not require refrigeration for stability. The clover-derived fusion protein induced, in injected rabbits, an immune response that recognized and neutralized the native antigen in modified neutral red cytotoxicity assays.

A comprehensive stability study was conducted by Smith *et al.* on HBsAg from soybean cells. Under optimum conditions, antigen stability was maintained for at least 1 month after isolation of the surface antigen, although excess detergent rendered the antigen susceptible to proteolytic degradation by contaminating plant enzymes. This proteolysis was counteracted by the addition of skimmed milk or its protein component, which stabilized the antigen for up to 2 months. Also, by altering the sodium ascorbate concentration or buffer pH, the proportion of HBsAg displaying monoclonal-reactive epitopes increased between 8- and 20-fold.

Although antigen stability will have to be investigated on a case-by-case basis, it is obvious that simple *in vitro* manipulations may prove valuable in increasing the immunogenicity and stability of plant-derived antigens. Castanon *et al.* investigated the minimal processing of a potato-derived rabbit hemorrhagic disease virus (RHDV) vaccine consisting of the VP60 gene product. Harvested potatoes were peeled, cut into pieces, lyophilized, powdered, stored and used in trials within 3 months of collection. Extracts made from the potato powder were used in rabbits as a subcutaneous primer and intramuscular booster. The rabbits immunized with the transgenic potato elicited specific antibody responses and were protected against challenge with virulent RDHV. Freeze-dried tomato fruit expressing LTB was orally immunogenic in mice, indicating the utility of the tomato system.

# CLINICAL TRIALS

The results of five clinical trials involving orally delivered plant-derived vaccines have been published. With the exception of LTB, each study used different recombinant antigens, and each reported the stimulation of serum antibodies, and in some cases, mucosal antibodies.

## Enterotoxic *E. coli* and *Vibrio cholerae*

Two of the most widely spread and well-studied enterotoxigenic forms of bacterial diarrhea are ETEC and *Vibrio cholerae.* The toxins they produce, labile toxin (LT) and cholera toxin (CT) respectively, are very similar in primary sequence, structure, and mechanism of action. They are homologous multi-subunit proteins in which the non-toxic B subunit mediates $GM_1$ ganglioside binding, and thus are candidates for vaccines that can neutralize toxin activity. The B-subunits of these toxins, LTB and CTB, are among the most potent oral immunogens known.

Oral delivery efficiently causes the accumulation of specific serum (lgG,lgA) and mucosal (slgA) antibodies. Both LT and CT also function as mucosal adjuvants, stimulating antibody production against co-delivered antigens. The ganglioside-binding activity of the LTB pentamer is required both for its mucosal immunogenicity and for the adjuvanticity of the holotoxin. Tacket *et al.* performed the first human clinical trial with a transgenic plant-derived vaccine. Fourteen volunteers ingested either 100 g of transgenic potato, 50 g of transgenic potato, or 50 g of non-transgenic potato. The LTB content of the tubers varied between 3.7 and 15.7 $\mu g\ g^{-1}$ of tuber weight and the doses were given on days 0, 7 and 21. Volunteers reported only a few instances of minor side effects (nausea, cramps or diarrhea), and the raw potato was well tolerated overall. Ten of 11 volunteers who ate the potatoes expressing LTB developed at least 4-fold increases in levels of toxin- neutralizing serum IgG against LTB ; none of the volunteers who ate the placebo potatoes showed any increase in anti-LTB antibodies. Five of 10 volunteers showed at least 4-fold increases in anti-LTB IgA, detected in stool samples. These data compared favourably with an earlier study in which volunteers were challenged with $10^9$ ETEC cells (Tacket *et al,* unpublished). This study was significant because it was the first ever to examine an edible plant vaccine in humans, and showed great potential for this new strategy. A more recent clinical study using LTB expressed in processed corn seed produced similar results to the potato study.

## Norwalk Virus

Tacket *et al.* reported a clinical trial performed using transgenic potato tubers carrying the gene for Norwalk virus capsid protein (NVCP), which assembled into Norwalk virus like particles (VLPs). Twenty adult volunteers ingested either two or three doses each of 150 g of raw transgenic potato tuber containing 215-750 μg NVCP (expression was variable). Only about 50% of the NVCP subunits assembled into VLPs in the potato cells, possibly as a result of low recombinant protein concentration. Thus, the effective amount of administered potato vaccine was 125-375 μg per dose. Unassembled subunits are likely to be much less stable in the gastrointestinal tract and thus less immunogenic. However, 19 of 20 subjects in the experimental group showed significant increases in the numbers of IgA-antibody forming cells (AFCs), ranging from 6-280 per $10^6$ peripheral blood mononuclear cells (PMBC), and 6 of 20 subjects in this group developed increases in IgG AFCs. Four volunteers shincreases in serum IgG anti-NVCP liters, four showed increased serum IgM and six showed increased IgA in their stool samples (a 17-fold mean increase). Although the antibody responses were less impressive than those obtained with LTB, the study showed that a plant-derived non-replicating antigens other than LTB and CTB could stimulate human immune responses after oral delivery.

Insect cellderived 250-μg doses of purified NVLP showed more effective seroconversion. Thus it is likely that part of the potato-delivered NVCP was unavailable for uptake in the gastrointestinal tract. More recent studies in transgenic tomato fruits with a plant-optimized NVCP gene resulted in higher expression levels and more potentimmune responses in mice fed freeze-dried tomatoes (X. Zhang and H.S. Mason,unpublished results). A clinical trial is planned with dried tomato powder formulated in gelatin capsules (D. Kirk, H.S. Mason, and C. Tacket, trial investigators) in order to evaluate safety and immunogenicity.

## Hepatitis B Virus

There is one published report concerning the immunogenicity in humans of orally delivered HBsAg expressed in plants. Two of three volunteers who ate two 150-g doses of transgenic lettuce (containing -1-2 μg HBsAg per dose) developed a modest protective (>10 IU $I^{-1}$) serum antibody titer after the second dose. The serum antibody titers declined rapidly after 4 weeks, probably due to the very low antigen dosage. However, the study showed that presumably naive subjects could be seroconverted by the oral delivery of plant-expressed HBsAg. In the US, a clinical trial was carried out at the Roswell Park Cancer Institute, Buffalo, NY, to test the oral immunogenicity of recombinant potatoes expressing HBsAg.

The study was limited to volunteers who had previously been vaccinated and seroconverted with the standard injectable yeast-derived HBsAg. The trial involved 33 volunteers who ate either two (days 0 and 28) or three (days 0, 14 and 28) 100-g doses of HBsAg potato tubers containing approximately 1 mg HBsAg per dose, while a group of 10 volunteers ate nontransgenic potatoes only. The potato HBsAg vaccine boosted serum IgG antibody titers in more than half of the volunteers (Thanavala, Mason and Arntzen, unpublished results), which suggests that the oral delivery of plant-produced HBsAg could be a viable delivery system for an HBV boosting vaccine.

## Rabies Virus

Yusibov *et al.* delivered to 14 human volunteers spinach expressing epitopes from the rabies virus glycoprotein and nucleoprotein fused to the coat protein of alfalfa mosaic virus (AlMV). Five of the fourteen subjects had previously received a conventional rabies vaccine. Three of those five and all nine of the initially naive subjects displayed significant antibody responses to the rabies virus. None of the control individuals showed a significant elevation in rabies-specific antibodies. The conventional rabies vaccine was administered to the nine initially naive subjects seven days after completing the oral vaccination. Three of these volunteers produced neutralizing antibodies against rabies virus, although none of the five control subjects did. This study showed a clear indication that the orally delivered rabies vaccine has potential as an oral booster for the conventional rabies vaccine.

# ISSUES AND CHALLENGES

## Development and Licensing of Plant-derived Vaccines

Conventional pharmaceutical development has been estimated to costs roughly between $100 million and $800 million per product, and takes over 12 years. The development of conventional vaccines may be somewhat less costly, although new recombinant DNA-derived subunit vaccines produced in fermentation-based systems are likely to be similar to protein pharmaceuticals in development costs. At present, none of the major pharmaceutical companies is directing funding towards the development of plant-derived vaccines for infectious diseases. This may reflect :

- doubts about the potential for significant return on investment;
- uncertainties in the regulatory processes;
- limited human clinical trial data that establish required dosages, timing of delivery, and evaluation of possible adverse immunological effects;
- a lack of personnel with sufficient expertise in plant biology.

As a result, the opportunity to produce vaccines in plants represents a classic example in which the reliance on market forces for the development of health products is failing. Participation of both the public sector and the non-profit sector will be essential to provide leadership and investment support to unlock the potential of plant-derived vaccines. A principal justification for the public sector promotion of plant-based vaccines is the significant favourable characteristics of this technology for the manufacture of vaccines against rare and neglected diseases. Developing new pharmaceuticals for these diseases is not a high priority for the major drug companies due to the low profit margins.

## Confronting GM Food Issues

The plant-based production of vaccines is a potentially transformative technology, but the use of a similar technology for agricultural biotechnology has stimulated significant public debate, especially focused on genetically modified foods (GM foods). Knowledgeable public debate is valuable, but the debates about GM foods have not always been based upon scientific considerations, and consequently the debates have become polarized.

Because plant-derived pharmaceuticals are not intended for use as food products, the crops that produce them must have special stewardship to ensure containment. In addition, the proteins they produce will have to be separated and purified in processing facilities dedicated to that purpose. The use of crop species that are currently part of the food supply for the production of oral vaccines will necessitate genetic separation from the food supply as an essential parameter for production of these materials. It is likely that global health organizations such as the Pan-American Health Organization (PAHO) and WHO will plan an essential role in the process of transferring the technology on a global scale as new regulatory frameworks are implemented.

## Edible Vaccines

In 2003, the UN Human Poverty Index showed that 1.2 billion of the developing world's 4.8 billion people were living on less than $1 per day, while a further 2.8 billion were living on less than $2 per day. The global poverty level, and the incidence of disease, is increasing dramatically every year. The World Health Organization estimates that approximately 170 million people worldwide are infected with hepatitis C virus (HCV), with 3-4 million new cases each year, and that more than one third of the world's population is infected with hepatitis B virus (HBV). In Asia, the prevalence of chronic hepatitis B and C is very high (about 150 million and 110 million people infected with HBV and HCV, respectively). A large majority of HCV-infected patients have severe liver cirrhosis and currently there is no vaccine available for thisdisease.

In addition, the rising cost of treatment for severe illnesses calls for the more economical production of therapeutic proteins. For example, the annual demand for insulin-like growth factor 1 (IGF-1) per cirrhotic patient is 600 mg (1.5-2 mg per day) and the cost of IGF-1 per mg is $30,000. Furthermore, the current annual cost of interferon therapy for viral hepatitis is $26,000 (Cowley & Geoffrey, Newsweek, April 22, 2002). It is evident that agricultural scale production of therapeutic proteins and vaccines is necessary to meet this large demand at an affordable cost. Plants have been used in medicine for many centuries but it is now possible to exploit plants as bioreactors for the production of human therapeutic proteins. The use of plants to produce biopharmaceuticals has several advantages, including the ability of plants to supply large quantities of therapeutic proteins to patients in a cost efficient manner.

In addition, plant-derived products are less likely than those derived from animal cells to be contaminated with human pathogenic microorganisms, since plants do not act as hosts for human infectious agents. Plants offer a number of advantages over other transgenic systems. For instance, the use of a cell fermenter or bioreactor is unnecessary and the need for post-translational modification and purification can be eliminated, thus lowering the production costs. The need for fermentation facilities has seriously limited the introduction of new thera peutic proteins in the past.

However, it has been estimated that one tobacco plant is able to produce more recombinant protein than a 300-liter fermenter of *E. coli* (Crop-Tech, VA) reflecting tobacco's biomass production of 40 tons of leaf fresh weight per acre (based on multiple mowings per season) and the production of up to one million seeds per plant. Consequently, the quantity of recombinant protein that can be harvested is only limited by the number of hectares that can be planted with transgenic crops. Furthermore, human proteins can be expressed without codon optimization, and fully assembled proteins can be produced since post-translational modifications such as glycosylation and the formation of disulfide bonds are possible in plant cells, especially within particular cellular compartments.

## HUMAN PROTEINS IN PLANTS

The first human protein produced in plants – growth hormone – was expressed as a fusion with the *Agrobacterium tumefaciens* nopaline synthase gene product. It was not until 1989 that the expression of antibodies in tobacco demonstrated the ability of plants to produce therapeutic proteins for human use. Transgenic tobacco, maize, soybean and alfalfa plants are expected to yield over 10 kg of therapeutic protein per acre, and this should reduce production costs by approximately 90% compared to other systems. However, with the exception of enzymes like phytase in tobacco (14% total soluble protein (TSP), and eubacterial glucanase in *Arabidopsis thaliana* (26% TSP, expression levels are generally less than 1% TSP in nuclear transgenic plants, especially for human therapeutic proteins. Such examples include the B subunit of *Escherichia coli* heat-labile enterotoxin (LTB) (0.01% TSP), hepatitis B virus envelope surface protein (0.01% TSP), human cytomegalo-virus glycoprotein B (0.02% TSP, transmissible gastroenteritis coronavirus glycoprotein S (0.06% TSP), human serum albumin (0.02% TSP), human protein C (0.001% TSP), human epidermal growth factor (0.0001% TSP), erythropoietin

(0.026% TSP), and human interferon-β (0.000017% of fresh weight). Therefore, strategies are required to increase expression levels, allowing commercial exploitation of the therapeutic and human proteins mentioned above. Such strategies have been the focus of transgenic plant research over the past several years.

## THE TRANSGENIC CHLOROPLAST SYSTEM

The world of plant genetic engineering research was revolutionized when it was reported that *Bacillus thuringiensis* (Bt) Cry2Aa2 protein could accumulate to a level of 46.1% TSP in transgenic tobacco chloroplasts. This was not only the highest level of recombinant protein ever achieved in plants, but a complete bacterial operon was expressed successfully for the first time resulting in the formation of stable Cry2Aa2 crystals. These results were possible because native bacterial genes were expressed at levels several hundred-fold higher in the chloroplast compartment than in nuclear transgenic plants. Chloroplast transformation not only results in higher protein expression levels compared to nuclear genetic engineering, but also provides several other advantages. Even though transgenic chloroplasts may be present in pollen, foreign genes do not escape to other crops because chloroplast DNA in not passed into the egg cell.

In most crop plants, plastid genes are inherited uniparentally in a strictly maternal fashion. Although pollen from such plants contains metabolically active plastids, the plastid DNA itself is lost during the process of pollen maturation and hence is not transmitted to the next generation. Also, the chloroplast can be a good place to store proteins, or their biosynthetic products, that might be harmful to the host plant and adversely affect its physiology if allowed to accumulate in the cytoplasm. This has been demonstrated in the case of the cholera toxin B subunit (CTB), a candidate oral subunit vaccine for cholera, because it accumulated in large quantities within transgenic plastids but was non-toxic to the plant; in contrast, even very small quantities of LTB were toxic when expressed in the cytosol. Similarly, trehalose, which is used in the pharamaceutical industry as a preservative, was very toxic when it accumulated in the cytosol but was non-toxic when it was compartmentalized within plastids. Transgenes are integrated into the spacer regions of the chloroplast genome by homologous recombination using chloroplast DNA flanking sequences. This allows site-specific integration and thus eliminates the position effects that are frequently observed in nuclear transgenic plants. As a result, it is not necessary to screen numerous putative transgenic lines to select those with high-level transgene expression.

All chloroplast transgenic lines express the same level of foreign protein, within a range of physiological variations. Yet another advantage is the lack of transgene silencing in chloroplast transgenic plants, which is a serious concern in nuclear transformation. It has been shown that there is no transcriptional gene silencing in chloroplast transgenic lines despite the accumulation of transcripts at a level 169-fold higher than in nuclear transgenic plants. Similarly, there is no transgene silencing at the translational level, despite the accumulation of foreign protein at levels up to 47%of the total plant protein in chloroplast transgenic lines. Purification costs can account for much of the expense involved in biopharmaceutical production. In the case of insulin, for example, chromatography alone accounts for 30% of the production costs and 70% of the set-up costs.

It is estimated that the oral delivery of properly folded and fully functional biopharmaceuticals in plant tissues could potentially reduce production costs by 90%. For such oral delivery to be successful, antibiotic selection should be avoided and the biopharmaceutical protein must be expressed in edible parts of the plant that require no cooking or processing. Antibiotic-free selection using a gene of plant origin has been developed recently". Alternatively, antibiotic resistance genes can be eliminated using direct repeats or theCre-*liox*P system. Another requirement for oral delivery is the ability to express foreign proteins in plastids that are present in non-green tissues. One such example is the expression of an antibiotic resistance gene *(aadA)* in tomato chromoplasts. More recently, stable and highly efficient plastid transformation has been achieved in the non-green tissues of carrot plants. Using carrots as the source of an edible vaccine or therapeutic protein provides many advantages, such as the uniform source of cell culture (which is one of the essential requirements for producing therapeutic proteins, i.e. a homogeneous single source of origin), rapid division, the large biomass using bioreactors, and direct delivery of a precise dose of vaccine antigens or biopharmaceuticals.

Furthermore, when carrots are used as the delivery vehicle, cooking is unnecessary and this preserves the structural integrity of therapeutic proteins during consumption. Most importantly, carrot plants do not flower in the first year. Therefore, there are no flowers or reproductive structures during the harvest and this provides complete gene containment, preventing outcross via pollen and gene flow via seeds. One of the most important features of the transgenic chloroplast system is that chloroplasts are able to carry out the processing of eukaryotic proteins, including the correct folding of subunits and the formation of disulfide bridges. Functional assays showed that chloroplast-derived cholera toxin B subunit (CTB) binds to the intestinal membrane $G_{M1}$-ganglioside receptor, confirming correct folding and disulfide bond formation in the plant-derived CTB pentamers. Formation of disulfide bonds within a single polypeptide, resulting in a fully functional human therapeutic protein, has also been demonstrated for human somatotropin (growth hormone), interferon-α and interferon-γ.

Human therapeutic proteins as small as 20 amino acids or as large as the anthrax protective antigen (83 kDa) have been expressed in transgenic chloroplasts. The ability to express operons or multi-gene cassettes allows the engineering of foreign pathways and the expression of multi-subunit therapeutic proteins. The successful assembly of a monoclonal antibody in transgenic chloroplasts has demonstrated the ability of this system to produce complex multi-subunit proteins and has also indicated the presence of chaperones required for the assembly of such proteins within plastids. These observations usher in a new era for the production of therapeutic proteins via chloroplast genetic engineering. The rest of this chapter provides examples of therapeutic proteins expressed in transgenic chloroplasts.

## Chloroplast-derived Human Antibodies

The most widely studied therapeutic proteins produced in plants include monoclonal antibodies for passive immunotherapy and antigens for use as oral vaccines. Antibodies against dental caries, rheumatoid arthritis, cholera, *E. coli* diarrhea, malaria, certain cancers, Norwalk virus, HIV, rhinovirus, influenza, hepatitis B virus and herpes simplex virus have been produced in transgenic plants. However, the anti- *Streptococcus mutans* secretory antibody for the prevention of dental caries is the only plant-derived antibody currently in Phase II clinical trials. Until recently, most antibodies were expressed in tobacco, potato, alfalfa, soybean, rice and wheat. It has been estimated that for every 170 tons of harvested tobacco, 100 tons represents harvested leaves. A single hectare could thus yield 50 kg of secretory IgA.

**Table 6.3. The production of therapeutic proteins for human or animal health in transgenic chloroplasts**

| *Biopharmaceutical proteins / Vaccine* | *Gene* | *Site of Integration* | *Promoter* | *5′/3′ regulatory elements* | *% tsp expression* | *Laboratory* |
|---|---|---|---|---|---|---|
| Elastin derived polymer | EG121 | *tml / tmA* | Prrn | T7gene10/ TpsbA TpsbA | ND | Daniell |
| Human somatotropin | HST | trnV/rpsl2/7 | Prrn[a], PpsbA[b] | T7gene10[a] or psbA[b]/Trpsl6 | 7.0%[a] and 1.0% [b] | Monsanto |
| Cholera toxin | *CtxB* | *trnl / tmA* | Prrn | ggagg/T*psb*A | 4% | Daniell |
| Antimicrobial peptide | MSI-99 | *tml / tmA* | Prrn | ggagg/T*psb*A | 21.5-47% | Daniell |
| Insulin like growth factor | *JGF*-1 | *trnl / tmA* | Prrn | PpsbA/T*psb*A | 33% | Daniell |
| Interferon alpha 5 | *INfα5* | *tml / tmA* | Prrn | *PpsbA / TpsbA* | ND | Daniell |
| Interferon alpha 2b | *INFα2B* | *trnl / tmA* | Prrn | PpsbA/TpsbA | 19% | Daniell |
| Human Serum Albumin | *Has* | *tml / tmA* | Prrn[a], PpsbA[b] | ggagg[a], psbA[b]/ T*psb*A | 0.02 %[a], 11.1%[b] | Daniell |
| Interferon gamma | *IFN-g* | *rbcL / accD* | *PpsbA* | PpsbA/TpsbA | 6% | Reddy |
| Monoclonal antibodies | *Guy's 13* | *tmI / trnA* | Prrn | ggagg/T*psb*A | ND | Daniell |
| Anthrax protective antigen | Pag | *tml / tmA* | Prrn | *PpsbA / TpsbA* | 4–5% | Daniell |
| Plague F1~V fusion antigen | CaFl~VLcrV | *tmF1 / trnA* | Prrn | *PpsbA / TpsbA* | 14.8% | Daniell |
| CPV VP2 protein | *CTB-2L21*[a], *GFP-2L21*[b] | *tml / trnA* | Prrn | *PpsbA / TpsbA* | 31.1%[a], 22.6%[b] | Daniell/ Veramendi |

Furthermore, it has been estimated that the cost of antibody production in plants half that in transgenic animals and 20 times lower than in mammalian cell cultures Based on the chloroplast's ability to produce bioactive and fully assembled proteins, a codon-optimized and humanized gene encoding a chimeric monoclonal antibody (IgA/G, Guy's 13) under the control of a specific 5'-untranslated region, was used successfully to synthesize and assemble monoclonal antibodies in transgenic tobacco chloroplasts, including the formation of disulfide bridges. Guy's 13 was developed to prevent dental caries, which is caused by *Streptococcus mutans* Daniell and Wycoff confirmed integration of the chimeric antibody gene into the chloroplast genome using PCR and Southern blot hybridization analysis. Protein expression was also confirmed by western blot analysis. Expression of the individual heavy and light chains as well as the fully assembled antibody was shown indicating the presence of chaperones for proper protein folding and enzymes for formation of disulfide bonds within transgenic chloroplasts.

## Chloroplast-derived Biopharmaceuticals

### *Human Serum Albumin*

Human serum albumin (HSA) is currently obtained by the fractionation of blood serum. It accounts for 60 % of the total protein in blood serum and it is the most widely used intravenous protein. Because of the low expression of HSA in nuclear transgenic plants (approximately 0.02% TSP it would be desirable to produce this biopharmaceutical protein in transgenic chloroplasts. To achieve higher expression levels, the HSA gene was placed under the translational control of three different regulatory elements: a Shine-Dalgarno sequence (SD), the 5′ *psbA* region, and the cry2Aa2 UTR. In seedlings transformed with the SD-HSA construct, HSA accumulated to a level of only 0.8% TSP, while 1.6% TSP was achieved using the 5' *psbA* control region, and 5.9% TSP was achieved using the cry2Aa2 UTR. On the other hand, HSA reached a maximum of 0.02%, 0.8% and 7.2% TSP in transgenic potted plants when the transgene was regulated by the SD sequence, *cry2Aa2* UTR and 5′ *psbA* sequence, respectively. This demonstrated that excessive proteolytic degradation could reduce overall yields unless compensated by enhanced translation Furthermore, purification of HSA was possible by centrifugation, due to the formation of inclusion bodies within transgenic chloroplasts.

Inclusion bodies were precipitated by centrifugation and easily separated from the majority of cellular proteins present in the soluble fraction by a single centrifugation step, which may eliminate the need for expensive affinity columns or chromatographic techniques. Because the 5′ *psbA* region is light dependent, the accumulation of HSA protein should be increased after continuous illumination. Fernandez-San Millan *et al.* monitored HSA accumulation by ELISA. A maximum of 50 h of continuous light produced an HSA yield of 11.1% TSP in mature leaves. This is the highest expression of H S A demonstrated so far and is 500-fold higher than previous reports of HSA expression in the leaves of nuclear transgenic plants.

Due to the high local concentration of the protein, inclusion bodies were formed and this protected it from proteolytic cleavage. Another positive result from this study was the knowledge gained about the regulatory elements (SD, 5′ *psbA* region and the ***cry2Aa2*** UTR). These serve as a model system for enhancing the expression of foreign proteins that are highly susceptible to proteolytic degradation and provide advantages during purification through the use of current purification techniques for inclusion bodies.

## *Human Insulin-like Growth Factor-1*

Insulin-like growth factor-I (IGF-I) is a naturally occurring, single chain polypeptide with three disulfide bonds. It is produced in the liver and has a molecular mass of 7649 Da. IGF-I is not only critical for the growth of muscle and other tissues, but its therapeutic potential currently is being evaluated in diabetes, IGF-I induced neuroprotection, and in the promotion of bone healing. This therapeutic protein is commercially valuable because cirrhotic patients require 600 mg of IGF-1 per year at a cost of 530,000 per mg. In the past, IGF-1 has been expressed in *E. coli* but the mature protein cannot be produced in this system because disulfide bonds do not form in the *E. coli* cytoplasm. Since IGF-I has a high eukaryotic codon content, chloroplast codon optimization was performed by Ruiz and a synthetic IGF-1 gene was created to further increase expression levels.

Integration of the IGF-1 gene into the tobacco chloroplast genome was achieved through homologous recombination and was confirmed using PCR and Southern blot analysis. Western blotting and chemiluminescence techniques were used to verify the high levels of IGF-I expression in transgenic chloroplasts. ELISA tests were used to quantify IGF-1 in transgenic chloroplasts containing the native and synthetic *IGF*-1 genes, revealing levels as high as 32%

TSP. Most importantly, these observations showed that the chloroplast translation machinery is quite flexible, unlike the bacterial translation machinery, which could only translate the synthetic chloroplast-codon optimized *IGF-1* gene.

## Human Interferon (IFNα2b)

Malignant carcinoid tumors, a symptom of carcinoid syndrome, presented a therapeutic challenge for many years until the introduction of interferon treatment in 1982 by Oberg and Eriksson. They reported that 47 out of 111 patients (42%) treated with interferon-α (median dose of 6 mega-units (MU) of interferon-α, five times weekly) demonstrated a significant biochemical response and 15% demonstrated more than 50% reduction in tumor size. Survival analysis demonstrated a median survival period of only 8 months in the group of patients treated with chemotherapy, compared to 80+ months ($p < 0.001$) in the groups treated with interferon-α. However, alpha interferons have other therapeutic uses, such as the inhibition of viral replication and cell proliferation, enhancement of the immune response, and most recently, the treatment of patients suffering from West Nile virus.

The Food and Drug Administration first approved a specific subtype of interferon-α (IFNα2b) in 1986 for the treatment of hairy cell leukemia in the United States and the recombinant IFNα2b now on the market is produced using an *E. coli* expression system. In plants, nuclear transformation has resulted in very low expression levels (0.000017%) TSP in transgenic tobacco and rice. Due to necessary *in vitro* processing and purification, the average cost of treatment is $26,000 per year. The injection of IFNα2b causes side effects such as fatigue, weight loss and anemia due to diarrhea, as well as flushing and bronchoconstriction. Treatment with alpha interferons became less popular when it was found that up to 20% of patients produce anti-IFNα antibodies, which reduce the effectiveness of the treatment because IFNα2b aggregates with human serum albumin in the blood. However, oral administration could be beneficial in the treatment of various infectious diseases. In the Daniell laboratory, rerambinant IFNα2b containing a polyhistidine purification tag and a thrombin cleavage site was expressed in transgenic tobacco chloroplasts, for use as an oral therapeutic.

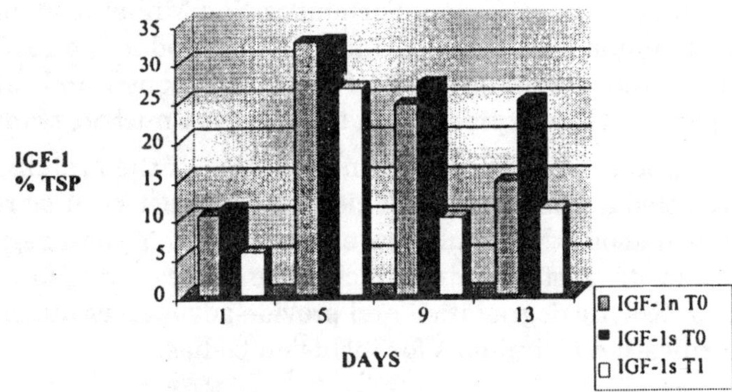


**Fig. 6.2. Expression of IGF-1 in transgenic chloroplasts after continuous light exposure for 13 days. IGF-1 expression is shown as a percentage of the total soluble protein. IGF-1 n is the native gene and IGF-1s is the chloroplast codon-optimized gene.**

For comparison, the gene cassette was integrated into the chloroplast genome of cv. Petit Havana and into a low-nicotine variety of tobacco, LAMD-609. Western blots performed using a monoclonal antibody to detected monomers and multimers of IFNα2b in both tobacco varieties, and disulfide bond formation was confirmed. Integration of the transgene into both chloroplast genomes was confirmed by Southern blot analysis.

In the Petit Havana transgenic lines, chloroplast genome homoplasmy occurred in the first generation, and this corresponded to the highest level of IFNα2b expression. ELISAs revealed IFNα2b levels of up to 18.8% TSP in Petit Havana and up to 12.5% TSP in LAMD-609. IFNα2b activity was confirmed by the ability of the recombinant protein to protect HeLa cells against the cytopathic effects of encephalomyocarditis virus (EMC) and through the identification of interferon-induced transcripts. Chloroplast derived IFNα2b was found to have the same activity as commercially produced IntronA. The mRNA levels of two genes induced by IFNα2b (2'-5' oligoadenylate synthase and *STAT-2)* were tested by RT-PCR using primers specific for each gene.

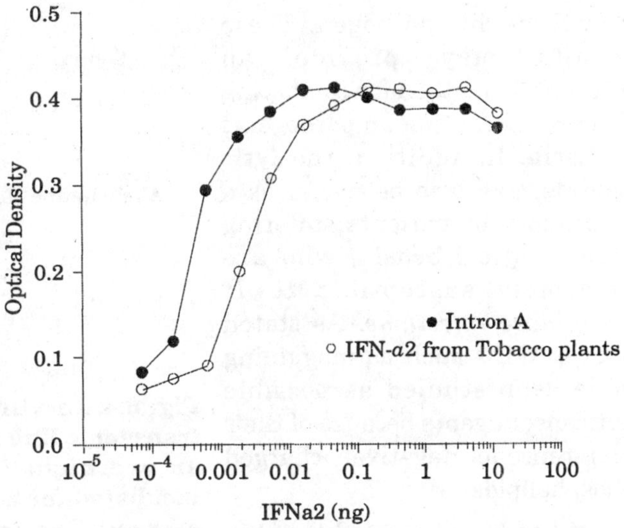

**Fig. 6.3. Demonstration of IFNα2b functionality by the ability of IFNα2 b to protect HeLa cells against the cytopathic effect of encephalomyocarditis virus (EMC). Chloroplast derived IFNα2b was as active as commercially produced Intron A.**

Chloroplast-derived IFNa2b induced the expression of both genes in a manner similar to commercial IFNα2b. These levels of expression and functionality are ideal for IFNα2b purification and for the further use of oral IFNα2b delivery in pre-clinical studies.

## *Anti-Microbial Peptides (AMPs): MSI-99*

Due to the increasing number of drug-resistant bacteria arising from the misuse of antibiotics and anti-microbial agents, new strategies to combat or prevent different human pathogens need to be developed. Research on defense peptides secreted from the skin of the African clawed frog *(Xenopus laevis)* has resulted in the discovery of magainin, an amphipathic α-helix-containing peptide, which has been investigated as a broad-spectrum topical agent, a systemic antibiotic, a wound healing agent, and an anticancer drug. Magainin has an affinity for negatively charged phospholipids in the outer leaflet of the prokaryotic membrane. A sequence encoding the magainin analog MSI-99, a synthetic lytic peptide, was integrated into the chloroplast genome of tobacco variety Petit Havana and expressed at high leyels (21.5-47%TSP)The minimum inhibitory concentration of MSI-99 was investigated.

Based on total inhibition of bacterial and fungal cells, MSI-99 was most effective against P. *syringae,* requiring only 1 ug per 1000 bacteria. The amount of antimicrobial peptide required to kill bacteria was used to estimate the level of expression in transgenic plants, since the lytic activity of antimicrobial peptides is concentration dependent. The effectiveness of the chloroplast-derived lytic peptide was tested *in vitro,* using a multi-drug resistant Gram-negative bacterium, *Pseudomonas aeruginosa,* which is an opportunistic pathogen of plants, animals and humans. Cell extracts prepared from $T_r$generation plants resulted in 96% inhibition in

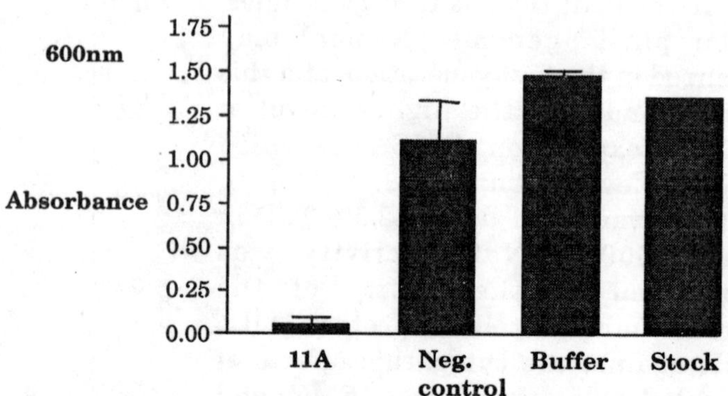


**Fig. 6.4. In vitro bioassay of $T_1$ generation chloroplast transgenic line against P. *aeruginosa*. Bacterial cells from an overnight culture were diluted to $A_{600}$ 0.1-0.3 and incubated for 2 hours at 25°C with 100 μg of total protein extract. One ml of LB was added to each sample and incubated overnight at 26 °C. Absorbance was recorded at 600 nm. Data was analyzed using a GraphPad Prism.**

growth of this pathogen. These results may provide an alternative method for combating drug-resistant human pathogenic bacteria. In addition, the lytic peptide may also be useful as a treatment for patients suffering from cystic fibrosis, who are extermely susceptible to *P. aeruginosa infections.* As stated above, AMPs such as magaining have been studied as possible anticancer agents because of their preference for negatively charged phospholipids.

It has been reported that the outer leaflet of melanoma and colon carcinoma cells contain 3-7 fold higher levels of phosphatidylserine than normal cells. Previous studies have shown that analogs of magainin were effective against hematopoietic cancer, melanoma, sarcoma and ovarian teratoma lines. The transgenic chloroplast system could thus provide sufficient expression levels for the commercially feasible production of AMPs as anticancer drugs and as antimicrobial agents.

## Chloroplast-derived Vaccine Antigens

Edible vaccines are currently being developed for a number of human and animal diseases, including measles, cholera, foot and mouth disease, hepatitis B and C. Many of these diseases ar likely to require booster vaccinations or multiple antigens to induce and maintain protective immunity. This problem could be solved, given the ability of plant to express more than one transgene, allowing the delivery of multiple antigens for repeated inoculations. Other advantages of plant-based vaccines include the reduced neeed for medical personnel and sterile injection conditions, heat stability, as explained by Webster et al. it is unlikely that an edible vaccine would lead to oral tolerance because it is achieved by dose-specific oral antigen delivery. To explain the concept of oral therapeutic protein delivery, hepatitis B surface antigen (HBsAg) was used for oral therapeutic protein delivery, hepatitis B surface antigen (HBsAg) was used for oral immunization.

No primary immune response was detected after two 300-μg doses of yeast-derived HBsAg. Howerver, a primary response began after two servings oftrangenic potatoes containgin 85-300μg of HBsAg. The main advantage of expressing biopharamaceuticals in plants is bioencapsulation. since the proteins are presented inside plant material, degradation occurs only slowly and a larger quantity of the protein can survive proteolysis. For this reason, less of the recombinant protein is needed to achieve the same theropeutic effect as a non-orally delivered protein.

## *Cholera Toxin B Subunit (CTB)*

The first step in developing a highly expressed edible vaccine containing an adjuvant is the successful expression of CTB in transgenic plants. Since CTB has previously been expressed in nuclear transgenic plants at levels of 0.01% TSP (leaves) and 0.03% TSP (tubers), expression of this adjuvant should increase if the correspinding gene is integrated into the chloroplast genome. Integration of an unmodified CTB-coding sequence into the chloroplast genome, confirmed by PCR and Southern blot analysis, resulted in the accumulation of CTB as funcitional oligomers, at levels of up to 4.1% TSP in transgenic chloroplast. $C_{MI}$-Ganglioside binding assays confirmed that chloroplast-derived CTB binds to the intestinal membrane receptor of cholera toxin establishing unequivocally that chloroplasts are capable of forming disulfide bridges during the assembly of foreign proteins. These results and the stable transformation of plastids in edible plant parts confirm the feasibility of producing in large quantities edible vaccines, of adjuvant-antigen fusion proteins, capable of eliciting an immune response in humans. In has been often shown that an adjuvant increases the immune response and in turn raises the antibody titer.

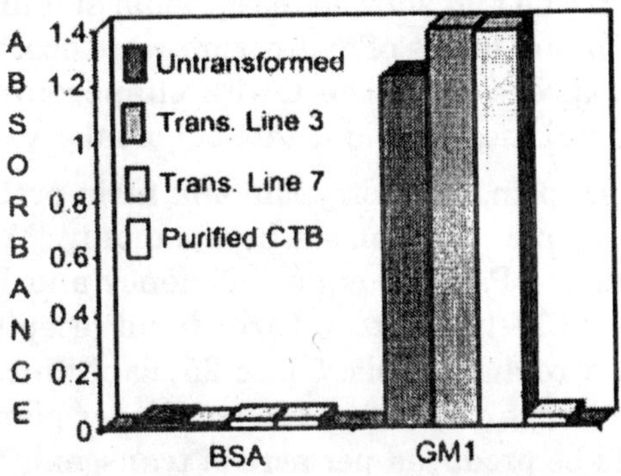


**Fig. 6.5. CTB-$G_{M1}$-ganglioside binding ELISA assay. Plates, coated first with $G_{M1}$-ganglioside and bovine serum albumin (BSA), respectively, were irrigated with total soluble plant protein from chloroplast transgenic lines (3 and 7) and 300 mg of purified bacterial CTB. The absorbance of the $G_{M1}$-ganglioside-CTB-antibody complex in each case was measured at 405 nm. Total soluble protein from untransformed plants was used as the negative control.**

## *Bacillus anthracis Protective Antigen*

The anthrax bioterrorist attacks that followed the events of September 11$^{th}$ 2001 resulted in a renewed interest *Bacillus anthracis,* the causative agent of this disease. Research has focused on the development of better vaccines than the one currently available. It has been estimated that the aerosolized release of 100 kg of anthrax spores upwind of Washington DC would cause mortalities of 1,30,000-30,00,000. Nonetheless, wild-type *Bacillus anthracis* is susceptible to conventional antibiotics, including penicillin, oxyfloxacin and ciprofloxacin. The problem lies not with the bacterial infection itself, but with three proteins released by the bacteria-protective antigen (PA, 83 kDa), lethal factor (LF, 90 kDa) and edema factor (EF, 89 kDa) -known as anthrax toxins. The Centers for Disease Control list *Bacillus anthracis* as a category A biological agent and estimate the cost of an anthrax attack to exceed $26 billion per 100,000 exposed individuals. Concerns regarding vaccine purity, the need for multiple injections, and the limited supply of PA, underscore the urgent need for an improved vaccine.

Therefore, in the Daniell laboratory, the PA gene *(pag)* was inserted into a chloroplast vector along with the *psbA* regulatory signals for enhanced translation. As in other investigations, integration of transgenes into the chloroplast genome was confirmed by PCR and Southern blot analysis. Crude plant extracts contained up to 2.5 mg of full length PA per gram of fresh leaf tissue and showed exceptional stability (several months in stored leaves or crude extracts). Maximum levels of expression were observed in mature leaves under continuous illumination. Co-expression of the ORF2 chaperonin from *Bacillus thuringiensis* did not increase PA accumulation or fold it into cuboidal crystals in transgenic chloroplasts.

Trypsin, chymotrypsin and furin proteolytic cleavage sites present in PA were protected in transgenic chloroplasts because only PA 83 was detected. Both CHAPS and SDS detergents extracted PA with equal efficiency and PA was detected in the soluble fraction. Chloroplast-derived PA bound to anthrax toxin receptor, heptamerized, and bound to lethal factor, resulting in macrophage lysis. Up to 25 μg of functional PA per ml of crude extract was observed. With an average yield of 172 mg of PA per plant, 400 million doses of vaccine (free of contaminants) could be produced per acre of transgenic tobacco, using a low yielding experimental cultivar in a greenhouse, which could be further enhanced 20-fold in commercial cultivars cultivated in the field (with multiple harvests).

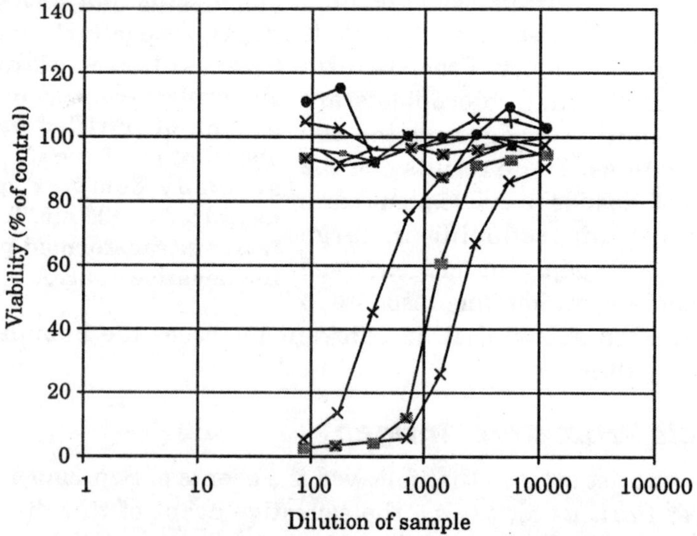


**Fig. 6.6. Macrophage cytotoxic assays of plant extracts. Supernatant samples were tested from TI-generation of chloroplast transgenic line pLD-JWl (proteins were extracted in buffer containing no detergent and MTT was added after 5 hours). pLD-JWl (extract stored for 2 days); pLD-JWl (extract stored for 7 days); PA 5 μg ml$^{-1}$; —x— Control wild type (extract stored for 2 days); Control wild type (extract stored for 7 days; Control wild type no lethal factor (LF) (extract stored for 2 days); —•— Control wild type no LF (extract stored for 7 days); Control pLD-JWl no LF (extract stored for 2 days); Control pLD-JWl no LF (extract stored for 7 days).**

## *Yersinia pestis F1~V Fusion Antigen*

*Yersinia pestis,* the Gram-negative bacterium responsible for plague, is also listed as a category A biological agent by the Centers for Disease Control. *Y. pestis* is naturally transmitted from one animal host to another either directly or via a flea vector. This pathogen has been used for centuries as a biological warfare agent and it raises concerns today as one of the microorganisms with potential for use against civilian or military populations. If used as a

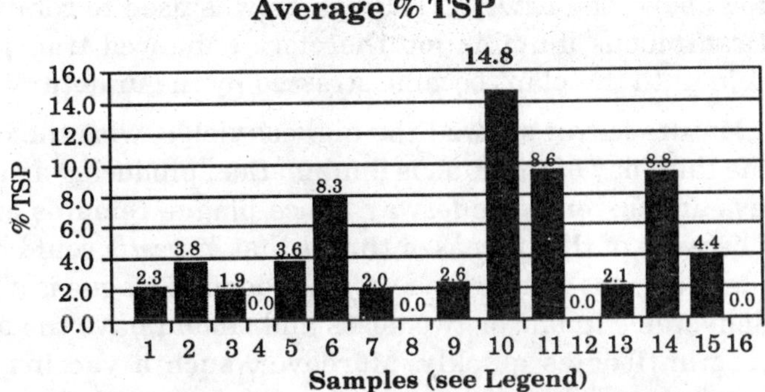


**Fig. 6.7. Enzyme linked immunoassays, with protein yields expressed as % TSP and μg of F1V per gram of fresh leaf material. For the continuous illumination experiment, leaf material was sampled on days 0, 1, 3 and 5. Young, mature, and old samples were taken for each experiment. The figure shows average levels of F1V antigen for: Samples 1-4 (DayO: (1) young, (2) mature, (3) old, and (4) wild type leaf samples); Samples 5-8 (Day 1: (5) young, (6) mature, (7) old and (8) wild type leaf samples)); Samples 9-12 (Day 3: (9) young, (10) mature, (11) old and (12) wild type leaf samples); Samples 13-16 (Day 5: (13) young, (14) mature, (15) old &. (16) wild type leaf samples).**

warfare agent, the pneumonic form of plague would be the most probable outcome of infection and is devastating due to the rapid onset of the disease, the high mortality rate, and the rapid spread of the disease. If a person with pneumonic plague is not treated with antibiotics within 18 hours of the onset of symptoms, he or she will die. Therefore, immunization against aerosolized plague has been an area of interest in current research. The current, killed whole unit vaccine is only slightly effective against the bubonic form of plague and not immunoprotective against pneumonic and septicemic'

Several subunit vaccines have been evaluated for immunogenicity and protective efficacy against *Y. pestis.* CaFl and LcrV are the two most favourable choices. F1 is a capsular protein located on the surface of the bacterium with anti-phagocytic properties. The V antigen is a component of the *Y. pestis* Type III secretion system and it may form part of an injectosome. Expression of these antigens as fusion proteins in *Salmonella typhimurium* has been shown to be immunogenic in mice, but has not proven to be safe. However, when these antigens are produced in *E. coli,* they have been shown to be safe and to provide protection against subcutaneous and aerosol challenges of *Y. pestis* in mice. Expressing subunit vaccines in chloroplasts is very promising for three reasons: (*i*) subunit vaccines, which do not express active toxins, are safe and do not multiply; (*ii*) bacterial and many viral genes have a high AT content allowing for high-level expression in the chloroplast; and (iii) oral delivery of vaccines yields high mucosal IgA liters along with high systemic IgG titers, enabling the immune system to fight off infectious agents at important entry points (e. g. the lungs and vagina). Investigations in the Danielllaboratory demonstrated expression of the F1 and V antigens in transgenic chloroplasts as a recombinant F1~V fusion protein, consisting of the F1 protein fused at its carboxyl terminus to the amino terminus of the entire V antigen.

The gene encoding the fusion proteins was obtained from the U.S. Army Medical Research Institute of Infectious Diseases (USAMRIID) and integrated into the chloroplast genome of Petit Havana and LAMD-609 varieties using a universal transformation vector. Expression levels of F1~V in the two varieties were compared by western blotting and ELISA, after continuous illumination for up to five days. For the continuous illumination analysis, leaf material was sampled on days 0, 1, 3 and 5, and samples were taken from young, mature and old leaves. As stated above, the *psbA* 5' UTR, which was used to control the fusion gene, is light regulated, and continuous illumination therefore enhanced translation. This led to a buildup of F1~V protein until the plant became stressed by the unnatural light period.

Mature leaves showed the highest yields, with the largest amount of F1~V accumulating on the third day of continuous illumination, producing an average yield of 14.8% TSP. Functional assays are currently underway. Since plague remains endemic in some regions of the world, and because of the increased threat that *Y. pestis* could be used as a biological warfare agent, the development of improved vaccines against plague is a high priority. The ideal vaccine should be deliverable in one or two doses and should have the ability to produce high-titer and long-lasting antibodies quickly. Moreover, such a vaccine should protect against aerosolized transmission of *Y. pestis*.

## *Canine Parvovirus (CPV) VP2 Protein*

Canine parvovirus (CPV) infects dogs and other Canidae such as wolves, coyotes, South American dogs and Asiatic raccoon dogs, producing haemorrhagic gastroenteritis and myocarditis. CPV-1 was the first strain of canine parvovirus to be discovered, and was initially described in 1967. It did not pose much of a threat except to newborn puppies. However, the CPV-2 strain, which appeared in the US in 1978, appeared to be a mutated form of the feline parvovirus (more commonly known as feline distemper virus). Infected animals shed CPV-2, which is hardy in the environment, in large numbers, and this led to rapid worldwide dissemination. Attempts to shield puppies from exposure were completely futile. A second mutant, CPV-2 a, was recognized in 1979 and was found to be more aggressive.

The vaccine was in short supply and many veterinarians had to use feline distemper vaccine as a substitute, since it was the closest one available. The most common form of the virus today is CPV-2 b and young animals are commonly vaccinated with an attenuated whole virus vaccine. Advances in molecular biology have lead to the identification of CPV antigens that are capable of eliciting a protective immune response. A linear antigenic peptide (2L21) from the VP2 capsid protein (amino acids 1-23) of CPV was selected and its DNA sequence was successfully introduced into tobacco chloroplasts. The 2L21 synthetic peptide, chemically coupled to a KLH carrier protein, has been extensively studied and has been shown to protect dogs and minks against parvovirus infection very effectively. This peptide was expressed in nuclear transgenic plants as an N-terminal fusion protein with β-glucuronidase (GUS) but expression levels were inadequate. Therefore, the 2L21 peptide, fused either to the cholera toxin B subunit (CTB) or the green fluorescent protein (GFP) was expressed in chloroplast transgenic plants, and accumulated to levels up to ten-fold higher than those previously reported in nuclear transformation.

The expression levels were dependent on plant age. Both young and senescent plants accumulated lower amounts of proteins than mature plants, showing that the time of harvest is important when scaling up the process of protein production. The maximum level of CTB-2L21 was 7.49 mg $g^{-1}$ fresh weight (equivalent to 31.1% TSP) and that of GFP-2L21 was 5.96mg $g^{-1}$ fresh weight (equivalent to 22.6% TSP). The inserted epitope was detected with a CPV-neutralizing monoclonal antibody, indicating that the epitope is correctly positioned at the C-terminus of the fusion proteins. The resulting chimeric CTB-2L21 protein retained an ability to form pentamers, possessed the $G_{M1}$-ganglioside binding characteristics of the native CTB, and induced antibodies that were able to recognize the VP2 protein of CPV following the intraperitoneal immunization of mice. These results show that plant derived CTB-2L21 recombinant protein is immunogenic via intraperitoneal administration, as shown by its ability to induce a humoral response that cross-reacts with the native VP2 protein.

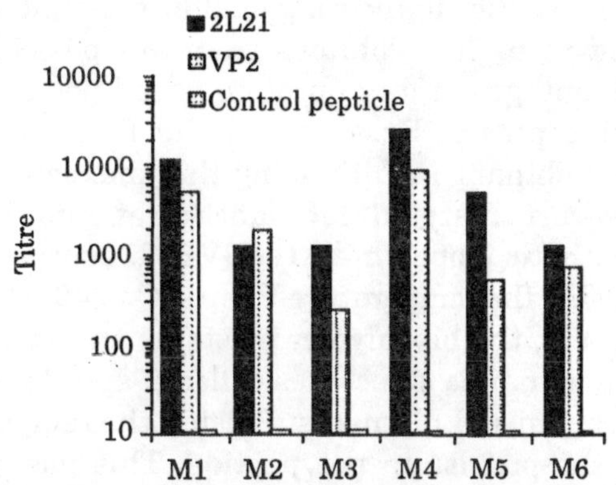


**Fig. 6.8. Titers of antibodies at day 50 induced by plant-derived CTB-2L21 recombinant protein. Balb/c mice were intraperitoneally immunized with leaf extract from CTB-2L21 transgenic plants. Animals were boosted at days 21 and 35. Each mouse received 20 µg of CTB-2L21 recombinant protein. Individual samples of mouse serum were titrated against 2L21 synthetic peptide,VP2 protein and a control peptide (amino acids 122-135 of hepatitis B virus surface antigen). Titers were expressed as the highest serum dilution lo yield twice Ihe absorbance mean of preimmune sera. M1-M6: mice 1 lo 6; 2L21: epitope from the VP2 protein of the canine parvovirus; CTB: cholera toxin B; VP2: protein of the canine parvovirus that includes the 2L21 epitope.**

Additional experiments are underway to check the ability of this fusion protein to induce specific immune responses after mucosal delivery. This is the first report of an animal vaccine epitope expressed in transgenic chloroplasts.

## PURIFICATION STRATEGIES

The purification of therapeutic proteins expressed in transgenic systems is a major challenge. For example, insulin produced in *E. coli* is purified by chromatography, which accounts for 30% of operating expenses and 70% of set up costs. Therefore, new strategies for biopharmaceutical purification are currently being investigated. To overcome the high cost of purification, protein-based polymers have been produced in *E. coli* through genetic engineering. For example, a synthetic polymer gene encoding the polypeptide $(GVGVP)_{121}$ was overexpressed in *E. coli* to the extent that polymer inclusion bodies occupied 80-90% of the cell volume. The inverse temperature transition property of this polymer makes it an ideal fusion protein for the purification of therapeutic proteins. The protein is soluble in water at temperatures below 25 °C, but the polymer aggregates into a more-or-dered, viscoelastic state called a coacevate at 37 °C.

Indeed, it is known that polypeptides, proteins, protein-based polymers and any polymers in which the hydrophobic and polar residues are in the correct balance, will fold and assemble (i.e. become more ordered) as the temperature increases and will disassemble and unfold (i.e. become less ordered) as the temperature decreases. This inverse temperature transition property makes it easier and cheaper to harvest polymers in aqueous solutions simply by increasing the temperature and this avoids cumbersome purification procedures and the use of enzymes and organic solvents which may alter the quality of the polymer or any biopharmaceutical protein. The idea of producing biopolymers in plants instead of bacteria seems very appealing since a further decrease in production costs could be achieved.

Genetic engineering of plants should allow the production of polymers at much lower costs and in higher volumes than are possible with microbial fermentation. Because of these advantages, a large protein based polymer $(GVGVP)_{121}$ was expressed in transgenic tobacco chloroplasts. Research in the Daniell laboratory confirms the possibility of producing recombinant insulin using this inexpensive purification approach and offers the prospect of low-cost treatment for diabetic patients. The proinsulin gene was fused to a gene encoding the synthetic biopolymer $(GVGVP)_{40}$, expressed in *E. coli* and the fusion protein was purified by raising the temperature from 4 °C to 42 °C, utilizing the inverse temperature transition property. At 4°C, the biopolymer exists as an extended molecule, but when incubated at 42 °C it folds into dynamic structures called β-spirals that further aggregate by hydrophobic association to form twisted filaments. Raising the temperature to 42 °C does not affect the tertiary structure of the proinsulin polypeptide. This has been demonstrated previously by the production of human insulin in bacteria using a temperature-responsive promoter. Production of the recombinant protein was induced by temperature shift from 30 °C to 42 °C with, out any adverse effects on protein stability. The fusion protein was purified twice using the inverse temperature transition property of the biopolymer, and the protein was analyzed by SDS-PAGE, western blotting and immunodetection using mouse anti-human proinsulin primary antibody.

Negative staining of the SDS-polyacrylamide gels with 0.3 M $CuCl_2$ confirmed the presence of the proinsulin-polymer fusion protein. Copper-stained gels with a black background appear to have dark bands against a light, semiopaque background when illuminated obliquely from above. The same gel was then stained with Commassie R-250, which would stain only the proinsulin but not the polymer. Commassie R-250 does not stain polymer proteins without a fusion, since the sulphonic acid groups on the dye only form ion pairs with lysine and arginine, which are not present in $(GVGVP)_{40}$. Since Commassie R-250 staining was successful, this confirmed the presence of the fusion protein. Thus, the polymer-proinsulin fusion protein has been confirmed by western blot, and two different staining methods.

# CHAPTER 7 Pharmaceutical Proteins in Plants

Molecular farming or *pharming*™is the production of pharmaceutically important and commercially valuable proteins in plants. Since the first production of a pharmaceutical protein in transgenic tobacco a broad range of proteins with potential medical applications has been expressed in plants. These range from small peptides such as enkephalins to complex, multisubunit molecules such as secreted antibodies of the IgA type. In this chapter we will address the decisions that must be made during the production and marketing of such proteins. This begins with the choice of production host, which can range from cellular green algae to a variety of crop species. For pharmaceutical applications, it is necessary to work under current good manufacturing practice (cGMP) conditions and the use of plant suspension cultures must therefore be considered as an alternative to whole plants.

Another crucial consideration is the design of the expression cassette, which mus optimized to achieve high-level accumulation of functional recombinant protein. Intrinsic factors such as how the protein is processed in the plant need to be taken into account, as well as the ability of the host species to carry out specific forms of posttranslational modification. When sufficiently high levels of expression and protein accumulation are achieved, efficient downstream processing protocols must be developed to insure product quality and the economic feasibility of production. As the demand for safe, recombinant pharmaceutical proteins continues to expand, the market potential of plant-produced recombinant proteins is considerable. Molecular farming can produce iccombinant proteins at a lower cost than traditional expression systems based on microbial or animal cell culture, and without the risk of contamination with human pathogens.

## PLANT SPECIES USED

Many species can be exploited for molecular farming with tobacco being the most widely used host plant to date. The advantages of tobacco plants include the high biomass yield (which can reach 50 – 100 tonnes $ha^{-1}$ in a high-density population depending on the cultivar), the ease of stable transformation either by co-cultivation with *Agrobacterium tumefaciens*, or transiently by infiltration with transgenic agrobacteria or transfection With viral vectors. Another benefit is that tobacco is not used as a food or feed crop, ensuring that a transformed line expressing a highly potent drug will not contaminate food resources by outcrossing or during the processing steps. A drawback of this species is its content of nicotine and other alkaloids. Although there are cultivars available with reduced alkaloid content it is necessary to remove all traces of these toxic compounds during downstream processing, especially if the recombinant protein is intended for clinical applications.

An additional disadvantage is the limited shelf life of transgenic leaf material after harvest, a problem that can be solved in other plant species by targeting the desired protein to natural storage organs like seeds or tubers. For tobacco leaves, downstream processing must commence immediately after harvest to ensure the stability of the recombinant protein, although Fiedler *et* al. reported no losses of scFv (single chain fragment variable) antibody specificity or antigen binding capacity in dried leaves three weeks after harvest. In the quest to find other plants that are suitable as bioreactors, various monocotyledonous and dicotyledonous species have been tested. These include corn, rice and wheat, alfalfa, potato, oilseed rape, pea, tomato and soybean.

The major advantage of cereal crops is that recombinant proteins can be directed to accumulate in seeds, which are evolutionar specialized for storage and thus protect proteins from proteolytic degradation. Recombinant proteins are reported to remain stable in seeds for up to five months at room temperature and for at least three years at refrigerator temperature without significant loss of activity. In addition, the seed proteome is less complex than the leaf proteome, which makes purification quicker and more economical. lt has been shown recently that the yields of recombinant protein in transgenic plants vary according to intrinsic factors such as the developmental stage and extrinsic factors such as the climate. Such factors also affect the precise nature of post-translational modification, particularly glycosylation.

Elbers and colleagues reported that the glycosylation profile, expressed as the ratio of complex to high mannose type glycans, increased with the age of tobacco leaves. Therefore, it is necessary to consider measures that ensure protein homogeneity, e.g. restricting expression to defined plant organs or specific compartments. Alternatively, it is possible to use an inducible expression system to restrict protein expression to a defined time point or period. Examples of such production methods include the postharvest system developed by Croptech and the amylase-based system for controlled expression in suspension cells.

## ALTERNATIVE EXPRESSION SYSTEM

Plant cell cultures have not been used as frequently as intact plants for molecular farming but there are several examples that demonstrate their suitability as hosts for the production of recombinant proteins. Free cell suspension is regarded as the most suitable cultivation system for large-scale biotechnology applications, but tentative experiments have also been performed with hairy roots and shooty teratomas. Examples of foreign proteins successfully expressed in suspension cells include monoclonal antibodies and their derivatives, cytokines, hormones and enzymes. The tobacco suspension cell line BY-2 appears to be a particularly suitable host cell line because of its high growth ratejtfl]. Like plant tissues, BY-2 suspension cells can be transformed easily by cocultivation with *A. tumefaciens*. In addition to this tobacco cell line, soybean cell culture has been used for the production of hepatitis B virus surface antigens and rice suspension cells have been employed for the expression of human α-1-antitrypsin and human lysozyme.

In contrast to the tobacco transformation process, gene transfer to rice is often performed by particle bombardment. Table 6.2 provides an overview of pharmaceutical proteins that have

**Table 7.1. Pharmaceutical Proteins Produced in Transgenic plant**

| *Protein* | *Regulatory elements* | *Additional elements of the construct* | | *Species/Tissue* | *Accumulation level* |
|---|---|---|---|---|---|
| α creatine kinase, MAK33 mAb (Fab) | Ca MV 35S promoter/ different termination sequence (ocs and rbcS) | LP of 2S2 storage protein of Arabidopsis *thaliana* | Fd and k chain with 3'ocs | *A. thaliana* (leaves) | 3.6% of TSP |
| | | | Fd chain with 3'ocs, κ chain with 3'rbcS | *A thalina* (leaves) | 6.5% of TSP |
| | | | Fd chain with 3'ocs, κ chain-KDEL with 3'rbcS | *A. Thaliana* (leaves) | 5.9 % of TSP |
| α human creatine kinase (fab, lgG) | CaMV 35S promoter/ ocs terminator | LP of the 2S2 storage Protein of A. *thaliana* | Fab | *Nicotiana tabacum* (callus) | 0.128% of TSP |
| | | | IgG | *N. tabacum* (callus) | 0.477% of TSP |
| | | | Fab | *A. thaliana* (callus) | 0.188% of TSP |
| | | | IgC | *A. thaliana* (callus) | 0.217% of TSP |
| α substance P (single-domain Ab) | CaMV 35S promoter/ nos terminator | *pel*B LP | | *Nicotiana benthamiana* | 1 % of TSP |
| IG of 38C13 mouse B Cell lympoma (scFv) | Viral vector TTO1A, a hybrid fusion of TMV and tomato mosaic virus, N-terminal: rice α-amylase LP | | | *N. benthamiana* (leaves) | 12.3–30.2 μg $g^{-1}$ FW |
| α Strerococcus mutans abhesin (IgA, IgG) | Ca MV 35S promoter/ nos terminator | Native mouse immunoglobulin LP The γ chain constant region was replaced by an α chain constant region | | *N. tabacum* | IgG: 7.7. α g $mL^{-1}$ Plant extract IgG1/A: 1.5 μg $mL^{-1}$ plant extact lgG2/A:2.1 μg $mL^{-1}$ plant extract |
| α *Streptococcus mutans* adhesin (slgA-G) | CaMV 35S promoter/ Nos terminator | Native mouse immunoglobulin LP, four transgenic lines expressing either the k, the hybrid lgA-G heavy chain, the murine J chain or the secretory component were crossed to obtain a sIgA experessing line | | *N. tabacum* | 200-500 μ $g^{-1}$ FW |
| α human (IgG)) | CaMV 35S promoter/ T7 terminator | 5'murine LP of HC and LC LC anc HC separrately transformed | | *Mddicago sativa* | 1% TSP (full-size IGG) |

**Table 7.1 Continued....**

| *Protein* | *Regulatory elements* | *Additional elements of the construct* | *Species/Tissue* | *Accumulation level* |
|---|---|---|---|---|
| α HCV-2 glycoprotein B (IgG) | CaMV 35S promoter/ nos termineator | LP of tobacco extensin | *Glycine max* | Not mentioned |
| Human hemoglobin (α + β) | CaMV enhanced 35S promoter/ CaMV 35S | Transit peptide of small subunit of Rubis CO of *Pisum sativum* | *N. tabacum* | 0.05% of seed protein |
| EGF (synathetic gene) | terminator CaMV 35S promoter/ nos terminator | Codon-optimized for high ,evel peptide expression in Escherichia coli | *N. tabcum* (leaves) | 0.001% of TSP |
| Human glutamic acid decarboxylase (hGAD) | CaMV enhanced 35S promoter/ TEV5′ UTR/nos terminator | | Low alkaloid tobacco *Salanum tuberosum* | 0.4% of TSP of leave or tuber |
| Isoform of hGAD (hGAD65) | CaMV 35S promoter/ nos terminator | | *N. tabacum* (leaves) | 0.01–0.04% of TSP |
| | | | *Daucus carota* (tap root) | 0.01% of TSP |
| | | | *N. Tabacum* (leaves) | 0.04% of TSP |
| | | $GAD67_{1-87}$ (rat)/$hGAD65_{88-585}$ Fusion | | 0.19%ofTSP |
| | Expression mediated by potato virus X infection | | *N. benthamiana* (leaves) | 2.2% of TSP |
| Human collagen Pra α (1) chain | CaMV 35S promoter/ CaMV 35S terminator | Pathogenesis-related protein (PR) LP or native IP | tobacco | >100 μg $g^{-1}$ tissue |
| Hepatitis B surface antigen | CaMV enhanced 355S promoter/ TEV 5′UTR/nos terminator | | *N. tabacum* (leaves) | 66 μg $g^{-1}$ TSP |
| | Patatin promoter | | *S. tuberosum* (tuber) | 1.1 μg $g^{-1}$ FW |
| | CaMV 35S promoter/TEV 5′ UTR/nos terminator | | | 0.33 μg $g^{-1}$ FW |
| | | Soybean vegetative storage protein LP (vspαS) | | 0.8 μg $g^{-1}$ FW |
| | | vspαS plus putative vacuolar targeting signal (vspαL) | | 2.4 μg $g^{-1}$ FW |
| | | Transit peptide of small subunit of RubisCO | | n.d. |
| | | KDEL | | 1.25 μg $g^{-1}$ FW |
| | CaMV 35S promoter/ΩTMV 5′UTR/nos terminator | | | 0.33 μg $g^{-1}$ FW |

**Table 7.1 Continued....**

| *Protein* | *Regulatory elements* | *Additional elements of the construct* | *Species/Tissue* | *Accumulation level* |
|---|---|---|---|---|
| | CaMV35S promoter/ Ω TMV 5'UTR/nos B terminator | | | 6.5 μg $g^{-1}$ FW |
| | CaMV/35S promoter/TEV 5'UTR/vsp B gene terminator | | | 16.0 μg $g^{-1}$ FW |
| Binding subunit of heat-labile enterotoxin B | CaMV35S promoter/TEV 5'UTR/vsp B terminator | With and without KDEL | *N. tabacum* (leaves) | 5 μg $g^{-1}$ TSP (without KDEL) 14 μg $g^{-1}$ TSP (with KDEL) |
| | | | *S. tuberosum* | 30 μg $g^{-1}$ TSP (without KDEL) 110 μg $g^{-1}$ TSP (with KDEL) |
| Cholera toxin B subunit | CaMV35S promoter nos terminator | Kozak seqence, KDEL | *Lycopersicon esculentum* | Leaves: 0.02% of TSP<br>Fruits: 0.04% of TSP |
| IL-10 | CaMV 35S promoter/ nos terminator | Native LP, with and without KDEL | *N. tabacum* | 800 ng $g^{-1}$ TSP (without KDEL) 55 μg $g^{-1}$ TSP (with KDEL) |
| IL-2 | Patatin promoter | Native LP | *S. tuberosum* | 115 units $g^{-1}$ FW |
| Enkephalins | Promoter and terminator of 2S1 albumin gene of *A. thaliana* | Integrated in 2S1 albumin of *A. thaliana* and flanked by tryptic cleavage sites | *A thaliana* (seed)<br>*Brassica napus* (seed) | 2.9% of TSP (200 nmol $g^{-1}$)<br>50 nmol $g^{-1}$ |
| T84.66 α CEA (scFv) | Legumin A seed specific promoter/Ω sequence TMV 5'UTR | Light chain LP of mAb24, KDEL | *P. sativum* (seed) | 9 μg $g^{-1}$ FW |
| T84.66 α CEA (diabody) | CaMV enhanced 35S promoter/ CHS 5'UTR/Pw 3'UTR of TMV/ | Light chain LP of mAb24, and without KDEL<br>CaMV 35S terminator | *N. tabacum* (leaves) | 1 – 5 μg $g^{-1}$ (apoplast/transient)<br>4 – 12 μg $g^{-1}$ (ER/ transient)<br>0.44 – 0.93 μg $g^{-1}$ (apoplast/stable)<br>3 – 9.3 μg $g^{-1}$ (ER/stable) |

**Table 7.1 Continued....**

| *Protein* | *Regulatory elements* | *Additional elements of the construct* | *Species/Tissue* | *Accumulation level* |
|---|---|---|---|---|
| Aprotinin | Maize ubiquitin promoter/ pinII terminator | Maize ubiquitin 5′intron, barley α-amylase LP | Extracellular matrix of embryo (maize seeds) | In T5: 0.35% of extractable seed proteins |
| Human serum albumin | CaMV enhanced 35S promoter/ AIMV RNA synthetic leader sequence/nos terminator | Native or PR-S leader peptide | *S. tuberosum* (leaves) | 0.02% of TSP |
| Hirudin | Oleosin promoter/ nos terminators | Fused to native oleosin | *B. napus* | 1% of seed protein |
| β-Interferon | CaMV 35S promoter/ CaMV 35S terminator | Native LP | *N. tabacum* | 170 ng $g^{-1}$ FW |
| α human Rhesus D (IgG) | CaMV 35S promoter (LC, HC and tandem construct)/E9 PolyA terminator of *P. sativum* | Murine immunoglobulin LP | *A. thaliana* | Cross of LC + HC: 0.6% of TSP, tandem : 0.3% of TSP |
| | *mas*1′2′ dual promoter (bidirectional)/nos and E9 terminator | | | Bidirectional: 0.12% of TSP |
| Dog gastric lipase | CaMV enhanced 35S promoter/ CaMV 35S terminator | LP of rabbit gastric lipase precursor | *N. tabacum* (leaves; | 7% of acid extractable proteins |
| | | LP of sweet potato sporamin precursor | *N. tabacum* (leaves) (vacuole) | 5% of acid extractable proteins |
| α Hepatitis B surface antigen (scFv) | CaMV 35S promoter/TMV Ω 5′UTR/nos terminator | | *N. tabacum* (cytosol) | n.d. |
| | | Sporamin LP | *N. tabacum* (apoplast) | 0.31 % of TSP |
| | | Sporamin pre-pro-peptide LP | *N. tabacum* (vacuole) | 0.032% of TSP |

**Table 7.1 Continued....**

| *Protein* | *Regulatory elements* | *Additional elements of the construct* | *Species/Tissue* | *Accumulation level* |
|---|---|---|---|---|
| | | Sporamin LP, KDEL | *N. tabacum* (ER) | 0.22% of TSP |
| α Hepatitis B surface antigen (scFV) | CaMV 35S promoter/TMV Ω 5'UTR/nos terminator | KDEL | *N. tabacum* (root exudates) | 630 – 760 ng $g^{-1}$ dry weight root/day |
| | | N-terminal ER LP from calreticulin | | |
| Human interferon (hIF-α2b and hIF-α8) | CaMV 35S promoter/ nos terminator) | Native LP of HIF-α2b and hIF-α8 resp. | *S. tuberosum* | 560 IU $g^{-1}$ tissue |

Abbreviations used in table 6.1: AIMV: alfalfa mosaic virus; CaMV: cauliflower mosaic virus; CEA: carcinoembryogenic antigen; CHS: chalcone synthase; Fab: antigen binding fragment; Fd: variable domain and first constant domain of the heavy chain; FW: freshweight; HC: heavy chain; hIF: human interferon; HSV: herpes simplex virus; IL: interleukin; LC: light chain; LP: leader peptide; mAb: monoclonal antibody; mas: mannopine synthase; n.d.: not detectable; nos: nopaline synthase; ocs: octopine synthase; *pel*B: pectate lyase; pin II: potato proteinase inhibitor II; PR-S: pathogenesis related protein S; Pw: pseudoknot wildtype of TMV; rbcS: ribulose-1, 5-bisphosphate carboxylase small subunit; RubisCO: Ribulose-1, 5-bisphosphate Carboxylase Oxigenase; scFv: single chain Fragment variable; TEV: tobacco etch virus; TMV: tobacco mosaic virus; TSP: total soluble protein; UTR: untranslated region; vspB: vegetative storage protein B.

been produced in plant cell cultures so far and the expression cassette elements used in each case. Possible contamination by chemical or biological substances is one of the most important concerns when producing pharmaceutical proteins. Plant cell cultures ensure the production of the desired protein in a controlled, sterile and sealed environment and can be adapted to cGMP conditions. Therefore, the risk of contamination is minimized and the production conditions can be modified more easily in a contained reactor than in the field.

Another advantage is the ability to freeze plant suspension cells in liquid nitrogen. So that master and working cell banks can be established, aprerequisitefor cGMP procedures. Although plant cell culture is not as cost effective as plant cultivation in the open field, it will become an economical process if higher protein yields can be achieved. The cultivation medium of plants is chemically defined, consisting of a carbon source, minerals, vitamins and phytohormones. Furthermore, it is protein-free and relatively inexpensive. In contrast, animal cells often require complex supplements such as fetal calf serum and/or expensive growth factors, although serum-free cultivation is possible in case of Chinese hamster ovary (CHO) cells. When trying to increase the yields of recombinant protein inplant suspension cultures, one should consider optimizing the nutrient supply and including product stabilizing agents.

Sharp and Doran showed that antibody accumulation in hairy root cultures was improved by increasing the dissolved oxygen tension to 150% air saturation and that loss of the antibody could be minimized by inhibiting protein transport in the secretory pathway with the antibiotic brefeldin A. The beneficial effects of stabilizing solutes were demonstrated among others by Magnuson *et al.* who achieved a 35-fold increase in scFv yield by adding polyvinylpyrrolidone (PVP) to the culture medium. Other proteins were stabilized to a lesser degree by supplementing the medium with the protease inhibitor bacitracin, bovine serum albumin (BSA), gelatin or dimethylsulfoxide (DMSO).

Fischer *et al.* reported a transient threefold increase in recombinant antibody levels after supplementing cell cultures with amino acids prior to harvest. All these studies show, however, that the benefits obtained by certain expression cassette elements or defined culture conditions cannot be generalized for all types of pharmaceutical proteins. These factors have to be investigated and optimized individually for every product. In order to facilitate purification of a recombinant protein from plant cell culture, the recombinant protein can be directed to the extracellular space (apoplast and culture medium) by attaching an appropriate signal sequence or using a native signal sequence if one is present. Only about 100-1000 endogenous proteins are thought to be secreted from cultured plant cells, so the purification of a recombinant protein from the culture medium is much easier than purification from whole cell extracts.

The plant cell wall is a natural barrier to protein secretion. Proteins >20 kDa tend to be retained although higher-molecular-weight proteins such as monoclonal antibodies are secreted. Secretion can be further enhanced by the addition of DMSO. Although cell cultures produce a much lower amount of biomass than plants cultivated in the open field, tobacco suspension cells have been cultivated at volumes of up to 20 $m^3$ and references therein).

## PROCESSING STEPS IN PLANTS

In order to make molecular farming commercially profitable, recombinant proteins must be produced at a sufficiently high yield and in an active form. It has become clear that, for

**Table 7.2. Pharmaceutical Proteins Produced in Plant Suspension Cultures**

| *Protein* | *Regulatory elements* | *Additional elements of the construct* | *Species* | *Accumulation level* |
|---|---|---|---|---|
| Human $\alpha_1$ antitrypsin | Promoter of rice a amylase gene *RAmy3D*/terminator of *RAmy3D* | LP of rice a amylase gene *RAmy3D* | *Orza sativa* | 4.6–5.7 mg $g^{-1}$ dry cell weight |
| | | Additional first intron of *RAmy3D* | | No significant increase |
| | | LP of rice α amylase gen *RAmy3D* | | 18.2–24.2 mg $g^{-1}$ dry cell weight |
| Human erythropoietin | CA MV 35S Promoter/CaMV35S terminator | Native LP | *N. tabacum* cv By-2 | 26ng$^{-1}$ TSP |
| α hepatitis B surface antigen-(scFv) | CaMV 35S Promoter/ nos terminator | Sporamin LP | *N. Tabacum* cv Havana SR | 1 mg $L^{-1}$ in culture supernatant and 5 mg $kg^{-1}$ wet weight from cell extracts |
| hIL-12 | CaMV enhaced 35S promoter | Native LP | *N tabcum* cv Havana SR | 175μ $L^{-1}$ |
| hIL-2, hIL-4 | CaMV 35S promoter/ T7 terminator | Native LP | *N. Tabacum* NT-1 | 0.1 mg $L^{-1}$ (hIL-2) and 0.18 mg $L^{-1}$ (hIL4) |
| Human Serum albumin | CaMV enhyanced 35S promoter/ AIMV RNA synthetic leader sequence/nos terminator | Native or PR-s LP | *N. tabacum* | 250 μg $g^{-1}$ TSP |
| Bryodin 1 | CaMV 35S promoter/ nos terminator | Extensin LP | *N. tabacum* NT-1 | 30mgL$^{-1}$ |
| hG-CSF | CaMV enhanced 35S promoter/ TMV Ω 5'UTR/nos terminator | | *N. tabacum* | 105 μg $L^{-1}$ extracellular |
| GM-CSF | CaMV35Spromoter/TEV5' UTR/T7 terminator | | *N. tabacum* cv BY-2 | 150 μg $L^{-1}$ intracellular and 250 ug $L^{-1}$ extracellular |
| | CaMV enhanced 35 S promoter/ TMV Ω 5'UTR/nos terminator | Native GM-CSF LP | *N. tabacum* cv Havana SR | 180 ug $L^{-1}$ extracellular, after adding of gelatin increase to 738 μgL$^{-1}$ |

Abbreviations used in table 2: AIMV: alfalfa mosaic virus; BY-2: *Nkotiana tabacum* cv Bright Yellow 2; CaMV: cauliflower mosaic virus; FW: freshweight; GM-CSF: granulocyte macrophage colony stimulating factor; hG-CSF: human granulocyte colony stimulating factor; hIL: human interleukin; LP: leader peptide; nos: nopaline synthase; PR-S: pathogenesis related protein S; RAmy 3D: rice amylase gene promoter 3D; scFv: single chain Fragment variable; TEV: tobacco etch virus; TMV: tobacco mosaic virus; TSP: total soluble protein; UTR: untranslated region.

high-level protein accumulation, the stability of transgene expression can be as important as the expression level itself. The quantity of protein is determined by the rate of protein synthesis, assembly as well as proteolytic degradation. Processes that influence the production of a stable and functional end product in plants and plant cell suspension cultures include transgene integration, transcriptional and translational activity, posttranslational modification. The first hurdle is stable transgene integration and expression in the genome of the transgenic plants or transformed cells, since transgenes are often silenced either immediately after integration or in subsequent generations.

A novel and very interesting method for dealing with this problem in transient gene expression is the use of tomato bushy stunt virus P19 protein. This is a known suppressor of posttranscriptional gene silencing (PTGS) and transient expression of a range of proteins was increased 50-fold in the presence of p19. Another approach is the inclusion of scaffold attachment regions (SARs) or matrix attachment regions (MARs) within the expression construct. Such elements, which attach to the nuclear matrix and may help divide interphase chromatin into functional domains, have been shown to enhance transformation efficiency to achieve stable gene expression and to increase transgene expression levels up to 650-fold although the level of enhancement appears to be dependent on the promoter. Increased transcription levels are assumed to result in increased protein synthesis. One approach to reach this goal is to raise the transgene copy number by the use of amplification-promoting sequences derived from a spacer sequence of tobacco ribosomal DNA. Posttranscriptional processes such as capping, splicing and polyadenylation are important for high protein yields, and it is also important to maximize mRNA stability. The synthesis of a protein may be hampered by suboptimal translation initiation or termination sites. Lukaszewicz *et al.* studied the context sequence of the initiation codon and Sawant *et al.* were able to increase the expression level of a reporter gene by inserting a sequence commonly found in highly expressed plant proteins.

Other problems can include the presence of cryptic splice or polyadenylation sites, and unfavourable codon bias (i.e. clusters of rare codons). If problems of this type are encountered, the construction of a synthetic gene can be considered. The artificial synthesis of a whole gene is expensive, but codon optimization at the 5′ end of the open reading frame can be sufficient to enhance gene expression. Batard *et al.* demonstrated an increase in wheat P450 enzyme activity in transgenic tobacco after optimization of the first 111 bp of the wheat open reading frame. This corresponds to 7% of the entire open reading frame. An useful alternative to nuclear transformation is the direct transformation of plastids, since plastid genomes are not affected by epigenetic silencing and they have a codon bias different to that of nuclear genes.

By creating transplastomic plants, very high levels of recombinant proteins have been obtained. In the case of pharmaceutical proteins, 7% total soluble protein (TSP) has been achieved for human somatotropin and 10% TSP has been achieved for human serum albumin (HSA). This exceeds the levels obtained through nuclear transformation, although the production of homoplastic plants is unfortunately time consuming and has only been achieved for a few species. Posttranslational modifications can have dramatic effects on the accumulation of a recombinant protein. If proteins are not properly cleaved, pro-proteins will accumulate instead of mature proteins.

For example, it has been shown that one of the causes for improper processing of soybean glycinins in potato tubers is the absence of the necessary proteases in the host plant. Some proteins need specific modifications to become active, but these are not carried out in plants. These include the addition of certain sugar molecules or the hydroxylation of proline for the production of native-like collagen in tobacco plants. Differences in glycan structures between plants and mammals (e.g. the presence of xylose and fucose residues in plants) have raised concerns regarding the potential immunogenicity of recombinant proteins, especially those used as human therapeutics. In order to obtain an antibody with a glycan profile similar to the mammalian type, Bakker *et al.* expressed human β-1, 4 galactosyltransferase in tobacco plants and crossed them with plants expressing a murine antibody. The result was a plantibody with partially galactosylated N-glycans. The folding of polypeptide chains and the assembly of multiple subunits are critical requirements when complex and multimeric proteins such as full size antibodies are expressed in plants.

Correct folding of the protein may be favoured by retention in the endoplasmic reticulum (ER), which can be achieved by adding a C-terminal K/HDEL ER-retention signal. This prolongs the exposure of the protein to endogenous chaperones like BiP (binding protein). Bouquin *et al.* expressed the heavy and light chains of an antibody individually in *Arabidopsis thaliana* plants. Strongly expressing parental lines were selected and crossed to obtain a hybrid line producing the full-size antibody. The multimeric antibody accumulated to higher levels than either of the individual chains in the parental lines, indicating that assembly of the mature protein had a stabilizing effect in comparison to the individual subunits. Expressing recombinant proteins as N-terminal fusions with ubiquitin is another strategy that can help to achieve proper folding.

Additionally, this is an elegant way to obtain proteins that do not start with a methionine residue, since the ubiquitin moiety is cleaved off by endogenous ubiquitin-specific proteases. If the desired protein is prone to proteolytic degradation, the use of appropriate targeting signals to direct the protein to different cellular compartments (e.g. the ER, chloroplast, vacuole or apoplast) or the use of tissue specific promoters (e. g. the hordein promoter for protein expression in barley grains) could be considered. Targeting is especially important if the recombinant protein is toxic to the production host. For example, targeting the non-pharmaceutical protein avidin to the cytosol in transgenic tobacco plants was toxic, but plants were regenerated successfully when this molecule was targeted to the vacuole. The expression of a secreted version of avidin in maize *(Zea mays)* resulted in partial or complete male sterility

## IMPROVED PROTEIN YIELDS

We will focus on recombinant antibodies (rAbs) and their derivatives (e.g scFvs, dia-bodies) in this section because these classic plant-produced pharmaceuticals have been expressed in many diverse production systems using a variety of regulatory elements. However, even in the case of this extensively studied class of proteins, direct comparisons between systems are difficult to make, since different genes, regulatory elements and production hosts have been used. Some investigators have compared the expression of a particular rAb in different plant species, or have targeted the antibody to different tissues or subcellular compartments in one particular host. Generally, such comparisons have revealed significant variation in expression and accumulation levels. Stoger *et al.* expressed the scFv antibody T 84.66, which recognizes the carcinoembryogenic antigen (CEA), and systematically compared several plant systems, regulatory elements and subcellular targeting signals.They concluded that the major factor determining overall accumulation levels was the stability of the antibody, which could be controlled in part by subcellular targetig.

In agreement with others, the highest levels of active antibody were obtained by targeting the protein to the secretory pathway using its native signal peptide, and then retaining the antibody in the ER lumen using a KDEL tag. Presumably because of the favourable biochemical environment, the lack of proteases and the presence of chaperones, rAbs accumulate to higher levels in the ER than in any other compartment. Furthermore, this strategy can limit the extent to which core glycans are replaced by plant-specific carbohydrates like α(1,3)-fucose and β(1, 2)-xylose.

Without ER retention, antibodies are secreted to the apoplast where accumulation *i* levels are typically 10 – 100-fold lower than in the ER. The cytosol appears to be an inappropriate compartment for antibodies, resulting in very low accumulation levels. An exception is described by De Jaeger *et al.* who targeted a phage display-derived scFv to the cytosol and achieved a level of 1% TSP. Stoger *et al.* also observed that between species, the amounts of scFv per unit fresh weight were in the same range, and did not correlate with the total protein content in the plant. For example, even though pea is a much more proteinaceous crop than rice, the amounts of antibody measured as a percentage of TSP were considerably leach species, individual promoters resulted in distinct, tissue-dependent accumulation patterns. The cauliflower mosaic virus (CaMV) 35S promoter, for example, led to high-level accumulation in callus and leaves whereas the maize ubiquitin-1 promoter was the best choice for producing recombinant proteins in cereal seeds even though it is not in itself seed-specific.

The lack of such comparative studies for proteins other than rAbs makes_it_ difficult to generalize an optimal expression strategy for all proteins. List recombinant proteins expressed in plants and provide details of the production system, promoters and other regulatory elements used in each case.

## DOWNSTREAM PROCESSING

When choosing a plant expression system, downstream processing should be taken into account since this will contribute a substantial proportion of the overall costs. A cost distribution for the production of β-glucuronidase in maize has been made by Evangelista *et al.* This is not a pharmaceutical protein, but the principles of cost analysis should hold true anyway. It was stated that protein extraction and purification would account for 40% and 48%, respectively, of the total annual operating costs, assuming a product purity of 83%. When higher purity is needed, as for pharmaceutical proteins, downstream processing could account for an even larger proportion of the total costs. Various strategies have been proposed to simplify downstream processing. We have already mentioned the use of tissue-specific promoters or targeting signals that allow protein secretion or accumulation in a particular organelle.

A useful example is protein targeting to oil bodies. which is used in combination with the oleosin-partitioning technique for protein isolation from oilseed crops. For new products that have yet to undergo clinical testing, the addition of an affinity tag can be considered to simplify protein isolation. Such tags have unique properties, which allow fusion proteins to be isolated from cellular extracts or media by affinity to particular ligands.

Alternatively, the tag may confer a novel physical property on the protein. An example is the synthetic elastin-like polypeptide (ELP) tag. When a temperature transition occurs, the tags fold into spirals that aggregate. These aggregates can easily be isolated by centrifugation. However, if such tags are used, they must not adversely affect the function of the protein. Importantly, the presence of affinity tags on pharmaceutical proteins may result in non-compliance with regulatory issues. A minimalist extraction buffer is often recommended for protein purification since most additives provide only a marginal improvement in yields, but will increase costs significantly.

The removal of phenolics, for example by tangential-flow ultrafiltration/diafiltration is an important step that should be carried out as early as possible in the purification procedure, since these molecules can become covalently linked to amino acid side chains and can oxidize certain residual groups. When using plant cell cultures for the production of biopharmaceuticals, targeting the protein to the apoplast (and from there to the culture medium) is beneficial for later purification. Plant cell culture media are protein-free, which makes the purification of the product relatively easy. Unfortunately, secreted recombinant proteins are often unstable in the cell culture medium. One strategy to overcome this problem was presented by James *et al* in form of an affinity chromatography bioreactor. By initiating protein purification while the cells are still growing, the exposure of the protein to destabilizing agents in the culture medium is minimized.

## PLANT-DERIVED PHARMACEUTICALS

The production of pharmaceutical proteins in plants has clear advantages over traditional systems in terms of cost-efficiency and product safety, since there is no risk of contamination with human pathogens. Furthermore plants are much less likely than mammalian cells to be affected by the expression of certain human proteins, such as growth factors and cell cycle inhibitors. Therefore, plants provide a strategic complement to existing microbial and animal production systems. Antibodies account for more than 20% of all biotechnology-derived molecules, and the market is expected to reach $5 billion by 2005. More than 250 companies are working on over 700 therapeutic antibodies, of which 220 are in clinicaltrials. These numbers demonstrate the ongoing commercial expansion of pharmaceutical protein production and indicate a growing demand for production capacity.

It has been estimated that plant-derived antibodies can be produced in a 250 $m^2$ greenhouse at about one tenth of the cost of hybridoma-derived antibodies. Planet Biotechnology Inc. presents an even rosier picture, estimating that the cost of plant-derived antibodies will be 5 % of the cost of antibodies produced in animal cells. This calculation, however, was based on yields of 500 $\mu g\ g^{-1}$ leaf fresh weight, a rather ambitious level for antibody production in plants. Once again, the greatest proportion of the cost reflects downstream processing and purification. When purification is needed, it is estimated that a recombinant protein must be expressed at greater than 0.01% of fresh weight or 1% of TSP for molecular farming to be commercially feasible. To date, several antibodies with therapeutic potential have been produced in plants. Among them are the Guy's 13 antibody that recognizes *Streptococcus mutans* (the causative agent of dental caries), humanized antibodies that recognize viral antigens (herpes simplex and respiratory syncitial virus), an antibody against CEA and a tumor-specific vaccine for the treatment of lymphoma.

Although none of these antibodies is produced commercially at the current time, the lymphoma scFv has completed phase I clinical trials and the Guy's 13 antibody has reached phase II clinical trials in the USA *(www.planetbiotechnology.com/products.html)*. Regarding other pharmaceutical proteins, enterotoxin B and aprotinin have both reached phase I trials, while gastric lipase is undergoing clinical phase IIa trials in France and in Germany *(www.meristem-therapeutics.com/GB/intro.htm)*. Although no plant-derived pharmaceutical protein is yet available on the market, two technical proteins (avidin and β-glucuronidase) are successfully produced in corn for commercial purposes.

## CONTAINMENT STRATEGIES

The issue that concerns people the most, with respect to the safety of plant biotechnology, is the possibility that transgenes could spread in the environment by out-crossing. Methods to contain foreign genes within transgenic plants are reviewed by Daniell and include maternal inheritance, male or seed sterility and induced transgene excision before flowering. Chloroplast genetic engineering takes advantage of the fact that the chloroplasts of most crops are maternally inherited. Male sterility can be achieved by crossing plants that produce the pharmaceutical protein of interest with plants that are male sterile or through the use of genetic methods to prevent the production of functional pollen.

Seed sterility can be achieved using Monsanto's Terminator Technology™ or the recoverable block of function systern. The Cre-loxP svstein has been used for transgene excision. Another strategy is the cultivation of transgenic plants in greenhouses. Although this increases production costs considerably, it could still be the cheapest solution overall. Khoudi *et al.* calculated the cost of 1 g of purified antibody produced in greenhouses, and found that it was only 10% of the cost of hybridoma technology. A further advantage of molecular farming in greenhouses is that the product itself would be contained, a desirable outcome in the case of biopharmaceuticals that are potentially harmful, capable of persisting in the environment or of accumulating in non-target organisms. Such safety issues came to the fore after the ProdiGene fiasco in late 2002, prompting the US Department of Agriculture in March 2003 to tighten its permit conditions relating to the field-testing of plants engineered to produce pharmaceutical and industrial compounds.

Another means to improve safety standards is the use of inducible promoters, which could be used to delay protein expression until the crop has reached the factory. This would also eliminate protein degradation problems caused by the delay between harvesting and processing. Selectable markers are an essential part of the transformation construct, since they allow the selective propagation of transformed cells.

Antibiotic resistance is the most widely used selectable phenotype, but there are concerns that resistance genes could be transferred into bacteria and might spread in the environment. For example Mercer *et al.* reported that oral bacteria can be transformed with naked DNA in human saliva. The persistence of DNA in the field was reviewed recently by Dale *et al.* who stated that there is no evidence for the incorporation of functional plant (trans)genes into the meat products of animals fed with genetically-modified feed. Nevertheless, the use of such markers has been legally challenged in Europe, and the commercial use of antibiotic resistance markers will be prohibited from 2005 (EU Directive 2001/18/EC released 21/03/2001). Other types of marker gene may be used, such as metabolic markers that allow the use of an alternative carbon source or growth-regulation markers that express enzymes involved in the biosynthesis of plant hormones. Another approach is the use of removable markers. For a review on these selection procedures see. Also, a recent publication proves the feasibility of marker-free transformation, which must be regarded as the best possible technique to prevent an unintended spread of antibiotic resistance genes.

## CONCLUDING REMARKS

Numerous issues must be taken into account in order to make molecular farming a commercial success. These issues include the yield and quality of the target protein, downstream processing methods, storage and purification, transgene containment and the cost of production. Currently, there are few examples of plant-derived pharmaceutical proteins that have been produced in a manner that addresses such concerns. We are facing a growing demand for therapeutics and diagnostics, but almost the entire capacity for cell-based production is already in use, leaving little room for the estimated 50-60 new protein pharmaceuticals that will be approved over the next 6-7 years. A shift to plant-bioreactors could therefore become a necessity. Although molecular farming still faces a number of challenges, one can clearly see the benefits and possibilities of the technology. Further research is needed to meet these challenges and facilitate the cost-effective and safe production of pharmaceuticals in plants.

CHAPTER

# 8 Identification of Neurons

As initiators of immune responses, dendritic cells (DCs) play a critical role in the defense against infectious agents. They are uniquely able to activate naive T cells, which are required for the initiation of adaptive immune responses to invading organisms. Optimal DC function requires appropriate development from bone marrow precursors, migration to peripheral lymphoid and nonlymphoid surveillance locations, maturation (including antigen processing/ presentation, costimulatory molecule expression, cytokine production) after encounter with pathogens and migration to lymph nodes and/or spleen to interact with T cells.

Langerhans cells (LCs), an epidermal DC resident population, are one well studiea example of a peripheral DC subset that participates in antigen transport to and T-cell activation in peripheral LN. In addition, some DCs are able to suppress or down-regulate immune responses and thus serve an immunoregulatory role. Ethanol exposure has the potential to interfere with many of these DC functions and thereby increase the organism's susceptibility to infection. The aim of this chapter is to briefly review what is known about the effect of EtOH exposure on DC function, and to outline methodologies that are applicable to studying DC function following chronic EtOH exposure in vivo, in murine models.

## Approaches to Studying DC Function After EtOH Exposure

It is impossible to precisely mimic in vitro the daily variability of EtOH exposure levels, local EtOH metabolite concentrations, or the microenvironment in which DCs normally reside in the intact animal. Thus, it is most relevant to study the effects of in vivo EtOH administration on DC function rather than exposing freshly isolated DC, or those generated from untreated animals, to EtOH in vitro. In some circumstances, no in vivo methods exist to adequately address the DC function under consideration. In these cases, in vitro experiments that measure events initiated in vivo or within a few hours of culture preparation (during which time DCs might be expected to maintain the deficits induced by EtOH in vivo) are reasonable surrogates.

## Effect of Chronic EtOH Exposure on DCs

Very few studies have addressed the effect of chronic in vivo EtOH exposure on DC function. Some clues about the effect of chronic EtOH on DC function can be had from studies of unpurified "APC" function from the spleen. Macrophage phagocytosi is reduced by EtOH exposure. Ag presentation and immunostimulatory cytokine production of APC from EtOH mice is also reduced and defects are observed in the afferent phase of DTH in EtOH mice. EtOH mice also show decreased serum levels of inflammatory cytokines produced by DC and macrophages, (such as interleukin (IL-6), tumor necrosis factor-α, and IL-1.

Antigen-presenting cell (APC) functions can be ascribed to a number of different cell types, including monocytes, macrophages, B cells, and DCs. Only one study has specifically investigated the effects of chronic in vivo EtOH exposure on DC function and demonstrated decreased allostimulatory capacity and lower levels of costimulatory molecule (CD86) and major histocompatibility complex (MHC) class II expression on EtOH exposed DC, particularly after inflammatory (CpG) stimulation.

# METHODS

Because of the advantages of an in vivo model of EtOH administration for determining pathophysiologically relevant effects on DC function, this chapter focuses on protocols that rely on this approach. All techniques have been performed after in vivo EtOH administration using the Meadows-Cook model described in the chapter by Coleman, et al. in this volume.

# MATERIALS

## Identification and Isolation of DCs and LCs

### *Purification of CD8α+ and CD8α– DCs from the Spleen*

1. Dense BSA Solution: 186mL of phosphate-buffered saline (PBS), 29 mL of 1 N NaOH in 65 mL of water. Carefully layer 106 g of bovine serum albumin (BSA; Intergen, Purchase, NY) onto the surface of the solution, cover beaker with foil and place at 4°C overnight. Check density by measuring refractory index; correct BSA density of 1.08g/mL corresponds to an index of 1.384 – 1.385 at 25°C. To reduce the refractive index by 0.0005 add 5mL of PBS or to increase the index by 0.0005, add 5g of BSA. The pH of the solution should be 7.0 – 7.4. Filter-sterilize the solution using a self contained filtering unit. Dense BSA may be replaced by Nycoprep 1.077A (cat. no. 10122311).
2. Nycodenz buffer: 9g NaCl, 0.6055g Tris, 0.22368 g KC1, 0.11167g ethylene diamine tetraacetic acid (EDTA); bring to 1.12L with deionized (DI) water.
3. Nycodenz stock solution: add exactly 15.275g of Nycodenz powder to 50mL of DI water. Filter and keep in the dark at 4°C.
4. MACS buffer: PBS, 0.5% BSA, 500 mM EDTA.
5. HT-2 Media: Iscove's DMEM IX 439mL, 10% fetal calf serum (PCS) 50mL, L-glutamine (0.292mg/mL $H_2O$) 5mL, Na Pyruvate (0.11 mg/mL $H_2O$) 5mL, gentamycin sulfate (0.5 mg/mL) 0.5 mL, 2-ME (0.05 m*M*) 500μL.
6. Anti-CD8α + Microbeads (Miltenyi, Auburn, CA, cat. no. 120 – 000 – 298).
7. Anti-CD11c+ Microbeads (Miltenyi, cat. no. 130 – 052 – 001).
8. LS Columns (Miltenyi, cat. no. 130 – 042 – 401).
9. Collagenase.

9. Hanks' Balanced Salt Solution (HBSS, 10X) without sodium bicarbonate, calcium and magnesium.
11. Complete media: Iscove's DMEM IX, 439 mL, 10% FCS (final concentration) 50mL, L-glutamine (29.2mg/mL $H_2O$ stock cone.) 5mL, Na pyruvate (100 m*M* stock cone.) 5mL, Pen/Strep (10μL/mL, 10μg/mL stock cone.) 5mL, nonessential amino acids (100X stock cone.) 5 mL, HEPES, 5 mL (1*M* stock concentration), 2-ME (89.3 m*M* stock cone.) 500 μL.

## Isolation of LCs for Flow Cytometry from Mouse Ears

1. Trypsin (Amresco, Solon, OH; cat. no. 0458 – 25G). Prepare 0.5% solution in PBS fresh on the day of the assay.
2. Balanced salt solution (BSS): 10x BSS stock solution 1 (10g dextrose, 0.6g $KH_2PO_4$, 3.58g $Na_2HPO_4{\bullet}7H_2O$ [1.89g anhydrous], 0.1 g phenol red powder or 20mL of 0.5% phenol red solution), 10 x BSS stock solution 2 (1.86g $CaCl_2{\bullet}2H_2O$ [1.41 g anhydrous], 4g KC1, 80g NaCl, 2g $MgCl_2{\bullet}6H_2O$ [1.04g anhydrous], 2g $MgSO_4$ $\bullet 7H_2O$ [1.74 g anhydrous]). To make a working solution, combine 50mL of stock solution #1 with 50 mL of stock solution #2 and bring to 500 mL with deionized water.
3. Ficolite-LM (Atlanta Biologicals, Lawrenceville, GA; cat. no. 140650).
4. Staining buffer: 475 mL BSS, 25 mL bovine calf serum-defined, 0.5 g $NaN_3$
5. 60-mm Petri dish (Falcon; cat. no. 351007).
6. Fine forceps (VWR; cat. no. 25607–856 for straight, 25607–890 for curved).
7. 70-μm filter (Falcon; cat. no. 352350).
8. Glass microscope slides with frosted ends.
9. Dissecting microscope.

## Purification of Pulmonary DC Subsets by FACS

1. Iscove's DMEM Ix.
2. Collagenase.
3. DNAse 1.
4. 0.5 M EDTA.
5. Dulbecco's modified Eagle's medium (DMEM).
6. 03 – 200/48 Nitex Mesh (Sefar, Kansas City, MO; cat. no. F03A0020048I043A02).
7. MACS buffer: PBS, 0.5% BSA, 500 mM EDTA.
8. Polypropylene round-bottom tubes (Falcon, Franklin Lakes, NJ; cat. no. 352063).
9. CDllc MACS Beads (Miltenyi Biotec, Auburn, CA; cat. no. 120 – 000 – 322).
9. FCS.
11. Cell Strainer (BD, Bedford, MA; cat. no. 352350).
12. LS Column (Miltenyi Biotec; cat. no. 130 – 042 – 401).

## Migration of DCs and LCs

### *Immunofluorescent Staining of LCs in Epidermal Sheets*

1. 0.5*M* ammonium thiocyanate: add 19 g of ammonium thiocyanate to 500mL of 0.1*M* phosphate buffer (pH 6.9) or PBS. Store at 4°C.
2. Blocking buffer: 1 mL of goat serum (Invitrogen, Carlsbad, CA; cat. no. PCN5000), 1 mL of 10X PBS, 8mL of deionized water. Mix, store at 4°C.
   (*a*) Alternate blocking buffer: 8mL of 100m*M* TBS, pH 7.3 (6.06 g of Tris base, 400 mL of deionized $H_2O$. Adjust pH to 7.3 with concentrated HC1. Adjust volume to 500 mL, then add 0.9 g of NaCl), 1 mL of rat antimouse FcγRII/III (clone 2.4G2), ImL of goat serum, 0.2g of BSA (Amresco, Solon, OH; cat. no. 0332-100G), 5μL of Tween 20.
3. Acetone.
4. Biotinylated or PE conjugated anti-mouse MHC class II.
5. Avidin-Cy5 (Jackson ImmunoResearch Laboratories, Inc., West Grove, PA; cat. no. 016–170–84) or Avidin-Cy3 (Southern Biotechnology Associates Inc. Birmingham, AL; cat. no. 7100–09L) if a biotinylated antibody is chosen in step 4.
6. Petri dishes (60 × 15 mm, BD Falcon, San Diego, CA; cat. no. 351007).
7. 24-well tissue culture plate (Corning Inc. Life Sciences, Acton, MA, Costar; cat. no. 3524) or 96 well plate (Costar cat. no. 3596).
8. Fine forceps.
9. Vecta Shield Slide Mounting Media (Vector Labs, Burlingame, CA cat. no. H-1000).
9. Microscope slides (Surgipath Medical Industries, Richmond, IL; cat. no. 00240).
11. Slide cover slips, 22 x 22 mm.
12. Clear fingernail polish.
13. Drugstore depilatory, *e.g.*, Nair.
14. Dissecting microscope.
15. Fluorescent microscope.

### *FITC Tracking of LCs and Dermal DCs to Lymph Nodes*

1. BSS.
2. Liberase blendzyme 3 (Roche Diagnostics, Chicago, IL; cat. no. 1814176)/DNase (Roche Diagnostics, cat. no. 104 159) working solution.
   (*a*) Stock Liberase solution: dissolve Liberase in RPMI 1660 + L-glutarnine (Invitrogen, Carlsbad, CA; cat. no. 1 1875) at 1 mg/mL
   (*b*) Stock DNase solution: dissolve DNase in RPMI at 4000 U/mL
   (*c*) Working solution: 25 μL of Liberase stock solution, 10 μL of DNase stock solution, 20μL of bovine calf serum, 945 μL of RPMI. Make fresh on day of use, and keep on ice.

3. Fixative for mouse cells: 12.5 mL of 10 × PBS, 2.7 mL of 37% formaldehyde solution, 200 mg of $NaN_3$. Bring to a final volume of 100mL.
4. n-butyl phthalate.
5. Acetone.
6. FITC.
7. Cytofix/Cytoperm solution (Becton Dickinson, San Diego, CA; cat. no. 554722).
8. Perm/Wash Buffer 10 x (Becton Dickinson, cat. no. 554723). Dilute to Ix before use, according to manufacturer's instructions.
9. Polyclonal unconjugated rabbit antimouse langerin (Imgenex, San Diego, CA; cat. no. IMG-5102A).
9. Cy5-conjugated AffniPure goat antirabbit IgG (H + L) (Jackson Immuno-Research, West Grove, PA; cat. no. 111 – 175 – 144).
11. Anti-CD 11c (and other antibodies to cell surface antigens of interest).
12. 12-well tissue culture tray (Corning Inc. Life Sciences, Acton, MA; cat. no. Costar 3512).
13. 18- and 23-gage needles.
14. Glass vials.
15. 1-mL syringes.
16. Scalpels (1 per lymph node).

### *Assessment of rDC Migration from the Lungs to the Lymph Nodes/ by i.n. CFSE Labeling*

1. 25 *mM* CFSE stock (Molecular Probes, Eugene, OR, cat. no. C1157): 25 mg dissolved in 1783.6μL of dimethyl sulfoxide (DMSO; stock solution).
2. Iscove's DMEM IX.
3. Anti-CD11c APC mAb (Caltag, Carlsbad, CA; cat. no. MCD11c05).
4. FACS buffer: 2% FCS, 0.02% sodium azide in PBS.

## Antigen Presentation by DCs

### *DC Activation of Naive CD8 T Cells In Vitro*

1. CD8α+ Microbeads (Miltenyi, Auburn, CA, cat. no. 120-000-298).
2. Anti-CD8β mAb biotinylated (Biolegend, San Diego, CA; cat. no. 118004).
3. Strepavidin PerCP (BD, San Jose, CA; cat. no. 554064).
4. CFSE (25*M* in DMSO; Molecular Probes, Eugene, OR; cat. no. C1157).
5. Bovine calf serum.
6. Complete medium: Iscove's DMEM IX, 439mL; 10% PCS (final concentration), 50mL; L-glutamine (29.2mg/mL $H_2O$ stock cone.), 5mL; Na pyruvate (100m*M* stock cone.), 5mL; Pen/Strep (10μL/mL, 10μg/mL stock cone.), 5mL; nonessential amino acids (100X stock cone.), 5mL; HEPES, 5mL (1*M* stock cone.); 2-ME (89.3 m*M* stock cone.), 500μL.

## *Influenza Infection of DCs*

1. Infect 1 × $10^6$ purified DC/mL at the desired MOI (5, 10, etc.) by adding virus diluted in Iscove's media without serum to cultured cells.
2. Incubate cells and virus on ice for 30min to allow virus to attach to the cells and then continue the incubation at 37°C for 30min to allow infection.
3. Centrifuge cells at 460 g (1600rpm) for 10min at 4°C. Remove any remaining virus in the supernatent and resuspend cells in complete media containing serum.
4. Incubate for 18 h at 37°C.

## *CDs T Cell Activation,*

1. Purify influenza specific HA Clone-4 CD8+ T-cells from the spleens of naive Clone-4 mice using CD8∞ microbeads.
   (*a*) Prepare single-cell suspension (spleen screen, slides, etc.) in Iscove's DMEM.
   (*b*) Centrifuge at 460 g (1600rpm) for 10min at 4°C and resuspend cells in MACS buffer at 90μL/$10^7$ cells.
   (*c*) Add 5μL CD8α+ beads per $10^7$ cells and incubate at 4°C for 10 min (refrigerate not on ice). Add 2mL of Iscove's DMEM.
   (*d*) Centrifuge at 460 g (1600rpm) for 10 min at 4°C and resuspend cells in MACS buffer at 10 × $10^7$ cells/mL.
   (*e*) Prewet column with 3 mL of MACS buffer solution.
   (*f*) Place cells over column (max of 5 lungs per column) and collect negative fraction.
   (*g*) Wash with 9 mL of MACS buffer.
   (*h*) Remove column from magnet and elute positive fraction with 6mL of media (2 mL of gravity, 2 mL of gentle plunge, 2 mL of forceful).
2. Centrifuge cells at 460 g (1600rpm) for 10min at 4°C and resuspend at 10 × $10^7$ T-cells/mL in Iscove's DMEM.
3. Label purified T cells with 1.5 μM CFSE. Incubate 10 min at 37°C. At the conclusion of the incubation, quench CFSE by adding an equal volume of calf serum.
4. Purify desired DC populations by MACS or FACS sorting and then infect as described previously.
5. Coincubate purified T cells and DCs at a 5:1 ratio (*i.e.* 100,000 T celsl:20,000 DCs) in 200μL of complete media at 5% $CO_2$ at 37°C for 48-72 h.
6. Harvest cells, stain for CD8 β with anti-CD8β biotinylated (clone:CD8b) and strepavidin PerCP and measure CFSE dilution by flow cytometry.
7. Method can be combined with staining for T cells activation markers such as CD25, CD69, etc.

## *DC Activation of Antigen-Specific T-Cell Proliferation in Vivo*

1. OTT or OTII mice as T-cell source (Jackson Laboratory, Bar Harbor, ME cat. no. 003831 or 004194)

2. C57B1/6 mice as DC source (donor) and as recipient of cell transfers.
3. 5mM stock solution of CFSE (5-(and -6) carboxyfluorescein diacetate succinimidyl ester, Molecular Probes, Eugene, OR; cat. no. c1157) in DMSO (store at –20°C, stable for 6mo.).
4. 10% PCS in PBS.
5. Super McCoy's medium: 85 mL of McCoy's 5A medium modified Ix with L-glutamine (Invitrogen, Carlsbad, CA; cat. no. 16600–082); 2mL of FCS; 2mL of essential amino acids 50x (Invitrogen, cat. no. 111–30–051); 1 mL of nonessential amino acids 100x (Invitrogen, cat. no. 11140); 1 mL of sodium pyruvate 100x (Invitrogen, cat. no. 11360); 1 mL of MEM vitamins 100x; 2mL of penicillin/streptomycin/glutamine 50 × (Invitrogen, cat. no. 5140 – 122); 50μL 0.1 *M* 2-mercaptoethanol.
6. Liberase/DNAse working solution.
7. OVA 258-264 peptide (for OTI assays) or OVA 323-339 peptide (for OTII assays) stock solution, 2mg/mL in sterile distilled water (Bio-Synthesis, Lewiston, TX; cat. no. 51023 –10).
8. 0.5% BSA (Amresco, Solon, OH; cat. no. 0332 – 100G) in PBS.
9. Anti-CD11c microbeads (Miltenyi Biotec, Auburn, CA; cat. no. 130 – 052 – 001).
9. Anti-mouse CD4 and CD8 antibodies, conjugated with a fluorochrome other than FITC.
11. 60-mm Petri dish (Falcon; cat. no. 351007).
12. Glass microscope slides with frosted ends.
13. ImL of insulin or tuberculin syringes.
14. 1-mL syringe with 23-gage needle.
15. AutoMACS separator

## Allostimulatory Assay of DC Function

1. BALB/c mouse as source of responder T cells.
2. C57B1/6 mouse as source of DCs in first stimulation.
3. BALB/c, C57B1/6 and C3H mice as sources of DCs in restimulation.
4. Super McCoy's medium.
5. 0.5% BSA (Amresco, Solon, OH; cat. no. 0332-100G) in PBS.
6. Liberase/DNase working solution.
7. Fico-Lite LM (Atlanta Biologicals, Lawrenceville, GA; cat. no. 140650).
8. Anti-CD 1 Ic microbeads (Miltenyi Biotec, Auburn, CA; cat. no. 130–052–001).
9. Anti-biotin microbeads (Miltenyi Biotec, cat. no. 130–090–485).
9. Biotinylated anti-B220, DX5, GD11c, GD11b, and Ter1 19 antibodies.
11. $^3$H thymidine (GE Healthcare, Piscataway, NJ; cat. no. 13-TRA120).
12. 96 well plate (Corning Inc. Life Sciences, Acton, MA; Costar, cat. no. 3596).
13. T25 culture flasks (Costar, cat. no. 430639).
14. 60-mm Petri dish (Falcon cat. no. 351007).

15. Glass microscope slides with frosted ends.
16. 1-mL syringe with 23-g needle.
17. AutoMACS separator
18. Scintillation counter.
19. Cell harvester (Cambridge Technologies, Cambridge, MA) and associated cell harvesting supplies.

# METHODS

## Identification and Isolation of DCs and LCs

### *Purification of Mature Bulk Splenic DCs*

1. Prepare collagenase solutions in HBSS with $Ca^{2+}$ and $Mg^{2+}$. (Dilute to 100U/mL. Also make 5 mL of 400U/mL dilution).
2. Collect 10 spleens (on Petri dish). Inject 1-2 mL of 100U/mL collagenase solution into each spleen.
3. Place the injected spleens into Petri dish. Collect the cells that were released in a 50-mL tube containing 10mL of ice-cold HBSS and keep on ice.
4. Add 6-8 mL of collagenase solution (100 U/mL) into the dish. Tease the spleen into very small pieces. Transfer released cells to collection tube and rinse twice more with 10mL fresh collagenase (100U/mL). Pipet up and down vigorously.
5. Add the high concentration collagenase solution (400U/mL) and incubate at 37°C for 30min.
6. Pipet up and down vigorously and mash dissociated tissue through the strainer.
7. Pool all the cells, filter through cell strainer and centrifuge at 460 g (1600rpm).
8. Resuspend cell pellet in dense BSA (no foaming!) Transfer 5mL of BSA containing cells from spleens into a 15-mL tube.
9. Overlay with 2mL of $Ca^{2+}$ $Mg^{2+}$-free HBSS. Spin for 15min at 2192 g (3300 rpm) without braking.
9. Collect low-density fraction (all of overlaid medium and upper 1 mL of BSA) with Pasteur pipet. Transfer into a 15-mL tube and fill with $Ca^{2+}$ $Mg^{2+}$-free HBSS, mix well, and centrifuge at 460 g.
11. Wash cells again in culture media and centrifuge at 300 g.
12. Plate low-density cells into 60-mm tissue culture dish in complete culture medium for 90min (37°C).
13. Rinse off nonadherent cells using a 10-mL pipet with prewarmed media (HT-2).
14. This wash must be done thoroughly. Pay attention to rims of the dishes. Aspirate the remaining media.
15. Repeat wash.
16. Culture adherent cells overnight (37°C).

17. Next day, collect nonadherent cells (>80% DCs, depends on thoroughness of wash steps 12–13).

## *Purification of Mature CD8α and CD8α⁺ DCs from Splenocytes by AutoMACS*

1. Follow purification protocol for purification of bulk DCs through differential adherence.
2. Harvest bulk DCs off of the differential adherence plates (wash quadrants 3 times/2 washes pre-warmed media). Spin at 460 g (1600 rpm) 10 min 4°C.
3. Remove supernatant, and resuspend cells in 30μL of anti-CD8α (Ly-2)-Microbeads. Bring to 3 mL with MACS buffer.
4. Incubate at 4°C, for 15min (refrigerate, not ice). Spin 460 g (1600rpm) 10min 4°C.
5. Resuspend cells in 3mL of MACS buffer, and run on autoMACS under "posselds" program.
6. Positive select the CD8α population with anti-CD11c (N418)-Microbeads by repeating steps 3-5, and keep the GD11c⁺ fraction.

## *Manual MACS Purification of Mature CD8α and CD8a⁺ DCs from Splenocytes*

1. Follow purification protocol for purification of bulk DCs through differential adherence.
2. Harvest bulk DCs off of the differential adherence plates (wash quadrants 3 times/2 washes with pre-warmed media). Spin 460 g (1600rpm) for 10min at 4°C.
3. Remove supernatant, and resuspend cells in 30μL of anti-CD8α (Ly-2)-Microbeads. Bring to 3 mL with MACS buffer.
4. Incubate at 4°C (refrigerator, no ice), for 15min. Spin at 460 g (1600rpm) for 10 min at 4°C.
5. Prewet LS+ column with 3 mL of MACS buffer solution during the last 3 min of spin.
6. Resuspend cells in 3 mL of MACS buffer, and pass cells through LS column, wash with 9 mL of MACS buffer.
7. Remove column from magnet and elute CD8α+ cells with 6mL of MACS buffer (2mL gravity, 2m L gentle plunge, 2mL forceful).
8. Repeat steps 3–6 on CD8α⁻ or cells.
9. Pool CD8α⁺ cells, and then positive select the CD8α population with anti-CD11c Microbeads. Repeat steps 3 through 6, and keep the CD11c⁺ fraction.

## *Isolation of LCs for Flow Cytometry from Mouse Ears*

1. Obtain ears from the mouse. With forceps, split each ear into dorsal and ventral halves.
2. Lay the ears in a 60-mm Petri dish so that the epidermal sides are facing up.
3. Pour 0.5% trypsin in PBS over the ears to submerge the ears in the Petri dish and place them in a 37°C water bath on a platform for 60 min.

4. Rinse the ears in BSS. Under the dissecting microscope, with fine forceps carefully peel off the epidermis taking care to preserve it as a sheet.
5. Prepare a single-cell suspension of the epidermis by gently rubbing it between the frosted ends of two microscope slides. Rinse the slides with BSS.
6. Pass the primary suspension through a 70-µm filter.
7. Transfer the suspension to 15-mL conical centrifuge tube.
8. Fill the tube with BSS. Centrifuge at 400 g at 4°C for 7min.
9. Resuspend the pellet by adding 3 mL of staining buffer, underlayer with 3 mL of Ficolite.
9. Centrifuge at 1700 g at room temperature for 20min.
11. Harvest cells from interface, and place them in a new 15-mL centrifuge tube.
12. Fill the tube with BSS. Centrifuge at 400 g, at 4°C for 7 min.
13. Resuspend the pellet. These cells are ready for study by flow cytometry.

## *Purification of Pulmonary DC Subsets by FACS*

All steps and media to be used in a sterile manner.

1. Remove lungs from naive or infected mice.
2. Split the lungs into groups of two lungs/tube in Iscove's media.
3. Digest lungs with 1 mg/mL of collagenase and 0.02mg/mL of DNase 1 in 7.5 mL for 15 min at 25°C in rotator at 18 g (300rpm).
4. Prepare single cell suspension (spleen screens, slides, etc.). Note: remove debris by passing through Nitex mesh.
5. Continue incubation for another 10min at 25°C in rotator.
6. Add 50µL of 0.5 M EDTA, mix by inverting tubes, and incubate at room temperature for 5 min.
7. Centrifuge at 460 g (1600rpm) for 10min at 4°C.
8. Remove a small amount of cells to count.
9. Enrich for bulk DC by MACS.
   - (*a*) Add 50 µL of anti-CD11c+ microbeads per $10^8$ cells and incubate at 4°C for 10 min. (Do not incubate on ice as this will cap Ab-beads on cells leading to contaminating cell populations.)
   - (*b*) Centrifuge at 460 g (1600 rpm) for 10 min at 4°C.
   - (*c*) Resuspend cells in MACS buffer at $10 \times 10^7$ cells/mL.
   - (*d*) Prewet LS+ column with 3 mL MACS buffer solution.
   - (*e*) Place cells over column (max of five lungs per column) and collect negative fraction.
   - (*f*) Wash with 9 mL of MACS buffer.
   - (*g*) Remove column from magnet and elute positive cells with 6 mL media (2 mL gravity, 2mL gentle plunge, 2mL forceful.

9. Resuspend in Sterile Iscove's + 10% PCS at 1 × $10^7$ cell/mL.
11. Stain enriched DC with anti-CD11b FITC (1μL/1 × $10^6$ cells), anti-CD11c APC 1μL/2 × $10^6$ cells), anti-I-A/I-E PE (1μL/2 × $10^6$ cells) mAbs.
12. Incubate cells for 30 min on ice.
13. Centrifuge at 460 g (1600 rpm) for 10 min at 4°C.
14. Resuspend in Sterile Iscove's + 10% BCS at 1 × $10^7$ cell/mL.
15. Run all cells through a cell strainer into new FACS tubes before sorting.
16. Sort into collection tubes of 1 mL of DMEM + 10% FCS.
17. Sort on CD11c + cells, then gate the aDC (CD11c+, I-A/I-E+, CD11b-), iDC (CD11c+, I-A/I-E+, CD11b+), and avelolar Macs (CD11c+, I-A/I-E inter, CD11b moderate (autofluorescent)).

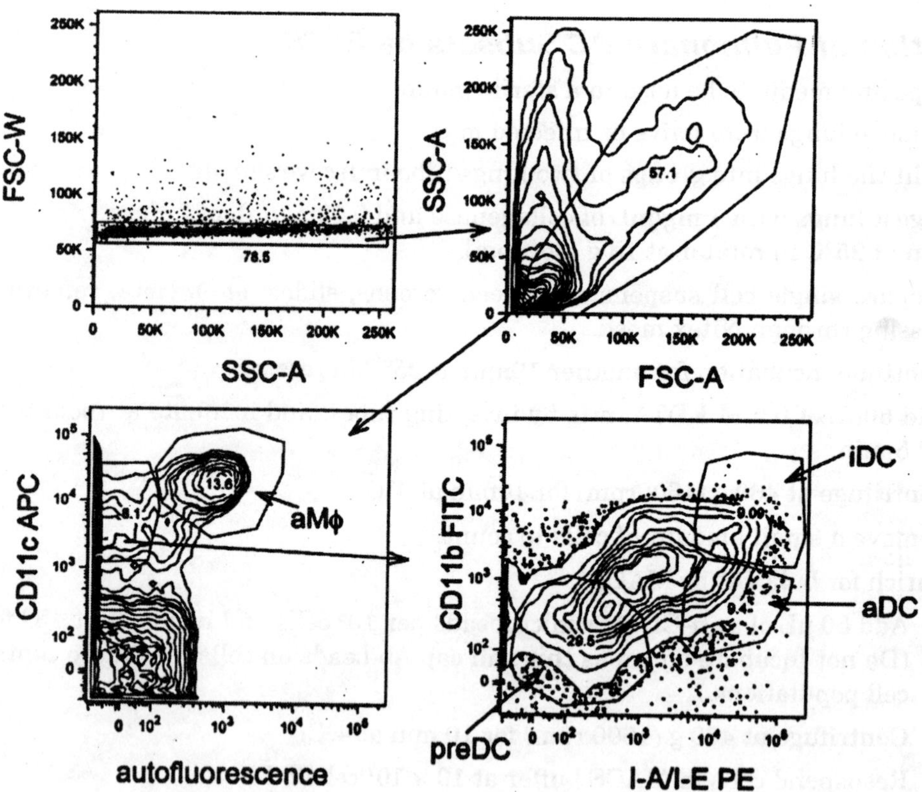


**Fig. 8.1. Gating strategy for FACS purification of pulmonary dendritic cells. Cells are first gated upon 311 a singlet gate (upper left) and then on a large live cell gate (upper right). Next, autofluorescent⁺ CD11c cells are sorted (alveolar macrophages) away from the autofluorescent$^{neg}$ CD11c cells (lower left). The autofluorescent$^{neg}$ CD11c⁺ cell population contains interstitial DC (CD11b⁺I-A/I-E⁺), airway and alveolar DC (CD11 b⁻ 1 A/I-E⁺), and precursor DC (CD11b⁻I-A/I-E⁻) cells (lower right).**

## Migration of DCs and LCs

### Preparation and Fixation of Epidermal Sheets

1. Obtain ears from the mouse. With forceps, split each ear into dorsal and ventral halves. Lay the ears in a Petri dish so that the epidermal sides are facing up.
2. Apply depilatory (*e.g.*, Nair) to the ears for 20 – 30 s. Wash with PBS.
3. Submerge the ears in 0.5 M ammonium thiocyanate (in the Petri dish) and place them in the 37°C water bath on a platform for 15-20 min.
4. Rinse the ears with PBS. Under the dissecting microscope, with fine forceps, carefully peel the epidermis off as a sheet.
5. Fix the epidermis in ice-cold acetone for 10-20min. Transfer to PBS for rehydration.

### Staining the Epidermis for LCs

1. Incubate the tissue for 1 h at room temperature in 500 αL of blocking buffer in a well of a 24-well tissue culture plate or in 200mL of blocking buffer in a 96-well ELISA plate.
2. Aspirate the blocking buffer.
3. Incubate with 500 μL of biotinylated anti-MHC class II or MHC class II PE direct conjugate at appropriate titer, overnight at 4°C protected from light.
4. Wash three times in PBS by filling and emptying the well.
5. Incubate with 500 μL of Avidin-Cy5 or Avidin-Cy3 for 1 h at room temperature.
6. Wash three times in PBS.
7. Mount on a slide with assistance of fine forceps and the dissecting microscope. Do not use a bright light source during this step.
8. Cover completely with Vecta Shield (about 10-20μL) and a cover slip. Protect from light.
9. Let dry for 1 h and then seal the four sides of the cover slip to the glass slide with fingernail polish.
9. Store at 4°C until microscopic examination for up to 3 d. For longer storage, freeze slides at –20°C.
11. Examine the slides using a fluorescent microscope. PE fluoresces in the Cy3 channel.

### FITC Tracking of LCs and Dermal DCs to Lymph Nodes

All cell washes in this procedure are performed at 4°C at 400 g for 7 min. 0.5% FITC solution preparation :

1. Weigh out 0.001 g of FITC into a glass vial.
2. Add 100 μL of acetone and 100 μL of n-butyl phthalate. Vortex intermittently for 5 min.

### Animal Application

1. At desired time, (*e.g.*, 12 h before sacrifice) apply 25 μL of the FITC solution to one side of the mouse's abdomen. Apply 25 μL of vehicle alone to the opposite side as a control.

2. Isolate each inguinal LN. Place each LN individually in 200 – 400 μL of liberase/DNase working solution. Keep on ice. A 12-well tissue culture tray works well and facilitates identification of individual LNs.
3. Using a scalpel, individually mince each LN in the liberase/DNase working solution.
4. Incubate at 37°C for 30 min.
5. After digestion, prepare a single cell suspension from the LN by passaging the fragments repeatedly in and out of a syringe, initially with an 18-gauge needle followed by a 23-gauge needle.
6. Wash the cell suspension in BSS. Resuspend the cells in Perm/Wash buffer as desired (300–500 μL is recommended for each LN) and perform a cell count.

## *Intracellular Staining*

1. Dilute only langerin and goat anti-rabbit secondary antibodies to the appropriate concentration for staining in Perm/Wash buffer.
2. Perform surface staining for at least CD11c (to identify DC) and other antigens of interest as usual. Remember to leave the Cy5 channel open for langerin.
3. At final wash, remove as much buffer as possible and resuspend the stained cells in 50 – 200 μL of Cytoperm/Cytofix solution (it is used undiluted from the manufacturer). Use 50μL for 0.6 × $10^6$ cells, and use 200 μL for 3 × $10^6$ cells. Vortex to mix and incubate at room temperature for 5 min.
4. Wash cells twice in Perm/Wash buffer. Add the diluted anti-langerin antibody (from step 1) and incubate at room temperature for 20 min.
5. Wash cells twice in Perm/Wash buffer. Add goat anti-rabbit IgG and incubate cells at room temperature for 20 min.
6. Wash cells twice in Perm/Wash buffer and transfer cells into mouse fix.

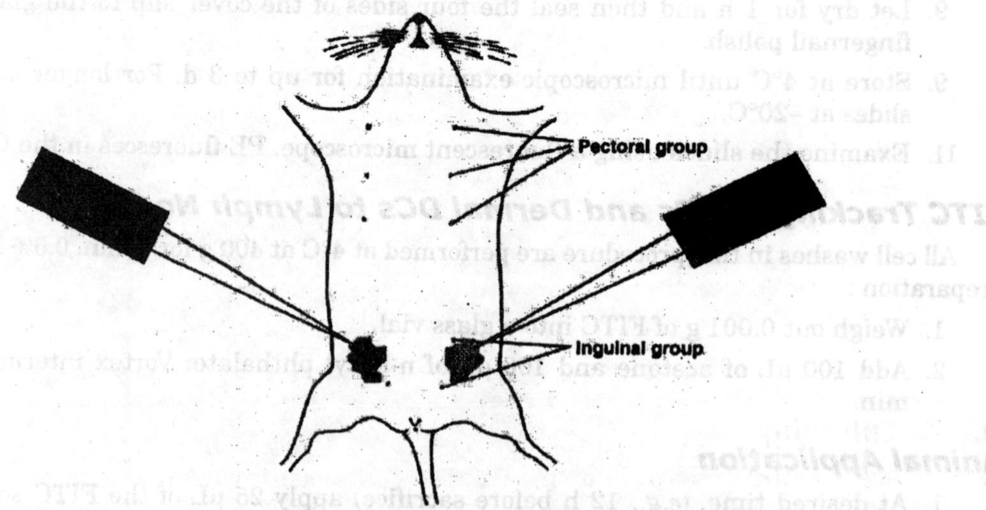


**Fig. 8.2. Location of FITC/vehicle control application.**

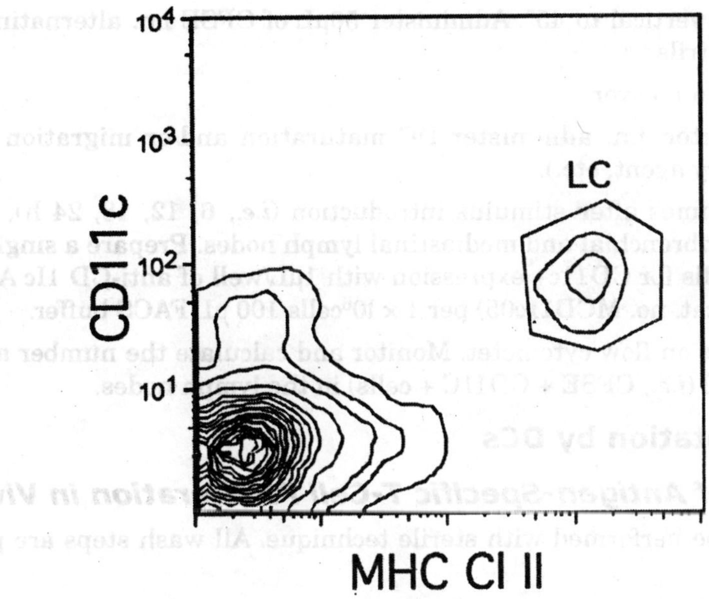


**Fig. 8.3. Results of flow cytometric identification of LC. The gated population of CDl lc+ MHC Class ll + cells identifies LC as approximately 3% of the cell suspension.**

## *Alternative Shortened Staining Procedure*

**This shortened procedure can only be used if the surface antigens are not sensitive to the fixation/permeabilization step.**

1. **Dilute all antibodies in Perm/Wash buffer.**
2. **After counting the cells, pellet them and remove the buffer. Depending upon how many cells are in the pellet, resuspend the cells in Cytoperm/Cytofix buffer (for 3 × $10^6$ cells, use 200 µL). Incubate at room temperature for 5 min.**
3. **Wash the cells twice with Perm/Wash buffer. Aliquot cells into staining tubes (if not already done) and add antibodies to surface antigens and langerin. Vortex to mix. Incubate cells for 20 min at room temperature.**
4. **Wash the cells twice with Perm/Wash buffer and add fluorochrome-conjugated avidin (if biotinylated antibodies have been used) and the goat anti-rabbit IgG. Incubate at room temperature for 20 min.**
5. **Wash cells twice with Perm/Wash buffer and transfer cells into mouse fix.**

## *Assessment of rDC Migration from the Lungs to the Lymph Nodes by i.n. CFSE Labeling*

1. **Prepare working CFSE solution from stock by diluting to an 8 *mM* solution with Iscove's DMEM Media (16 µL of CFSE + 34 µL media /mouse.**
2. **Anesthetize 1-4 mice.**

3. Hold mouse vertical to 45°. Administer 50μL of CFSE i.n. alternating back and forth between nostrils.
4. Allow mice to recover.
5. Six hours later, i.n. administer DC maturation and/or migration stimulus (virus, inflammatory agent, etc.).
6. At various times after stimulus introduction (*i.e.,* 6, 12, 18, 24 h), remove the lung draining peribronchial and mediastinal lymph nodes. Prepare a single-cell suspension and stain cells for CD11c+ expression with 1μL/well of anti-CD 1Ic APC mAb (Caltag/ Invitrogen, cat. no. MCD11c05) per 1 × $10^6$cells/100 μL FACS buffer.
7. Analyze cells on flow cytometet. Monitor and calculate the number and percentage of recruited DC (*i.e.,* CFSE + CD11C + cells) in the lymph nodes.

## Antigen Presentation by DCs

### *DC Activation of Antigen-Specific T-Cell Proliferation in Vivo*

All steps must be performed with sterile technique. All wash steps are performed at 4°C for 7 min at 400 g.

### *Day 1*

1. Prepare.a single-cell suspension of splenocytes from OTI or OTII mice, wash in PBS, count, and resuspend at 5 × $10^7$/mL in PBS.
2. Mix 5m*M* CFSE stock solution: cell suspension 1:10 (to achieve a final concentration of 5μ*M* CFSE), and incubate at 37°C for 10 min.
3. Wash three times in cold (4°C) 10% FCS in PBS.
4. Inject 5 × $10^7$ CFSE-labeled splenocytes into each recipient C57B1/6 mouse intravenously using an insulin syringe.

### *Day 2*

1. Remove spleen from donor C57B1/6 mice (EtOH or $H_2O$ treated), being careful not to disrupt the capsule. Place into a 60mm Petri dish. Inject 1 mL of Liberase/DNase working solution into the spleen, and incubate 20min in a 37°C water bath, supported by a platform.
2. Prepare a single-cell suspension of splenocytes in the Petri dish by rubbing the spleen between the frosted ends of two microscope slides, and aspirate the suspension through a 1 mL syringe with a 23-g needle to break up remaining cell clusters.
3. Collect the cells in a 15-mL centrifuge tube. Wash with 0.5% BSA in PBS, and count the cells.
4. Label with GDI 1c+ microbeads as follows:
   (*a*) Resuspend the cells in 400 μL of 0.5% BSA in PBS/$10^8$ cells.
   (*b*) Add 100 μL of anti-CD lie microbeads per $10^8$ cells, mix well, and incubate at 4°C (in the refrigerator, not on ice) for 15 min.

(*c*) Wash with 0.5% BSA in PBS.

(*d*) Resuspend cells in 500 μL of 0.5% BSA in PBS/$10^8$ cells.

5. Purify the DC on the autoMACS following the manufacturer's instructions (using "POSSELD" selection mode).
6. Wash the DCs once in 0.5% BSA in PBS and resuspend in Super McCoy's medium at $5 \times 10^6$/mL.
7. Pulse the purified DC with OVA 258-264 (for OTI) or OVA 323-339 (for OTII) stock solution to achieve a final concentration of 100n*M*.
8. Incubate at 37°C for 2 h, wash in PBS, and resuspend at $5 \times 10^7$/mL in PBS.
9. Transfer $5 \times 10^6$ peptide-pulsed DC intravenously into each recipient using insulin syringe.

### Day 3

1. Prepare a single cell suspension of recipient splenocytes as in step 1 of day 1.
2. Stain with anti-CD8 (OTI) or anti-CD4 (OTII) using a fluorochrome conjugate other than FITC (this channel is needed to collect the CFSE fluorescence).
3. Run samples on flow cytometer.

## Allostimulatory Assay of DC Function

All steps in this procedure must be performed under sterile conditions. All washes are performed in 0.5% BSA in PBS, 400 g at 4°C for 7min unless otherwise noted.

### Preparation of T Cells

1. Prepare a single-cell suspension of splenocytes from BALB/c mouse. Wash and resuspend in 3 mL of 0.5% BSA in PBS:
2. Underlayer the cell suspension with 3 mL of Fico-Lite LM. Centrifuge the cells at 1700g for 20min at room temperature. Remove the cells from the interface. Wash and resuspend the cell pellet in 1 mL of 0.5% BSA in PBS.
3. Incubate splenocytes with appropriate concentrations of biotinylated anti-B220, DX5, CD11c, CD11b, and Ter1 19 antibodies for 20min on ice.
4. Wash three times. Resuspend cell pellet in 80 μL of 0.5% BSA in PBS per $10^7$ cells. Add 20μL of anti-biotin microbeads per $10^7$ cells. Mix well and incubate for 15 min at 4°C in the refrigerator (not on ice).
5. Deplete the antibody-coated cells on the autoMACS according to the manufacturer's instructions using the "DEPLETE" mode.
6. Wash the cells, resuspend in Super McCoy's medium, and count.

### Preparation of DC for Initial Stimulation

1. Purify DCs from C57B1/6 mouse spleen by using anti-CD11c microbeads on the autoMACS according to steps 1-5.

2. Wash the cells. Resuspend the purified DCs in Super McCoy's medium, and count the cells. Put DCs and T cells into a T25 culture flask at a 1:10 ratio of DC to T cells (the density is generally around 1.1 × $10^6$cells/mL, 10mL per flask). Incubate in 5% $CO_2$ at 37° C.

### *Restimulation of T Cells*

1. On day 3 of culture, harvest the cells. Wash 3 times and resuspend in 400mL of 0.5% BSA in PBS per $10^8$ total cells.
2. Deplete the DCs from the cell suspension. Use anti-CD11c microbeads to label the DCs as in step 1 of "Preparation of DCs for Initial Stimulation" in this protocol, but run the autoMACS in "DEPLETE" mode to remove rather than purify the DCs from the cell suspension.
3. Resuspend the re-purified T cells in Super McCoy's medium and place into a new T25 culture flask. Rest these cells for 2 d by incubating in 5% $CO_2$ at 37°C.
4. On day 3, purify DCs from the spleens of C57BL/6, C3H or BALB/c mice as in step 1 of "Preparation of DCs for Initial Stimulation" in this protocol.
5. Resuspend the cells in Super McCoy's medium, count, and place into culture in a 96-well plate at graded doses (*e.g.*, 1000, 5000, and 10,000 cells) with 1 × $10^5$ T cells/well.
6. Incubate in 5% $CO_2$ at 37°C for 3 d.
7. Measure T-cell proliferation by adding 1μCi of $^3$H-thymidine/well and incubating for an additional 18 h.
8. Harvest the cells using a cell harvester and measure the radioactivity incorporated into the divided cells with a scintillation counter using standard methods.

## NOTES

1. Liberase and DNase stock solutions are stable at 4°C for 1 wk or at –20°C for at least 2 mo.
2. DC purification can also be done using the LS Column (Miltenyi Biotech, Auburn, CA; cat. no. 130–042–401) as described in Subheading 3.1.5
3. CD11c+, CD8α-fraction can be further enriched for CD11c + CD4 + DC by using the Miltenyi Biotec Multisort Release Kit to remove anti-CD11c Microbeads, stain with anti-CD4 Microbeads and positively select CD4+ cells.
4. Generally approx 1-3% of the cell suspension is LC, as identified by GD11c+ MHC Class 11 + cells by flow cytometry.
5. Negative fraction can be used for unstained and comps (be sure to lyse RBC using HBSS).
6. Anti-CD11c (clone N418) PE or biotin conjugates have also been used in our hands to label LCs in the ear. In general the biotinylated antibodies give brighter staining with less background than the direct conjugates. PE has a tendency to photobleach readily.
7. Avoid excessive handling of the epidermal sheets as they are quite friable.
8. Avidin-Cy3 is brighter than avidin-Cy5 in our hands. The latter is useful primarily if multicolor staining is performed.

9. Antigens that have been tested for sensitivity and found to be unaffected are epitopes recognized by the following antibodies: 4B12. (CCR7), C34–3448 (CCR5), G1-1 (CD86), 1C10 (CD40), and N418 (CDllc).
9. Performing the permeabilization in bulk will aid in the cell recovery if that is a concern.
11. LCs are identified in LN as FITC+langerin+, and dermal DC as FITC+CD1 1c+ langerin.

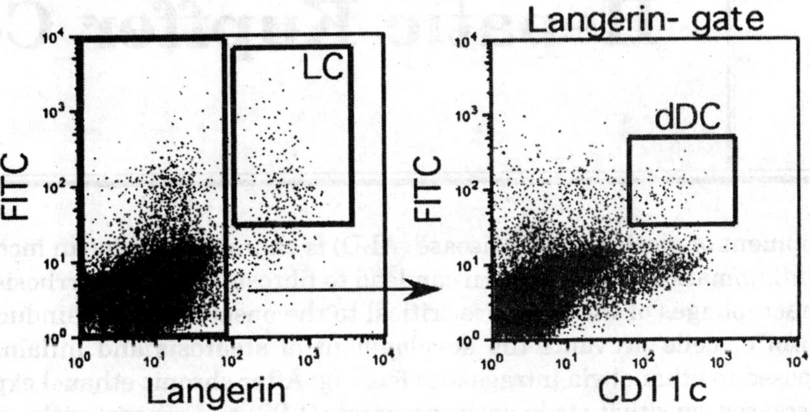


**Fig. 8.4. Gating strategy for LC and dDC in draining LN following FITC skin painting. Recently immigrated LC are identified as langerin + FITC +, and dDC are langerin-CD11c + FITC +.**

12. CFSE solution will be cloudy with some precipitate. However, it will still label cells.
13. Mouse must be anaesthetized enough to allow inhalation of CFSE without coughing or swallowing it.
14. Be careful not to get onto hands, skin, or other areas of mouse or those DCs will label and you will observe CFSE + DC in the appropriate non-lung draining lymph nodes.
15. There is substantial bleed-over of CFSE into the FL2 channel. Careful compensation is required.
16. To measure T cell cytokine production, you can choose to skip the CFSE labeling steps.
17. Supernatants can be used for ELISA.
18. Intravenous injections are easily accomplished by accessing the retro-orbital plexus immediately behind the medial aspect of the eye. Insulin (or tuberculin) syringes work best because there is no void volume left in the syringe (all the cells are transferred).
19. Restimulation with C57B1/6 DC measures the allogeneic response of the T cells, restimulation with BALB/c DC measures the syngeneic response, and restimulation with C3H DC measures the third party (primary) response.

CHAPTER

# 9 Hepatic Kupffer Cells

The development of alcoholic liver disease (ALD) is characterized by an increase in hepatic steatosis and inflammation, which in turn can lead to fibrosis and then cirrhosis. Kupffer cells, the resident macrophages of the liver, are critical to the onset of ethanol-induced liver injury. Ablation of Kupffer cells prevents the development of steatosis and inflammation in rats chronically exposed to ethanol via intragastric feeding. After chronic ethanol exposure, Kupffer cells exhibit increased sensitivity to lipopolysaccharide (LPS) and consequently exhibit enhanced LPS-stimulated inflammatory cytokine production. Tumor necrosis factor-α (TNF-α) production, resulting from Kupffer cell activation, has been shown to be a critical component in the development of ALD. Recent studies have shown that increased sensitivity to LPS after chronic ethanol exposure also leads to an increased production of reactive oxygen species (ROS).

Isolation of Kupffer cells from rats fed an ethanol-ontaining diet provides sufficient material for studies using cellular and molecular biology approaches aimed at understanding the effects of ethanol on Kupffer cell function. For example, isolation of mRNA from Kupffer cells from pair- and ethanol-fed rats has shown that LPS-stimulated increases in TNF-α expression can be inhibited by adiponectin. In this chapter, we will focus on methods used for isolating Kupffer cells from animals fed the Lieber-DeCarli ethanol diet and their pair-fed controls, as well as how to isolate mRNA from these primary Kupffer cell cultures.

## MATERIALS

### Surgical Portion of the Procedure

Materials needed for the surgical portion of the Kupffer cell isolation can be found in Table 9.1. Sources used and catalog numbers associated with them are also included in the table.

### Elutriator

Materials needed for the elutriation portion of the Kupffer cell isolation can be found in Table 9.2. Also included are the sources used and the catalog numbers associated with them.

### Isolation and Culture of Kupffer Cells

Materials for the isolation and culture of the Kupffer cells can be found in Table 9.3 along with the sources used and catalog numbers.

**Table 9.1. Materials of Surgical Portion of Procedure**

| *Material* | *Vendor* | *Catalog number* | *Concentration or amount needed* |
|---|---|---|---|
| 1. Autoclaved water | | 6% | |
| 2. Hydrogen peroxide | | | |
| 3. 8-channel peristaltic pump | Rainin | 7103-058 | |
| 4. Four-"butterfly needles" (Blood Collection Infusion Set 19 gauge ¾ with 12" tubing) | BD Biosciences | 364919 | |
| 5. Waterbath at 39°C | | | |
| 6. Hanks-(Hank's Balanced Salt Solution (w/o $Ca^{2+}$, $Mg^{2+}$, $NaHCO_3$) | Mediatech | 55–021-PC | 250mL/Rat |
| 7. HEPES | Fisher | BP310–500 | 10mM |
| 7. EGTA | Sigma | E4378 | 1mM |
| 9. Williams E | Sigma | W4125 | 250mL/rat |
| 10. Collagenase type IV (from *Clostridium histolyticum*) | Sigma | C5138 | 50mg/rat |
| 11. 9" Pasteur Pipets | | | |
| 12. 95% $O_2$ 5%, $CO_2$ gas mixure | Praxair | mmoxcd5-k | |
| 13. Heparin | Sigma | H3393 | 1000U/mL |
| 14. Ketamine (100mg/mL) | Henry Schein | 995-2949 | |
| 15. Xylazine (20mg/mL) | NLS Animal Health | 105650 | |
| 16. Acepromazine (10mg/mL) | Henry Schein | 356-7290 | |
| 17. Eight 1-mL syringes | Fisher | 14–823–2F | |
| 17. Four 26-gauge 5/8-inch needles | Fisher | 14–826-6A | |
| 19. Four 21-gauge needles | Fisher | 14-826C | |
| 20. Kimwipes | | | |
| 21. Buttonhole twist thread (2 colors- ~5") | | | |
| 22. Two surgical pans (9 × 13 baking pan) | | | |
| 23. Two racks (cookie coling racks must fit over pan) | | | |
| 24. Two pairs of sharp/sharp 5 1/2" sciessors | Fisher | 13–808–4 | |
| 25. One pair of blunt/sharp 5 1/2" sciessors | Fisher | 13–808–2 | |
| 26. One pair of blunt-point curved 5" forceps | Fisher | 08–887 | |
| 27. One pair of curved eye dressing 4 1/2" forceps | Henry Schein | 101-7739 | |
| 27. Four 15-mL centrifuge tubes | Sardstedt | 52 554 205 | |
| 29. Four divided Petri dishes | Fisher | 08–757–110 | |

Item 21 can be autoclaved in a small box and items 24-27 can be autoclaved together in an autoclaveable box.

**Table 9.2. Materials for Elutriation Procedure**

| | *Material* | *Vendor* | *Catalog number* | *Concentration or amount needed* |
|---|---|---|---|---|
| 1. | Beckman Avanti J25 Centrifuge | Beckman | | |
| 2. | JE6B Elutriation Rotor and accessories | Beckman | | |
| 3. | Masterflex L/S Variable Speed Pump | Cole Parmer | EW-77521-40 | |
| 4. | Masterflex L/S Easey Load II Pump Head | Cole Parmer | EW-77200-60 | |
| 5. | Autoclaved water | | | |
| 6. | Hydrogen Peroxide | | | 6% |
| 7. | Catalase | Sigma | C9322 | 10–15mg/ 100mL |
| 7. | Gey's Balanced salt solution | | | 1L |

If working with endotoxin, autoclave and bake all glassware before use.

## METHODS

The following method is designed for the isolation of Kupffer cells from male Wistar rats that were allowed free access to the Lieber-DeCarli liquid diet (7) containing 36% of calories from ethanol or pair-fed an isocaloric diet that substituted maltose dextrins for the ethanol. The following procedure has been adapted from methods originally developed by the Friedman and Roll, Knock *et. al.,* and Munthe-Kaas *et. al.*

Kupffer cells are isolated as previously described *(2)*. In brief livers are perfused with 0.05% collagenase, and the resulting suspension of liver cells is treated with 0.02% pronase for 15min at 12°C. The resulting cell suspension is centrifuged 3 times at 50 g for 2min, and the supernatant collected after each centrifugation. The pooled supernatant is then centrifuged at 500 g for 7min to collect nonparenchymal cells. Kupffer cells are then purified by centrifugal elutriation *(2)*. The following protocol is for isolation of Kupffer cells from 2 pair-fed and 2 ethanol-fed animals; however, the protocol can be scaled up or down according to the quantity of Kupffer cells required.

### Setting up the Peristaltic Pump and Perfusion

All eight channels of the peristaltic pump must be used for isolating Kupffer cells from four rats. Set up the Rainin peristaltic pump per the manufacturer's instructions. It is necessary to check the flow rates of the peristaltic pump before beginning the isolation. For perfusion of two rats, set the flow rate to 25mL/min and to 50mL/min for four rats. To sterilize the tubing, first run sterile water through the pump, and then run approx 200 mL of 6% hydrogen peroxide ($H_2O_2$) through the pump. Finally, run more sterile water through the pump until all peroxide bubbles are gone from the tubing. While sterilizing the pump, thaw 1 × 10-mL aliquots of 10 mM HEPES and 1 mM ethylene glycol-bis (2-amnioethylether)- N, N, N', N'-tetraacetic acid (EGTA), and bring 1L of Hanks (pH 7.4) and Williams E ~(pH 7.4) to 37°C.

**Table 9.3. Materials for Isolation and Culture of Kupffer Cells**

| *Material* | *Vendor* | *Catalog number* | *Concentration or amount needed* |
|---|---|---|---|
| 1. Waterbath at 37°C | | | |
| 2. 12°C Shaking Waterbath | | | |
| 3. Beckman Tabletop Centrifuge | | | |
| 4. Hanks+(Hank's Balanced Salt Solution (with $Ca^{2+}$, $Mg^{2+}$ w/o $NaHCO_3$) | Sigma | H6136 | 10mL/rat |
| 5. CMRL Medium 1066 | Invitrogen | 11530-037 | |
| 6. Antibiotic-Antimycotic | Invitrogen | 15240-062 | |
| 7. L-glutamine | Invitrogen | 25030-081 | 2mM |
| 7. Fetal bovine serum certified | Invitrogen | 16000-044 | 10% |
| 9. Pronase | Roche | 165 921 | 30mg/2 rats |
| 10. DNase I | Roche | 104 159 | 405mg/rat |
| 11. Gey's Balanced Salt Solution | | | |
| 12. Nycodenz | Accurate Chemical | ACN1002424 | 11% and 24.6% |
| 13. 3% Acetic acid | | | |
| 14. Trypan blue | | | |
| 15. Thirty-50 mL centrifuge tubes | Sardstedt | 62 547 205 | |
| 16. Six-15 mL centrifuge tubes | Sardstedt | 52 554 205 | |
| 17. Two-500 mL polypropylene flasks | Fisher | 10–041–10D | |
| 17. Two-250 mL polypropylene flasks | Fisher | 10–041–10C | |
| 19. Two plastic funnels | Fisher | | |
| 20. One pair straight dissecting dressing 4- 1/2″ forceps | Fisher | 13–812–39 | |
| 21. One pair curved dissecting VWR 4- 1/2″ forceps | VWR | 25717-005 | |
| 22. Two 3 × 3″ pieces of Coarse mesh (132 μm) | Sefar Americo (Tetko) | 03–132–40 | |
| 23. Two 3 × 3″ pieces of Fine mesh (224 μm) | Sefar Americo (Tetko) | 03–225–42 | |
| 24. Two-60-mL syringes | Fisher | 13–689–8 | |
| 25. Two 18-gauge needles | Fisher | 14–826–5G | |
| 26. 5-3/4″ Pasteur pipets | | | |
| 27. 9" Pasteur pipets | | | |
| 27. Rubber bulb | VWR | 56311–062 | |
| 29. 1.5-mL Eppendorf tubes | | | |
| 30. Pipet tips (0.5 – 10 μL, 10 – μL, 10 – 100 μL, 100 – 1000 μL) | | | |
| 31. Sterile reagent reservoirs | Fisher | 07 200 127 | |
| 32. Sterile tissue culture-treated plates (96-well, 24-well, 6-well, chamber slides) | | | |

Items 17 – 18 should be autoclaved and items 20 – 23 can be autoclaved together in an autoclaveable box. If working with endotoxin, autoclave and bake all glassware before use.

When the media are at 37°C, oxygenate each medium with a mixture of 95% $O_2$ and 5% $CO_2$ for approx 10min, and then add 10mM HEPES and 1 mM EGTA to the Hanks-, mix thoroughly and run through the peristaltic pump. Once the Hanks- is through, stop the pump until you are ready to perfuse the livers. Once the Williams E has been oxygenated, add 200 mg of collagenase, mix by swirling and place in the water bath.

## Setting up the Elutriator

Before using the elutriator to isolate cells, it is necessary to check the flow rates of the peristaltic pump at the desired rate of centrifugation. Before pumping cells through the elutriator, the tubing should be autoclaved. Set up the elutriation chamber and tubing and cool to 4°C. To prepare the elutriator, turn the pump to the 45 mL/min flow rate and run sterile water through the pump and chamber (~100 – 200 mL). Next, recirculate 100 mL of 6% $H_2O_2$ through the tubing and chamber for approx 5min, with the centrifuge running at 50*g*.

Make sure the pressure gauge stays under 20 psi at all times. Stop the centrifuge, recirculate 100mL of catalase (10 – 15 mg/100mL) through the tubing, and restart the centrifuge at 24*g* for approx 5 more minutes or until all of the peroxide bubbles are gone. Stop the centrifuge again, and then rinse out the tubing and chamber with sterile water. Once the tubing and chamber have been sterilized, run Gey's Balanced Salt Solution (pH 7.4) (GBSS) through the system (let it run all the way through to remove any air bubbles) and then stop the pump until needed.

## Setting Up the Tissue Culture Room

It is helpful to set up needed cell culture supplies before beginning the perfusion or while the perfusion is in progress. Place the following in the cell culture hood:

2 – 30-mg aliquots of pronase in a 50-mL centrifuge tube

2 – 4.5-mg aliquots of DNase I in a 15m-L centrifuge tube

1 – 500-mL bottle of CMRL with antibiotic/antimycotic and 2 mM L-glutamine (CMRL).

Fill each 50-mL centrifuge with 50 mL of CMRL and each 15-mL centrifuge with 3 mL of CMRL.

## Anesthesia

Rats are weighed and the dose of anesthetic is calculated based on diet and body-weight. In our experiments, we use a mixture of xylazine (20 mg/mL), ketamine (100mg/mL), and acepromazine (10mg/mL), which expires after 1 mo.

| | | |
|---|---|---|
| Ketamine: | 150mg (100mg/mL) | 1.5mL |
| Xylazine: | 30 mg (20mg/mL) | 1.5mL |
| Acepromazine: | 5 mg (10 mg/mL) | 0.5 mL |
| | | 3.5 mL |

In chow-fed animals, we use a dose of 0.12mL/100g bodyweight. Ethanol-fed animals on the Lieber-DeCarli ethanol diet receive a dose of 0.075 mL/100 g bodyweight, whereas their pair-fed controls receive 0.12mL/100g bodyweight. Animals receive an intraperitoneal injection in the lower middle abdomen (right or left of midline is fine). Allow 5 – 10 min for the animals to reach complete sedation. If there is no movement when the foot or base of the tail is pinched, the animal is completely anesthetized.

## Surgery

1. Once the animals are anesthetized, wipe the fur with a Kimwipe wet with water (do not use ethanol) to get rid of any dander, loose fur, or dirt.
2. Using the sharp/blunt scissors and blunt-point curved forceps, peel the abdominal fur by pulling up on the fur at the lower part of the abdomen and cutting up towards the thoracic area, removing enough fur to have a clear surgical field.
3. Remove loose fur from the scissors and animals by wiping with another wet Kimwipe.
4. Using the same sharp/blunt scissors and blunt-point curved forceps, make a horizontal cut into the lower part of the abdomen at midline. Carefully insert the scissors into the opening (do not to puncture the bladder) and make a smooth, V-shaped cut up both sides of the abdomen. Stop the cut at the diaphragm of the animal. Do not puncture the diaphragm.
5. Using a Kimwipe, push the organs to the side and out of the abdominal cavity and place a 15-mL centrifuge tube under the thoracic area to better expose the portal vein.
6. Place two loose ligatures (two different colors of autoclaved buttonhole twist thread) around the hepatic portal vein using the curved eye dressing forceps.
   (*a*) Insert the forceps in the connective tissue near the hepatic portal vein, go underneath the vein and around to the other side. Then, grasp onto one end of the thread and gently pull it through. Pulling too hard on the thread can twist the hepatic portal vein and also cut off circulation to the liver.
   (*b*) After pulling both ligatures through, tie a loose knot in preparation for inserting the perfusion "butterfly" needle and lay the ends of the string out so they are easy to pick up.
7. Inject 0.1 mL of heparin using a 26-gauge, 5/8-inch needle into the posterior vena cava. After the injection, press down on the injection site to stop the bleeding with a Kimwipe, and wait 1 min before proceeding.
   (*a*) If you would like to harvest other tissues (skeletal muscle, adipose), this would be the most appropriate/ideal time to do so. The animal will still be alive, and the area for liver perfusion is ready.
   (*b*) To take a blood sample from the animal, use an 18-gauge needle (smaller needles will shear red blood cells resulting in a hemolyzed plasma/serum sample). Insert the needle into the posterior vena cava and pull back slowly. Try to minimize bleeding by applying pressure to the vena cava. The perfusion should be started as soon as possible after collecting the blood.
7. For the perfusion, start the peristaltic pump at the 25mL/min flow rate, insert the 19-gauge, 3/4-inch perfusion "butterfly" needle behind the rear ligature into the hepatic portal vein and past the front ligature. Tighten the knots of the two ligatures to secure the needle.

9. As each perfusion needle is secured, cut through the diaphragm and cut the portion of the posterior vena cava in the pericardial cavity. This releases the perfusate and allows it to drain into the pan below.
10. First, perfuse with the Hanks- medium. Stop the pump just before all the media is in the tubing, place the tubing into the William's E + collagenase and start the pump again. Do not allow bubbles to get into the liver.

## Isolation of Kupffer Cells

1. After the perfusion, using the second pair of sharp/sharp scissors and the blunt-point curved forceps, remove the liver. Take the forceps and grasp onto a portion of the diaphragm near the liver, at midline. Cut a small "tag" of diaphragm to hold onto while cutting the connective tissue that attaches the liver to other organs. When removing the liver, be careful not to puncture the intestines or stomach, as they are a source of potential bacterial contamination. Once each liver has been removed, place it into a 100mm divided Petri dish. The remainder of the procedure is performed in a laminar flow hood under sterile conditions.
2. Rinse each liver with approx 10 mL of Hanks+.
3. Add 15 mL of CMRL to each liver. Using the straight and curved dissecting forceps, tease each liver slightly.
   (*a*) If the perfusion was successful, the liver will come apart with little effort.
   (*b*) If the liver is hard to tease apart, the perfusion did not work well and Kupffer cell yields will be low.
4. Once the livers have been teased apart, carefully pour the cell suspension from each Petri dish into its appropriately labeled (P or E) 500-mL flask through the plastic funnel. Rinse each Petri dish with 20 mL of CMRL adding the rinse to the flask. Then, add the 50 mL of CMRL + pronase to the flask.
   (*a*) If hepatocytes are required, for experiments, remove them prior to adding the pronase.
5. Pour the 3mL of CMRL + DNase I into each flask and cover with aluminum foil. At this point, there should be approx 100mL of media in each flask.
6. Shake the 500-mL flasks for 15 min at 12°C in a shaking water bath (100rpm).
7. After shaking, filter each suspension through the coarse mesh (224 μm) into each 250-mL flask.
7. Next, filter the suspension from each 250 mL flask through the fine mesh (132μm) into 4 x 50-mL centrifuge tubes labeled P or E.
9. Centrifuge at 500 g for 7 min (4°C).
   (*a*) This step pellets all cell types: hepatocytes, red blood cells, Kupffer cells, etc.
10. Carefully aspirate off most of the supernatant (leave 1–2mL over the cells as the pellet is very loose).

11. Resuspend each of the pellets from two tubes into 10-15mL of CMRL with antibiotic/ antimycotic, 2mM L-glutamine, and 10% fetal bovime serum (FBS; CMRL 10% FBS) and bring each one to 50 mL. Repeat for the other two tubes, and then the other treatment group.
12. Centrifuge at 50 g for 2min.
13. Carefully pour approx 25–30mL of the supernatant from each tube into two sterile centrifuge tubes labeled PA (for pair-fed tube A) or EA (for ethanol-fed tube A) and place them in ice.
    (*a*) The volume of media poured off will depend on the size of the pellet. At this step the pellet is very soft and easily disrupted. Pour off as much supernatant as possible without displacing the pellet. The nonparenchymal cells are in the supernatant.
14. Resuspend the pellet in the remaining media in each tube and take the volume up to 50mL using CMRL 10% FBS.
15. Repeat steps 12 – 14 two more times.
    (*a*) This step is to separate the Kupffer cells from hepatocytes. After repeating there will be 6 × 50-mL centrifuge tubes for each treatment group.
16. Centrifuge all 12 centrifuge tubes at 500 g for 7 min to pellet nonparenchymal cells.
17. Aspirate off most of the supernatant (the pellet is soft and easily disrupted).
17. Combine the pellets of the six centrifuge tubes from each treatment group into one centrifuge tube with GBSS and then bring the volume up to 50mL.
19. Once the cells are resuspended in 50 mL of GBSS, load the cells into a 60-mL syringe with an 18-gauge needle attached.
    (*a*) First carefully remove the plunger from the syringe, turn it over, and set aside to avoid contamination. Next place the 60-mL syringe (without plunger) into an empty 50-mL centrifuge tube.
    (*b*) Carefully, but quickly, pour the cell suspension into the syringe, insert the plunger (some of the suspension will be lost into the centrifuge tube), and invert the syringe so the needle is facing upward.
    (*c*) Get rid of as many bubbles as possible and push out any air left in the syringe. Pour the lost cell suspension into a reagent reservoir. Invert the syringe and collect the remaining cells by pulling back on the plunger.
    (*d*) Again, remove all the air from the syringe. Replace the cap of the needle onto the syringe and place the syringe in ice.

It is now time to separate the Kupffer cells from other cell types within the cell suspension in the elutriation chamber. Depending on the centrifuge, make sure that the correct rotor is selected on the instrument panel (JE6B) and check that the centrifuge is at 4°C.

20. Turn on the peristaltic pump allowing the GBSS to flow through the tubing into the elutriator and out into the collection bucket. Run the elutriator for approx 1 min and then turn the pump to the 10mL/min flow rate.
    (*a*) Some bubbles may have collected in the tubing near the pressure gauge. Allow these bubbles to pass out of the elutriator into the collection bucket.
21. Once the bubbles are gone, start the centrifuge at 24 g. If there are any bubbles left in the chamber allow them to pass out of the elutriator. If no bubbles are present and the pressure gauge is reading close to 0 psi, increase the speed to 96 *g*, check again for bubbles and high pressure, and increase the speed finally to 602 g.
22. Load the cells into the elutriation chamber at 10mL/min. Take one cell-loaded syringe from the ice and invert it to make sure the cells are in suspension. Remove the 18-gauge needle from the end and discard. Invert the syringe and twist it into place on the stopcock making sure it is secure. Turn the valve on the stopcock so the pump starts to pull the cell suspension into the tubing. Once the cells have started to move into elutriator tubing, place the outlet tubing into a 50-mL centrifuge tube.
23. Allow the majority of the cell suspension to be taken into the tubing, return the stopcock to its original position allowing the GBSS to rinse the cells through the elutriation chamber. Remove the syringe, discard it, and allow the first 50-mL centrifuge tube to fill.
24. Change the pump setting to 21 mL/min, and collect another 50 mL of cells; this will elute off any remaining red blood cells, hepatic stellate cells, and endothelial cells.
25. Change the pump setting to 45 mL/min and collect 2 × 50mL of cells; this will elute off the Kupffer cells.
26. Once you have collected the Kupffer cells, return the outlet tubing to the collection bucket and stop the centrifuge. The cells that were not eluted out of the chamber will wash out into the collection bucket at this time.
27. Allow 100-200 mL of GBSS to wash the elutriation chamber before repeating steps 21 – 26 for the other cell suspension.

    If there is debris left in the chamber, disassemble the chamber and clean out the debris. If left inside the chamber, it may increase the pressure during the next round of elutriation.

27. Centrifuge the Kupffer cells (2P and 2E 50-mL centrifuge tubes) at 500 g for 7 min.
29. Under the tissue culture hood, prepare the nycodenz (ND) gradient for the final purification step.
    (*a*) Add 4mL of 24.6% ND into 2 × 15-mL centrifuge tubes for each treatment group,
    (*b*) Slowly layer (drop by drop initially) 4mL of 11% ND onto the –24.6% ND. An interface will be visible between the layers if done correctly.
30. Take the pelleted cells from step 28, aspirate off the GBSS, and resuspend in 10 mL of fresh GBSS.
31. Take 5 mL of cells and slowly layer them onto each ND gradient.

32. Centrifuge the gradients at 1644 g for 15 min WITH NO BRAKE. Using the brake will disrupt the separation of cells within the gradient.
33. After spinning, pair-fed Kupffer cells will be suspended in the 24.6% ND. Anything that pellets at the bottom should be discarded. With cells from rats that were ethanol-fed, two layers of cells can be seen within the gradient. The top layer, within the 11% ND, will be fatty hepatocytes. The bottom layer within the 24.6% ND, is Kupffer cells. Any cells that pelleted through the gradient should be discarded.
34. Under the tissue culture hood, take a 9-inch Pasteur pipet and attach a rubber bulb. Squeeze the rubber bulb and carefully insert the Pasteur pipet through the ND to the Kupffer cells and carefully-remove the cells within the 24.6% ND layer.
35. Remove the Pasteur pipet and place the collected cells into a new 50-mL centrifuge tube. Repeat step 34 to remove all of the Kupffer cells from the gradient.
36. Take the 50-mL centrifuge tube with the Kupffer cells (step 35) and fill to 50mL with GBSS.
37. Centrifuge the Kupffer cell suspensions at 500 g for 7 min.
37. Aspirate off the GBSS and resuspend the pellet with 10 mL of CMRL 10% FBS.
39. Into a 1.5-mL sterile Eppendorf tube, pipet 70 μL of 3% acetic acid, 10 μL of trypan blue, and 20 μL of Kupffer cells for counting.
40. Resuspend the cells to your desired density for plating. In our lab, we resuspend the cells to $2 \times 10^6$ cells/mL.
    (*c*) We use a plating density of $6.25 \times 10^5$ cells/cm$^2$ or 0.625 mL/cm$^2$, maintaining a consistent plating density across different types of cell culture plates.
41. Wash within 1 h of plating to remove any nonadherant cells.

## Isolation of RNA From Primary Kupffer Cells

Analysis of gene expression in isolated Kupffer cells from pair- and ethanol-fed rats is integral to our understanding of the mechanisms that contribute to the pathology associated with chronic ethanol exposure. Using real-time reverse transcriptase polymerase chain reaction (rtPCR), the effects of ethanol on the expression of several genes can be examined rapidly, even with a small number of Kupffer cells. The brief protocol below describes a method to isolate total RNA and synthesize cDNA from rat Kupffer cells for gene expression analysis by real-time rtPCR.

1. Seed six wells of a 96-well plate at a density of $6.25 \times 10^5$ cells/cm$^2$ for each experimental condition.
2. The RNeasy Micro kit (Qiagen, Valencia, CA) is used to isolate total RNA from Kupffer cells. The total number of cells per experimental condition is not to exceed $5 \times 10^5$. Fifty-eight microliters of the lysis buffer, provided with the kit and supplemented with 2-mercaptoethanol by the user, is added directly to each well using a multi-channel pipet. This results in a total of 355 μL of total lysis buffer for a maximum of $5 \times 10^5$ cells. Wells of the same experimental condition are combined into a single DNase-RNase-free 1.5-mL microcentrifuge tube before nucleic acid precipitation.

3. The remainder of the RNA isolation, including on-column DNA digestion, proceeds according to the manufacturer's instructions.
4. RNA yield is small (250-500 ng/sample) when compared with isolating RNA from, for example, whole liver. Therefore, quantify the RNA and test purity using the least amount of sample possible. The NanoDrop instrument (NanoDrop Technologies, Wilmington, DE) is ideal as it only requires 1 μL of undiluted RNA per sample.
4. We routinely reverse transcribe the RNA immediately to maximize the amount of sample for subsequent analyses. Total RNA is reversed transcribed using Ambion's RETROscript kit (Austin, TX) according to the manufacturer's instructions, using random decamer as primers according to manufacturer's instructions.
5. Finally, cDNA can be diluted 1:2 with sterile, RNase- DNase-free water. This will only increase threshold cycle time by one, but will double the starting material and therefore maximize the number of genes one is able to examine per sample.

## NOTES

1. We make use of two surgical pans (similar to a 9 × 13 baking pan), and two wire mesh racks on which to place the four animals. One wire mesh rack is placed on each baking pan, and one pair-fed and one ethanol-fed animal are placed on top of each wire mesh rack. The purpose of this set up is to allow the 2L of perfusate (1L of Hanks- and 1L of Williams E) to drain from the animal into the surgical pan. This allows for a more sterile field to be maintained around the liver because it is not covered with perfusate and also allows for easy clean up of the area after the perfusion is complete.
2. It may be helpful when first learning the perfusion procedure to insert the butterfly needle into the hepatic portal vein before turning on the peristaltic pump. However, starting the pump before inserting the needle generally allows for a better perfusion of the liver.

   If the needle is inserted before starting the pump, many red blood cells may be left at the edges of the liver, which are hard to separate from the Kupffer cells.
3. It is also important to test different batches of collagenase before starting experiments. We use Collagenase type IV from Sigma (Catalog C5138). If you plan on doing multiple experiments, it is a good idea to call the Reserves Department at Sigma and request a 100-mg tester batch of a lot of collagenase with sufficient reserves to completer all of our experiments. Chow-fed control animals can be used to test the new batch of collagenase. It is important to look at the success of the perfusions, as well as, the yield of Kupffer cells. If there is a problem with either of these, it is likely that the lot of collagenase used will not work for other animals. Call Sigma for another tester batch, if necessary.
4. Centrifuging the cells at 4°C and keeping them in ice, starting at Subheading 3.6., step 13, is important for good quality and yield of Kupffer cells.
5. When taking the ethanol cells off the ND gradient, be careful not to remove the fatty hepatocytes. One trick is to insert the Pasteur pipet just past the fatty hepatocyte layer

and then gently squeeze out any remaining air from the rubber bulb. This will push any cells that were in the tip of the pipet out before inserting the pipet down to the Kupffer cell layer.

6. The 3% acetic acid will cause any remaining red blood cells to lyse, and the trypan blue will allow for enumeration of the viable Kupffer cells.
7. The total number of cells from which RNA can be isolated using the RNeasy micro columns is $5 \times 10^5$. If you exceed this number of cells, it is likely that RNA yields will suffer in quantity as well as quality. Therefore, determine the number of cells per well and ensure the total number of cells per experimental condition does not exceed $5 \times 10^5$. Cell counts can be done immediately before experiment initiation. If the cell culture medium is to be changed before initiation of the experiment, count cells after the media change. If needed, adjust experimental protocol accordingly (*i.e.,* use fewer wells/experimental condition).
7. DNase I used for the on-column DNA digestion step of the RNeasy micro procedure can be dispensed in 10 μL aliquots in 600 μLof RNase- DNase-free tubes and stored at –20°C until use. One of these aliquots is used per sample during RNA isolation.
9. It is also possible to perform the reverse transcription reaction without prior RNA quantification. This is due to the fact that the reagent concentrations for the reverse transcription reaction, using Ambion's RETROscript kit, are suitable for up to 2 μg of RNA, far above what is typically isolated using the small numbers of Kupffer cells described here. Input sample amounts can be normalized within and between samples by comparing the expression of a house keeping gene to the gene of interest.

# CHAPTER 10 Hepatic Abnormalities

It is widely recognized that a key component in the development of alcohol-induced liver disease is the diminished capacity of liver to generate and maintain sufficient levels of ATP. This resulting decrease in bioenergetic capacity is thought to contribute to decreased hepatocyte viability and ultimately the pathological changes that occur in alcohol-dependent hepatotoxicity. The importance of mitochondrial dysfunction in this disease process has long been appreciated due to chronic alcohol-mediated defects on the oxidative phosphorylation system, which leads to decreased ATP synthesis.

Moreover, evidence indicates that these chronic alcohol-induced alterations in mitochondria structure and function also contribute to increased production of reactive oxygen and nitrogen species in mitochondria. This increase in mitochondrial oxidant production could result in post-translational oxidative modifications and inactivation of mitochondrial macromolecules, particularly proteins, thereby further compromising mitochondria function in the chronic alcohol consumer. Although the mechanisms responsible for alcohol-dependent mitochondrial dys-function have been studied, the impact of chronic alcohol consumption on the overall content of mitochondrial proteins, that is, the global mitochondrial proteome, has only recently been investigated.

Early studies by Cunningham and colleagues demonstrated that chronic alcohol consumption decreases the synthesis of the 13 mitochondrial-encoded proteins that are components of complexes I, III, IV, and V *(6,7)*, as the result of defects in mtDNA and a decrease in functional mitochondrial ribosomes. It is important to note, however, that there are well over 600 proteins that comprise the mitochondrial proteome, and that close to 100 of these are components of the oxidative phosphorylation system, most of which are encoded by the nuclear genome. Because very little information is available on the effects of chronic alcohol consumption on the total mitochondrial proteome, especially effects on nuclear encoded proteins, we have used two complementary proteomics approaches, two-dimensional isoelectric focusing (IEF)/SDS-PAGE and blue native-PAGE (BN-PAGE) to improve our understanding of the impact these changes have on mitochondrial functionality and their contribution to the development of alcohol induced liver pathology.

## MATERIALS

### Isoelectric Focusing

1. Buffer for rehydration of IEF gel strips: 7*M* urea, 2*M* thiourea, 2% CHAPS, 0.5% *N*-dodecyl-β-D maltoside (*i.e.*, lauryl maltoside), and 0.002% bromophenol blue. Store at –20°C in 1.0-mL aliquots.

2. Equilibration buffer for IEF gel strips: 6M urea, 2.0% sodium dodecyl sulfate (SDS), 0.375*M* Tris-HCl, 20% glycerol, and 0.002% bromophenol blue, pH 8.8 (with HCl). Store at –20°C in 2.0-mL aliquots. Before using, warm to 37°C to redissolve urea into solution.
3. DTT stock solution: 1*M* DTT in water. Store at –20°C in 50-μL aliquots. Do not refreeze and reuse DTT aliquots for future experiments.
4. Ultrapure, low melting-temperature agarose (Aqua*Por*™LM, National Diagnostics, cat. no. EC-204) for sealing IEF gel strips onto SDS-polyacrylamide gel electrophoresis (PAGE) gels: 1.0% agarose (*w/v*) in IX SDS-PAGE running buffer. Make fresh on day of experiment.
5. The Invitrogen ZOOM IPG runner (cat. no. ZM0001) is used to perform IEF in combination with Invitrogen ZOOM strips (cat. no. ZM0011, pH 3 – 10), ZOOM IPG Runner Cassettes (cat. no. ZM0003), and the ZOOM Dual Power Supply (cat. no. ZP10002). Please refer to manufacturer's manual for additional details on setup and use.
6. The ampholyte carriers used for IEF are Ampholines Electrophoresis Reagent (Sigma, cat. no. A5174, pH 3 – 10). Other carrier ampholines or ampholytes can be substituted.

## Blue Native-polyacrylamide Gel Electrophoresis

### *One-dimensional Blue Native-polyacrylamide Gel Electrophoresis*

1. Cathode buffers: a) High-blue cathode buffer: 50 mM tricine, 15mM BisTris, and 0.02% Coomassie brilliant blue G-250, pH 7.0. b) Low-blue cathode buffer: 50 mM tricine, 15 mM BisTris, and 0.002% Coomassie brilliant blue G-250, pH 7.0. Store at 4°C.
2. Anode buffer: 50 mM BisTris, pH 7.0. Store at 4°C.
3. 3X gel buffer: 1.5M aminocaproic acid, 150 m*M* BisTris, pH 7.0. Store at room temperature.
4. Extraction buffer: 0.75 M aminocaproic acid, 50 mM BisTris, pH 7.0. Store at 4°C.
5. Coomassie brilliant blue G-250 suspension: 0.5 M aminocaproic acid and 5% Coomassie brilliant blue G-250, pH 7.0. Store at 4°C.
6. Lauryl maltoside solution: 10% (w/v) *N* dodecyl-(β-D-maltoside in water. Store at –20°C.
7. Molecular weight standards for 1-D BN-PAGE: High-molecular weight native marker kit (Amersham Biosciences, cat. no. 17–0445–01). Dissolve contents of 1 vial into 200 μL of extraction buffer, 25μL lauryl maltoside, and 12 μL of Coomassie brilliant blue. Store at –20°C.

### *Two-dimensional Blue Native-polyacrylamide Gel Electrophoresis*

1. Cathode buffer: 100m*M* Tris, 100m*M* tricine, and 0.1% SDS, pH 8.25. Store at room temperature.
2. Anode buffer: 200m*M* Tris, pH 8.9 with HCl. Store at room temperature.
3. 2-D BN-PAGE Gel buffer: 3*M* Tris-HCl and 0.3% SDS, pH 8.45 with HCl. Store at room temperature.

4. SDS/β-Mercaptoethanol solution: 20μL of β-mercaptoethanol, 200μL of 10% SDS into 1780μL of ultrapure $H_2O$. Make fresh on day of experiment.
5. Agarose solution for sealing 1-D BN-PAGE gel strips onto SDS-PAGE gels: 100 mg of low melting temperature agarose (Aqua Por™LM, National Diagnostics, cat. no. EC-204), 1 mL 10% SDS, 100μL of β-mercaptoethanol into 9.0 mL of ultrapure $H_2O$. Make fresh on day of experiment.

## Total Protein Staining for Gels

1. Coomassie blue stain: 0.3 g of Coomassie blue R-250, 100mL of glacial acetic acid, 250mL of isopropanol, and 650mL of ultrapure $H_2O$. Use 100mL per one mini-gel. Stain gel overnight and do not reuse stain. Gels are destained the next day using 10% glacial acetic acid solution. A small piece of a paper towel is added to the gel container to absorb the Coomassie blue as it "leaches" from gel. Change blue-stained paper towels every 2 – 4 h until clear background on gels is achieved.
2. SYPRO® Ruby Protein Gel Stain (Invitrogen, cat. no. S-12000): After electro-phoresis, fix gel in 100 mL of a 40% methanol and 10% glacial acetic acid solution for 1 h. After fixation step, discard fix solution and incubate gel with 60 mL SYPRO Ruby solution overnight. After staining gel, discard SYPRO Ruby solution and destain gel in a 10% methanol/7% glacial acetic acid solution for 1 – 2 h before visualizing gels.

   It is recommended that gels are destained for at least 16 – 24 h using multiple changes in destain solution to minimize background (*i.e.,* decrease speckling) of gels. A number of imaging platforms have been validated for visualizing SYPRO Ruby-stained gels. Please refer to the SYPRO Ruby product information sheet provided by Invitrogen to obtain the excitation sources and emission filters for optimal visualization of protein in gels.

## METHODS

It is hypothesized that chronic alcohol-induced alterations to the mitochondrial proteome negatively affects mitochondrial bioenergetics and signaling, which contribute, in part, to the development of liver pathology. To address this question, our laboratories have begun to profile the mitochondrial proteome using both conventional 2-D IEF/SDS-PAGE and BN-PAGE. Recent advancements in the field have now made it possible to identify post-translational modifications to mitochondrial proteins by ROS/RNS, identify defects in the assembly of multi-protein complexes in mitochondria, and separate the more hydrophobic respiratory complex proteins of the inner membrane using newly refined gel electrophoresis methods. A scheme depicting our strategy for detecting alterations to the mitochondrial proteome in models of chronic alcohol-induced liver injury.

Briefly, mitochondria isolated from the liver of control and alcohol-fed animals can be subjected to conventional 2-D IEF/SDS-PAGE and BN-PAGE. The high-resolution protein "maps" generated via these electrophoresis methods can then be analyzed for changes in protein abundance or a variety of well-defined ROS/ RNS-mediated post-translational modifications using immunoblotting techniques. Finally, proteins of interest can then be identified using several mass spectrometry techniques. Proteomic studies from our laboratories are considered significant in that novel changes in several key energy metabolism pathways, including fatty acid oxidation, TCA cycle, and oxidative phosphorylation system were found to be altered in mitochondria isolated from animals exposed to ethanol chronically. These alterations were in response to alcoholic-mediated metabolic stress and correlated to those pathways with a direct impact on the etiology of disease. A detailed discussion of these techniques is provided in the sections to follow.

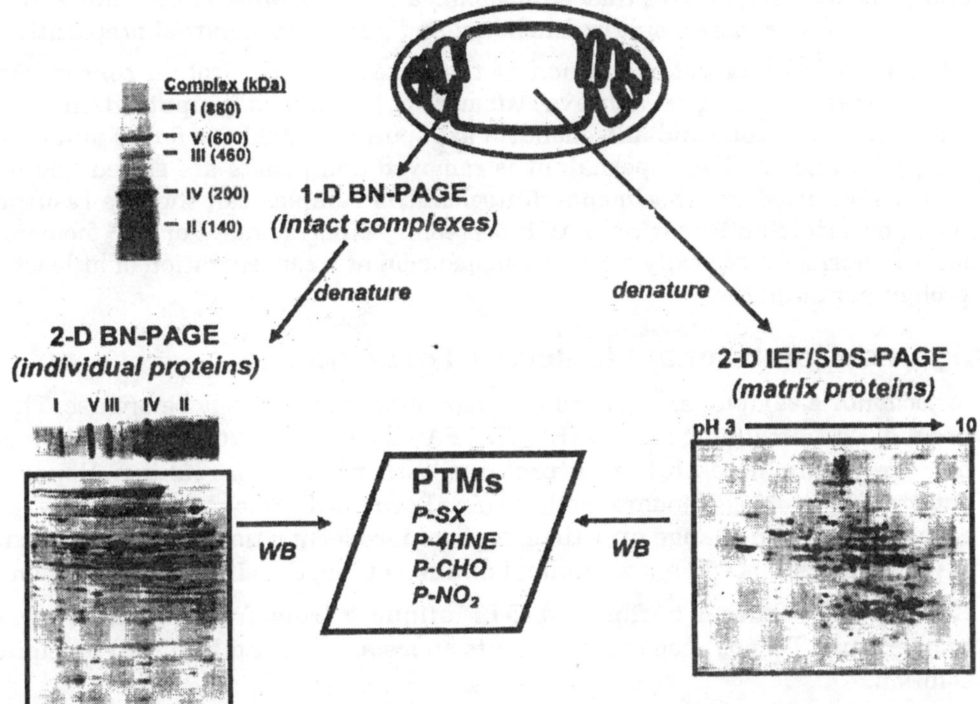


**Fig. 10.1. Analysis of the mitochondrial proteome using 2-D IEF/SDS-PAGE and BN-PAGE proteomics. 2-D IEF/SDS-PAGE is used to generate high resolution maps of the soluble matrix proteins (right panel), whereas 1-D and 2-D BN-PAGE is used to define changes to respiratory chain proteins (left panels). The right-side panel illustrates a 2-D IEF/SDS-PAGE separation of proteins from whole liver mitochondria using the Invitrogen IPGRunner® System in combination with pH 3-10 IEF gel strips, whereas the left side panels demonstrate the BN-PAGE separation of inner membrane proteins. Roman numerals are used to identify the five oxidative phosphorylation complexes in the BN-PAGE gels. During nondenaturing 1-D BN-PAGE, the five complexes remain intact (first step) whereas under the denaturing conditions used for 2-D BN-PAGE the individual proteins of each complex are resolved by size (second step). Thus, those proteins that comprise each complex are aligned vertically within the gel. Note that both gel platforms can be subjected to image analysis to look for changes in protein abundances or subjected to western blotting (WB) to reveal post-translational modification (PTMs) by reactive oxygen and nitrogen species, as well as electrophilic lipids, as a number of antibodies have been developed against these modifications. P-SX, protein thiol modifications; P-4HNE, 4-hydroxynonenal protein adduct; P-CHO, protein carbonyls; P-$NO_2$ 3-nitrotyrosine protein modification.**

## Preparation of Mitochondrial Samples

1. Mitochondria are prepared from fresh liver tissue by standard differential centrifugation techniques and should exhibit tightly coupled respiration with respiratory control ratios

in the range of 4 – 8. Moreover, mitochondrial protein yield should average 25 – 30 mg of mitochondrial protein/g wet weight liver and cytochrome *c* oxidase and citrate synthase activities should be determined as additional markers of mitochondria purity and yield. It is imperative that mitochondria used for proteomics studies are isolated by methodologies that result in functional and pure mitochondrial preparations.

After isolation and determination of the mitochondrial protein concentration, the appropriate volume to achieve 1.0 mg of protein is pippetted into a 1.5-mL microcentrifuge tube and mitochondria are spun at 15,000 × g for 10 min to form a 10 mg protein pellet. The supernatant is removed and pellets are stored and dried at – 80°C before used in experiments. Mitochondria samples can then be resuspended in the appropriate buffer for BN-PAGE proteomic experiments. For IEF focusing studies mitochondria are typically stored in suspension at a concentration of at least 10 mg of protein per milliliter.

## Running One-dimensional Isoelectric Focusing Gels

1. Mitochondria samples are thawed and kept on ice immediately before use. The optimal protein concentration for a 2-D IEF/SDS-PAGE gel is 50 – 200 mg of protein per gel to achieve maximal resolution of proteins with minimal streaking. Repeat protein determination using standard methods (*i.e.,* Lowry or Bradford protein assay) as protein concentration can change over time. A protein concentration of no less than 5 mg/mL is preferred as this will allow minimal dilution of reagents in IEF rehydration buffer.
2. Remove rehydration buffer and DTT aliquots from freezer and thaw to room temperature. Do not keep these reagents on ice as urea and DTT will precipitate from solution.
3. Into 1.0 mL of rehydration buffer add 10 μL of pH 3 – 10 ampholines, 10 μL of 200 mM tributylphosphine (BioRad, cat. no. 163 – 2101), and 40 μL of 1M DTT. Mix well and keep at room temperature.
4. IEF strips can be rehydrated overnight with a total volume of 150 – 160 μL containing the mitochondrial sample. The following example conditions are given below for the rehydration of strips with buffer and a protein sample that has a concentration of 10 mg/mL. Samples are allowed to extract in rehydration buffer for at least 30 min at room temperature before loading samples onto IEF gel strips. Gently vortex samples every 5 – 10 min during extraction.

| *Protein amount* | *Quantity of sample* | *Quantity of rehydration buffer* |
|---|---|---|
| 50 μg | 5μ L | 155μ L |
| 100 μg | 10μ L | 150μ L |
| 200 μg | 20μ L | 140μ L |

5. Following the procedure described by the specific manufacturer for the instrumentation being used, slowly load 160 μL of sample onto the strip. Samples will load along the entire lane by capillary action. Pipet slowly to minimize the introduction of bubbles

into the gel lane as this could result in uneven rehydration of the IEF strips and loss of sample. Load all samples into cassette lanes before removing IEF gel strips from freezer.

6. After loading samples, remove the IEF strip pack from the freezer and keep strips on ice to minimize condensation. For each sample, carefully remove an IEF strip from the plastic backing, holding it with clean forceps by the marked negative (–) end. Holding the IEF strip from the (–) end and with the gel side up, gently slide the IEF gel strip into the cassette well until the strip reaches the (+) end of the cassette.
7. After rehydration of the IEF gel strips is complete, assemble the electrophoresis apparatus according to the manufacturer's instructions.
8. Perform IEF gel electrophoresis using the following conditions: 175 V for 20 min, ramp to 2000 V for 45min, 2000 V for 30 min, ramp down to 500 V for 30 min, and 500 V for 2 h. As IEF proceeds, the bromophenol blue dye front will migrate from the top to the bottom of the IEF strips and a green/blue/yellow band will appear at the anode (+) after IEF.
9. After IEF, seal cassettes in plastic wrap and freeze at –80°C until ready to perform second dimension SDS-PAGE.

## Running 2-D IEF/SDS-PAGE Gels

1. For second dimension SDS-PAGE gels a wide range of equipment is commercially available that is compatible with the first dimension IEF. Homogenous (10 or 12%) or gradient (8 – 15%) acrylamide gels can be used for 2-D SDS-PAGE. It is important that spacer plates are selected that provide sufficient width for inserting the IEF gel strips.
2. Remove IEF gel strips from freezer and allow IEF gel strips to thaw for a few minutes before beginning the equilibration step. For equilibration, prepare 1.0-mL equilibration buffer per IEF gel strip. Remove equilibration buffer and DTT aliquots from freezer and thaw at room temperature. To 1.0 mL of equilibration buffer, add 50 µL of 1 *M* DTT.
3. To remove the IEF gel strips from cassette, remove plastic covering gently and using forceps place each strip into a 15-mL conical tube, gel side up, and cover with 1.0 mL of equilibration buffer. Gently rock strips with equilibration buffer for 15 min. During the equilibration step, prepare the agarose solution and keep warm. The aga-rose solution is used to "seal" the IEF gel strips to the top of SDS-PAGE gel.
4. After equilibration, remove strips from conical tubes, cut small plastic end from the anode (-) side, rinse strips in IX SDS-PAGE running buffer, and slide the IEF gel strip between the plates of the SDS-PAGE gel. Gently pipet the warm agarose solution over strip to "seal" the IEF gel strip on top of the SDS-PAGE gel. Do not allow bubbles to form between the IEF gel strip and the top of the SDS-PAGE gel. Slide a "tooth" from a 1.5-mm gel comb between glass plates at the (–) end of the gel strip. This will serve as the "well" for molecular weight markers. Gels are placed upright for 15 – 30 min to allow the agarose to set before beginning electrophoresis.

5. Once the agarose has set, gently remove the gel comb "tooth" and load broad range molecular weight markers to the well. Assemble the gel apparatus per the manufacturer's instructions, fill inner and outer chambers with standard 1 X SDS-PAGE gel electrophoresis running buffer, and run gels for 1 – 2 h at 100 V or until the dye front reaches the bottom of gel.
6. After electrophoresis, gels can be stained for total protein using conditions described above (Coomassie Blue or SYPRO Ruby) or subjected to western blotting. Typically, we stain the 2-D IEF/SDS-PAGE gels with SYPRO Ruby stain to increase sensitivity of low abundance proteins. An example of a 2-D IEF/SDS-PAGE gel stained with SYPRO Ruby stain is presented in Fig. 10.1.

## Running 1-D BN-PAGE Gels

1. BN-PAGE is a specialized method of electrophoresis that facilitates the high-resolution analysis of membrane protein complexes in their native state from biological membranes. A detailed discussion of the development and theory of BN-PAGE for analysis of the oxidative phosphorylation system proteins is presented in the seminal papers of Schagger and von Jagow.
2. For BN-PAGE, the procedure more recently described by Brookes et al. is used with the 1-D BN-PAGE, typically performed with 5 – 12% or 5 – 16.5% gradient gels with 1.5-mm spacer plates. Recipes for gels are as follows

| | *Resolving Gel solutions* | | | *Stacking gel solution* |
|---|---|---|---|---|
| | *Light-5%* | *Heavy-12%* | *or Heavy-16.5%* | *4%* |
| Protogel | 0.66mL | 1.60mL | 1.91mL | 0.60mL |
| Water | 1.97mL | 0.56mL | – | 2.40mL |
| 3XGel Buffer | 1.34mL | 1.34mL | 1.16mL | 1.50mL |
| Glycerol | 26.0µ | 26.0µL | 23.0µL | 70.0µL |
| TEMED | 4.0µL | 4.0µL | 3.50µL | 9.0µL |

3. On the day of the experiment, remove the 1.0 mg mitochondrial pellets from the –80°C freezer and keep on ice. Add 100 µL of extraction buffer and 12.5 µL of 10% *N*-dodecyl-β-D-maltoside to each 1.0 mg of mitochondrial pellet. Gently resuspend mitochondrial protein by pipetting to extract and dissociate the protein. Place samples on ice for 30 – 60 min with gentle vortexing every 5 – 10 min. After this extraction step, centrifuge samples at 15,000 × g for 5 min at 4°C to pellet any non-dissolved material, remove supernatant to a fresh microcentrifuge, and then determine the protein concentration of the extract. A typical protein concentration of sample prepared by the procedure described here is 5 – 8 mg/mL.
4. After determining the protein concentration in each sample, add 6.3µL of the cold Coomassie brilliant blue G-250 suspension (Solution #5,) to each tube of mitochondrial extract and gently vortex. Load samples immediately onto gel for 1-D BN-PAGE.

5. After assembling the gel apparatus per manufacturer's directions, fill wells of the stacking gel with cold Hi-Blue cathode buffer, and load 75 – 250μg of protein per lane. Fill the inner (*i.e.* upper) buffer chamber with cold Hi-Blue cathode buffer and the outer chamber with cold anode buffer Electrophoresis is performed in the cold room at 40V for 1 h or until samples have migrated into the resolving gels. At this time the cathode buffer is changed to cold Low-Blue cathode buffer and electrophoresis is continued for an additional 3 – 4 h or until dye front reaches the bottom of the gel.

6. At this point, gels can be stained with Coomassie blue to visualize the content of the intact oxidative phosphorylation complexes in samples or gels can be processed for 2-D BN-PAGE by the procedures described below. A representative 1-D BN-PAGE gel is shown in the upper left panel of Fig. 10.1.

## Running 2-D BN-PAGE Gels

1. To resolve the individual polypeptides that comprise each oxidative phosphorylation complex, the intact gel lane containing all five complexes is cut from the 1-D gel, rotated 90°, and laid on top of a Tris-Tricine/SDS-PAGE gel. The recipe for preparing one, 1.5-mm thick 2-D BN-PAGE gel is as follows leaving a 1.0-cm gap at the top of the gel.

| | |
|---|---|
| 2-D BN-PAGE Gel Buffer | 2.98 mL |
| Protogel | 2.98 mL |
| $H_2O$ | 2.31 mL |
| Glycerol | 0.72 mL |
| 10% AMPS | 60.0 μL |
| TEMED | 6.00 μL |

2. To apply the 1-D BN-PAGE gel strip on to the top of the SDS-PAGE resolving gel, raise-up the gel plates in the clamps and place the top, that is, a 1– to 2-cm gap on an angled hot plate or heat block. Pour approximately 4 – 5 mL of hot agarose into the gap at the top of the gel sandwich, and using the back plate as a "staging area" and the hot agarose as a "lubricant", gently slide the gel lane down between the plates until the strip lays on top of the SDS-PAGE resolving gel. Once the gel is in place, insert a single "tooth" of a gel comb on the end to serve as a well for molecular weight markers.

3. Remove the gel plate assembly from the hot plate and allow the agarose to set for 20 – 30 min. Excess agarose is then removed from the top of the gel with a scalpel blade and then the gel is overlaid with a thin layer of the SDS/β-mercaptoethanol solution every 5 – 10 min for 30 – 45 min to denature proteins.

4. Fill the inner chamber with Cathode buffer (Solution #1,) and the outer chamber with the anode buffer load molecular weight markers into well, and run gels at 30V for 45 min followed by 110V for 1.5 to 2 h.

5. After electrophoresis, gels can be stained for total protein using the conditions (Coomassie Blue or SYPRO Ruby) or subjected to western blotting. Typically we stain 2D BN Page gels with Coomassie Blue stain.

## Image Analysis of Gels

Instructions for image analysis of gels are detailed in the following publications from our laboratories. In brief, scanned TIFF images for 1-D and 2-D BN-PAGE gels can be analyzed using either Scion Image Beta 4.02 (Scion Corp) or Quantity One (BioRad Laboratories) using the directions provided for each software program. For 2-D IEF/SDS-PAGE image analysis, gels are scanned, saved as TIFF files, and analyzed for differences in protein abundance between control and ethanol samples using PDQuest Image Analysis software (BioRad Laboratories).

## Mass Spectrometry for Identification of Proteins

Proteins found by image analysis to change in relative abundance in response to a treatment are cut from gels and subjected to matrix-assisted laser desorption ionization time-of-flight (MALDI-TOF) mass spectrometry by standard methodologies as described in to identify proteins. The peptide masses identified using MALDI-TOF mass spectrometry are entered into the MASCOT search engine (see www.matrixscience.com) and the NCBI data base is searched to match the tryptic peptide fingerprint with a parent protein.

## ANALYSIS DISEASES

Mitochondrial ribosomes (mitoribosomes) are responsible for the translation of 13 proteins coded for by the mitochondrial genome. These proteins (ND1, ND2, ND3, ND4, ND4L, ND5, ND6, Cyt b, COI, COII, COIII, ATPase6, and ATPase8) are required for the successful assembly of the respiratory complexes NADH dehydrogenase, ubiquinol: cytochrome c oxidoreductase, cytochrome c oxidase, and ATP synthase, respectively. Accurate assembly and subsequent activity of mitoribosomes is therefore essential for the maintenance and correct functioning of the electron transport chain.

The rat liver mitochondrial ribosome has a molecular mass of approximately 3.57MDa, dimensions of 26.2 nm × 23.6 nm, and a sedimentation coefficient of 55S. It has a chemical composition of 75% protein and 25% RNA and comprises a large 39S subunit (LSU) and a small 28S subunit (SSU). Subunits are assembled inside the mitochondrial matrix from 12S (SSU) and 16S (LSU) rRNA, encoded for by the mitochondrial genome, and ribosomal proteins, encoded for by the nucleus and imported into the mitochondrion. One of the earliest observations seen in the liver during chronic ethanol feeding is decreased mitochondrial protein synthesis.

Although a number of potential molecular mechanisms may explain this decrease, one specific mechanism involves the impaired assembly of mitoribosomes. Investigations conducted by the Cunningham lab have revealed that mitoribosomes isolated from young male rats (200 g) fed ethanol (Lieber-DiCarli diet) for 31 d exhibit decreased sedimentation rates through sucrose density gradients when compared with their paired controls. Further, sedimentation velocity experiments reveal a significant decrease in the average sedimentation coefficient for the intact ethanol mitoribosome (52.2, ethanol-fed; 53.7, control) as well as for the small ribosomal subunit (27.0, ethanol-fed; 28.3, control).

In addition, mitoribosomes isolated from ethanol-fed animals exhibit significant decreases in their translational diffusion coefficients ($1.02 \times 10^{-7} cm^2 s^{-1}$, ethanol-fed; 1.10, control) and larger hydrodynamic diameters (42.1 nm, ethanol-fed; 39.1, control). These ethanol-elicited alterations in the physicochemical properties of the hepatic mitoribosome are accompanied by significant amounts of disassociation of the intact 55S monosome (8 – 14%) into its constituent subunits. Further, in vitro studies on the translation capacity of mitoribosomes have revealed that while total translation activity is depressed by 29% in the ethanol-fed animals relative to the controls, no difference is detected between the two treatment groups in the activity of the intact monosomes which suggests that the decrease in translation activity observed in ethanol-fed animals is caused by either increased disassociation or decreased association of functional ribosomes rather than impaired activity of intact monosomes.

Whether this means that ethanol feeding induces increased disassociation or decreased association of mitoribosomes in vivo or whether it causes a more relaxed and loosely assembled monosome which becomes disassociated upon isolation has yet to be determined. The sedimentation properties rnitochondrial ribosomes are affected not only by the size of the ribosomes but also by their shape. Any ethanol-elicited alterations in the ratios of constitutive mitochondrial ribosomal proteins (MRPs) may lead to a change in the shape of the ribosome and consequently a decrease in sedimentation rates. Additionally, altered MRP levels may lead to formation of inaccurately assembled ribosomes that are easily dissociated into subunits or unable to bind to functionally important docking sites on the inner mitochondrial membrane (IMM). Further, it is unclear as to whether any of the MRPs are susceptible to the kinds of protein damage commonly incurred during chronic ethanol feeding, that is oxidative modification, nitrosative damage and formation of adducts such as acetaldehyde (AA), malondialdehye-acetaldehyde (MAA), and the hydroxyethyl radical (HER). Ethanol-elicited modifications of MRPs may result in an alteration in their binding to other MRPs, rRNA or IMM components, as well as alter their rates of degradation. In addition to altering the overall structure of the mitochondrial ribosome, ethanol feeding has been shown to significantly decrease the levels of a specific population of MRPs. The reasons for the ethanol-elicited decreases in a specific population of MRPs as opposed to a global down-regulation are unclear; however, a recent proteomics study by Chevallet et al. has demonstrated the selective depression of nine MRPs upon depletion of mitochondrial DNA (mtDNA) in 143B osteosarcoma cells. Our laboratory has shown that mtDNA depletion is a common occurrence during chronic ethanol feeding.

In this chapter, we detail methodologies for the isolation of mitochondrial ribosomes from hepatic mitochondria, determination of their activities, extraction of MRPs from the ribosomes and two-dimensional nonequilibrium pH gradient electrophoretic (NEPHGE) analysis of MRPs. Separated proteins can then be excised from the gels and subjected to mass spectrometry or transferred to nitro-cellulose membranes and probed for oxidative and nitrosative modifications by western blotting.

## MATERIALS

### Isololation of Mitochondrial Ribosomes from Mitochondria

1. Isolation buffer: 0.25 *M* sucrose and 2 *mM* HEPES, pH 7.4 (pH with *6M* KOH). Store at 4°C.
2. RC DC Protein Assay Reagents (BioRad, cat. no. 500 – 0120).
3. Digitonin (solid, free of water-insoluble constituents, Sigma cat. no. D-141).

4. Buffer A: 0.26M sucrose, 40mM KC1, 15mA/ $MgCL_2$, 14mM Tris-HCl, pH 7.5, and 0.8 mM ethylene diamine tetraacetic acid (EDTA); Store at 4°C. Just before use add β-mercaptoethanol to a final concentration of 5 mM.
5. Buffer B: 100m*M* KC1, 20 m*M* $MgCL_2$, and 20 mM triethanolamine pH 7.5. Store at 4°C. Just before use, add β-mercaptoethanol to a final concentration of 5 m*M*.
6. Puromycin dihydrochloride (solid, Sigma, cat. no. P 7255).
7. Magnesium-deficient buffer B: buffer B without the $MgCl_2$ but including 1 m*M* EDTA.

## Ribosome Activity Assay

1. Ribosome activity buffer: 50 m*M* KC1, 50 m*M* Tris-HCl, pH 7.8, and 20 m*M* $MgCl_2$. Store at 4°C.
2. Transfer RNA (from baker's yeast, Type, X., Sigma cat. no. R9001).
3. Pyruvate kinase (from rabbit muscle, Type VII, cat. no. P7768).
4. Phosphoenol/pyruvate (Sigma, cat. no. P 7127).
5. ATP (Sigma, cat. no. A 7699) and GTP (Sigma, cat. no. G 5884).
6. L-[4-$^3$H]Phenylalanine (Amersham, cat. no. TRK204-250UC).
7. Econo-Safe (RPI, cat. no. 111175).

## Preparation of Soluble Mitochondrial Translation Factors

1. Hypotonic buffer: 20m*M*Tris-HCl, pH 7.5, 20mM $MgCl_2$, and 40mM KCl Store at 4°C. Just before, use add β-mercaptoethanol to a final concentration of 5 mM.
2. 2*M* KCl.
3. Sephadex G-25 (Amersham, cat. no. 17 – 0031 – 02).

## Extraction of Ribosomal Proteins from Pelleted Mitochondrial Ribosomes

Both MRP extraction procedures produce similar 2DE profiles.

## Lithium Chloride-Urea Procedure

This procedures is used in the lab when purified ribosomes or subunits are first pelleted; starting material usually two livers.

1. Ribosomal protein solubilization solution: *6M* urea, 3*M* LiCl, 50m*M* KCl, and 5m*M* β-mercaptoethanol, pH 3.5 – 5.5, with 3N HC1. Can be stored at –20°C in 500 μL aliquots.
2. 3N HC1.
3. 20% (w/v) trichloroacetic acid (TCA).
4. Ethanol/ether (1:1 vol/vol).

## Acetic Acid-Acetone Procedure

This procedure is used when MRPs are extracted directly from sucrose density gradient fractions without prior sedimentation of purified ribosomes or subunits; starting material can be one liver.

1. Glacial acetic acid.
2. 1*M* magnesium chloride.
3. Acetone.

## 2D-Electrophoretic Analyses of Mitochondrial Ribosomal Proteins (NEPHGE)

1. 40% Biolyte 3/10 ampholyte (BioRad, cat. no. 163 – 1113) and 40% Biolyte 5/7 ampholyte (BioRad, cat. no. 163 – 1153).
2. Ribosomal protein lysis buffer: 9.5 M urea (ultrapure™urea, Invitrogen, cat #. 15505 – 035), 2% (w/v) Triton X-100, 2% ampholytes (1% Biolyte 3/10 ampholyte, 1% Biolyte 5/7 ampholyte). Store at –20°C as 500 μL aliquots.
3. 30% acrylamide/bis solution, 29, 1 ratio (BioRad, cat. no. 161 – 0156). Store at 4°C.
4. 10% (w/v) Triton X-100.
5. 10% (w/v) ammonium persulfate (Sigma, cat #. A3678). Make up fresh every 2 wk.
6. TEMED (BioRad, cat. no. 161 – 0801).
7. Equilibration buffer: 10% (w/v) glycerol, 2.3% (w/v) sodium dodecyl sulfate (SDS), 0.002% bromophenol blue, and 62.5 mM Tris-HCl, pH 6.8.
8. 0.5% (w/v) agarose (ultrapure™agarose, Invitrogen, cat. no. 15110 – 019) in equilibration buffer.
9. Dithiothreitol (DTT), solid (BioRad, cat. no. 161 – 0610).
10. Possible staining procedures: SYPRO® Ruby protein gel stain (Invitrogen, cat no. S-12000), Brilliant Blue R staining solution (Sigma, cat. no. B6529) and SilverQuest™Silver Staining kit (Invitrogen, cat. no. LC6070).
11. Destaining solution (for Brilliant Blue R staining only): 30% (v/v) ethanol and 10% (v/v) acetic acid.

## EQUIPMENT

1. Nitrocellulose filter disks (Millipore).
2. Gel extrusion needles (BioRad, cat. no. 165 – 1944).
3. Appropriate ultracentrifuge, rotors and centrifuge tubes.
4. Microcentrifuge.
5. UV plate reader or spectrophotometer.
6. Scintillation counter.
7. Gradient fractionator (e.g. Labconco, Auto-Densiflow).
8. Liquid nitrogen.
9. Probe sonicator (*e.g.*, Fisher Sonic Dismembrator 300, with Microtip).
10. Lyophilizer.
11. Glass tubes for NEPHGE (*e.g.*, 7 mm OD, 5 mm ID, 15 cm length).

12. Electrophoresis unit (*e.g.*, Hoefer Scientific Instruments, GT1).
13. Image analysis system and suitable software.
14. Microwave (optional).

# METHODS

## Isolation of Mitochondrial Ribosomes from Mitochondria

This procedure provides sufficient yields of mitochondrial ribosomes for sedimentation profile analyses and, when used in conjunction with silver staining, 2D/SDS-PAGE mini-gel analyses of ribosomal proteins. If 55S ribosomal activity is to be determined or 2D electrophoretic analysis of ribosomal proteins using larger gels is intended we suggest using two rat livers as the starting tissue.

1. Isolate hepatic mitochondria using standard procedures.
2. Perform a protein assay on the isolated mitochondria. We use the Lowry protein assay but any detergent compatible assay will work just as well. Digitonin needs to be added to the mitochondria at a ratio of 0.11 mg of digitonin per milligram of mitochondrial protein. Dissolve a suitable amount of digitonin in mitochondrial isolation buffer such that when the digitonin solution is added to the mitochondria it not only gives the required detergent to protein ratio but that it also dilutes the mitochondrial protein concentration down to 20 mg/mL.
3. Let the digitonin solution cool down before adding it to the mitochondria on ice. Stir the mitochondrial suspension gently for exactly 15 min.
4. Dilute the suspension down 10-fold with ice-cold isolation buffer, allow to continue stirring for a further 10s, then centrifuge immediately at 12,000 g for 10 min.
5. Collect the pelleted mitoplasts and resuspend them in isolation buffer to wash them. Re-pellet the mitoplasts and resuspend them in buffer A. Mitoplasts can be stored overnight at –70°C.
6. Perform a protein assay on the mitoplasts and dilute them down with buffer A to a concentration of 20 mg/mL. Lyse them with Triton X-100 (2% w/v final concentration).
7. Clarify the suspension at 12,000 g for 45 min. Remove the supernatant and layer it onto 15ml cushions of 1M sucrose and 1% (w/v) Triton X-100 dissolved in buffer B. Centrifuge at 200,000 g) for 4 h

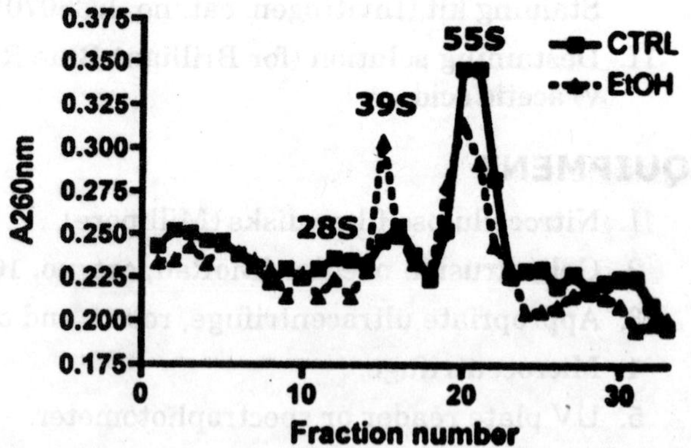


Fig. 10.2. Effects of chronic ethanol feeding (12mo) upon the sucrose density gradient profiles of mitochondrial ribosomes and their subunits isolated from a single control and alcoholic liver. Sucrose cushion purified hepatic mitochondrial ribosomes were separated through 10-30% sucrose gradients into monosomes and subunits. The $A_{260\ nm}$ of the 50-μL fraction aliquots was monitored using a 96-well fluorescent plate-reader (in UV mode).

(a 50.2Ti is a good choice of rotor). The pellets consist of mitochondrial ribosomes and some mitoplast membrane fragments.

8. Decant the supernatant, resuspend the crude mitochondrial ribosomal pellets in 1 ml of buffer B and layer them on top of a 10 – 30% linear sucrose gradient (30mL) in buffer B and centrifuge at 53,000 g for 14 h (SW27 rotor). Fractionate the gradients and monitor the $A_{260\ nm}$ of the fractions for ribosomal RNA. We detect the $A_{260\ nm}$ of 50-μ.L aliquots of our gradient fractions using a 96-well fluorescent plate reader that is also capable of reading U.V. Alternatively, fractions can be monitored spectrophotometrically. Absorbance peaks of free ribosomal particles corresponding to the small (28S) and large (39S) ribosomal subunits and the intact 55S monosome should be detected (see Fig. 10.1). A small peak is also detected at the bottom of the gradient. This corresponds to a population of ribosomes that may be attached to inner mitochondrial membrane fragments via unique attachments, for example, nascent polypeptide chains being inserted into the inner mitochondrial membrane.
9. If the aim is to isolate mitochondrial ribosomal proteins without any contaminating nascent polypeptides or, alternatively, if it is required to strip from the IMM those ribosomes attached via newly synthesized polypeptides, then incubate the crude ribosomal pellet in 1 ml buffer B containing 1 mM puromycin for 15 min at 37°C before layering on top of the sucrose gradients.
10. Fractions corresponding to the intact ribosomes or the ribosomal subunits can be centrifuged at 230,000 g for 5 h (60Ti rotor) and stored as a pellet at –70°C. Alternatively, the proteins associated with the intact 55S ribosome or the respective subunits can be subjected directly to electrophoresis and detected by silver staining.
11. If it is desired to separate 55S mitochondrial ribosomes into their 28S and 39S, small and large ribosomal subunits, respectively, resuspend the pelleted 55S monosomes or crude ribosomal pellets in magnesium-deficient buffer B and sediment them through 10 – 30% sucrose gradients made up in the same buffer.

## Ribosome Activity Assay

1. Incubate 2 – 5 pmol of 55S mitochondrial ribosomes in ribosome activity buffer supplemented with (final concentrations) 500 mg/mL tRNA, 100 mg/ml pyruvate kinase, 2 mM phosphoenolpyruvate, 5 mM ATP, 0.3 mM GTP, 0.2 mg/mL translation factors and 10 μCi [$^3$H]phenylalanine in a final volume of 50 μL.
2. Remove 5 μL of the reaction mixture and spot onto a nitrocellulose filter disk. This represents time zero. Start the reaction by adding poly (U) to a final concentration of 1 mg/ml, remove 5 μL aliquots at suitable time points up to 1 h (we find that radiolabel incorporation is linear for this time period) and spot them onto nitrocellulose filter disks (1 cm$^2$).
3. Incubate the filter disks in ice-cold 10% TCA (w/v, 2mL per 1 cm$^2$ filter) for 30 min with gentle swirling to stop the reaction and to precipitate the protein onto the filter.
4. Wash the disks in fresh ice-cold 10% TCA for 10 min with gentle swirling.
5. Place the disks in 5% TCA and heat to 90°C for 15 min.

6. Quantify the radioactivity present on the filters by adding 5 μL of a scintillation cocktail of choice (Econo-Safe works well for us). Express the phenylalanine polymerization as pmol [$^3$H] phenylalanine/mg RNA/pmol mitochondrial ribosomes.

## Preparation of Soluble Mitochondrial Translation Factors

1. Resuspend mitochondria in hypotonic buffer at a concentration of 5 mg/mL and incubate on ice for 15 min to allow the mitochondria to swell.
2. Subject the mitochondria to three cycles of freeze-thawing using liquid nitrogen then sonicate (we use 4 × 30s pulses at 50%; to disrupt the mitochondria.
3. Clarify the mitochondrial suspension at 16,000 g for 20 min at 4°C. Remove the supernatant and adjust it to 500 mM KCl.
4. Centrifuge at 200,000 g for 4 h at 4°C to pellet mitochondrial ribosomes.
5. Pass the supernatant over a Sephadex G-25 column (to remove endogenous amino acids) and collect the void volume. This contains the mitochondrial translation factors. Lyophilize the translation factors and store at –70°C.

## Extraction of MRPs from Pelleted Mitochondrial Ribosomes

### *Lithium Chloride-Urea Procedure*

1. Resuspend 55S mitochondrial ribosomes or mitoribosomal subunits in 150 μL of ribosomal protein solubilization solution in a 1.5 mL microcentrifuge tube.
2. Place a micro stir bar in the centrifuge tube (the stir bar will stand on its end) and stir the ribosomes overnight at 4°C (make sure the sample is stirring and it is not the tube that is spinning).
3. Pellet the RNA at 200,000 g (we use a TLA-100 rotor with 200 μL capacity tubes) for 1 h. Remove the supernatant and reserve. Re-extract the RNA pellet by adding another 150μL of ribosomal protein solubilizing solution. Stir for 6 h and pellet the RNA at 200,000 g for 1 h. Remove the supernatant and combine it with the supernatant from the first extraction.
4. The RNA pellet should contain rRNA and any mRNA that remains associated with the ribosome.
5. Precipitate the ribosomal proteins overnight at 4°C by adding 2 volumes of 20% TCA. Centrifuge at 10,000 g in a microcentrifuge for 15 min. Remove the supernatant and wash the pellet in 1:1 (v/v) ethanolether.
6. Allow to air-dry for 5 – 10min before resuspending the ribosomal proteins in an electrophoresis buffer of choice. We suggest using ribosomal protein lysis buffer supplemented with DTT to a final concentration of 1% (w/v) just before use.

### *Acetic Acid-Acetone Procedure*

We have used this procedure on as little as 0.1$A_{260\ nm}$ units of mitochondrial ribosomes.

1. Collect fractions from sucrose density gradients corresponding to intact 55S monosomes or ribosomal subunits.

2. Add magnesium chloride to a final concentration of 100 m*M*.
3. Add 2 volumes of glacial acetic acid. Invert a couple of times. Leave for 45 min on ice.
4. Pellet precipitated RNA at 5000 *g* for 10 min.
5. Add 5 volumes of ice-cold acetone. Invert a couple of times. Leave at −20°C overnight.
6. Pellet precipitated protein at 5000 g for 10 min.
7. Wash pellets with ice-cold acetone. Re–pellet protein at 5000 g for 10 min.
8. Remove supernatant and allow pellet to air-dry (approximately 10min).
9. Resuspend pellet in ribosomal protein lysis buffer supplemented with DTT to a final concentration of 1% (w/v) just before use.

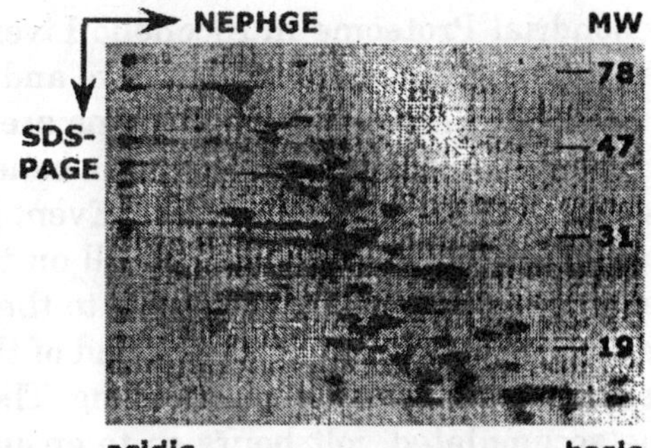


**Fig. 10.3. Two-dimensional (NEPHGE/SDS-PAGE) proteomic analysis of mitochondrial ribosomal proteins extracted from intact 55S monosomes (modified from). Proteins were extracted from the 55S monosome using lithium chloride and urea separated in the first dimension by NEPHGE (pH 3-10) and in the second dimension by 12% SDS-PAGE. Proteins were stained with Coomassie Brilliant Blue.**

## 2D-Electrophoretic Analyses of Mitochondrial Ribosomal Proteins (NEPHGE)

Early isoelectric focusing methodologies in which ampholytes were used to form pH gradients suffered from gradient drift. The development in recent years of immobilines (buffering acrylamide derivatives that contain either a free carboxylic acid group or a tertiary amino group) has allowed the formation of stable pH gradients in the range of pH 3 – 12. These immobilized pH gradients (IPGs) are reproducible and insensitive to disturbances from sample components. We have used the ZOOM® IPGRunner™ system from Invitrogen in conjunction with silver staining to analyze the protein content of the large ribosomal subunit from control and ethanol mitochondrial ribosomes. Figure 10.1 shows a sucrose density gradient profile of mitochondrial ribosomes isolated from control and ethanol-fed animals. Ethanol-fed animals show decreased levels of intact 55S ribosomes but an increase in the level of 39S ribosomal subunits. This is suggestive of an ethanol-elicited increase in ribosome dissociation or a decrease in ribosome association. The 39S peaks from Fig. 10.1 were extracted using the acetic acid-acetone procedure and separated in the first dimension by isoelectric focusing using IPGs (ZOOM strips, Invitrogen) of range pH 6 – 10 followed by second dimension SDS-polyacrylamide gel electrophoresis (PAGE; Fig. 10.2). A significantly greater amount of ribosomal protein was extracted from the 39S peak in the ethanol-fed animals when compared with the paired control. This greater yield of ribosomal protein reflects the increase in 39S subunit seen in Fig. 10.1 and suggests that the increase in 39S fraction as measured by $A_{260nm}$ is the result of increased levels of intact 39S subunits and not due to increased levels of mRNA and rRNA associated with the 39S fraction.

The methodologies involved in using Invitrogen's ZOOM® IPGRunner™system are described in detail in the chapter "Proteomic Approaches to Identify and Characterize Alterations to the

Mitochondrial Proteome in Alcoholic Liver Disease" by Bailey et al. in this volume. We will therefore focus here on an alternative and excellent method of separating ribosomal proteins, that is, NEPHGE. Ribosomal proteins are, for the most part, extremely basic in charge. The 39S ribosomal subunit for example is believed to comprise 48 proteins with pis ranging from 6.6 to 12.3 with an average of 9.6. Even, if these proteins were to be focused using IPGs, it would be difficult to include them all on the same 2D/IEF/SDS-PAGE gel. During NEPHGE proteins are not necessarily focused to their pIs but instead migrate at different rates across the gel due to their charge. At the end of the run, some proteins will have focused to their pIs while the others will still be migrating. The protein pattern produced on .the gel is dependent on the accumulated volt hours so to ensure reproducibility this must remain constant for a specific population of proteins. The method described below is intended for use with the larger 2D system, that is, 15-cm tube gels and 16cm × 16cm SDS-PAGE slab gels which we find to be excellent for resolving the complete population of ribosomal proteins. Mini-systems such as the Mini-PROTEAN® Tube Cell Module from Bio-Rad can be used but we find the larger set-up easier to manipulate.

1. Rinse out the glass tubes with water followed by acetone. Allow to air dry.
2. Stopper one end of the tubes with parafilm.
3. To make 10 mL of gel mix, place 5.5 g of urea, 1.33 mL of 30% acrylamide mix, 2 mL of 10% Triton X-100, 1.97mL of $dH_2O$, 0.25 mL of Biolyte 3/10 ampholyte (40%), and 0.25 mL of Biolyte 5/7 ampholyte (40%) into a 125-mL side-arm flask.
4. Heat slightly to dissolve urea.
5. Add 30 µL of AMPS.
6. De-gas for 1 min.
7. Add 21 µL of TEMED.
8. Pour tube gels using a suitable syringe to approx 1 cm from the top. Overlay with $dH_2O$. Leave to polymerize for 1 – 2 h. To test whether the gel mix has polymerized successfully we suggest sucking up any remaining gel mix, after the tubes have been poured, into a plastic pipet. Once it has polymerized inside the pipet it is usually safe to assume that it has also polymerized inside the tubes.
9. Fill the lower reservoir of the electrophoresis apparatus with 0.02 *N* NaOH. Remove the parafilm from the bottom of the tubes. Place the tubes inside the tube holder according to apparatus instructions.
10. Load ribosomal protein sample solubilized in ribosomal protein lysis buffer (50 µL is a good volume) onto polymerized gel. For Coomassie blue staining of gels, we suggest 200µg per tube gel, and for silver staining of gels we suggest 50 µg. We suggest using the RC DC Protein Assay (BioRad, cat. no. 500 – 0120) to quantify sample protein levels. This assay is compatible with a number of components of the ribosomal protein lysis buffer, for example, ampholytes and thiol reducing agents that would normally interfere with protein determination.
11. Overlay with ribosomal protein lysis buffer diluted 1 to 4 leaving a small space at the top.

12. Fill the tube to the top with upper reservoir solution, that is, 0.01 M phosphoric acid.
13. Run electrophoresis with the polarity reversed *i.e.* red lead from apparatus into black port on power pack, black into red. Run at 400 V for 1 h and 30 min depending upon the desired separation profile. This corresponds to 600 Vh.
14. Extrude gel from glass tube using a syringe with a long thin gel extrusion needle. Fill the syringe with $dH_2O$ and insert it gentle inside the glass tube down the side of the gel. The needles recommended can easily be slid inside the tubes to just over half way. Pipet water gently while sliding the needle around the outside of the gel. Then withdraw the needle, turn the glass tube around and reinsert the needle into the other side of the tube. Gently pipet water as before. On withdrawing the needle from the tube this time, the gel inside should slide easily out. If nol, determine which end of the gel is sticking and reinsert the needle into that side and pipette. We suggest placing a sheet of parafilm or aluminum foil on the lab bench to catch the tube gel should it slide out unexpectedly during extrusion.
15. Place tube gel into 15mL conical tube and add 5mL of equilibration buffer supplemented with DTT (0.5% w/v) just before use. Incubate with gentle rocking for 30 min.
16. Pour an SDS slab gel (8–20% resolving gel works well, 5% stacking gel) according to standard procedures. Assemble the electrophoresis apparatus according to manufacturer's instructions.
17. Add a small amount of 1% agarose dissolved in equilibration buffer containing 0.5% (w/v) DTT to the top of the gel to serve as a platform for the tube gel to sit on. Once the agarose has set, lay the tube gel on top. Allow some room at one end to add molecular weight markers. We normally embed a small volume of markers in about 20µL of 1% agarose in equilibration buffer, containing 0.5% DTT in the bottom of a 1.5mL centrifuge tube. After the agarose has set pop out the markers plug and sit it next to the tube gel on the agarose platform.
18. Pipet more of the agarose/equilibration buffer solution over the top of the tube gel and markers plug to hold them in place.
19. Run the gel at 25 mA for 7 h or until the Bromophenol blue has migrated to the bottom of the gel.
20. Detect ribosomal proteins using a suitable staining procedure. We have successfully used Coomassie Brilliant Blue R250, SYPRO® Ruby protein gel stain and silver staining procedures. All these staining systems are compatible with the excision of protein spots from the gels, tryptic digestion of the proteins and analysis by mass spectrometry.
21. Alternatively, the proteins can be transferred from the gel to nitrocellulose membranes and analyzed for specific protein modifications by western blotting.

## Analysis of Proteins Separated by 2DG

A number of current systems and software products are available for protein analyses in 2D gels. Our laboratory possesses a Kodak ImageStation 440CF with Kodak Digital Science ID & 2D Software. This is convenient for protein staining via Coomassie Brilliant Blue, SYPRO® Ruby and silver.

## NOTES

1. Adjust the pH of these solutions at 4°C because gels are run in the cold room (4°C).
2. The 3X gel buffer can be stored at room temperature. However, the pH must be adjusted at 4°C because gels are run in the cold room (4°C).
3. These gel volumes can be proportionally increased to prepare a resolving gel that has an extended length. This is typically done if the 1-D BN-PAGE gels are to be run overnight to improve the resolution (*i.e.,* separation) of the complexes. Similarly, when gels are run overnight the cathode buffer can be switched to a "blue-free" buffer, that is, 50 mM Tricine, 15mM BisTris, pH 7.0, with no added coomassie blue G-250. This helps to minimize the high blue background when 1-D BN-PAGE gels are destained for imaging and densitometry purposes.
4. For the separation of large membrane protein complexes it is recommended that the concentration of the coomassie blue G-25.0 in the sample be one quarter of the lauryl maltoside concentration for electrophoresis. Therefore, based on the given stock solutions and protocol for the extraction of inner membrane proteins this works out to be approximately 6.3μL of a 5% (w/v) coomassie blue G-250 solution. Addition of a molar excess of dye to protein also helps to remedy one problem of native electrophoresis, that is, the tendency of membrane proteins to form aggregates in the presence of detergents. However, the problem of aggregation is minimized by the inclusion of the coomassie blue G-250. This dye binds to the hydrophobic regions on the proteins surface, which induces a negative surface charge thereby reducing protein aggregation. In addition, the anionic nature of coomassie blue G-250 induces the needed "charge shift" on proteins such that they will migrate to the anode at pH 7.5.
5. A high level of magnesium ions (15 – 20 mM) is necessary to ensure that mitochondrial ribosomal subunits remain associated during the isolation procedure.
6. To dissolve digitonin, place it in a suitable amount of buffer in a glass beaker and heat it up in a microwave. Watch for the buffer to just begin to boil then stop the microwave and remove the beaker.
7. Digitonin is a detergent and is being added in a precise amount to the mitochondria to remove the outer mitochondrial membrane and in doing so eliminate the possibility of contamination by cytoplasmic ribosomes. If it is left in contact with the mitochondria for too long it will begin to permeabilize the inner mitochondrial membrane.
8. Solubilization of the mitoplasts occurs immediately and is accompanied by a darkening of the mitoplast suspension as matrix contents are released.
9. Puromycin is an antibiotic compound that binds to the ribosome as an analogue of aminoacyl-tRNA and arrests the formation of nascent polypeptide chains. The polypeptides chains are then liberated from the ribosome. If using puromycin be warned that the compound also has an absorbance at 260 nm and this will mask the presence of the 28S subunit. If the 28S subunit is required we suggest removing the puromycin by pelleting the crude ribosomes through a second sucrose cushion before layering them onto the sucrose gradients.

10. We have found that complete dissociation of rat liver hepatic mitochondrial ribosomes is achieved by incubating them in a buffer containing no $Mg^{2+}$ rather than low levels of $Mg^{2+}$ as is suggested in some publications. The inclusion of 1 mM EDTA into the $Mg^{2+}$-deficient buffer B ensures a complete absence of $Mg^{2+}$.
11. On the basis of the molecular weight of the 55S ribosome 1 $A_{260\ nm}$ unit approximates to 40 pmol.
12. It is necessary to remove nucleic acid from protein samples that are to be used in isoelectric focusing or NEPHGE analyses because it can bind to the proteins and cause artifactual migration and streaking.
13. Air bubbles can inhibit gel polymerization.
14. Make sure that no air bubbles are present at the bottom of the tubes as they sit in the lower reservoir solution. If there are, they can easily be removed by blowing reservoir solution across the bottom of the tubes using a glass pipet that has been heated gently and curved at its narrow end.
15. It is very important to pipette the agarose solution gently to avoid the formation of bubbles. These can impede electrophoresis and cause distortions of the protein patterns. If air bubbles form, suck them out with a glass pipette. Also make sure that the marker plugs have no air bubbles present.

CHAPTER 11

# Detecting

Early free radical researchers were for the most part chemists (*e.g.*, Henry John Horstman Fenton of the "Fenton" reaction. However, free radical formation was considered for the most part an ex vivo phenomenon. Gershman and Gilbert first proposed that radiation-induced cell killing might be mediated by oxygen free radicals, based on the observation that the pattern of toxicity of radiation was quite similar to that hyperoxygen in vivo. In the same year, Harman published the "free radical hypothesis of aging". This strength of this hypothesis can be demonstrated by the simple fact that it has endured for almost 50 yr.

It was generally thought that free radical formation in the cell was limited to either external stimuli (such as radiation), or random events and did not occur under "normal" circumstances. A major paradigm shift in free radical research came in 1969, when Fridovich and McCord described the function of SOD as a catalytic reducer of $O_2^-$ to $H_2O_2$. On the basis of these results, the concept that oxidants are produced by the cell under normal conditions gained hold. The discovery of enzymes that normally produce prooxidants (in contrast to electron leakage from other enzyme systems), such as NAD(P)H oxidases, xanthine oxidases, and myeloperoxidases, further strengthened the case that prooxidants are regularly found in vivo.

Discoveries of catalytic functions of the glutathione peroxidases, glutaredoxin peroxidases, thioredoxin peroxidases, and catalases indicated that hydroperoxides are also kept in close check in vivo. It became clear that balance between prooxidants and antioxidants is critical for survival and functioning of aerobic organisms. An imbalance favoring prooxidants and/or disfavoring antioxidants, potentially leading to damage, was coined as oxidative stress by Sies. Reactive oxygen and nitrogen species (ROS and RNS, respectively) can be products of normal cellular metabolism and have potentially beneficial effects (*e.g.*, cytotoxicity against invading bacteria); indeed, there are enzymes in which the key function is to produce ROS/RNS, such as nitric oxide synthases (NOS), NA(D)PH, oxidases, and myeloperoxidases.

Oxidative stress is proposed to be critically involved in numerous diseases, including alcohol-induced liver disease. In 1966, DiLuzio was the first to characterize lipid peroxidation after chronic exposure to alcohol a finding confirmed by others. After the introduction of the spin trapping technique, in vitro and in vivo, evidence for a free radical derived from ethanol, the α-hydroxyethyl free radical was presented. A general problem of studying prooxidants in vivo is that, because of their inherent reactivity, they generally cannot be measured directly. One is therefore often left with measuring products of the reaction of these molecules with other molecules (*e.g.*, oxidative or nitrative modification of proteins).

This indirect detection of "footprints" of ROS/NOS makes it difficult to conclusively identify the parent oxidant. However, coupled with in vivo modulation of free radical production, one can obtain relatively clear data supporting the involvement of specific free radical species involvement in alcohol-induced tissue damage. This chapter will summarize some techniques that, have been used in the study of oxidative stress in alcohol-induced liver injury. Immunohistochemical detection of 4OH-nonenal- and 3-nitrotyrosine-modified proteins will be detailed.

In the past, immunohistochemical techniques have served largely as qualitative measures of changes but have the advantage of illuminating the location and zonation of the haptene of interest. However, with the advent of digital photography and image analysis, one can now quantitate treatment-related changes in the tissue of interest. Other advantages of these techniques are that they are readily performed in most basic research laboratories and do not require pretreatment of the animals to attain a signal. The latter point implies that historical samples may be analyzed.

Although the methods described herein are based liver, these immunohistochemical techniques have been employed successfully in most tissue types with minor modifications. Although immunohistochemical analysis of ROS/RNS-modified proteins is a useful technique, careful attention must be paid to minimize the effect of variability and artifacts. Obvious possible sources of in vivo variability (*e.g.*, gender, age, dose, etc) should clearly be avoided. Furthermore, variability caused by the processing and staining of the tissue itself should also be minimized. This ex vivo variability is attributable predominantly to the staining itself; it is therefore imperative that all samples to be analyzed are stained simultaneously to avoid interassay inconsistencies. Reasonable measures should also be taken to avoid all possible sources of ex vivo variability, including harvesting and processing of the tissue for immunohistochemistry.

## MATERIALS

All solutions should be of analytical grade or purer. Nanopure water is used for preparing all solutions. Glassware should be freed of possible detergent residue by prerinsing with nanopure water.

### Generation of an In Vivo Positive Control for Staining

1. Laboratory safety equipment (gloves, mask, protective eyewear, etc.).
2. Lipopolysaccharide (LPS) from *Escherichia coli* (O55:B5; Sigma).
3. Sterile saline: 0.9 % sodium chloride.
4. Needles for injection.
5. Borosilicate glass tubes (12 × 75 mm; Fisher).

### Harvesting of Tissues for Immunohistochemistry

1. Laboratory safety equipment (gloves, mask, protective eyewear, etc.).
2. Rodent anesthetic (*e.g.*, sodium pentobarbital, ketamine/xylazine, etc.).

3. Tools for harvesting tissue.
4. Sterile saline.
5. 10% neutral-buffered formalin (Fisher).
6. 10-mL syringes with plastic infusion needles (16 gage or larger).
7. Razor blades.
8. Plastic tissue cassettes.

## Immunohistochemistry

1. Laboratory safety equipment (gloves, mask, protective eyewear, etc.).
2. Ventilated work area (*e.g.*, chemical fume hood).
3. Magnetic stir plate capable of maintaining a slow stir rate (<100rpm) and a 1-inch magnetic stir bar (Fisher).
4. >3X glass staining dishes (45x75x32 mm; "Coplin jars") and slide holders (Wheaton).
5. Xylene (Fisher) or less toxic alternative (*e.g.*, CitriSolv; Fisher).
6. 100% and 95% ethanol (Aaper).
7. Phosphate-buffered saline with 0.5% Tween-20 (PBS) (10x stock) 1.37*M* NaCl, 27 *mM* KC1, 100mM $Na_2HPO_4$, 18mM $KH_2PO_4$ (adjust to pH 7.4), and 5% Tween-20.
   (*a*) Autoclave before storage at room temperature. Sodium azide (0.02%) may be added to the stock to help prevent contamination. Prepare working solution by dilution of one part with nine parts water.
8. 10 mM sodium citrate (pH 6).
9. 30% hydrogen peroxide solution (Fisher).
10. Bovine serum albumin, fraction V (BSA; Sigma).
11. Clear nail polish or Dako Pen (Dako).
12. Slide rack (Fisher). A slide storage box (Fisher) works quite well for this purpose.
13. Rabbit polyclonal anti-4OH-nonenal antibody (Alpha Diagnostic); Rabbit polyclonal anti-nitrotyrosine antibody (Upstate).
14. En Vision, HRP-based staining system (Dako). It is not required, but this system usually gives a good signal to noise ratio and contains both the HRP-conjugated secondary antibody and DAB solutions. Alternatively, HRP-conjugated goat antirabbit antibodies (Vector) and diaminobenzidine (DAB) solution (Dako) may be purchased separately.
15. Aqueous hematoxylin (Dako).
16. Mounting media (Permount; Fisher) and coverslips (25 × 50 × 0.15 mm; Fisher).

## Digital Image Acquisition and Analysis System

1. Microscope: Eclipse E600 with a 10 × ocular Plan Fluor series objectives (4 ×, 10×, 20×, 40x and 100 ×) (Nikon)
2. Camera: Spot RT slider BW/color digital CCD camera (Diagnostic Instruments)

3. Imaging software: MetaMorph image-acquisition and analysis software (Molecular Devices).

## METHODS

Both staining protocols are relatively similar and have shared methods. The in vivo methods require prior approval by the local Institutional Animal Care and Use Committee (IACUC).

### Generation of an in Vivo-Positive Control for 4OH-Nonenal and 3-Nitrotyrosine

It is important to have a positive in vivo control for the staining. In our experience, injections of hepatotoxic doses of LPS generate suffiecient signals of both 4OH-nonenal and 3-nitrotyrosine to serve as positive controls in Fig. 11.1 contains representative photomicrographs (10 × 10 magnification) of a liver from a mouse 24 h after injection with LPS (10mg/kg intraperitoneally) and immunostained for 4OH-nonenal. The supplier also recommends additional positive and negative controls for 3-nitrotyrosine staining. These can be generated by following manufacturer instructions.

1. Prepare an LPS-injectable solution in sterile saline (injection volume 1-2mL/ kg). Solution must be made in borosilicate glass tubes (LPS binds to plastic and Teflon). *Note:* LPS is highly toxic care should be taken when handling this compound to avoid exposure.
2. Inject the rodent (rat or mouse) with LPS intraperitoneally. This laboratory uses a dose of 10 mg/kg LPS (intraperitoneally) for both species. However, because of batch-to-batch variability in LPS potency, the dose should be titrated in the individual laboratory to achieve serum alanine aminotransferase (ALT) activity of 200–300 IU/L 24 h after LPS injection.
3. Harvest liver 24 h after LPS injection.

### Harvesting Tissue for Immunohistochemistry

1. Anesthetize the rodent according to your institution's IACUC-approved protocol.
2. Perform a laparotomy to fully-expose the liver.
3. Exsanguinate the animal by cutting the aorta just before entering the heart.
4. Flush the liver/animal (Optional). Flush the liver with a 10-mL syringe filled with saline by inserting into the portal vein just below the liver. Follow with a 10-mL syringe filled with 10% neutral buffered formalin. This not only helps minimize possible ex vivo haptene formation but also helps retain the architecture of the organ. If other organs are also to be harvested, exsanguination, flushing, and fixing of the whole animal may be achieved by cardiac puncture and perfusion. A larger syringe (*e.g.,* 50 mL) will be required.
5. Cut thin (2- to 4-mm thick) transverse sections of liver (or other organ of interest).
6. Place in tissue cassettes (labeled with pencil) and fix at 4°C for >48 h.
7. Paraffin-embed tissue, section (5- to 6-μm thick), and mount on microscope slides.

## Deparaffinization and Rehydration of Tissue Slides

At least one sample should be processed in duplicate in this procedure for primary antibody control. Normally, the positive in vivo control is selected for this control.

1. Slide racks for the Coplin jars have 10 slots. They may hold up to 20 slides if placed "back-to-back" (*i.e.,* tissue side facing out). The jar should be filled with liquid to completely cover the slides (~225 mL). With a stir bar in the bottom of the jar, it is placed on stir plate (set to the lowest RPM). Incubate in each of the following solutions for 5 min :
   (*a*) Xylene (or alternate solution); repeat once
   (*b*) 100% ethanol

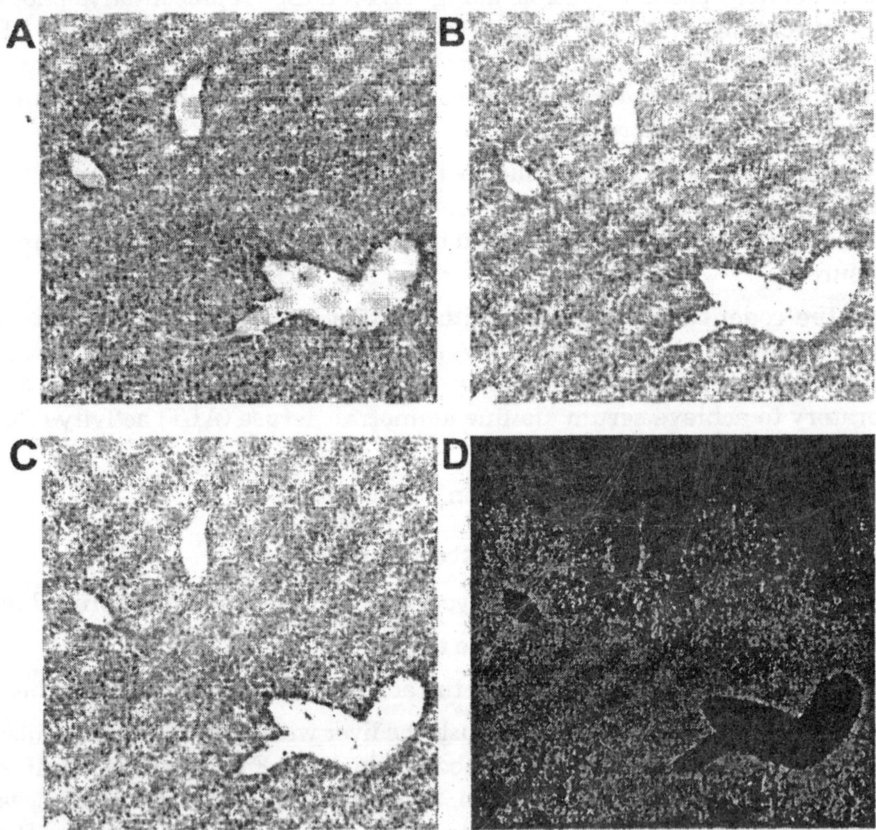


**Fig. 11.1. Effect of LPS on 4OH-nonenal adduct accumulation in mouse liver as determined by quantitative immunohistochemistry. Representative photomicrographs are shown depicting 4OH-nonenal immunostaining of a liver section (100x) of a mouse 24 h after injection with LPS (10 mg/kg intraperitoneally). The same microscope region is shown before (A) and after (B) white balancing of the image-analysis software. Note that the acellular space in (A) is light yellow instead of white. (C) is the same as (B), but after the blue exposure is increased 10%. Note that the contrast between the DAB chromogen (brown) and the hematoxylin counterstain (blue) is enhanced by digitally increasing the blue exposure. (D) shows the microscope region from (C), determined by the computer to be within the threshold range for positive staining.**

(*c*) 95% ethanol

(*d*) 70% ethanol

(*e*) 30% ethanol

(*f*) Nanopure water

2. Create a buffer dam around tissue (optional). While carefully avoiding drying of the section (do one slide at a time), paint a circle around the tissue section with the Dako pen (or nail polish), carefully avoiding touching the tissue with the solution (>3 mm away from the section). This buffer dam decreases the likelihood of the sample drying-out during antibody incubations. Place samples in a new Coplin jar containing fresh water.
3. Perform antigen demasking (optional). This procedure is strongly recommended for 3-nitrotyrosine immunohistochemistry.

   (*a*) Place slides in Coplin jar containing citric acid solution.

   (*b*) Remove metal handle from slide rack. Place jar in a glass dish of water and microwave on high for 2.5 min (700-watt oven). Warning, buffer will get hot; avoid scalding.

   (*c*) Let cool for 3 min. Refill Coplin jar with citrate buffer, if needed.

   (*d*) Repeat steps b-c three times. Let cool in citrate buffer for 20 min.

   (*e*) Rinse sections in Coplin jar containing nanopure water for 5 min. Repeat once.
4. Rinse the samples in a Coplin jar containing PBS-Tween for 5 min. Repeat once.

## Peroxidase and Protein Block of Tissue Sections

1. Make 250 mL of 3% hydrogen peroxide solution diluted in PBS-Tween. Pour into Coplin jar.
2. Incubate the slides for 5 min in peroxide solution.
3. Rinse the samples in a Coplin jar containing fresh PBS-Tween for 5 min. Repeat once.
4. Make PBS-Tween containing 1% BSA (PBA). You will need approx 250 μL per slide.
5. Remove the slides from the Coplin jar one at a time. Pour off excess solution and pipette PBA into buffer dam around tissue. Place on slide rack.
6. Incubate slides for 1 h at room temperature. Keep rack covered and hydrated by placing wet paper towels in the bottom of the rack. Add fresh solution as needed to avoid sample drying.
7. Proceed immediately to immunohistochemistry.

## Immunohistochemical Detection of Haptene Signals

1. While samples are blocking, prepare the primary antibody solution by diluting the antibody in PBA (~250 μL/slide). Historically, a 1:500 and 1:50 dilution of the 4OH-nonenal and 3-nitrotyrosine antibodies, respectively, were effective. Dilutions may vary on antibody batch.

2. Shake off excess PBA from slides one at a time and replace with primary anti-body solution. The primary antibody control slide will receive PBA without antibody.
3. Incubate on slide rack (covered and hydrated) for 30 min (4OH-nonenal) or an hour (3-nitrotyrosine). Incubation times may need to be altered to improve signal to noise ratio. Rinse the samples in a Coplin jar containing fresh PBS-Tween for 5 min. Repeat once. Shake off excess PBS-Tween from slides one at a time and replace with secondary antibody solution. Place on slide rack and incubate for 15 min (if using the Dako EnVision kit). Dilutions and incubation times for secondary antibodies from other, suppliers should be determined by manufacturer's suggestions.
4. Rinse the samples in a Coplin jar containing fresh PBS-Tween for 5 min. Repeat once.
5. Prepare DAB working solution according to manufacturer's instructions. Place a piece of white paper in the bottom of the slide rack. If the slide rack does not have a cover, make a cover with aluminum foil (DAB is light sensitive).
6. Apply DAB solution to each slide. Cover slides and incubate for 8 – 10 min. Use the in vivo positive control to determine the length of staining. When the brown DAB precipitate appears to not increase in this slide, stop the incubation. Transfer slides to a Coplin jar containing nanopure water.
7. Rinse Coplin jar by placing under a gentle tap of water for 6 min.
8. Transfer slide rack to a Coplin jar containing hematoxylin. Incubate in solution for 30 seconds while gently moving the slide rack up and down.
9. Return slide rack to Coplin jar containing water. Rinse under tap for 6 min.

## Dehydration and Coverslipping of Slides

1. Incubate the slide rack in each of the following solutions for 3-5 min.
   (*a*) Nanopure water
   (*b*) 30% ethanol 100% ethanol
   (*c*) 70% ethanol 95% ethanol
   (*d*) 95% ethanol 70% ethanol
   (*e*) 100% ethanol
   (*f*) Xylene (or alternate solution); repeat once
2. Keep the slides in the second xylene solution until ready to be coverslipped. Place a thin bead of Permount on one long edge of a coverslip. Gently role the slide onto the coverslip. Capillary action will pull the pieces of glass together. Avoid bubbles over the tissue sections. Should bubbles occur wash off coverslip with xylene solution and recoverslip. Let the slides dry in the fume hood for at least 24 h.

## Image Analysis of Immunohistochemistry

An absolutely critical point in performing quantitative image-analysis of immunohistochemistry is to ensure that all samples to be compared are analyzed simultaneously. It is recommended that before performing image analysis that the slides be critically analyzed under the microscope for any variation in chromogen signal not owing to the in vivo condition. Furthermore, image-analysis should be viewed as a means to quantitate differences between treatment groups, not to create.

Specifically, the differences should be readily observable with the eye under the microscope. For microheterogeneous tissues, such as the liver, a low power magnification (*e.g.*, 100 ×, ocular and objective) is usually sufficient for analysis. For macrohetergeneous tissues (*e.g.*, specific brain regions), a higher magnification may be required to focus on the area of interest. Should staining be located in a specific area within a tissue (for example, the perivenous area of the liver lobule), higher magnification photomicrographs may be also justified. It is recommended, how-ever, that initial attempts be performed at lower magnifications, as this reduces the risk of investigator bias in acquiring the pictures. It is also a good idea to blind analysis of the tissues to reduce the risk of investigator bias.

1. Power-up microscope, camera and image-analysis computer. Turn-on the microscope light source to ~80% of full brightness and allow it to sit for at least 20 min. Most bright-field bulbs are unstable light emitters until they are sufficiently warmed-up. A simple way to validate that the bulb is stable is to take five consecutive photos of the same microscope region and validate that they are consistent in brightness, contrast and color.
2. While the system is warming-up, ensure that the microscope is optimized for taking pictures at the selected magnification (*e.g.*, adjusting the diaphragm height and width). These techniques are routine to experienced microscopists.
3. Make sure that the image analysis software is set to manually determine exposure settings. Otherwise, the software may adjust the exposure between pictures, causing variations in staining color and intensity.
4. Using the image-analysis software, calibrate the exposure settings for the micro-scope field. First, move the microscope stage to a region of the slide that does not contain the tissue (but does have the coverslip) and defocus the image. Microscopes often have a gradient of light intensity across the field. Most image-analysis programs have a function to account for this gradient and normalize this gradient in captured photomicrographs (in MetaMorph, the command is "get flatfield". Next, in the same area, determine optimal exposure settings for the digital camera (*i.e.*, "white balance"). Fig. 11.1A,B compare the same microscope field before (A) and after (B) appropriate white balancing.
5. After performing flatfield and white balance commands, the photomicrographs captured by the software should be similar to those seen with eye. Validate this by capturing a representative image and comparing the digital picture with the image seen with the eye. If necessary, manually adjust the RGB exposure settings.
6. Enhance contrast with digital filtering (optional). Image analysis is much more sensitive to contrast than the human eye. Immunostaining may look very clear by manual analysis but may be fraught with noise during digital analysis. This noise can be alleviated by enhancing the digital contrast. For example, slightly increasing the blue exposure on the RGB setting will enhance the contrast between the DAB chromogen and the hematoxylin counterstain. However, care should be taken if this filtering

approach is performed that artifactual signals are not introduced. Fig. 11.1C is the same microscope region as Ib, but after the blue exposure has been increased 10%.

7. Once steps 1 through 6 are completed, it is imperative that the neither the microscope nor the exposure settings are altered during the image capture phase. Select and capture 10 microscope fields from each section (assuming 10x10 magnification). The regions should be randomly selected with the caveat that staining artifacts (*e.g.,* edges, folds, tears, etc.) should be avoided. Morphologic variations found in the tissue (*e.g.,* very large vessels in the liver lobule) may also be avoided. However, due diligence should be paid to avoid investigator bias in selecting regions. Should a higher magnification be used to focus on a specific region of the tissue, it is recommended that the number of pictures captured be increased.
8. Repeat this step for all tissue sections to be analyzed.
9. Select an image file with strong positive staining for the haptene. Use this image to determine the color, intensity, and saturation ranges for the image-analysis (called "threshold" in MetaMorph). It is recommended that a "conservative" range be initially attempted to reduce noise in the sections Fig. 11.1D shows the microscope region from Fig. 11.1C determined by the computer to be within the threshold range for positive staining.
10. Pool the summary area data from each tissue section to determine the percentage of microscope field positive for the haptene signal. Use this value to determine group means.

## In Vivo Modulation of Free Radical Production Using Genetically Modified Mice

The inherent reactivity of ROS/RNS makes it very difficult to directly measure them under in vivo conditions. Even electron spin resonance spectroscopic analysis of free radicals usually requires a spin-trap to stabilize the radical signal long enough for detection. Therefore, although assessment of changes in ROS/RNS by the techniques described above are useful, these 'footprints' do not give any information on the type(s) and source(s) of reactive species produced that caused these changes.

To date, most work has employed chemical inhibitors of suspected prooxidant enzymes. The effect of these drugs in vivo may not necessarily be specific to inhibition of the enzyme in question. Therefore, caution should be used when basing conclusions on work with pharmacologic inhibitors. One advance towards this end in the alcohol field has been the ability to address key questions at the molecular level by using genetically-altered mice (*e.g.,* knockout/transgenic). Such strategies have their own potential pitfalls, such as induction of compensatory mechanisms in knockout mice, which may cause false-positive or negative results.

## NOTES

1. To decrease ex vivo variability, our group always harvests sections for immunohistochemistry from the same liver lobes (left lateral and right medial) and the same location with the lobe (transverse section from the middle of the lobe) to avoid any potential geographic variation in the haptene signals.

2. There are many possible iterations of this system, but the emphasis should be on a system that is able to take high-quality and –resolution brightfield images. Important features of the system include flat-field correction, white balance, and manual control of red/green/blue (RGB) exposures.
3. Windows-compatible personal computer. Most new PCs will have enough speed and RAM to handle MetaMorph. A large hard drive is recommended to accommodate the large microscope files.
4. Microscope sections must be of consistent high quality (e.g., few if any folds or tears). Should the laboratory lack a practiced hand at cutting and mounting tissues, it is highly recommended that the tissues be prepared and processed by a professional histopathology service.
5. A number of newer and more rapid methods to deparaffinize and rehydrate microscope slides are available that use less reagent volumes than Coplin jars. However, it has been the experience of this author that deparaffinization/rehydration can be inconsistent and incomplete with these alternate approaches. These techniques are therefore not recommended for quantitative immunohistochemistry.
6. The Dako En Vision kit supplies a peroxidase block reagent. However, more consistent blocking has been achieved by our group using freshly diluted concentrated hydrogen peroxide in a Coplin jar.
7. A critical component of quantitative immunohistochemistry is consistency, to avoid intra-assay variability. Careful attention should be paid so that the incubation steps for each sample are similar, especially in steps that require you to work with one sample at a time. A simple solution is to measure the length of time required to add reagent to each sample.
8. The following are general guidelines for using genetically altered mice in experiments. These guidelines are derived from the Banbury conference on genetic background in mice and have been adopted by many scientific journals:
   (*a*) Genetic background of mouse strain must be clearly reported.
   (*b*) Common strains of mice should be used to aide reproducibility by other laboratories.
   (*c*) The genetically modified strain should be backcrossed into a common back-ground strain.
   (*d*) The strain should be backcrossed into the wild-type strain at least four times to avoid genetic variation at other sites besides the locus of interest.
   (*e*) Genetically modified line should not be maintained by crossing homozygotes, but rather maintained by backcrossing heterzygotes against the wild-type strain. This avoids the potential of genetic drift of the modified strain away from the background strain.

Results with modified strain should be validated with multiple generations. Further, weight-of-evidence can be increased by comparing pharmacologic targeting of the same protein with the knockout strain.

CHAPTER

# 12 Defensive Mechanism

The roots of immunoanalytical methods are found in the 19th century. By 1890 von Behring and Kitasato demonstrated the neutralising characteristics of bacterial antitoxins. In an effort to identify the origin of the most common bacterial diseases at the turn to the 20th century, the Widal test to verify typhoid (1896) or the Wässerman test to confirm syphilis (1906) antibodies were used as analytical tools. Subsequently, the rapid progress in biosciences, together with a deeper understanding of the basic principles of the immune system, led to a vast increase in different methods taking advantage of the specific interaction between antibodies and antigens. It is beyond the scope of this chapter to describe all the laboratory methods, that rely on the antigen-antibody interaction. But a few methods that are currently used will be described in more detail.

Since the analysis of insulin by Berson and Yalow 623, several different immunoassays for determination of a wide variety of drugs, marker substances in blood, and biological proteins have been developed. To date, many assays are commercially available as kits being used routinely in automated clinical laboratories. When a kit cannot be purchased, part of this chapter should help to establish an immunoassay in your lab. Additionally, immunoanalytical methods become more and more important in our understanding of how cells function.

The development of flow cytometers makes multiparametric analysis of single cells possible. Furthermore, confocal laser scanning microscopic imaging of tissues helps us to visualize processes occuring within cells and tissues. Both techniques use immunoflourescence and provide for a powerful tool in life sciences in the future.

## STRUCTURE AND CHARACTERISTICS OF ANTIBODIES

The antibodies are glycoproteins that belong to the group of immunoglobulins. Primarily, they are secreted by plasma cells in response to an immunogen. Each plasma cell secretes about 2000 antibody molecules per second, amounting to about 20% (w/w) of blood. Because blood can be sampled easily, antibodies provide for a powerful analytical tool as a result of their specific interaction with antigens. Structurally, antibodies are Y-like proteins composed of four polypeptide chains, which stick together by three disulfide-bridges and hydrophobic interactions. Each molecule contains two identical heavy chains, each composed of about 440 amino acids and two identical light chains, each containing 210 to 230 amino acids.

There are five types of different heavy chains termed as α-chains, γ-chains, δ-chains, ε-chains, or μ-chains. Accordingly, five classes of antibodies, namely IgA, IgG, IgD, IgE, and IgM, are distinguished (Tab. 12.1). These isotypes of antibodies can contain two identical types of light chains, either κ-chains or λ-chains. The heavy and light chains are made up of domains with sequence homologies comprising about 110 amino acids, which fold into parallel sheets linked by an intradomain disulfide bond. Thus, light chains contain two domains and heavy chains contain four domains.

**Table 12.1. Isotypes of Immunglobulins. IgM and IgA Additionally Contain Joining Peptides to Connect the Y-units. IgA also Exhibits a Secretary Protein.**

| *Isotype* | *heavy chain* | *light chain* | *molecular formula* | *mass (kD)* | *valency* | *function* |
|---|---|---|---|---|---|---|
| IgG | γ | κ or λ | $\gamma_2\kappa_2$ or $\gamma_2\lambda_2$ | 150 | 2 | Secondary response |
| IgM | μ | κ or λ | $(\mu_2\kappa_2)_5$ or $(\mu_2\lambda_2)_5$ | 950 | 10 | Primary response |
| IgA | α | κ or λ | $(\alpha_2\kappa_2)_{1-3}$ or $(\alpha_2\lambda_2)_{1-3}$ | 180–500 | 2, 4, 6 | Secretory Ig |
| IgD | δ | κ or λ | $\delta_2\kappa_2$ or $\delta_2\lambda_2$ | 175 | 2 | Not known yet |
| IgE | ε | κ or λ | $\varepsilon_2\kappa_2$ or $\varepsilon_2\lambda_2$ | 200 | 2 | Allergy Ig |

The amino-terminal domains of the heavy and light chains mediate antigen binding and are variable, whereas the remainder are constant regions. Each variable domain is composed of four framework regions and three hypervariable regions. The short loops of the hypervariable region of each heavy and light chain are referred to as complementary determining regions (CDR). Each CDR comprises 5–10 amino acids and protrudes, forming the site of interaction with the antigen. The area of interaction between the epitope, which represents the deteminant on the surface of the antigen, and the paratop which is the complementary structure formed by six CDRs of the antibody, was found to be about 1.6 × 0.6 nm.

The interaction between antigen and antibody is reversible and is in equilibrium with the free components. Thus, the precise fitting is a prerequisite for non-covalent interactions such as hydrogen bonds, van der Waals forces, hydrophobic bonds, and sometimes ionic interactions, which stabilize the immune complex. The high variability of antigen-binding site derives from about 45,000 different variable domains of the heavy chain, which are genetically encoded by 4 J(joining)-regions, 15 D(diversity) regions, and 250 V(variable) regions, which is extended by three kinds of combination.

The variable domain of the light chains is encoded by 250 V-segments and 4 J-segments, resulting in about 3000 different variable domains of the light chain. Linking the variability of heavy and light chains results in at least $1.35 \times 10^8$ antibodies with different specificity. Additionally, somatic mutations contribute to the diversity of antibody specificity. Exposition to proteolytic enzymes demonstrates the highly folded structure of antibodies. Upon treatment with papain, the heavy chains are cleaved N-terminally to the interconnecting disulfide bond at amino acid 224, yielding three fragments. Two single fragments with paratops at the N-terminus for antigen binding are known as Fab-fragments, corresponding to the arms of the Y. The third fragment that corresponds to the base of the Y was isolated by crystallisation and is known as Fc-fragment. This part of the immunglobulin is not involved in antigen-binding but plays a key role in immune regulation, *e.g.*, activation of the complement system and macrophages.

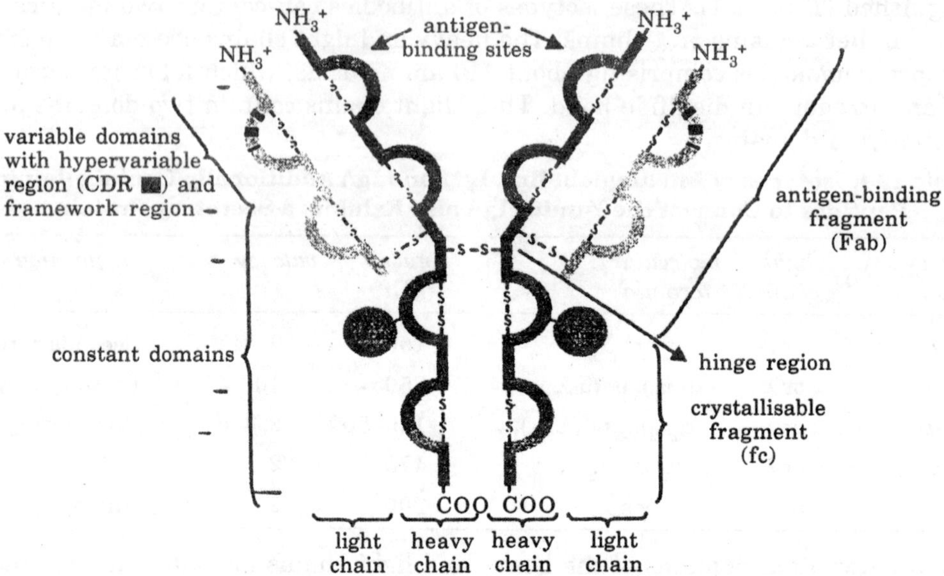


**Fig. 12.1. Scheme of IgG-molecule**

The region between the Fab and Fc-fragments is known as hinge-region, which provides for flexibility of the Fab-fragments. Lateral and rotational movement facilitates the binding of two determinants of different conformations of a certain antigen. Especially as a reagent in immunoanalytical techniques, the product of pepsin treatment becomes more and more important. Pepsin attacks the immunoglobulin at C-terminal positions below the heavy chain connecting disulfide bond at amino acid 234. Thus, a divalent fragment referred to as $F(ab)_2$-fragment is obtained, which possesses two antigen binding sites, but the Fc-portion is fully digested by pepsin. This $F(ab)_2$-fragment is a useful substitute for intact immunglobulins, as the Fc-fragment is known to interfere in many techniques, resulting in higher background staining or higher signal/noise ratio.

There are three terms closely related to each other that describe utility of antibodies in practice. Affinity describes the exactitude of stereochemical fit of the epitope of the antigen to the paratope of the antibody. Thermodynamically, it indicates the strength or energy of the interaction. Deriving from the law of mass action, it is described by an association constant $K_a$ indicating the ratio of bound and unbound antigen ($mol^{-1}$). Thus, the affinity constant is the reciprocal value of that concentration of a certain antigen, where half of the antigen is bound to the paratopes. In practice, this is the amount of antigen-antibody complex at equilibrium. High affinity antibodies will bind larger amounts of antigen in a shorter period of time than low affinity antibodies. Consequently this term $K_a$ is independent from the number of paratopes of an antibody molecule.

To characterize the overall strength of the antigen-antibody interaction, the term avidity is used. It includes affinity, valency of the antibody, and arrangement of the interacting compounds. Assuming that the affinity of an antigen to the paratope of IgG and IgM is equal, the avidity of IgM is five times higher than that of IgG because of the pentameric IgM. Thus, avidity is also expressed by an association constant K for a particular test system. When an antigen binds to an antibody, this interaction is termed to be specific. But some antigen determinants are shared between molecules, especially closely related ones. Thus, cross-reactivity describes the fact that common epitopes on different antigens are bound by the same antibody. On the other hand, cross-reactivity also can derive from binding of structurally different epitopes by the same antibody, especially in antisera.

# FROM ANTIGENS TO ANTIBODIES

## Basic Considerations

Antibodies used as reagents in immunoanalytical methods are host proteins that are produced by the immune system in response to foreign large molecules in the body. To provoke formation of antibodies by B-cells and cellular immune response mediated by T-cells, an immunogen is required. According to the clonal selection theory by Sir MacFarlane Burnet, the immunogen binds to well-suited antibodies located at the surface of virgin B-cells, which determine specificity of the antibodies finally obtained. The immunogen is degraded within the virgin B-cell, and its fragments are presented at the surface as a complex with class II proteins. This complex is recognised by T-helper cells, which induce differentiation of B-cells to antibody-secreting plasma cells.

For these reasons, epitope binding, degradability, and mediation of the B-cell – T-cell interaction require a minimum molecular weight of the immunogen of about 5 kDa. In this context, it has to be considered that animals usually used for immunization have developed some self-tolerance, eliminating well-suited B-or T-cells. Thus, the immunogen or carrier protein should be foreign to the immunized species.

# ANTIGEN PREPARATION

## Large Molecules

According to the requirements of epitope binding, degradability, and mediating interaction of B-and T-cells, high-weight proteins should elicit an immune response, which usually increases with a higher degree of conformation and structural rigidity. The antigenic sites at the surface of proteins molecules can be made up of different domains of amino-acid side chains, which are distant in sequence but close in space. In contrast to these conformation-dependent determinants, fibrous proteins possess sequiential determinants comprising up to six amino-acid residues.

Probably all proteins are immunogenic, but individual proteins may differ markedly in immunogenicity. Self-aggregation of proteins usually is associated with a slight change in its specificity, but with increased immunogenicity. On the other hand, large molecules can be made more immunogenic by denaturation. Heating or heating in the presence of sodium dodecyl sulphate will increase immunogenicity, but, concurrently, specificity will be altered because hidden epitopes will become accessible. Polysaccharides can act as immunogens in purified form, but only in humans and mice, not in rabbits or guinea pigs. Additionally, antibodies against nucleic acids are described, but they are elicited only in humans suffering from the autoimmune disease lupus erythematodes or in animals with similar diseases.

### *Small molecules*

In contrast to large molecules, small molecules including drugs, steroids, or peptides do not elicit an immune response. As first shown by Landsteiner exemplified by 2,4-dinitrophenol, low-molecular-weight compounds termed as haptens have to be coupled to carrier proteins to provide for class II protein-T-cell binding, which results in formation of antibodies reacting specifically with the dinitrophenyl group.

**Table 12.2. Reagents for conjugation of haptens to proteins**

## Hapten with Carboxylic Group–carrier Amino Group

*Mixed anhydride method:* Activation of the tributylammonium salt of the hapten in anhydrous milieu with isobuytlchloroformate to yield the rather unstable mixed anhydride. Coupling at alkaline pH, adjustment of pH necessary.

**Notes :** Racemisation omitted; no side products to be removed; isolation of activated hapten not necessary.

*Carbodiimide method:* Activation with DCC in anhydrous or EDAC in aqueous milieu at pH 4-5, coupling at pH 8.

**Notes :** Works well; probable formation of autopolymers; remove excess cross-linker; use water-miscible anhydrous solvents for activation such as DMF, THF, or dioxane.

*Active ester method* (carbodiimide/N-hydroxy-succinimde method): Activation and coupling is performed at neutral pH.

**Notes :** Works well; isolation of activated intermediate recommended; thiol-, hydroxyl-, and amino groups can react as well.

## Hapten with Amino Group

*Imidoester method :* Homobifunctional cross-linker, activation and coupling at pH 9.

**Notes :** Isolation of activated intermediate required to omit polymerization, design of spacer-length possible (dimethyl-adipimidate 0.86 nm, dimethyl-pimelimidate 0.92 nm, dimethyl-suberimidate 1.1 nm).

*Glutardialdehyde :* Homobifunctional cross-linker, activation and coupling at alkaline pH.

**Notes :** Isolation of intermediate is recommended; reagent is a polymer; sometimes reduction by sodium-borohydrid is required for stability.

*SPDP (N-succinimidyl-3-(2-propyldithio)-propionate:* Heterobifunctional, hapten activated at pH 7–9, coupling to thiol-containing carrier after conversion of amino- to thiol-residues with 2-iminothiolan or cleavage of carrier disulfide bonds with dithiothreitol.

**Notes** : Mild conditions; degree of substitution measured by appearance of 2-thiopyridon.

## Hapten with Hydroxyl-group Carrier with Amino Group

*Hemisuccinoylation :* Hapten converted to carboxylderivative by reaction with succinic anhydride in anhydrous milieu followed by use as carboxylated hapten.

**Notes :** Spacer is introduced.

*Divinylsulfone :* Derivatisation of hydroxyls at pH 11, coupling at pH 9.

**Notes :** Inactivation of excess reagent by addition of glycine.

*Periodate oxidation :* Mildly alkaline conditions required for cleavage and oxidation of cis-diol partial structures, coupling of aldehyde at pH 9.

**Notes :** Stability is increased by reducing the aldimin with sodium borohydride.

To gain better access to virgin B-cell antibodies, antennary exposition of the hapten by covalent linkage via spacer molecules at the surface of the carrier molecule is favoured. The most common carrier proteins are serum albumins of various species, preferably the well-characterized bovine serum albumin or high immunogenic keyhole limpet hemocyanin from the mollusc megathura crenulata, but oval-bumin and thyroglobulin or fibrinogen also are described. For conjugation of hapten and carrier protein, different protocols are described in the literature, but the use of heterobifunctional cross-linkers and two-step mechansims is highly recommended to avoid formation of homopolymers and precipitation of the hapten-carrier complex.

The amount of hapten coupled per molecule of carrier protein can be determined only after rigorous purification of the artificial antigen by dialysis, gel permeation chromatography, or precipitation with ammonium sulphate or any other useful method. Most simply, the degree of substitution can be determined by UV-difference spectroscopy of the carrier protein and the conjugate. Appearance of an additional peak at the maximum absorbance of the hapten in the spectrum of the conjugate in comparison to the carrier protein is a qualitative confirmation of successful coupling. When the solutions are carefully prepared by weighing accurate amounts of the lyophilised conjugate and the carrier protein and the solutions are adjusted to the same concentration, the hapten content of the conjugate can be estimated quantitatively.

An alternative is an indirect assay by determining the number of free amino groups of the carrier protein prior to and after coupling of the hapten. The difference points to the number of modified amino residues, which corresponds to the number of hapten molecules attached. A rather time-consuming method is the 2, 4-dinitro-2-fluorobenzene test, originally applied by Sanger for determination of insulin.

A simple and rapid alternative is the trinitrobenzene sulfonic acid test, which avoids total hydrolysis and extraction steps of dinitrophenylated amino acids. But it should be considered that non-covalent interactions also can contribute to hapten-fixation on the carrier protein, which amounts in the case of bovine serum albumin to about 2 to 3 hapten molecules. While bovine serum albumin has 59 lysine residues, only 35 are accessible at the surface for coupling at neutral pH. A conjugation rate of about 10 hapten molecules per carrier molecule is sufficient to obtain not only serum albumin-specific anti-bodies but also hapten-specific ones

## Immunization

Before an immunization protocol is started, the legal responsibilities should be considered. In the countries of the European Union, as well as in the USA, a personal and a project license is required for all immunization and bleeding procedures. The choice of the animal species to be immunized greatly depends on the facilities. Most commonly, rabbits of 4 to 6 months of age are used, as about 25 ml of serum is yielded from a single bleed. Guinea pigs, which are often hard to bleed, rats, or hamsters yield 1 to 2 ml, while mice yield only 200 μl. For rabbits, two animals should be used as a minimum, whereas in the case of rodents three to six animals are recommended.

It is self-evident that the animals should be held under proper environmental conditions preventing any stress. Prior to immunization of an animal, the antigen is mixed with adjuvants.

Because adjuvants enhance the humoral immune response and activate unknown sectors of the immune system, immunization never should be done without adjuvants. Heat-killed bacteria containing lipopolysaccharides *(Borde-tella pertussis)* or muramyldipeptides *(Mycobacterium tuberculosis)* stimulate lymphokine production resulting in strong local inflammatory reaction, which mobilizes macrophages.

On the other hand, aluminium salts precipitate the antigen and mineral oils form water-in-oil emulsions. Both result in formation of a depot, which prevents rapid degradation of the antigen and guarantees sustained nonspecific stimulation of the immune response. Most commonly, the primary injection is given as an emulsion of the immunogen in complete Freund's adjuvant containing killed *Mycobacterium tuberculosis* and mineral oil, which provokes aggressive and persistent granulomas. But the boosts are administered as a mixture with incomplete Freund's adjuvant containing mineral oil and emulsifier only. As a rough basic rule, in rabbits about 0.5 to 1 mg antigen in adjuvant is administered, preferably subcutaneously, on multiple sites at low volumes for the primary immunization, which corresponds to 50 to 100 μg antigen for mice. 'When only small amounts of poor immunogens are available, the antigen might be injected in rabbits in the poplietal lymph node at the lower leg to enhance the probability of an appropriate immune response. In this latter case, however any adjuvant must be omitted. The dose of the following boost injections about three to four weeks after the first injection might be one-third to one-half that of the first immunization. For deeper insight into immunization procedures, there is some useful literature available.

## Sampling Serum and Storage

After the first injection of the immunogen, the primary response results in exponential increase of antibody secretion occuring within 30 days. In the majority of cases, the antibody concentration in the blood, termed as liter, reaches a maximum within 9 to 11 days so that the first test bleed can be collected. When using rabbits, the animal is placed in a small box and a small distal area of the rear ear is rinsed with a small amount of 70% ethanol.

When the marginal ear vein is not apparent, it might be dilated by gently rubbing or warming with a lamp, but avoid use of toluene or xylene. The area, which should be incised or opened by insertion of a thick injection needle, is covered with Vaseline. The blood is collected by droping in a clean test tube. When the blood flow stops by clotting, the incision area should be wiped with sterile warm water. After collecting usually 5 ml, but 20 ml at the maximum, the needle is removed and gentle pressure is applied to the cut for half a minute. Check to be sure that the blood flow has stopped. Before screening this first test bleed for presence of specific antibodies, the blood is allowed to coagulate for one hour at 37 °C.

The clot is removed with a Pasteur pipette or by simply rotating a skewer made from wood followed by centrifugation for 10 min. at 13,000 rpm to remove any remaining insoluble material. The clot might be allowed to retract overnight to yield an additional few milliliters of serum. Additionally, the serum can be heated to exactly 56°C for 10 to 20 min. to inactivate the complement proteins. These proteins can interfere with some immunoanalytical techniques by binding to the antigen-antibody complex. The serum can be stored for many years at –20 °C or below. When stored at 4°C for weeks, it is recommended to add 0.02% sodium azide or thiomersal at 1:10,000. But be aware of the fact that sodium azide is highly sonous and can interfere with some immunoanalytical techniques. Test bleeds can be taken from mice in a similar way or from rats by incision the tail veil.

The test bleed usually yields a volume of 200 to 400 μl and can be treated as described above. The first test bleed predominantly contains IgM and only few IgG, but antibody formation declines with time. About one month after the first immunization, another exposure to the immunogen provokes the secondary immune response, which yields a stronger response by activating memory cells. Moreover the antibodies of the secondary response belong predominantly to the IgG-type and often exhibit higher affinity, which might even be increased by the number of boosters.

The titer should be monitored by appropriate methods such as enzyme immunoassays or simple immunodiffusion assays. When a good titer has developed, regular boosts are preformed every 6 weeks, and up to 50 ml blood (rabbit) is collected 10 days after the booster injection. Through the course of time, affinity, specificity and cross-reactivity are altered.

The concentration of the antigen in the animal decreases with time so that lymphocytes producing low-affinity antibodies are not stimulated further, whereas those secreting high-affinity antibodies are still stimulated. By boost injections with low doses of antigen, the threshold value for stimulation remains high, and only those lymphocytes remain to be stimulated that produce high-affinity antibodies. The decrease in specificity with time probably results from early formation of antibodies with low affinity. Each determinant of the antigen provokes formation of specific antibodies with different lag times. Thus, shortly after administration of the immunogen the number of antibodies with different specificity is low, but it increases with time. Additionally, cross-reactivity increases by time. In response to different determinants located at the surface of an antigen, a higher number of different antibodies is formed. Thus, it is likely that certain antibodies also recognize antigens with similar conformation.

## Polyclonal Antibodies

According to the clonal selection theory, a clone of B-cells produces antibodies of only one specificity. When an animal is immunized with an antigen, a random number of clones is activated and, consequently, a polyclonal antiserum is obtained. This strategy of the immune system to produce polyclonal antibodies yields important advantages in terms of affinity and specificity. A major advantage of polyclonal antibodies is the formation of large insoluble immune complexes with polyvalent antigens. At a molar ratio of 4 Mol antigen mol antibody, the multivalent antigens and the bivalent antibodies forms a three-dimensional network, which is opaque and can be determined quantitatively. Additionally, polyclonal antibodies are usually well-suited for cell staining, immunoblotting, and immunoassays with labeled antigens, but it is difficult to perform immunoassays with labeled antibodies. In the latter case, use of monoclonal antibodies is recommended. When usmg antibodies as a reagent tor immunoanalytical techniques, it should be considered that there is a batch-to-batch variation of polyclonal antibodies, even those deriving from the same animal.

The antibodies are heterogenous with regard to specificity, isotype composition, and optimum binding conditions so that each assay has to be adapted. Additionally – with exception of specific pathogen-free animals – the antiserum contains antibodies against all antigens to which the animal was ever exposed. Thus, the antiserum contains at most 20% to 30% immunglobulins, of which about 10% are specific antibodies against the desired immunogen.

## Monoclonal Antibodies

In 1984 G. Kohler and C. Milstein were awarded with the Nobel Prize for development of the hybridoma technique, which yields monoclonal and, consequently, monospecific antibodies. The rationale for this technique is to fuse antibody-secreting lymphocytes from the spleen, which produce polyclonal antibodies, with immortal myeloma cells. Thus, the capability of antibody production deriving from the B-cells and the immortality deriving from the B-lymphocyte tumor cells are joined. In order to render the polyclonal antibodies in the supernatant of the fused cells monoclonal, certain antibody-secreting cells are selected by cloning.

The growing hybridomas are diluted and grown in microplate wells as far as the antibodies secreted might stem from the same fused, antibody-secreting cell. Production of monoclonal antibodies requires mice, usually the inbred strain BALB/c, for immunization and plasmacytoma cells such as the non-secreting, HAT-sensitive P3-X63-Ag8.653 cells. Additionally, good tissue culture facilities including a 5% $CO_2$/95% air incubator, a sterile workbench, and an inverted microscope for visual control of the cultures are needed. In the following some basic items for production of monoclonal antibodies are given.

There are many different protocols for production of monoclonal anti-bodies; for detailed information; the reader is referred to the literature. Per immunogen, a minimum of three RALB/c mice is immunised by subcutaneous injection of about 25 μg immunogen in complete Freund's adjuvant at multiple sites at the back. Four weeks later, half of the amount of immunogen mixed with incomplete Freund's adjuvant is administered intraperitoneally. About one week after the booster injection, the animals are sacrified and the spleen is removed under aseptic conditions.

Using a stainless sieve, the spleen is homogenised, and clotted cells of the connective tissue are removed by centrifugation. Erythrocytes are lysed by osmotic shock, and the spleen cells are recovered in the pellet after centrifugation. The myeloma cells P3-X63-Ag8.653 represent a variant of BALB/c plasma-cytoma cells that do not secrete antibodies and that are sensitive to HAT medium. They should be in the logarithmic phase of cell growth. While fusion of the cells in early work was achieved by inactivated Sendai virus, PEG 4000 is commonly used today. Fusion of the lymphocytes with myeloma cells by PEG requires strict compliance with the selected procedure with regard to time and amounts of reagents added.

After fusion, the cell suspension contains hybridoma cells, non-fused lymphocytes, and myeloma cells. Whereas lymphocytes are mortal and lose their viability during further propagation, the myeloma cells are immortal. In order to eliminate the myeloma cells, the cell suspension is cultured in a selective culture medium (HAT medium) containing hypoxanthin, aminopterin, and thymidin. The myelomn cells suffer from a deficiency of the enzyme hypoxanthin-guanine-phosphoribosyl transferase (HPGRT). Thus, they are not able to synthesize DNA in presence of the HAT medium and lose viability. In contrast, in hybridoma cells the enzyme deficiency is compensated py the parent lymphocytes and they still grow. About two weeks later, the cells are cultured in HAT medium and cells from dense cultures are propagated further on feeder layers containing macro-phages from the peritoneum of mice.

They are obtained by rinsing the abdominal cavity of adult mice with sterile saccharose-solution. Upon propagation of the cell suspension containing hybridomas on feeder-layers, non-growing hybridomas (99%) are degraded by the macrophages, whereas the growing hybridomas (1%) take advantage of the growth factors provided by the macro-phages. Different methods are available to select the hybridomas secreting relevant antibodies.

Most commonly, enzyme-linked immunosorbent assays are used. After coating the wells of a microplate with antigen and washing, an aliquot of the supernatant of the hybridomas is added and washed after incubation. The antigen-bound antibodies are detected with goat-anti-mouse IgG/IgM labeled with horseradish peroxidase. After removal of excess second antibody by washing and addition of substrate solution the absorption of the cleaved, stained substrate is determined photometrically after stopping the enzyme reaction, by addition of acid. According to the results of screening the hybridoma-containing wells for antibody secretion, relevant hybridomas are cloned by limited dilution rather than by the agar technique.

**Table 12.3. Characteristics of Polyclonal Antisera and Monoclonal Antibodies**

| *Characteristics* | *Polyclonal antisera* | *Monclonal antibodies* |
|---|---|---|
| Purity of immunogen | Important | Not important |
| Time and expense | Low | Initially high |
| Specific antibodies | 0.1–1.0 mg/ml | 5-25 mg/mL in culture supernatants mg range in bioreactors |
| Nonspecific antibodies | About 10 mg/mL (> 80%) | None in serumfree supernatants |
| Specificity | Recognizing all determinants of immunogens the animal | Recognizing a single determinant of the imunogen |
| Affinity | Mixture of high and low Affinity antibodies | Hingh or low-according to the screening assay |
| Avidity | High | Low |
| Cross-reactivity | Yes | Usually not, but cannot be excluded |
| Antigen-precipitation | Yes | No |
| Isotypes and subtypes of immunoglubulins | Typical spectrum | One isotype and subtype |
| Physical stability | Usually high | Differs markedly |
| Utility for: | | |
| Cell staining | Good | antibody-dependent, but excellent with pooled antibodies |
| Immunopreciptation | Good | Antibody-dependent, but excellent with pooled antibodies |
| Immunoblots | Good | Antibody-dependent, but exellent with pooled antibodies |
| Immunoaffinity Purification | Poor | Antibody-dependent, poor with pooled antibodies |
| Immunoassays with Labeled antibody | Difficult | Good, excellent with pooled antibodies |
| Labeled antigen | Good | Antibody-dependent, but excellent with pooled antibodies |

Usually limiting dilution is done in three subsequent series, ensuring that the antibodies in the supernatant derive from fusion of a single B-cell and a myeloma cell. A problem sometimes encountered with production of monoclonal antibodies is loosening of chromosomes. Thus, a significant number of initially positive clones can be lost. This can be prevented by omitting overgrowth of the hybridomas in the wells, but this phenomenon occurs as the hybridomas become monoclonal. The hybridoma cells can be propagated continuosly *in vitro,* and the supernatant contains homogenous antibodies, which recognize only one or a few closely related antigens.

In contrast to polyclonal antibodies, monoclonal anti-bodies are homogenous molecules with regard to specificity, affinity, and isotype. Additionally, in contrast to polyclonal antisera, the predominant protein in the supernatant of hybridomas is the specific antibody that might be used as a homogenous reagent for nearly all immunoanalytical techniques especially for antibody-labeled techniques. Because of their monospecificity, monoclonal antibodies exhibit only poor antigen-precipitating properties as single reagents. Thus, indirect measurement of antigen-antibody-binding by use of a label is possible, which in turn represents the more sensitive assays.

## IMMUNOASSAYS

Immunoassays take advantage of the specific interaction between antigen and antibody. They obey the law of mass action:

$$\text{antibody} + \text{antigen} \leftrightarrow \text{antigen} - \text{antibody complex.}$$

At equilibrium, the antigen-antibody complex is formed as described by the association rate constant K1 and concurrently dissociated to yield again its free components according to the dissociation rate constant K2. But when at a constant amount of one free component the amount of the second free component is increased, the equilibrium of the interaction is driven toward complex formation. Thus, immunoassays enable detection and quantification of antigens and antibodies. Formerly, non-labelled techniques such as immunoprecipitation, agglutination or light scattering were used, but they still suffered from low sensitivity in the micromolar range.

Today labelled immunoassays are used that exhibit at least 1000 fold higher sensitivity. Antibodies or antigens are decorated with labels that impart measurable signals such as radioactivity, fluorescence emission, or enzymic activity. The extent of complex formation is then determined by measuring the amount of labelled antigen or antibody. Depending on the assay design, the signal intensity quoted is directly or inversely proportional to the analyte of interest. In order to describe the quality of an immunoassay, the terms sensitivity and accuracy are used. In a narrower sense, sensitivity corresponds to the change in response per amount of reactant, whereas sometimes the term is used to characterize the detection limit of the assay. Accuracy relates the amount of substance determined by the assay to the true value.

### Design of an Immunoassay

The criteria to classify immunoassays can be as follows: (1) determination of the antigen or antibody, (2) labelling of antigen or antibody, (3) homogenous or heterogenous assay, and (4) competitive or non-competitive assays.

## Homogeneous Assays

Homogenous methods are based on changes in enzyme activity, which is mediated by formation of an immune complex. Consequently, modulation of enzyme activity reflects the degree of the immunochemical reaction resulting in enhancement or inhibition. Thus, measurement of the degree of complex formation is carried out without any physical separation of bound and free compounds.

Enzymes used as labels in homogenous assays are lysozyme, malate dehydrogenase, and β-galactosidase. An assay design using enzyme-labelled antigen or analyte is referred to as enzyme multiplied immunoassay technique (EMIT) and was developed for determination of antiepileptic drugs, cardioactive drugs, anti-asthmatic drugs, and drugs of abuse. The EMIT assay system requires an active enzyme-hapten-conjugate. Upon binding of hapten-specific antibodies, the activity of the conjugate is inhibited by inducing or preventing changes in conformation of the enzyme as necessary for enzyme activity.

Thus, free hapten as analyte and hapten-enzyme conjugate compete for antibody binding. In the presence of excess conjugate, the enzyme activity is directly proportional to the concentration of the hapten. Generally, homogeneous methods for detection of small or large analytes are less sensitive than heterogeneous assays. A major drawback of homogeneous assays is reproducible and time-consuming preparation of well-characterised conjugates.

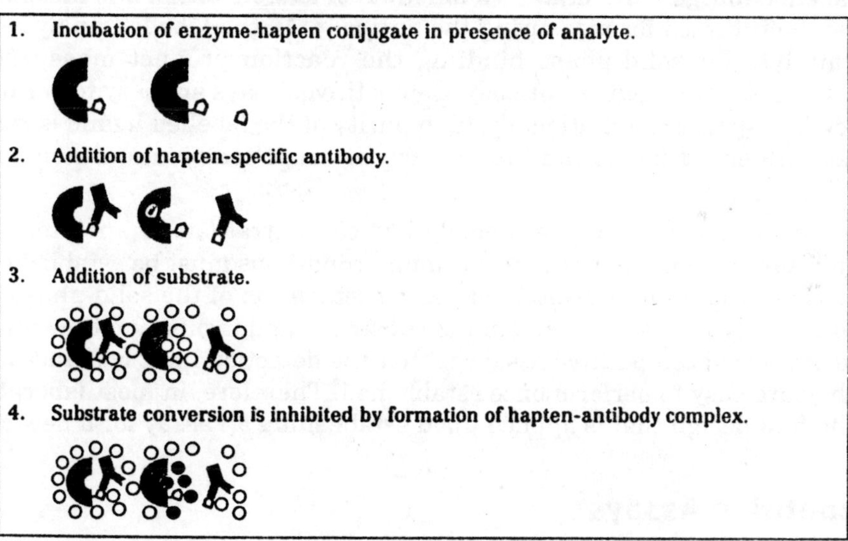


**Fig. 12.2. Scheme of EMIT using enzyme-hapten conjugate, free hapten as analyte, and specific antibodies for competitive binding.**

In order to modify the enzyme activity by complex formation, the hapten needs to be exposed to guarantee unrestricted interaction with the antibody. On the other hand, a separation step as a source of imprecision is omitted. When a homogeneous method is established, the assay still consists of mixing and measuring facilitating automatisation.

## Heterogeneous Assays

In contrast to homogeneous methods, in heterogeneous systems the activity of the labelled ligand is not altered upon formation of the antigen-antibody complex. Thus, most radioimmuno-

and enzymeimmuno-techniques require separation of bound and free-labelled antibody or antigen in order to be able to discriminate between bound and free-labelled ligand, which in turn allows quantification of the analyte of interest. Usually the activity of radiolabelled reactants is not affected by complex formation, whereas the activity of enzyme-labelled immunoreactants can be altered upon binding to the complementary antigen or antibody depending on the site of coupling the enzyme.

In heterogenous enzyme immunoassays, usually horse radish peroxidase, alkaline phosphatase, (β-galactosidase, or glucose-oxidase are used as labels. Methods for separation include use of solid-phase-bound antigen or antibody as well as second antibody precipitation. When enzyme-labelled antigen or antibody is used for detection of the reciprocal solid-phase-bound reactant, this is referred to as enzyme-linked immunosorbent assay (ELISA). The amount of the bound or free fraction is determined by enzyme-driven conversion of a colourless or non-fluorescent substrate to a intense coloured or fluorescent substrate, which is quantitated by photometry or fluorimetry.

### Competitive Assays

In competitive assays, the competition in binding to a fixed antibody between a constant amount of labelled antigen and unknown amounts of sample antigen is measured. Of course, this assay can be reversed for measuring the antibody. As the labelled analyte competes with unlabelled analyte for solid-phase binding, the reaction product measured is inversely proportional to the test antigen or antibody. Competitive assays are easy to perform but require high accuracy in dispensing. Additionally, high purity of the labelled ligand is required, but the result is less influenced by contaminants such as cross-reactive and non-specific binding compounds.

However, the measurement of extremely low concentrations of analytes can be affected by non-specific adsorption. Furthermore, optimum conditions must be established for each step of the assay. This includes optimisation of the concentration of the solid-phase ligand and the labelled ligand, the type as well as amount of substrate, and incubation time and temperature. The main advantage of competitive assays is that the detection limit is found in the picogram-range and they are easy to perform once established. Therefore, in most laboratories this type of assay is the first design that is applied upon establishing an assay for a new compound to be quantified.

### Non-competitive Assays

These are often called immunoradiometric assay (IRMA) in the case of radioactive tracers or immunoenzymometric assays. In this assay excess of immobilized antibody is incubated with standard antigen or sample antigen. Formation of the antigen-antibody complex is measured in a second step. When multivalent antigens are measured, the antigen can be detected by a second, labelled antibody. Because at least three layers of immunoreactants corresponding to antibody-antigen-antibody are built up, this technique is called sandwich immunoassay. Since the measured signal derives from the amount of antigen-bound, labelled second antibody,

1. Absorb specific antibody to the solid phase.
2. Remove excess of unbound antibody by washing.
3. Saturate remaining binding sites by incubation with blocking buffer.

1. Adsorb specific antibody to the solid phase.
2. Remove excess of unbound antibody by washing.

3. Saturate remaining binding sites by incubation with blocking buffer.
4. Remove excessive blocking agents by washing.

5. Add test antigen in the presence of constant amounts of enzyme-labelled antigen.

6. Remove excess reactants by washing.
7. Add substrate solution and incubate for a certain period of time at appropriate temperature.

8. Stop enzyme activity and quantify converted substrate, which is inversely proportional to the test antigen.

**Fig. 12.3. Competitive ELISA with solid-phase antibody using enzyme-labelled antigen for quantification of test antigen.**

4. Remove excessive blocking agents by washing.
5. Add test antigen in the presence of constant amounts of enzyme-labelled antigen
6. Remove excess reactants by washing.
7. Add substrate solution and incubate for a certain period of time at appropriate temperature.
8. Stop enzyme activity and quantify coverted substrate, which is inversely proportional to the test antigen.

there is a direct proportionality between analyte and signal quoted. Conversely, this design is also applicable to immobilized antigen, and the antigen-antibody complex can be detected by a second, heterologous-labelled antibody, which recognizes the Fc-part of the first antibody. The results obtained by sandwich ELISA are comparable to those of radio-immunoassays in terms of sensitivity. Additionally, optimisation steps are required as described above.

1. Adsorb specific first to the solid phase.
2. Remove excess of unbound antibody by washing.
3. Saturate remaining binding sites by incubation with blocking buffer.
4. Remove excessive blocking agents by washing.

1. Adsorb specific first antibody to the solid phase.
2. Remove excess of unbound antibody by washing.

3. Saturate remaining binding sites by incubation with blocking buffer.
4. Remove excessive blocking agents by washing.

5. Incubate with multivalent, unlabelled antigen.
6. Remove unbound antigen by washing.

7. Add excess of second, enzyme-labelled antibody.

8. Remove unbound reporter antibody by washing.
9. Add substrate solution and incubate under appropriate conditions.

10. Stop enzyme activity and quantify converted substrate, which is directly proportional to the concentration of the antigen.

**Fig. 12.4. Sandwich ELISA for determination of polyvalent antigens with second enzymelabelled, antigen-specific antibody.**

5. Incubate with multivalent, unlabelled antigen.
6. Remove unbound antigen by washing.
7. Add excess of second, enzyme-labelled antibody.
8. Remove unbound reporter antibody by washing.
9. Add substrate solution and incubate under appropriate conditions.
10. Stop enzyme activity and quantify converted substrate, which is directly proportional to the concentration of the antigen.

## PRACTICAL CONSIDERATIONS

### Antigen and Antibody

It is highly desirable to have pure antigen at one's disposal. When two monoclonal antibodies or affinity-purified antibodies are available, which recognize two independent epitopes of the antigen, the sandwich assay is appropriate to detect even impure antigen preparations. To perform competitive assays for detection of antigen, pure or partially purified labelled antigen is required. On the one hand, the solid support is coated with antiserum or culture supernatant, and subsaturating amounts of a mixture containing labelled and non-labelled antigen is allowed to compete for binding to the antibody.

On the other hand, the antigen is immobilised and a mixture of antigen and labelled antibody is added. Consequently, high amounts of antigen inhibit binding of the labelled antibody to the immobilized antigen. If pure antigen is not available, the wells are coated with crude antigen preparation followed by detection with saturating amounts of labelled antibody. For this reason and because of possible interference with compounds of the antiserum or the biological matrix containing the analyte, it is sometimes necessary to purify and concentrate the antibodies, especially polyclonal anti-sera. This can be done most simply by ammonium sulphate precipitation. IgG usually precipitates at 33% to 40% ammonium sulphate saturation. This can be performed simply by mixing equal volumes of antiserum and PBS-buffer, followed by addition of two volumes of saturated ammonium sulphate solution.

After stirring for 30 min. the pellet is collected by centrifugation and washed with half-saturated ammonium sulphate solution. Finally, the pellet is dissolved in PBS (one volume or less) and dialysed against PBS. Thus, the antibody preparation is purified and concentrated. An alternative method is to adsorb serumproteins on a cation-exchange matrix. The IgG is neutral or slightly positively charged at pH 6.5, whereas most of the serum proteins are negatively charged. Consequently, the serum proteins are bound to the DEAE-matrix, whereas IgG passes the column or is detected in the supernatant of the batch. Large amounts of ion exchanger are needed and the preparation is diluted.

Furthermore, the pH should be adjusted to 8 as soon as possible to prevent denaturation of the immunoglobulin. Much more comfortable is isolation of IgG by purification with protein A immobilised on beads that are commercially available. Because the Fc-proportion of IgG is bound to the surface protein of *Staphylococcus aureus* at pH 8, all contaminants are removed by washing the column. The IgG is dissociated from the beaded carrier by elution at pH 2. Thus, pure and concentrated IgG is yielded by elution in counter-current flow, but the pH of the eluate should be adjusted to 8.5 immediately. Only minimum amounts of the commercially available matrix are necessary, as the IgG-binding capacity of protein A is high (about 20 mg IgG/mg protein A). For preparation of monospecific antibodies from polyclonal antisera, affinity chromatography on immobilized antigen is performed.

This rather time-consuming method requires synthesis of an affinity matrix containing covalently bound antigens via 6C-12C spacers. The antiserum is bound to the matrix at physiological pH and, after removal of impurities by washing the column, the monospecific antibodies are recovered by lowering the pH, enhancing the ionic strength of the buffer, or addition of chaotropic agents. Again, physiological conditions in the eluate should be restored as soon as possible. In general, the sensitivity of immunoassays increases with purity of antibody and antigen preparation.

## Diluents

The diluent should provide optimum conditions for formation of the antigen-antibody complex. Furthermore, extremely low concentrations of analytes are applied, as they are susceptible to surface adsorption on the reaction vessels and denaturation. Whereas monoclonal antibodies and enzyme labels are more sensitive to ionic conditions and temperature, for polyclonal antibodies a pH near neutrality and an ionic strength of 0.15 is recommended, usually PBS pH 7.4. Addition of bovine serum albumin up to 1% minimizes adsorption to the surface of the vessel and stabilizes antigens and antibodies, especially in highly diluted solutions.

However, in assays for haptens addition of albumin should be avoided, as it binds many drugs and hormones. Washing buffers can contain up to 0.5% Tween 20 to enhance solubilisation of non-specifically adsorbed's compounds. Inclusion of bacteriostatic agents, *e.g.*, 0.02% sodium azide, enhances storage stability of antibody solutions but sometimes hardly interferes with enzymic detection. To improve storage stability of concentrated antibody solutions upon freezing, addition of glycerol up to 50% is recommended. Prior to the assay, the solution should be highly diluted with working buffer.

## Solid Supports

A solid matrix with immobilised antigen or antibody greatly facilitates manipulation during immunoassays. Formerly, beads were used as carriers, but today 96-well microplates are used in labs routinely for many purposes. Sometimes the reactants are immobilised covalently, but non-covalent interactions are used more frequently. In the latter case, the solid support can be made from polyvinylchloride or polystyrene, each exhibiting a protein-binding capacity of up to 300 ng/cm$^2$. Because polyvinylchloride microplates are not translucent, they are not suited for enzyme immunoassays measuring conversion to coloured substrate.

Translucent polystyrene microplates with flat bottoms are recommended for enzyme immunoassays, as they are commercially available with high and low protein-binding capacity. Infrequently, protein-A-Coated beads serve as solid support. As protein A interacts with the Fc-fragment of the IgG-molecule, the site-directed immobilisation of the antibody renders the antigen-combining sites freely accessible. Furthermore, so-called "magneto-beads" are commercially available and contain a core made from iron, which facilitates separation by magnetic fields omitting centrifugation.

## Detection

There are three ways of detection in immunoassays: (1) direct detection using labelled antigen or antibody; (2) indirect detection using a secondary labelled analyte, most frequently a second antibody recognizing the Fc-fragment of the first antibody; and (3) use of the biotin-avidin system comprising detection of biotinylated antigen or antibodies with enzyme-tagged or radiolabelled avidin. Direct detection requires fewer steps in the assay procedure and background problems are minimal. However, pure analyte is required and needs to be conjugated covalently with the label. In the case of radioimmunoassays, labelling procedures are well established, but for enzyme immunoassays this can be crucial to do in the lab.

Additionally, direct detection is less sensitive than the indirect way. The main advantage of indirect detection is commercial availability of labelled antibodies, which might be used to detect a wide range of antibodies with different specificity. The interaction between the glycoprotein streptavidin and the vitamin biotin is characterised by a dissociation constant of $10^{-15}$ mol/L, which corresponds to a half life of 160 days and approaches covalent interactions. Streptavidin consists of four identical subunits, each binding one molecule of biotin. Thus, an enhancement effect can result from the multivalency of streptavidin, which might be amplified further by a multilayer design using avidin-biotin complexes.

Reagents for biotinylation of proteins containing spacer molecules are commercially available, and avidin is easily coupled to enzymes. The type of the label needs to be carefully selected prior to the experiment. Radiolabelling is easy to perform by the chloramin-T iodination or by the lactoperoxidase method, but safety regulations concerning use, storage, and disposal need to be fulfilled. In radioimmunoassays low amounts of radioactivity are necessary, about 10 to 100 μCi/μg in the case of exhibiting a half life of 60 days, and about 50,000 cpm/well are required for reliable results. Consequently, danger to health is minimal if handled correctly, but special equipment is a prerequisite. In comparison, labelling with enzymes is safe, but it is emphasized that some cross-linking agents and substrates are hazardous.

The equipment necessary for enzyme immunoassays is inexpensive and the reagents exhibit a long shelf-life. Additionally, enzyme immunoassays allow a wide variety of assay designs, and detectability is comparable to that of radio-immunoassays, which is supported by the amplification effect of enzymes and by use of fluorogenic substrates. Usually 5 to 50 ng enzyme-labelled antibody well yields a colour development by substrate conversion corresponding to an absorption of 0.2 to 2.0. In practical terms, direct visualisation of the result on the microplate is possible.

For further enhancement of detectability, fluorogenic substrates might be used. A high quantum yield provides for the best detection limit, a large Stoke's stift should minimise the interaction between excitation and emission, and naturally occuring fluorophores in the sample should not interact. Quenching effects need to be characterised for each individual assay design. A major disadvantage of enzyme immunoassays is the possibility of interference of enzyme activity with compounds of the biological matrix of the sample, which might be reduced by selection of an appropriate assay design.

## PROTOCOLS

### Quantification of Cortisol by ELISA

**Protocol 1 :** Antigen preparation

1. A solution containing 5.5 mmol cortisol and 22 mmol succinic anhydride in anhydrous pyridine is refluxed under protection from light and humidity for 4 h.
2. The solvent is completely removed by evaporation under vacuum, and the oily residue is dissolved in chloroform followed by extraction with distilled water. The extraction is repeated three times and the organic phase is dried with anhydrous sodium sulphate.

   Formation of cortisol-21-hemisuccinate is confirmed by thin-layer chromatography on KGF254 in chloroform/methanol/water (65+30+6, v/v/v) indicated by hRf 49 as compared to cortisol with hRf 65. IR-spectroscopy and mass spectroscopy confirm these results.

3. The antigen is prepared by the mixed anhydride method.

   Under a hood, a solution containing 1.6 mmol cortisol-21-hemisuccinate in 32-ml anhydrous dioxane is cooled to 13 °C and 1.6 mmol tri-N-butylamine is added under stirring. After addition of 1.6 mmol isobutylchloroformate, the solution is incubated for 30 min. followed by addition of a solution containing 880 mg BSA in a mixture of 26 ml distilled water pH 10 and 17.2 ml dioxane. Immediately, the pH is adjusted to 7.5-8.5 by addition of 1 M NaOH and maintained for the next 2 h.

4. The BSA-cortisol conjugate is dialysed extensively against distilled water followed by lyophilisation. For immunization of mice, the conjugate is further purified by chromatography on Sephadex G-25 using TRIS-buffer pH 8.6 for elution to remove even traces of the free immunosuppressive drug.

The degree of substitution is determined by UV-difference spectroscopy. At the absorption maximum of cortisol at 246 nm, the difference in absorption of accurately equimolar solutions of conjugate and carrier protein is determined and calculated from the calibration curve of cortisol to be 19.4 mol cortisol/mol carrier protein.

**Protocol 2 :** Immunization and preparation of monoclonal antibodies

1. Three Balb/c mice were immunised with 100 μg antigen emulsified in 1 ml complete Freund's adjuvant for the first immunization and 50 μg antigen in 1 ml incomplete Freund's adjuvant for the subsequent two boosters.
2. Fusion of isolated spleen cells and $2 \times 10^8$ myeloma cells was performed with PEG as described above, following a protocol reported in the literature. The primary hybridoma cultures (one drop of fused cells in 2 ml medium) were grown in 24-well culture dishes as described above, yielding 75 growing clones.
3. After reaching confluency, the supernatant was screened for cortisol-specific antibodies yielding six positive clones, which were propagated further.
4. Cloning was done by limited dilution. The primary dilution series was done in 96-well microplates at a density of $10^4$ cells/well in 200 μl HAT medium. The secondary dilution series was done accordingly from positive clones.
5. Screening of the supernatants resulted in eight cortisol-antibody secreting clones. These clones were further propagated in 24-well microplates.

**Protocol 3 :** ELISA protocol

1. The wells of a 96-well microplate were coated with 100 μl antigen solution δ (0.003% in PBS containing calcium and magnesium) overnight at 4 °C.
2. After washing three times with 100 μl PBS each, the surface of the wells was saturated by incubation with 100 μl 1% BSA-solution in PBS for 2 h at room temperature.
3. After washing three times with 100 μl PBS each, a mixture of 50 μl hybridoma-supernatant and 50 μl cortisol in PBS (serial dilution, 5–200 ng) was added and incubated for 2 h at 37 °C.
4. The wells were washed three times with PBS. After addition of 100 μl goat-anti-mouse-IgG-conjugated peroxidase, the microplate was incubated for 2 h at 37 °C.

5. The supernatant was removed and the wells were washed again three times with PBS. 100 μl of freshly prepared substrate solution consisting of 10 mg o-phenylendiamin hydrochloride, 9 ml 0.11 M aqeous disodium-hydrogen-phosphate, 1 ml 0.5 M aqueous citric acid, and 10 μl 30% $H_2O_2$ were added.
6. After incubation for 30 min. under protection from light substrate, conversion was stopped by addition of 100 μl 3 M sulfuric acid. Immediately, the absorption was read at 450 nm using a microplate reader.

The samples were assayed in triplicate at the minimum, including blanks (well-filled with PBS), negative controls (as above, but without addition of cortisol-antibody and cortisol), and positive controls (as above, but without addition of cortisol).

Cross-reactivity of the antibody with structurally related steroids can be determined by the same assay, but add the steroid of interest instead of cortisol.

## IMMUNOFLUORESCENT ANALYSIS

Because of the fascinating facilities concerning the characterisation of cells, immunofluorescence techniques in combination with flow cytometric analysis have gained more and more interest over the past years. Application of fluorescent-labelled monoclonal antibodies directed against specific proteins or polysaccharides located at the cell surface or with some limitations-selected cytoplasmic, as well as nuclear, ligands offers the possibility to obtain direct information on the state of activation, proliferation, and differentiation of the cells.

In comparison to microscopical methods, the advantage of flow cytometry is the rapid quantification of the amount of a particular stain per cell, which offers a good statistical evaluation because of the large number of cells usually assessed. The objective of this chapter is to give an overview of the basic principles of flow cytometric measurements and to provide readers with basic considerations concerning immunofluorescence techniques. Upon understanding the important parameters, the specificity and sensitivity of the staining methodology can be optimized for a particular application.

### Principles of flow Cytometry

The basic components of a flow cytometer consist of a fluidic system to transport the cells across the microscopic field; an optical system for illumination and detection; and the electronics for light collection, data management and control. The most important part of the *flow system* is the flow chamber, where the microscopic observation takes place. An important parameter for successful flow cytometric analysis is the hydrodynamic focusing of the cells in this flow chamber providing for individualisation and correct positioning of the cells under investigation. Furthermore, the sample suspension is injected into a particle-free "sheath fluid", and the diameter of the sample flow is reduced dramatically at the injection point of the flow chamber to guarantee serial alignment of the individual cells during optical analysis. The optical system consists of three parts, the first is the *illumination optics,* which represents the light source for the flow cytometry and is made up of either a conventional lamp or a laser, generally an argon ion laser. Via prisms and mirrors, and if necessary monochromatic filters, the illuminating light is directed to the liquid stream and focused by several lenses to the point of microscopic observation. The other two components belong to the detection system of the flow cytometer.

The *forward scatter collection optics* is a microscope with low numerical aperture observing the liquid stream from opposite the illuminating light. The "forward scatter" (FS) collects the light scattered by the cells in the flow stream in the range of 2° to 20° off the axis of the illuminating light. The scattering signal measured by the FS corresponds to the diameter and, there-fore, the size of the cell under investigation. The second detection component includes the fluorescence and side scatter collection optics, comprising a second microscope with a long working distance and high numerical aperture, which analyses the light scattered perpendicular to the illumination beam as well as the liquid stream.

The "side scatter" (SS) signal is dependent on the surface characteristics and the granularity of the measured cells, whereas the fluorescence detectors collect the light emitted by fluorescent compounds associated with the analyse cells. The electronic system is necessary for the correct collection and processing of the light signals gained. Conversion of optical into electrical signals is done by semiconductor photodiodes or/and photomultiplier tubes.

To analyse two or more fluorescent compounds with overlapping emission spectrum (*e.g.,* fluorescence isothiocyanate and phycoerythrin) simultaneously, the flow cytometer should be equipped with an electronic fluorescence compensation network for correction of fluorescence overspill and proper calculation of the different fluorescence signals. For data calculation, either a linear or a logarithmic amplifier can be used, depending on the dynamic range of the parameter under investigation. In addition, a trigger circuit provides for correct separation between signals and electronic noise or debris by setting a threshold level for one parameter (*e.g.,* forward scatter). Each analog pulse acquired by the photomultiplier tubes that exceeds this trigger threshold is converted into a digital signal and transmitted to a computer for further storage and data evaluation.

## Data Evaluation

Today, state-of-the-art flow cytometers offer the possibility to analyse five parameters per cell simultaneously: forward and side scatter as well as up of three fluorescence light parameters. Normally, the intensity of the scattered light is analysed on the basis of linear amplification, but the intensity of immunofluorescence is analysed upon logarithmic amplification over four decades. Thus, a single gain setting can be used for data acquisition allowing direct comparison between relative fluorescence intensities of various cells. For evaluation a multitude of data plots is available. Usually a two-dimensional dot plot is used to characterize the scattering properties of a cell population by plotting the FS signal *versus* the SS signal.

One-dimensional histograms are best suited for illustration of the intensity of fluorescence staining emphasising the quantitative aspects of flow cytometry. In order to exclude non-specifically stained particles (*e.g.,* dead cells, debris) or conglomerates of cells from analysis, cells for data evaluation can be preselected by setting lower and upper thresholds for certain parameters ("gating").

Statistical calculation of data is done by the cytometer software, making the. evaluation of immunofluorescence assays rather simple. In this statistical evaluation, the cell populations are usually described by the relative mean and the coefficient of variation (CV) of a particular fluorescence parameter as well as the number of cells in the respective population.

The mean is defined as the arithmetic mean in the case of linear scale and geometric mean for logarithmic scale. The CV is calculated as half the width of the distribution at 0.6 times maximum height. Usually the distribution of fluorescence intensities on a logarithmic scale varies symmetrically around the mean because of both biological variation (*e.g.*, size of cells, density of target antigen and staining antibody) and variation of measurement (*e.g.*, focusing, orientation). Absolute quantification of the amount of antigen molecules per cell is difficult because of the necessity to include a reference method for calibration of the relative immunofluorescence.

## Basic

The area of immunofluorescence covers the staining of cellular structures with antibodies and other specific ligands directly or indirectly labelled with fluorescent dyes. During an immunofluorescence analysis, all the steps prior to the flow cytometric measurement, including preparation of cells, staining, and fixation, are most critical.

## Preparation of Cells

In order to obtain appropriate data from flow cytometric analysis, attached or connected cells have to be individualized prior to staining and measurement. Preparation of a single-cell suspension is done easily by repeated re-suspension in the case of natural single cells derived from blood, bone marrow, or non-adherent cultured cells. Adherent growing cells or cells from tissues can be separated mechanically or enzymatically either prior to or after the staining procedure.

Upon use of enzymes for disintegration of cells, care has to be taken in order to ensure that the antigen of interest or the attachment of the immune label is not affected by the dissociation procedure. Recent protocols often include pre-enrichment steps, especially in the case of blood samples, *e.g.*, the depletion of erythrocytes for analysis of leukocytes. Among these procedures, density gradient centrifugation and lysis are used frequently. Pre-enrichment of rare cells is advantageous because of the reduction of staining volume and elimination of cells that could be stained unspecifically, such as dead cells, resulting in improved discrimination of the cells of interest.

## Staining of Cells

The staining procedure can be performed either directly or indirectly. In the case of direct staining, the fluorescence-label is attached covalently to the antibody against the specific cellular antigen under investigation and used for analysis. Indirect staining is done by incubating the cells with an unconjugated primary antibody followed by detection of the antibody with a secondary, fluorescent-labelled, usually polyclonal anti-antibody. Direct staining is preferred because of easier handling and better control of the staining parameters.

Especially in routine use and for multiparametric analysis using a combination of various markers in one staining step, control and handling are important to consider. Prior to use the labelled primary antibodies of interest have to be purified. Indirect staining methods facilitate application of the primary antibody without any previous labelling and purification. A further advantage of indirect staining protocols is the possibility to improve the sensitivity of detection by increasing the number of secondary antibodies bound to one primary antibody, thus increasing the number of fluorescence markers per specific antigenic determinant.

But particularly in multiparameter analysis, control of proper indirect staining is not trivial and requires highly purified isotype-specific secondary antibodies or complete blocking of free binding sites of the secondary antibody prior to staining of the next parameter to avoid any cross-reactions. To combine high specificity and high sensitivity, an indirect staining protocol using primary antibodies conjugated to haptens such as digoxigenin or biotin can be followed. In these systems the detection is pursued very specifically by fluorescence-labelled secondary antibodies directed against the hapten or streptavidin, while the sensitivity is controlled by the number of marker molecules per mol secondary antibody/streptavidin. In order to avoid any loss of or damage to cells, the number of steps in the staining protocol should be kept to a minimum.

Additionally, reproducible results can be expected only if the concentrations of cells and antibodies used for detection are standardized. In so doing it is important to apply concentrations that are as low as necessary for predominance of specific high-affinity antigen-antibody interactions over low-affinity cross-reactions but are high enough to ensure that staining is as bright as possible to obtain good quality of data. The optimum concentration of cells depends mainly on the nature of cells and the antibodies used for detection and should be determined in preliminary studies.

The lower limit amounts to approximately $10^5$ cells in 10 μl providing for flow cytometric analysis of $10^4$ cells. Thus, 10 μl is the minimum volume necessary for proper staining in order to avoid significant changes in the concentration of the staining antibody during the washing procedure. Usually staining volumes of 20 to 100 μl containing 0.7 to $5 \times 10^6$ cells are used. All these concentrations-refer to viable positive cells; additional non-stainable cells can be neglected.

However, for flow cytometric measurement of rare cells the number of negative cells has to be considered upon optimization of the staining protocol. To stain larger numbers of cells, it is better to increase the volume of the cell suspension than the concentration of cells. To obtain results that guarantee adequate statistical evaluation, the optimum concentration of staining reagents should be determined by titration. For that matter, it should be kept in mind that too-low concentrations resulting in poor discrimination between positive and negative cells are as problematic as too-high concentrations resulting in non-specific staining of negative cells or subpopulations due to low affinity cross-reactions.

Titration can be performed by flow cytometric analysis of a mixture of positive and negative cells stained with varying concentrations of the fluorescent-labelled antibody using a standard protocol. By plotting the mean cell-associated fluorescence intensity against the concentration of the staining reagent, optimum conditions in an antibody concentration exist where the difference in cell binding between positive and negative cells is highest. If there are any problems with cross-reactions, it might be useful to apply lower concentrations. Higher concentrations are recommended in the case of low antigen density. The incubation with the antibody should be kept as short as possible because high-affinity antigen-antibody reactions occur very fast.

Extending the incubation time may lead to enhanced low-affinity cross-reactions and non-specific adsorption. Since with most staining reagents about 90% of the maximum staining is achieved within 5 min., doubling the time for safety reasons results in a standard staining time of 10 min. For immunofluorescent staining of intracellular components, a prolonged staining period after fixation and permeabilisation of the cells is necessary. To guarantee appropriate diffusion of the staining antibodies into the cytoplasm to reach the antigens of interest, staining times of about 1 h are used. To avoid any problems deriving from cell physiology, staining of live cells is done at an incubation temperature of 4 °C. At this temperature, the metabolism of the cells is minimized and the fluidity of the cell membrane is reduced, resulting in a preferred binding of the staining reagents to the cell surface.

In the case of staining cytoplasmatic parameters of fixed and permeabilized cells, the temperature during staining should be raised to 37 °C in order to enhance diffusion of the staining antibodies. Staining of live cells at higher temperatures may lead to altered results that are due to modulated expression of the antigenic receptors (*e.g.*, lymphocytes can "cap" their receptors with the staining antibodies, throw off the cap into the medium, and then appear "negative"). Blocking cell physiology by addition of 0.003% sodium azide in the staining medium may reduce these problems.

## Washing and Fixation of Cells

Prior to flow cytometric analysis, unbound staining antibodies have to be removed by washing. Therefore, the cell suspension is spun down at 1000 rpm followed by removal of the supernatant and resuspending the cells in fresh medium or isotone buffer. Because under even optimum conditions 10% of the cells are lost per washing step, washing should be reduced to a minimum. Applying an adequate low concentration of reagents during staining can help to restrict washing of the cells to one or two steps, particularly in direct staining protocols. In the case of indirect staining, proper washing of the cells is very important to improve staining results and to avoid any reaction between primary and secondary antibody in solution.

In these protocols, the specificity of the antigen-antibody interaction often is improved by addition of 1% BSA to the washing solution for blocking of residual non-specific binding site. For routine analysis cells usually are fixed prior to flow cytometry using para-formaldehyde (2% in PBS) to avoid time-dependent alterations of the cell staining as well as the risk of infections, thus standardizing analysis parameters. However, like all other manipulations, fixation of cells may lead to unpleasant side effects. Especially for analysis of rare cells, discrimination between rare positive cells and non-specifically stained dead cells is crucial.

After fixation it is very difficult to distinguish whether a cell was dead before staining or not. In such cases it is more convenient to analyse live cells without fixation immediately after the staining procedure in the presence of propidiumiodide, gating out dead cells according to the scatter and the propidiumiodide fluorescence. In contrast, staining of cytoplasmatic or nucleic substructures requires fixation and permeabilisation of cells because of the inability of antibodies to penetrate the cell membrane of live cells. For this reason, cells are fixed either by addition of ice-cold methanol (10 min. at –20°C) and subsequent rehydratisation in PBS or with para-formaldehyde (2% in PBS, 10 min). In the case of para-formaldehyde, fixation has to be followed by an extra permeabilisation step using Triton X-100 (0.1% in PBS, 10 min.).

## Measurement and Data Evaluation

Prior to flow cytometric analysis, the sample usually amounting to 50 to 200 μl has to be resuspended in 1 ml Cell Pack. Amplification of the fluorescence signals during flow cytometry should be adjusted in such a way that the auto-fluorescence signal of unlabelled cells is put in the first decade of the 4-decade log range. For each measurement between 5000 and 10,000 cells of interest are accumulated. Generally data evaluation is done by using a forward *versus* side scatter plot for gating and the relative mean of the particular fluorescence parameter for further calculation, as described above in detail.

## Staining and Flow Cytometric Analysis

**Protocol 4 :** Direct staining of surface antigens

1. Prepare 50 µl of a single-cell suspension containing $1 \times 10^6$ cells.
2. Add 50 µl of the specified fluorescence-labelled antibody solution and incubate for 10 min. at 4°C. The adequate concentration of antibody depends on antigen density on target cell, fluorescein/protein ratio of the labelled antibody, concentration and nature of cells, etc., and has to be determined by titration in a preliminary study.
3. Add 100 µl PBS and spin down the cell suspension at 4°C and 1000 rpm. Discard 150 µl of the supernatant and resuspend the cells using 150 µl of PBS. This washing step can be repeated once. Finally the volume is adjusted to 100 µl.
4. If fixation is required, add 100 µl of a para-formaldehyde solution (4% in PBS) and incubate for 10 min. at r.t.
5. Dilute the sample with 1 ml Cell Pack and analyse by flow cytometry, accumulating 5000-10,000 cells per measurement.

**Protocol 5 :** Indirect staining of surface antigens

1. Prepare 50 µl of a single cell suspension containing $1 \times 10^6$ cells.
2. Add 50 µl of the specified primary antibody solution followed by an incubation for 10 min. at 4°C. The adequate concentration of antibody: depends on antigen density on target cell, fluorescein/protein ratio of the labelled antibody, concentration and nature of cells, etc., and has to be determined by titration in a preliminary study.
3. Add 100 µl 1% BSA in PBS and spin down the cell suspension at 4°C and 1000 rpm. Discard 150 µl of the supernatant and resuspend the cells in 150 µl of 1% BSA/PBS. This washing step can be repeated once. The volume should be adjusted to 50 µl.
4. Add 50 µl of the corresponding fluorescence-labelled secondary antibody solution and incubate again at 4°C for 10 min. For considerations on the concentration of antibody, see above.
5. Repeate the washing procedure as described above.
6. If fixation is required, add 100 µl of a para-formaldehyde solution (4% in PBS) and incubate for 10 min. at r.t.
7. Dilute the sample with 1 ml Cell Pack and analyse by flow cytometry accumulating 5000-10000 cells per measurement.

**Protocol 6 :** Staining of cytoplasmatic or nucleic tructures

1. Prior to staining, the single-cell suspension has to be fixed and permeabilized:
2. By addition of ice-cold methanol (70%) and incubation for 10 min. at –20°C followed by rehydratisation at 4°C.

A. By addition of para-formaldehyde in PBS (final concentration 2%) and incubation for 10 min. at r.t.

B. Use of para-formaldehyde for fixation requires additional permeabilisation of the cells using 0.1% Triton X-100 in PBS and incubation for 10 min. at r.t.

3. Staining is performed as described above, but incubation time is extended up to 1 h.
4. During washing the cells should be incubated after resuspension for about 15 min. to guarantee proper removement of antibody and blocking of residual binding sites.

All in all, immunofluorescence techniques in combination with flow cytometric analysation offers exciting possibilities to determine specific antigens at the surface of cells or even cytoplasmatic and nucleic structures. The advantages of flow cytometric analysis are based mainly on the ability to determine cell-associated fluorescence intensities at the single-cell level and the rapid assessment of a large number of cells leading to excellent statistical data. Thus, flow cytometry is best qualified to improve characterisation of cells in all areas, ranging from basic immunology and molecular biology to clinical medicine and development of new strategies for pharmaceutics, as well as discrimination between even very close subsets of mammalian cells as applied to detect and monitor certain diseases.

Confocal laser scanning microscopy has been established within the last 15 years as a valuable tool for obtaining high-resolution images and three-dimensional reconstructions for a variety of biological specimens. Convential light microscopy creates images with a depth of field of 2 to 3 μm, and the real information in the focal plane is overlayed with "out-of-focus" information from optical planes above and below. This is especially a problem in fluorescence miroscopy where "out-of-focus" fluorescence creates a diffuse haze around the objects of interest. The aim of confocal light microscopy is to use only the information of the plane that is in focus and to divide three-dimensional objects into multiple optical slices. These images can then be digitized and reconstructed into three-dimensional objects using modern computer technology.

## Components of a Confocal Laser Scanning Microscope

Conventional epifluorescence microscope: Objective (oil or water immersion).

stage, focus motor (for optical slicing).

Confocal unit: laser, excitation filter, illumination pinhole, scanning unit (oscillating mirrors), dichroic mirror, detection pinhole, reflection filter, photo-multiplier.

## Optical Principle

The principle of a confocal laser scanning microscope is shown in Fig. 12.5. A laser light beam from a strong laser source is expanded to make optimar use of the optics in the objective. This improves spatial resolution by the objective compared to that of conventional light sources. To generate a two-dimensional image, the laser light beam is scanned across the specimen. This is achieved by an *x-y* deflection mechanism (reflection from galvanometer-driven oscillating mirrors). This enables the scanning laser beam to focus by an objective lens on a very small area of the fluorescent specimen termed the excitation light. Using a high numerical aperture objective lens, the crossover of the conical beam of laser light is approximately 0.2 μm in diameter. The fluorescent specimen emits light of a higher wavelength that has to pass the objective lens, the oscillating mirrors, the reflecting dichroic mirror, an excitation light filter, and a confocal pinhole and ultimately is focused onto a photodetector. The emitted light from the fluorescent specimen is "de-scanned" by the same oscillating mirrors scanning unit. The reflecting dichroic mirror or beamsplitter allows only the emitted light up to a defined wavelength to pass through and reflects light with higher wavelengths (which includes all the excitation light coming back from the specimen). The confocal pinhole is constructed so that it eliminates all light from "out-of-focus" planes. This is especially important when evaluating thick specimens. A "confocal spot" is the term used for the area that is detected by the laser beam.

To generate a two-dimensional image of a small area of the specimen or "optical section", a raster sweep of the specimen is done at one particular focal plane. As the laser scans across the specimen, the analog light signal is detected by the photomultiplier and converted into a digital signal. The combination of scans creates a pixel-based image (256 × 256 pixels, 512 × 512 pixels, or 1056 × 1056 pixels) that is displayed on a computer monitor. The relative intensity of the fluorescent light emitted from the laser hit point on the fluorescent specimen corresponds to the intensity of the resulting-pixel in the image (8-bit grayscale). Colour can be added to the 8-bit grayscale images to define reacting proteins in a double or triple labeling and prior to three-dimensional reconstruction to highlight regions of interest.

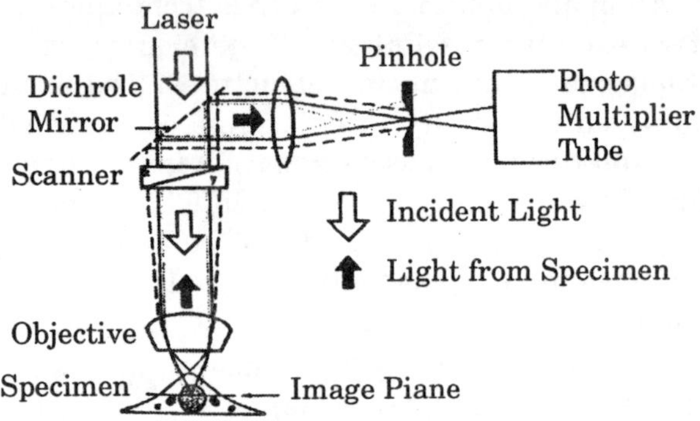


**Fig. 9.5. Confocal laser scanning microscope.**

Most frequently, assigning colours by thresholding is used to make images look clearer. This is done by assigning various intensity ranges to different colours. The fluorescence intensity of each assigned colour is ramped from dark to light to reflect fluorescence intensity, termed pseudocolor. The colours are usually shown in an 8-bit colour look-up table (LUT), which is included in the image legend.

## Pinhole and Confocality

The confocal microscope produces an optical slice through a specimen of defined thickness as a consequence of the pinhole. The numerical aperture of the lens influences the thickness of the optical slice: Oil and water-immersion lenses are most suitable and can generate slices as thin as 1 μm. The diameter of the pinhole inversely correlates with confocality. For example, a large pinhole increases the signal but decreases the axial resolution of a plannar object and can result in a toss of confocality resulting in conventional microscopy.

A small pinhole limits brightness. In practice, microscope users often open up the size of the prinhole to increase the signal from a weekly fluorescent or scattering object. For each confocal microscopy specimen, it is necessary to determine an optimal pinhole diameter for slice thickness and brightness.

## Fluorophores

Specimens are stained using antibodies labeled with fluophores for imaging with CLSM. This technique is called immunofluorescence histochemistry. However, other techniques using direct staining with specifically reacting substances like phalloidins also can be used. Fluorescent probes are available for a broad range of biomolecules, permitting the use of the CLSM to image macromolecular structures (proteins, lipids, carbohydrates, and nucleic acids) and physiological ions (*e.g.*, calcium, protons) for static fluorescent specimens, as well as for dynamic quantitative imaging and measurements. The choice of fluorophores used for immunofluorescence is critical because auto-fluorescence intrinsic from the tissue can mimic the appearance of the fluorophores. To exclude auto-fluorescence interfering with the fluorophore emission, the unstained specimens are examined with the same set of filters for excitation that are used with the specific fluorophore.

Auto-fluorescence is usually the highest, with excitation between 400 and 450 nm, and decreases with increasing wavelengths Additionally, the choice of fluorophores depends on their excitation/emission wavelength, the laser source, and the filter sets available. Immunofluorescence can be performed by either direct or indirect methods. Direct immunofluorescence involves the conjugation of the primary anti-body with a fluorophore such as fluorescein or rhodamine. For indirect immunofluorescence, the primary antibody is detected by a secondary fluorescence-labeled antibody that is directed against the first antibody.

This method is more commonly used because it is easier to label secondary antibodies than primaries and it is more sensitive, *i.e.*, more than one molecule of the secondary antibody can bind to the primary antibody. The specificity of the staining antibodies for an indirect staining protocol is especially important for multi-colour immunofluorescence because of cross-reactivity. Each individual secondary antibody must recognize only one primary antibody and must not cross-react with intrinsic proteins in the specimen or with other secondary antibodies. Unspecific staining of the tissue by a secondary antibody can be determined by staining the tissue in the absence of the primary antibody.

To ensure that a secondary antibody recognizes only the primary antibody, it is necessary to determine whether the secondary antibody binds to the primary antibody. To avoid cross-reactivity, it is ideal to use secondary antibodies that have been raised in species different from the species in which the primary antibodies have been derived. Some of the important considerations for choosing a laser for a CLSM system are cost, wavelength of emission, output power at each wavelength, efficiency, and stability. Laser lines below 400 nm are expensive to achieve, therefore, only a few CLSM systems with such lasers are available commercially.

**Table 12.4. Commonly used Laser Sources.**

| *Source* | | *Excltation lines* | |
|---|---|---|---|
| He Ne | 543 nm | 633 nm | |
| Argon lon 1 | 488 nm | 514 nm | |
| Argon lon 2 | 458 nm | 488 nm | 529 nm |
| Argon Krypton | 488 nm | 568 nm | 647 nm |

Use of filters in a CLSM system: Gray filters are used for the attenuation of the laser power, and single lines are selected by excitation filters. Ideally, these filters should have minimal loss of laser intensity and should permit the detection of the fluorophores around their maximal emission wavelength. For multi-labelling experiments, fluorophores should have only minimal overlap in their excitation/emission wavelengths. To eliminate channel "crosstalk", it is necessary to use sequential excitation and imaging of each fluorophore to spectrally isolate their fluorescence emission. This method requires special filter combinations based on each of the fluorophore's excitation and emission wavelengths and is time-consuming because the specimen must be scanned several times with different excitation and emission wavelengths.

## Anti-fade Reagents

Because of the high intensity and focus of the laser beam, the labeled specimens undergo significant bleaching/fading. The bleaching effect is more severe than in conventional epifluorescent microscopy, where the entire specimen is observed under lower-power excitatory light. This results in accelerated degradation of fluorophores and makes longer observation periods impossible. Several factors influence fading, among them the fluorescence intensity, pH, and base solution of the embedding medium. Antifade reagents can slow down the degradation process of the fluorophore, resulting in longer periods of observation, fluorimetry, and pattern recognition.

**P-phenylenediamine (PPD) :** This is one of the most effective antifade reagents, but it is photo/thermo sensitive and cannot be used for *in vivo* studies because of its toxic effects. The optimal PPD antifade mixture consists of 90% glycerol, 10 PBS, PPD concentrations between 2 mM and 7 mM, pH 8.5–9.0 [51].

**N-propylgallate (NPG) :** NPG is non toxic and photo- and thermostable. Though less effective as an antifade, it can be used for *in vivo* studies. Optimal working concentrations: glycerol base, 3 mM–9 mM NPG.

## Standard Protocols for Labeling

*Direct labeling :* Staining F-Actin with fluorescent phallotoxins (protocol from Molecular Probes, modified).

## Preparation of Stock Solution

Fluorescent phallotoxins are available from Molecular Probes, The Netherlands or Eugene, Oregon. The content (5 μg) is dissolved in 1.5 ml methanol (200 units/ml). Molecular Probe defines one unit as the amount of material used to stain one microscopic slide of fixed cells. This stock solution can be stored at 4°C. Prior to staining, the necessary volume is taken from the stock solution and transferred to a test tube, evaporated, and then dissolved again in the appropriate buffer immediately before use.

**Protocol 7 :** Procedures for staining slides

This procedure has yielded consistent results in most instances. Modifications may be necessary for particular specimens. For cells grown on glass coverslips (for each coverslip):

1. Fix coverslips or slides in 4% paraformaldehyde in PBS pH 7.3 for 3 min. and then 10 min. at room temperature.
2. Wash several times with PBS.
3. Place the coverslip in a glass petri dish and permeabilize the cells with ice-cold acetone.
4. Air dry sample.

5. Evaporate 5 μl of the stock solution in a small test tube and dissolve it again in 150 μl of PBS. Place on the coverslip for 20 min. at room temperature.
6. Wash twice rapidly with PBS.
7. Mount the coverslip on a slide with the cell side down in FluoroSave.
8. Seal the edges of the coverslip with rubber cement.
9. Specimens prepared in this manner retain actin staining for at least 5 to 6 months when stored in the dark at 4°C.

## Indirect Labeling

**Protocol 8 :** Method for immunofluorescence on coverslips or slides

1. Fix coverslips or slides in 4% paraformaldehyde in PBS pH 7.3 for 3 min. and then for 10 min. at room temperature.
2. Wash several times with PBS.
3. To stain intracellular proteins, the membrane must be permeabilized prior to the labeling by rapid dehydration/rehydration using different alcohol concentrations (70%, 80%, 90%, 100%, 90%, 80%, 70%)
4. Wash once with PBS.
5. Block nonspecific staining by incubating in PBS + 0.1% bovine serum albumin (BSA) and normal serum from the host species of the secondary antibody, for example 5% natural goat serum (NGS) for 1 h at room temperature. To detect intracellular antigens, it is necessary to add 0.05% Saponin to this buffer and keep the BSA and Saponin through the procedure.
6. Incubate the primary antibody at room temperature between 1 and 3 h (depending on the characteristics of the primary antibody). Use approximately 100 ml per coverslip, more for slides. Wash several times with PBS, BSA, and Saponin if permeabilized.
7. Incubate with fluorescent-labeled second antibody for 1 h at room temperature in the dark to protect fluorescence (add 5% NGS to the secondary antibody solution).
8. Wash several times with PBS and once with distilled water.
9. Mount in Fluorosave™(embedding and antifade reagent) or other embedding media containing antifade agents.

***Preparation of solutions***

- **4% paraformaldehyde in PBS :** Heat 8 g paraformaldehyde in 100 ml PBS until dissolved (must be prepared in a fume hood). Remove particles using a paper filter in a glass funnel. Adjust to the original volume (100 ml), then mix 1:1 with 2x PBS to obtain a 4% paraformaldehyde solution in PBS. (This fixative works best when used fresh.)
- **Phosphate buffered Saline (PBS), 10x :** Dissolve 2.0 g KCl, 2.0 g $KH_2PO_4$,1.0 g $MgCl_2.6H_2O$, 80,0 g NaCl, 14.34 g $Na_2HPO_4 \cdot 2H_2O$, and 1.0 g $CaCl_2$ (add last) in 1000 ml distilled water, adjust pH to 3. Before diluting PBS to use as 2x or Ix solution, adjust pH to 7.2 with 1M NaOH.

### *Multi-labeling*

For multi-labeling of fixed cells with antibodies, the basic protocol must be performed twice. During the labeling, the coverslips must stay in PBS + 0.1% BSA + 5% NGS. It is important to use the right combination of primary and secondary antibodies. If primary antibodies of similar origin (*e.g.*, mouse) are used, cross-reactions can occur during the detection with secondary antibodies. This can be avoided with appropriate blocking steps. If possible, use direct staining with directly labeled probes, such as Texas red-X-Phalloidin or DAPI. List of primary and secondary antibodies used in the multi-labeling experiment to stain the activation of cytoskeletal proteins:

1. Labeling: Anti-Phosphotyrosine, polyclonal, rabbit IgG (#06-427, Upstate Biotechnology); secondary antibody: FITC (Fluorescein isothiocyanate) anti rabbit IgG (Ex = 496/Em = 516).
2. Labeling: Anti-paxillin, monoclonal, mouse IgG (#P17920, Transduction Laboratories); secondary antibody: TRITC (Tetramethylrhodamine isothiocyanate) anti-mouse IgG (Ex = 543/Em = 571).
3. Labeling: direct, for cytoskeletal F-actin: Texas red-X-Phalloidin, (Ex = 594/ Em = 608).

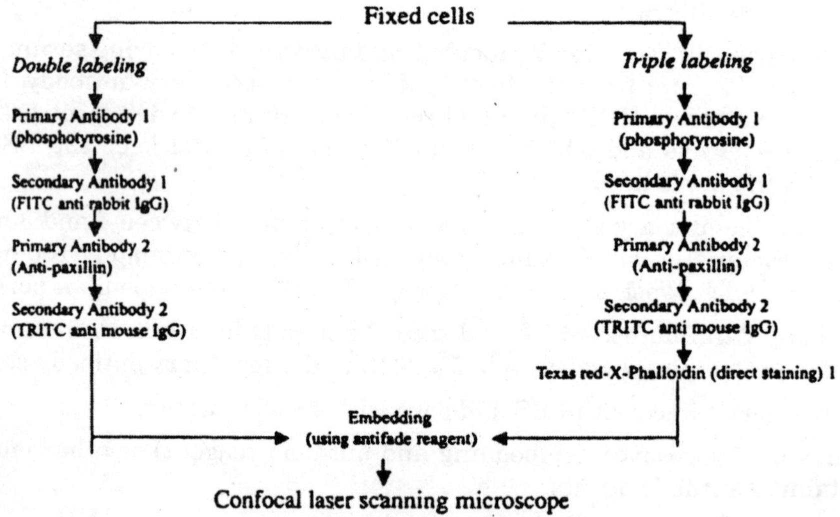


## CONCLUSIONS

CLSM for multicolour immunofluorescence is a complex task involving detailed knowledge of immunohistochemistry, confocal microscopy operation and instrumentation, and image presentation and analysis. The development of CLSM systems with multi-wavelength excitation and powerful microcomputer systems allows confocal imaging to be used with triple-labeled specimens.

CHAPTER

# 13

# Biochemistry of Nucleic Acid

Both the replication and transcription of DNA are complex processes. Although the basic chemistry is relatively simple many enzymes and other proteins are required. In part this reflects organizational and topological problems associated with the huge amount of DNA present as a single molecule within a chromosome.

## THE TOPOLOGY AND ENVIRONMENT OF DNA

Although we can isolate DNA in the form of simple double helical fragments, the topology of natural DNA is always more complex. Covalently closed circular DNA such as that in plasmids, mitochondria, and bacterial chromosomes is supercoiled and bound to proteins.

The DNA of chromosomes and bacteriophage particles is folded further into more compact forms. For example, the chromosome of *E. coli* contains DNA about 1.5 mm in length folded within a cell that is only 2 μm long. The diploid length of DNA in a 20 μm cell of a human is about 1.5 meters. At the time of cell division human DNA must all be replicated and packaged into chromosomes, 23 pairs in each cell. The density of the compacted DNA varies. A bacterial nucleoid may contain 10–30 mg/ml of DNA.[2] In chromatin of a eukaryotic nucleus there may be 200 mg / ml of DNA and in nucleosomes 330–400 mg / ml. The tightly compacted head of the T4 bacteriophage contains 520 mg/ml.

### DNA in Viruses

In the simplest filamentous DNA viruses such as M13 the DNA is coated by a helical protein sheath, as it is extruded from a cell. The sheath is peeled off as the virus enters another cell. However, in the large tailed phage, which contains ~ 160 kb of polynucleotide chains, the DNA is closely packed within the heads. In a model that seems to accommodate most

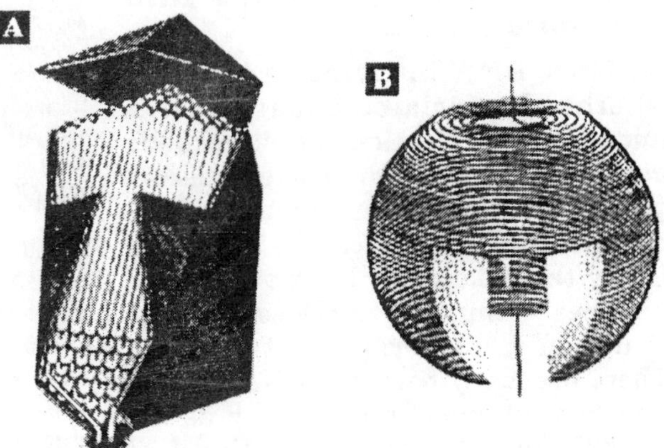


**Fig. 13.1 Possible ways of packing DNA into the heads of bacteriophage particles. (A) Spiral-fold. (B) Concentric shell model.**

experimental results, the DNA rod bends sharply into a series of folds, which are laid down around the long axis of the head in spirally arranged shells. The end of the DNA that enters the phage head first appears to be located in the center with successive shells of DNA around it. In the large bacteriophage G the DNA appears to be folded to form 12 icosahedrally arranged pear-shaped rings in the corners of the capsid. The 2.0 nm diameter double helical segments of DNA lie roughly parallel and are separated by only 0.5 – 1.0 nm of solvent,[6] which contains cations such as the polycation of sper-midine. Another possibility is that the DNA may be wound as on a spool of thread. The DNA chains have an external diameter of ~2 nm with 0.6 nm additional for a hydration layer. In a phage head the adjacent parallel chains are 2.6 – 2.7 nm apart. Thus, the packing is very tight. Even so, capsids tend to be only about half filled with DNA. An exception is provided by the tobacco mosaic virus in which the RNA genome is held precisely by protein subunits, which dissociate to release the RNA during infection.

## Bacterial Chromosomes and Plasmids

Most DNA in living organisms, whether bacteria or eukaryotes, is underwound. That is, the superhelix density is about –0.05 or one supercoil per 200 base pairs. In eukaryotes this negative supercoiling can be accounted for by the winding of DNA around the histones within the nucleosomes. The situation in bacteria is not as clear. There are many bacterial DNA-binding proteins but one of them known as HU is particularly abundant: In E. *coli* it exists as mixed dimers and tetramers of 9.5 (α) and 9.2 (β) kDa subunits. Each HU tetramer can bind~ 60 bp of DNA. There are about 60,000 HU monomers per cell, enough to coat ~ 20% of the genome.

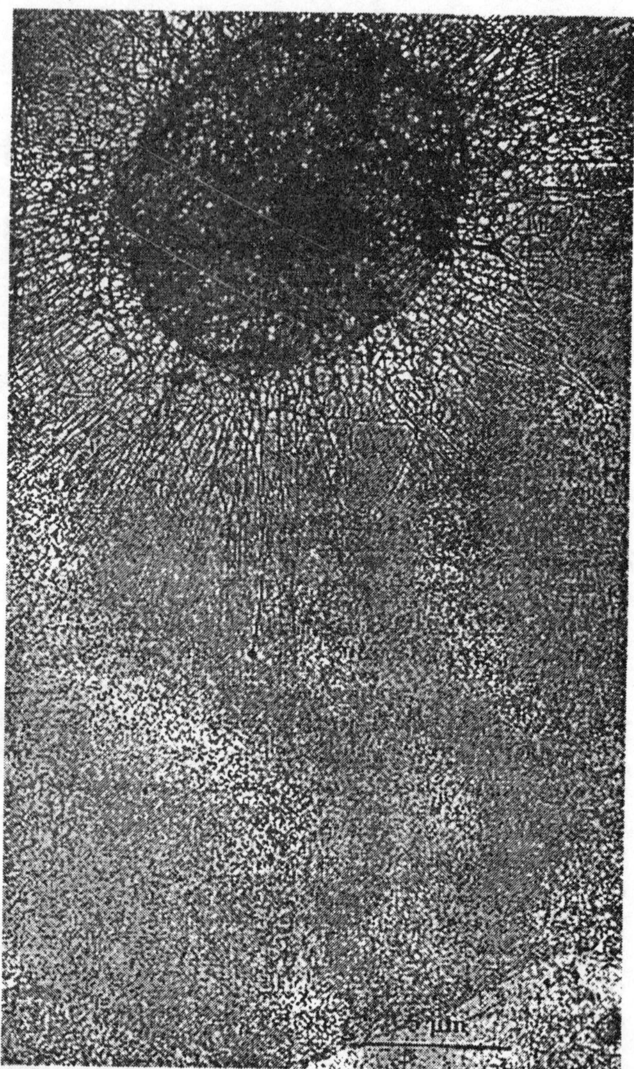

**Fig. 13.2. Electron micrograph of a bacterial nucleoid. The DNA is usually contained within the 'cage', but has been spread, using Kleinschmidt's procedure, to yield a surrounding skirt. The cage contains a protein network which includes elements of the cytoskeleton, enclosing the residual nuclear substructures. The denser fibrils radiating from this cage disappear to nuclease digestion and are probably aggregates of DNA fibers which merge with individual DNA strands at the extremities of the skirt. These strands are highly supercoiled, indicative of intact DNA.**

Possible modes of interaction with DNA have been proposed on the basis of the X-ray structure of HU. Binding to HU causes the DNA to be more tightly wound, introducing additional negative supercoils. The resulting unreleased torsional stress may be important in the functioning of the DNA. Binding is strongest to four-way junctions and to DNA with nicks and gaps or to structures induced by supercoiling. Other basic histonelike proteins may also bind to the DNA. However, there are no structures that resemble eukaryotic nucleosomes. If bacterial cells are lysed under certain conditions, e.g., in 1 M NaCl or in the presence of a "physiological" 5 mM spermidine, the entire bacterial chromo-some can be isolated.

The DNA in these isolated chromosomes retains some torsional tension that, however, can be relaxed by nicking with nucleases or by γ-irradiation. However, a single nick relaxes the DNA very little. The explanation appears to be that the DNA is held by proteins of the nucleoid matrix in a series of loops. A single nick relaxes just one loop. On this basis there are 43 ± 10 loops per genome with ~ 100 kb of DNA per loop. A 136-residue protein H-NS is involved in condensation of bacterial DNA. It may act as a scaffolding protein, but it also functions in controlling transcription.

## Protamines, Histones, and Nucleosomes

Within bacterial cells the negatively charged phosphate groups of the DNA are neutralized to a large extent by the positively charged polyamines, by cations such as $K^+$ and $Mg^{2+}$, and by basic proteins such as HU. Within the mature heads of sperm cells of fish, the tightly packaged DNA is neutralized by the *protaniinea,* small ~5-kDa proteins rich in arginine. Similar basic proteins are found in mammalian sperm. However, within most eukaryotic cells, the charges on DNA are balanced principally by a group of basic proteins, first isolated and named *histon* by Kossel in 1884.

There are five classes of histones, which range in molecular mass from ~ 11 to 21.5 kDa.

$H_1$ (including $H_1$°, $H_5$; lysine-rich "linker" histone)

$H_2A$, $H_2B$ (moderately lysine-rich) ⎤<br>
⎬ "Core" histones<br>
$H_3$, $H_4$ (arginine-rich) ⎦

All of the core histones share a conserved 65-residue *histone fold.* The arginine-rich histones have a strongly conserved amino acid sequence, histone $H_4$ from pea seedlings differing from that of the bovine thymus by only two amino acids. On the other hand, the lysine-rich $H_1$ is almost species-specific in its sequence. Differentiated tissues contain at least seven variant forms of histone $H_1$ including proteins designated $H_1$°, Hit, and $H_5$. The N-terminal 25-40 amino acids of the core histones are positively charged and highly conserved. The 135-residue histone $H_3$ of calf thymus carries a net charge of +18 within the first 53 residues. This is probably the portion that binds to DNA. On the other hand, the carboxyl terminal end is hydrophobic and only slightly basic.

Histones undergo substantial amounts of *micromodification* including phosphorylation, acetylation, and methylation. Mono-, di-, and tri-methyllysine residues may be present. The core histones all undergo acetylation on specific lysyl side chains. Nuclear histone acetyltrans-ferases transfer acetyl groups from acetylCoA and hydrolytic deacetylases may remove them. The amount of acetylation varies during different stages of the cell cycle, suggesting a regulatory role. Acetylation sites in $H_3$ and $H_4$ are highly conserved in all eukaryotes. A small fraction of histone $H_2A$ undergoes phosphorylation and dephosphorylation continuously, but $H_1$ and $H_3$ are phosphorylated and dephosphorylated at specific stages of the cell cycle.

Phosphorylation of $H_1$ has been thought essential for "condensation" of chromatin, the folding into the tightly packed chromosome structures. However, more recent experiments point to the N terminus of histone $H_2B$ as the required site of phosphorylation for chromosome condensation. In addition, histone $H_1$ may interact with membrane lipids. Histone $H_2A$ exhibits the greatest heterogeneity and appears to function in regulation of transcription, of gene silencing, and of repair of double-strand breaks in DNA. Each of the histones appears to be regulated separately. In animal chromatin ~10% of histone $H_2A$ and small fractions of $H_2B$ and of tissue-specific histones are covalently linked to ubiquitin. However, this monoubiquitination may not be related to proteolytic degradation. Archea contain histones that dimerize and bind DNA to form nucleosomes.

***Nucleosomes.*** An early idea of the function of histones was that they serve as gene repressers. To some extent this view is still valid. However, the large quantity of histone and uniform distribution over the DNA suggested some other role. This was. clarified when electron micrographs showed that chromatin fibers form *nucleosomes,* regular repeating structures resembling beads on a string. The same structure is seen in the "minichromosomes" formed from virus SV40.

Two molecules each of histones $H_2A$, $H_2B$, $H_3$, and $H_4$ form the core of the nucleosome around which ~146 bp of dsDNA is coiled into approximately two negative, left-handed toroidal superhelical turns. Digestion of chromatin by nucleases causes rapid cleavage into ~200-bp fragments and slower cleavage to 146 ± 20-bp fragments. This suggested that ~ 200-bp segments of DNA are folded around a histone octomer, contracting the 68 nm extended length of relaxed B-DNA into a 10-nm nucleosome. A short linker region of variable length, up to 80 bp, lies between the nucleosomes. The fifth histone, $H_1$ (or $H_5$ in some species), may bind to this linker DNA.

A nucleosome with bound $H_1$ is sometimes called a *chromatosome***.** The superhelix density of ~0.05 observed for DNA extracted from eukaryotic cells is just equal to one negative superhelix turn per nucleosome. For example, the number of nucleosomes seen in the minichromosome of Fig. 13.3 matches the numbers of supercoils in the SV40DNA. If there are two negative supercoils per nucleosome, the DNA in the nucleosome must be wound more tightly than in relaxed DNA (10.0 bp per turn instead of the 10.6 of relaxed DNA).

NMR data suggest that within the nucleosome tne regular base pairing in the DNA may be partially disrupted and that someparts of the histone have a high degree of mobility. Although nucleosomes are distributed rather evenly along the DNA of a cell, there are some DNA sequences that favour nucleosome formation. The resulting *positioned nucleosomes* are often found in the vicinity of gene promoters, enhancers, and other control sequences. It seems likely that positioning of nucleosomes is related to control of transcription or of other activities. The same basic structure for chromatin has been found in animals, fungi, and green plants.

***The linker histones.*** Anucleosome is pictured in Fig 13.3A as if held in a compact configuration by the binding ~ of histone $H_1$ at a position that marks a pseudo twofold axis that lies in the plane of the nucleosome. However, this is only one of several possible locations for linker histones of the $H_1$, $H_5$ family. The structure of the linker histones is somewhat different from that of core histones. They have an 80-residue globular domain with long N-terminal *and* C-terminal chains, both of which are rich in basic residues and evidently available for binding to DNA.

Perhaps they bind both to the DNA entering the nucleosome and to that leaving the nucleosome, reducing electrostatic repulsion of those two parts of the DNA superhelix. Another possible location for histone $H_1$ or $H_5$ is above the histone surface and *inside* the DNA loop. A third suggested location for the globular linker core is *between* the two turns of the DNA strand. While one function of the linker histones may be to stabilize mononucleosomes, they may also play a role in compaction of the DNA into the 30-nm fibers universally seen in nuclei of cell Histone $H_1$ can also be regarded as a general represser, holding chromatin tightly folded and preventing transcription.

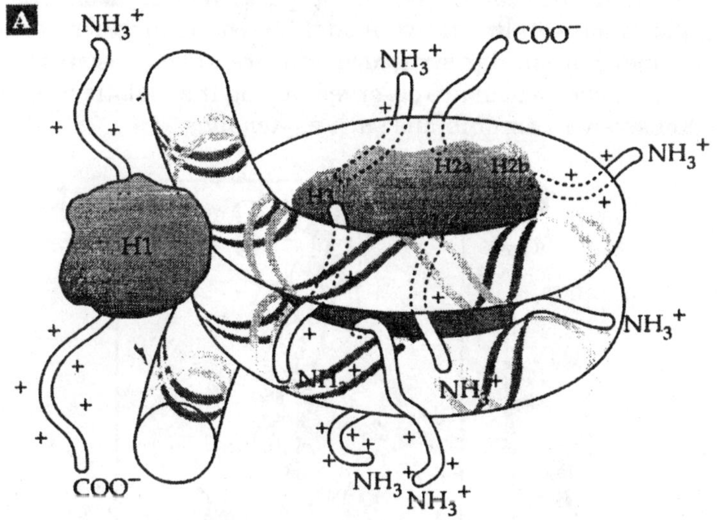

**Fig. 13.3(A) A nucleosome is formed when dsDNA, shown schematically as a tube, wraps roughly twice around a histone octomer. This complex. or nucleosomal core particle, includes at its center two copies of histone $H_3$, two of histone $H_4$, and two $H_2A$–$H_2B$ dimeric pairs, one of which is not visible, Ends of each histone molecule are thought to protrude like tails from the core, ready to interact with other molecules. In many organisms, histone $H_1$, portrayed at left in one possible position, helps to anchor DNA to the core and promotes further compaction of the DNA into a 30-nanometer fiber.**

The possible roles of acetylation, phosphorylation, methylation, ubiquitination, and other modifications of histones in controlling transcription, replication, and DNA repair are receiving increasing attention. *Active chromatin,* where transcription is occurring, has an altered nucleosome structure and increased susceptibility to nuclease action. It appears to be less tightly packed than inactive chromatin and to contain regions called *hypersensitive sites* that are accessible to nucleases or chemical modification reagents.

There seems to be a direct link between increased acetylation of histones and enhanced initiation of transcription by RNA polymerase II. Conversely, deacetylation is associated with repression of transcription. Both histone acetylase and deacetylase activities have been found in transcriptional regulators. The observation that $H_1$ becomes phosphorylated during the initiation step of mitosis suggests another control mechanism for its represser functions. Several multiprotein complexes that "remodel" chromatin have been identified. These complexes contain ATP-dependent *helicase* activities that open up DNA for transcription.

***Folding of nucleosome chains; chromosomes.*** Electron microscopy shows that chromatin is packed into the nucleus largely as fibers of ~30- to 36-nm diameter. Thoma *et al.* proposed that the fibers could be formed by winding the string of nucleosomes into a simple one-start helix with six nucleosomes per turn and a pitch of ~11 nm. The $H_1$ molecules would be close together in the center. Its zigzag pattern of adjacent nucleosomes would generate a two-start helix. Another alternative model envisions larger solenoids with interdigitated nucleosomes. A somewhat similar structure appears to be present in the specialized eukaryotic *lampbrush chromosomes* which are observed during the meiotic prophase of oocytes. They have been studied intensively in amphibians such as *Xenopus*.

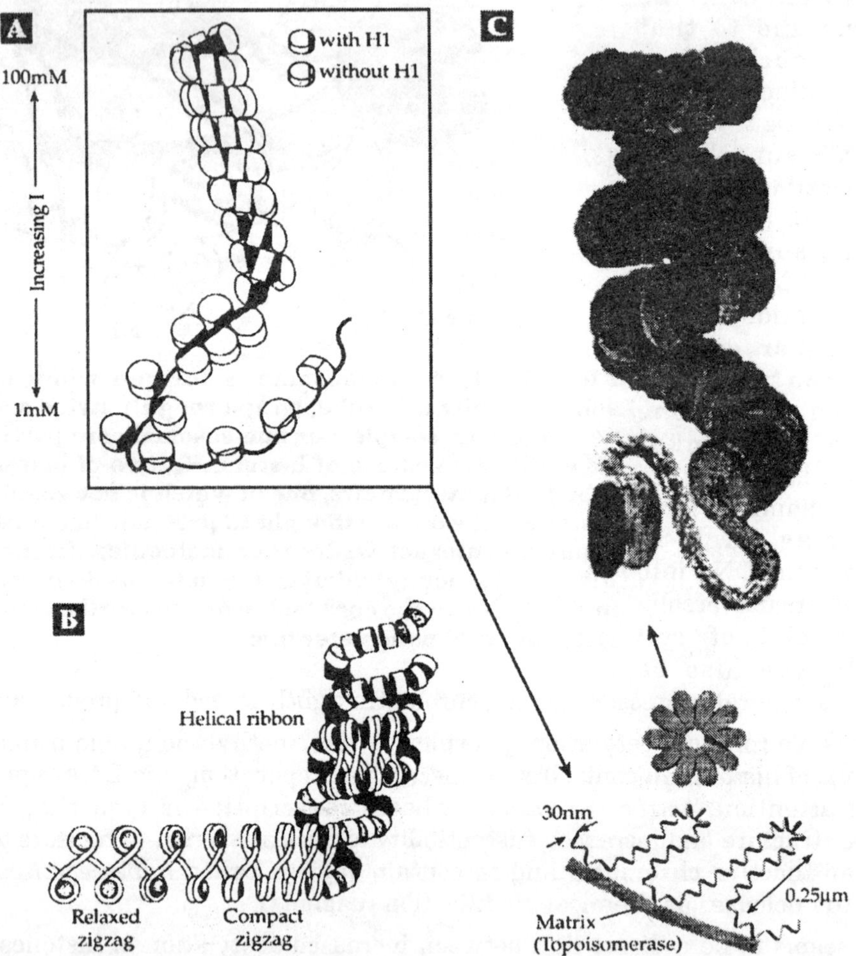


**Fig. 13.4 (A, B) Two possible models of the 30-nm chromatin fiber. (A) Thoma *et al.* (B) Woodcock *et al.* The fully compacted structure is seen at the top of each figure. The bottom parts of the figures illustrate proposed intermediate steps in the ionic strength-induced compaction. (C) Possible organization of the DNA within a metaphase chromosome. Six nucleosomes form each turn of a solenoid in the 30-nm filament as in (A). The 30-nm filament forms -30 kb-loop domains of DNA and some of these attach at the base to the nuclear matrix that contains topoisomerase II. About ten of the loops form a helical radial array of 250-nm diameter around the core of the chromosome. Further winding of this helix into a tight coil ~ 700 nm in diameter, as at the top in (C), forms a metaphase chromatid.**

A lampbrush chromosome is actually a homologous pair of chromosomes, each one in turn consisting of two closely associated chromatids. The chromosomes are highly expanded, and about 5% of the DNA is extended in the form of ~4000 perfectly paired loops visible with an electron microscope. Each loop consists of ~50 μm or ~ 150 kb of extended DNA. No evidence of any breaks in the DNA is seen, a fact that supports the belief that a single DNA molecule extends from one end of the chromosome to the other through all of the loops. Like the puffs of polytene chromosomes, which may have a similar structure, lampbrush chromosomes appear to be actively engaged in transcription.

Approximately 3% of the DNA may be functional in producing mRNA that is accumulated within the oocyte and is used as a template for protein synthesis during early embryonic development. A different arrangement is present in metaphase chromosomes, which appear as two dense parallel sister chromatids of ~ 700-nm diameter. The DNA must be highly folded.

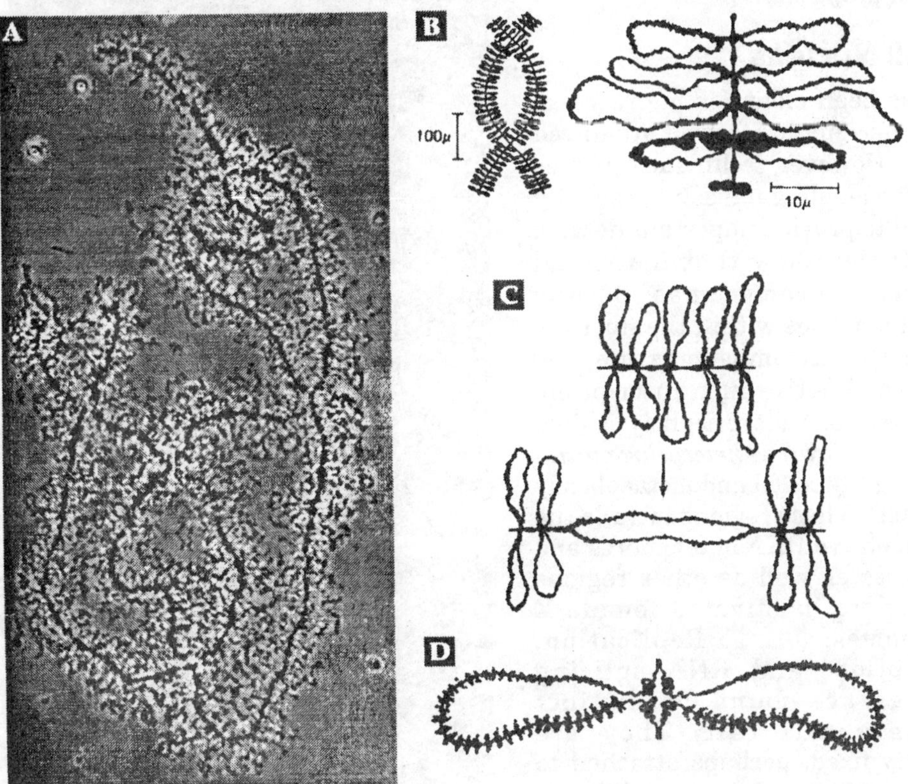


**Fig. 13.5. (A) Photomicrograph of a lampbrush chromosome from the nucleus of an oocyte of the newt *Triturus*. (BD) Diagrammatic views of lampbrush chromosomes. (B) The two homologous chromosomes (left) are held together by two chiasmata. A portion of the central chromosome axis (right) shows that two loops with identical morphology emerge at a given point, evidence that each chromosome has already split into two chromatids. (C) Accidental stretching of a chromosome reveals the continuity of the loop axis with the central axis. (D) A single loop pair, showing the single DNA molecules on which RNA chains (indicated by fuzzy shading) are being transcribed.**

In the model shown in Fig. 13. 5C the 30-nm fiber is folded into ~ 30-kb loops, each one formed from ~25 turns of the 30-nm helix. The loops then form a helical array 250 nm in diameter with ~ ten loops (300 kb) per turn. This helix is further wound into a tight helix of ~700-nm diameter. A single turn of this helix may contain as much as 9 Mb of DNA. About four hundred coils (an average of 18 coils per human haploid chromosome) could accommodate the entire genome.

A group of five proteins, some of which are designated *SMC* (structural maintenance of chromosomes) proteins, form a complex called *cohesin.* SMC proteins, large multidomain proteins found in all eukaryotes, are also present in bacteria. Interphase chromatin must be much less tightly packed and contains regions in which large loops, e.g., of 20–120-kb size, are uncoiled enough to allow transcription factors and other proteins to locate their target sequences. Many models of interphase chromatin have been proposed.

## The Cell Nucleus

It has been clear for many years that the nucleus has a well-organized structure. However, techniques such as fluorescent labeling have only recently been used to provide important details. These studies show that individual interphase chromosomes occupy discrete territories within the nucleus. Parts of the chromosomes may be unfolded and active in transcription, while others are more tightly coiled. Some parts, known as *heterochromatin,* are very, tightly coiled and metabolically almost inert. These regions include the highly repetitive DNA of telomeres and centromeres as well as other regions and complete inactivated female X chromosomes. Replication, transcription, and RNA splicing complexes are found at distinct locations within cells. They are apparently fixed, perhaps attached to the inner nuclear membrane, while the DNA passes through the complexes.

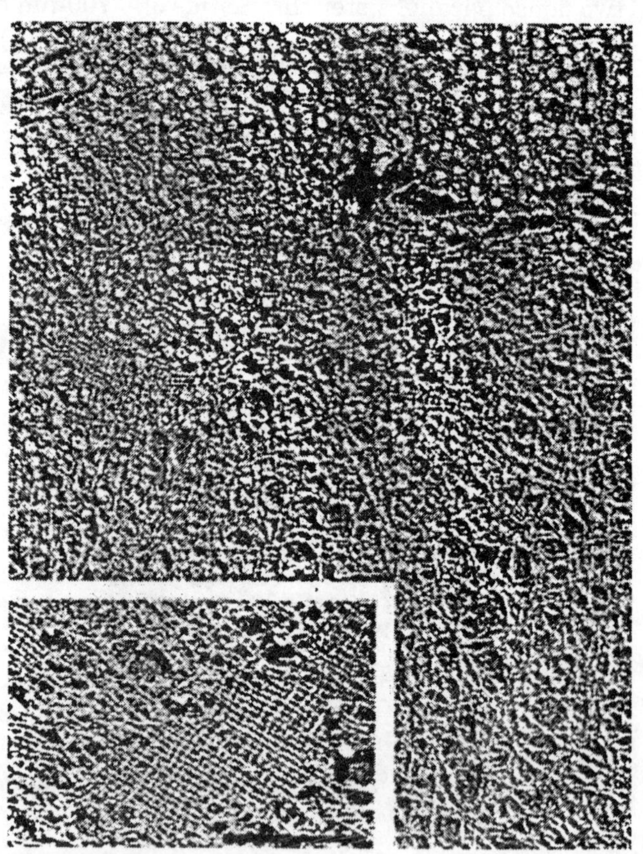

**Fig. 13.6. Native nuclear lamina of *Xenopus* oocytes. Freeze-dried metal-shadowed nuclear envelope extracted with Triton X-100, revealing the nuclear lamina meshwork partially covered with arrays of nuclear pore complexes. Inset, relatively well-preserved area of the meshwork of nearly orthogonal filaments from which pore complexes have been mechanically removed.**

***The nuclear matrix.*** The lipid bilayers, the histones and other soluble proteins, and the DAN can all be removed from nuclei by extraction and enzymatic digestion. An in soluble residue, **the nuclear** matrix, is left. Largely protein in nature, this matrix is spread throughout the nucleus. Remnants of the membranes remain in the form of proteins that were in or along the bilayer. The nucleolus is clearly defined. The DNA appears to be bound to the nuclear matrix proteins. A specific 320-kb piece of *a Drosophila* chromosome has been mapped and used to locate nontranscribed scaffold (or matrix) attachment regions of DNA bound to matrix proteins. These were found at intervals of 26-112 kb, the intervening loops containing up to five or more genes. A 120-kDa protein together with topoisomerase II may be components of a *nuclear scaffold* that constrains the loops of DNA. The scaffold may also provide locations for the complexes of proteins involved in replication and other processes.

The *matrix attachment regions* (MARs) may also act as *insulators* that shield promoters for transcription within certain loops from control elements such as *enhancers* that may be present in adjacent loops. At least one nuclear matrix component becomes phosphorylated and moves to the nuclear poles during mitosis.

***Other nonhistone nuclear proteins.*** Polyacrylamide gel electrophoresis revealed more than 450 components in HeLa cell nuclei. Most are present in small amounts of <10,000 molecules per cell and are not detectable in cytoplasm. Among the more acidic proteins are many enzymes including RNA polymerases. There are also gene repressors, hormone-binding proteins, protein kinases, and topoisomerases.

Among the six most abundant nonhistone nuclear proteins in the rat are the cytoskeletal proteins myosin, actin, tubulin, and tropomyosin. A group or' small (<30 kDa) proteins, the *high mobility group (HMG)* proteins, can be extracted from chromatin with 0.35 M NaCl. Two pairs, HMG–1 + HMG–2 R (renamed *–HMGB*) and HMG–14 + HMG–17 (renamed *HMGN*) are present in nuclei of all mammals and birds. HMG–14 and HMG-17 appear to be concentrated where active transcription is occurring. HMG-17 undergoes a variety of modifications including acetylation, methylation, phosphorylation, ubiquitination, glycosylation, and ADP-ribosylation, suggesting that it may assist in regulation of transcription. Other members of the HMG family bind more specifically to certain DNA sequences or structures.

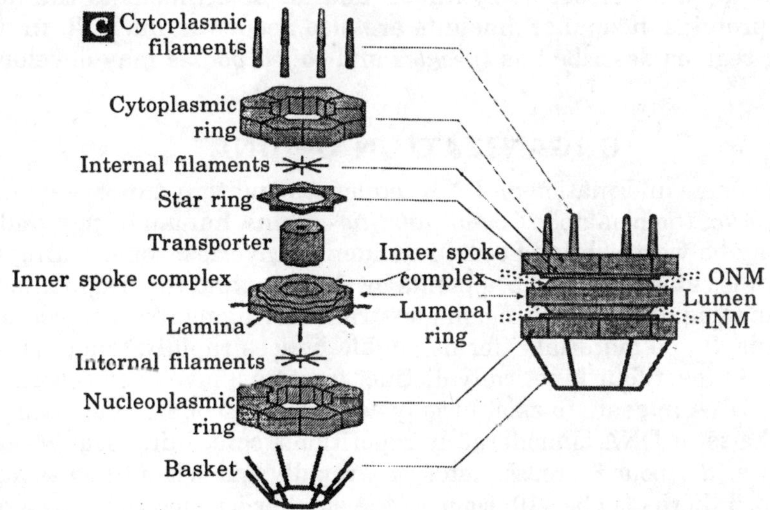


**Fig. 13.7. Electron micrographs of nuclear pores from a Xenopus oocyte.**

***Nuclear membrane, pores, and lamina.*** The nuclear membrane consists of two bilayer membranes separated by a 40– to 60–nm perinuclear space. Both of the membranes and the intervening space are penetrated by large proteinaceous *nuclear pores.* The two membranes are fused together around" the pores. On the inside of the innermost nuclear membrane is a relatively insoluble meshwork of intermediate f ilaments, the *nuclear lamina.* It acts as a scaffold for tfie pores and may also interact with chromatin. Three proteins, the *lamins* A, B, and C, are the major components of the nuclear lamina. The A and C lamins, which differ only at their C termini, are homologous to cytoplasmic intermediate filament proteins and may exist as a network of coiled-coil polymers. Lamins are phosphorylated, and the lamina is disassembled during the prophase of mitosis, apparently under the influence of the cdc2-cyclin B complex and reappears in telophase at the end of mitosis. Nuclear pores consist of octameric rings of protein subunits with a complex structure and an outside diameter of 120 nm and an inside diameter of –80 nm.

A pore may consist of as many as 80 to 100 different proteins and have a mass of ~125 MDa. The pores are sometimes seen to be blocked by -35 nm granules, perhaps pre-ribosomes. Transport through nuclear pores occurs in both directions. Numerous proteins enter the nucleus where many bind, to RNA and are then exported as ribosomal subunits. Transfer RNAs and messenger RNAs must also be exported. A family of nuclear transporters known as *importing* and *exporting* mediate the movement of protein and RNA through the pores. They depend upon the G protein Ran, which also functions in spindle formation, and the hydrolysis of GTP. Large conformational changes may be linked to the transport cycle and also have been observed in response to changes in the $Ca^{2+}$ level.

***The nucleolus.*** This organelle is a dynamic structure, which breaks down and reforms in each mitotic cycle. It is organized around clusters of genes for the 28S, 18S, and 5.8S ribosomal RNA subunits. Both chromatin and ribonucleoproteins that transport the rRNA out to the cytoplasm are present. The major nucleolar protein *nucleolin* is characterized by its glycine-rich C terminus, which contains seven repeats of GGRGG and also contains $N^{\gamma}N^{\gamma}$-dimethylarginine. Nucleolin may control transcription of the DNA that carries the rRNA genes. It appears to be required for ribosome synthesis and for attachment to the nuclear matrix. A variety of other proteins in smaller amounts are also needed in nucleoli. In addition to the true nucleoli other regions described as *speckles* and *coiled bodies* may develop around RNA splicing centers.

## ORGANIZATION OF DNA

A vast amount of new information on DNA sequences and structures is available for yeast, *Arabidopsis, Drosophila,* the nematode *Caenorhabditis elegans,* human beings, and other species. Nevertheless, it may be worthwhile to consider older discoveries, some dating back 20 years or more. When DNA is cut into ~10-kb fragments by shearing and the fragments are denatured by heat, the renaturation of the resulting single-stranded fragments upon cooling takes place in two or more steps. Some material reforms double helices rapidly, whereas other material is slow to renature. At least four kinetically distinct fractions have been recognized: (1) About 70% of mammalian DNA appears to exist largely as single copies, i.e., with unique sequences. (2) About 20% of the total DNA is moderately repetitious, containing sequences that may be present $10^3$–$10^5$ times. (3) About 5% reassociates very rapidly and is identified as highly repetitive *satellite DNA* of which there may be ~$10^6$ copies. (4) A smaller fraction contains long palindromic sequences.

As a consequence, the single-stranded pieces can fold back almost instantaneously to form hairpin structures. These differences also show up in density gradient centrifugation in CsCl, whether on large pieces of DNA in the presence of ligands such as $Ag^+$ or on smaller restriction fragments which have often been studied by polyacrylamide gel electrophoresis. If there are ~35,000 human genes, protein or RNA coding sequences must occupy only 2–3% of the genome. The pufferfish *Fuga rubripes* probably has almost as many genes as we but only –13% as much DNA. In further contrast the newt *Triturus cristatus* has six times as much DNA as a human.

The compact genome of the green plant *Arabidopsis thaliana* occupies only 120 Mb. In contrast are the 415–, 2500–, and 5300-Mb genomes of rice, maize, and barley, respectively. What is the function of all of the apparently noncoding DNA present in some organisms? It is often viewed as *"junk,"* whose only function may be to facilitate evolutionary changes in the genome. However, there is doubtless important undiscovered information in these regions. It has been very hard to determine sequences, in part because of the large amount of highly repetitive DNA. Some regions have been *"unclonable"* in prokaryotic systems because of the presence of transposon-like sequences or *"kinkable"* elements (TG•CA steps), palindromes, etc.

## Repetitive DNA

Rapidly renaturing DNA fragments often have a different base composition than the bulk of the DNA and, consequently, often separate as small satellite bands upon centrifugation in a CsCl gradient. Satellite DNA is usually associated with regions of the chromosome that do not unravel in telophase as does the bulk of the DNA. Satellite DNA usually consists of short highly repetitive sequences, which occur in large clusters of up to 100 Mb of DNA, often near centromeres or telomeres or on a Y chromosome. The DNA of a satellite band from the kangaroo rat contains the sequence 5'-GGACACAGCG-3' repeated so often that it accounts for 11% of the entire DNA of the cell. Longer repeating sequences -170 bp are also often present as are *microsatellites* of 2– to 5-base-pair repeats. At least 30,000 microsatellite loci are present in the human genome.

***Centromeres.*** The attachment of spindle fibers to chromosomes depends upon the segments of DNA known as centromeres to organize the attached kinetochores. In the yeast *Saccharomyces cerevisiae* a 120-bp region containing three short conserved sequences is present in the centrosomes of all ten chromosomes. This may fold into a distinctive looped structure. Human centrosomes are large and complex, but the DNA is highly repetitive, giving rise to *α-satellite DNA*. Sequences such as $(TGGAA)_n$ are repeated many times. Such sequences can form self-complementary looped structures containing some unpaired guanines that intercalate and stack between sheared G • A pairs.

Complex regional centromeres involving kilobases or megabases of DNA have also been identified in fission yeasts, *Drosophila,* and green plants. The great variability indicates that centromeric sequences undergo rapid evolution. This may be related to the fact that of the four cells formed by female meiosis only one becomes an egg. A series of unique *centromeric proteins* (CENP-A to CENP-G) bind to the DNA sequences of centromeres and direct the formation of the kinetochores. Even for the simpler centromere of budding yeasts, kinetochores have a complex structure. The CENP proteins were first identified as autoantigens in sera of patients with the autoimmune disorder *scleroderma.*

***Telomeres.*** The DNA sequences at the chromosome ends have a TG-rich strand, such as the *(TTGGGG) (Tetrahymena)* and the $(TTAGGG)_n$ of both human and trypanosome chromosomes. The complementary DNA strand is CA-rich. The *S. cerevisiae* telomers have ~350 base pairs containing the sequences ($TG_{1-3}/C_{1-3}A$) as well as one or more copies of a 6.7-kb nonrepetitive sequence and other elements. In many species the repetitive telomeric sequences have 3' poly(G) tails at the ends of the DNA molecules. These tails are able to form G quartet structures. A variety of telomere-binding proteins have been isolated. Some of these bind to G quartet structures and some, such as RAP1 of yeast, to double-stranded telomere repeat regions. Special problems associated with replication of telomeres are discussed in Section C,8.

***Short interspersed sequences** (SINES).* Much of the reiterated DNA is present in repeated segments 100–500 bp in length that lie between 1– to 2–kb segments of unique DNA. The best known example is the human ***Alu*** family, socalled because it contains a site for cleavage by restriction enzyme *Alu*I. Sequences of the *Alu* family also exists in other primates and in rodents. The ~300-bp *Alu* sequence is reiterated over 500,000 times in the human genome with various sequence alterations, but an 80–90% homology. This sequence consists of two similar ~130-bp segments called the *"left monomer"* and *"right monomer."* The right monomer contains a 31-bp insertion and the left end carries a poly(dA) sequence.

In addition there is a short 7– to 20-bp sequence, which is variable between different *Alu* sequences but is directly repeated at each end of a given *Alu* sequence. The *Alu* sequence has strong homology with the 7S RNA that is part of the signal recognition particle involved in transport of newly synthesized peptide chains across the membranes of the ER. *Alu* sequences are transcribed into hnRNA, the precursors to mRNA. Some *Alu* sequences are present in intervening sequences (introns) within genes anizf others are in noncoding sequences between genes.

Sharp suggested that specific proteins in the nucleus may bind to the *Alu* sequences, preventing hnRNA from leaving the nucleus before it has been processed to remove introns and other sequences absent in mRNA. However, the presence of the poly(dA) regions and the direct terminal repeats suggests that the whole *Alu* sequence is *pseudogene* derived from a *retroposon,* a type of transposon that originated from an RNA molecule. *Alu* and other SINES contain an RNA polymerase III promoter and are transcribed. Active retroposons form reverse transcripts (cDNAs) that can be integrated at various points in the genome. One theory is that retroposons have no biological function but have invaded the genome at random locations. Many SINES and other families of repetitive sequences have been characterized by the presence of a restriction enzyme cleavage site in each copy of the sequence. *Eco*RI cuts the previously mentioned α-satellite DNA.

A 319-bp reiterated sequence in the human genome surrounds a *Hint* site, and Sau3A cleaves a 849-bp sequence with ~1000 copies per haploid genome. In the hermit crab 30% of the genome consists of repeated sequences, one 156-bp unit occuring ~7 million times. Many identical 14-bp GC-rich inverted repeats are present. *Neurospora* and yeast both contain many copies of the GC-rich palindromic sequence 5'-CCCTGCAGTACTGCAGGG-3', which contains the two underlined *Pstl* sites. Some repetitive sequences such as the **CAT** family and the *homeodomain sequence* are found in the control regions preceding the 5' ends of groups of genes that are regulated coordinately. Some repetitive sequences seem to be unstable in the genome and may be excised and lost or may increase in number during aging.

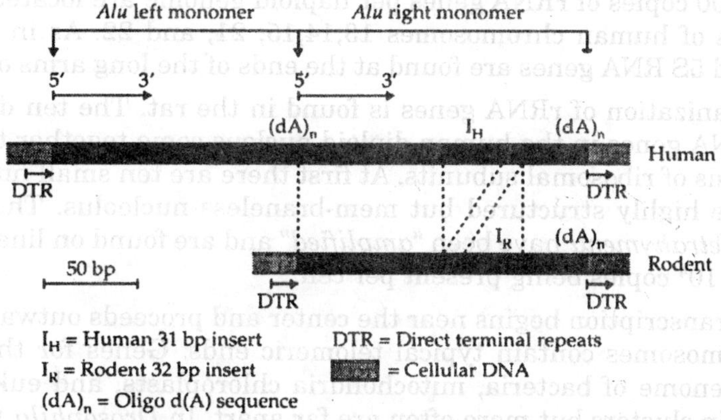


**Fig. 13.8. Structure of human and rodent *Alu* sequences in DNA**

Repeated DNA sequences have apparently originated not only with 7S RNA but also from mRNA or tRNA molecules. A characteristic SINE in rodent DNA is known as the ID (identifier) sequence because it was once thought to be a marker for genes transcribed in neural tissues. Rat DNA contains ~130,000 copies of the ID sequence. Other *clustered repeats* are frequently found between genes that are present in large numbers. These include the genes for ribosomal RNA, tRNA, small nuclear RNAs, and histones.

***Long interspersed repeat sequences (LINES).*** These moderately repetitive sequences may be several kb in length. Just one type seems to be abundant in each mammalian species. The human LI sequence is over 6 kb in length. Like the *Alu* sequence it has a poly(dA) sequence at the 3'-end, but it does not contain short terminal direct repeats. These interspersed repeat sequences usually contain genes which may be functional, but many of the copies contain pseudogenes or genes that are randomly truncated, having lost a segment from the 5' end. These sequences, too, seem to have been dispersed throughout the genome by retroposons.

Human chromosome 22, sequenced in 1999, contains within its 33.4 Mb of euchromatin at least 545 genes, 134 pseudogenes, and 8043 L1 sequences (9.7% of the DNA), as well as 20,188 *Alu* sequences (16.8% of the DNA) and many other interspersed repeats. Repetitive DNA sequences are less common in prokaryotes, but they do exist. For example, many dispersed and clustered repeating units are present in *Halobacterium* DNA.

## Genes for Ribosomal RNA and Small RNA Molecules

Most genes are present as one copy each per hap-loid genome. However, there are many copies of the genes for ribosomal RNA and tRNA. In *Xenopus* DNA there are ~450 repeats of the 28 S and 18 S rRNA genes on one chromosome and ~24,000 copies of the 5 S RNA genes at the ends of the long arms of most of the chromosomes. Nontranscribed spacer regions lie between the repeats of the 28S and 18S gene pairs, as can be seen with the electron microscope in Fig. 13.9. The gene for 5 S RNA has a high GC content. A-denaturation map of the DNA shows easily melted regions separated by shorter 120-bp sequences, apparently of high GC content and presumably coding of the 5 S RNA. The easily denatured AT-rich spacers are ~630 bases long.

Using restriction enzymes much of this DNA can be cut into segments that contain repeats within repeats. One 15-unit polynucleotide contains the sequence $A_4CUCA_3CU_3G$ repeated about

30 times. About 200 copies of rRNA genes per haploid genome are located at the constrictions in the short arms of human chromosomes 13,14,15, 21, and 22. As in *Xenopus,* clusters of tandemly repeated 5S RNA genes are found at the ends of the long arms of most chromosomes.

A similar organization of rRNA genes is found in the rat. The ten different chromosome ends carrying rRNA genes in the human diploid nucleus come together to form the nucleolus the site of synthesis of ribosomal subunits. At first there are ten small nucleoli, but fuse these to form the single highly structured but mem-braneless nucleolus. The rRNA genes of the macronucleus of *Tetrahymena* have been *"amplified"* and are found on linear 21-kb palindromic molecules, about $10^4$ copies being present per cell.

Initiation of transcription begins near the center and proceeds outward in both directions. These short chromosomes contain typical telomeric ends. Genes for the tRNAs are spread throughout the genome of bacteria, mitochondria chloroplasts, and eukaryotic nuclei. They sometimes occur in clusters but more often are far apart. In *Drosophila* there are probably at least 600 tRNA genes, many occurring in pairs of opposite polarity, i.e., as inverted repeat sequences. The genes for small nuclear RNAs U1-U6 are organized in a variety of ways. The ~ 100 human U1 genes occur on a single chromosome, perhaps organized in a tandem array and interspersed with as many as 10,000 defective pseudogenes.

## Other Gene Clusters and Pseudogenes

Closely related structural genes often occur in clusters. Among these are clusters of *immunoglobulin* genes and clustered genes for the α and β *globins,* which encode the protein sequences for the hemoglobins. The human α globin gene cluster occupies about 30 kb on chromosome 16 and the β globin genes 60 kb on chromosome 11. The α1 and α2 genes encode identical peptides, while the related ξ gene encodes the corresponding subunit of embryonic hemoglobin. The β cluster includes, in addition to the adult gene, a pair of fetal globin genes ($\gamma^G$ and $\gamma^A$ ) differing by only one amino acid (Gly vs Ala) at position 136, the embryonic ε chain, and the minor adult δ chain. Besides the functional genes the globin cluster contains *pseudogenes,* which are given the prefix ψ in fig 13.10. These are nonfunctional genes, which appear to encode peptides homologous to the known globins.

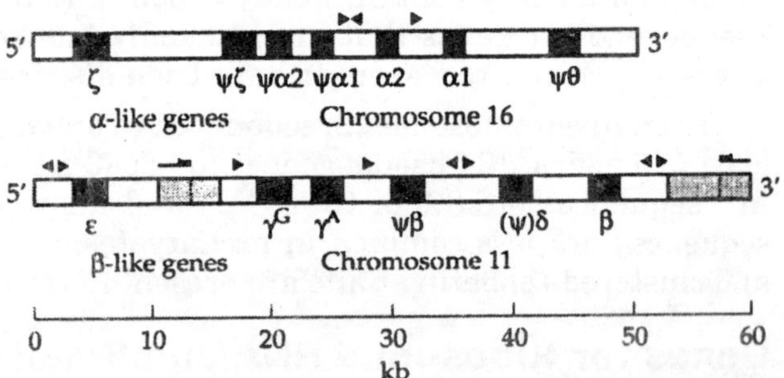


**Fig. 13.9. Organization of the globin genes on human chromosomes 11 and 16. The composition of the various embryonic, fetal, and adult hemoglobins is also indicated. Closed boxes indicate active genes and open boxes pseudogenes. The triangles (→) indicate *Alu* repetitive sequences and their orientation. The shaded boxes indicate Kpn repeat sequences and the half-arrows (←) their respective orientation. The Kpn sequence between the ε and $\gamma^G$ genes in fact consists of two tandemly linked Kpn repeats.**

However, they contain mutations that prevent expression. For example, deletion of a single nucleotide near the beginning of the pseudogene will scrambl: the genetic message by changing the reading frame. Globin pseudogene ξψ has a nonsense mutation at codon 6, ψα1 contains a whole set of mutations that prevent transcription and translation, while ψα2 is mutated almost to the point of being unrecognizable as a globin relative. Do pseudogenes have a physiological function?

The presence of pseudogenes in close proximity to functional genes suggests that there may be an as yet unrecognized controlling function. On the other hand, they may be relics of evolution. If a gene family arises by duplication and mutation of duplicated copies, we may expect to find gene copies that have not been selected as useful but are still present in the genome. There are regions of the genome that seem to be largely single copy DNA with few reiterated sequences or gene clusters. While some gene families exist as clusters others are dispersed throughout the genome.

## Introns, Exons, and Overlapping Genes

A major difference between prokaryotes and eukaryotes is the presence of intervening sequences (*introns*) between the coding sequences (*exons*) in eukaryotes. Introns are especially numerous in higher organisms. For example, the gene for the myosin heavy chain present in rat embryonic skeletal muscle encodes a 1939-residue peptide but occupies a length of 22 kb of DNA. The gene is split into 41 exons, whose transcribed RNA must be cut and spliced at 40 places to form the mRNA. By comparison the corresponding gene from the nematode *Caenorhabdatis elegans* is far less fragmented.

It is not clear why some genes have so many introns and others so few. The very compact chromosomes of viruses and most bacteria do not contain introns. However, they are sometimes present in mitochondrial and chloroplast genes. A surprise, which was first recognized in viral RNA and DNA, is that genes sometimes overlap. For example, two proteins, one long and one shorter, are synthesized starting at the same point in the RNA genome of phage Qβ.

In the DNA of phage ϕX174 the third nucleotide of the stop signals for some genes are also the first nucleotides for the start signal for translation of the next gene.' Pairs of genes have been found in which one of the genes of the pair is found completely inside the other gene but is translated in a different reading frame. A fourbase overlap between dihydrofolate reductase and thymidylate synthase has been found in the DNA of phage T4. A transposable DNA insertion sequence in *E. coli* encodes two genes, one of which is contained within the other and which is transcribed from the opposite strand of DNA.

The double-stranded RNA of a reovirus produces two peptides from the same sequence using two different AUG initiation codons in different reading frames. Overlapping obviously limits severely the mutational alterations that are allowable. Perhaps it is for this reason that eukaryotic genes seem to be dispersed in widely separated locations and overlap is rare. However, the human type IV collagen genes for the αl and α2 chains are encoded on opposite strands with their 5' flanking regions overlapping. Some introns contain genes. Chlorarachniophyte algae contain a multimembraned chloroplast thought to be a vestigial remnant of an endosymbiont. Its 380-kb genome consists of three short chromosomes that encode overlapping genes and contain the shortest introns (18–20 base pairs) known.

## DNA of Organelles

Mitochondrial DNA (mtDNA), varies in size from <6 kb to 367 kb. The small circular mtDNA of animals is extremely efficiently packed with genes for tRNAs, rRNA, and a small number of protein subunits. However, the 78-kb yeast mtDNA contains many long AT-rich spacers as well as long introns, some of which contain genes for splicing enzymes. Otherwise, it is similar to animal mitochondrial DNA. Mitochondrial genomes of higher plants are much larger; that of maize is a 570-kb circle, which contains both direct and inverted repeat sequences. Recombination between these sequences is apparently the origin of smaller incomplete circular genomes that are also present.

Plant mtDNA appears to contain the genes present in animal mitochondria plus additional genes. For example, one encodes a 5S RNA. The most unusual mtDNA is found in hemoflagellates such as *Trypanosorna* and *Leishmania.* The single large mitochondrion or *kinetoplast* located near the base of the flagellum contains a network of catenated circular DNA molecules (Fig. 9 .16). The genetic makeup is similar to that of other mtDNA, but the rRNA genes are unusually short. Chloroplast DNA varies in size from 120 to 160 kb

## Methylation of DNA

As mentioned in already a significant fraction of the pyrimidine and purine bases in DNA is methylated. One function of such methylation in bacteria, to protect against the action of restriction endonucleases. For example, the gene for the well-known *EcoRl* endonuclease is carried in *E. coli* cells by an R factor. This plasmid also carries (just 29 base pairs away) the gene for a 326-residue *$N^6$-adenine methylase.* This enzyme uses S-adenosylmethionine (AdoMet) to methylate the two adenines (marked by asterisks) in the six-base-pair recognition sequence 5'-G A A* T T C / 3'-C T T A* AG converting them to $N^6$-methyladenines ($m^6A$). Other DNA methyltransferases place methyl groups on N-4 of cytosine or on C-5 of cytosine.

The latter utilizes a mechanism illustrated in Eq. 13.4. Such enzymes are components of both type II and type I restriction-methylation systems. However, most of the $m^6A$ in the E. *coli* chromosome arises from action of a different methylase, one that recognizes the palindromic sequence 5'-GATC and methylates adenines in both chains. This *DNA adenine methylase,* a product of the *dam* gene, plays an important role in mismatch repair, transposition, regulation of transcription, and initiation of DNA replication. The same methylase regulates at least 20 genes induced during infection by *Salmonella typhimurium.* Some of these genes are essential to virulence. A similar methyl-transferase appears to control differentiation of the stalked *Caulobacteria* cells.

***CG doublets.*** The only modified base commonly found in eukaryotes is 5-methylcytosine, which upon deamination becomes thymine. Most methylation occurs when C is followed by G. Usually 60–90% of all 5'–CG sequences (CpG sequences) in eukaryotic DNA are methylated. However, the fraction of methylated cytosine varies from almost zero for *Drosophilia, Caenorhabditis,* and *Saccharomyces* to as much as 30% in higher plants. When CG pairs are methylated on both cytosines, an interesting result arises upon replication of the DNA. The methyl groups don't prevent replication, but only one of the DNA strands in each daughter duplex is methylated.

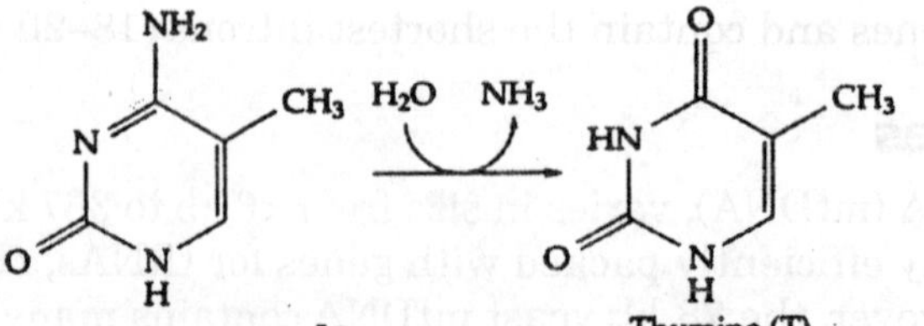

$H_2O$ $NH_3$

5-Methylcytosine ($m^5C$) → Thymine (T)

However, additional methylation of DNA occurs within three hours of DNA replication by a *maintenance methyltrans-ferase* that recognizes methylated CpG sequences in the old DNA strand and methylates the cytosine in the 3′-GpC of the newly synthesized strand. Mammalian cells contain at least three different DNA (cytosine-5) methyltransferases. Enzymatic *demethylation* which converts $m^5C$ residues back to cytosine, may also occur. It has been difficult to demonstrate enzymatic demethylation of 5-methylcytosine has been reported.

The observed loss of methylation during embryonic development if may be a result of loss of the maintenance methylase. Most CpG doublets are found in large *"islands"* of several hundred bases to about *2* kb in length, which are unusually rich in G + C. These islands lie near the 5' ends of ~ 60% of all human genes and near origins of replication. They are also found in regions of the DNA that are compacted into heterochromatin. The genes in methylated regions tend to be *"silent"* i.e., they are not actively transcribed. Demethylation may permit transcription. The degree of methylation of the CG doublets is variable both in position within a chromosome and with stage of development. Cytosine methylation is essential to embryonic development, and mice lacking the maintenance DNA methytransferase are developmentally retarded and die at mid-gestation.

DNA replication

Hemimethylated DNA

Maintenance methylation

Normal methylation pattern retained

Not all CG sequences are methylated and various patterns of DNA methylation are generated at different stages of development by rounds of methylation and by the action of demethylases. The maintenance methylase ensures that a stable methylation pattern persists until altered by new rounds of methylase or demethylase action. A parallel is found in the prokaryotic *Canlobacter* in which three chromosomal sites are fully methylated in swarmer cells, become hemimethylated in stalked cells, and are fully methylated again just prior to cell division. DNA methylation affects transcription, either directly by preventing the binding of transcription factors or indirectly via a series of binding proteins specific for methylated CG doublets.

In early stages of embryonic development there is very little methylation, but some genes are quickly silenced as methylation takes place. Heterochromatic regions including inactivated

X chromosomes are heavily methylated. However, additional alteration in chromatin is required for complete silencing of genes. Recent studies indicate that an abundant mammalian protein binds to the methylated DNA along with a *historic deacetylase.* The latter acts on acetylated mstones to free lysine side chains, which may interact in an inhibitory manner with the DNA. In female mammalian cells most of the genes on one of the two X-chromosomes are completely inactivated.

DNA methylation plays a major role in this process. A perfect correlation has been observed between 5'-methylation of cytosines in CpG islands and inactivation of X-chromosome genes. Methylation may also play a role in recombination and repair. Methylation of DNA decreases with increasing age. It increases as a result of oncogenic transformation of cells.

***Imprinting.*** With the exception of X-linked genes each person has two copies of each gene, one of maternal and one of paternal origin. Both copies of most of these genes are expressed. However, a few of the genes receive from one parent or the other an *imprint,* a mark that distinguishes the parental origin. Such imprints are maintained in cells through embryonic development but are erased in embryonic gonads to allow for a new imprint in the germ cells. Imprinting depends upon DNA methylation, and all imprinted genes show the presence of differentially methylated regions.

## REPLICATION

Following the discovery of the double helix and the enthusiasm that it engendered many people thought that the synthesis of DNA was simple. The nucleptide precursors would align themselves along separated DNA template strands and perhaps spontaneously snap onto the growing chains. In fact, replication is a complex process that requires the cooperative action of many different gene products and perhaps an association with membrane sites. The matter is made more complex by the fact that some of the enzymes involved in replication are also required in the processes of genetic recombination, in repair of damaged DNA molecules, and in defensive systems of cells.

### Early Studies

That the DNA content doubles prior to cell division was established by microspectrophotometry. It was clear that both daughter cells must receive one or more identical molecules of DNA. However, it was not known whether the original double-stranded DNA molecule was copied in such a way that an entirely new double-stranded DNA was formed or whether, as we now know to be the case, the two chains of the Original molecule separated. The latter is called *semiconservative* replication, each of the separated strands having a new complementary strand synthesized along it to form the two identical double-stranded molecules. The first definitive evidence for semiconservative replication was reported by Meselson and Stahl in 1958.

Cells of *E. coli* were grown on a medium containing isotopically pure $^{15}NH_4^+$ ions as the sole source of nitrogen. After a few generations of growth in this medium the DNA contained exclusively $^{15}N$. Then the cells were transferred abruptly to a medium containing $^{14}NH_4^+$ and were allowed to grow and to double and quadruple in number. At various stages DNA was isolated and subjected to ultracentrifugation in a CsCl gradient. Small but easily detectable differences in density led to separation of dsDNA molecules containing only $^{15}N$ from those containing partly $^{15}N$ and from those containing only $^{14}N$. At the beginning of the experiment only DNA containing entirely $^{15}N$ was present. However, after one generation of growth in the $^{14}N$-containing medium, the density of *all* the DNA was such as to indicate a content of one-half $^{14}N$ and one-half $^{14}N$.

After a second generation of growth half of the DNA atill contained both nitrogen isotoped in equal quantity, whereas half contained only $^{14}N$, exactly the result expected for semiconservative replication.

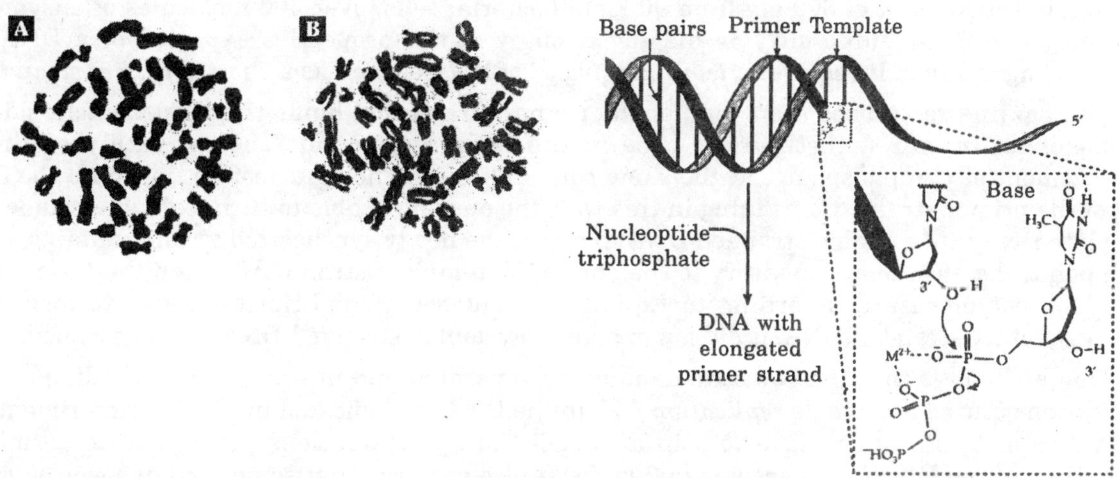


**Fig. 13.10 (A) Human chromosomes after one replication in the presence of 5-bromodeoxyuridine (BrdU). Both chromatids of each chromosome contain BrdU in one strand of the DNA duplex and normal thymidine in the other. (B) After two replications in the presence of BrdU one chromatid of each chromosome contains BrdU in both strands of the duplex and stains strongly with a special differential staining procedure. The other chromatid contains only normal thymidine in one strand of the duplex and is not stained.**

***Autoradiogrpahy.*** Later a technique of direct autoradiography of DNA using $^3H$-labeled thymidine was applied by Cairns. Cells of *E. coli* were grown on a medium containing the radioactive thymidine for various times but typically for 1 h (~2 generations). The cells were then ruptured, the DNA was spread on a thin membrane filter, and autoradiograms were prepared. When the DNA contained [$^3H$]thymidine the exposed trace in the autoradiogram could be followed around the entire 1.1–1.4 mm circumference of the spread DNA molecule.

Molecules partially labeled and in the process of replication could also be identified. After two hours of growth in the presence of $^3H$-labeled thymidine about half of the bacterial DNA was fully labeled, but half contained regions that were labeled only half as heavily. They presumably contained $^3H$ label in a single strand and, therefore, represented unreplicated regions. All of the molecules had undergone one round of replication with tritiated thymidine to yield lightly labeled molecules; parts of the molecules had not completed the second round. The more heavily labeled regions were interpreted as fully replicated.

The shapes of the *"replication figures"* suggested that DNA is synthesized in a continuous manner starting from one point and continuing around the circular molecule at a constant rate. Although subsequent experiments (considered in Section 4) show that replication is usually *bidirectional,* the experiments of Cairns were important because they introduced a technique for direct visualization of replication *in vivo.*

***The chemistry of DNA polymerization.*** What are precursors of DNA? Early experiments showed that the nucleoside [$^3$H]thymidine was efficiently incorporated into DNA, but for energetic reasons it seemed unlikely that thymidine was an immediate precursor. Evidence favouring the nucleoside triphosphates was provided in 1958 when Arthur Kornberg identified a DNA polymerase from *E. coli.* Kornberg's enzyme, now known as *DNA polymerase I,* was isolated in the amount of 600 mg from 90 kg of bacterial cells (over 400 molecules of enzyme per cell). The 928-residue enzyme displayed many of the properties expected of a DNA-synthesizing enzyme. It requires a *template strand* of DNA as well as a shorter *primer strand.*

The enzyme recognizes the 3' end of the primer strand and binds the proper nucleoside triphosphate to pair with the next base in the template strand. Then it catalyzes the displacement of pyrophosphate, at the same time linking the new nucleotide unit onto the 3' end of the primer strand. Continuing in this way, the enzyme is able to turn a single-stranded template DNA into a double-stranded DNA in which the newly synthesized strand contains, at each point, the base complementary to the one in the template strand. Although the action of the DNA polymerase I, according to Eq. 13.3 provided a straightforward way to form a complementary strand of DNA, it did not explain how double-stranded DNA could be copied.

One problem is that the two strands must be separated and unwound. If unwinding and replication occured at a single *replication fork* in the DNA, as indicated by Cairns' experiment, the entire molecule would have to spin at a speed of 300 revolutions per second to permit replication of the *E. coli* chromosome in 20 min. It also required that some kind of a *swivel,* or at least a nick in one chain, be present in the chromosome as indicated in the following drawing. Another problem was posed by the fact that the two chains in DNA have opposite orientations. Thus, at the replication fork one of the new chains might be expected to grow by addition of a new nucleotide at the 3' end, while the other chain would grow at the 5' end. *If so, there should be two DNA polymerases,* one specific for polymerization at each of the two kinds of chain end. Nevertheless, despite intensive search the only DNA polymerases found added new units only at the 3' end.

***Discontinuous replication and RNA primers.*** In 1968, Okazaki reported that during the time that replication of DNA is taking place bacterial cells contain short fragments of DNA. These are now called *Okazaki fragments* or *replication fragments.* A second development was the discovery of the enzyme *DNA ligase* which is able to join two pieces of DNA to form a continuous chain. These two discoveries provided an explanation for the lack of a second kind of DNA polymerase. One strand, the *leading strand,* of the replicating DNA could be synthesized continuously in the 5' to 3' direction while the other strand, the *lagging strand,* would have to be synthesized in segments (the *Okazaki fragments*), which could then be joined by the DNA ligase.

In 1971 Brutlag *et al.* reported that initiation of synthesis of the DNA of phage M13 in E. *coli* required formation of a short segment of RNA as a primer. It was subsequently shown that *with few exceptions priming by RNA synthesis is always required for replication.* Newer studies have revealed great complexity in the mechanisms of replication. As for synthesis of any polymer there are distinct steps of *initiation, elongation,* and *termination.*

Topological problems associated with the unwinding and rewinding of the double helices and with disconnection of catenated circles and untying of knots are solved with the aid of special enzymes, the *helicases* and *topoisomerases*. Replication requires both DNA and RNA polymerases, a ligase, and ancillary proteins, some of whose functions aren't yet clear. Many of these proteins associate to form large multiprotein complexes, which are given names such as *primosome* (for priming) and *replisome*. Many bacteriophages and plasmids also replicate within cells of *E. coli* utilizing bacterial proteins as well as proteins encoded in the viral or plasmid genome.

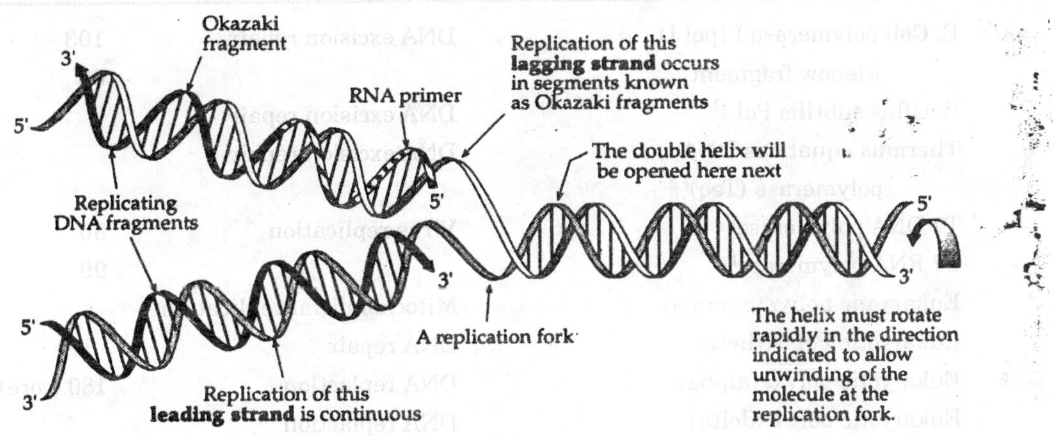


**Fig. 13.11.**

## DNA Polymerases

Immediately after its identification DNA polymerase I was generally assumed to be the principal chain elongation enzyme. However, an *amber* mutant of *E. coli deficient* in DNA polymerase I (gene *polA.*; Fig. 13.4A) synthesized DNA normally. This finding stimulated' an intensive search for new polymerases. Two were found: DNA polymerases II (gene *pol* B) and III. Both are present in amounts less than 25% of that of DNA polymerase I. Both have properties similar to those of polymerase I, but there are important differences.

By now DNA polymerases have been isolated from many organisms, many genes have been cloned and many sequences, both of bacterial and eukaryotic polymerases are known. Comparisons of both sequences and three-dimensional structures, suggest that the polymerases belong to at least six families. These include the RNA-dependent DNA polymerases known as *reverse transcriptases* as well as some *RNA polymerases*. Some of the polymerases exist as single polypeptide chains, while others function only as large complexes. In every case a two-metal ion catalytic mechanism with in line nucleotidyl transfer, appears to be used by the enzymes. Two-metal ion catalysis is also observed for phosphatases and ribozymes.

***Exonuclease activities, proofreading, and editing,*** DNA polymerase I not only catalyzes the growth of DNA chains at the 3' end of a primer strand but also, at about a 10-fold slower rate, the hydrolytic removal of nucleotides from the 3' end (*3'- 5' exonuclease activity*). The same enzyme also catalyzes hydrolytic removal of nucleotides from the 5' end of DNA chains. This latter 5'- 3' *exonuclease activity,* the DNA polymerase activity, and the 3'-5' exonuclease activity all arise from separate active sites in the protein. DNA polymerases II and III do not catalyze hydrolysis from the 5' end. Treatment with proteolytic enzymes cuts a 323-residue piece containing the 5'-3' exonuclease from DNA polymerase I leaving a larger C-terminal piece known as the *Klenow fragment.* This fragment retains the polymerase activity as well as the 3'-5' exonuclease activity and is widely used in genetic engineering.

**Table 13.1. Families of DNA Polymerases.**

| *Class* | *Name* | *Function* | *Molecular mass (kDa)* |
|---|---|---|---|
| A | E. Coli polymerase I (pol I) | DNA excision repair | 103 |
| | Klenow fragment | | 68 |
| | Bacillus subtilis Pol I[b] | DNA excision repair | |
| | Thermus aquaticus DNA polymerase (Teq)[c.d] | DNA excision repair | |
| | T7 DNA polymerase[e] | Virus replication | 80 |
| | T7 RNA polymerase[f.g.] | | 99 |
| | Eukaryotic poly γ (gamma) | Mitochondirial replication | |
| | Eukaryotic Pol θ (theta) | DNA repair | |
| B | Eukaryotic poly α (alpha) | DNA replaction | 180 (core) |
| | Eukaryotic poly δ (delta) | DNA replaction | |
| | Eukaryotic poly ε (epsilon) | DNA replaction | |
| | Eukaryotic poly ζ (zeta) | Bypass synthesis | |
| | Bacteriophage T4 DNA Pol + accessory proteins[h,i] | DNA replaction | 43 |
| | E. coli Poll II[j] | | 90 |
| C | Bactrial DNA Pol III[p] + accessory proteins | DNA replication | ~ 900 (holoenzyme) |
| D | Euryarchaeotic Pol II | | |
| X | Eukaryotic DNA Pol β[n,o] | DNA repair | 39 × 2 |
| | Eukaryotic Pol λ (lambda) | Base excision repair | |
| | Eukaryotic Pol μ (mu) | Non-homologous end-joining | |
| | Eukaryotic Pol σ (sigma) | Sister chromated cohesion | |
| Y | E.coli UmuC protein | SOS response | |
| | Eukaryotic Pol η (eta, XP-V, RAD 30) | Bypass synthesis | |
| | Eukaryotic Pol ι (iota) | Bypass synthesis | |
| | Eukaryotic Pol κ(kappa) | Bypass synthesis | |
| Reverse transcriptase family | | | |
| | HIV reverse transcriptase[k,1] | | |
| | Telomerase[m] | | |
| | RNA-dependent RNA polymerses | | |

In the Klenow fragment the large C-terminal domain contains the polymerase. The N-terminal domain contains the 3',5'-exonuclease activity, which is thought to fu lfill a *proofreading* and *editing* function. The polymerase acts at the 3' end of the growing DNA chain. Before moving on to the next position, the enzyme verifies that the correct base pair has been formed in the preceding polymerization event. If it has not, the exonuclease action removes the incorrect nucleotide and allows the polymerase to add the correct one. Thus, each base pair is checked twice, first before polymerization and then after polymerizations. A puzzle was the fact that structural studies indicate that the editing center is over 3 nm away from the catalytic center in Pol I.

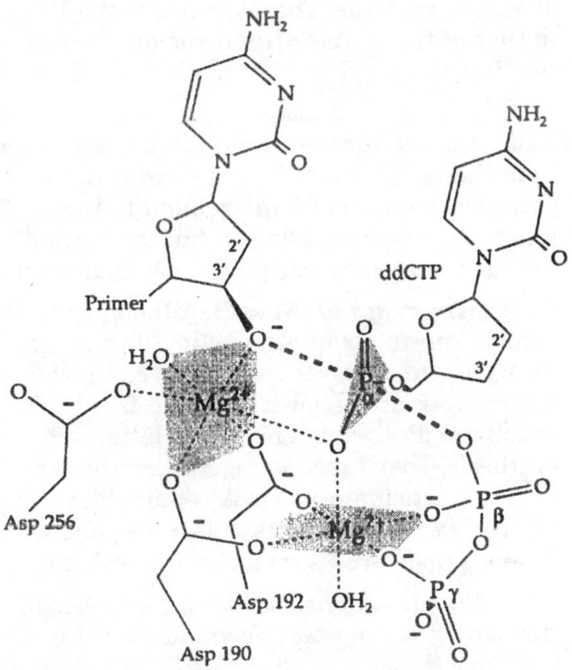

When the catalytic center *"identifies"* a nucleotide triphosphate as able to form a proper Watson-Crick nucleotide pair, it catalyzes the formation of the new nucleotide linkage. Then it releases the 3'-chain end, which sometimes *"melts"* and over a 10- to 100-ms time interval is able to reach over into the exonuclease site where the newly added nucleotide may be hydrolytically removed. However, if the newly formed nucleotide is properly paired, it will be less likely to melt, and the new nucleotide will be more likely to be retained. DNA polymerase I and other related polymerases utilize *processive mechanisms,* moving from one site to the next without diffusing away from the DNA. The schematic picture shown in fig. 13.14 also indicates how the 5'-3' exonuclease activity can come into play, when the polymerase reaches the end of a single-stranded gap.

**Fig. 13.12. Proposed mechanism and transition state structure for the synthetic nucleotidyl transfer activity of DNA polymerase β (and other DNA polymerases). The chain-terminating inhibitor dideoxy CTP is reacting with the 3'-OH group of a growing polynucleotide primer chain. This -OH group (as $-O^-$) makes an in-line nucleophilic attack on $P_\alpha$ of the dideoxy-CTP. Notice the two metal ions, which interact with the phospho groups and which are held by three aspartate side chains. Two of the latter, Asp 190. and Asp 256, are present in similar positions in all of the polymerases. The active centers for the hydrolytic 3'-5' and 5'-3' exonuclease activities of some of the polymerases also appear to involve two-metal catalysis and in-line displacement.**

***Other Class A polymerases.*** The *Thermus aquaticus (Taq)* polymerase is best known for its widespread use in the polymerase chain reaction. Like *E. coli* I the enzyme is a large multidomain protein. The structure of the catalytic domains of the two enzymes are nearly identical, but the *Taq* polymerase has poor 3'-5' editing activity. The enzyme has been carefully engineered to improve its characteristics for use in the PCR reaction. Some bacteriophage encodetheir own DNA polymerases. However, they usually rely on the host cell to provide accessory proteins. The sequence of the DNA polymerase from phage T7 is closely homolo-gous to that of the Klenow fragment and the 3D structurs are similar. The 80-kDa T7 polymerase requires the 12-kDa thioredoxin from the host cell as an additional subunit.

It has been genetically engineered to improve its usefulness in DNA sequencing. About 45% of the sequence of the *RNA polymerase* Encoded by phage T7, which transcribes RNA from the phage DNA, is also similar to that of the Klenow fragment. Sequences of these DNA polymerases are distantly related to those of reverse transcriptases. The 136-kDa polymerase y functions in mitochondria but is encoded in a nuclear gene. It is the only DNA polymerase that is inhibited by antiviral nucleotide analogs such as AZT.

***Polymerases of Class B.*** Although E. *coli* polymerase II is a member of this family, relatively little is known about its function. It may participate in DNA repair in the "SOS" response. The catalytic subunits of the major *eukaryotic* DNA polymerases as well as of archaeal DNA polymerases are members of the B family. Eukaryotic cells contain at least 13 DNA polymerases which are desig-nated by Greek letters. Polymerases α, β, and ε are essential for nuclear DNA synthesis. They function together with accessory proteins in pri-mase and replisome complexes. Others participate in DNA repair. Phage T4 encodes a DNA polymerase much used in the laboratory because of its ability to polymerize using a long single-stranded template. It also depends upon accessory factors provided by the bacterial host.

***Other DNA polymerases.*** *Reverse transcriptases* synthesize DNA using an RNA template strancr. Tney are best known for their function in retroviruses. The HIV reverse transcriptase is a heterodimer of 51- and 66-kDa subunits. The larger subunit contains a *ribonuclese H* domain. The enzyme is a prime target for drugs such as AZT and others. A different reverse transcriptase is found in all eukaryotic cells *telomerase,* an enzyme essential for replication of chromosome ends. Reverse transcriptases have also been found in rare L1 sequences that are functioning retrotransposons. The ~335-residue catalytic subunit of *eukaryotic polymerase* β , which has a DNA repair function, is the simplest known DNA polymerase. The active enzyme is as small as 38 kDa. It lacks proofreading and is less accurate than the eukaryotic replicative polymerases. Structural analysis revealed a folding pattern related to that of a *nucleotidyltransferase superfamily* that includes enzymes such as terminal deoxynucleotidytransferase and the glutamine synthase adenylytransferase. However, its active site is similar to that of other DNA polymerases.

***DNA polymerase III.*** This Class C enzyme is the major bacterial polymerase for DNA replication. In its complete holoenzyme form it can synthesize new DNA strands at rates as high as ~1 kilobase $s^{-1}$ without dissociation from a template. A genetic approach has provided important information about DNA replication. A series of temperature-sensitive mutants of E. *coli* unable to carry out DNA synthesis were obtained. From these mutations, genes *dnaA, F, G, I, J, K, L, P, I, X,* and Z were identified and located at various points on the chromosome map. Genes C and D map close together at 89 min, and it now appears that they are one gene. Gene F encodes a ribonucle-otide reductase.

The functions of genes *A-E, G, X,* and Z are discussed in the following sections. None of these genes code for DNA polymerase I, but gene *'dnaE* was identified as that of DNA polymerase III, which is now known to be the major polymerase in bacterial DNA replication. To obtain rapid error free synthesis it must be combined with a number of other subunits to form *polymerase III holoenzyme* (or replisome). Ten different subunits, some as two or more copies, form the holoenzyme.

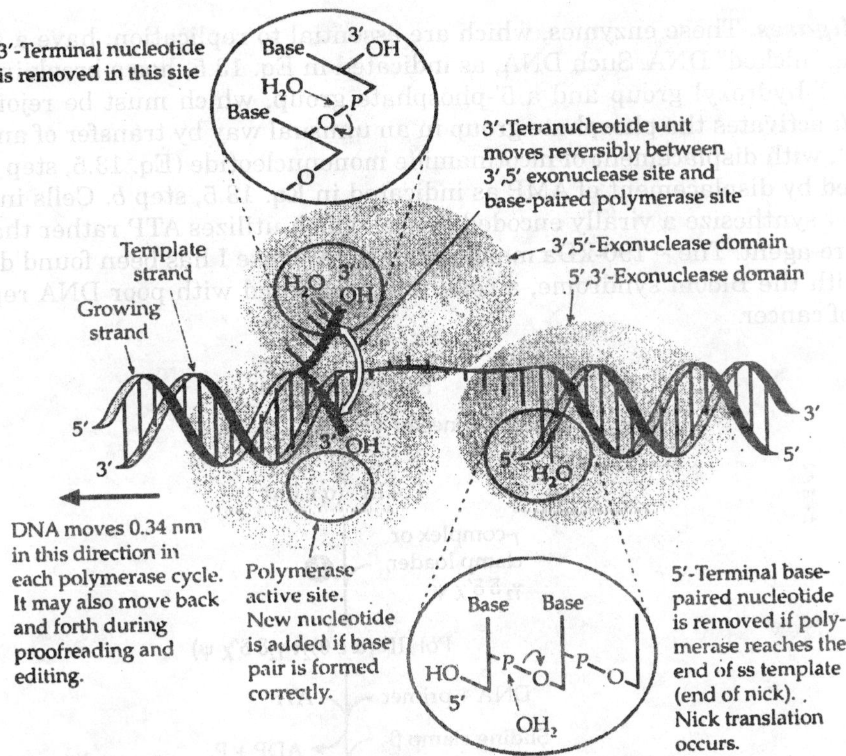


**Fig. 13.13. Schematic representation of DNA polymerase action on a nicked strand of DNA in which the nick has been enlarged. At the catalytic center new nucleotide units are added at the 3' end of a growing strand. At the 3'-5' exonuclease site the 3' terminal nucleotide may be removed hydrolytically. This will happen to the greatest extent if the nucleotide is poorly paired in the duplex. At the 5'-3' exonuclease site nucleotides are hydrolytically removed from the 5' end of a strand in the chain.**

The polymerase catalytic center is in the 132-kDa a subunit. The 27.5-kDa ε subunit contains the 3'-5' exonuclease editing activiry. Mutation in its gene *(dnaQ)* leads to a high spontaneous mutation rate in bacteria. Together with the θ subunit, α and ε form the polymerase III core. This complex has polymerase activity and improved proofreading ability but is still unable to act rapidly, accurately, and processively. Full catalytic activity requires at least the additional β, γ, δ, and τ subunits. The presence of the i subunit causes the core complex to dimerize to form Pol III' can add only about ten nucleotides to a growing DNA chain before it dissociates. The presence of the $\beta_2$ dimer, known as the *processivity factor* or *sliding clamp,* is essential For highly processive polymerization. The β protein forms a ring around the duplex DNA and interacts with the polymerase clamping it tightly to the DNA. Putting the $\beta_2$ clamp onto the DNA is an ATP-dependent process that first involves binding of the *γ–complex* or *clamp loader* ATP-dependent opening of the $\beta_2$ ring, and insertion of the DNA duplex (step *c*). This complex may form a replisome structure that acts simultaneously on the leading and lagging strands. The *Bacillus subtilis* replisome appears to contain two different catalytic (a) subunits, perhaps one for each strand.

## Other Replication Proteins

***DNA ligases.*** These enzymes, which are essential to replication, have a specific function of repairing "nicked" DNA. Such DNA, as indicated in Eq. 13.5, has a break in one strand and contains a 3'-hydroxyl group and a 5'-phosphate group, which must be rejoined. The ligase from *E. coli* activates the phosphate group in an unusual way by transfer of an adenylyl group from $NAD^+$, with displacement of nicotinamide mononucleotide (Eq. 13.5, step *a*). The reaction is completed by displacement of AMP as indicated in Eq. 13.5, step *b*. Cells infected by bacteriophage T4 synthesize a virally encoded ligase, which utilizes ATP rather than $NAD^+$ as the activating re-agent. The ~ 190-kDa mammalian DNA ligase I has been found deficient in some patients with the Bloom syndrome, a condition associated with poor DNA repair and a high incidence of cancer.

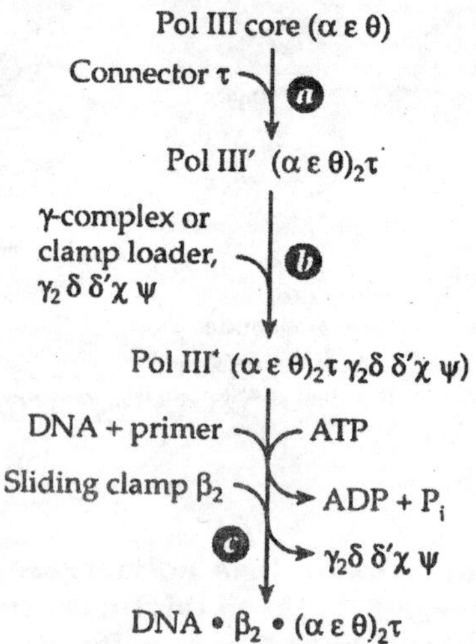


***Single-strand binding proteins.*** Genetic analysis of replication of the DNA of phage T4 within cells of *E. coli* revealed that at least five genes of the virus are required. One of these, gene 43 specifies the T4 DNA polymerase, while gene 32 codes for a single-strand binding protein, also known as the DNA unwinding, melting, or helix-destabilizing protein. It has a greater affinity for ssDNA than for dsDNA and binds to a length of ssDNA causing unwinding of the double helix and exposure of the purine and pyrimidine bases of the template strand. The protein is required for replication, genetic recombination, and repair of DNA. Similar proteins are encoded in the genomes of many viruses. The 87-residue single-strand binding pro-tein encoded by gene 5 of phage M13 forms a dimer, which completely coats newly synthesized ssDNA preventing the DNA polymerase system of the host bacteria from converting it into dsDNA. The polynucleotide chain binds into a groove in the protein with one tyrosine intercalated between the DNA bases.

The *E. coli* single-strand binding protein, another helix-destabilizing protein that is usually called simply **SSB,** is a tetramer of 18.5-kDa subunits. It is essential to DNA replication. About 35 nucleotides may bind to each tetramer. The situation is not as clear in eukaryotes where DNA is largely coiled around histones in the nucleosomes. Several single-strand binding proteins have been identified, but the need for SSB proteins in eukaryotic nuclear replication is uncertain. A human mitochondrial SSB resembles that of *E. coli.*

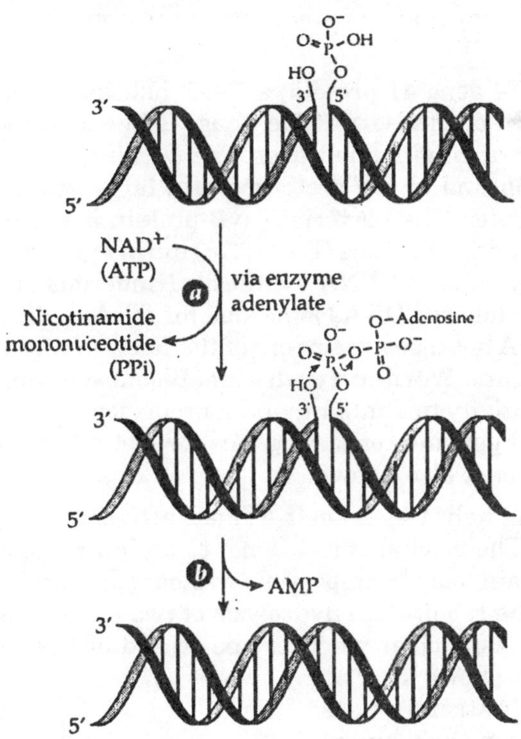


***Helicases and tapoisomerases.*** Cells of *E. coli* contain at least 12 DNA-dependent ATPases that cause unwinding of DNA at the expense of hydrolysis of ATP. The activity of these **helicases** is essential to replication, repair, and recombination of DNA in all organisms. The primary replicative helicase of *E. coli,* which unwinds DNA ahead of the replication fork, is encoded by gene *dnaB.* The active form is a hexamer, which exists in at least two conformational states. A segment of ~ 20 nucleotide units of DNA binds to one hexamer. Helicases have ATPase activity, and the dnaB hexamer contains six ATP-binding sites. However, only three of them may be occupied. Many helicases have a hexameric ringlike structure; that of *Bacillus subtilis* is seen clearly in electron micrographs. Three-dimensional structures are known for some. Although these enzymes may bind to duplex DNA, they also bind to and move along single-stranded DNA in either the 5'→3' or 3'→5' direction. The directionality of a helicase can be determined by annealing two small pieces of ssDNA to the 5' and 3' ends of a longer strand of ssDNA. A 5'→3' helicase will translocate along the long ssDNA and displace the oligonucleotide annealed at the 3' end of the strand, while a 3'→5' helicase will displace the oligonucleotide annealed at the 5' end.

The dnaB protein is a 5'→3' helicase. However, the first helicase identified, the product of the *E. coli rep* gene, is a 3'→5' helicase. It is one of the host proteins needed for the propagation of phages such as ϕX174 and M13. It catalyzes the unwinding of the double-stranded replication forms of these viruses. It binds to a stretch of ~ 20 nucleotides in a single-stranded region of nicked DNA. The hydrolysis of ATP moves the enzyme along the bound strand in a 3′ to 5′ direction opening up the DNA at *a* replication fork. Another E. *coli* 3′→5′ helicase is protein *priA* (also called n'), a component of some primasome structures involved in replication of viruses.

The bacteriophage T4 gene 41 protein, a 5′→3′ helicase, functions together with the gene 61 primase in replication of that virus. The phage T7 gene 4 protein and virus SV40 large T antigen are also hexameric ringlike helicases. The *E. coli* protein **RecQ** is required for various aspects of recombination and is the prototype of a large group of helicases present in both prokaryotes and eukaryotes. The bacterial ruvB protein is a hexameric helicase that propels branch migration in Holliday junctions (Fig. 13.28) during genetic recombination, while helicase rho is required for termination of RNA synthesis. Numerous eukaryotic helicases have been identified and purified. Helicase DNA2 is needed for DNA replication in nuclear extracts from yeast. Human 3′→5′ DNA helicases, members of the RecQ family, are defective in some patients with **Bloom syndrome** and **Werner Syndrome.** Bloom syndrome causes growth retardation, immunodeficiency, sensitivity to sunlight, and a predisposition to skin cancers and leukemias. The yeast *(S. cerevisiae)* genome contains genes for 134 different proteins that are prob-ably helicases. RNA helicases are also known.

A characteristic of all helicases is their ATPase activity, which apparently provides energy for "melting" the DNA. The mechanisms are not clear, but rapid separation of the stacked and hydrogen-bonded base pairs may be impossible without some assistance from an ATP-dependent process. In the case of the rep protein hydrolysis of two molecules of ATP seems to be required to melt one base pair. It isn't clear whether one strand of DNA passes through the center of the oligomeric ring, as shown in Fig. 13.15, or whether both strands pass through. Helicases vary in their amino acid sequences, but they all possess several characteristic **signature sequences** including the Walker A and B motifs found in other ATPases, in sequences related to the *E. coli* **recA** proteins and in synthases.

While helicases can cause the unwinding of linear DNA duplexes, they cannot alter the linking number of covalently closed circular double-stranded DNA. The latter is the function of **topoisomerases,** which have been found in organisms from bacteria to higher plants and animals and may also be encoded by viruses. There are two basic types of DNA topoisomerase. Those of type I change the linking number in steps of 1. One way that this might be accomplished is for an enzyme to nick

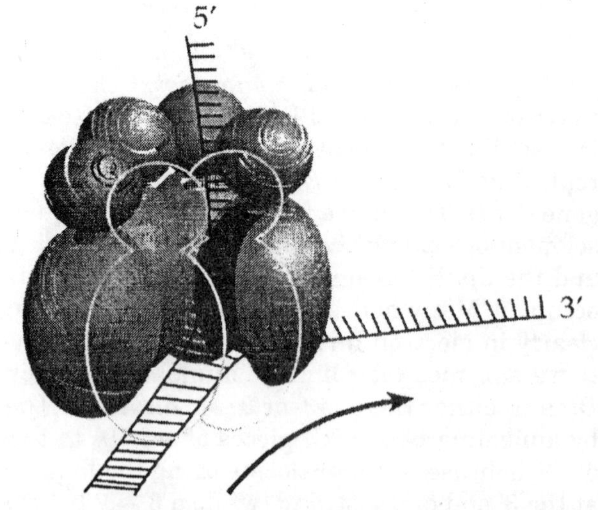


**Fig. 13.14. Model of DnaB hexamer proposed by** Jezewska ***et al.*** **The arrow indicates the direction of movement of the DNA relative to the position of the helicase.**

one strand in the DNA allowing one of the cut ends to swivel around the unbroken strand, then to reseal the chain. However, it was found that topoisomerases can also cause catenation or decatenation of circular duplex DNA as long as at least one of the reacting DNA molecules is nicked. This observation suggested that a topoisomerase binds to a single-stranded region at a nick and cuts the chain but does not release the ends. This permits either a single strand or a duplex to pass through the broken strand, which is then resealed.

Topoisomerases of type I usually act most rapidly on negatively supercoiled DNA. They relax it by decreasing the number of negatively supercoiled turns one at a time. Negative supercoiling presumably facilitates binding of the enzyme to a single-stranded region by unwinding of the duplex. No ATP or other obvious source of energy is needed by type I topoi-somerases. The chain cleavage involves a simple nucleophilic displacement by an –OH group of a tyrosine side chain (Tyr 319 of *E. coli* topoisomerase I), which attacks a phosphorus atom in the DNA chain (Eq. 13.6). The result is covalent attachment of the enzyme to the

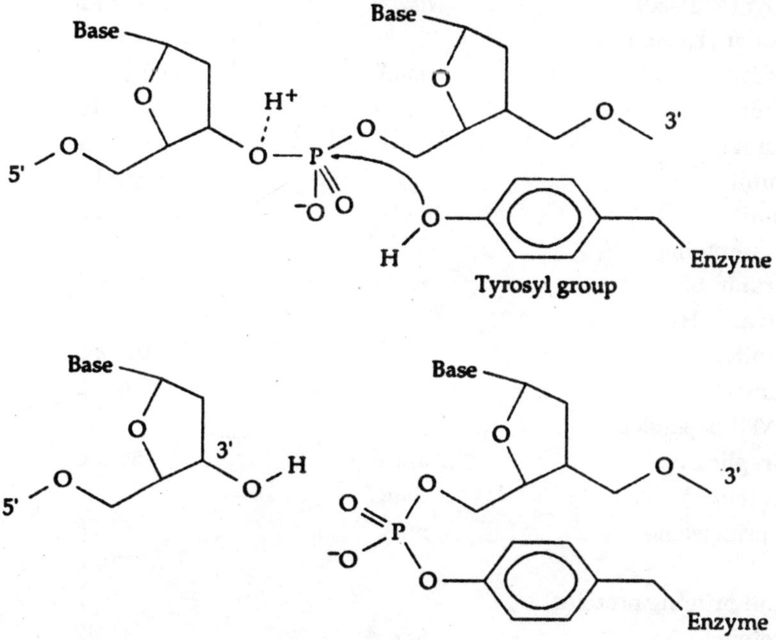

Base Base O H+ O O 3' 5' O O—P O O O Enzyme H Tyrosyl group

Base Base O 3' O H O 3' 5' O O O P O O Enzyme

5'-end of the cut strand in type IA topoisomerases and at the 3'-end in type IB topoisomerases. After the passage of the other strand through the gap in the cut strand of a type IB topoisomerase (Fig. 13.16) the free 3'-OH oxygen atom (or 5'-OH) attacks the phosphorus atom in the phos-photyrosine diester to reform the chain and release the enzyme. X-ray diffraction studies show that both the Type IA E. *coli* topoisomerase I and the human type IB topoisomerase are large proteins with holes of appropriate diameter for a DNA double helix. As illustrated in Fig. 13.16B, the protein may open to allow a double helix to enter and occupy a suitable position for cleavage of one chain by the active-site Tyr 319. Topoisomerases are metalloenzymes, usually functioning best with $Mg^{2+}$. *E. coli* topoisomerase I also contains 3–4 tightly bound $Zn^{2+}$ ions.

**Table 13.2. Some Proteins of DNA Replication *E. coli***

| *Name* | *Gene* | *Mass of monomer (kDa)* | *Map location (min) Fig. 9.4* |
|---|---|---|---|
| Ploymerase I (also has 3′–5′ and 5′–3′ exonulcease, and RNase activities | *polA* | 103 | 87 |
| Polymerase III | | | |
| Core | | | |
| α subunit (polymerase) | *dnaE* | 130 | 4 |
| ε subunit (3′–5′ exonuclease) | *dnaQ* | 27.5 | 5 |
| θ subunit | *hol E* | 10 | |
| Sliding clamp, $\beta_2$ | *dnaN* | 37 × 2 | 83 |
| Connector τ (ATPase) | *dnaX* | 71 | |
| Clamp loader (*Y* complex) | | | |
| γ subunit | *dnaX* | 47.5 × 2 | 11 |
| σ subunit | | 35 | |
| σ′ subunit | | 33 | |
| *Z* subunit | | 15 | |
| ψ subunit | | 12 | |
| DNA binding proteins | | | |
| Single-strand, SSB | *ssb* | 18.5 × 4 | 92 |
| Double-strand, HU | | | |
| α subunit, | | 9.5 × 2 | |
| β subunit | | 9.5 × 2 | |
| Helicases (ATP-dependent) | | | |
| Primary replicative | *dnaB* | 52 × 6 | 92 |
| Dna C protein | *dnaC* | | 99 |
| PriA(n′), primosome | *priA* | 76 | 88 |
| Rep | | 76.4 | 85 |
| Initiation and priming proteins | | | |
| Dna A protein | *dna A* | 52 | 83 |
| Primase (an RNA polymerase) | *dnaG* | 60 | 67 |
| PriB (n) primosome | *priB* | | 96 |
| PriC (n″) primosons | *priC* | | |
| DnaT (primosome assembly) | *dnaT* | | 99 |
| Ribonuclease HI | *rnhA* | | |
| DNA ligase | *lig* | 75 | 52 |
| Topoisomerases | | | |
| Type I | *top A* | | 28 |
| Type II, DNA gyrase ($\alpha_2\beta_2$) | | | |
| Subunit α | *gyrA* | | 97 |
| Subunit β | *gyrB* | | 90 |

Topoisomerases of type II change the linking number by 2 in either the positive or negative direc-tion and hydrolyze ATP in the process. The best know n of these is the *E. coli* **DNA gyrase,** an $\alpha_2\beta_2$ dimer of 97-kDa ($\alpha$) and 90-kDa ($\beta$) subunits. The enzyme catalyzes the ATP-dependent introduction of negative supercoils into DNA. It also relaxes negatively supercoiled DNA slowly in the absence of ATP. Type II topoisomerases are found in all organisms. They are encoded by some bacteriophage such as T4 and by plasmids. However, most differ from bacterial DNA gyrase in not coupling DNA supercoiling to ATP hydrolysis. They require ATP but like topoisomerase I cause a relaxation of the supercoiling. Strands of one segment of DNA (called the "gate" or **G-segment)** are cleaved by the enzyme with staggered cuts four base pairs apart. Another segment of DNA (the "transport" or **T-segment)** is men passed through the gate and is thought to be released from a second gate in the complex (Fig. 13.16C). The enzyme subunits bind through phosphotyrosine linkages as in Eq. 13.5 to the 5' phospho groups of the two cleaved chains, while the subunits bind ATP and may like tiny muscles twist the DNA. Topoisomerases II are large dimeric proteins. The subunits of yeast topoisomerase II (Fig. 13.16C) are 1200-residue multidomain proteins. Mechanisms of DNA cleavage by types I and II topoisomerases appear to be related. However, the ATP-dependent conformational changes involved in a two-gate mechanism are unique to topoisomerases H. The bacterium *Sulfolobus* contains a type I topoisomerase that is called **reverse gyrase** because it utilizes ATP to introduce *positive* supercoils into DNA. This is in contrast to gyrase, a type II topoisomerase that introduces negative supercoils.

Type II topoisomerases are essential and func-tion in replication, DNA repair, transcription, and chromosome segregation at mitosis. Yeast with a *top2* mutation dies during mitosis with hopelessly entangled daughter chromosomes. A fluorescent antibody to eukaryotic topoisomerase II binds to chromosomes, probably at the bases of the radial loops present during mitosis, and also to centrosomes Higher organisms contain more than one topoisomerase II. Their specific functions are uncertain, but one appears to be to unknot entangled chromosomal DNA. In the crowded conditions of a cell nucleus topoisomerase can also cause inadvertent *formation* of knots.

The functional role of topoisomerases of type I is less clear. Staining with fluorescent antibodies to the enzyme has revealed its presence in the transcriptionally active "puffs" of polytene chromosomes and in centromeres of mitotic cells. A current hypothesis is that in *E. coli* class I topoisomerases act to relax negatively supercoiled strands of DNA behind transcription complexes, while gyrase acts to generate superhelical twists, which favour opening of the duplex ahead of transcription complexes. Transcription of a supercoiled rRNA gene *in vitro* is diminished by the selective topoisomerase I inhibitor **camptothecin,** one of a group of antitumor drugs directed against topoisomerase of both types I and II.

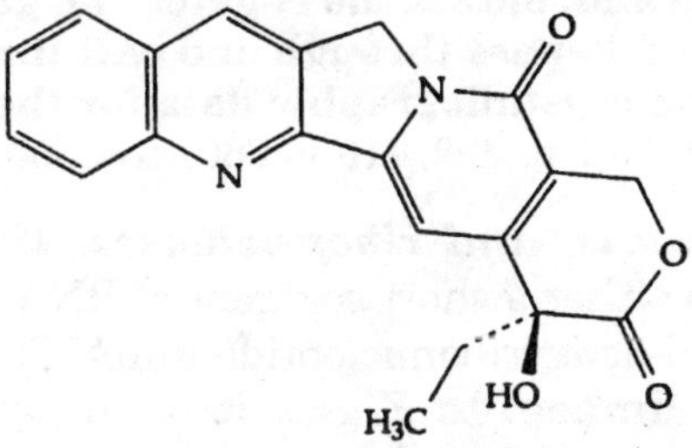

N N O O O HO $H_3C$

**Camptothecin, a topoisomerase I inhibitor**

In the autoimmune diseases scleroderma and systematic lupus erythematosus antigens to nuclear proteins or nucleic acids ate present in the blood. Many patients with severe scleroderma have an antibody against topoisomerase I.

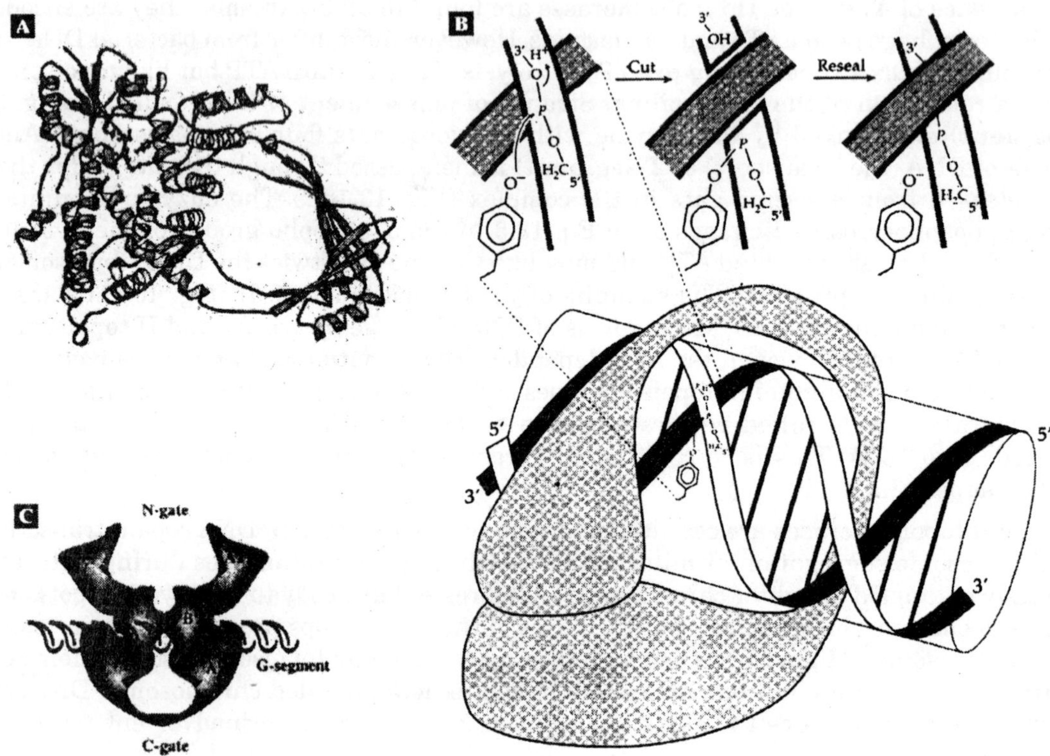


**Fig. 13.15. (A) Ribbon drawing of a large 67-kDa fragment of the 97-kDa (864-residue) *E. coli* topoisomerase I showing the position of the activesite tyrosine 319 and an adjacent arginine. (B) Schematic diagram indicating a way in which topoisomerases of type 1 may pass one strand of DNA through another. The protein is shown binding to a single strand of a DNA duplex. This binding is facilitated by negative supercoiling. The enzyme then cuts the same strand by means of a nucleophilic displacement on a phosphorus atom using a tyrosinate side chain. The other cut end is held noncovalently by the enzyme, while the second strand passes through the gap. Then the gap is resealed by a reversal of the cleavage reaction. (C) Schematic model of a type II topoisomerase bound to a G-segment of DNA. This double helix is cut and another double helical strand, the T-strand, enters the N-gate. The gate then closes, and the central gate opens to allow the T-strand to pass through and exit through the C-gate. The shapes of the three domains are based on crystallographic data for the yeast enzyme. The ATPase, B', and A' domains consist of residues 1 to 409, 410 to 660, and 660 to 1200, approximately.**

***Primuses, initiator proteins, and ribonucleases.*** The priming segment needed for initiation of DNA replication is either a short segment of RNA or an oligonucleotide containing a mixture of ribonucleotide and deoxyribonucleotide units. The enzyme forming the primer is an RNA polymerase called **primase.** In *E. coli* it is encoded by gene *dna*G. Under some circumstances other RNA polymerases can act as the primase. Bacteriophages and plasmids may also encode primases (Table 13.1). For example, gene 61 of phage T4 (Fig. 13.2) encodes a primase, which together with the T4 helicase forms the priming particle or **primosome.** The phage T7 gene 4 encodes a 63-kDa mutlifunctional protein that is both a primase and a helicase. The primase active site is on the out-side of the hexameric ring. Additional proteins representing products of genes *dnaA* and C are also required for initiation of replication in *E. coli.* Several molecules of the dnaA **initiator protein** bind to a specific DNA **origin** sequence and participate in assembling a primosome that also contains the hexameric dnaB helicase and, transiently, protein dnaC. Replication of some single-stranded phages, such as $\phi$X174, in E. *coli* also require the *E. coli* priA, priB, and priC proteins (Table 13.2). For successful completion of replication chaperonins the products of genes *dnaJ* and *grp* E are needed as is a ribonuclease that digests the primer segments after they have been used as replication initiators.

## Replication of Bacterial DNA

The basic mechanisms of replication implied in Eq. 13.3 seems to be universal, but several questions had to be asked. "Is replication initiated at a fixed point or points in a chromosome?" and "Does replication occur in one direction only or do two forks form at the point of origin and travel in opposite directions?" To answer these questions both genetic methods and electron microscopy have been employed.

***Directions of replication.*** One technique for establishing the direction of replication in *E. coli* was to insert a $\lambda$ prophage at the *att* site and DNA from phage Mu–1 at a variety of other sites around the chromosome. Phage Mu–1 was especially useful because it can be integrated at many different locations within the well-mapped genome. Integration within a gene inactivates that gene (an addition mutation) and allows the localization of the Mu prophage. Bacteria were prepared containing both $\lambda$ and Mu-1 prophage, the latter at various sites. The bacteria were also auxotrophic for certain amino acids. Because of this, replication could be stopped by amino acid starvation. The bacteria usually completed any replication cycle in progress and then stopped. When the missing amino acids were added, replication began, again starting from the replication origin. Bromouracil, which enters DNA in place of thymine, was added at the same time. Consequently, the newly synthesized DNA strands were denser than the parent strands. After various times of replica-tion the newly formed strands were separated by centrifugation in a CsCl gradient and were tested for hybridization with both A, and Mu-1 DNA. Since the cells did not all begin replication at the same time after addition of amino acids, a variety of lengths of newly replicated DNA were present. Nevertheless, from the observed ratios of Mu-1 DNA to *K* DNA for the various strains it was possible to map the progress of replication beginning at an **origin** or/C near gene *ih* at 74 min. Replication was found to progress bidirectionally around the chromosome arid to terminate between genes *trp* and *his* at ~25 min.

The use of autoradiographic methods confirmed bidirectional replication. Strains of amino acid auxotrophs with small nucleoside triphosphate pools were used. The addition of amino acids after starvation led to initiation of replication with only a 6-min lag. The cells were labeled with [$^3$H]thymidine, and after the replication forks had moved a short distance from the origin of replication the cells were given a pulse of "superhot" [$^3$H]thymidine. Using autoradiography it was possible to observe the clearly bidirectional replication forks. Replication in other bacteria is also bidirectional.

***Origins of replication.*** Replication of the *E. coli* chromosome begins and proceeds bidirectionally from its defined origin *oriC*. Replication of linear phage T7 is also bidirectional and begins at a point 17% of the way from one end. In mammalian mitochondrial DNA the origin of replication for the H-strand is in the D-Ioop but that for the L-strand is 2 / 3 of the way around the circular chromosome within a cluster of tRNA genes. The single-stranded circular DNA genomes of Ff, 0X174, and G4 phage also have distinct origins for initiation of replication to give RF circles.

Most origins have quasipalindromic nucleotide sequences, perhaps so that DNA can be looped out from the main duplex as is shown in Fig. 13.18A and B. The lengths of on sequences vary, as does the com-plexity of their possible folding patterns. Plasmids have been constructed which not only contain the *E. coli* origin, but are dependent upon that origin for their own replication. Study of those plasmids indicate that a 245-bp *oriC* sequence is essential. This sequence, which is shown in Fig. 13.18, contains several repeated oligonucleotides including 11 GATC sequences, which are sites of adenine $N^6$-methylation, and four "9-mers" (commonly known as dnaA "boxes") with the consensus sequence

$$\mathrm{TTAT}^{A}_{C}\mathrm{CA}^{A}_{C}\mathrm{C}$$

These appear in both 5' to 3' and 3' to 5' orientations, allowing them to form two base paired "stems." .In addition, there are three direct repeats of a 13-residue consensus sequence 5'-GATCTNTTNTTTT, which form an AT rich duplex. Other bacterial replication origins often follow a pattern similar to that in Fig. 13.18. However, the origin for *Mycoplasma genitalium* has been hard to detect. Replication origins of archaea have character-istics similar to those of bacteria and of organelles.

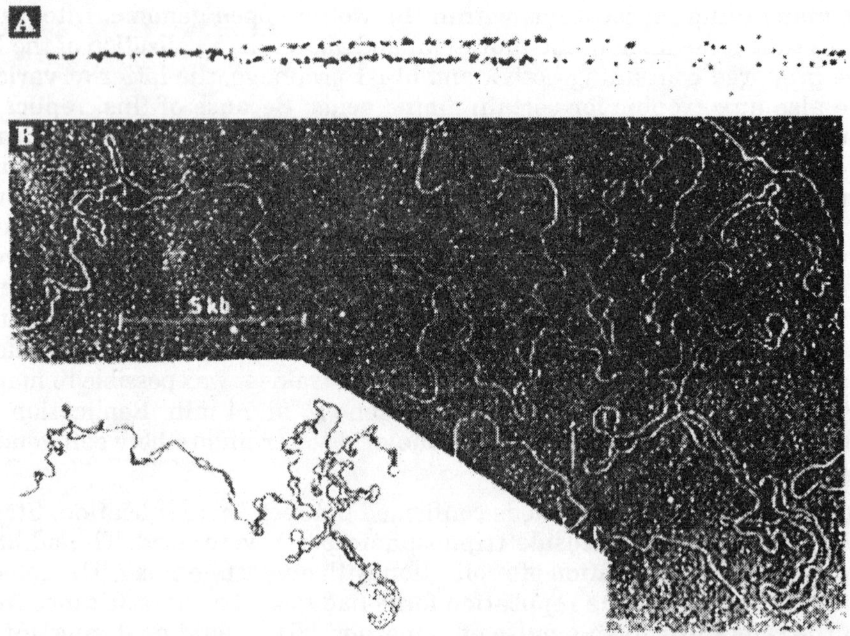


**Fig. 13.16. Bidirectional replication. (A) Replication forks in the *E. coli* chromosome. The autoradiographic pattern was produced by a chromosome that initiated replication with [$^3$H]thymine (5 Ci/mmol) and was subsequently labeled with [$^3$H]thymidine (5 ci/mmol) for 6 min. The total length of the grain track is 370 µn. (B) Fragment of replicating chromosomal DNA from cleavage nuclei of *Drosophila melanogaster*. The DNA, which was spread in the presence of formamide, contains several "eyes" formed where the DNA has been replicated.**

***Priming and initiation of DNA synthesis.*** The first step in initiating a new round of replication and a new cell cycle in *E. coli* appears to be the binding of the initiator protein, dnaA, to the 9-mers in the negatively supercoiled origin. The resulting complex is visible in an electron microscope and may consist of a core of up to 20–40 molecules of dnaA protein with the DNA wrapped around them. ATP is also required and is bound to the protein where it hydrolyzes slowly. Similar initiator proteins are found in many bacteria and are encoded by some viruses and plasmids. In archaea, as well as in humans, the initiator proteins seem to combine functions of the *E. coli* dnaA protein with those of SSB.

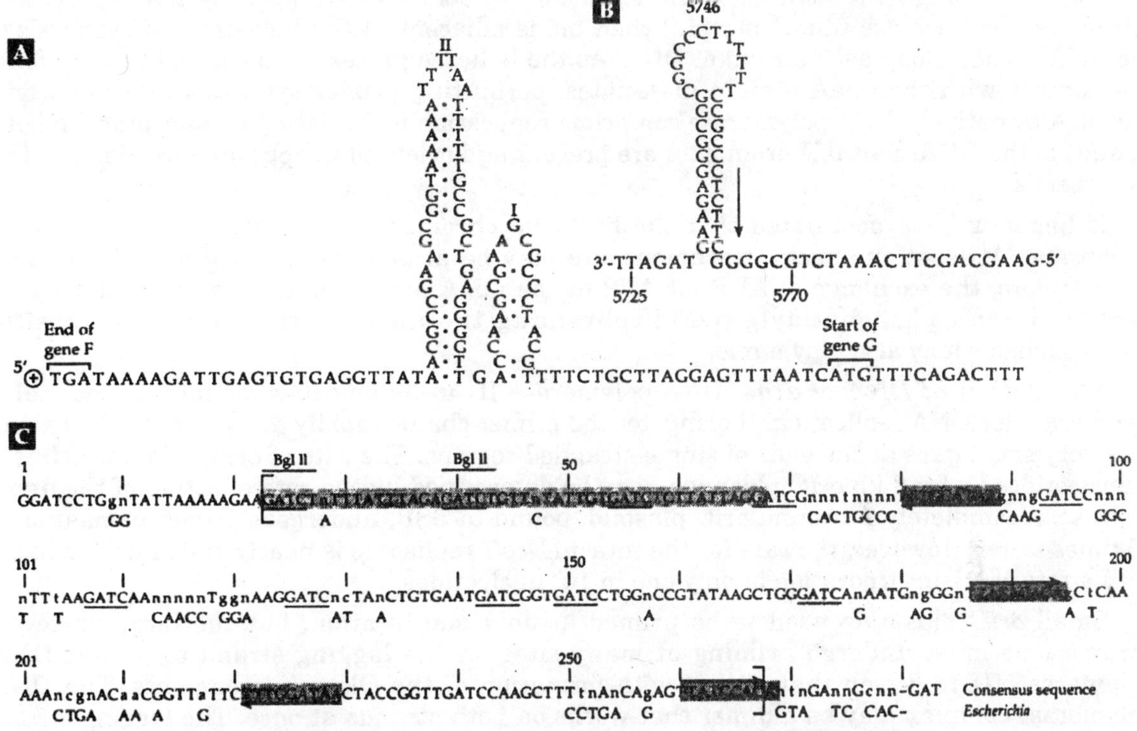


**Fig. 13.17 Nucleotide sequences and proposed secondary structures for some origins of replication. (A) Bacteriophage ϕX174. The ends of the neighbouring genes F and G are shown. (B) Human mitochondrial DNA; origin of replication of the L strand. The sequence shown is for the H strand. The arrow indicates the direction of replication. (C) Bacterial *ori* region. The consensus sequence of origins of replication that will function in *E. coli.* Derived from sequences from *E. coli, Salmonella typhimurium, Enterobacter aerogenes, Klebsiella pneumoniae,* and *Erwinia carotovora.* In the consensus sequence a large capital letter means that the same nucleotide is present in all six origins; a lowercase letter means that the same nucleotide is present in three of the five origins; n means that any of the four nucleotides may be present; represents a deletion, G being present here in the *Enterobacter* origin. Underlined are *dam* methylation sites. Green shaded sequences, three 13-mers in a region that undergoes easy opening of the duplex for insertion of replication proteins; gray boxes with arrows, four 9-mers that are specific sites for binding of replication protein A. Brackets enclose a 245-bp minimum origin.**

Following the binding of the dnaA protein the hexameric helicase dnaB is loaded onto the adjacent DNA in the region of the 13-residue repeated sequences, which are labeled 1, 2, and 3. As shown in this figure introduction of the helicase is assisted by protein dnaC, which forms a complex with the helicase. Additional binding of protein HU and a temperature of >30°C are essential for tight binding of the proteins. Although the dnaC protein is needed for formation of the preprinting complex, it dissociates after the dnaB protein becomes firmly bound. One dnaB hexamer binds to each single strand of the DNA duplex with opposite orientations. With the dnaB helicase in place on each strand, this ATP-driven enzyme processes along the DNA, unwinding the duplex in both directions. (Perhaps it may be more accurate to say that the *DNA moves throtigh the helicase.)* The resulting "bubble" is held open by binding of SSB tetramers. The primase (dnaG protein) then binds adjacent to the helicase and synthesizes the RNA primer along each strand of DNA. As the helicase processes to the right in the fork, the complex with the dnaA protein dissociates, permitting primer synthesis into the origin region. Alternatively, RNA polymerase can prime replication by initiating transcription on both strands of the DNA. Suitable promoters are present and oriented in opposite directions on the two strands.

It has long been postulated that the bacterial chromosome is attached to the plasma membrane. At least one such attachment site may be at or near the origin of replication. Furthermore, the exchange of ADP for ATP in the dnaA protein is catalyzed specifically by cardiolipin and phosphatidylglycerol containing the unsaturated oleic acid. Inositol polyphosphates may also play a role.

***Elongation of DNA chains.*** DNA polymerase III in its holoen-zyme form is the major polymerase for DNA replication. It elongates the primer chains rapidly and processively leaving only very small gaps at the ends of single-stranded regions. The rate of elongation, which is ~ 3 nucleotides / s for 8 kb *oriC* plasmids, may be determined by the rate of action of the dnaB helicase. A completely unwound oriC plasmid, bound to SSB, undergoes primer elongation ~ 10 times faster. However, the rate for the intact *E. coll* replisome is nearly 1000 nucleotide $s^{-1}$ with a rate of misincorporation of only one in $10^9$ nucleotides.

Small *oriC* plasmids need to be primed at only one location, but the large bacterial chromosome must undergo priming at many sites on the lagging strand to permit DNA polymerase III to act on that strand with formation of the Okazaki fragments. The DNA polymerase complex may be a dimer that works on both strands at once. The lagging strand may be looped out to allow it to lie parallel with the leading strand. The appearance of the electron micrograph in Fig. 13.20B supports this suggestion. However, the manner in which the lagging strand can be shifted to bring the next primed initiation site to the replication complex is not clear. DNA polymerase III holoenzyme itself may be organized as an asymmetric oligomer that operates on both strands in a complex such. The primase, either the dnaG protein or the primosome used by phage $\phi$X174 (Section 5), may synthesize the RNA primers on the lagging strand.

***Termination of replication.*** *As* each Okazaki fragment is completed along the template for the lagging strand, the RNA primer piece is digested out, replaced by DNA, and the nick sealed by action of DNA ligase. Ribonuclease H, which is found in both bacteria and eukaryotic cells, specifically degrades the RNA component in these RNA-DNA hybrid regions. In bacteria another mechanism for primer removal is available. The 5' – 3' exonuclease activity of DNA polymerase I will cut out the RNA segment, while the 5' – 3' polymerase activity of the same enzyme will fill the gap.

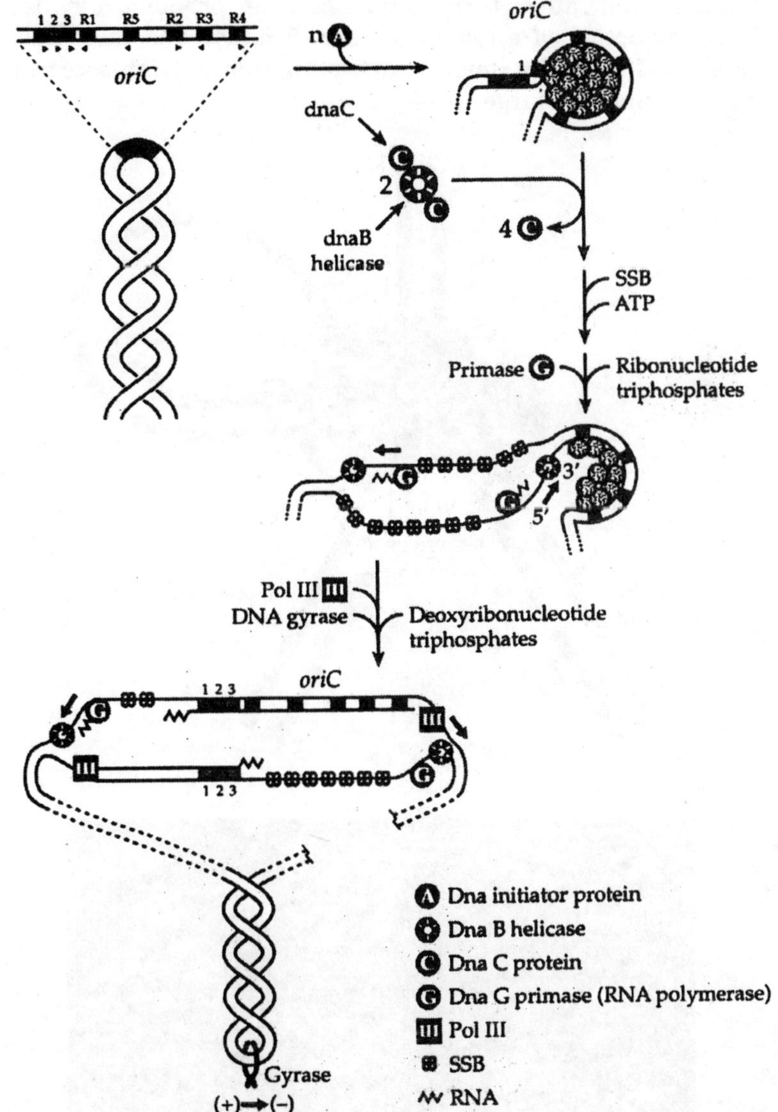


**Fig. 13.18. Hypothetical scheme for initiation of bidirectional replication in *E. coli*. The closed boxes R1 through R5 represent the 9-residue recognition sequences for the *E. coli* dnaA protein. The open boxes 1, 2, and 3 represent the three 13-residue repeats, possible sites for binding of the dnaB-dnaC protein complex.**

Replication of *oriC* plasmids may occur by simply allowing replication of the leading strands at both replication forks to continue all the way around the circle. However, in *E. coli* bidirectional replication continues only until the two replication forks converge. This can occur anywhere between two **terminators,** Tl and T2, located at 28.1 min and 35.6 min. The terminators slow replication in the counterclock-wise and clockwise directions, respectively. A gene *(tus)* near Tl encodes a **terminator utilization substance,** a DNA-binding protein that associates with Tl and T2 and causes termination. Another problem may be the separating of catenated DNA circles by action of a topoisomerase. Finally, it is essential to **partition** the original chromosome and its replica, one to each daughter cell. This requires at least three other gene products including one large 170-kDa protein.

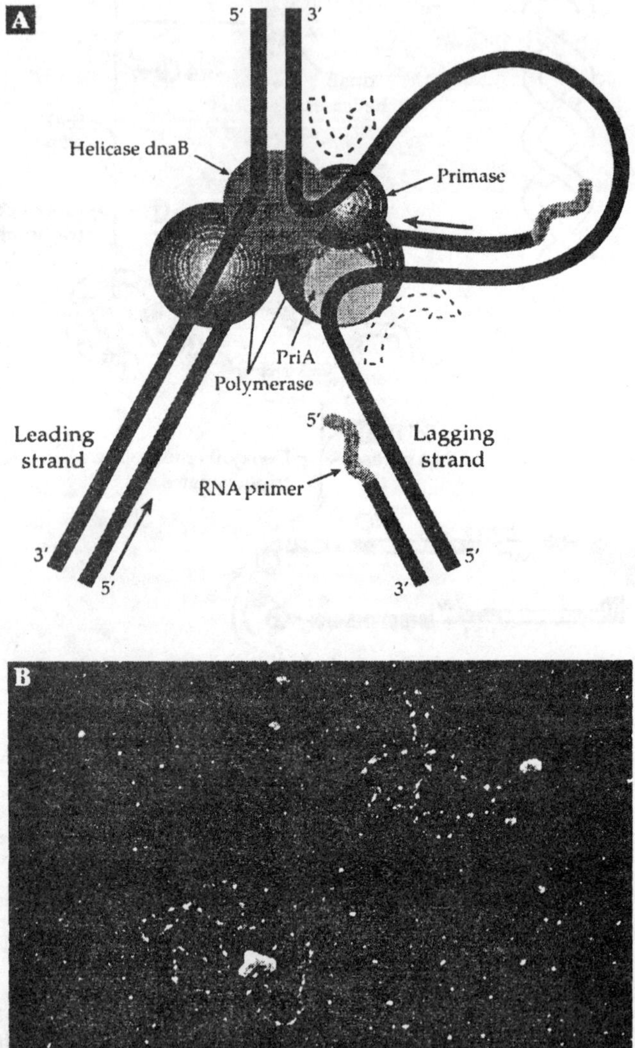


**Fig. 13.19. (A) Hypothetical replisome for concurrent replication of leading and lagging strands by a dimeric polymerase associated with helicase dnaB and a primosome. Open arrows indicate directions of movement of DNA, which is forming a loop as the polymerase fills a gap to complete an Okazaki fragment. The primase will then form a new primer and a new loop. (B) Electron micrograph of the primosome bound to covalently closed $\phi$X174 duplex replicative form. These enzymatically synthesized duplexes invariably contain a single primosome with one or two associated small DNA loops.**

## The Replication of Viral DNA

The replication of viral DNA usually depends upon the genes of both the .host organism and the virus. For example, *ts* mutations in the *E. coli* genes *dnaB, D, E, F,* and G lead to a loss in ability to support growth of phage λ as well as loss of ability to reproduce under conditions where the *ts* gene products are inactivated. However, the phage can replicate in E. *coli* with mutated genes *dnaA* and C because phage λ encodes its own initiator proteins by genes *O* and P (marked on the gene map in Fig. 13.4). In addition to these two proteins, seven *E. coli* proteins are required to initiate replication at the lambda origin *ori* A, and to complete replication. The *E. coli dna*B helicase and the dnaC protein are needed. The chaperonins dnaj, dnaK, and GrpE are also necessary for replication of phage A, and other viruses. As we have seen (Section 2) many viruses contain genes specifying their own DNA polymerases and primases, which function in cooperation with host proteins. For many dsDNA viruses the origins of replication, priming reactions, and chain elongation processes closely resemble those of *E. coli*.

In contrast, the first step in replication of the filamentous Ff viruses (fl, fd, and M13) or the small icosahedral ϕX174 or G4 is conversion of the single-stranded closed circular DNA molecules of the infecting virus particles into circular double-stranded **replicative forms** (RF). This occurs as the DNA enters the bacterial cell and is accomplished entirely by the enzymes of the host cell. Phage G4 DNA, whose replication has the simplest known requirements, contains a tight hairpin region at its origin. The rest of the DNA must be coated by SSB for replication to occur. The hairpin resists melting and serves as a binding site for *E. coli* primase. This is the only known case in which no other priming proteins are needed. Primase synthesizes up to a 29-ribonu-cleotide primer after which DNA polymerase III holoenzyme copies the rest of the chain.

Replication of the closely related ϕX174 is more complex. It requires assembly of a **primosome** made up of at least seven host proteins : dnaB, dnaC, primase (dnaG), and proteins priB, A, C (n, n′, n″), and dnaT (i), Table 13.2. The 76-kDa helicase priA (n′) may locate the **primosome assembly sequences,** which are ~70 nucleotides in length, and displace SSB from them. These sequences can adopt secondary structures with a pair of hairpin loops. Kornberg and associates suggested that the same kind of primo-some formed at these sites in the ϕX DNA may participate in replication of the lagging strand of the chromosomal DNA. If so, helicase priA presumably functions in the replisome on one strand and dnaB helicase on the other as depicted.

In the second and third stages of replication ϕX174 RF molecules are themselves replicated and are then used for synthesis of new viral (+) strands. At both stages a virally encoded **gene A protein,** which has endonuclease activity, nicks the duplex. Cutting the (+) strand it leaves a free 3″-OH on DNA residue 4305, while the 5′-phospho group of residue 4306 becomes covalently attached to a tyrosyl residue in the A protein. The free 3′-OH serves as the primer for a **rolling-circle synthesis** (Eq. 13.7). As a new viral strand is synthesized along the complementary (–) strand as a template, the original viral DNA (+) strand is displaced (Eq. 13.7) as a single-stranded tail.

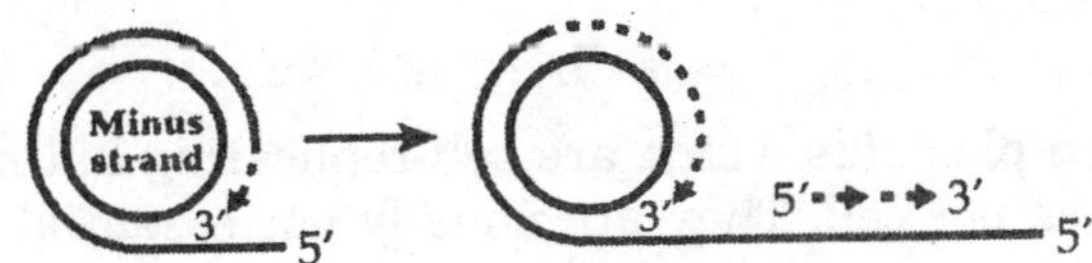

A strand complementary to the single-stranded tail is then formed in segments. A complete turn of the circle produces a viral strand twice the normal length. Cleavage by the endonuclease activity of the gene A protein and closure of the circle completes the replication. The displaced (+) strand can be cut off and either incorporated into a progeny phage or converted into another RF circle. The A protein, attached to the 5'-terminus of the (+) strand, is involved in either case. It can participate in repeated sequences of initiation and termination of viral (+) strand synthesis. Once double-stranded circles are formed, they undergo several replications to give additional RF circles, which serve as templates for the synthesis of many single strands of viral (+) DNA, which are incorporated into the mature viruses. This synthesis of additional RF circles requires transcription of some viral RF genes.

In the final stage of replication the single-stranded (+) chains formed by the rolling-circle mechanism are packaged into phage particles. The gene 5 single-strand binding protein of M13 coats the DNA chains as they are formed, evidently preventing their conversion to RF circles. In the case of $\phi$X174 the new single-stranded DNA circles are packaged as they are synthesized to form complete icosahedral virus particles.

Replication of the larger tailed viruses, which have many genes, is complex and varied. The lytic phage A. resembles the smaller viruses in using the host replica-tion enzymes. In the final stages a rolling-circle mechanism is utilized to form **concatemers** consisting of linear DNA duplexes with numerous successive copies of the viral DNA. The ssDNA that is formed in the rolling circle is converted to dsDNA as it is formed. Finally a **terminase** cleaves the DNA at specific cos sites, using staggered cuts, to form cohesive ends. However, there are uncertainties. The linear dsDNA enters an empty preformed procapsid, apparently pumped in an ATP-dependent fashion, perhaps by a rotating portal ring.

## Packaging or Viral Genomes

The construction of intact virus particles from the genomic DNA and protein subunits is often a complex process. It is simplest for the small filamentous ssDNA viruses (Fig. 13.7). The subunits are synthesized as soluble proteins, which enter the cell membrane, then lose their leader sequences. As the viral DNA coated by the viral gene 5 ssDNA-binding protein enters the membrane, the binding protein is replaced by the coat subunits.

The process is somewhat more complex for the icosahedral viruses. In the $\phi$X, G4, $\alpha$3 family the icosa-hedral procapsid is constructed with the aid of both internal and external **scaffolding** proteins as is illustrated in Fig. 13.28. In phage that replicate via concatameric dsDNA the terminase that cleaves the DNA also interacts in a precise way with the packaging apparatus of the prohead. For the tailed phage the ringshaped oligomeric headtail connector, together with an ATPase, may function as a rotatory pump to feed the DNA into the prohead. This has been demonstrated for phage $\phi$29. In some cases the terminase produces new phage DNA of unit size, but in other cases, e.g., with phage T4, the DNA may be cut more randomly when the head is full or when another piece is needed to fill it. After the DNA is packaged the virus capsids usually expand and become stronger.

## Plasmids

Most bacteria contain plasmids which are self-replicating but stably maintained at well-defined numbers of copies per cell. They are usually not essential to the cell but may carry

traits such as anti-biotic resistance or toxin formation that benefit the. bacterial host. A plasmid always carries in its DNA an orign (on) of replication and a gene, usually designated *rep,* for an initiator protein. It usually encodes other proteins as well but may depend largely on host pro-teins for replication. Plasmids may use the *oriC* copied from the bacterial host's DNA, the origin from phage λ, yeast autonomously replicating sequences (ARSs), or other origins. Replication of the small 6.6-kb plasmid ColEl, which is present at ~20 copies per cell, depends entirely on the host-cell replication machinery· However, the control of copy number depends upon synthesis of **antisense RNA** and its reaction with the plasmid DNA. Similar copy number control is used by the larger ~ 100-kb resistance factor **R plasmids.** Some plasmids use replication systems very similar to those of viruses such as ϕX174, often using rolling-circle replication. However, the plasmids lack the proteins for virus coat formation and maturation.

The F factor plasmids are large 100-kb circular DNA molecules containing ~ 60 genes, about 20 of which encode proteins involving transfer of DNA into another bacterial cell F plasmids display strict copy number control with only 1-2 copies per host chromosome. The controls lie in a region known as the partition locus, which resembles regions of the host chromosomes that are involved in partition of the bacterial genome. They have repetitive sequences suggesting a similarity to centromeres of eukaryotic chromosomes.

## Chromosome Ends

The T-odd bacteriophages Tl, T3, T5, and T7 are medium-sized phage with linear duplex DNA genomes. Replication of linear DNA in these and in many other genomes presents a problem. Even if the RNA primer segment is made at the very 3' end of the template strand, there will be a gap in the final replicated strand when the primer is digested out. Since there is no known enzyme that will add to the 3' end of a chain, this gap will remain unfilled. The problem is solved by **terminal redundancy,** the presence of a common 260-nucleotide sequence at both ends. Several daughter DNA molecules with gaps at the 5' ends can be joined by their cohesive ends to form a long **concatamer.** DNA polymerase I fills any gaps, and the chains can then be ligated and cleaved at different points to generate complete 5' ends.

Another mechanism, which is utilized by some single-stranded parvoviruses, obviates the need for an RNA primer by use of a palindromic sequence to form a hairpin loop (Eq. 13.8).

Yet another solution to this problem is used by some viruses. **Phage ϕ29** of *Bacillus subtilis* primes the replication of its 19,285 bp dsDNA at both ends by a **terminal protein,** which is linked covalently through its Ser 232 –OH group to dAMP. The 3'–OH of the deoxyadenosyl group primes the DNA replication. In a similar fashion replication of the eukaryotic **adenoviruses,** whose genome is a 35– to 36-kb linear DNA duplex, starts at the ends and is primed by one residue of dCMP covalently attached through a 5'-phosphodiester linkage to a serine side chain in a 80-kDa **preterminal protein.** It substitutes for the RNA oligonucleotides that prime most DNA synthesis. The dCMP pairs with guanine at the 3' terminus of the template strand and provides the initiating 3'–OH group. During the replication the preterminal protein is cleaved to the 55-kDa **terminal protein,** which remains covalently attached, one molecule at the 5' end of each strand. The genome can be replicated *in vitro by* five proteins: the virally encoded preterminal protein, DNA polymerase, DNA-binding protein, and two cellular transcription factors that bind in the adenovirus origin region.

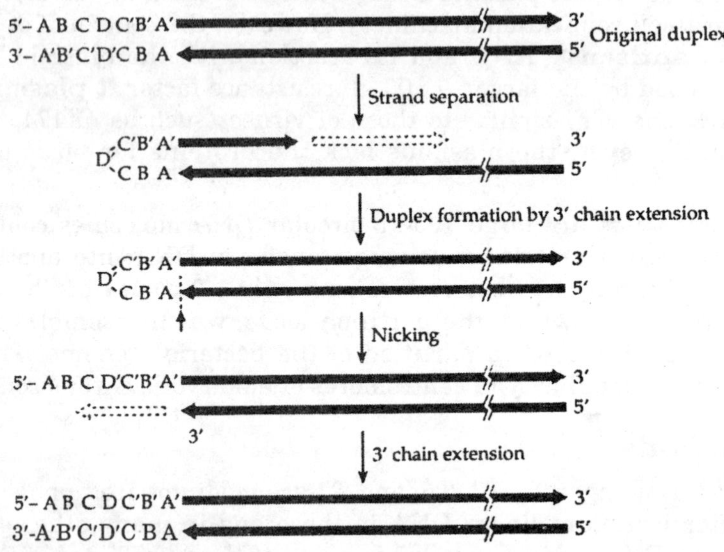


**Fig. 13.20. Role of terminal proteins.**

The chromosome end problem is solved in an-other way in eukaryotes. As discussed in Section B,1, **telomeres** (chromosome ends) contain repeated sequences of variable length. One DNA strand is always G-rich. For example, in human cells the sequence 5'–(TTAGGG),, –3', where *n* may be ~20, oc- curs at the 3' ends of the G-rich strands. The other strand, whose 5' end is at the telomere end, is C-rich and has the complementary sequence 3'–$(AACTCCC)_H$ – 5'. The 3' end of the G-rich strand is always longer by 12–16 nucleotides than the end of the C-rich strand. This 3' extension may fold back to form non-Watson-Crick structures that apparently involve G-quartets. The shorter 5' end is thought to result from the need for a short RNA primer during replication. When replicated in the normal fashion the full G-rich leading strand will be formed, but the C-rich lagging strand will be 8-12 nucleotides short, when the RNA primer is digested away. A result of this is that human somatic cells gradually lose telomeric repeats. However, in tumor cells, germline cells, and unicellular organisms the enzyme **telomerase** prevents this telomere loss.

Telomerase is a reverse transcriptase that copies the DNA sequence of the telomeric repeats from a small **guide RNA** that is part of the enzyme. The first telomerase studied was the relatively abundant enzyme from *Tetrahymena.* It contains a 159-nucleotide RNA with the sequence 5'–C AACCCC AA–3' at positions 43–51. This sequence is complementary to the 5'–TTGGGG–3' repeat sequence of the *Tetrahy-mena* telomeres. A 127-kDa human protein contains a similar guide RNA with the sequence 5'-CUAACCCUACC–3', which is complementary to the human telomere repeat sequence. Telomerases evidently allow the cell to elongate the telomere 5'-ends using the guide sequence and the reverse transcriptase activity of the telomerase. Any number of additional repeats may be added to the 5' ends. The shortened 3' ends can also be lengthened in the next round of replication. The control of telomerase must be important. The enzyme is active in early embryonic cells and some stem cells. However, most normal cells have little or no telomerase activity and lose telomere length throughout their lifespan with eventual growth arrest. On the other hand, excessive telomerase activity may induce cancer. Certain mutations in the telomerase guide RNA can cause greatly increased telomerase activity. The control of telom-erase is still poorly understood but involves specific telomere binding proteins.

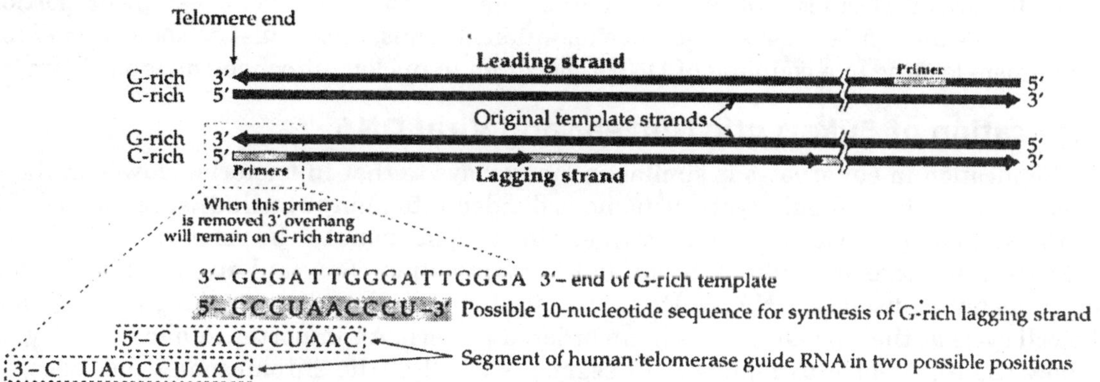


**Fig. 13.21. Aspects of telomere synthesis. The end of the chromosome and the 5' end of the C-rich strand is at the left. This is the template for replication of the leading G-rich strand (green). The primer lies far back in the chromosome. The C-rich strand is replicated in segments from several RNA oligonucleotide primers, one of which lies at the 5'-terminus. This first primer is removed by RNase activity leaving a 12–16 nucleotide 3'-overhang. The telomerase guide RNA can hybridize with the 3'-end of the G-rich strands providing a template that allows additional growth of the G-rich strand and extension of the C-rich strand also in the next replication.**

## Mitochondrial and Chloroplast DNA

Replication of the ~ 16-kb mammalian mtDNA begins with RNA priming within a small **displacement loop** or D-loop. One daughter strand, the heavier or H strand, starts to grow on the primer. As it does, the parental H strand is displaced and the D-loop is enlarged. The H-strand grows until -70% of the parental H strand has been displaced and the L-strand origin is uncovered. Then a new light L strand is laid down to form the second daughter duplex. The rate of formation of the new L strand is only 10 nucleotides / s, an hour being required to complete the process. The DNA formed is initially relaxed, another 40 min being needed to introduce the 100 superhelical turns present in the finished chromosome.

The kinetoplast DNA of trypanosomes consists of thousands of catenated circular DNA molecules. Among these the smaller minicircles always contain the sequence GGGGTTGGTGTA at their origins of replication. The minicircles are individually removed from the mass prior to replication. The two progeny circles are then both recatenated into the mass.

Chloroplasts contain large 120– to 169-kb circular genomes encoding about 100 proteins. A characteristic feature of most chloroplast DNA is the presence of long inverted repeat sequences (10,058 bp in the liverwort, 25,339 bp in tobacco). These are separated by 19,813 and 81,095 bp single copy regions in the liverwort and by similar sized regions in tobacco. Plastid DNA exists as a mixture of monomeric molecules with smaller amounts of dimers, trimers, and tetramers.

Ethidium bromide inhibits the replication of chloroplast DNA and causes partial degradation of existing DNA in chloroplasts without interfering with replication of DNA in the nucleus. The effect is similar to that of the same drug on mitochondrial DNA.

However, cells of *Chlnmydomonas* treated with ethidium bromide are able later to regenerate their chloroplast DNA. This result has been interpreted to mean that there may be one or a few "master copies" of chloroplast DNA in specially protected locations. The result should also be considered in relationship to the following observation. Although nuclear and organelle DNA molecules replicate at different times in the cell cycle, constant proportions of the organelle and nuclear DNA tend to be maintained. Thus, there must be some kind of control mechanism leading to a coupling of DNA replication in nuclei, mitochondria, and chloroplasts.

## Replication of Eukaryotic Nuclear and Viral DNA

Replication in eukaryotes is similar in many ways to that in bacteria. However, the ~ $10^6$ kb of DNA in a typical eukaryotic genome is divided into many **replicons,** segments of DNA 30-150 kb in length, each having its own origin of replication. In the relatively small ~ 14-Mb yeast genome there are –400 replicons, but in mammalian DNA and also in plant DNA there are probably thousands. DNA synthesis is initiated at different times during the S-phase of the cell cycle at the various origins in an ordered pattern. An important unanswered question is how the cell is able to replicate all segments of all of the chromosomes just once before entering mitosis.

Many of the proteins of eukaryotic replication are closely related in sequences and functions to those of bacteria. There are initiator proteins analogous to *E. coli* dnaA. Eukaryotic polymerases a and 5 and possibly *E* are essential for replication. Polymerase a, which is inhibited specifically by the fungal metabolite **aphidocolin,** is a complex of a ~170-kDa DNA polymerase core, an RNA-synthesizing primase consisting of 58- and 49-kDa subunits, and a 70-kDa subunit

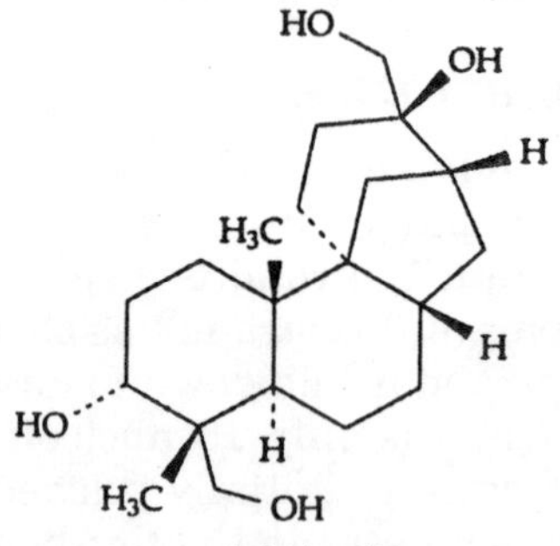

Aphidocolin

of uncertain function. The complex makes an RNA-DNA primer consisting of ~10 nucleotides of RNA and ~30 of DNA. This pol α primer is replaced early in replication by the highly processive polymerase δ and, perhaps, under some conditions by polymerase ε. The ringlike processivity factor or "clamp" that is provided in *E. coli* by protein dnaB is called the **proliferating cell nuclease antigen** (PCNA) in eukaryotes. It is loaded on to the DNA by a **clamp loader,** the 5-subunit replication factor C (RFC). As in *E. coli* an SSB type protein known as **replication protein A (RPA)** is also essential. PCNA is not only essential to eukaryotic replication but is also required for recombination and repair. The poloα/primer primes leading strand synthesis initially but then switches to replication origins on the lagging strand where, together with other proteins, it primes the formation of the Okazaki fragments. Polα/primer is also a logical participant in the control of the initiation of the S-phase of the cell cycle.

***Eukaryotic Viruses.*** Investigation of viruses provided the first insights into eukaryotic DNA replication. Most of the factors needed for replication of the DNA of adenoviruses, simian virus 40 (SV40), and polyomavirus within animal cells are supplied by the host. Replication of the 5-kb SV40 DNA, whose DNA forms typical nucleosomes, appears to be an excellent model for eukaryotic replication in general. The single SV40 origin of replication is a 64-bp sequence containing the 5–bp sequence GAGG C four times as pairs of inverted repeats. These are recognized by the 95-kDa virally encoded initiation protein which also has helicase activity and is known as the T antigen. A nearby 17-bp sequence containing only AT pairs is presumably the region of entry of the host cell's polymerase a/primer. Single-stranded DNA regions are coated with the replication protein A. After the primer is formed, the RFC complex loads the sliding clamp PCNA, and polymerase a is replaced by polymerase 8 on both leading and lagging DNA strands permitting highly processive bidirectional chain elongation. Topoisomerase activity is required to decatenate the replicated chromosomes. Since SV40 DNA forms typical nucleosome, its replication is thought to mimic chromosomal replication quite closely. The more complex herpes simplex virus HSV–1 has a 153-kb genome, a linear DNA duplex. It has –75 genes and encodes its own DNA polymerase, origin-binding protein, SSB, and other proteins needed for replication within eukaryotic cells.

***Artificial chromosomes.*** Another approach to understanding eukaryotic replication, similar to the use of *ori*C plasmids in *E. coli,* is to study **autonomously replicating sequences** (ARSs) and plasmids and **artificial chromosomes** made from them. ARS sequences were first found in the budding yeast *S. cerevisiae.* Plasmids containing an ARS, whose core consensus sequence is 5′-(A/ T)TT TAT(A/G)TTT(A/T), replicate autonomously during S-phase. Such plasmids have been genetically engineered, providing them with telomeres and some kind of functional centromere, to form artificial chromosomes. **Yeast artificial chromosomes** (YACs) have become extremely important as cloning vehicles and they also serve as important tools for studying eukaryotic replication and its control. They can be cultured in yeast cells or can be transferred into animal cells, etc.

As mentioned human centromeres are rich in the repetitive a-satellite DNA. By joining a-satellite DNA-containing fragments of the X-chrotnosome to cloned telomeric DNA, human **minichromosomes** have been created. These have been developed into **human artificial chromosomes,** which may be practical vehicles for gene transfer in human therapy.

***Replication of nuclear DNA.*** The budding yeast *Saccharomyces cerevisiae* has permitted the most detailed picture of DNA replication in a eukaryote. The complete genome sequences are known and the ARSs have been physically mapped. For example, in chromosome VI there are nine origins that differ in frequency of initiation and which replicate sequentially during the S-phase of the cell cycle. The initiation **(replicator)** regions surround the 11-bp consensus sequence of the ARSs, each occupying at most –150 bp. However, in metazoa and even in the fission yeast *Schizosaccharomyces potiibe* the ARSs range from 500 to 1500 bp in length. These origin regions frequently overlap the **promoter** sequences, which control initiation of transcription. This association with transcription origins has also been observed in metazoan cells, where replication origins are often clustered. However, there is no sequence homology between the ARSs of S. *cerevisiae* and replication origins in other species, even those of S. *pombe.*

The study of replication in yeast ARSs and artificial chromosomes has revealed that initiation of replication requires not only an initiator protein but a complex of six proteins that form an **origin recognition complex (ORC)**. This complex, which is essential to initiation of replication, may be joined by additional proteins in a **prereplication complex.** At least some of the ORC proteins have their homologous counterparts in metazoa, suggesting a highly conserved initiation machinery.

One or more of the proteins that bind to the ORC may constitute a **license** to replicate. The licensing concept states that when replication occurs the license is destroyed and the origin involved cannot initiate replication again without a new license. The **replication licensing factor** (LRF) is postulated to be unable to pass through the nuclear membrane. It can only reach the replication origin after the S-phase has concluded and mitosis has taken place. At this time the membrane has been disrupted. A second signal, the **S-phase promoting factor** (SPF), cannot act without an intact nucleus and a license in place. This system ensures that DNA is replicated only once per cell cycle. Among the proteins involved in the licensing is a group of **minichromosome mainte-nance (MCM) proteins,** so-named because of their importance to replication of ARSs and artificial chromosomes. Six of these proteins (MCM2–MCM7) can form a hexameric complex with one subunit of each type as well as other complexes, e.g., $(MCM4,6,7)_2$. The latter acts as an ATP-dependent helicase. A somewhat simpler MCM complex is found in archaea. Some of the cell cycle proteins including Cdc6 and the protein kinases Cdc7 and Cdc28 as well as other proteins, are also required for regulation of replication origins. Proteins homologous to those of yeast have been identified in humans and other eukaryotes. Licensing of replication involves association of the MCM helicases with each ORC during the G1 phase of the cell cycle. Binding of the initiation factor Cdc6/18 and of a recently discovered loading factor Cdt-1 apparently completes the licensing. Once licensing has occurred both cdc6/18 and Cdt-1 can dissociate from the DNA. Removal of Cdt is facilitated by its binding to another protein **Geminen,** found first in the frog *X. laevis*. The ORC complexes may remain at the origins. It has been estimated that a yeast cell contains –400–600 molecules of the very stable ORC, about one ORC per replication origin. However, a large excess of MCM proteins may be present. Their concentrations may regulate the number of ORC molecules that associate with DNA. Replication of DNA is not the only aspect of cell growth. For example, as DNA is replicated histones must be synthesized and assembled. This synthesis occurs during the S-phase and is tightly coupled to replication.

Initiation of replication in metazoans is still confusing. Almost any piece of DNA will be replicated if introduced into a *Xenopus* egg, where initiation appears to occur just once at a random position. However, in differentiated tissue the origins of replication seem to be fewer in number and more specifically located. A possible explanation is that high concentrations of ORC and MCM proteins in the embryo may lead to many relatively nonspecific origins and a replicon size of ~7 kb. The lower concentrations of these auxiliary factors in somatic cells may lead to fewer but more specific origins with a replicon size of ~170 kb. Three distinct mammalian origins have been studied in detail. One is in the β globin locus (Fig. 13.10). A second is near the dihydrofolate reductase gene. A third, which is activated early in S-phase, is at the 3' end of the lamin B2 gene. The latter has been localized to a 500-bp region. These findings suggest that the replicon concept, as developed for yeast, may be generally applicable.

Replication of the intact genome of *Drosophila* has been studied in rapidly dividing nuclei by electron microscopy. The replication rate in these nuclei is ~300,000 bases/s, but it has been estimated that replication forks in animal chromosomes move no faster than ~50 bases/s. Thus, we would anticipate at least 6000 forks, or one fork per 10 kK bases. Indeed, this number of forks has been observed. They occur in pairs with many short regions containing single-stranded DNA as if one strand at the fork is replicated more rapidly than the other as in mitochondrial DNA. The arrangement of the ssDNA regions at the two forks in a pair suggests bidirectional replication. However, replication forks are rarely seen in higher eukaryotes, but extensive regions of single-stranded DNA are often visible. Benbow and associates suggested that in higher eukaryotes the strands of duplex DNA may be separated throughout a whole looped domain of DNA. Replication could then occur with initiation at many points along each strand.

Replication reactions are similar in bacteria and eukaryotes, but some details differ. In eukaryotes at least two DNA polymerases, α and δ, are required. In budding yeast polymerase ε is also essential. Both polymerases δ and ε may replicate separate strands at the replication fork. Processing of Okazaki fragments also differs from that in bacteria, where either RNase H or the 5′ to 3′ exonuclease activity of DNA pol I removes the RNA primer (Fig. 13.14). This exo-nuclease activity is lacking in eukaryotic polymerases. Replication primers are removed in a two-step process by **RNase HI,** which makes an endonucleolytic attack that removes all but one nucleotide residue of the primer in a single piece, leaving a 5′-phospho group on the remaining ribonucleotides. That residue is removed by a 5′ to 3′ exonuclease designated RTH-1 nuclease (Eq. 13.9). This is a homolog of the yeast RAD27 protein. A polynucleotide kinase may then phosphorylate the 5' end of the DNA fragment.

Another difference between bacterial and eukaryotic replication is the presence of nucleosomes in eukaryotes. Some evidence suggests that nucleosomes may open and close to allow replication forks to pass through. Studies of SV40 minichromosomes indicate that passage of the replication machinery does destabilize nucleosomes, which must be partially reconstructed about 260 nucleotides past the elongation point. Another factor is the variable extent and location of modifications to histones, in particular to the H3 and H4 his tone tails. A code has been proposed according to which certain modifications would favor transcription or mitosis, while lack of modification would silence the genes.

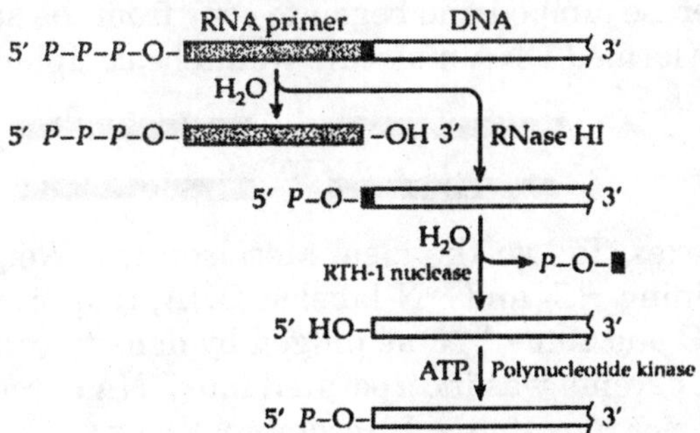


Much of the control of replication is at the initiation stage. Growth factors and other mitogenic stimuli acting at the plasma membrane can stimulate expression of such nuclear proteins as those encoded by the protooncogenes *c-myc, c-myb,* and *c-fos.* These may initiate a regulatory cascade (Fig. 13. 13) and trigger mitosis. As indicated in the 34-kDa protein kinase encoded by fission yeast gene *cdc2* (budding yeast CDC28) is essential for progression of the cell cycle through the Gl phase into mitosis (Eq. 13.3). A single oscillation in this kinase activity induced by a B-type cyclin can promote both replication and mitosis. However, in *S.cerevisiae* there are 14 different cyclinlike proteins, and their individual functions are not clear. The signal that is sent to the ORCs is likewise unclear. However, theoretical models involving Eq. 13.3 and many additional components have been proposed. Multiple phosphorylations may occvir, some on the RPA initiator protein. Many proteins required for replication, including DNA polymerase and primase, are associated with the nuclear matrix. The nuclear membrane may also be important in controlling replication.

## INTEGRATION, EXCISION, AND RECOMBINATION OF DNA

The exchange of genetic information between chromosomes, plasmids, and viruses occurs in many ways, which are described collectively as recombination. Mutants of E. *coli* deficient in recombinational ability often have defects in genes designated *recA, B, C,* etc. (for recombination), or *ruvA, B,C* ... (for resistance to ultraviolet light). Some of these mutants are unusually sensitive to ultraviolet light because of their inability to repair damage to DNA. Several of the recombination enzymes are used for repair of ultraviolet damage and of double-strand breaks in DNA arising from other causes. In eukaryotes recombination occurs during meiosis. Many viruses, including phage, also carry genes for their own general recombination systems. In addition, the DNA of some viruses, such as the temperate phage *"k,* undergoes recombination with the host DNA. This can happen during the processes of integration of the viral DNA into the host genome or excision of the viral DNA during the lysogenic cycle of replication. Recombination occurs around specific sites in the chromosomes of both the virus and its host and is called **site-specific recombination.** Genetic recombination is essential to the development of genomic diversity, to the survival of a species, and to evolution.

### Recombination Mechanisms

How can the homologous regions of two different DNA duplexes be brought together? As illustrated schematically in Eq. 13.10, the strand exchange must occur at exactly the same point in each duplex. An early attempt to explain this postulated a "copy choice" mechanism of replication. It was assumed that replication occurred along one DNA strand up to some random point at which the polymerase jumped and began to copy from the second of a pair of homologous chromosomes. The newly formed DNA molecule would be complementary to different parts of

both parental DNA duplexes. To test the idea, Meselson and Weigle infected *E. coli* with two strains of phage *X* containing $^{13}$C- and $^{15}$N-labeled DNA, respectively. Recombinant DNA was found to contain some $^{13}$C and some $^{15}$N, as judged by density gradient centrifugation. It was clear that DNA from both parents was incorporated into DNA of recombinant progeny, a finding that ruled out the copy choice hypothesis. If recombination occurs instead by enzymatic cutting of two homologous duplex DNA molecules followed by rejoining, how is it possible to avoid inactivation of genes by addition or deletion of genetic material? Recombination cannot depend upon the random action of a nonspecific enzyme with random rejoining. Yet, general recombination can occur at any point and with a roughly constant frequency throughout the DNA chain. The explanation of these facts lies in the occurrence of base pairing between at least some short homologous regions of strands of the two different DNA duplexes.

***The Holliday recombination intermediate.*** In 1964, Holliday suggested a recombination process that would give rise to characteristic H-shaped intermediates. Recom-bination could be initiated at special points on the duplexes, recognizable by a recombination enzyme. A short amount of unraveling would be followed by strand exchange with the two broken strands being rejoined by a ligase. The crossover points would then migrate up or down the chains as the two helices turned about their own axes. Long regions of heteroduplex DNA could be gen-erated in this way, and the process could be terminated at a random distance from the starting point, accounting for the observed unifor-mity of genetic recombination events. Chain cleavage and rejoining of two of the strands would terminate the process. If these were the same strands broken in the initiation event (cleavage at points *aa'*), genes lying outside the heteroduplex region would not be recombined, but cleavage of the other chains (at points *bb"*) would lead to their recombination. Intermediates of the type predicted by the Holliday model were soon observed by electron microscopy. Three-dimensional structures have been determined by X-ray crystallography and have been studied by atomic force microscopy.

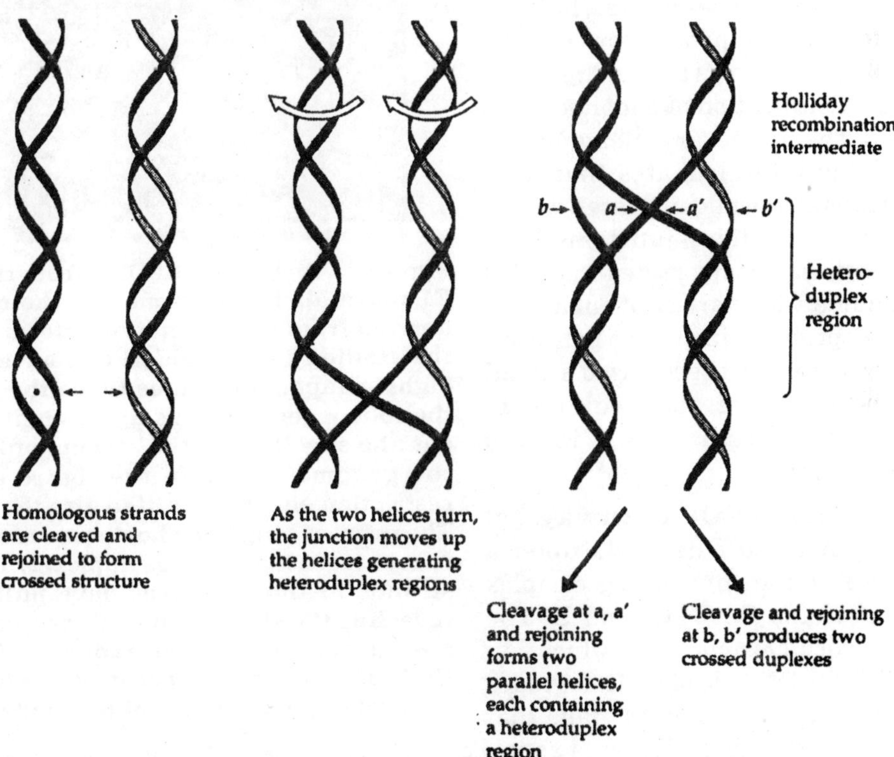


**Fig. 13.22. A recombination mechanism involving single-stranded exchanges. After Holliday.**

The cross-stranded structure can be formed with all base pairs in both duplexes intact. All that is required is formation of a nick in each of the two polynucleotide chains and a rejoining of the backbones across the close gap between the duplexes. This model also accounts for the cutting of the two crossed strands at exactly equivalent points to terminate the process. Various mechanisms of recombination exist, and most make use of the key

Holliday recombination intermediate. Such four-way junctions can arise in several ways. For example, a 3' or 5' single-stranded tail in a piece of dsDNA can "invade" another dsDNA that has a homologous sequence as indicated in Eq. 13.11. In this drawing a 5' tail has been displaced during repair of a gap in one strand. The resulting D loop may be trimmed out and a new connection made to give the Holliday intermediate. The cleavage points a, a′, b, b′ marked in Eq. 13.11 correspond to those in Fig. 13.22. Holliday junctions may also be formed in stalled replication forks and must be removed to allow replication and transcription to continue. The existence of the Holliday intermediate has been supported not only by electron micrographs such as that of Fig. 13.23 but also by the identification of endonucleases that carry out the necessary cleavages of synthetic Holliday intermedi-ates that have been made artificially. Endonucleases with a high specificity for Holliday junctions have been found in bacteria, among proteins encoded by viruses, and in a wide variety of eukaryotic cells. Additional proteins including helicases, DNA-binding proteins, and specialized **strand exchange proteins** are also required to catalyze the individual steps in the recombination process.

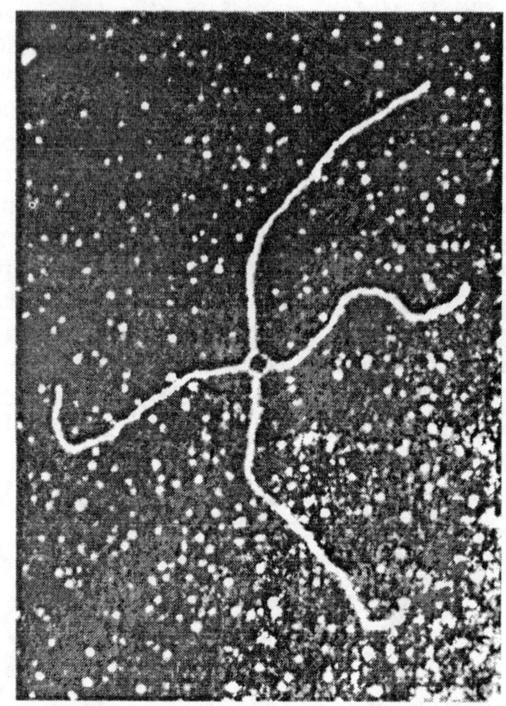

**Fig. 13.23. A chi form of DNA from the colicin E1 plasmid. These forms are thought to be derived from recombination intermediates of the Holliday type, which appear as "figure eight"-shaped molecules twice the length of the colicin genome. This figure eight form was cut at a specific site that occurs only once in the genome (twice in the figure eight) by restriction enzyme EcoRl to give the chi form. The pairs of long and short arms are believed to represent homologous duplexes. The single strands in the crossover have pulled apart revealing the strand connections clearly. Such a structure would be expected from the Holli-day intermediate, e.g., if one of the two vertical duplexes were rotated end over end.**

The main **RecBCD pathway** of recom bination in *E. coli* depends upon a dsDNA nuclease and an unwinding complex consisting of proteins RecB, C, and D. The complex is a powerful *exonuclease,* which can digest the ends of a DNA duplex. It degrades the 3' ends most rapidly, leaving 5'-tails that can invade other homologous duplexes as in Eq. 13.11. The RecBCD complex is also an ATP-dependent helicase, which unwinds the DNA, preparing ssDNA for reaction with the strand-exchange protein **RecA.** The RecB-CD complex also functions to completely degrade foreign dsDNA such as that from invading bacteriophages. Why doesn't it also degrade the genomic DNA of the *E. coli* cell in which it functions? The answer lies in an eightbase DNA sequence, a recombination "hot spot" called chi ($\chi$): 5-GCTGGTG G-3'. This $\chi$ sequence occurs 761 times in the leading strands for DNA replication in the *E. coli* genome. When the RecBCD complex reaches a $\chi$ sequence, when approaching it from the 3' end, the enzyme stops its exonuclease action by inactivating the nucleolytic activity of the D subunit and promotes recombination about five-to ten-fold as fast as at other sites.

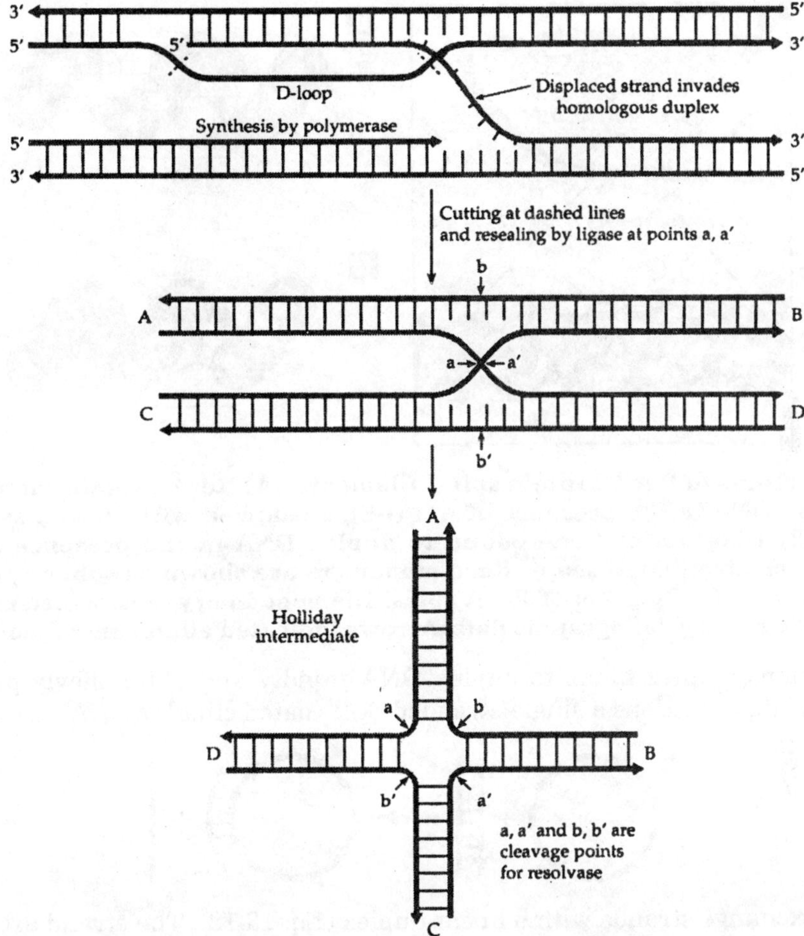


***RecA and other strand-exchange proteins.*** The 352-residue product of the *E. coli RecA* gene is a multi-functional **recombinase,** which is required both for recombination and also for DNA repair. In its repair function the RecA protein acts as a DNA-dependent protease that cleaves a number of repressers in response to damage to DNA. It has a quite different role in recombination where it (1) brings a piece of single-stranded DNA (an end or a gap) together with a duplex; (2) locates homologous sequences; and (3) forms a synaptic complex in which strand exchange can occur. Electron microscopic observations show that the RecA protein binds to either single-stranded or duplex DNA in a cooperative manner to form long rodlike spiral filaments. Measurement of the lengths of RecA protein-covered duplexes shows that the DNA is underwound and stretched by about 50%. It contains ~18 nucleotides per turn. Similar filaments are formed with single-stranded DNA, 3–4 nucleotides being bound per RecA protein monomer. Formation of this ssDNA complex, which may be regarded as an initiation complex for recombination, requires MgATP and is faciliated by prior coating of the DNA with SSB protein. The RecA protein subunits are added in the 5' – 3' direction of the DNA, and SSB is displaced in the process.

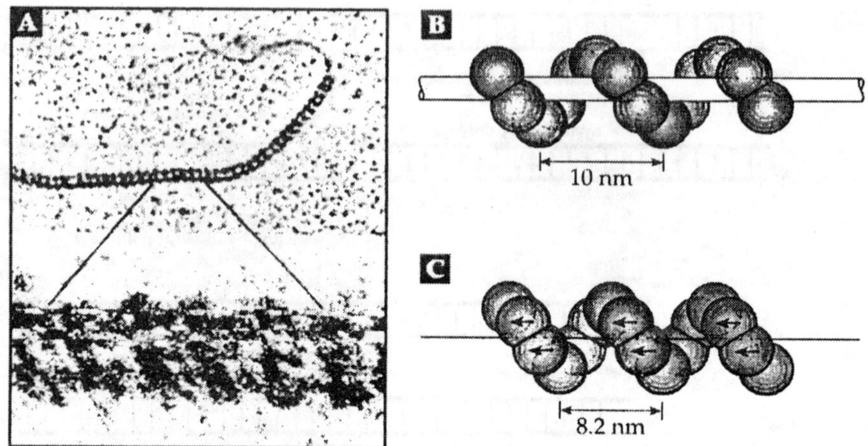


**Fig. 13.24 Structures of RecA protein spiral filaments. (A) RecA protein filament formed on circular duplex DNA in the presence of ATP(γ-S), shadowed with Pt and seen by electron microscopy. (B) Diagram of RecA bound to duplex DNA in the presence of ATP(γ-S), as determined by electron microscopy. RecA monomers are shown as spheres, but their exact shape is unknown. (C) Diagram of RecA spiral filament in crystals of RecA protein free of DNA, based on X-ray crystallographic data. Arrows indicated alignment of monomers.**

The initiation complex binds to duplex DNA rapidly and more slowly promotes strand exchange. In related reactions a single-stranded SSB-coated circular DNA will bind to RecA

protein, then exchange strands with a linear duplex (Eq. 13.12). The strand exchange requires ATP and advances in the 5' to 3' direction along the original single strand at the rate of a few bases / s. Strand exchange can also occur between two duplexes if there is a suitable gap in one strand, e.g., as is illustrated in Eq. 13.13.

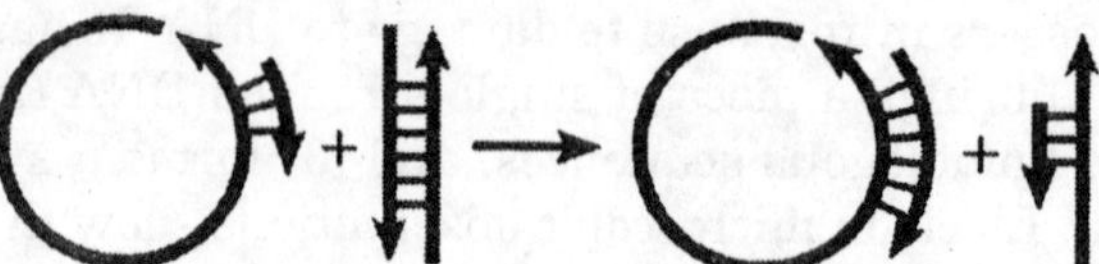

A possible mechanism of strand exchange is illustrated in Figs. 13.25 and 13.26. The RecA protein has binding sites that can accommodate nucleotides from two DNA molecules, one single-stranded and the other a duplex. It may also accommodate two DNA duplexes. As shown in Fig. 13.25, the RecA protein could test the hydrogen-bonding between many base pairs at once in a search of homologous regions. The two DNA chains would have to either slide past each other or repeatedly dissociate and reassociate until a homologous region was found. Then strand exchange could occur. The single strand may be wound into the major groove of the duplex to form an interwound triplex. The matching of hydrogen-bonding atoms may be an attractive way of searching for homology, but the actual search seems to substitute speed for precision. Base substitutions are quite permissive. The need for precise hydrogen-bonding has not been demonstrated, and the exact recognition mechanisms in homologous recombina-tion remain uncertain. Whole chromosomes must be aligned and checked rapidly in the homology search.

Many proteins similar to the RecA protein and with similar functions have been found. These include the products of gene *uvsX* of phage T4, the β protein of phage lambda, the yeast RAD51 and human RAD51 proteins, a meiosis-specific human homolog of the RecA protein, and corresponding proteins from plastids of higher plants. Both the UvsX protein of phage T4 and human RAD51 protein yield strands of coated DNA similar.

***Processing the Holliday junction.*** Completion of the recombination process requires "resolution" of the Holliday intermediate by endonuclease action followed by ligation and perhaps by gap repair. The major recombination pathway in £. *coli* employs a binding protein, a nuclease, and a helicase encoded by genes *RuvA, B,* and C. **RuvA** is a DNA binding protein specific for symmetric Holliday junctions. **RuvB** is a closely associated ATP-dependent helicase. On the basis of genetic and X-ray crystallographic evidence it is now evident that some of the functions previously attributed to RecA are carried out by the RuvABC complex. As indicated, RuvA binds to the Holliday junctions, holding it in the symmetric square configuration in which branch migration is possible. Two molecules of the oligomeric RuvB helicase apparently rotate the DNA, causing branch migration and movement of the DNA through the RuvAB complex. Under some circumstances a different helicase, encoded by *E. coli* gene **RecG,** moves Holliday junctions in the opposite direction. In addition to RuvA a variety of other Holliday junction-binding proteins are known. These include p53 and the nuclear HMG proteins.

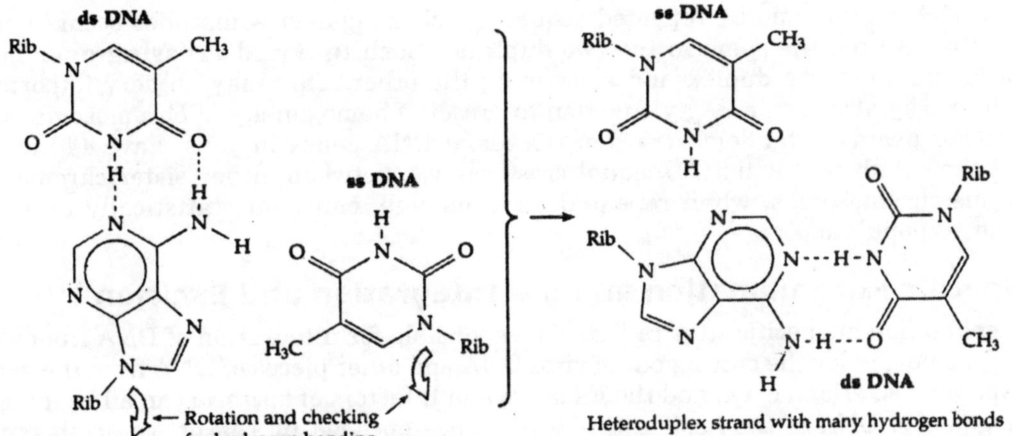

**Fig. 13.25 A possible mechanism for homologous pairing of an ssDNA with a duplex DNA and strand exchange. The ssDNA (right) binds together with a hydrogen-bonded duplex (left). The RecA protein rotates the bases into the heteroduplex configuration, where hydrogen bonds may be formed in many of the base pairs.**

**RuvC** is an endonuclease that is highly specific for Holliday junctions. It is a **resolvase** that cuts at either points a,a′ or b,b′ of Eq. 13.11 to form either "patched" or "spliced" recombiiiant DNA. Similar resolvases process bacteriophage DNA and have also been found in yeasts and in mammals. All are dimeric metal ion-dependent proteins.

## Nonreciprocal Recombination and Unequal Crossing-Over

The phenomenon of **gene conversion** or **nonreciprocal recombination** was first recognized in genetic studies of fungi for which the four haploid miotic products can be examined individually (tetrad analysis). Instead of the normal Mendelian ratio of 2 : 2 for the gene distribution in the progeny at a heterozygous locus a ratio of 3 : 1 is sometimes ob-served . One of the recombinant chromosomes appears to have been altered to a parental type. A reasonable mechanism by which this can occur arises from the fact that heteroduplex regions, which are present in recombination intermediates, contain defects in base pairing. One strand of the heteroduplex will have a base that does not properly pair with the base in the other strand, or will have an extra base that loops out from the heteroduplex. Since cells contain repair mechanisms that search for defects and carry out a repair process, there is a likelihood that one strand in the heteroduplex region will be altered to restore perfect base pairing, thus causing the observed gene conversion. "Flanking" genetic markers outside of the heteroduplex region are unaffected by gene conver-sion, and during meiosis crossing-over between these markers occurs in about 50% of gene conversion events as would be predicted from the model of Fig. 13.21. Data from yeast show that nonreciprocal re-combination during meiosis may also result from double-strand breaks and gap formation followed by repair synthesis using both strands of the homologous chromosome as templates.

Recombination is not limited to meiosis but can occur between homologous chromosomes during mitosis, during the $G_1$ period preceding mitosis, or even during the $G_2$ period. Certain mutations in yeast abolish meiotic recombination but have much less effect on mitotic recombination. Thus, the two processes are not identical. It has been suggested that mitotic recombination is utilized to maintain sequence homogeneity between repeated eukaryotic genes.

Since DNA contains many repeated sequences, crossing-over sometimes occurs between locations that are not the same in the two duplexes. Such **unequal crossing-over** has the effect of lengthening one duplex and shortening the other. This may be very important in evolution. It may also, surprisingly, function to preserve homogeneity of chromosomes within a species. For example, tandem arrays of ribosomal RNA genes in yeast have 140 identical copies of their 9-kb repeat unit. Unequal crossing-over between either sister chromatids or homologous chromosomes, when repeated often enough, can lead statistically to a highly homogeneous population.

## Site-Specific Recombination and the Integration and Excision

Recombination at specific sites in DNA is responsible for integration of DNA from viruses into the genome and for the cutting out of viral DNA and other pieces of DNA from the genome. The temperate bacteriophage λ, and the F factors and R factors of bacteria can all be integrated into the genomic DNA of the host in this way. Genes encoded by the phage or plasmid are required. In the case of phage λ the viral genes *int* and *xis* are required for integration and excision, respectively. These are not the same as the enzymes of the *rec* loci of the bacterium or the general recombination genes *exo* and *bet* of the phage. In addition, both integration and excision require an *E. coli* protein called integration host factor (IHF), a DNA bending protein resembling the DNA-binding HU.

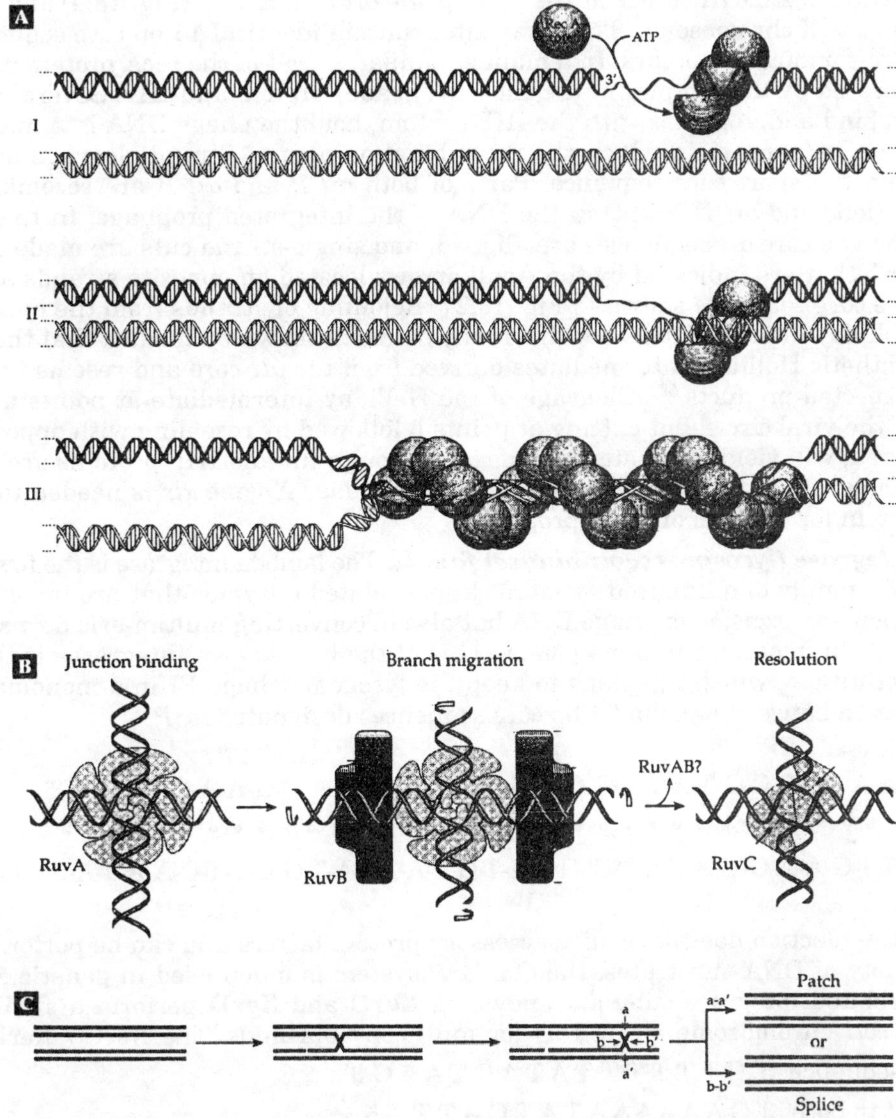


**Fig. 13.26 (A) Model for genetic recombination proposed by Howard-Flanders *et al.* (I) RecA protein binds cooperatively to the single strand in a gapped duplex to form an initiation complex in preparation for pairing. (II) The initiation complex binds to the intact duplex, making transient contacts until a homologous site is reached. For clarity, the initiation complex is drawn with only a few protein monomers, but in reality it is likely to extend over hundreds or thousands of nucle-otides. (III) When homologous contacts are made and the strands become paired locally, the initiation complex acts as nucleus for further cooperative binding, which extends the RecA spiral filament around all three or perhaps all four interacting strands. (B) Arrangement of the proteins and DNA during three stages of recombination catalyzed by the RuvABC system. The two RuvB hexameric rings are shown in cross section with the DNA passing through their centers. (C) Scheme of DNA rearrangement during homologous recombination in *E. coli***

Integration of λ, DNA occurs at the ~25-bp site *att B* in *E. coli* (Fig. 13.4) and the ~240-bp site *att P* in the X chromosome. These two sites contain identical 15-bp core sequences within which the re-combination occurs. In a manner similar to that of the recA protein a homologous region is located by the complex of the Int **(integrase)** protein and IHF. Several molecules of Int protein bind and, together with the IHF protein, hold the phage DNA in a nucleosomelike structure (an **intasome)** in which the recombination occurs. Strand cleavage and rejoining occur within the short core sequence. Parts of both *att P* and *att B* are recombined to give sites *att L* (left) and *att R* (right) in the DNA of the integrated prophage. In the integration complex the two core *att* sequences are aligned, and single-strand cuts are made at one of the points *a* or *b* that are indicated by the small arrows located on opposite strands and seven bp apart in the core sequence shown in Fig. 13.27. Rejoining of strands from the opposite duplex yields a Holliday intermediate. That this really occurs is shown by the fact that the Int protein cleaves synthetic Holliday intermediates derived from the *att* core and reseals the strands to give the expected products.[581] Cleavage of the Holliday intermediate at points a will lead to excision of the viral circle, but cutting at points b followed by resealing with opposite strands, as is observed, will yield integrated prophage. Although Int and IHF proteins are sufficient to pro-mote integration, the **excisionase** encoded by phage *X* gene *xis* is needed together with the Int protein for excision of the *X* prophage.

***The integrase (lyrosine recotnbinase) family.*** The lambda integrase is the first recognized member of a family of a hundred or more closely related enzymes that are involved not only in integration and excision of phage DNA but also in converting multim-eric forms of bacterial and plasmid chromosomes into monomers. One of the best known integrases is the 38.5-kDa **Cre recombinase,** which functions to keep the lysogenic phage PI in a monomeric form by re-combination between pairs of 34-bp core sequences designated *loxP.*

17' 1 1' 17'
5'-A-T-A-A-C-T-T-C-G-T-A-T-A-G-C-A-T-A-C-A-T-T-A-T-A-C-G-A-A-G-T-T-A-T-3'
• • • • • • • • • • • • • • • • • • • • • • • • • • • • • • • • • •
3'-T-A-T-T-G-A-A-G-C-A-T-A-T-C'-G-T-A-T-G-T-A-A-f-A-T-G-C-T-T-C-A-A-T-A-5'
1'

Since the reaction doesn't require accessory protein factors and can be performed *in vitro* with a variety of DNA substrates, the *Cre-loxP* system is much used in genetic engineering. A pair of related integrase subunits known as **XerC** and **XerD** perform a similar function for the *E. coli* chromosome as well as for multicopy plasmids. The XerC / XerD system is

**Core** sequence 5' C T T T T T T A T A C T A A G 3'
present in both 3' G A A A A A A T A T G A T T C 5'
*att* P and *att* B b

atypical because it utilizes a pair of integrase subunits rather than just one. However, the basic chemistry is the same for the entire family. As with the λ, integrase the XerC / XerD com-plex acts on a pair of identical core sequences that are aligned in an antiparallel fashion. Active sites in all of the integrases contain the conserved amino acid sequence Arg-His-Arg-Tyr. All of these residues are essential for catalysis. Staggered cuts are made sequentially in the core sequences, e.g., at points a and b in Fig. 13.27 and adjacent to the 5'-terminal thy-midylate residues in Fig. 13.28A. A transesterification reaction forms 3'-phosphotyrosyl linkages from the cut DNA to the integrase protein. The freed 5'-OH groups

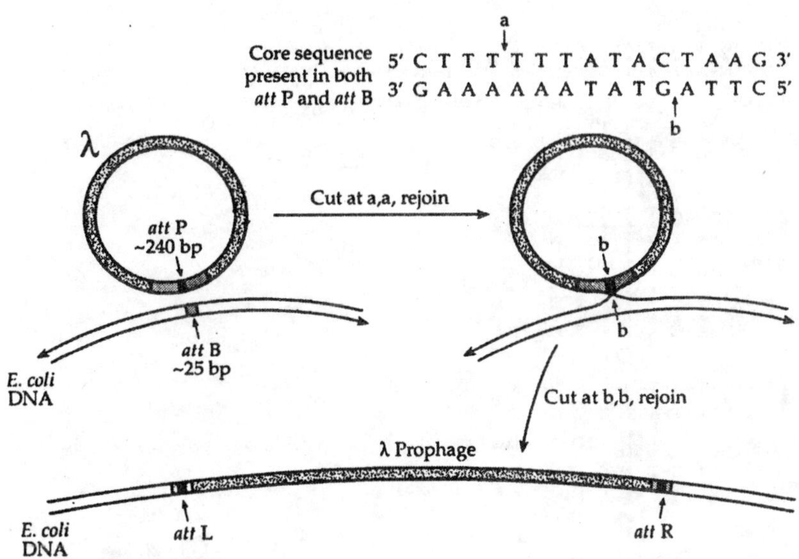


**Fig. 13.27 Integration of the phage λ genome into the E. *coli* chromosome at site affB. The same recognition sequences are present at *attB* and *attP.* These are cut at points *a* with rejoining to give the structure at center right. This is cleaved at points *b* with rejoining to give the integrated prophage.**

on the other cut end fold back and recombine with the bound 3' ends of the second duplex to generate a Holliday junction. In the mechanism proposed in this figure, three nucleotide units are involved in the folding back. The hydrogen bonds of their initial base pairs are broken, and new bonds are formed. This "base swapping" process is the equivalent of a short branch migration and verifies homology of the two recombination sites.

The Holliday junction can isomerize readily be-tween two forms. In one a base-stacked double helix runs between ends a and b in Fig. 13.28B and another runs between ends c and d. The core hexanucleotide sequences lie between the marked XerC and XerD cleavage sites. In the other isomer (lower drawing) one helix has ends a and d and the other b and c. Following the isomerization the XerD active sites act in two transesterification reactions (not shown but analogous to those in steps *a* and *b)* with base swapping of the trinucleotides at the cut ends. This generates the two separate recombinant duplexes. These might be two circular chromosomes or plasmids formed by recombination from a double length chromosome or plasmid. The previously men-tioned A, amd Cre recombinases appear to act by closely similar mechanisms.

Tyrosine recombinases of the lambda family also function in eukaryotes. Best known is the **FLP (Flip) recombinase,** which is encoded by the 2-µm plasmid of *Saccharomyces cerevisiac* and is thought to function in amplifying the number of plasmid copies. The 6.3-kbp plasmid contains a unique DNA sequence that lies between two 599-bp repeats in inverted orienta-tion. Embedded in each repeat is an FLP recombina-tion target (FRT) sequence, which is recognized by the plasmid recombinase. Each FRT segment includes inverted repeats 13 bp in length with an 8-bp spacer between them. As with other integrase systems the 8-bp spacer or **strand exchange region is asymmetric** and establishes the orientation of the recombination sites. The role of the recombinase is to invert one of the 599-bp repeats with respect to the other. This switches replication of the plasmid to a rolling circle pattern.

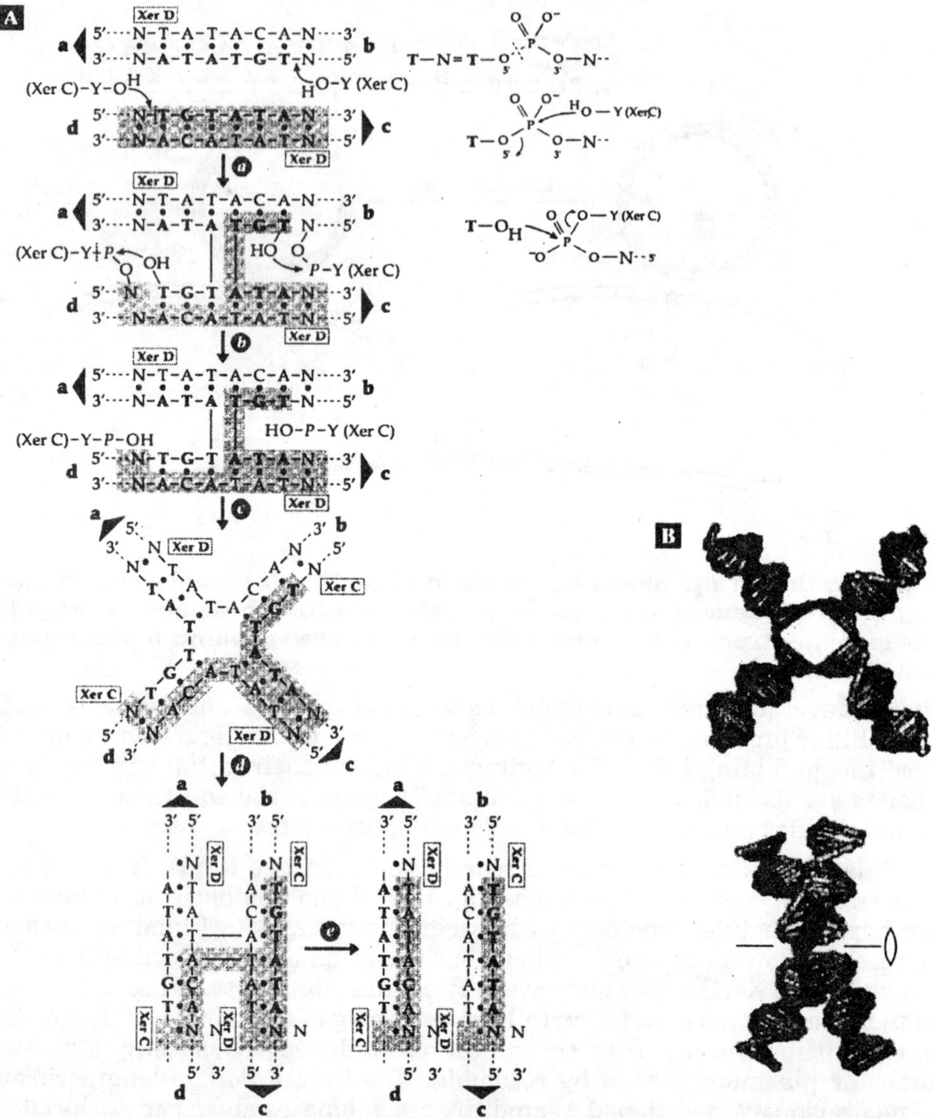


**Fig. 13.28 (A) Action of the integrase XerC / XerD on a pair of *E. coli* sites, containing the central six bp sequences TATACA/ATATGT, which are shown in an antiparallel orientation. In step *a* the active site tyrosine hydroxyl groups (Y-OH) of a pair of XerC subunits carry out transesterification reactions on the 5'-terminal thymidylate residues of the central hexa-nucleotide sequences to yield 3'-phosphodiester linkages to the XerC tyrosines. In step *b* the cut 5'-ends, containing free thymidylate 5'-OH groups, fold back to form new base pairs, and the strands are resealed in a second transesterification. This generates a Holliday junction, which *isomerizes* (steps *c* + *d*) to an isome'ric species that is acted on by a second pair of transesterification steps *(e)* that are catalyzed by protein XerD, again with folding back of the central trinucleotides. (B) The two isoforms of an antiparallel stacked X Holliday junction are shown. These can be reached from the symmetric square form shown in Fig. 13.28 and schematically in (A) by folding into X-conformations in which all base pairs are stacked either in pairs a,b and c,d or a,c and b,d.**

5′-G-A-A-G-T-T-C-C-T-A-T-T-C-T-C-T-A-G-A-A-A-G-t-A-T-A-G-G-A-A-C-T-T-C

3′-C-T-T-C-A-A-G-G-A-T-A-A-G-A-G-A-T-C-T-T-T-C-a-T-A-T-C-C-T-T-G-A-A-G

FLP recombination target (FRT) sequence

***The resolvase invertase family and invertible DNA sequences,*** A second large family of recombi-nases act by cleaving a target DNA sequence hydrolytically leaving a free 3'-OH end (Eq. 13.14, step *a*). This free end then attacks a phosphodiester linkage in a second strand of DNA, cleaving that strand with an in-line nucleophilic displacement (step *b*). Active sites usually contain a characteristic cluster of aspartate and glutamate (DDE) side chains, which probably act together with a metal ion, perhaps as in Fig. 13.13. Enzymes in this resolvase/ invertase family act either to resolve cointegrates in transposon action (next section) or to invert DNA sequences.

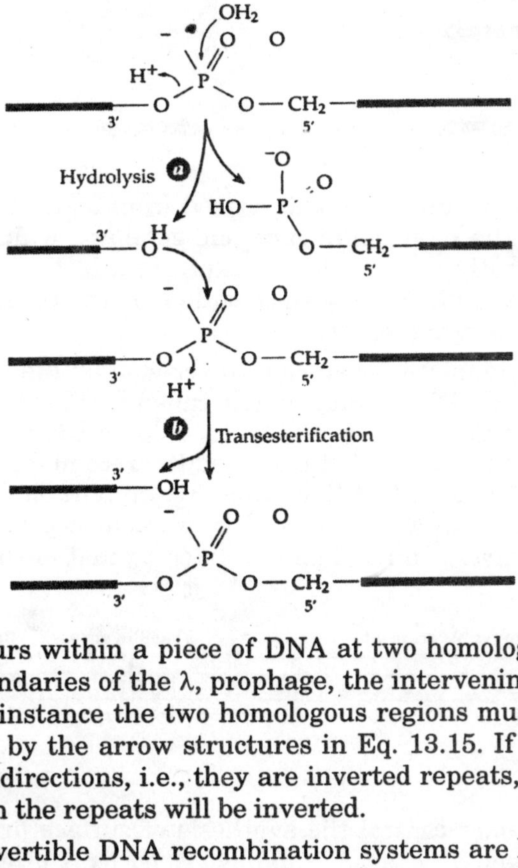

If recombination occurs within a piece of DNA at two homologous sites such as the *attL* and *attR* sites at the boundaries of the λ, prophage, the intervening DNA will be excised as a circular particle. In this instance the two homologous regions must be repeated in the same direction, as is indicated by the arrow structures in Eq. 13.15. If the homologous sequences are oriented .in opposite directions, i.e., they are inverted repeats, excision will not occur but the piece of DNA between the repeats will be inverted.

A number of such invertible DNA recombination systems are known and are some-times used to control specific genes. For example, many strains of *Salmonella* have two types of flagella, which are composed of flagellin subunits encoded by genes H1 and H2 respectively. On rare occasions an individual bacterium switches from one flagellar "phase" to the other. This occurs by recombination, as in Eq. 13.16, between two 26-recombination sites *hixL* and *hixR*, each of which contains a 14-bp inverted repeat. The 993-bp invertible segment encodes a recombinase gene called *hin* and a promoter, i.e., an mRNA initiation site, which has a specific

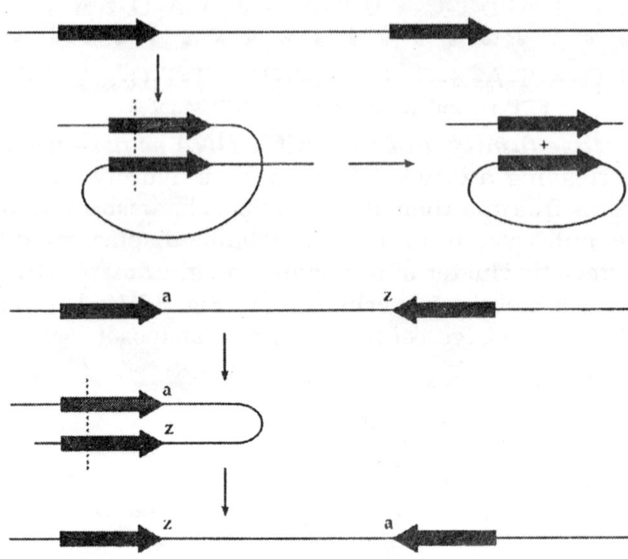


orientation. In one orientation mRNA is transcribed from a short operon that includes the right inverted repeat IRR, the $H_2$ flagellin gene and gene $rh_1$, which encodes a represser for flagellin gene $H_1$. Consequently, only gene $H_2$ is expressed. In the other orientation the RNA transcription is in the opposite direction so that neither $H_2$ nor $rh_1$ is expressed. However, $H_1$, which is located elsewhere, is expressed freely.

Two other proteins are required for efficient inver-sion by the Hin recombinase. A dimer of a 98-residue helix-turn-helix DNA binding protein called Fis (factor for inversion stimulation), a relative of protein HU, must bind to an enhancer, a 65-bp DNA segment. Binding of Fis to the enhancer helps to hold the supercoiled DNA and the recombinase in a correct orientation for reaction. Protein HU is also needed. The same Fis protein binds to an enhancer for a 3-kbp invertible DNA sequence, which controls alternative host preferences for bacteriophage Mu. The chemistry of the inversion reaction is related to that of the replicative transposons discussed.

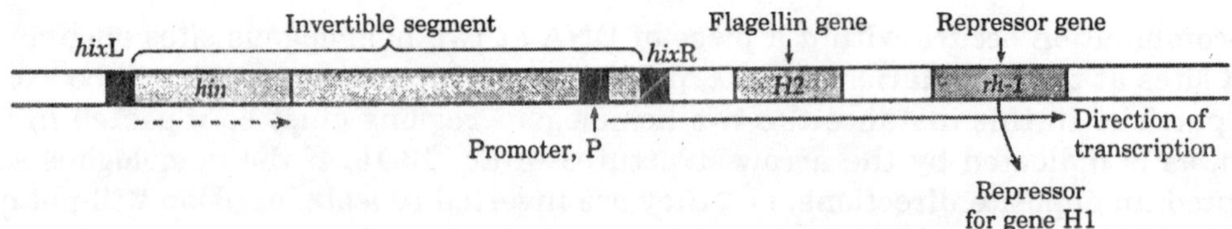


Microorganisms sometimes control the synthesis of surface proteins using segments of invertible DNA. The pathogenic bacterium *Campylobacter fetus* utilizes DNA rearrangements to allow one of a large family of surface layer (S-layer) proteins to be formed. The yeast FLP recombinase, mentioned in the preceding section, also inverts the sequence flanked by the 599-bp repeats.

## Transposons and Insertion Sequences

The first evidence that some genes can move from one location to another within the genome came from studies of *Zea mays* by Barbara McClintock in the late 1940s. She concluded that the variegated kernels found in some colored maize were a result of **controlling elements,** which could move from place to place turning on or inhibiting expression of various genes including some of those determining anthocyanin pigment formation. Two of these systems have been studied especially intensively: the **"activator (Ac)-dissociation (Ds)"** system, discovered by McClintock and the "enhancer (En)-inhibitor (I)" system, discovered by Peterson and independently by McClintock.[603] Each contains two genetic elements (segments of DNA) of which Ac and En are autonomous, i.e., they can move by themselves. Ds and I, however, cannot move unless the other element of the pair is also present. Both Ds and En have now been cloned and sequenced.

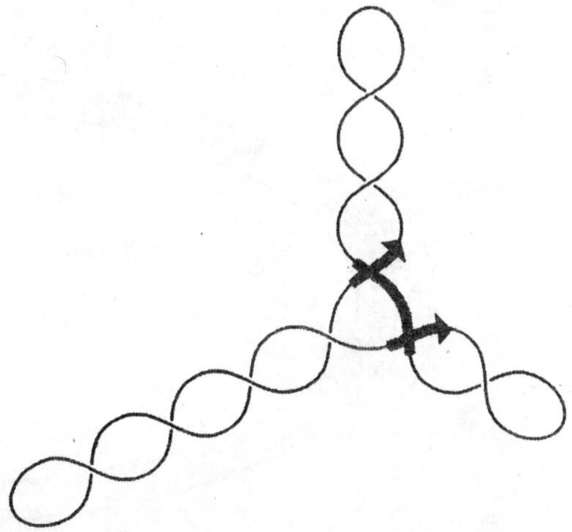

**Fig. 13.29. Formation of a synaptic complex of a super-coiled circular DNA containing the sites *gix* (green), which pass over and under the enhancer (gray). The recombinase Gin and the enhancer-binding Fis form a synaptic complex with DNA in this form as seen directly by electron microscopy.**

The general importance of transposable genetic elements was not appreciated by most molecular biologists until about 20 years after McClintock's discoveries. Then several moveable **insertion sequences** (Is elements) were found in enteric bacteria. Like the controlling elements of maize these small (0.8–1.4 kb) sequences can move and insert themselves at many points in the genome, often inactivating genes which they enter. The *E. coli* genome contains eight copies of IS1, and five of IS2, as well as several others. Most species of *Shigella* contain more than 40 copies of IS1, Mobile elements similar to those present in maize also exist in archaea.

Following the discovery of the IS elements it was found that transposable elements named **transposons** could transfer resistance to antibi-otics between bacteria. All of these transposable elements have inverted repeat sequences at the ends. For example, IS1 contains the following sequence at both ends but with opposite orientation as if in a palindrome. Some complex

C C
CAATT T TGAT

transposons have an IS sequence at each end. For example, Tn10 contains IS10, a relative of IS1, at both ends. These provide the characteristic inverted repeat termini. Fig. 13.30 shows a schematic drawing of a bacterial drug resis-tance plasmid containing IS1, IS2, and IS10 as well as transposons Tn3, Tn10, and Tn55. The two IS1 elements surround the large resistance determinant, which can be transferred as a block. To be autono-mously mobile a transposon must contain a **transposase** that enables it to be transferred. It usually carries other genes as well and may also contain one or more signals for transcriptional regulation such as promoters.

The chemistry of transposition is more complex than that of simple site-specific recombination. Trans-position can occur at many sites in a genome, and no homology with

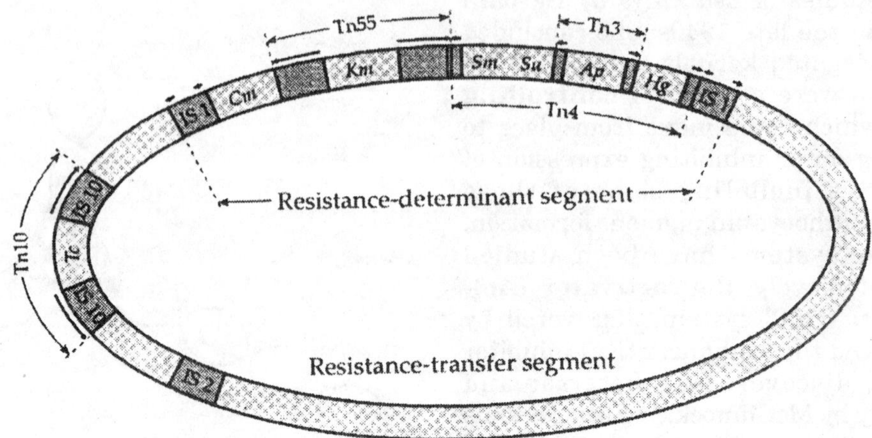


**Fig. 13.30 Transposons in an antibiotic-resistance plasmid. The plasmid appears to have been formed by the joining of a resistance-determinant segment and a resistarke-transfer segment; there are insertion elements *(IS1)* at the junctions, where the two segments sometimes dissociate reversibly. Genes encoding resistance to the antibiotics chloramphenicol *(Cm)*, kanamy-cin *(Km)*, streptomycin *(Sin)*, sulfonamide *(Su)*, and ampicillin *(Ap)* and to mercury *(Hg)* are clustered on the resistance-determinant segment, which consists of multiple transposable elements; inverted-repeat termini are designated by arrows pointing out-ward from the element. A transposon encoding resistance to tetracycline (Tc) is on the resistance-transfer segment. Transpo-son Tn3 lies within Tn4. Each transposon can be transferred independently.**

the transposon termini is required. Transposition is accompanied by *duplication of a short sequence of the recipient DNA* exactly at the ends of the transposon. Usually 5, 9, or 11 base pairs are duplicated as in the hypothetical example of Eq. 13.17. This happens because staggered cuts 5, 9, or 11 bp apart in the recipient DNA are made during recombination. These are indicated by the small arrows in Eq. 13.17. Transposons causing 5-bp duplications are Tn3, Tn7, γδ, phage Mu of *E. coli,* and Tyl of yeast; IS1, Tn10, and Tn5 of E. *coli* cause 9-bp duplications. When a transposon of one major group moves to a new location, the original copy remains. In this case transposition involves a combination of replication and site-specific (for the transposon) recombination. As a consequence, a circular DNA molecule containing the transposon will often react with a second circular DNA to form a large circle, a **cointegrate,** which con-tains two copies of the transposon. However, another group of transposons utilize a "cut-and-paste" mechanism that doesn't require extensive DNA duplication.

5′-C-A-C-G-A-T-C-T-T-- Recipient DNA
3′-G-T-G-C-T-A-G-A-A--

Transposon

5′-C-A-C-G-A-T-C Transposon C-G-A-T-C-T-T—
3′-G-T-G-C-T-A-G Transposon G-C-T-A-G-A-A—

Duplicated pentamer (27-17)

***Cut-and-paste (nonreplicative) transposons,***The transposases of *E. coli* Tn5, Tn7, and Tn10 act by hydrolytically cutting both strands of duplex DNA at the transposon ends leaving the phospho groups attached to the 5' cut ends, as is depicted in detail in Eq. 13.14, step *a*. The two 3' ends then carry out transesterification reactions, as in Eq. 13.14, step *b*. These two steps are used to nick both strands of the DNA carrying the transposon and to join them to a target DNA sequence to give a branched intermediate (Eq. 13.18, step *a)*. Nonreplicative transposons apparently cut off two arms, e.g., A and B, and heal up the small gaps by repair synthesis, leaving the transposon in a new location between C and D. The gap repair accounts for the duplication of the end sequences of the cut target DNA.

***Replicative transposons.*** In 1979 Shapiro proposed the mechanism illustrated in Fig. 13.31 for replicative transposons. The two inversely repeated segments (green) at the ends of the transposon are aligned with the recipient DNA whose ends are la-beled C and D. In fact, the recombining DNA mole-cules must be supercoiled. Staggered cuts are made in the recipient DNA at points *a* and *b,* which are 5, 9, or 11 bp apart, depending upon the specific recombinase. Nicks are also made in the transposon ends. The 3' ends from the transposon are resealed with the 5' ends from the recipient DNA (step *a)* to give a structure that in effect has two replication forks. Replication (step *b)* yields the cointegrate, which contains two copies of the transposon as indicated. In a third step (step c) recombination between the two integrated transposons yields a copy of the original transposon-containing donor and the recipient DNA, which now also contains a copy of the transposon.

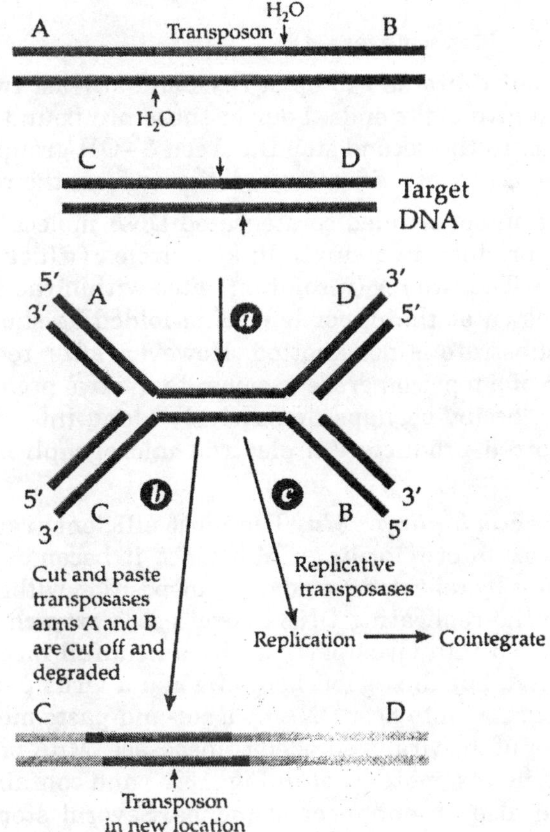

When a transposon reacts with another part of the same DNA circle there are two possibilities. The piece of DNA lying between the transposon and the recipi-ent site may be excised as a circle containing a copy of the transposon. Alternatively, there will be inversion of that sequence as well as replication of the transposon.

The closely related transposons Tn3 and γδ are understood best. They contain not only a transposase gene but also a **resolvase** gene. The transposase carries out the recombination reactions that yield the cointegrates, while the resolvase catalyzes site-specific recombination between the two transposons in the cointegrate to complete the transposition. Several subunits of the resolvases bind the two **resolution (res) sites** of the supercoiled DNA in a parallel orien-tation with the DNA supercoiled as in Fig. 13.29. However, no enhancer-binding protein is needed, and the two *res* sites must be supercoiled. Purified y§ resolvase uses hydroxyl groups of the serine-10 side chains as the nucleophiles to cleave the DNA by displacement at specific *res* sites to give transient enzyme-bound phosphodiester linkages.

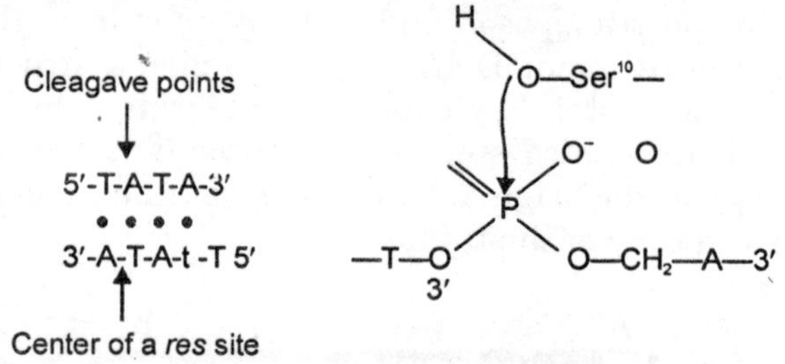

Cleagave points
5′-T-A-T-A-3′
• • • •
3′-A-T-A-t -T 5′
Center of a *res* site

H
O—$Ser^{10}$—
$O^-$ O
P
—T—O
3′
O—$CH_2$—A—3′

The synaptic complex contains 240 bp of DNA and at least two resolvase dimers. All four DNA chains are cut to give eight ends. Four of these are bound to the serine side chains in phosphodiester linkage. In the second step the freed 3′–OH groups react with the bound ends of the other duplex via a transesterification reaction to form the recombinant chains.

The resolvases act on supercoiled cointegrated DNA molecules that contain two directly repeated res sites to produce two singly linked circles (which are still supercoiled) each containing one res site. The two *res* (resolution) sites within the transposons are aligned, the open circle of DNA shown at the upper left being folded as shown in the lower part of the drawing. .The DNA substrate is not knotted. However, after recombination it is catentated and will require action of a topoisomerase to separate the two products. Occasionally additional recombina-tion events occur, perhaps processively along inter-wound double helices. This produces various knotted products. An electron micrograph of one of these is shown in Fig. 13.17.

***The temperate bacteriophage Mu,*** The most efficient transposon known is the 37-kb genome of the **mutator bacteriophage** Mu. Once it becomes integrated into a bacterial chromosome, it replicates by repeated rounds of transposition within the host bacterium. During the lytic cycle some of the replicating DNA is excised as extrachromosomal circles of various sizes, which are packaged into virus particles by a headfull mechanism. The circles contain copies of some host DNA, but this is left behind when a virus particle infects a new cell. The phage Mu DNA is integrated into host DNA by a cut-and-paste mechanism in a transpososome that contains a tetramer of the virally encoded transposase (MuA protein) bound to a supercoiled DNA. The transpososome resembles that in Fig. 13.29 and contains cleavage sites at the ends of the transposon and also an enhancer sequence. Several steps involving conformational alterations occur in the transpososome. One is an ATP-dependent action of a second Mu-encoded protein MuB. During the lytic cycle replicative transposition predominates.

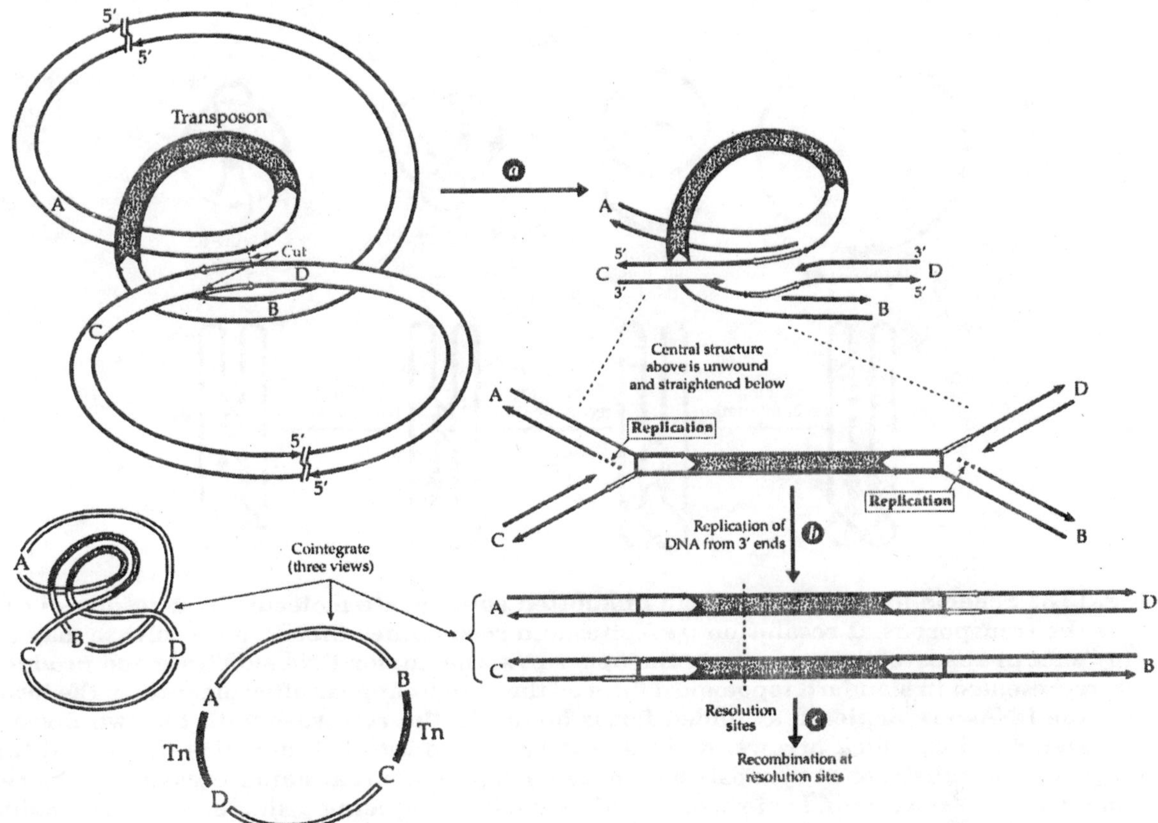


**Fig. 13.31 Scheme for integration of a transposon (stippled duplex) present in a piece of DNA with ends A and B into another piece of DNA with ends C and D and containing a suitable recognition sequence (open bars). Inversely repeated sequences in the transposon are shown as solid bars with a direction arrowhead. Arrowheads point toward 5' strand ends. Cleavage and rejoining at points a, a and b, b yield an intermediate with two replication forks. Replication through the transposon yields one unchanged DNA segment with ends A and B and a transposon inserted into the other DNA segment. If A is continuous with B, a cointegrate structure is formed.**

***Some other transposons,*** Transposons have a variety of biological functions. For example, haploid cells of the yeast *S. cerevisiae* exist as one of two mating types *a* or α. The mating type is established by trans-position of one of two "cassettes" of genes from two different "silent" locations to a location from which they can be expressed.

One of the best known eukaryotic transposons is the P element of *Drosophila,* which transposes only within the germ line cells of developing embryos, somatic cells being unaffected. It belongs to the same family as Barbara McClintock's Ac element of maize and the *Tel* family of nematodes.[627] P ele-ments use a nonreplicative cut-and-paste method of transposition. The 87-kDa transposase protein requires GTP and $Mg^{2+}$ for activity. It was only in the past few decades that P elements have been found in *D. melanogaster.* They may have entered this fruit fly from another species, possibly transferred by a mite.

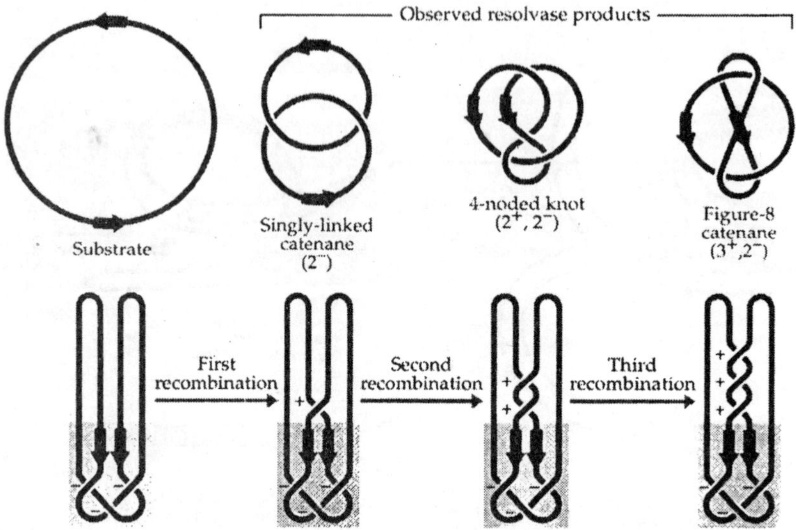


**Fig. 13.32 Scheme for resolution of an unknotted cointegrate molecule by a resolvase that cuts the transposons at resolution (*res*) sites and recombines them. The resolvase may act only once or repeatedly as shown. In the upper row, the duplex DNA substrate and products are represented in standard topological form as they might appear after nicking. In the lower row the DNAs are depicted as folded forms bound to the resolvase with the two directly repeated *res* sites (thick arrows) dividing the substrate into two domains (thick and thin regions). The substrate at synapsis has three (–) supercoils that entail crossing of the two domains. Successive rounds of recombination, each introducing a single (+) interdomainal node, are drawn in the lower row. Bound resolvase maintains the three synaptic supercoils. After dissociation from the resolvase at any stage, the product supercoil nodes either cancel with ones of opposite sign or are removed by subsequent nicking. The node composition is indicated in parentheses.**

A second *Drosophila* transposon called **mariner** typifies the *mariner* / *Te*1 transposon superfamily, which also contains members from nematodes, other invertebrates, fishes, amphibia, and possibly human beings. These transposons encode a transposase containing a D, D, D or D, D, E motif but no other proteins. They contain short ~30-bp terminal inverted repeats and become inserted into host TA sequences. Movement of some repetitive sequences of the LINE and SINE families within the human genome may be assisted by *mariner* transposons.

The maize transposon *ac* is widely used as a means of inactivating genes and placing a "tag" that can be used to map the gene and to permit it to be cloned and sequenced. Although initially of use only in maize the method has been extended to other plants, and genetically engineered transposons have allowed it to be utilized in animals.

A different kind of transposition controls self-sterility in maize. The cause of self-sterility in one strain has been traced to the presence of two linear episomes called S–1 (614 kb) and

S–2 (514 kb) within mitochondria. These have inverted terminal 208 bp repeats. On rare occasions they recombine with the circular mtDNA converting it to a linear form with the episomes covalently linked to one end. The change is accompanied by reversion to fertility.

As described already, mammalian DNA contains many **retrotransposons** (retroposons) that lie within short direct repeats characteristic of transposons. However, they contain a poly(A) tail at the 3' end, an indication of their relationship to RNA transcripts, and are discussed in.

## Other Causes of Genetic Recombination

Because their integration sites in host DNA do not depend upon homology with the transposon ends, transposition is sometimes called **"illegitimate recombination."** Although no homology is required, there are prefered sites for integration. For example, Tn10 is transposed most readily into certain "hot spots" in the *Salmonella* DNA among which is the sequence 5'-GC$^{m5}$CAGGC. Illegitimate recombina-tion can also be induced by other processes that in-volve DNA chain cleavage, e.g., by topoisomerases. Whenever a DNA chain breaks, it must be repaired, a process that often also involves recombination. Recombination is often observed to occur between direct repeat sequences, which are a major cause of instability in the genome.

Under some circumstances selected segments of the genome are **amplified** by repeated replication of a gene or genes. Amplication of specific genes occurs in viruses, in bacteria where it may provide for adaptation to conditions of stress, and in eukaryotes. In oocytes of amphibia, such as *Xenopus,* excess DNA accumulates around the nucleoli and later breaks up to form 1000 or more separate nucleoli. As many as 3000 copies of the rDNA (which forms a dis-tinct satellite band upon centrifugation) may be present. Much of this DNA exists as extrachromosomal rings containing 1–20 rDNA units. Using these genes as many as $10^{12}$ ribosomes per oocyte are synthesized.

*Tetrahymena* contains only one set of rRNA genes per haploid genome in its diploid **micronucleus.** Following sexual conjugation the chromosomes of the micronucleus undergo multiple replications to form polytene chromosomes containing thousands of copies. However, about 5% of the resulting DNA is ex-cised as linear pieces with characteristic inverted repeat sequences and 3' single-stranded tails:

$$5'\text{--}\overrightarrow{C_4A_4C_4A_4C_4}\text{ - - -//- - -}G_4T_4G_4T_4G_4T_4G_4T_4G_4\text{- - }3'$$
$$3'\text{--}G_4T_4G_4T_4G_4T_4G_4T_4G_4\text{ - - -//- - -}\overleftarrow{C_4A_4C_4A_4C_4}\text{- -}5'$$

These tails are joined by a protein to form the circles, which are segregated in the **macronucleus.** The other 95% of the amplified DNA is degraded. At the next meiosis the macronucleus is discarded entirely, and a new one is formed at the next diploid stage.

Toxic drugs often cause cells to amplify genes that help resist the drug. This can be a major problem in the chemotherapy of cancer. For example, a culture of human leukemia cells grown in the presence of in-creasing concentrations of methotrexate increased its level of dihydrofolate reductase 240–fold. The cause is an increase in the number of copies of a chromosomal region containing the gene. Cancer cells tend to amplify oncogenes such as *c-myc*.

There are several mechanisms for gene amplification. The formation and breakup of polytene chromosomes in *Tetrahymena is* one. The circular copies of rDNA in *Xenopus* and many other species are generated by a rolling circle mechanism similar to that in Eq. 13.7. If each circle excised from the original chro-mosomes contains an origin of replication, many copies can be formed. Another possible mechanism is replication of a local region of DNA several times followed by excision of pieces of the DNA. A mechanism that may give rise to homogeneously staining regions is unequal crossing-over repeated several times within a gene cluster. Transposition can cause excision of DNA in a circular form, which can be amplified by a rolling-circle mechanism.

## DAMAGE AND REPAIR OF DNA

A characteristic of living things is their high degree of mutability. Harmful mutations take a toll of human life at an early age, and the very high incidence of cancer in older persons is largely a result of the accumulation of somatic mutations. Mutations are a major factor in aging and are continuously introducing new genetic defects into the population. Mutations can be described as **base substitutions, deletions, or additions.** Base substitution mutations are classified as **transitions,** in which a pyrimidine in one strand is replaced by a different pyrimidine. In the complementary strand a purine is replaced by the other purine, e.g.,

$$C \bullet G \rightarrow T * G \xrightarrow{\text{Replication}} T \bullet A \quad C \bullet + \quad G \qquad (27\text{-}19)$$

$$A \bullet T \rightarrow \underset{\text{Mispair}}{G * T} \longrightarrow \underset{\text{Mutation}}{G \bullet C} \quad A \bullet + \quad T \qquad (27\text{-}20)$$

In a *transversion* a purine in one chain is replaced by a pyrimidine, while the pyrimidine in the complementary chain is replaced by a purine:

$$C \bullet G \rightarrow A * G \longrightarrow A \bullet T \quad C \bullet + \quad G \qquad (27\text{-}21)$$

Here the central asterisks designate mispairs and the green shade marks mismatched bases and also the resulting mutant base pair formed in the next replica-tion cycle.

### Causes of Mutations

DNA can be damaged in many ways. Sponta-neous hydrolysis of the glycosidic bonds between nucleic acid bases and the deoxyribose to which it is connected cause the loss of ~ $10^5$ purines and pyrimi-dine rings per day from the DNA in a mammalian cell. About 100 residues per day of cytosine are deaminated by such agents as nitrite or bisulfite to form uracil. Like thymine, uracil will pair with adenine causing C • G → T • G a transition mutation as in Eq. 13.19. A few adenines per day are also deaminated to form hypoxanthine. Oxidized bases are formed by attack of HO• radicals and other species of reduced oxygen. Alkylating agents from the external environment as well as *S*-adenosylmethionine carry out slow, nonspecific alkylation of purine and pyrimidine bases. Polycyclic aromatic hydrocarbons and other carcinogens are converted to metabolites that alkylate DNA, and alkylated bases often mispair during replication. Ultraviolet light induces formation of photohydrates, pyrimidine dimers, and other photochemical products. X-rays and gamma rays cleave nucleic acid bases and break chromosomes. Natural radioactivity has a measurable effect. Mistakes caused by mispairing or misalignment are made during replication and recombination and even during repair of DNA. Errors in replication are especially numerous in highly repetitive DNA sequences. Some errors probably arise as a result of tautomerization and others from incorporation of uracil in place of thymine. Thus, to keep its DNA in repair a cell must continuously deal with missing bases, wrong bases, altered bases or sugars, pyrimidine dimers and other crosslinkages, deletions, and insertions.

## Fidelity of Replication

During DNA replication in *E. coli* only one mistake is made on the average during polymerization of $10^9$–$10^{10}$ nucleotides. The rate varies among different sites in the genome. In eukaryotes the error rate may be only 1 in $10^{-10}$ base pair or less per generation. Early workers often attributed the specificity in base pairing and the resultant high precision of replication entirely to the strength of the two or three hydrogen bonds formed together with the stabilization provided by the adjacent helix. However, the Gibbs energy of formation of the base pairs is small, and the additional energy of binding to the end of an existing helix is insufficient to account for the speci-ficity of pairing. Thus, according to Eq. 13.30 a difference in $\Delta G$ of binding between the correct nucle-otide and an incorrect one of 11 kJ/mol would give an error rate of 1 in 100.

***The role of polymerases.*** The polymerase enzymes play a major role in ensuring correct pairing during replication, transcription, and protein synthesis. RNA and DNA polymerases are large molecules. The binding site on the enzyme can completely surround the double helix. Water will be excluded as the enzyme folds around the base pairs. This may have the effect of greatly increasing differences in the Gibbs energies of binding and thereby enhancing selectivity. It has traditionally been assumed that formation of proper hydrogen bonds is essential for the base selection. A hypothetical active site is portrayed in Fig. 13.33. A guanine ring of the template strand of DNA is portrayed at the point where the complementary DNA strand is growing from the 3' end. The proper nucleoside triphosphate must be fitted in to form the correct GC base pair before the displacement reaction takes place to link the new nucleotide unit to the growing chain. Let us suppose that the enzyme possesses binding sites for the deoxyribose unit of the template nucleotide and for the sugar unit of the incoming nucleoside triphosphate and that the two binding sites are held at a fixed distance one from the other. As indicated in Fig. 13.33, some group H-Y might also be present at each binding site to hydrogen bond to the nitrogen or oxygen indicated by the heavy

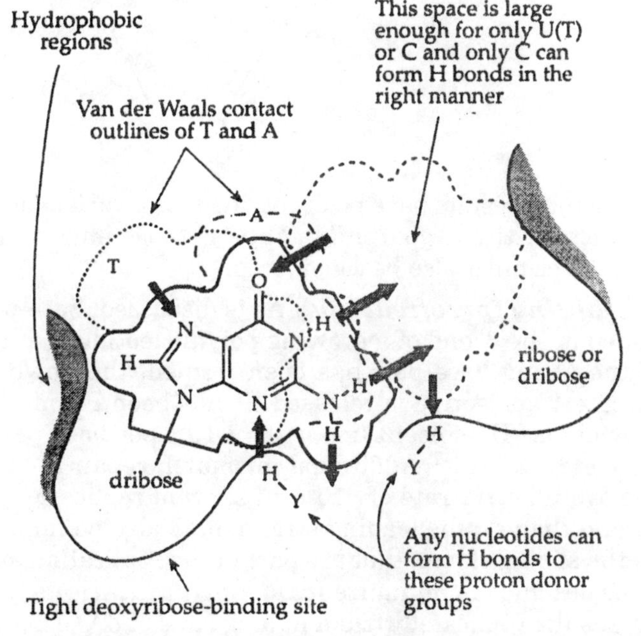


**Fig. 13.33 Selecting the right nucleotide for the next unit in a growing RNA or DNA chain. A deoxyguanosine unit of the template chain is shown bound to a hypothetical site of a DNA polymerase.**

green arrows. All four of the bases could form hydrogen bonds of this type in the same position. Hydrophobic interactions could provide additional stabilization. With such an arrangement the correct nucleoside triphosphate could be selected no matter which one of the four bases occupied the binding site on the left side of the figure. (The outlines of the thymine and adenine rings have been drawn in with dotted and dashed lines, respectively.) If a purine is present on the left side, as is shown in the drawing, there is room on the right side only for a pyrimidine ring. Thus, A and G are excluded, and the choice is only between C and U (or T). However, U will be excluded because the'dipoles needed to form the hydrogen bonds point in the wrong direction. These dipolar groups are hydrated in solution. They are unlikely to give up their associated water molecules unless hydrogen bonds can be formed within the base pair. Not only would a molecule of U (or T) be unable to form the stabilizing hydrogen bonds within the vacant site but also the electrostatic repulsion of the like ends of the dipoles would tend to prevent associa-tion. This would lower the affinity of the polymerase for mispaired bases. Verification that the proper base pair has been formed could be accomplished by using its tautomeric properties to sense an electronic displacement through the hydrogen-bonded network. Considerable experimental evidence suggests, however, that hydrogen bonding may not be important to the selectivity of DNA polymerases. For example, the triphosphate of the following analog of thymidine is efficiently incorporated by *E. coli* DNA polymerase I into DNA opposite the thymine base;pair partner adenine.

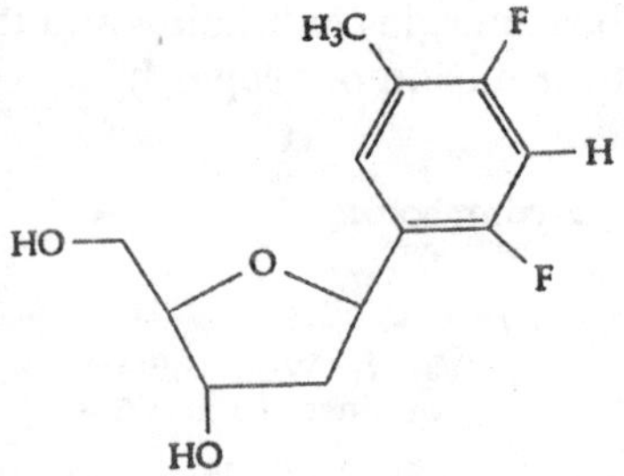

$H_3C$ F H HO O F HO

Structural analog of thymidine

Similar results have been obtained for a variety of other analogs with poor hydrogen bonding character-istics. Apparently *shape* is most important. Binding strengths in transition state structures must also be considered.

***Editing (proofreading).*** As is discussed, base-pairing is checked after a new monomer is added at the 3' end of a growing polynucleotide chain, as well as before polymerization occurs. If the wrong base pair has been formed, the newly created linkage is hydrolyzed, and the incorrect nucleotide is released. It has been estimated that in E. *coli* the error rate for DNA polymerase III holoenzyme is $\sim 1 \times 10^{-7}$ per base pair of which proofreading by the ε subunit may provide $\sim 10^{-2}$. Additional mismatch repair reduces the error rate 200- to 300-fold to give the overall error rate of $<10^{-9}$. Fidelity of replication, which seems to be higher on the lagging strand than on the leading strand, may also be improved by operation of other proteins, such as the sliding clamp, that are part of the replication apparatus. Replicative proteins also appear to be designed to minimize **frameshift mutations** that could arise by slippage of the template versus the replicated strand in long runs of dA•dT pairs.

## Repair of Damaged DNA

A final check of the fidelity of replication is made after a new strand has been formed. Mismatched base pairs are identified, and the incorrect nucleotides are cut out and replaced by correct ones. Some of the thymine dimers created by the action of light are also repaired photochemically by photolyases. **Photoreactivation** was the first DNA repair process recognized. However, most thymine dimers in human DNA must be excised and replaced. Loops, gaps, and double-stranded breaks are also sensed, and appropriate corrections are made. If necessary the sequence of a badly damaged segment of DNA may be copied from the DNA in a sister chro-matid before mitosis is completed or from the homologous chromosome.

Much of our understanding of DNA repair comes from investigations of mutant strains of *E. coli,* which display an elevated incidence of mutations. On this basis, the "mutator" genes *dam, mutD, mutH, mutL, mutS, mutU* (also called *uvrD),* and *mutY* were implicated in DNA repair. An additional series of genes, *uvrA, B, C,* and *D,* were identified by their participation in resistance to ultraviolet radiation damage and were also recognized as being involved in DNA repair. Many corresponding genes were found later in the yeast, *Drosophila,* and human genomes. The study of human DNA repair has also been facilitated by the recognition of a group of human inherited defects (Box 9-A). There are 130 known human DNA repair genes.

***Methyl-directed and other mismatch repair.*** In E. *coli* the methylase encoded by gene *dam,* within ~7 min of replication, methylates adenine at N-6 in the sequence GATC, which occurs at many locations. This methylation provides a label for the template strand in a newly replicated duplex. The GATC sites in the newly synthesized strand will not yet be methylated, and repair enzymes recognize it as the strand on which to carry out excision repair. An other system, which may function during recombination, acts on fully *dam* methylated DNA. Since eukaryotes do not use methylation of GATC sites to distinguish the old template and newly replicated DNA strands, other mechanisms must operate. Ten proteins are required for methyl-directed mismatch correction. Of the required *mut* genes, *mutD* (also called *dnaQ)* was found to be the structural gene for the proofreading subunit e of DNA polymerase III. *MutH, L,* and *S* as well as ATP are also needed, as is a helicase, an exonuclease, DNA pol III, and a ligase. Homodimers of proteins MutS and MutL preferentially bind to DNA with mismatched base pairs such as Pur-Pur (e.g., A•G) or Pyr•Pyr (less well recognized). Some of these mismatched pairs may contain modified bases. After **recognition** by the MutS•MutL complex the endonuclease MutH is activated and cuts the DNA chain at the nearest unmethylated GATC sequence. An exonuclease then degrades the chain past the mismatched base leaving a gap that must be filled by the action of DNA polymerase and DNA ligase. This **long-patch mismatch repair** (MMR) is utilized also by eukaryotes, using proteins homologous to E. *coli* proteins MutS and MutL. However, methylation is not involved. In both yeast and human beings there are at least six MutS homologs, some of which have functions other than repair. For example, in yeast MutS- and MutL-related proteins are essential for normal levels of meiotic crossing-over. Defects in human mismatch proteins may cause hereditary non-polyposis colorectal cancer.

***Excision repair.*** The *E. coli* mismatch repair is a type of excision repair. However, a different **nucleotide excision repair** system (NER) is utilized by all organisms from bacteria to human to remove a variety of defects. These include thymine dimers, photohydrates, oxidized bases, adducts of cisplatin, mutagens derived from polycyclic aromatic compounds, and poorly recognized OC mismatched pairs. In *E. coli* this excision repair process depends upon proteins encoded by genes ***UvrA,* B, C,** and **D** and also DNA polymerase I and DNA ligase. A dimer of protein UvrA forms a complex with helicase UvrB.

$$(UvrA)_2 \xrightarrow{UvrB} (UvrA)_2 \cdot UvrB \xrightarrow{DNA} UvrB \cdot DNA \xrightarrow{UvrC} UvrB \cdot C \cdot DNA \xrightarrow{-UvrB,\ UvrC} UvrB \cdot C \cdot DNA$$

The helicase, driven by ATP hydrolysis, may move along the DNA chain with its associated UvrA protein in search of defects. When one is located UvrB binds tightly. UvrA then dissociates, and the nuclease UvrC binds and cleaves the DNA chain in two places as in the following scheme. One is at the fourth or fifth phosphodiester linkage in the 3' direction from the defect. The other is at the eighth phosphodiester on the 5' side. The resulting gap is filled by a DNA polymerase and DNA ligase. Another helicase, encod-ed by *UvrD,* is also required.

Cut Cut Cut

5'-TpTpTpTpTpTpTpTpTpTpTpTpTpTpTpTpTpTpTpT-3'

Defect

3'-pppppppppppppppppp-5'

UvrBC preincision complex

Both yeast and human cells have similar but more complex systems of nucleotide excision repair. The dual incision steps require at least six components. Several of these (XPA, C, F, G) have been found defective in various forms of the inherited disease **xeroderma pigmentosum,** in which there is a high incidence of UV-induced skin cancer. Replication protein A (Section C,10), a three-subunit ssDNA binding protein, is also essential. In eukaryotic excision repair the DNA is cleaved at the same position as in the bacterial preincision complex (preceding scheme), on the 3' side of the defect but further out on the 5' side, the excised polynucleotide being –24–32 nucleotides in length. Repair of the resulting gap in the DNA duplex is accomplished by synthesis mediated by DNA polymerase 5 or e and action of a DNA ligase.

One of the lesions removed by nucleotide excision is the thymine photodimer. In the fission yeast S. *pombe* an **alternative excision repair** system, special-ized for removal of thymine dimers and 6-4 photo-products, produces two-nucleotide gaps with 3'-OH and 5'-phospho-group ends. Alternative NER pathways are also employed by bacteria.

***Base excision repair (BER),*** The N-glycosyl linkages of the purine and pyrimidine bases to the deoxribose residues of the sugar-phosphate backbone of DNA are subject to spontaneous hydrolysis, one important source of damage to DNA. Similar hydrolytic reactions are catalyzed by **DNA glycosylases,** which remove many mismatched or damaged bases. At least seven enzymes of this type are present in cells of *E. coli.* One of them, a **uracil-DNA glycosylase,** hydrolyzes the glycosyl linkage wherever uracil has been incorporated accidentally in place of thymine or has been produced by deamination of cytosine. Acting via a nucleophilic displacement mechanism, it removes the uracil, leaving the DNA backbone intact but with an **apyrimidinic site in** one chain. The resulting apyrimidinic (apurinic sites) are recognized and cleaved by

**apurinic /apyrimidinic DNA endonucleases** (AP nucleases). These enzymes cut the DNA backbone, some leaving a 3′-phosphate group and a 5′-OH terminus and others a 3′-OH and 5′-phosphate. The resulting gap can be filled, for example by the action of DNA polymerase I and a ligase. To prevent the eliminated uracil from being converted to dUTP and reincorporated into DNA a **deoxyUTPase,** essential to both *E. coli* and yeast, hydrolyzes dUTP to dUMP and inorganic phosphate, decreasing the concentration of dUTP. Other enzymes, known as **DNA glycosylase/AP** lyases, use an amino group of an enzyme side chain as a nucleophile in a ringopening reaction that is followed by P elimination of the nucleotide base with formation of a Schiff base intermediate (green arrow, Eq. 13.23). An example is the bacteriophage T4 enzyme **T4 endonuclease V.**

Others include the *E. coli* **endonu-clease III,** encoded by the **Nth** gene, its yeast homolog **Nth-Spo**, and human DNA pol ι (iota). After the Schiff base is formed, the DNA back-bone is cut on the 3′ side of the abasic site by a second β elimination reaction. A δ elimination may also occur, cleaving the DNA chain on the 5′ side of the lesion. Alternatively, a hydrolytic cleavage is possible as is also indicated in Eq. 13.23. Either type of DNA glycosylase forms a single-nucleotide gap or a gap missing just a few nucleotides. This gap can also be filled by polymerase and ligase action.

Both NER and BER forms of excision repair remove a great variety of defects, many of which are a result of oxidative damage. Most prominent among these is **7,8-dihydro-8-oxoguanine** (8-OG), which is able to base pair with either cytosine (with normal Watson-Crick hydrogen bonding) or with adenine, which will yield a purine-purine mismatch and a C•G → A•T transversion mutation, a frequent mutation in human cancers.

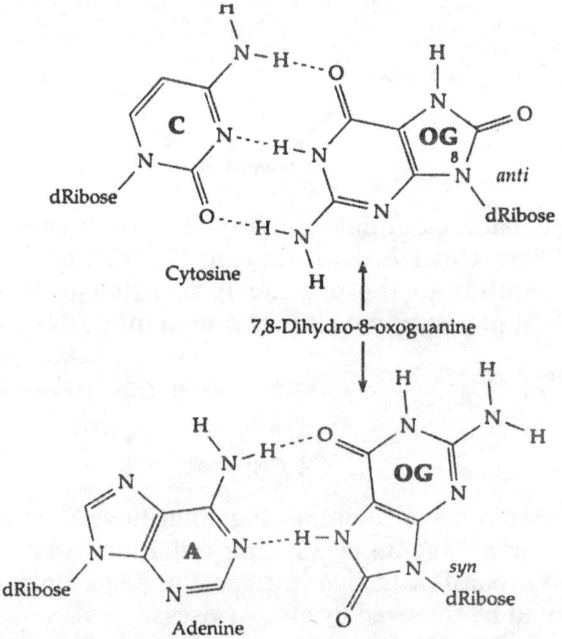

In *E. coli* three enzym,es protect against 8-OG DNA mismatches. A glycosyltransferase encoded by the *mutM (orfpg)* gene removes 8-OG from DNA. Because some A*OG base pairs may remain, a second glycosyltransferase (**MutY**) removes adenine from them. Interestingly,

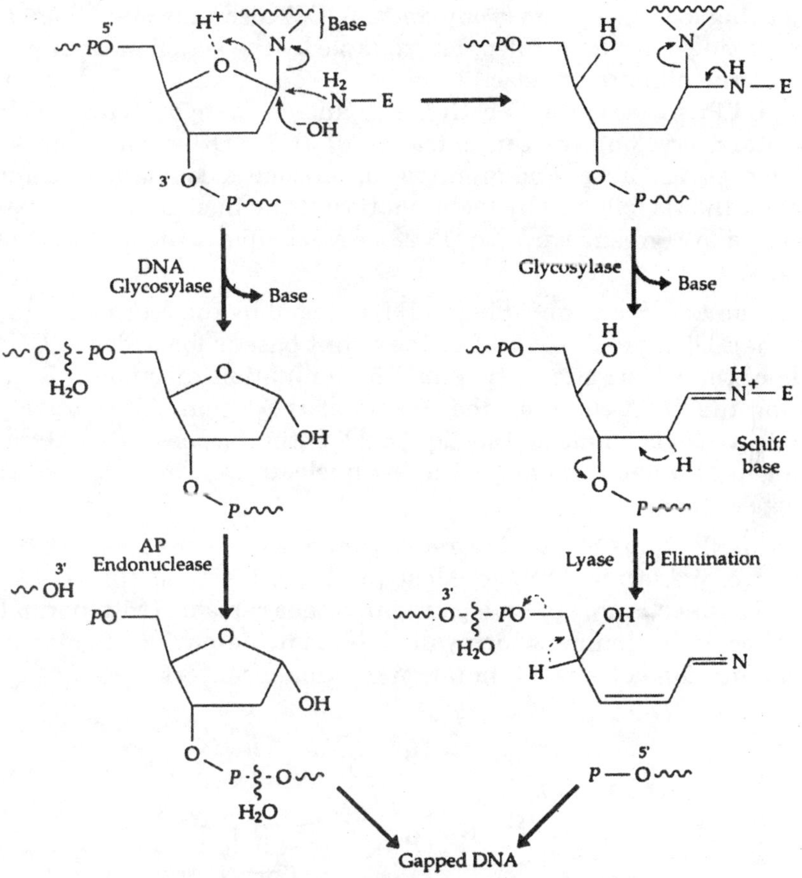

MutY contains an $Fe_4S_4$ cluster, essential to its function. Both yeast and human cells have a corresponding enzyme. The third *E. coli* enzyme **(MutT)** is a nucleoside triphosphate pyrophosphohydrolase, similar to the previously mentioned dUTPase. It preferentially hydrolyzes free 8-oxo-dGTP, preventing its incorporation into DNA.

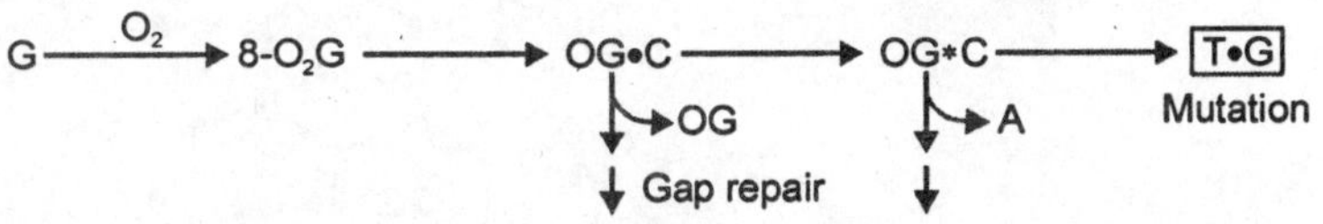

$$G \xrightarrow{O_2} 8\text{-}O_2G \longrightarrow OG{\bullet}C \longrightarrow OG{*}C \longrightarrow \boxed{T{\bullet}G}$$

OG•C → OG; Gap repair. OG*C → A. T•G: Mutation

Other DNA glycosylases remove thymine rings that have been converted to saturated or fragmented forms by oxidizing agents or ionizing radiation. Among these are 5,6-hydrated thymine and urea, which are still attached in ribosyl linkage. Purines such as hypoxanthine and 3-methyladenine can all be removed by glycosylases.

***Reactions of alkylated bases.*** The $O^6$ alkylation of guanine rings in DNA is highly mutagenic, presum-ably because mispairing with thymine will cause C•G → T•A mutations.

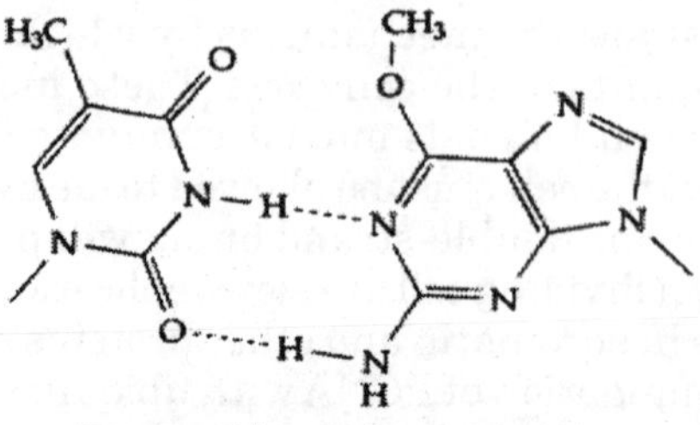

Thymine–O⁶-methylguanine pair

Both bacteria and higher eukaryotes have an enzyme-like **O$^6$-methylguanine-DNA methyltransferase,** whose synthesis may be induced by culture of cells in the presence of alkylating agents. This 354-residue $Zn^{2+}$-containing **Ada protein** acts as both acceptor and catalyst for transfer of the methyl group off from the O$^6$-methylguanine onto the sulfur atom of a cysteine side chain in the protein. The presence of alkylating agents also induces synthesis of DNA glycosylases that remove 3-methyladenine and other alkylated bases. 5-Methylcytosine, present in CpG islands in human DNA, will occasionally be hydrolyzed to thymine, giving T*G mismatches. The mismatched thymines are removed in cells of *E. coli* by a specialized endonuclease (**vsr** gene), which hydrolyzes the phosphodiester linkage preceding the mismatch. In eukaryotic cells a **thymine-DNA glycosylase** accomplishes the removal.

***Repair of double-strand breaks,*** Following replication both in bacteria and in eukaryotes, recombinational exchanges may occur between sister chromatid duplexes or between homologous pairs of chromosomes. A newly replicated chain may have gaps because of defects in the template strand from which it was copied. Copying from a sister duplex may permit assembly of a correct chromosome sequence and survival for the cell. Recombinational repair is also a major mechanism for preventing loss of genetic information from **double-strand breaks.** Such breaks are induced by ionizing radiation or chemical damage. They are often created at replica-tion forks, stalling DNA synthesis with potentially disastrous consequences to the cell. Repair of these breaks, which is required for completion of replication, depends upon a large complex of replication and recombination proteins. In *E. coli* this includes protein RecF. Repair of double-strand breaks by homologous recombination may be the most frequently employed type of DNA repair in bacteria, yeasts such as *S. cerevisiae* and also in mammals and other organisms.

A second repair pathway, called **nonhomologous endjoining (NHEJ),** is also utilized and may be relatively more important in higher organisms. The NHEJ mechanism repairs double-strand breaks caused by ionizing radiation and is also employed in specialized genomic rearrangements of developing B and T cells of the immune system. This V(D)J recombination is used to create the huge array of antibodies and antigen receptors required for immunological recognition and protection. NHEJ doesn't depend upon templates but simply rejoins duplex ends. Errors can be made if the ends have been damaged or if the wrong ends are joined. Several special proteins are required. One is a DNA-activated protein kinase (DNA-PK) and a heterodimeric **endbinding protein** with subunits **KU70** and KU80. The serine / threonine protein kinase ATM is also activated and phosphorylates protein p53 in response to $\gamma$-irradiation of cells. This is thought to be part of a signaling pathway for control of the cell cycle following DNA damage. Mitosis may be delayed while repair is completed. Alternatively, the damaged cell may be killed by apoptosis.

Subjects of current interest are the mechanisms by which completion of DNA synthesis is signaled at the various checkpoints in the cell cycle. These include points in addition to those marked in Fig. 13.15. At every point signals must accumulate to indicate that replication must be delayed to allow repair or that the cell must be allowed to die by apoptosis. Failure to accomplish repair may lead to cancer. Even one double-strand break will prevent the completion of mitosis. Among other problems faced by a dividing cell are slow replication and stalling at replication forks. A range of chemical alterations in chromatin and other proteins are associated with DNA damage and repair. Among these are conjugation of PCNA with ubiquitin and SUMO.

***The SOS response and translesion repair.*** If cells of bacteria or eukaryotes are heavily damaged by UV, X-ray irradiation, or mutagenic chemicals, an emergency or **SOS response** is initiated. In *E. coli* the two proteins specified by genes *lexA* and *recA* initiate the response. The *lexA* protein is a represser that prevents transcription of a group of SOS genes. It is thought that some product from damaged DNA activates the RecA protease activity. The activated RecA protein then cleaves the lexA protein allowing transcription of the SOS genes. The SOS response is transient but complex. It includes increased recombinational activity, alterations in repli-cation initiation, inhibition of nucleases, and induction of an **errorprone DNA synthesis.** The cell now replicates DNA more rapidly than normal but with an increased frequency of errors. For example, it will bypass thymine dimers and other defects, and it will continue the DNA synthesis even though incorrect bases have been put into the chain opposite the thymine dimer. Later the errors may be corrected by recombinational repair or by photoreactivation. In *E. coli* this translesional repair depends upon DNA polymerase III, the RecA protein, and two proteins encoded by genes *umuC* and *umuD.* These genes were recognized as providing resistance to mutations induced by UV radiation. Study of similar genes in yeast led to the discovery of two new and unusual DNA polymerases essential for translesional repair. DNA polymerase ζ (zeta), encoded by gene *REV3* and *REV7* subunits, bypasses abasic sites by inserting a dCMP residue into the growing DNA chain. Because C may be the wrong base the process is "errorprone." Polymerase η (eta) bypasses thymine photodimers by placing two consecutive dAMP residues in the growing strand. This is an errorfree process. Genes for human polymerases ζ and η have also been identified, and additional polymerases have been found in yeast and other organisms. The human XP-V gene for polymerase η is defective in xeroderma pigmentosa variant type. Polymerase ι may act sequentially with pol η to bypass highly distorting lesions. The very imprecise pol θ may also be used to bypass hard-to-remove lesions. Polymerase λ is thought to play a role in DNA repair during meiosis, while pol κ is needed in some way to provide cohesion between sister chromatids. Some of these DNA polymerases are very inaccurrate. It has been suggested that human pol i may participate in hypermutation of immunoglobin variable genes. Polymerase ϕ of yeast may be required for synthesis of ribosomal RNA.

***Poly (ADP-ribose).*** A eukaryotic peculiarity, which is not well understood, is the synthesis of poly(ADP-ribose) chains attached to many sites in nuclear proteins. Increased synthesis is observed following damage to DNA. The poly (ADP-ribose) polymerase binds to DNA near strand breaks or nicks and, using $NAD^+$ as a substrate, synthesizes the highly branched polymer attached to a small number of nuclear target proteins.

## MUTAGENS IN THE ENVIRONMENT

More than 500 new chemicals are introduced into the environment industrially each year. Some widely used drugs, e.g., **hycanthone**, are mutagenic. Powerful mutagens are present naturally in some foods. Others have been added through ignorance. Although many of these have now been removed, the problem cannot be ignored.

One way in which chemical compounds can induce base substitution mutation is through

their incorporation into the structure of DNA itself. Thus, 5-bromodeoxyuridine (or bromouracil) can replace thymidine in DNA, where it serves as an efficient mutagenic agent. 2-Aminopurine, an analog of adenine, pairs with thymine, just as does adenine when incorporated into DNA.

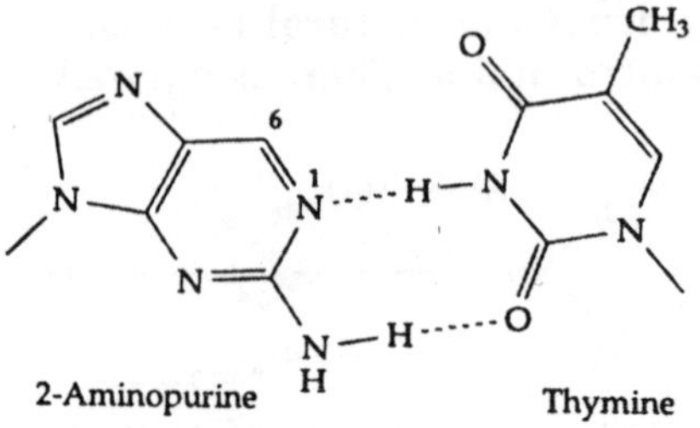

2-Aminopurine Thymine

However, if it protonates on N-l, it can pair with cytosine, causing mutation.

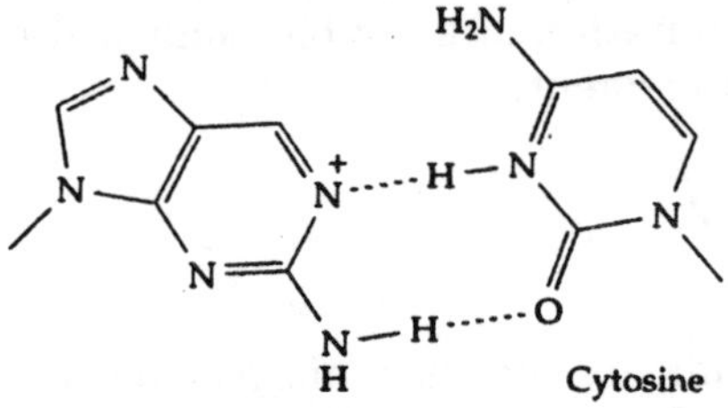

Cytosine

Another mutagenic base is $N^6$-hydroxylaminopurine.

Many alkylating agents are powerful mutagens. Alkylation can occur at many places in DNA, but the N-7 position of guanine is especially susceptible (Eq. 13.18). The resultant positive charge on the imidazole ring portion of the alkylated guanine causes hydrolysis of the N-glycosyl linkage and depurination. However, this may be a lethal rather than a mutagenic event. The previously mentioned methylation on O-6 of guanine is probably more important in inducing mutations.

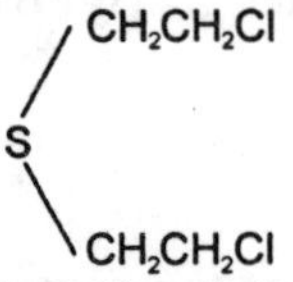

$S(CH_2CH_2Cl)_2$

bis-(2-chloroethyl) sulfide

Among the most biologically reactive alkylating agents are the nitrogen and sulfur "mustards" such as bis-(2-chloroethyl)sulfide. These toxic bifunctional compounds cause lethal crosslinking of DNA chains. The monofunctional half-mustards are mutagenicfaut are less acutely toxic. Another group of alkylating agents, the **nitrosamines**, are highly mutagenic. Much used in the laboratory as a mutagen is **N-methyl-N'-nitro-N-nitroso-guanidine:**

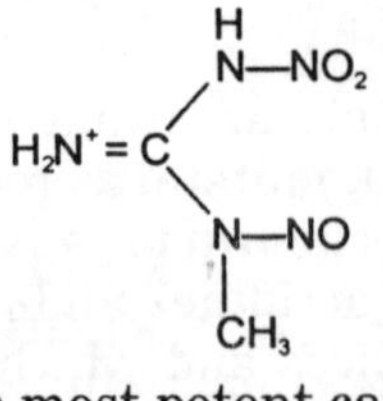

$H_2N^+=C(NH-NO_2)(N(CH_3)-NO)$

N-Ethyl-N-nitrosourea is one of the most potent carcinogens known. These compounds, as well as diazomethane and other related substances, probably act via a common intermediate: $CH_3N_2^+$ or $C_2H_5N_2^+$

Any secondary amine will react with nitrous acid to form a nitrosamine (Eq. 13. 25). Tertiary amines can also react with loss of one alkyl group. This can occur in the stomach, and the nitrosamines may be absorbed into the system. All plants contain some nitrate and some, such as spinach and beets, have large amounts. Bacon and other cured R' R meats contain both nitrites and nitrates, and many drugs and natural food constituents are secondary amines. Cigarette smoke also contains a nitrosamine. There is a possibility that these substances may induce human cancer.

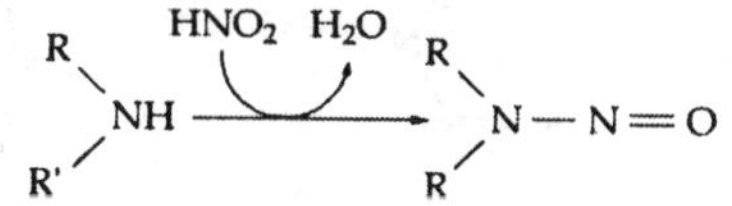

$$R(R')NH \xrightarrow{HNO_2 \quad H_2O} R(R)N-N=O$$

Nitrosamine

About 3% of residents of North America excrete feces that contain mutagenic unsaturated glyceryl ethers, which have been named fecapentaenes. They could be among the causative agents of colorectal cancer. Protonation of a double bond will produce a reactive carbocation, which may be an active alkylating agent.

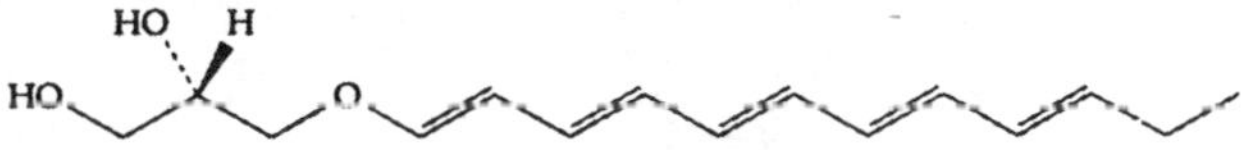

HO H
HO O

A fecapentaene

Many halogenated compounds are carcinogenic. Among these is 1,2-dibromoethane, which has been produced in the United States in quantities as great as $10^8$ kg/year. It has been widely used as a fumigant for foods, as an industrial solvent, and as an additive to gasoline. It is a **procarcinogen,** whose carcinogenic properties are expressed only after **metabolic activation.** One pathway of activation is reaction with glutathione to form a thioether, which can cyclize to a sulfonium compound. The latter alkylates nucleophilic groups in DNA. Other bifunctional electrophiles such as acrolein, malondialdehyde, vinyl chloride, and urethane are procarcinogens. Styrene, another procarcinogen, is converted to the carcinogenic styrene oxide by action of a cytochrome P450. Metabolites of glucose such as **pyruvaldehyde, methylglyoxal,** and other reactive aldehydes can also attack DNA.

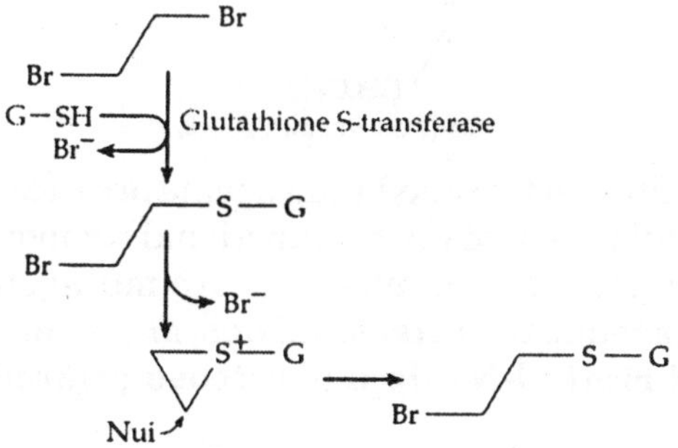

Br
Br
G−SH
$Br^-$
Glutathione S-transferase
S—G
Br
$Br^-$
$S^+$—G
Nui
S—G
Br

Less common than base pair exchanges are frame-shift mutations. A characteristic of such mutations is that they do not revert (back mutate) as readily as do base substitution mutations, and reversion is not induced by chemicals known to cause base substitutions. However, reversion of frame-shift mutations is induced by acridines and other flat molecules that are known to act as intercalating agents in DNA helices and which promote frame-shift mutations. They are especially effective in causing mutations of regions in which long repeated sequences of a single base such as $(A)_n$ or $(G)_n$ occur. For example, deletion of two base pairs from a "hot spot"

(site of frequent mutation) in the *Salmonella* histidine operon with the following sequence is induced by 2-nitrosofluorene and causes reversion of a (–1) histidine-requiring mutant: 5′-CGCGCGCG. Whereas simple intercalating agents are often not very mutagenic, compounds that are both an intercalating agent and an alkylating agent are especially potent. An example is the following compound, which contains a half-mustard side chain:

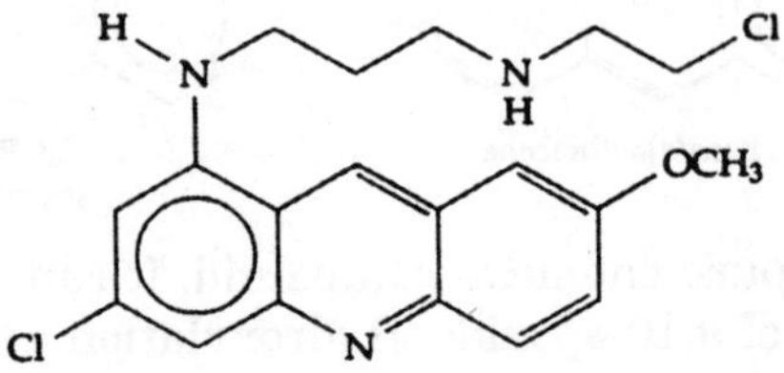

When the $CH_2Cl$ group of the side chain is replaced by $CH_2OH$, the compound is 100 times less mutagenic.

The carcinogenic **aflatoxins,** which are produced by *Aspergillus flavus,* may be present in infected pea-nuts and other foodstuffs. Like many other compounds that are carcinogenic or

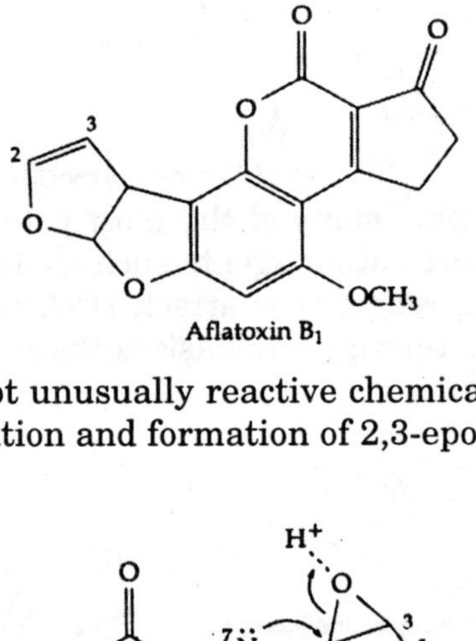

Aflatoxin $B_1$

mutagenic, the aflatoxins are not unusually reactive chemically. However, they are activated in the animal body by hydroxylation and formation of 2,3-epoxides. The latter may react with N-7 of guanyl residues in DNA.

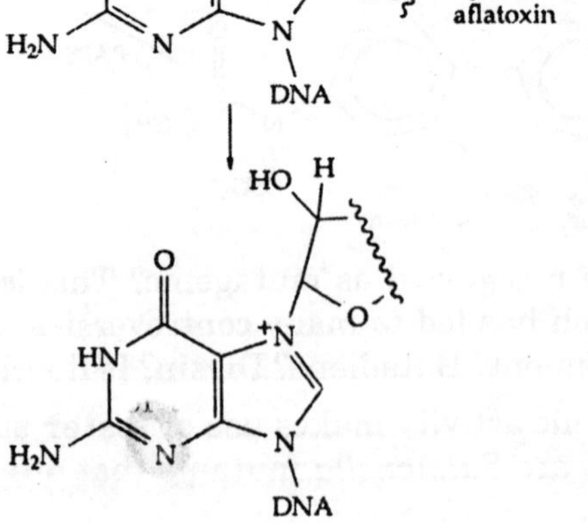

Epoxide of aflatoxin

Benz(*a*)anthracene is also converted in the body to a carcinogenic epoxide. Benzo(a)pyrene was isolated from coal tar in 1929 and in 1930 provid-ed the first demonstration

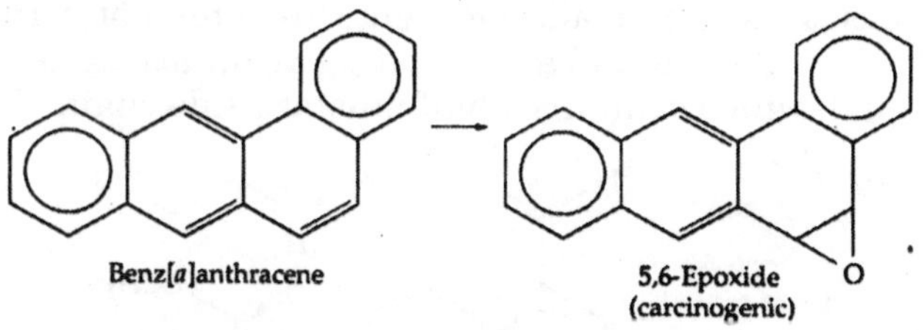

Benz[*a*]anthracene

5,6-Epoxide (carcinogenic)

of the carcinogenicity of a pure chemical compound. It can also be activated by conversion in the ER to the 7,8-dihydrodiol 9,10-epoxide. Hydroxylation activates 2-acetylaminofluorene to a carcinogenic N-sulfate, which reacts with guanine rings.

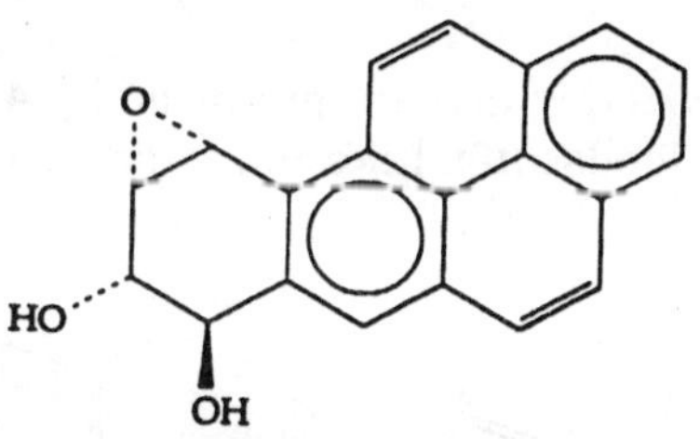

Benzo(*a*)pyrene diol epoxide

Although it seems incongruous, many of the most potent antitumor drugs are powerful mutagens. Among these are intercalating agents such as daunomycin, neocarzinostatin, and bleomycin. These are alkylating reagents or attack DNA in other ways. The fact that such compounds are in use for chemotherapy emphasizes the need for new approaches to cancer treatment.

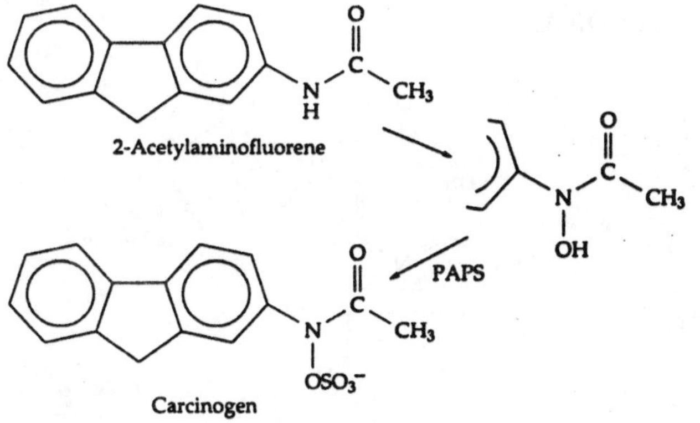

2-Acetylaminofluorene

PAPS

Carcinogen

How can compounds be recognized as mutagenic? This is an important question that is difficult to answer and which has led to many controversies. For example, how dangerous is formaldehyde in the environment? Butadiene? Dioxin? Is fluoride a carcinogen?

A quick test for mutagenic activity makes use of **tester strains** of bacteria developed by Ames and associates. These are *Salmonella* mutants that are unable to synthesize their own

histidine, but which can grow when a mutagenic agent produces a back mutation. One of the strains can be mutated by agents causing base exchanges, while the other three, which contain different types of frame-shift mutations, are affected differently by various mutagens. About $10^9$ bacteria are spread on a Petri plate, and a small amount of the mutagenic chemical is introduced in the center of the plate. Where back mutation has occurred, a colony of the bacteria appears. The strains all carry a mutation in the main DNA excision-repair system so that most mutations are not repaired, and the test is very sensitive. Addition of a liver homogenate plus an NADPH-generating system to the sample tested allows activation of many aromatic chemicals by hydroxylation. Feeding of mutagens to *Drosophila* eye color mutants permits testing for back mutation in a eukaryote.

The bacterial tests have been widely used and have been of great value. For example, they revealed that certain chemicals that were being used as flame retardants in children's clothing are mutagens, and that mutagens can be generated during cooking of meat and other foods. However, there are good reasons for using other methods of identifying mutagens together with the bacterial tests. Government regulatory agencies in the United States have relied largely on tests with rodents. Compounds are tested at very high doses with these short-lived animals. Extrapolation to human exposures at very low levels has been criticized. However, rodent tests have identified many true carcinogens. A broader range of tests are now in use. Direct monitoring of the accumulation of defects in animal and human cells is now possible. For example, in the **$^{32}$P-postlabeling technique** DNA is enzymatically digested to the nucleotide level, and the adducts with mutagens are labeled with $^{32}$P, separated, and their quantities measured. Laser-excited fluorescence from such adducts, GC / mass spectrometry, and immunological methods can also be used to identify DNA adducts. Careful vigilance is needed to keep our mutation rate at as low a level as possible.

## A DEFICIENCIES IN HUMAN DNA REPAIR

Problems with human repair systems account for several serious diseases. One is the rare skin condition **xeroderma pigmentosum** (XP), an autosomal recessive hereditary defect. Homozygous individuals are extremely sensitive to ultraviolet radiation and have a high incidence of multiple carcinomas. Defects in at least seven human genes, listed in the accompanying table, can cause XP. Most of the genes have been cloned and found to be homolougs of genes for nucleotide excision repair (NER). In rodents the *ERCC* genes have the same functions as do genes for resistance to ultravi-olet light in yeast *(RAD* genes). This information has elucidated the functions of the XP genes and the basis for this group of human diseases.

The XP-A protein (see table) is a zinc-finger-containing DNA-binding protein, which interacts with human replication protein A. It may also interact with proteins corresponding to the rodent ERCC1 and ERCC4 proteins to form a complex that recognizes DNA defects. Proteins XPC and XPE may be specialized DNA binding proteins for cyclobutane dimers and other photoproducts· The XPG protein is an endonuclease that probably cuts on the 3′ side of the DNA defect. In human cell lines with a variety of XP mutations the DNA problems have been corrected by transfer of the corresponding normal genes into the cells.

While defects in protein XPD often cause typical XP symptoms, some defects in the same protein lead to **trichothiodystrophy** (TTD, brittle hair disease). The hair is sulfur deficient, and scaly skin mental retardation, and other symptoms are observed. Like their yeast

counterparts (proteins RAD3 and RAD25), XPB and XPD are both DNA helicases. They also constitute distinct subunits of the human transcription factor TFIIHP, which is discussed in. It seems likely that XPD is involved in **transcription-coupled repair** (TCR) of DNA. This is a subpathway of the nucleotide excision repair (NER) pathway, which allows for rapid repair of the transcribed strand of DNA. This is important in tissues such as skin, where the global NER process may be too slow to keep up with the need for rapid protein synthesis. Transcription coupled repair also appears to depend upon proteins CSA and CSB, defects which may result in the rare **cockayne syndrome.** Patients are not only photosensitive but have severe mental and physical retardation including skeletal defects and a wizened appearance.

In a variant form of XP, designated XP–V, nucleotide excision repair is normal, but DNA replication is very slow. Postreplicational translesional repair (bypass repair) is also slow, and patients are cancerprone. The recently discovered DNA polymerase $\eta$ may be defective.

Children with **ataxia telangiectasia (AT)** have progressive neurological problems, a weak immune system, premature aging, and a high incidence of cancer. Their skin fibroblasts are deficient in the ability to repair X-ray damage, which causes many double-strand breaks: Apparently in this disease cells do not wait until DNA repair has been carried out after exposure to ionizing radiation but attempt to replicate the damaged DNA. The defective protein ATM (ataxia telangiectasia, mutated) has been identified as a large 370-kDa Ser/Thr protein kinase with a carboxyl terminal domain similar to phosphatidylinositol 3-kinase. It appears to play a crucial role in the cell cycle DNA damage check-points by participating in the detection of double-strand breaks and in delay of replication, while they are repaired by homologous recombination. Although AT is an autosomal recessive disease, women heterozygous for the ATM gene have an increased susceptibility to breast cancer. This observation led to the discovery that the proteins encoded by the well known breast cancer genes *Brcal* and *Brcn2* form a complex with ATM. Phosphorylation of BRCA1 by ATM may initiate a signaling pathway through the p53, c-Abl, and Chk2 proteins that cause cell cycle arrest. BRCA1 and BRCA2 also form a complex with protein RAD51, a RecA homolog necessary for homologous recombination. BRCA1 also may be essential to transcription-coupled repair.

There are two major forms of hereditary susceptibility to colon cancer. Familial adenomatous polyposis is caused by defects in the *APC* gene. The more common **hereditary non-polyposis colorectal cancer** (HNPCC), which includes many endometrial, stomach, and urinary tract tumors, results from defects in DNA mismatch repair. The proteins hMSH2 and hMSLl are homologs of the *E.coli* MutS and MutL.

Cells of patients with **Bloom syndrome** (BS) have many chromosome breaks and a high frequency of sister chromatid exchanges, perhaps in an effort to correct these breaks. The body is small but well-proportioned. A somewhat similar disease, the **Werner syndrome** (WS), is associated with premature aging. The Bloom's protein BLM and the WS gene product WRN are both helicases related to *E.coli* RecQ. Protein BLM colocalizes with replication protein A as discrete foci in the meiotic synaptonemal complex. Protein WRN also seems to be associated with DNA replication. Defects appear to increase homologous and illegitimate recombination. Both proteins may also function in transcription.

Many other diseases leading to a high incidence of cancer are known. Among them is the Nijmegan breakage syndrome, in which chromosomes are hypersensitive ro oreakage by

ionizing radiation. The gene has been identified by positional cloning, and its protein is apparently involved in repair of double strand breaksPP **Fanconi anemia, Gardner syndrome,** and hereditary **retinoblastoma** may also involve defects in DNA repair.

**Some Human Hereditary Defects of DNA Repair**

| *Human diseease* | *Human gene invloved* | | *Yeast gene* | *Function* |
|---|---|---|---|---|
| **Xeroderma pigmentosum (XP)** | | | | |
| XP – A | XPA | | RAD14 | DNA-binding, damage recognition |
| XP–B | XPB | ERCC3 | RAD25(SSL2) | DNA helicase |
| XP–C | XPC | | RAD4 | DNA-binding, thymine dimers |
| XP–D | XPD | ERCC2 | RAD3 | DNA helicase |
| XP–E | XPE | | RAD16 | |
| XP–F | XPF | | RAD10 | DNA nuclease complex |
| XP–G | XPG | | RAD2 | DNA nuclease |
| XP–V | XPV | | RAD30 | DNA polymerase η |
| **Cyckayne sundrome (CS)** | | | | |
| CSA | CSA | | | Interaction with helicase CSB and with TFIIH |
| CSB | CSB | | | DNA helicase |
| **Tricthiodystrophy (TID, brittle hair disease)** | | | | |
| | XPDE | | | |
| **Ataxia telangiectasia (AT)** | | | | |
| | ATM | | | Cell cycle delay for repair of ds breaks |
| **Human nonpolyposis colorectal cancer (HNPCC)** | | | | |
| | hMSH2 | MSH2 | Mismatch repair | |
| | hMSL1 | | MSL1 | |
| | GTBP | | | |

CHAPTER

# 14

# Cell Death

CYP2E1, an ethanol-inducible form of P450, metabolizes and activates many toxicologically important substrates, including ethanol, carbon tetrachloride, acetaminophen, and N-nitrosodimethylamine to more toxic products. Induction of cytochrome P450 2E1 by ethanol appears to be one of the central pathways by which ethanol generates a state of oxidative stress.

In the intragastric model of ethanol feeding, significant alcohol injury occurs. In these models, large increases in lipid peroxidation have been observed, and the ethanol-induced liver pathology has been shown to correlate with CYP2E1 levels and elevated lipid per-oxidation, and to be blocked by inhibitors of CYP2E1. Although there is disagreement of the role of CYP2E1 in alcohol liver injury, convincing evidence has been presented that CYP2E1 in the hepatocyte is a key source of reactive oxygen production and in DNA damage. Anapproach that our laboratory has used to try to understand basic effects and actions of CYP2E1 is to establish cell lines that constitutively express human CYP2E1 . HepG2 cell lines, which overexpress CYP2E1 , were established either by retroviral infection methods (MV2E1-9 cells, or E9 cells) or by plasmid transfection methods (E47 cells).

Hepatotoxins such as CC14 or acetaminophen were more toxic in E9 cells than control HepG2 cells validating the utility of the model to study CYP2E1-dependent toxicity. We have characterized the toxicity of ethanol, polyunsaturated fatty acids (PUFA) such as arachidonic acid (AA) and iron in the E9 and E47 cell lines. Concentrations of ethanol or AA or iron that were toxic to the CYP2E1 -expressing cells had no effect on control HepG2 cells not expressing CYP2Elor to HepG2 cells expressing a different P450, CYP3A4 (3A4 cells). Toxicity to CYP2E1 -expressing cells was found when glutathione (GSH) was depleted by treatment with 1-buthionine sulfoximine. Inhibitors of CYP2E1 prevented the toxicity by the aforementioned treatments. Antioxidants such as vitamin E, trolox, and catalase also prevented toxicity. The aforementioned treatments of CYP2E1-expressing cells with ethanol, AA, iron, or BSO resulted in an increase in oxidative stress to the cells

In other experiments, damage to mitochondria, e.g., decline in mitochondrial membrane potential, was found in CYP2E1-expressing cells treated with AA or iron or BSO. Please refer to Caro and others for recent reviews of this work and for CYP2E1 biochemistry and toxicology. This chapter describes protocols used with the HepG2 cells developed to express CYP2E1 to study CYP2E1-dependent toxicity, mitochondrial damage, and oxidant stress. Assays for CYP2E1

expression, for loss of cell viability, for apoptosis, for decline in mitochondrial membrane potential, for lipid peroxidation and generation of reactive oxygen species (ROS), and for activation of mitogen activated protein kinases (MAP kinases) when the CYP2E1-expressing HepG2 cells are challenged with pro-oxidants will be described herein.

## MATERIALS

Unless otherwise mentioned, solutions are prepared in distilled, deionized water and, except for buffers, prepared fresh.

### Cell Line Maintenance ana Storage

1. Dulbecco's Modified Eagle's Medium (DMEM, Sigma, St Louis, MO), supplemented with 10% fetal bovine serum (FBS, Sigma, St Louis, MO) and 1% penicillin-streptomycin-glutamine (GIBCO, Grand Island, NY).
2. Trypsin (0.25%) and ethylenediamine tetraacetic acid (EDTA; 1 *mM)* from GIBCO.
3. G418 antibiotic (GIBCO, Grand Island, NY).
4. Cell freezing medium, DMEM containing 20% FBS, 10% dimethyl sulfoxide (DMSO, Sigma), 1% penicillin-streptomycin-glutamine mixture.

### CYP2E1 Oxidation Activity

1. 1*M* $KH_2PO_4/K_2HPO_4$ buffer, pH 7.4.
2. 10 mM/p-nitrophenol (FLUKA) in water.
3. 10 mM NADPH (Sigma) in water.
4. 20 % (W/V) Trichloroactic acid (Sigma).
5. 10N NaOH.

### Cell Toxicity

1. MTT Test Solution, Thiazolyl' Blue Tetrazolium (Sigma), 3.3mg/mL in PBS, filter before use. N-propanol (Sigma).
2. LDH test solution: A. 50m*M* $KH_2PO_4/K_2HPO_4$, pH. 7.3; B. NADH, 4.1 mg/mL in solution A; C. pyruvic acid, 2.1 mg/mL in solution A. In 100mL LDH Test Solution: Solution A 94mL, Solution B 3mL, Solution C 3mL.
3. Trypan Blue Exclusion Test, 0.4 % Trypan Blue Solution (Sigma).

### Measurement of Malondialdehyde and ROS

1. Stock TCA-TBA-HCl reagent: 15% w/v trichloroacetic acid, 0.375 % w/v thio-barbituric acid, 0.25 N hydrochloric acid. 0.1m*M* butylated hydroxy toluene (BHT). This solution may be mildly heated to assist in the dissolution of the thiobarbituric acid.
2. Dichlorofluorescin diacetate: 5mA/ stock solution in DMSO (DCF-DA).

## Determination of GSH

1. Buffer A, 0.1 M sodium phosphate, pH 7.4, 5μ*M* EDTA, 0.6m*M* 5,5′-dithiobis (2- Nitrobenzoic acid) (DTNB, Sigma), 0.2m*M* NADPH (Sigma), pH 7.5.
2. Buffer B, 10U/mL glutathione reductase (Roche) in 0.1 *M* sodium phosphate buffer, pH 7.5.
3. 10% trichloroacetic acid (TCA).
4. Reduced-form GSH (Sigma).

## Apoptosis Analysis

1. Cell lysis buffer: 10mM Tris-HCl, pH 7.5, 10m*M* NaCl, 10m*M* EDTA, 100 μg/ mL proteinase K, and 0.5 % sodium dodecyl sulfate.
2. Chloroform/isoamyl alcohol (24:1) (Sigma), 0.1 μ.g/mL RNase (Sigma), μg/mL ethidium bromide (Sigma),
3. Rhodamine 123 (5 μg/mL final), propidium iodide (25 μg/mL final) (Sigma).

## *Caspase Activity*

1. Lysis buffer: 50m*M* Tris HCl, pH 7.4, 50m*M* β-glycerophosphate, 15mM $MgCl_2$, 15m*M* EDTA, 10 μ*M* phenylmethy sulfonyl fluoride; ImA/dithiothretol; 150 μg/mL digitonin.
2. Caspase 3—substrate II, AC-DEVD-AMC.
3. Caspase 8 substrate II, fluorogenic Z-IETD-AFC.
4. Caspase 9 substrate 1, fluorogenic AC-LEHD-AFC (CalBiochem).
5. Prepare 50m*M* stock solution in DMSO.
6. Reaction buffer: 100m*M* HEPES (pH. 7.5), 10% sucrose, 10m*M* dithiothreitol 0.5m*M* EDTA (Sigma).

## Mitogen- Activated Protein Kinase (MAPK) Phosphorylation

1. pp38/p38, p-JNK/JNK, p-ERK/ERK antibodies (Santa Cruz Biotechnology, Inc., Santa Cruz, CA).
2. p38 inhibitor SB203580, JNK inhibitor SP600125, ERK inhibitor U0126 (CalBiochem). Stock solutions of 5 mM dissolved in DMSO. Horseradish conjugated anti-IgG antibody (Sigma)
3. Enhanced chemiluminescence immune blot detection reagent (ECL, Amersham Biosciences Inc.).

## *Methods*

The HepG2 E47 cell line which constitutively expresses human CYP2E1 was established as follows. Human CYP2E1 cDNA, excised from a plasmid p91023(B)-2El (kindly provided by Dr. F. J. Gonzalez, National Cancer Institute, Bethesda, MD) was inserted into the EcoRI restriction site of pCI-neo expression vector (Promega, Madison, WI) in the sense and antisense

orientation to form the plasmids pCI-2E1 and pCI-as-2El. The empty pCI plasmid was used to prepare the control, non-CYP2El-expressing HepG2 C34 cells. Transfections of HepG2 cells were carried out with the use of LipofecAMINE reagent (Life Technologies, Grand Island, NY). Cells were then selected with 0.8mg/mL geneticin in the medium, colonies formed from the survivor cells were grown to large scale, and subjected to Western blot analysis and PNP oxidation activity assay to determine CYP2E1 activity. Positive clones were subjected to another two rounds of limited dilution screen-ing to create stable cell lines.

## *Cell Culture and Storage*

1. HepG2 E47 or HepG2 C34 cells are maintained in DMEM containing 10% FBS and 1% penicillin-streptomycin-glutamine mixture and 0.1 mg/mL G418. Cells are passaged when approaching confluence by splitting with trypsin/EDTA in 1:5 ratio to provide new maintenance cultures.
2. An aliquot of cells is centrifuged at 600g for 5min, the pellet is resuspended in the cell freezing medium at a concentration of $1 \times 10^6$ /mL and aliquoted into cryogenic vials in 1 mL each, and the vials are stored at –20°C for 2 hi and then at –80°C for 1 h and finally the vials are stored in a rack in liquid nitrogen.
3. When experiments are to be initiated, one or two frozen vials of cells are taken out from the liquid nitrogen and immediately put in a 37°C water bath for 5 min. Cells are centrifuged at 1000rpm for 5min, the supernatant is discarded and the cells are resuspended in DMEM in two 75-mL culture flasks. Cells are cultured for about 1 wk, and are now ready to set up new experiments.

## CYP2E1 Activity

1. When the culture of HepG2 E47 or C34 cells grown in the 75-mL flask approaches confluence, cells are harvested with a cell policeman and centrifuged 600g for 5min. The cell pellet is resuspended in 5mL of 0.15M KC1 and sonicated for 10s in an ice bath with a W-375 sonicator (50% duty cycle, output at 4 W) to obtain the cellular lysate.
2. Cell lysate is centrifuged at 9000 g for 15 min, and the supernatant is further centrifuged at 105,000 g for 1 h. Afterward, the microsomal pellet is collected.
3. The microsomal pellet is resupended in 200 mL of 0.15M KC1 with a miniho-mogenizer, and the protein concentration is determined by a protein assay kit from BioRad (Hercules, CA).
4. 100 µg of microsomal protein is incubated in 100- µL reaction buffer containing 0.1*M* $KH_2PO_4/K_2HPO_4$, pH 7.4, 0.4 *mM* p-nitrophenol, 1*mM* NADPH in a 37°C shaking water bath for 1 h.
5. At the end of this period, 30 µL of 20% TCA is added to stop the reaction and the mixture is centrifuged at 10,000 g for 10 min. The supernatant is collected and mixed with 10 µL of 10 *N* NaOH. The absorbance of the pink-yellow colour supernatant is determined in a spectrophotometer at 510nm, and the CYP2E1 activity is calculated using an extinction coefficient of $9.53 \times 10^5\ M^{-1}\ CM^{-1}$. The activity is expressed as pmoles/mg/min.

6. HepG2 E47 cell CYP2E1 protein content can be determined using a regular Western blot method with a CYP2E1 antibody.

## MTT Assay to Determine Cytotoxicity

1. HepG2 E47 and control C34 cells are seeded onto 24-well plates at concentrations of $1 \times 10^4$/1 mL DMEM / well. Cells are cultured for 24 h and the medium is replaced with fresh DMEM.
2. Arachidonic acid (30-60 µM, diluted with serum to prepare 5 *mM* stock solution, aliquot and keep in –20°C), BSO (0.1 mM), (prepare 0.1M stock solution with IX PBS), Ethanol (50-100mM) or iron (25 µM, prepared as Fe/nitrilotriacetate = 1:3 couple at concentration of 25 mM stock solution) are added to the cells for 24 h, or varying time points
3. Two hours before ending the treatment, 75 µL of MTT solution (tetrazolium salts) is added onto each well and continue to incubate at 37°C for 2h. The medium is then discarded followed by adding 1 mL of *n*-propanol to stop the metabolism, place in room temperature for 1 h in the dark.
4. The absorbance at 570 and 630 nm are recorded. The net $A_{570}$–$A_{630}$ is taken as the index of cell viability. The net absorbance from the wells of cells cultured with control medium is taken as the 100% viability value. The percent viability of treated cells is calculated by the formula $(A_{570} - A_{630})_{sample} / (A_{570} - A_{630})_{control} \times 100\%$.

## LDH Assay to Determine Cytotoxicity

1. HepG2 E47 or HepG2 C34 cells are seeded onto six-well plates at a concentration of 5 × 10V2mL DMEM/well for 24 h. The medium is replaced with fresh medium and treated.
2. At the end of treatment, the culture medium and the cells are collected separately.
3. The cells are resuspended in 0.5 mL of PBS and sonicated for 5 s in an ice bath (duty cycle 50%, output control 4W).
4. Mix 50 µL of medium or cell lysate with 950 µL of LDH reaction solution and measure LDH activity (A Absorbance per min) due to NADH formation by scanning in a spectrophotometer at 340 nm, 35°C for 1 min. The Cytotoxicity index is expressed as the ratio of $LDH_{medium}/LDH_{lysate + medium} \times 100\%$.

## Trypan Blue Exclusion Assay to Determine Cytotoxicity

1. HepG2 E47 or C34 cells are seeded onto six-well plates at a concentration of $5 \times 10^5$ 2mL DMEM per well for 24 h. The medium is replaced with fresh medium.
2. Treat the cells.
3. At the end of treatment, collect the medium and cells (trypsin/EDTA harvesting).
4. Collect cells by trypsinization and resuspend the cells in 1 mL of PBS and treat the cells with 0.1 mL of 0.05% trypan blue for 5min.

5. Count the excluding non-staining or staining cells in a hemocytometer under a light microscope. The cell toxicity is expressed as: Staining $_{total}$/Staining$_{total}$ + Excluding$_{total}$ × 100%.

## Lipid Peroxidation Analysis

1. HepG2 E47 or C34 cells are seeded onto six-well plates at a concentration of 5 × $10^5$ / well and cultured for 24 hours.Fresh medium is added and the cells are then treated with different toxins
2. Cells are harvested by trypsinization and resuspended in 0.5 mL of PBS containing 0.1 *mM* butylated hydroxytoluene (BHT) and sonicated for 10 s in an ice bath using an Ultrasonic Cell Disrupter (50% duty cycle, output at 6). Protein concentration of the lysate is determined as mentioned previously.
3. Mix 0.5 mg of lysate protein with 0.5 mL of trichloroacetic acid-thiobarbituric acid-HC1 solution (containing 0.1 *mM* BHT) and heat in a 100°C water bath for 1 h.
4. Centrifuge the reaction solution at 14,000rpm in a microcentrifuge for 10 min-utes and collect the supernatant.
5. The formation of TBA-reactive substances (TEARS) is determined by measuring absorbance at 535 nm and using an extinction coefficient of 1.56 × $10^5$M/cm to calculate malondialdehyde equivalents (nmoles/mg protein).

## Measurement of ROS Production

1. HepG2 E47 or C34 cells are seeded onto six-well plates at a concentration of 5 × $10^5$ / 2mL DMEM per well and cultured for 24 h. Replace with fresh medium and treat with different toxins as mentioned in Subheading 3.3., step 2.
2. One-half hour before ending treatment, 2′,7′-dichlorofluorescin diacetate is added at a final concentration of 2 μg/mL for 30min.
3. Cells are harvested by trypsinization and resuspended in 3mL I×PBS (avoid direct exposure to strong light).
4. Fluorescence is immediately determined in a Perkin-Elmer 650-105 fluorescence spectrometer using wavelengths of 490 nm for excitation and 525 nm for emission. Results are expressed as arbitrary units of fluorescence per milligram of protein.

## Measurement of GSH Levels

1. HepG2 E47 or C34 cells are seeded onto six-well plates at a concentration of 5 × 10 $^5$/ 2mL DMEM per well for 24 h and treated with different toxins.
2. Cells are scraped and harvested in 0.5mL Ix PBS and mixed with 0.5mL 10% TCA solution and vortexed briefly.
3. Centrifuge the cell mixture solution at 10,000 rpm for 5min and collect the supernatant.
4. In a 1-mL cuvette, add 900 μL of Buffer A, 100 μL of Buffer B and 50 μL of sample and determine absorbance at 412nm for 2min.
5. Prepare a standard curve with four different concentrations of reduced form of glutathione from 0 to 100 μM in 5% TCA and determine the absorbance at 412 nm for

2min as step 4 and evaluate the GSH levels (nmoles/mg protein) by comparing with the standard curve.

## Mitochondrial Membrane Potential Analysis

1. Mitochondria membrane potential is analyzed from the accumulation of rhodamine 123. Cells are seeded onto six-well plates at a concentration of 5 × $10^5$ well for 24 h and treated as detailed previously.
2. One hour before ending treatment, rhodamine 123 is added at a final concentration of 5 μg/mL for 1 h.
3. Cells are harvested by trypsinization and resuspended in 1 mL of PBS containing 25 μg/mL PI and the intensity of the fluorescence from rhodamine 123 and PI is determined by Fas-Scan Flow Cytometry. Cells with high mitochondrial potential and viable are mainly in the lower right quadrants of the PI vs rhodamine 123 curves.

## Apoptosis Determination by Flow Cytometry

1. E47 or C34 cells are grown in 6 well plates and treated with different toxins as mentioned previously.
2. Harvest the cells by trypsinization and fix with 70% ethanol for 15 min at room temperature.
3. Cells are washed with PBS and incubated with 50 μg/mL RNase A for 2h at room temperature.
4. Stain the cells with 50 μg/mL PI for 30 min and the fluorescence (excitation, 488 nm; emission, 575 nm) is measured with a flow cytometer. The cells undergoing apoptosis are calculated from the sub-G hypodiploid area programmed by the analyzer.

## Apoptosis Analysis by Determination of DNA Fragmentation (DNA Ladder)

1. E47 or C34 cells at concentration of 5 × $10^5$/2mL DMEM per well are grown on six-well plates and treated as detailed previously.
2. The cells are harvested and resuspended in 1 mL of'lysis buffer and incubated for 1 h at 50°C.
3. The lysis samples are then extracted with 2mL of phenol (neutralized with TE buffer, pH 7.5), followed by extraction with 1 mL of chloroform/isoamyl alcohol (24:1).
4. The aqueous supernatants are precipitated with 2.5 volumes of ice-cold ethanol plus 10% volume of 3M sodium acetate, pH 5.2, at –20°C overnight.
5. Centrifuge the samples at 13,000g for 10 min and the pellets are airdried and resuspended with 50μL of TE buffer, pH 7.5, supplemented with 0.1 μg/mL RNase A.
6. Samples are electrophoretically separated on a 1.5% agarose gel in 0.5 × TBE buffer containing 1 μg/mL ethidium bromide,at 100 V for 2h.
7. Pictures of the separated DNA on the gel are taken under UV transillumination

## Immunohistochemistry Analysis to Determine Nuclear DNA Fragmentation

1. Nuclear DNA fragmentation is determined by a DNA Fragmentation Detection Kit from CalBiochem. HepG2 E47 cells or C34 cells are grown on a cover slide in six-well plates and treated with different toxins as mentioned previously.
2. After treatment, the slides are fixed with 4% para-formaldehyde at room temperature for 15 min and are dehydrated by using ethanol (concentrations are from 70% to 95%).
3. Permeabilize the cells with proteinase K. Inactivate endogenous peroxidase by incubating the cells on the coverslide with 3 % $H_2O_2$.
4. The fragmented DNA is labeled with labeling solution of the kit and the labeled DNA is mixed with a conjugate solution provided by the kit.
5. Incubate the cells on the slide with 3, 3′-diaminobenzidine (DAB)/$H_2O_2$/urea solution (provide by the kit).
6. Nuclear DNA fragmentation is detected by stained dark-brown cell nuclei and the number of stained cells is counted under a light microscope.

## Caspase Activity

1. Cells are grown on six-well plates at the concentration of $5 \times 10^5$/2mL DMEM per well and treated with different toxins. Cells are scrapped and harvested.
2. The cell pellet is suspended in 300 µL of lysis medium and incubated for 30 min at room temperature.
3. The cell suspension is centrifuged at 200g for 5mins and the supernatant is collected.
4. 100µg of supernatant protein is added into 0.5 mL of the assaying solution containing 50 µ*M* caspase substrates (caspase 3- substrate II, AC-DEVD-AMC; casepase 8 substrate II or Granzyme B substrate II, Z-IETD-AFC; caspase 9 substrate I, fluorogenic, AC-LEHD-AFC respectively, CalBiochem) and incubated for 12h in a 30°C shaking water bath overnight.
5. Caspase activities are determined by measuring the. fluorescence associated with released AMC (7-amino-4-methyl-coumarin, excitation at 380nm, 'emission at 460 nm) or AFC (excitation at 400 nm, emission at 505 nm) in a spectrofluorimeter. Results are expressed as arbitrary units per 100(J.g cell lysate protein.

## Activation ofp38 MAPK

1. E47 or C34 cells are grown on six-well plates for 24 h, and fresh medium is added.
2. The cells are treated with either 60 µ*M* AA or 0.1 mAf BSO in the absence or presence of 5 µ*M* of the p38 MAPKJnhibitor SB203580 for 0.5, 1.0, 2.0, 4.0, 8.0, or 12h, respectively
3. Collect the cell pellets, resuspend in 200 µL of PBS and sonicate for 10s in an ice bath (50% duty cycle, output at 5 – 7 W).
4. Prepare a 8% SDS-polyacrylamide gel, loading 20 µg of each lysate protein onto the gel and carry out the gel electrophoresis for 2 h at 50 V.

5. The blotted membranes are incubated with pp38 MAPK antibody (Santa Cruz, 1:200 diluted) for 4h at room temperature (or overnight in the cold room).
6. Wash the blotted membranes with PBS containing 0.05% Tween 20 and incubate with anti-IgG antibody conjugated with horseradish peroxidase (Sigma, 1:10,000 diluted) for Ih.
7. Wash the membrane and develop the fluorescence using the enhanced chemilu-minescence immune blot detection agent. The arbitrary units of density of each lane are determined after scanning with computer software Image! (National Institutes of Health).
8. The blotted membranes are washed in a Re-Blot Plus Strong Solution for 15 minutes to wash out the previous bound pp38 MAPK antibody and re-incubate the membrane with p38 MAPK antibody, and repeat step 6 and 7 to determine the total p38 MAPK. The final pp38 MAPK activation levels are expressed as the ratio of pp38 MAPK/p38 MAPK.
9. Similar procedures can be carried out to determine the activation of JNK or ERK or Akt by use of specific antibodies recognizing these kinases.

## IMMUNOLOGICAL RESPONSES

Considerable efforts have been made to determine the mechanistic changes that are responsible for immune deficiency and autoimmunity in the chronic alcoholic. Alterations that could reflect regulatory disturbances involving some aspect of T-cell function include: (1) increased serum levels of IgA, IgG, and IgM; (2) reduced cell-mediated immunity such as a loss of DTH reactions; (3) reduced lymphocyte numbers, altered subset distributions, persistent T-cell activation, and increased CD4/CD8 ratio; (4) changes in cytokine levels, which can be acute or chronic. In work to characterize the T cells of chronic alcoholics, it was established that alcoholics have an increased percentage of activated CD8+ T cells, measured as HLADR surface expression.

Subsequent work by three- and four-colour flow cytometry showed that these patients also have a significant shift from a naive cell expression of the leukocyte common antigen (CD45RA+) toward a memory cell phenotype (CD45RO+). This change is present in both CD8+ and CD4+ T cells and is accompanied by a significant reduction of L-selectin and an increase in CD 1 Ib on the surface of CD8+ cells, and by a loss of L-selectin in some but not all patients' CD4+ T cells. Most alcoholics also have stably increased expression of the carbohydrate-rich marker, CD57. The CD57+ T cell subsets of both patients and controls respond to polyclonal stimulation through the TCR with a rapid burst of IFN-γ and TNF-∝ production, but not IL-4.

In addition, the CD57+ subset does not require a second signal for this production, whereas the CD57- subset does. These findings are consistent with the concept that this subset is a differentiated effector cell with cytotoxic potential and a TH1 immediate response cytokine profile. It has been demonstrated also that there is a substantial increase in rapid cytoplasmic IFNγ production in the T cells of chronic ethanol consuming mice, along with an increase in the memory phenotype in the activated T cells of these mice. Because there are also increases in IL-4 production, the findings do not exclude TH2 excess in the chronic ethanol mice, but it has not been possible to demonstrate any TH1/TH2 skewing in antigen-specific T cells of chronic ethanol mice.

Recently, it has been observed in our laboratory that in spite of the activated phenotype of the CD8+ T cells in these mice, the production of antigen-specific CD8+ cells during the primary response to the intracellular bacterium *Listeria monocytogenes,* is significantly reduced. The adequacy of both polyclonal and antigen-specific T-cell responses is of increasing interest as we attempt to understand the mechanisms of the 19 Polyclonal and Antigen-specific Responses of T Cells and T Cell Subsets immunodeficiency of chronic alcohol abuse. This chapter describes some of the basic techniques available for characterization of these responses.

# MATERIALS

## Isolation of T cells From Spleen, Thymus, or Lymph Node

1. Dulbecco's phosphate buffered saline (DPBS, Invitrogen, Carlsbad, CA).
2. RPMI-1640 media (Invitrogen, Carlsbad, CA).
3. Lysis buffer (lyses red blood cells in cell suspension): 155*mM* $NH_4Cl$, 1 *mM* Tris-HCl, pH 7.20-7.5. Mix 4.15 *g* of $NH_4Cl$ with 0.061 g of Trizma. Base (Sigma, St. Louis, MO.) to make 1L with reagent grade water. Adjust pH to 7.2-7.5. Filter to sterilize and store at room temperature (RT).
4. Trypan blue stain 0.4% (Invitrogen, Carlsbad, CA). Stain used to discriminate live cells from dead cells when counting cells in a hemocytometer.

## Column Fractionation of Spleen Cell Suspension

1. Magnetic Cell Sorting (MACS) Kit (Miltenyi Biotech, Auburn, CA). Kits contain biotinylated-antibody cocktail and streptavidin microbeads for cell separation. Negative-selection kits are available for: CD4+ cells (cat. no. 130-090-860), CD8+ cells (cat. no. 130-090-859), and B cells (cat. no. 130-090-862).
2. Column buffer (for the MACS separation system): 2mM ethylene diamine tetraacetic acid (EDTA), 0.5% bovine serum albumin (BSA) in DPBS. Place 1 liter of DPBS on stir plate with stir bar. Add 0.75 g of EDTA (Sigma, St Louis, MO) and stir until dissolved. Then add 5 g of biotechnology grade BSA (Amresco, Solon, OH) and stir until dissolved. Store at 4°C.
3. Columns and magnetic holders for cell separation (Miltenyi Biotech, Auburn, CA).
4. Biotinylated-anti CD45R/B220 antibody (clone # RA3-6B2) for B-cell depletion of splenocytes.
5. Streptavidin microbeads (Miltenyi Biotech, Auburn, CA) for B-cell depletion of splenocytes.
6. Lipopolysaccharide (LPS, Sigma, St Louis, MO). Purified from bacteria, lyophilized and sterilized by $\gamma$-irradiation.

## Flow Cytometry Staining

1. Flow cytometry wash buffer: PBS-0.1% azide (used for washing flow cytometry tubes). Add 1.Og of sodium azide (Fisher Scientific, Pittsburgh, PA) to 1 liter of DPBS. Store at RT.
2. Flow cytometry fixative (used to fix cells before flow cytometry): Mix 1 part 10% buffered formalin with nine parts flow cytometry wash buffer. Make fresh each week. Store at RT.

## Stimulation and Cytokine Response of Cells

1. Conjugation buffer (for immobilizing antibody on culture plates): Mix 6.055 g of Tris-HCl (Sigma, St. Louis, MO) with 0.9 mL of *2N* HCl (Sigma, St. Louis, MO) in 1 liter of reagent grade water. Adjust pH to 9.25-9.50. Filter to sterilize and store at RT.

2. Purified anti-CD3 antibody (clone #1445-2C11), (Becton Dickinson, Sparks, MD). Dilute antibody in conjugation buffer (1 μg/mL). Dilute antibody just before use. Allow antibody to conjugate to plate for 4 h at RT or overnight at 4°C.
3. Purified, azide-free, low endotoxin, anti-CD28 antibody (clone # 37.51; Becton Dickinson, Sparks, MD). Add 1 μg/mL to media in anti-CD3 positive control well after washing and before adding cells.
4. Culture medium - RPMI-10, Mix 500 mL of RPMI with L-glutamine, 50 mL of heat-inactivated fetal calf serum (HI-FCS, Invitrogen, Carlsbad, CA), 100 μL of 50 mM beta-mercaptoethanol (Sigma, St. Louis, MO), 100 μL of 50mg/ml Gentamicin (Invitrogen, Carlsbad, CA). Filter to sterilize and store at 4°C. Use 1.0 mL/well for a 24-well tissue culture plate.
5. BD Cytofix/Cytoperm™ Plus Kit (with BD GolgiPlug ™Protein Transport Inhibitor; Becton Dickinson, Sparks, MD).
6. Micro-porous tape (3 M, Minneapolis, MN).

## Antigen-Specific T Cells, Using Attenuated, L. Monocytogenes as a Model Organism

1. ACT A-*Listeria monocytogenes,* DPL 1942. The stock we used, was kindly provided by John Harty, University of Iowa for production of frozen stocks of attenuated Listeria.
2. Tryptic Soy Broth (TSB) Media for culture of, *L. monocytogenes:* Mix 30.0 g of BD Bacto™ Tryptic Soy Broth powder (Becton Dickinson, Sparks, MD) in 1 liter of reagent grade water. Warm to dissolve powder and autoclave. For plates, add 15.0 g of Bacto™ Agar (Becton Dickinson, Sparks, MD) per liter. Add streptomycin sulfate (Sigma, St. Louis, MO) to a final concentration of 50μg/mL.
3. *Listeria monocytogenes* Listeriolysin O molecule (LLO)-specific peptide (Bio-Synthesis, Inc., Lewisville, TX). The peptide sequence for the C57B1/6 CD4+ epitope used in our lab is NEKYAQAYPNVS (LLO AA# 190-201). For Balb/c the CD8+ epitope is GYKDGNEYI (LLO AA# 91-99).
4. Intracellular staining buffer for cytoplasmic staining. Flow cytometry wash buffer with 1% HI-FCS.
5. Sterile saline solution for injection of Listeria. 0.9% Sodium chloride, inj., USP, (Hospira, Inc., Lake Forest, IL).
6. Igepal for homogenization of Listeria-containing mouse organs. Igepal C630 (octylphenoxy)polyethoxyethanol (Sigma, St. Louis, MO). Igepal is a surfactant to help release Listeria from spleen/liver cells.
7. Variable speed tissue homogenizer to homogenize organs. Tissue Tearor Model 985-370 (Biospec Products, Inc., Bartlesville, OK).
8. P815 cell line - (ATCC # TIB-64™, cultured in RPMI-10).

## Adoptive Transfer

1. Pentamer wash buffer: 0.1% sodium azide, 0.1% biotechnology grade BSA in DPBS.

2. Pentamer fixation buffer: 1% HI-FCS, 2.5% formaldehyde (Fisher Scientific, Pittsburgh, PA) in DPBS.

## METHODS

## Isolation of T Cells From Spleen, Thymus, or Lymph Node

### *1. Organ Harvest*

Place animals in a $CO_2$ chamber until respiration ceases. Remove animals and perform cervical dislocation. Excise desired organs and place them in 5 mL of either buffer (DPBS) or culture medium (RPMI-10).

### *2. Mechanical Disruption*

Several alternative methods can be used :

(*a*) Transfer organs and buffer to a 60 × 15-mm Petri dish. Take two sterilized glass microscope slides with frosted ends. Place the organ on the frosted side of one slide and crush the tissue with the frosted end of the other. Spleens can be cut into two pieces to facilitate crushing.

(*b*) Place a cell dissociation sieve with a 60-mesh screen in a 100 × 15-mm Petri dish and transfer the organs and buffer to the sieve. Press the tissue through the screen using either a tissue grinder pestle or the plunger of a 10-mL syringe. The screen helps to sift out pieces of connective tissue. This method is especially effective when a number of organs are being pooled.

(*c*) Place a small volume of buffer in a glass tissue homogenizer. Grind tissue with the pestle to make a single-cell suspension.

### *2. Removal of Red Blood Cells*

Transfer the cell suspension to a 15-mL culture tube, rinsing with enough buffer to fill the tube. Pellet the cells for 5min at 250 RCF. Aspirate the media from the pellet. Resuspend the cell pellet in lysis buffer, using 1 ml of lysis buffer/organ and incubate at RT for 1-2 min. Rinse away the lysis buffer by adding approx 10 volumes of buffer or media.

### *3. Quick Spin*

Pull the chunks of connective tissue to the bottom of the tubes by centrifuging them until the force just reaches 70 RCF. Remove the tubes from the centrifuge and decant the supernatant into clean tubes. Pellet the cells in the new tube as before. Alternatively, cell suspensions .may be filtered through nylon mesh cell strainers (Becton Dickinson, Sparks, MD).

### *4. Count*

Resuspend the cells in a small volume (~1mL) of buffer or media. Determine the cell density by counting an aliquot of the cells diluted 1:100 with Trypan blue stain in a hemocytometer. The cells are now ready for culture as whole splenocytes or for further fractionation of the population. Save cells for lineage marker analysis if using flow cytometry ($10^6$ cells per tube).

## Column Fractionation of Spleen Cell Suspension

Cell suspensions may be fractionated to allow studies of purified populations or to enrich for rare cell types. The refinement of magnetic bead separations allows easy fractionation of T and B cells with 90%+ purity. Cells are isolated by either positive or negative selection. The positive selection methods can result in activation of the selected cells, which is often an undesirable effect in culture.

Briefly, cells are incubated with a biotinylated antibody cocktail followed by antibiotin microbeads that react with the antibodies, attached to the cells. Elution of the cell/bead suspension over a column in a magnetic field retains the labeled cells on the column allowing the un-labeled (negatively selected) cells to pass unimpeded. The column is then removed from the magnetic field and the labeled (positively selected) cells are eluted, if desired.

### *Column Purification of CD4+ T Cells by Negative Selection*

### 1. Antibody Incubation

Prepare single-cell suspension as discussed in Subheading 3.1. Suspend the cells in column buffer with 40mL of buffer for every $10^7$ total cells. Add 10mL of biotinylated-antibody cocktail from kit (Miltenyi cat. no. 130-090-860) for every $10^7$ total cells and incubate at 4°C for 10 min.

### 2. Microbead incubation

Add an additional 30µL of column buffer and 20 µL of anti-biotin microbeads (provided in kit) for every $10^7$ cells. Mix and incubate 15min at 4°C.

**3.** Wash cells in approx 10x volume of column buffer and pellet by centrifuging for 5 min at 250 RCF. Resuspend them in degassed column buffer with 500 µL of column buffer for every $10^8$ cells.

### 4. Magnetic column elution

Use one Miltenyi LS size column for every $10^8$ cells. Place the columns in the proper magnetic holder and pre-wet columns with 3 mL of de-gassed column buffer, discarding the eluate. Load $10^8$ cells (0.5 mL) on each column and let them run through, collecting the eluate. Wash the column four times using 3 mL of de-gassed column buffer in each wash. Pellet and resuspend the cells in the desired buffer, pooling tubes as needed. The negatively selected cells are then ready for culture or further treatment. Take aliquots of $10^6$ cells for lineage marker analysis if using flow cytometry to determine the purity of the cell fractionation. If positively selected cells are to be collected, remove the column from the magnet holder and elute the retained cells into a clean tube with four 3 mL.washes of column buffer. Otherwise, discard the used column.

### *Column Purification of CD8+ T Cells by Negative Selection*

Follow the same protocol as the CD4+ purification given above (Miltenyi Biotech Inc., cat. no. 130-090-859).

### *Column Purification of B Cells with Miltenyi B-Cell Kit; Overnight Activation*

1. Antibody incubation: prepare single-cell suspensions of splenocytes from two or three norm alm ice.Resuspend cells in colum n buffer with 40 µL of buffer and 10µL of antibody

cocktail from kit (Miltenyi Biotech Inc., cat. no. 130-090-862) for every $10^7$ cells. Incubate at 4°C for 10 min.

2. Microbead incubation: Add an additional 30 μL of column buffer and 20 μL of anti-biotin microbeads for every $10^7$ cells. Incubate 15 min at 4°C. Wash once with approx 10x volumes of column buffer. Resuspend cells in de-gassed column buffer at 500μL/$10^8$ cells.
3. Magnetic column elution: prepare column and elute cells.
4. Resuspend purified cells in RPMI-10 at $2 \times 10^6$ cells/mL and add 20 μg/mL LPS. Incubate overnight at 37°C, 5% $CO_2$.
5. The next day, wash cells twice with DPBS or RPMI to remove LPS. Resuspend in RPMI-10. Cells are now ready for peptide loading.

## *B-Cell Depletion of Splenocytes*

### 1. Antibody incubation

Prepare a single-cell splenocyte suspension as discussed in Subheading 3.1. Resuspend the cells in RPMI or DPBS without serum. Add biotinylated-anti-CD45R/B220 antibody at a concentration of 1 μg antibody/106 spleen cells. Incubate 20 minutes at 4°C.

### 2. Microbead incubation

Wash away excess antibody with approx 10x volumes of RPMI or DPBS by pelleting cells once. Add streptavidin microbeads (Miltenyi Biotech Inc., cat. no. 130-048-101) with 10 μL of beads for every 107 cells and incubate 15 min at 4°C to allow the attachment of labeled cells to the beads. Wash once with approx 10x volumes of column buffer. Resuspend cells in de-gassed column buffer at 500μL/108 cells. 3. Magnetic column elution : Prepare column and elute cells.

## Flow Cytometry Staining

Staining lineage markers on the cells is used to demonstrate both that the starting cell population was as expected and that any fractionation of the cells was successful. The markers used may vary from experiment to experiment. Markers to document the makeup of the donor animals would include markers for the various cell line-ages: T cells (e.g., CD3, CD4, CD8), B cells (e.g., CD19), macrophages and DCs (e.g., CD11b, CD11c), etc., as determined by the aim of the experiment. The same panel of stains is used on any purified or enriched cell suspension to demonstrate that the fractionation was successful.

### 1. Stain cells

Place $10^6$ cells in a 12 × 75-mm polystyrene tube in a volume of approx 100μL. Add the antibodies to the cells. Antibodies either can be from commercial sources or can be locally produced from hybridoma lines. Commercially produced antibodies may have a recommended working concentration; however, the optimum concentration of antibody should be determined by titrating the response of the cells used in any particular laboratory.

### 2. Incubate Stain

Hcubate cells with antibodies in the dark at either RT or 4°C for 15-20min.

### 3. Wash

wash to remove excess antibody with 1 ml of flow cytometry wash buffer. Pellet cells for 5 min at 250 RCF.

### 4. Secondary Antibody

If a biotin-labeled primary antibody is used, add the fluorescent-labeled-streptavidin conjugate in wash buffer at 1 μg/$10^6$ cells and incubate for 20 min. Wash as above.

### 5. Fix

Resuspend cells in 0.3 mL of flow fixative. Store at 4°C in dark until ready to analyze.

## Stimulation and Cytokine Response of Cells After Polyclonal Stimulation

The response of cells is measured by exposing, a controlled population of cells to defined stimuli in culture. Positive and negative control conditions are included to aid in determination of the cells' response to the test stimuli. The volume of cells to be used in culture is determined by the parameters being measured and by the number of cells available. Cytokine production provides one means to quantify the functional response of the cells being tested. Cytokine production can be quantified by measuring either the total (soluble) cytokines secreted or the intracellular cytokines.

Intracellular staining allows the identification of the cell type producing the cytokines within a given 4-6 hour period of stimulation. Brefeldin A prevents the release of cytokines from cells by blocking protein transport from the endoplasmic reticulum (ER) to the Golgi apparatus. The number and subset of cells producing cytokines can be identified using combined cell surface marker and cytoplasmic staining.

### 1. Preparation of the Culture Plate

Volumes (1.0mL/well) and concentrations given are for a 24-well culture plate. Prepare the plates by adding a polyclonal stimulator, such as anti-CD3 antibody, to the positive-control wells. Immobilized monoclonal anti-CD3 antibody (clone #145-2C11) activates T cells in culture in a polyclonal fashion providing a good positive control. Dilute the antibody to a working concentration of 1 μg/mL with conjugation buffer and pipet into (positive control) culture wells (1.0mL/well). Seal with Parafilm® (Pechiney Plastic Packaging, Menasha, WI) and place plate at 4°C overnight or keep at RT for 4h. Wash the wells.

Do not allow wells to dry out. After the final wash of the anti-CD3 wells, add 0.5 mL of culture medium to *all* wells that will have cells. If intracellular cytokine levels will be measured, use media containing 2μl/mL BD GolgiPlug™(BD Cytofix/Cytoperm™Plus kit, Becton Dickinson, Sparks, MD) and place 0.5 mL in the plates/wells for those samples. The final Brefeldin A concentration in the wells after the addition of cells will be 1 μg/mL. The cells being cultured can be either from whole organ preps (e.g., whole splenocytes), B-depleted or purified T cells (CD4+ or CD8+).

2. Adjust the concentration of cells to be cultured to 4.0 × $10^6$/mL and add 0.5 mL of the cell suspension to the wells (final concentration of 2 × $10^6$ cells/mL). Cover and seal lid to plate with microporous tape (3M, Minneapolis, MN.). Incubate for the desired time period. Cells and/or culture supernatants may be harvested for assay at the end of the incubation period.

3. For supernatant collection, pipet the liquid contents of each well into a centrifuge tube. Centrifuge the tubes to pellet any cells and then transfer the supernatants to tubes. Freeze at –80°C for cytokine ELISA or Cytokine Bead Array (CBA) analysis.
4. To collect cells resuspend them with a Pasteur pipet. Transfer the suspended cells to a centrifuge tube. Wash the well twice with the same volume of media to increase cell recovery. Centrifuge to pellet the cells and discard the supernatant. The cells are then ready for cell surface marker staining, cytoplasmic staining, or RNA/DNA isolation, etc.

## (Antigen-Specific T Cells, Using Attenuated, L. Monocytogenes as a Model Organism

Antigen-specific T cells can be induced in the experimental animal model for study of adaptive immunity in the experimental animal encompassing primary, secondary and memory responses. Inoculation of the experimental animal with an infectious organism (either virulent or attenuated) or with an isolated immunogenic molecule produces an antigen-specific response that can be measured later by stimulating cells in culture with the infectious organism, a specific molecule or a peptide epitope. Any cell preparation method discussed above may be used, depending on the aim of the experiment.

The cell preparation may be from either individual mice or pooled populations. Both the economics and logistics of cell population purification discourage processing a large number of samples. Listeriolysin O (LLO) T-cell peptide epitopes have been identified that can be used to measure T-cell specific responses. The peptides used in our lab are targeted to CD4+ or CD8+ T cells in C57B1/6 or Bal/6 mice, respectively. The specific peptide chosen must be titered with the cell system being studied to determine the optimum concentration for culture.

### *Inoculation*

ACT A - *Listeria monocytogenes* (DPL 1942), may be used to infect mice and elicit a measurable immune response to known specific peptide epitopes. ACT A - Listeria has been engineered so that the intracellular bacterium is unable to nucleate host cellular actin and therefore is incapable of moving from cell to cell to amplify the infection; it is thus less likely to kill the host.

1. Preheat a shaking incubator to 37°C. Prepare a 50-mL culture tube with 10 mL of TSB media containing 50 μg/mL streptomycin. Remove 1 mL of media and save to use as the blank in a spectrophotometer. Thaw a 1-mL aliquot of ACT A- Listeria and add it to the 9 mL of media in the culture tube. Incubate in the pre-warmed shaking incubator for approx 1 h. After 30 min, the $OD_{600}$ should be at least 0.01. Continue monitoring $OD_{600}$ until it is 0.06-0.1 (log phase). Remove the tube from the incubator and hold it at RT. An $OD_{600}$ of 1.0 = $10^9$ CFU/mL.
2. Calculate the amount of inoculum to be prepared. Example: Inoculation of 10 mice with 0.3-mL doses of $5 \times 10^5$ CFU Listeria requires 3.0 mL, but extra should be prepared. Prepare 6.0 mL (20 doses):

   20 doses × ($5 \times 10^5$ CFU/dose) = $1 \times 10^7$ CPU needed.

   A culture with an $OD_{600}$ = 0.06 contains $6 \times 10^7$ CFU/mL.

   ($1 \times 10^7$ CFU needed) / ($6 \times 10^7$ CFU/mL) = 0.167mL culture needed. Mix the 0.167mL of Listeria culture with 5.833 mL of sterile saline and inject mice intraperitoneally.

For intravenous infection, the number of CFU's of Listeria given the mice is reduced to $10^5$ CFU and the injected volume is decreased to 0.1 mL.

3. Prepare four or five 10-fold serial dilutions of the inoculum. Plate 20 µL of each dilution on separate TSB-strep agar plates to determine the actual dose of bacteria that was given to the experimental animals. Incubate plates overnight at 37°C and count the colonies the next day. (# colonies) × (dilution factor) × 50 = # CFU/mL.

### *Cell Preparation*

Harvest organs from Listeria-infected mice. Prepare singel-cell suspensions for culture from mice as discussed above. Prepare postitive and negative control wells, as well as the LLO-specific peptide wells, and add cells. Positive control wells contain anti-CD3 +/– anti-CD28 as needed. If measuring intracellular cytokines, add Brefeldin A (GolgiPlug™in kit) to the wells 4–6 h before harvesting the cells.

### *Culture*

Culture the cells in a 37°C, 5% $CO_2$ incubator for the time period determined to provide the desired response for the cells being tested (usually 6h;). A complete set of test/control wells is required for each time period used. After the time has elapsed, supernatants can be collected to measure soluble cytokines or cells may be collected from Brefeldin A-treated wells for measurement of intracellular cytokines. Freeze supernatants for soluble cytokine measurement, at –80°C until assayed. Stain cells for intracellular cytokine expression.

### *Antigen Presenting Cells (APCs)*

When culturing purified T cells (CD4+, CD8+ cells), APCs may be pre-loaded with specific peptide antigen. Either activated purified B cells or cell lines (e.g., P815 cells) may be used.

### 1. Whole Splenocytes

In a whole spleen preparation, the native APCs are present and functional and therefore no additional APCs are required. This is the starting point for determining optimum antigen concentrations for culture. Prepare whole spleen, single-cell suspensions from either pooled or individual animals. Add prepared cells to the test wells (negative control, positive control and peptide) with a final concentration of $2 \times 10^6$ cells/mL. The concentration of peptide must be titrated in the cell system being studied. Start with a range of serial dilutions (e.g. $10^{-4}$ *M*, $10^{-5}$ *M*, etc.) and determine the concentration that induces the optimum response. Culture the cells for the specified time and collect cells and/or supernatants for analysis.

### 2. Loaded B Cells

Prepare activated B cells, beginning.the day before, as discussed in Subheading 3.2.3., culturing (activating) them overnight with LPS. Wash two or three times with approx 10x volumes of.DPBS or RPMI to remove LPS. Adjust B-cell concentration to $2 \times 10^6$ cells/mL in RPMI-10 and add the specific peptide at the concentration found to be optimum when culturing whole splenocytes with the peptide. Incubate for 1 h at 37°C. Wash twice with 10x volumes of RPMI. Add loaded B cells to culture wells along with the purified T cells in a range of several T cell:B cell ratios such as 100:1 down to 1:1. No additional peptide is added to the wells.

## 3. P815 Cell line as APCs

P815 (ATCC # TIB-64™) is a mouse lymphoblast-like mastocytoma cell line that can be used as antigen presentation cells. They can be loaded with specific peptides in the same manner as loading purified B cells P815 cells, grown in suspension, are washed once with RPMI-10. The cell concentration is adjusted to $5 \times 10^6$ cells/mL and peptide added to the optimal concentration as determined by-titration in the cell system being used. Incubate for 1 h at 37°C. Wash once or twice with RPMI-10. The cells are then ready to add to the prepared culture plate. It is best to initially test a range of cell ratios as with the B cells. A negative control well is required in which nonpeptide-loaded P815 cells are added to the purified T cells. This is used to determine non-specific binding during analysis of the flow cytometry data.

## Intracellular Staining for Cytokines after Brefeldin A Treatment

This staining is a multistep process. *The cell* surface markers (CD4, CDS, etc.) must be stained first, then the cells are permeabilized to allow the antibodies against the cytokines to reach the cytoplasmic constituents of the cells. Figure 14.1 shows a plot of purified CD4+ cells stained for cytoplasmic interferon-γ (IFN-γ) and CD44+ expression. CD4+ T cells were isolated from C57B1/6 mice 7d after inoculation with. ACT A- Listeria, and incubated with LLO peptide for 6 h.

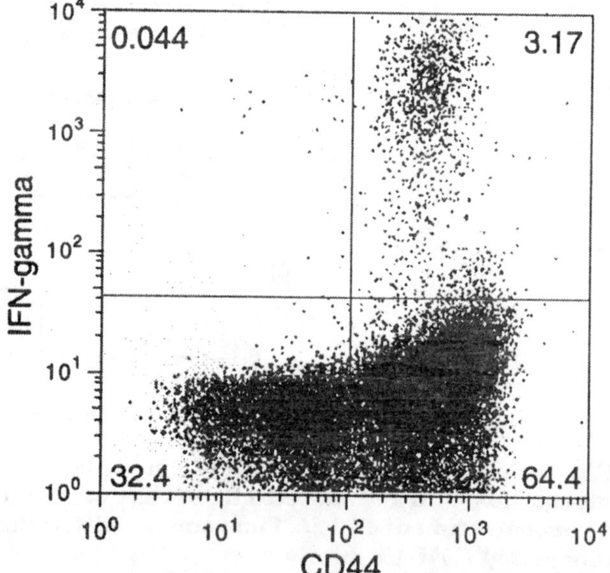


**Fig. 14.1. Pattern of cytoplasmic IFN-γ and surface CD44+ expression in a culture of purified CD4+ cells. CD4+ cells were isolated from C57B1/6 mice 7d after inoculation with ACT A- *Listeria monocytogenes* and cultured for 6h with B cells loaded with listeriolysin peptide (AA# 190–201) at a ratio of 1 : 1 with added GolgiPlug™, then permeabilized, stained and analyzed on a BD™FACS Calibur. The IFN-γ+ cells represent those responding specifically to the'peptide during the peak of the primary response. The numbers represent the percentage of CD4+ cells in each quadrant.**

1. Stain the surface markers after the final wash step, omit the fixation buffer. Instead, the cells are washed twice with 1.0 mL of intracellular staining buffer.
2. Resuspend the cell pellet in 250 μL of the Cytofix/Cytoperm solution provided in the BD Cytofix/Cytoperm™ Plus kit (Becton Dickinson, Sparks, MD.). Vortex the cells as the buffer is added to decrease clumping of the cells. Incubate 20min at 4°C.
3. Prepare the Perm/Wash buffer (approx 4.2mL/sample will be needed) by diluting the 10 x stock provided in the kit with reagent grade water. Wash the cells twice with 1.0 mL of diluted Perm/Wash buffer, vortexing the tube as the buffer is added.
4. During the second wash of the cells, prepare fluorescent-

labeled anti-cytokine solution by diluting the antibody in Perm/Wash buffer (50 μL of diluted antibody/sample). Vortex the cell pellet while adding the antibody and incubate in the dark at 4°C for 30 min.

5. Wash twice with 1.0 mL of Perm/Wash buffer. Vortex while adding buffer.
6. Resuspend in 250 μL of staining buffer and vortex well. Keep at 4°C in the dark until the samples can be run on a flow cytometer. They must be run within 24 h or signal could be decreased.

## Time Course of Response

Establish a response curve by measuring the cytokine response over time. A typical response curve for *Listeria monocytogenes* infection is shown in Fig. 14.2. Culture cells from Listeria-infected mice at various times after inoculation (e.g. daily, weekly, monthly) to measure immune response over time depending on the parameters being studied. Soluble (total) cytokines may be measured by either cytokine ELISA or by CBA assay (e.g. Becton Dickinson's BD™CBA or Bio-Rad's Bio-Plex™).

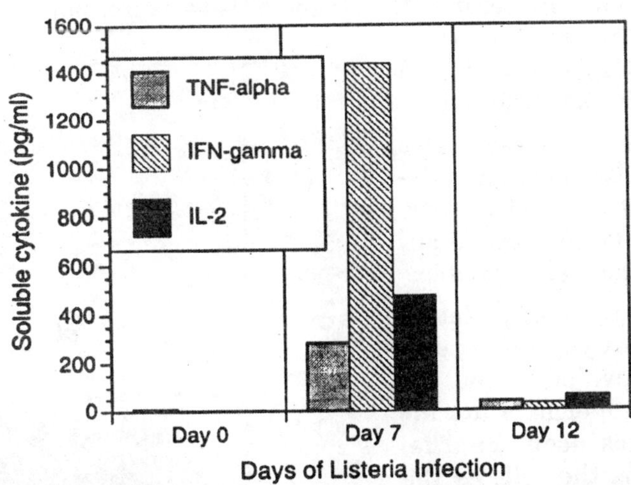


**Fig. 14.2. Time course of soluble cytokine production after ACT A- *Listeria monocytogenes* inoculation Unfractionated spleen cells cultured for 8h with listeriolysin peptide (190–201) at a concentration of $10^{-4}$ *M*. Time course with uninfected (Day 0), seven and twelve day Listeria inoculated C57B1/6 mice**

## Adoptive Transfer

Adoptive transfer of immunity can be used to study primary and secondary immune responses. Congenic strains of mice (e.g., C57 B1/6J [CD45.2+] and B6.SJL-Ptprc$^a$ Pep$^b$/BoyJ [CD45.1+]) allow one to discriminate between the responding cell types. Monoclonal antibodies are available to separate CD45.1+ cells (clone # A20) and CD45.2+ cells (clone # 104).

An immune response is induced in the donor mice. Then T cells are transferred to congenic recipient mice. Inoculation of the recipient animals produces cytokine responses from both donor (secondary response) and recipient primary response) T cells. Donor mice are inoculated with Listeria. T cells are isolated, usually after 28+ days, and transferred intravenously to congenic recipients. The recipient mice are then inoculated immediately with Listeria.

At the peak primary time for Listeria infection, 6–7 d later, spleen cells are cultured with and without Listeria-specific peptide and then the cytokine response is measured. Figure 14.3A illustrates the resolution of congenic cells in a whole spleen culture one day after adoptive transfer of CD4+ T cells. Staining for intracellular cytokine expression and gating on CD4+ cells allows the discrimination of cytokine-producing donor cells from cytokine-producing recipient cells as shown in Fig. 14.3B.

## *Adoptive Transfer Utilizing Congenic Strains of Mice*

1. Inoculate the donor mice (e.g., the CD45.1+ strain, B6.SJL-Ptprc[a] Pep[b]/BoyJ) with Listeria.
2. Isolate T cells from donor mice at the desired interval by column purification or B220-depletion. Resuspend cells in DPBS and inject intravenously or via retro-orbital sinus into recipient mice (e.g., the CD45.2+ strain, C57B1/6J). Adoptive transfers have been performed in our lab using up to a 1:1 ratio of donor to recipient splenic lymphocytes as starting material. Average cell recoveries after purification of donor cells allow the routine transfer of $1 \times 10^7$ CD4+ cells and $8 \times 10^6$ CD8+ cells. Fewer transferred cells may be preferred in some experimental protocols.

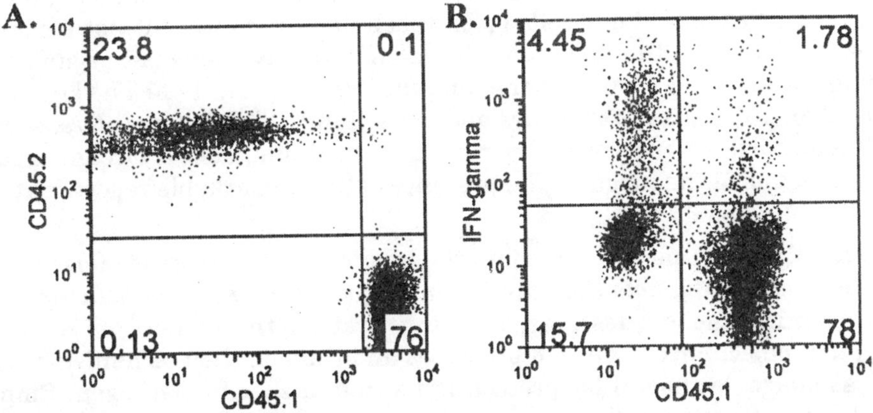


**Fig. 14.3. Patterns of congenic CD4+ cells after adoptive transfer. Donor (CD45.2+) CD4+ cells were transferred to recipient (CD45.1+) mice. Recipients were inoculated with Listeria 1 day later and organs harvested 6d after Listeria inoculation. Analyzed on BD™ FACS Calibur. (A) Whole splenocytes gated for total CD4+ cells. (B) Whole splenocytes cultured 6 hours with CD4+ LLO-specific peptide and GolgiPlug™. Stained for cytoplasmic IFN-γ. In this case the donor cells were obtained 58 d after Listeria inoculation. Thus, the donor haplotype (CD45.1⁻) response is the response of memory cells at the peak time of memory response (6d), whereas the recipient response (CD45.1+) is a primary response, and is one day before the usual peak primary response at 7 d after inoculation.**

3. Infect recipient mice. This can either be done the same day or the day following the cell transfer.
4. After 6d (peak memory response time), harvest organs and culture cells, either as whole spleen or purified T-cell fractions as described previously. Intracellular staining for cytokines combined with haplotype staining discriminate the response of donor cells from those of the recipient. Analysis requires gating for the transferred T cells (CD4+ or CD8+) followed by displaying anticytokine-labeled cells against either the donor or recipient haplotype.

### *Fluorescent-labeled MHC Class I Pentamers and MHC Class II Tetramers*

These are commercially available to identify antigen-specific T cells (CD8+ and CD4+ respectively) without the need for congenic strains of mice. Antigen-specific T cells are generated and cultured as discussed previously.

1. Collect $10^6$ cells to stain for flow cytometry. Wash with 1.0 mL and resuspend in 50 μL of pentamer wash buffer.
2. Add 10μL of fluorescent-labeled pentamer and incubate 10-15 min at RT in the dark.
3. Wash with 1.0 mL of pentamer wash buffer and add 50μL of pentamer wash buffer containing fluorescent-labeled anti-T-cell antibody.
4. Incubate on ice 20-30 min in the dark.
5. Wash twice with 1.0 mL of pentamer wash buffer and resuspend in 0.3 mL of pentamer fix buffer. Store the .samples in the dark at 4°C until, they can be analyzed by flow cytometry.

## DEFENSIVE SYSTEM

We are constantly exposed to an incredible diversity of bacteria, viruses, and parasites, many of which would flourish in our cells or extracellular fluids were it not for our immune system. Remarkably, we are often even able to defend ourselves against organisms that we have never before encountered. How does the immune system protect us? The key is our ability to produce more than $10^8$ distinct *antibodies* and more than $10^{12}$ *T-cell receptors,* each of which presents a different surface for specifically binding a molecule from a foreign organism and initiating the destruction of the invader. The presence of this remarkable repertoire of defensive molecules poses a challenge.

What prevents the immune system from attacking cells that express molecules normally present in our bodies; that is, *how does the immune system distinguish between nonself and self?* We shall examine these questions, focusing first on the structures of the proteins participating in the molecular recognition processes and then on the mechanisms for selecting cells that express molecules useful for protecting us from a specific pathogen. Emphasis will be on the modular construction of the proteins of the immune system — identifying structural motifs and considering how spectacular diversity can arise from modular construction.

## PRINCIPLES OF EVOLUTION

The immune system comprises two parallel but interrelated systems. In the *humoral immune response,* soluble proteins called *antibodies (immunoglobulins)* function as recognition elements that bind to foreign molecules and serve as markers signaling foreign invasion. Antibodies are secreted *by plasma cells,* which are derived from *B lymphocytes (B cells).* A foreign macromolecule that binds selectively to an antibody is called an *antigen.* In a physiological context, if the binding of the foreign molecule stimulates an immune response, that molecule is called an *immunogen.* The specific affinity of an antibody is not for the entire macromolecular antigen but for a particular site on the antigen called the *epitope* or *antigenic determinant.* In the *cellular immune response,* cells called *cytotoxic T lymphocytes* (also commonly called *killer T cells)* kill cells that display foreign motifs on their surfaces.

Another class of T cells called *helper T lymphocytes* contributes to both the humoral and the cellular immune responses by stimulating the differentiation and proliferation of appropriate

B cells and cytotoxic T cells. The celluar immune response is mediated by specific receptors that are expressed on the surfaces of the T cells. The remarkable ability of the immune system to adapt to an essentially limitless set of potential pathogens requires a powerful system for transforming, the immune cells and molecules present in our systems in response to the presence of pathogens. *This adaptive system operates through the principles of evolution, including reproduction with variation followed by selection of the most well suited members of a population.* If the human genome contains, by the latest estimates, only 40,000 genes, how can the immune system generate more than $10^8$ different antibody proteins and $10^{12}$ T-cell receptors? The answer is found in a novel mechanism for generating a highly diverse set of genes from a limited set of genetic building blocks.

Linking different sets of DNA regions in a combinatorial manner produces many distinct protein-encoding genes that are not present in the genome. A rigorous selection process then leaves for proliferation only cells that synthesize proteins determined to be useful in the immune response. The subsequent reproduction of these cells without additional recombination serves to enrich the cell population with members expressing a particular protein species. Critical to the development of the immune response is the selection process, which determines which cells will reproduce.

The process comprises several stages. In the early stages of the development of an immune response, cells expressing piolecules that bind tightly to self-molecules are destroyed or silenced, whereas cells expressing molecules that do not bind strongly to self-molecules and that have the potential for binding strongly to foreign molecules are preserved. The appearance of an immunogenic invader at a later time will stimulate cells expressing immunoglobulins or T-cell receptors that bind specifically to elements of that pathogen to reproduce — in evolutionary terms, such cells are selected for. Thus, the immune response is based on the selection of cells expressing molecules that are specifically effective against a particular invader; the response evolves from a population with wide-ranging specificities to a more-focused collection of cells and molecules that are well suited to defend the host when confronted with that particular challenge.

Just as Medieval defenders used their weapons and the castle walls to defend their city, the immune system constantly battles against foregn invaders such as viruses, bacteria, and parasites to defend the organism.

## ANTIGEN-BINDING UNITS

Antibodies are central molecular players in the immune response, and we examine them first. A fruitful approach in studying proteins as large as antibodies is to split the protein into fragments that retain activity. In 1959, Rodney Porter showed that *immunoglobulin G (IgG),* the major antibody in serum, can be cleaved into three 50-kd fragments by the limited proteolytic action of papain. Two of these fragments bind antigen. They are called $F_{ab}$ (F stands for fragment, *ab* for antigen binding). The other fragment, called $F_c$ because it crystallizes readily, does not bind antigen, but it has other important biological activities, including the mediation of responses termed *effector functions.* These functions include the initiation of the *complement cascade,* a process that leads to the lysis of target cells.

Although such effector functions are crucial to the functioning of the immune system, they will not be discussed further here. How do these fragments relate to the three-dimensional structure of whole IgG molecules? Immunoglobulin G consists of two kinds of polypeptide chains, a 25-kd *light (L) chain* and a 50-kd *heavy (H) chain* Figure 14.2). The subunit composition is $L_2H_2$. Each L chain is linked to an H chain by a disulfide bond, and the H chains are linked to each other by at least one disulfide bond. Examination of the amino acid sequences and three-dimensional structures of IgG molecules reveals that each L chain comprises two homologous domains, termed *immunoglobulin domains.* Each H chain has four immunoglobulin domains.

Overall, the molecule adopts a conformation that resembles the letter Y, in which the stem, corresponding to the $F_c$ fragment obtained by cleavage with papain, consists of the two carboxyl-terminal immunoglobulin domains of each H chain and in which the two arms of the Y, corresponding to the two $F_{ab}$ fragments, are formed by the two amino-terminal domains of each H chain and the two amino-terminal domains of each L chain. The linkers between the stem and the two arms consist of relatively extended polypeptide regions within the H chains and are quite flexible. Papain cleaves the H chains on the carboxyl-terminal side of the disulfide bond that links each L and H chain

Thus, each $F_{ab}$ consists of an entire L chain and the amino-terminal half of an H chain, whereas $F_c$ consists of the carboxyl-terminal halves of both H chains. Each $F_{ab}$ contains a single antigen-binding site: Because an intact IgG molecule contains two $F_{ab}$ components and therefore has two binding sites, it can cross-link multiple antigens. Furthermore, the $F_c$ and the two $F_{ab}$ units of the intact IgG are joined by flexible polypeptide regions that allow facile variation in the angle between the $F_{ab}$ units through a wide range. This kind of mobility, called *segmental flexibility,* can enhance the formation of an antibody-antigen complex by enabling both combining sites on an antibody to bind an antigen that possesses multiple binding sites, such as a viral coat composed of repeating identical monomers or a bacterial cell surface.

The combining sites at the tips of the $F_{ab}$ units simply move to match the distance between specific determinants on the antigen. Immunoglobulin G is the antibody present in highest concentration in the serum, but other classes of immunoglobulin also are present. Each class includes an L chain (either $\mu$, or $\nu$) and a distinct H chain. The heavy chains in IgG are called i chains, whereas those in immunoglobulins A, M, D, and E are called $\chi$, $o$, $\phi$, and $\gamma$, respectively. *Immunoglobulin M (IgM)* is the first class of antibody to appear in the serum after exposure to an antigen.

The presence of 10 combining sites enables IgM to bind especially tightly to antigens containing multiple identical epitopes. The strength of an interaction comprising multiple independent binding interactions between partners is termed *avidity* rather than affinity, which denotes the binding strength of a single combining site. The presence of 10 combining sites in IgM compared with 2 sites in IgG enables IgM to bind many mukivalent antigens that would slip away from IgG. *Immunoglobulin A (IgA)* is the major class of antibody in external secretions, such as saliva, tears, bronchial mucus, and intestinal mucus. Thus, IgA serves as a first line of defense against bacterial and viral antigens.

The role of *immunoglobulin D (IgD)* is not yet known. *Immunoglobulin E (IgE)* is important in conferring protection against parasites, but IgE also causes allergic reactions. IgE-antigen complexes form cross-links with receptors on the surfaces of mast cells to trigger a cascade that leads to the release of granules containing pharmacologically active molecules. Histamine, one of the agents released, induces smooth muscle contraction and stimulates the secretion of mucus. A comparison of the amino acid sequences of different IgG antibodies from human beings or mice shows that the carboxyl-terminal half of the L chains and the carboxyl-terminal three-quarters of the H chains are very similar in all of the antibodies.

Importantly, the amino-terminal domain of each chain is more variable, including three stretches of approximately 7 to 12 amino acids within each chain that are hyperyariable. The amino-terminal immunoglobulin domain of each chain is thus referred to as the *variable region,* whereas the remaining immunoglobulin domains are much more similar in all antibodies and are referred to as *constant regions.*

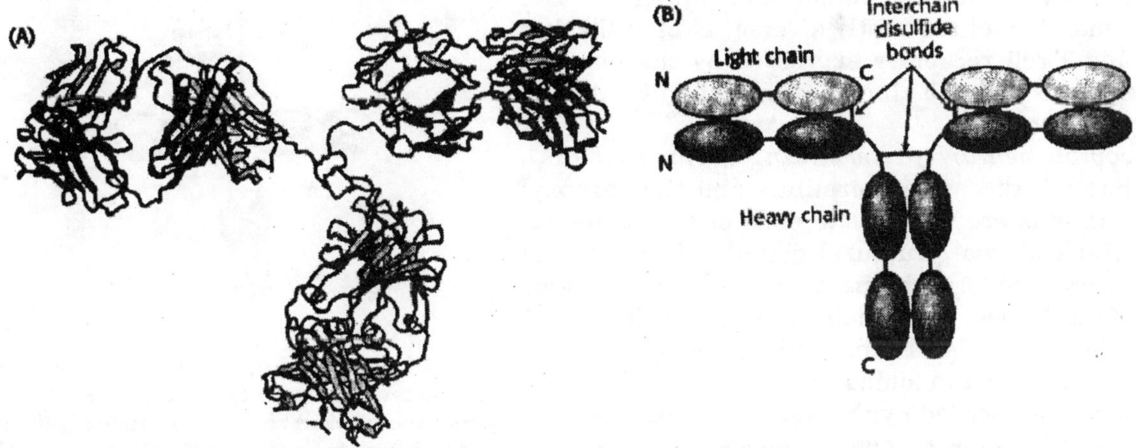


**Fig. 14.4. Immunoglobulin G Structure. (A) The three-dimensional structure of an IgG molecule showing the light chains the heavy chains. (B) A schematic view of an IgG molecule indicating the positions of the interchain disulfide bonds. N, amino terminus; C, carboxyl terminus.**

**Table 14.1 Properties of immunoglobulin classes**

| Class | Serum concentration (mg/ml) | Mass(kd) | Sedimentation coefficients(s) | Light chains | Heavy chains | Chain structure |
|---|---|---|---|---|---|---|
| IgG | 12 | 150 | 7 | $\mu$ or $\nu$ | $\iota \times 3$ | $\mu_2\iota_2$ or $\nu_2\,\iota_2$ |
| IgA | 3 | 180–500 | 7, 10, 13 | $\mu$ or $\nu$ | $\chi$ | $(\mu_2\chi_2)_n$ or $(\nu_2\,\chi_2)_n$ |
| IgM | 1 | 950 | 18 – 20 | $\mu$ or $\nu$ | o | $(\mu_2 o_2)_5$ or $(\nu_2\,o_2)_5$ |
| IgD | 0.1 | 175 | 7 | $\mu$ or $\nu$ | $\phi$ | $\mu_2\phi_2$ or $\nu_2\,\phi_2$ |
| IgE | 0.001 | 200 | 8 | $\mu$ or $\nu$ | $\gamma$ | $\mu_2\gamma_2$ or $\nu_2\,\gamma_2$ |

*Note:* $n$ = 1, 2, or 3. IgM and oligomers of IgA also contain J chains that connect immunoglobulin molecules. IgA in secretions has an additional component.

## HYPERVARIABLE LOOPS

An IgG molecule consists of a total of 12 immunoglobulin domains. These domains have many sequence features in common and adopt a common structure, the *immunoglobulin fold.* Remarkably, this same structural domain is found in many other proteins that play key roles in the immune system. The immunoglobulin fold consists of a pair of δ sheets, each built of antiparallel δ strands, that surround a central hydrophobic core. A single disulfide bond bridges the two sheets. Two aspects of this structure are particularly important for its function. First, three loops present at one end of the structure form a potential binding surface. Thes loops contain the hypervariable sequences present in antibodies and in T-cell receptors.

Variation of the amino acid sequences of these loops provides the major mechanism for the generation of the vastly diverse set of antibodies and T-cell receptors expressed by the immune system. These loops are referred to as *hypervariable loops* or *complementaritydetermining regions (CDRs).* Second, the amino terminus and the carboxyl terminus are at opposite ends of the structure, which allows structural domains to be strung together to form chains, as in the L and H chains of antibodies. Such chains are present in several other key molecules in the immune system. The immunoglobulin fold is one of the most prevalent domains encoded by the human genome — more than 750 genes encode proteins with at least one immunoglobulin fold recognizable at the level of amino acid sequence. Such domains are also common in other multicellular animals such as flies and nemotodes.

However, from inspection of amino acid sequence alone, immunoglobulin-fold domains do not appear to be present in yeast or plants. However, structurally similar domains are present in these organisms, including the key photosynthetic electron-transport protein plastocyanin in plants. Thus, the immunoglobulin-fold family appears to have expanded greatly along evolutionary branches leading to animals — particularly, vertebrates.

## SPECIFIC MOLECULES

For each class of antibody, the amino-terminal immunoglobulin domains of the L and H chains (the variable domains, designated $V_L$ and $V_H$) come together at the ends of the arms extending from the structure. The positions of the complementarity-determining regions are striking. These hypervariable sequences, present in three loops of each domain, come together so that all six loops form a single surface at the end of each

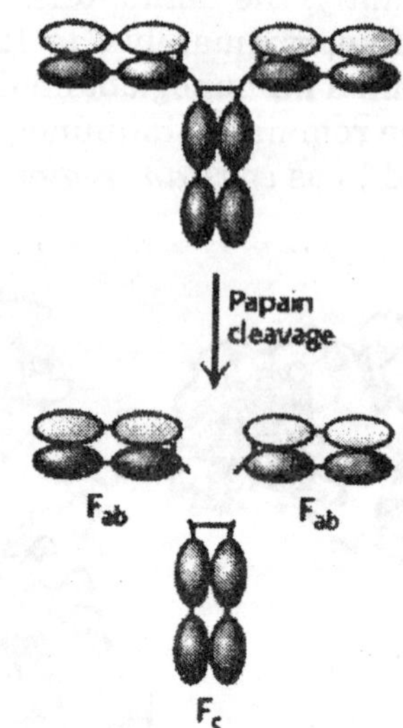


**Fig. 14.5 Immunoglobulin G Cleavage. Treatment of intact IgG molecules with the protease papain results in the formation of three large fragments: $F_{ab}$ fragments that retain antigen-binding capability and one $F_c$ fragment that does not.**

**Fig. 14.6. Antigen Cross-Linking. Because IgG molecules include two antigen-binding sites, antibodies can crosslink multivalent antigens such as viral surfaces.**

arm. Because virtually any $V_L$ can pair with any $V_H$, *a very large number of different binding sites can be constructed by their combinatorial association.*

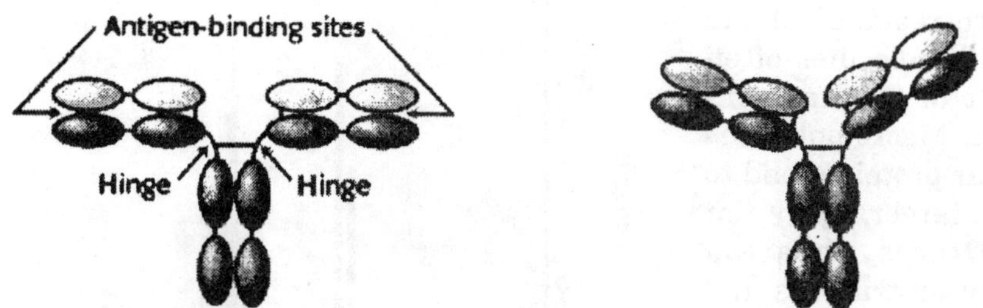


**Fig. 14.7. Segmental Flexibility. The linkages between the $F_{ab}$ and the $F_c$ regions of an IgG molecule are flexible, allowing-the two antigen-binding sites to adopt a range of orientations with respect to one another. This flexibility allows effective interactions with a multivalent antigen without requiring that the epitopes on the target be a precise distance apart.**

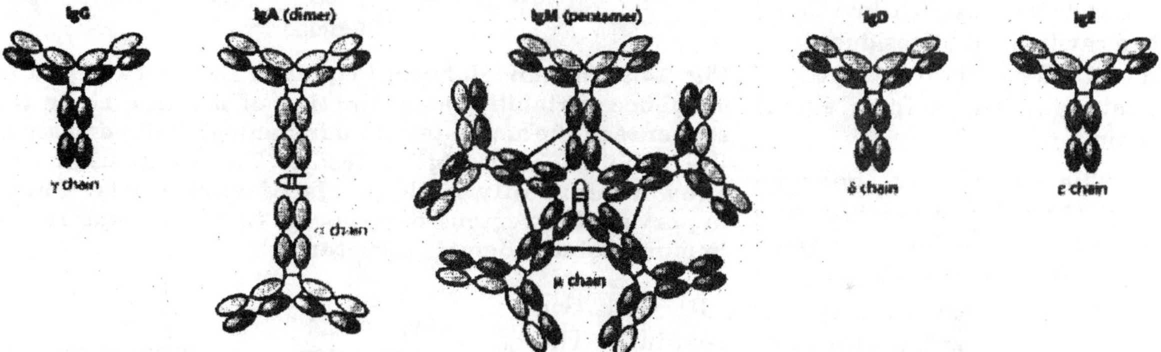


**Fig. 14.8. Classes of ImmunoGlobulin. Each of five classes of immunoglobulin has the same light chain (shown in yellowj-eefribined with a different heavy chain (τ, χ, o, ϕ, *or* μ). Disulfide bonds are indicated by green lines. The IgA dimer and the IgM pentamer have a small polypeptide chain in addition to the light and heavy chains.**

## X-Ray Analyses

The results of x-ray crystallographic studies of many large and small antigens bound to $F_{ab}$ molecules have been sources of much insight into the structural basis of antibody specificity. The binding of antigens to antibodies is governed by the same principles that govern the binding of substrates to enzymes. The apposition of complementary shapes results in numerous contacts between amino acids at the binding surfaces of both molecules.

Numerous hydrogen bonds, electrostatic interactions, and van der Waals interactions, reinforced by hydrophobic interactions, combine to give specific and strong binding. A few aspects of antibody binding merit specific attention, inasmuch as they relate directly to the structure of immunoglobulins. The binding site on the antibody has been found to incorporate some or all of the CDRs in the variable domains of the antibody. Small molecules (e.g., octapeptides) are likely to make contact with fewer CDRs, with perhaps 15 residues of the antibody participating in the binding interaction.

Macromolecules often make more extensive contact, interacting with all six CDRs and.20 or more residues of the antibody. Small molecules often bind in a cleft of the antigen-binding region. Macromolecules such as globular proteins tend to interact across larger, fairly flat apposed surfaces bearing complementary protrusions and depressions. A well-studied case of small-molecule binding is seen in an example of phosphorylcholine bound to $F_{ab}$. Crystallographic analysis revealed phosphorylcholine bound to a cavity lined by residues from five CDRs — two from the L chain and three from the H chain.

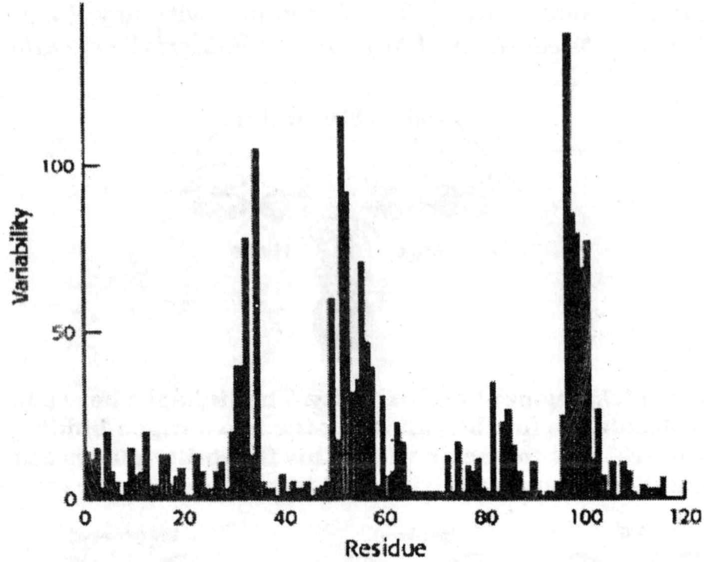


**Fig. 14.9. Immunoglobulin Sequence Diversity. A plot of sequence variability as a function of position along the sequence of the amino-terminal immunoglobulin domain of the H chain of human IgG molecules. Three regions (in red) show remarkably high levels of variability. These hypervariable regions correspond to three loops in the immunoglobulin domain structure.**

The positively charged trimethylammonium group of phosphorylcholine is buried inside the wedge-shaped cavity, where it interacts electrostatically with two negatively charged glutamate residues. The negatively charged phosphate group of phosphorylcholine binds to the positively charged guanidinium group of an arginine residue at the mouth of the crevice and to a nearby lysine residue. The phosphate group is also hydrogen bonded to the hydroxyl group of a tyrosine residue and to the guanidinium group of the arginine side chain.

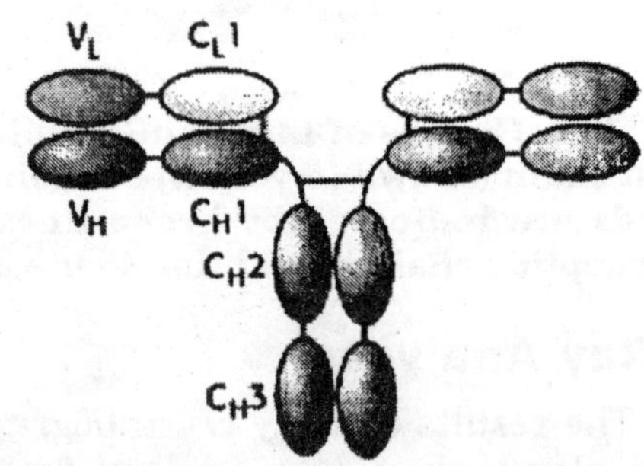


**Fig. 14.10. Variable and Constant Regions. Each L and H chain includes one immunoglobulin domain at its amino terminus that is quite variable from one antibody to another. These domains are referred to as $V_L$ and $V_H$. The remaining domains are more constant from one antibody to another and are referred to as constant domains ($C_L1$, $C_H1$, $C_H2$, and $C_H3$).**

Numerous van der Waals interactions, such as those made by a tryptophan side chain, also stabilize this complex. The binding of phosphorylcholine does not significantly change the structure of the antibody, yet induced fit plays a role in the formation of many antibody-antigen complexes. A malleable binding site can accommodate many more kinds of ligands than can a rigid one. Thus, induced fit increases the repertoire of antibody specificities.

## Antigens and Antibodies with Numerous Interactions

How do large antigens interact with antibodies? A large collection of antibodies raised against hen egg-white lysozyme has been structurally characterized in great detail. Each different antibody binds to a distinct surface of lysozyme. Let us examine the interactions present in one of these complexes in detail. This antibody binds two polypeptide segments that are widely separated in the primary structure, residues 18 to 27 and-116 to 129. All six CDRs of the antibody make contact with this epitope.

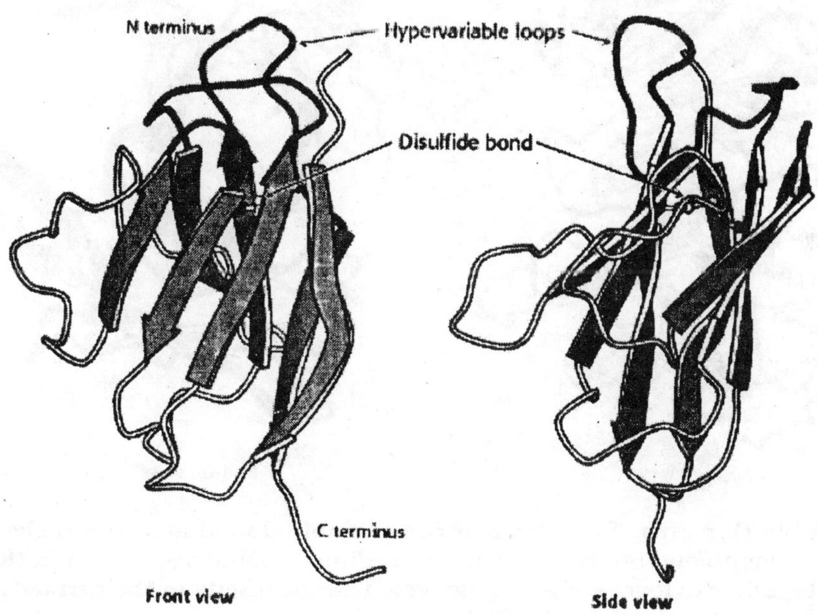


**Fig. 14.11. Immunoglobulin Fold. An immunoglobulin domain consists of a pair of δ-sheets linked by a disulfide bond and hydrophobic interactions. Three hypervariable loops lie at one end of the structure.**

The region of contact is quite extensive (about 30 × 20 Å). The apposed surfaces are rather flat. The only exception is the side chain of glutamine 121 of lysozyme, which penetrates deeply into the antibody binding site, where it forms a hydrogen bond with a main-chain carbonyl oxygen atom and is surrounded by three aromatic side chains. The formation of 12 hydrogen bonds and numerous van der Waals interactions contributes to the high affinity ($K_d$ = 20 nM) of this antibody-antigen interaction. Examination of the $F_{ab}$ molecule without bound protein reveals that the structures of the $V_L$ and $V_H$ domains change little on binding, although they slide 1 Å apart to allow more intimate contact with lysozyme.

## GENE REARRANGEMENTS

A mammal such as a mouse or a human being can synthesize large amounts of specific antibody against virtually any foreign determinant within a matter of days of being exposed to it. We have seen that antibody specificity is determined by the amino acid sequences of the variable regions of both light and heavy chains, which brings us to the key question: How are different variable-region sequences generated? The discovery of distinct variable, and constant regions in the L and H chains raised the possibility that the genes that encode immunoglobulins have an unusual architecture that facilitates the generation of a diverse set of polypeptide 'products.

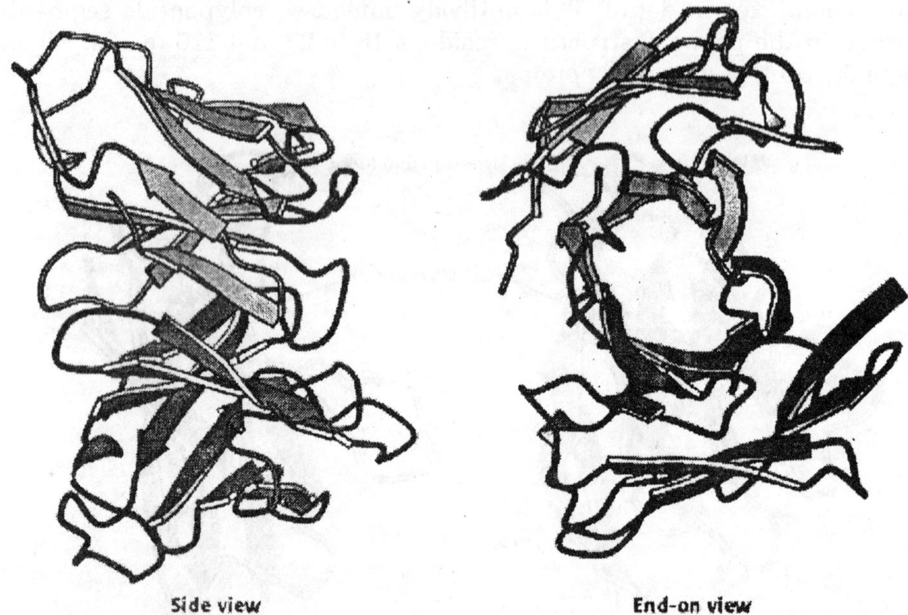


**Fig. 14.12. Variable Domains. Two views of the variable domains of the L chain and the H chain (blue); the complementarity-determining regions (CDRs) are shown in the figure. The six CDRs come together to form a binding surface. The specificity of the surface is determined by the sequences and structures of the CDRs.**

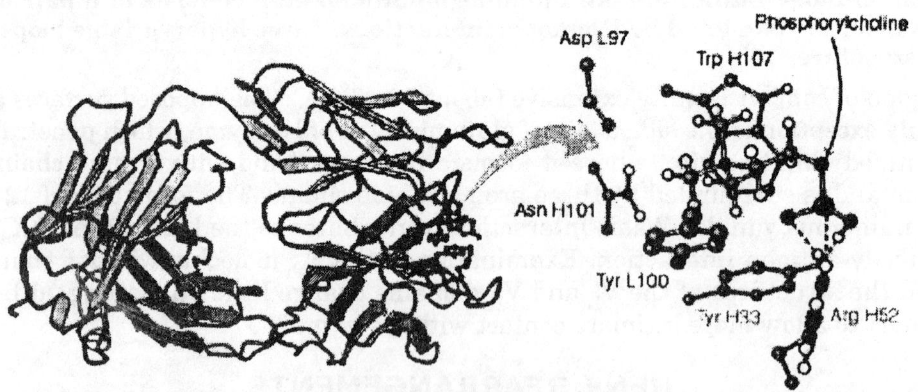


**Fig. 14.13. Binding of a Small Antigen. The structure of a complex between an $F_{ab}$ fragment of an antibody and its target in this case, phosphoryl-choline. Residues from the antibody interact with phosphorylcholine through hydrogen bonding and electrostatic and van der Waals interactions.**

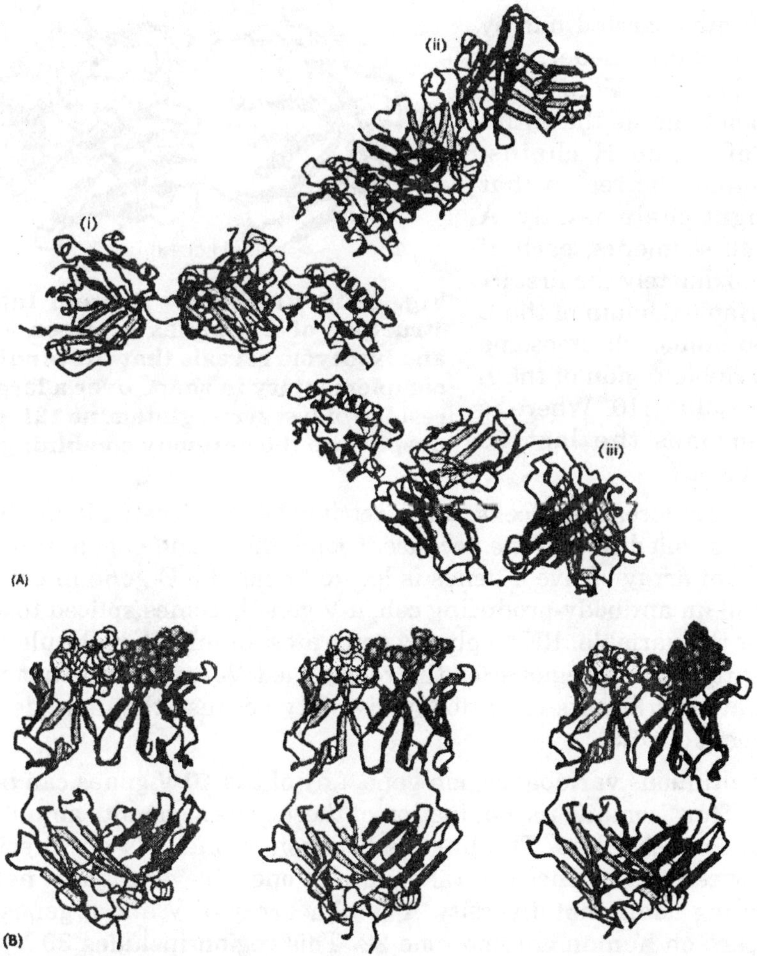


**Fig. 14.14. Antibodies Against Lysozyme. (A) The structures of three complexes (i,ii,iii) between $F_{ab}$ fragments and hen egg-white lysozyme shown with lysozyme in the same orientation in each case. The three antibodies recognize completely different epitopes on the lysozyme molecule. (B) The $F_{ab}$ fragments from part A with points of contact highlighted as space-filling models, revealing the different shapes of the antigen-binding sites.**

In 1965, William Dreyer and Claude Bennett proposed that multiple *V (variable) genes* are separate from a single *C (constant) gene* in embryonic (germ-line) DNA. According to their model, one of these V genes becomes joined to the C gene in the course of differentiation of the antibody-producing cell. A critical test of this novel hypothesis had to await the isolation of pure immunoglobulin mRNA and the development of techniques for analyzing mammalian genomes. Twenty years later, Susumu Tonegawa found that V and C genes are indeed far apart in embryonic DNA but are closely associated in the DNA of antibody-producing cells. Thus, immunoglobulin genes are rearranged in the differentiation of lymphocytes.

## Antibody Diversity

Sequencing studies carried out by Susumu Tonegawa, Philip Leder, and Leroy Hood revealed that V genes in embryonic cells do not encode the entire variable region of L and H chains. Consider, for example, the region that encodes the p. light-chain family. A tandem array of 40 segments, each of which encodes approximately the first 97 residues of the variable domain of the L chain, is present on human chromosome 2. However, the variable region of the L chain extends to residue 110. Where is the DNA that encodes the last 13 residues of the V region?

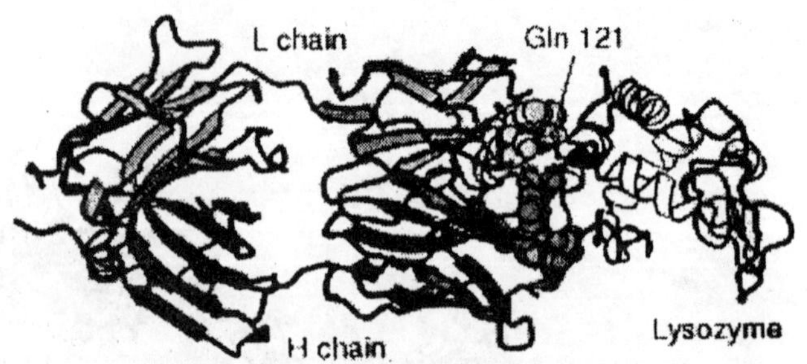


**Fig. 14.15. Antibody - Protein Interactions. The structure of a complex between an $F_{ab}$ fragment and lysozyme reveals that the binding surfaces are complementary in shape over a large area. A single residue of lysozyme, glutamine 121, penetrates more deeply into the antibody combining site.**

For L chains in undifferentiated cells, this stretch of DNA is located in an unexpected place: near the C gene. It is called the *J gene* because it *j*oins the V and C genes in a differentiated cell. In fact, a tandem array of five J genes is located near the C gene in embryonic cells. In the differentiation of an antibody-producing cell, a V gene becomes spliced to a J gene to form a complete gene for the variable. RNA splicing generates an mRNA molecule for the complete L chain by linking the coding regions for the rearranged VJ unit with that for the C unit J genes are important contributors to antibody diversity because they encode part of the last hypervariable segment (CDR3).

In forming a continuous variable-region gene, any of the 40 V genes can become linked to any of five J genes. Thus, somatic recombination of these gene segments amplifies the diversity already present in the germ line. The linkage between V and J is not precisely controlled. Recombination between these genes can take place at one of several bases near the codon for residue 95, generating additional diversity. A similar array of V and J genes encoding the v light chain is present on human chromosome 22. This region includes 30 $V_{\upsilon}$ gene segments and four $J_{\upsilon}$ segments. In addition, this region includes four distinct C genes, in contrast with the single C gene in the μ locus. In human beings, the genes encoding the heavy chain are present on chromosome 14. Remarkably, the variable domain of heavy chains is assembled from *three* rather than two segments. In addition to $V_H$ genes that encode residues 1 to 94 and $J_H$ segments that encode residues 98 to 113, this chromosomal region includes a distinct set of segments that encode residues 95 to 97.

These gene segments are called D for diversity. Some 27 D segments lie between 51 $V_H$ and 6 $J_H$ segments.

The recombination process first joins a D segment to a $J_H$ segment; a $V_H$ segment is then joined to $DJ_H$. A greater variety of antigen-binding patches and clefts can be formed by the H chain than by the L chain because the H chain is encoded by three rather than two gene segments. Moreover, CDR3 of the H chain is diversified by the action of terminal deoxyribonucleotidyl transferase, a special DNA polymerase that requires no template. This enzyme inserts extra nucleotides between $V_H$ and D. The *V(D)J recombination* of both the L and the H chains is executed by specific enzymes present in immune cells. These proteins, called *RAG-1* and *RAG-2,* recognize specific DNA sequences called *recombination signal sequences (RSSs)* adjacent to the V, D, and J segments and facilitate the cleavage and religation of the DNA segments.

## Combinatorial Association and Somatic Mutation

Let us recapitulate the sources of antibody diversity. The germ line contains a rather large repertoire of variable-region genes. For μ light chains, there are about 40 V-segment genes and five J-segment genes. Hence, a total of 40 × 5 = 200 kinds of complete $V_\mu$ genes can be formed by the combinations of V and J. A similar analysis suggests that at least 120 different v light chains can be generated. A larger number of heavy-chain genes can be formed because of the role of the D segments. For 51 V, 27 D, and 6 J gene segments, the number of complete $V_H$ genes that can be formed is 8262. The association of 320 kinds of L chains with 8262 kinds of H chains would yield $2.6 \times 10^6$ different antibodies. Variability in the exact points of segment joining and other mechanisms increases this value by at least two orders of magnitude. Even more diversity is introduced into antibody chains by *somatic mutation* — that is, the introduction of mutations into the recombined genes.

In fact, a 1000-fold increase in binding affinity is seen in the course of a typical humoral immune response, arising from somatic mutation, a process called *affinity maturation.* The generation of an expanded repertoire leads to the selection of antibodies that more precisely fit the antigen. Thus, nature draws on each of three sources of diversity — a germ-line repertoire, somatic recombination, and somatic mutation — to form the rich variety of antibodies that protect an organism from foreign incursions. .

## Oligomerization of Antibodies

The processes heretofore described generate a highly diverse set of antibody molecules — a key first step in the generation of an immune response. The next stage is the selection of a particular set of antibodies directed against a specific invader. How does this selection occur? Each immature B cell, produced in the bone marrow, expresses a monomeric form of IgM attached to its surface. Each cell expresses approximately $10^5$ IgM molecules, but *all of these molecules are identical in amino acid sequence and, hence, in antigen-binding specificity.* Thus, the selection of a particular immature B cell for growth will lead to the amplification of an antibody with a unique specificity.

The selection process begins with the binding of an antigen to the membrane-bound antibody. Associated with each membrane-linked IgM molecule are two molecules of a heterodimeric membrane protein called Ig χ-Ig-δ. Examination of the amino acid sequences of Ig-χ and Ig-δ is highly instructive. The amino of each protein lies outside the cell and corresponds to a single immunoglobulin, and the carboxyl terminus, which lies inside the cell, includes a sequence of 18 amino acids called an *immunoreceptor tyrosine-basedactivation motif (ITAM)*. As its name suggests, each ITAM includes key tyrosine residues, which are subject to phosphorylation by particular protein kinases present in immune-system cells. A fundamental observation with regard to the mechanism by which the binding of antigen to membrane-bound antibody triggers the subsequent steps of the immune response is that *oligomerization or clustering of the antibody molecules is required.* The requirement for oligomerization is reminiscent of the dimerization of receptors triggered by growth hormone and epidermal growth factor indeed, the associated signaling mechanisms appear to be quite similar.

The oligomerization of the membrane-bound antibodies results in the phosphorylation of the tyrosine residues within the ITAMs by protein tyrosine kinases including Lyn, a homolog of Src. The phosphorylated ITAMs serve as docking sites for a protein kinase termed spleen tyrosine kinase (Syk), which has two SH2 domains that interact with the pair of phosphorylated tyrosines in each ITAM. Syk, when activated by phosphorylation, proceeds to phosphorylate other signal-transduction proteins' including an inhibitory subunit of a transcription factor called NF-μB and an isofor of phpspholipase C. The signaling processes continue downstream to activate gene expression, leading to the stimulation of cell growth and initiating further B-cell differentiation. Drugs that modulate the immune system have served as sources, of insight into immune-system signaling pathways.

For example, *cyclosporin,* a powerful suppressor of the immune system, acts by blocking a phosphatase called *calcineurin,* which normally activates a transcription factor called NF-AT by dephosphorylating it.

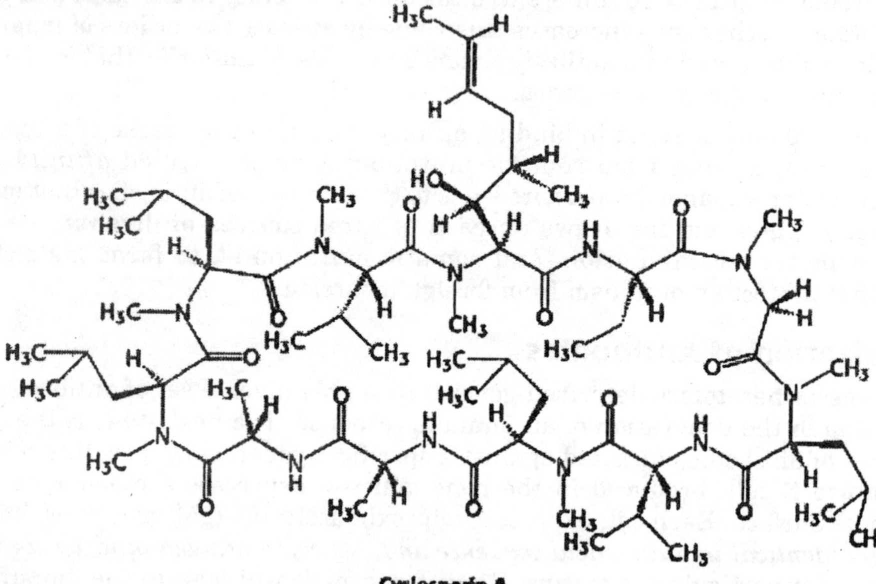

**Cyclosporin A**

The potent immune supression that results reveals how crucial the activity of this transcription factor is to the development of an immune response. Without drugs such as cyclosporin, organ transplantation would be extremely difficult because transplanted tissue expresses a wide range of foreign antigens, which causes the immune system to reject the new tissue. The role of oligomerization in the B-cell signaling pathway is illuminated when we consider the nature of many antigens presented by pathogens.

The surfaces of many viruses, bacteria, and parasites are characterized by arrays of identical membrane proteins or membrane-linked carbohydrates. Thus, most pathogens present multiple binding surfaces that will naturally cause membrane-associated antibodies to oligomerize as they bind adjacent epitopes. In addition, the mechanism accounts for the observation that most small molecules do not induce an immune response; however, coupling multiple copies of the small molecule to a large oligomeric protein such as keyhole limpet hemocyanin (KLH), which has a molecular mass of close to 1 million daltons or more, promotes antibody oligomerization and, hence, the production of antibodies against the small-molecule epitope. The large protein is called the *carrier* of the attached chemical group, which is called a *haptenic determinant.* The small foreign molecule by itself is called a *hapten.* Antibodies elicited by attached haptens will bind unattached haptens as well.

## Different Classes of Antibodies

The development of an effective antibody-based immune response .depends on the secretion into the blood of antibodies that have appropriate effector functions. At the beginning of this response, an alternative mRNA splicing pathway is activated so that the production of membrane-linked IgM is supplanted by the synthesis of secreted IgM secreted IgM is pentameric and has a relatively high avidity for multivalent antigens. Later, the antibody-producing cell makes either IgG, IgA, IgD, or IgE of the same specificity as the intially secreted IgM. In this switch, the light chain is unchanged, as is the variable region of the heavy chain.

Only the constant region of the heavy chain changes. This step in the differentiation of an antibody-producing.cell is called *class switching.* In undifferentiated cells, the genes for the constant region of each class of heavy chain, called $C_o$, $C_\phi$, $C_\tau$, $C_\gamma$, and $C_\chi$, are next to each other. There are eight in all, including four genes for the constant regions of ι chains. A complete gene for the heavy chains of IgM antibody is formed by the translocation of a $V_H$ gene segment to a $DJ_H$ gene segment. How are other heavy chains formed?

Class switching is mediated by a gene-rearrangement process that moves a VDJ gene from a site near one C gene to a site near another C gene. Importantly, *the antigen-binding specificity is conserved in class switching because the entire $V_H DJ_H$ gene is- translocated in an intact form.* For example, the antigen-combining specificity of IgA produced by a particular cell is the same as that of IgM synthesized at an earlier stage of its development. The biological significance of $C_H$ switching is that a whole recognition domain (the variable domain) is shifted from the early constant region ($C_O$) to one of several other constant regions that mediate different effector functions.

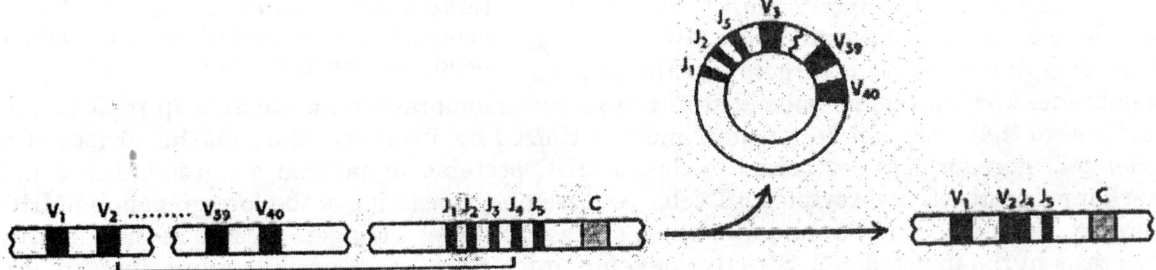


**Fig. 14.16. VJ Recombination. A single V gene (in this case, $V_2$) is linked to a J gene (here, $J_4$) to form an intact VJ region. The intervening DNA is released in a circular form. Because the V and J regions are selected at random and the joint between them is not always in exactly the same place, many VJ combinations can be generated by this process.**

## MAJOR-HISTOCOMPATIBILITY-COMPLEX PROTEINS

Soluble antibodies are highly effective against extracellular pathogens, but they confer little protection against microorganisms that are predominantly intracellular, such as viruses and mycobacteria (which cause tuberculosis and leprosy). These pathogens are shielded from antibodies by the host-cell membrane. A different and more subtle strategy, *cell-mediated immunity,* evolved to cope with intracellular pathogens. *T cells* continually scan the surfaces of all cells and kill those that exhibit foreign markings. The task is not simple; intracellular microorganisms are not so obliging as to intentionally leave telltale traces on the surface of their host. Quite the contrary, successful pathogens are masters of the art of camouflage.

Vertebrates have devised an ingenious mechanism—cut and display—to reveal the presence of stealthy intruders. Nearly all vertebrate cells exhibit on their surfaces a sample of peptides derived from the digestion of proteins in their cytosol. These peptides are displayed by integral membrane proteins that are encoded by the *major histocompatibility complex (MHC).* Specifically, peptides derived from cytosolic proteins are bound to *class I MHC proteins.* How are these peptides generated arid delivered to the plasma membrane? The process starts in the cytosol with the degradation of proteins, self proteins as well as those of pathogens. Digestion is carried out by proteasomes.

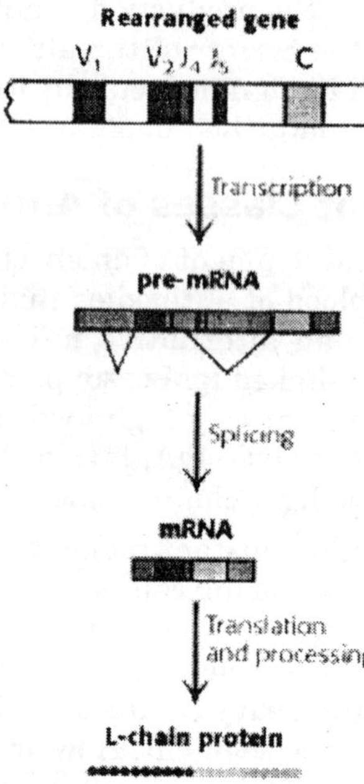

**Fig. 14.17. Light-Chain Expression. The light-chain protein is expressed by transcription of the rearranged gene to produce a pte-RNA molecule with the VJ and C regions separated. RNA splicing removes the intervening sequences to produce an mRNA molecule with the VJ and C regions linked. Translation of the mRNA and processing of the initial protein product produces the light chain.**

The resulting peptide fragments are transported from the cytosol into the lumen of the endoplasmic reticulum by an ATP-driven pump. In the ER, peptides combine with nascent class I MHC proteins; these complexes are then targeted to the plasma membrane. MHC proteins embedded in the plasma membrane tenaciously grip their bound peptides so that they can be touched and scrutinized by T-cell receptors on the surface of a killer cell. Foreign peptides bound to class I MHC proteins signal that a cell is infected and mark it for destruction by cytotoxic T cells. An assembly consisting of the foreign peptide-MHC complex, the T-cell receptor, and numerous accessory proteins triggers a cascade that induces apoptosis in the infected cell. Strictly speaking, infected cells are not killed but, instead, are triggered to commit suicide to aid the organism.

## Peptides Presented by MHC Proteins

The three-dimensional structure of a large fragment of a human MHC class I protein, *human leukocyte antigen A2 (HLA-A2),* was solved in 1987 by Don Wiley and Pamela Bjorkman. Class I MHC proteins consist of a 44-kd $\chi$ chain noncovalently bound to a 12-kd polypeptide called $\delta_2$ - *microglobulin.* The $\chi$ chain has three extracellular domains ($\chi_1$, $\chi_2$, and $\chi_3$), a transmembrane segment, and a tail that extends into the cvtosol. Cleavage by papain of the HLA $\chi$ chain several residues-before the transmembrane segment yielded a soluble heterodimeric fragment. The $\delta_2$-microglobulin subunit and the $\chi_3$ domains have immunoglobulin folds, although the pairing of the two domains differs from that in antibodies. The $\chi_1$ and $\chi_2$ domains exhibit a novel and remarkable architecture.

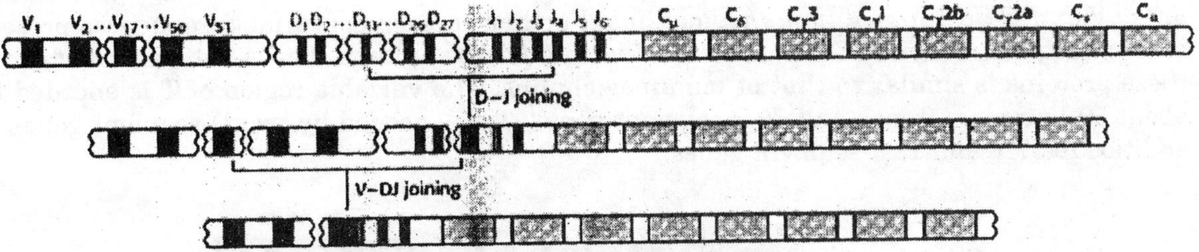


**Fig. 14.18. V(D) J Recombination. The heavy-chain locus includes an array of 51 V segments, 27 D segments, and 6 J segments. Gene rearrangement begins with D-J joining, followed by further rearrangement to link the V segment to the DJ segment.**

They associate intimately to form a deep groove that serves as the peptidc-binding site. The floor of the groove, which is about 25 A long and 10 A wide, is formed by eight 8 strands, four from each domain. A long helix contributed by the $\chi_1$ domain forms one side, and a helix contributed by the $\chi_2$ domain forms the other side. *This groove is the binding site for the presentation of peptides.* The groove can be filled by a peptide from 8 to 10 residues long in an extended conformation. As we shall see, MHC proteins are remarkably diverse in the human population; each person expresses as many as six distinct class I MHC proteins and many different forms are present in different people. The first structure determined, HLA-A2, binds peptides that almost always have leucine in the second position and valine in the last position.

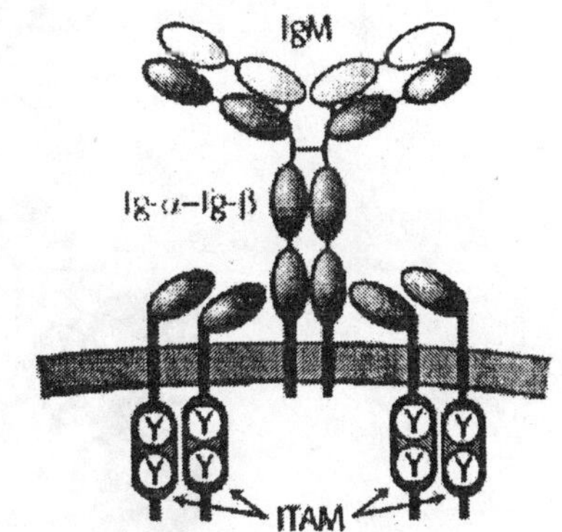


**Fig. 14.19. B–Cell Receptor, This complex consists of a membrane-bound IgM molecule noncovalently bound to two Ig-χ-Ig-δ heterodimers. The intracellular domains of each of the Ig-χ and Ig-δ chains include an immunoreceptor tyrosine-based activation motif (ITAM).**

Side chains from the MHC molecule interact with the arnino and carboxyl termini and with the side chains in these two key positions. These residues are often referred to as the *anchor residues.* The other residues are highly variable. Thus, many millions of different peptides can be presented by this particular class I MHC protein; the identities of only two of the nine residues are crucial for binding. Each class of MHC molecules requires a unique set of anchor residues. Thus, a tremendous range of peptides can'be presented by these molecules. Note that *one face of the bound peptide is exposed to solution where it can be examined by other molecules, particularly T-cell receptors.* An additional remarkable feature of MHC-peptide complexes is their kinetic stability; once bound, a peptide is not released, even over a period of days.

## Antibody-like Proteins

We are now ready to consider the receptor that recognizes peptides displayed by MHC proteins on target cells. The *T-cell receptor* consists of a 43-kd $\chi$ chain ($T_\chi$) joined by adisulfide bond to a43-kd δ chain $T_\delta$ . Each chain spans the plasma membrane and has a short carboxyl-terminal region on the cytosolic side. A small proportion of T cells express a receptor consisting of τ and ϕ chains in place of χ and δ. $T_\chi$ and $T_\delta$ , like immunoglobulin L and H chains, consist *of variable* and *constant* regions. Indeed, *these domains of the T-cell receptor are homologous to the V and C domains of immunoglobulins.* Furthermore, hypervariable sequences present in the V regions of $T_\chi$ and $T_\delta$ form the binding site for the epitope. The genetic architecture of these proteins is similar to that of immunoglobulins. The variable region of T is encoded by about 50 Vsegment genes and 70 J-segment genes. $T_\delta$ is encoded by two D-segment genes in addition to 57 V and 13 J-segment genes.

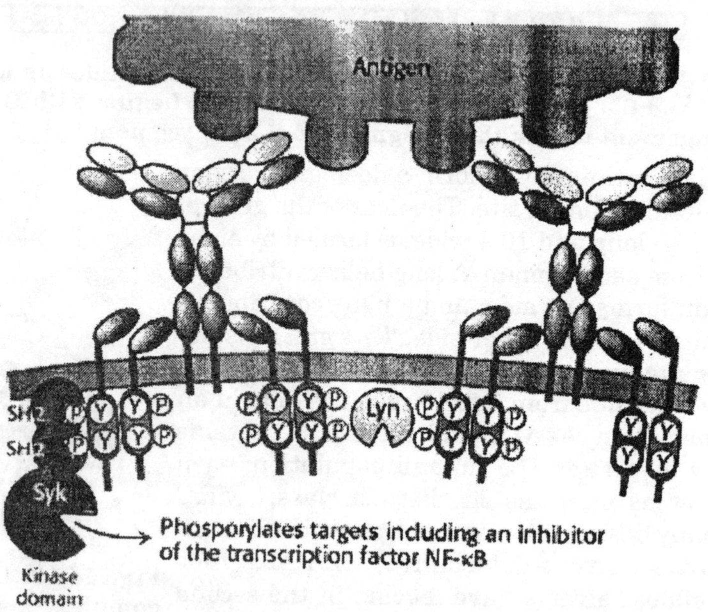


**Fig. 14.20. B-Cell Activation. The binding of multivalent antigen such as bacterial or viral surfaces links membrane-bound IgM molecules. This oligomerization triggers the phosphorylation of tyrosine residues in the ITAM sequences by protein tyrosine kinases such as Lyn. After phosphorylation, the ITAMs serve as docking sites for Syk, a protein kinase that phosphorylates a number of targets, including transcription factors.**

Again, the diversity of component genes and the use of slightly imprecise modes of joining them increase the number of distinct proteins formed. *At least $10^{12}$ different specificities could arise from combinations of this repertoire of genes.* Thus, T-cell receptors, like immunoglobulins, can recognize a very large number of different epitopes. All the receptors on a particular T cell have the same specificity. How do T cells recognize their targets? The variable regions of the χ and δ chains of the T-cell receptor form a binding site that recognizes a combined epitope-foreign peptide bound to an MHC prntein. Neither the foreign peptide alone nor the MHC protein alone forms a complex with the T-cell receptor. Thus, fragments of an intracellular pathogen are presented in a context that allows them to be detected, leading to the initiation of an appropriate response.

## Cytotoxic T Cells

The T-cell receptor does not act alone in recognizing and mediating the fate of target cells. Cytotoxic T cells also express a protein termed *CDS* on their surfaces that is crucial for the recognition of the class I MHC-peptide complex. The abbreviation CD stands for *cluster of differentiation,* referring to a cell-surface marker that is used to identify a lineage or stage of differentiation. Antibodies specific for particular CD proteins have been invaluable in following the development of leukocytes and in discovering new interactions between specific cell types. Each chain in the CD8 dimer contains a domain that resembles an.immunoglobulin variable domain CD8 interacts primarily with the relatively constant χ3 domain of class I MHC proteins. This interaction further stabilizes the interactions between the T cell and its target. The cytosolic tail of CDS contains a docking site for Lck, a cytosolic tyrosine kinase akin to Src. The T-cell receptor itself is associated with six polypeptides that form the CD3 complex. The ι, ϕ, and γ chains of CD3 are homologous to Ig-χ and Ig-δ associated with the B-cell receptor 53; each chain consists of an extracellular immunoglobulin domain and an intracellular ITAM region. These chains associate into CD3 ι, ϕ and CD3 ϕ γ 7 heterodimers. An additional component, the CD3 I chain, has only a small extracellular domain and a larger intracellular domain containing three ITAM sequences. On the basis of these components, a model for T-cell activation can be envisaged that is closely parallel to the pathway for B-cell activation. The binding of the T-cell receptor with the class I MHC-peptide complex and the concomitant binding of CDS from the T-cell with the MHC molecule results-in the association of the kinase Lck with the ITAM substrates of the components of the CD3 complex. Phosphorylation of the tyrosine residues in the ITAM sequences generates docking sites for a protein kinase called ZAP-70 (for 70-kd zeta-associatedprotein) that is homologous to Syk in B cells. Docked by its two SH2 domains, ZAP-70 phosphorylates downstream targets in the signaling cascade. Additional molecules, including a membrane-bound protein phosphatase called CD45 and a cell-surface protein called CD28, play ancillary roles in this process. T-cell activation has two important consequences.

First, the activation of cytotoxic T cells results in the secretion of *perform.* This 70-kd protein makes the cell membrane of the target cell permeable by polymerizing to form transmembrane pores 10 nm wide. The cytotoxic T cell then secretes proteases called *granzymes* into the target cell. These enzymes initiate the pathway of apoptosis, leading to the death of the target cell and the fragmentation of its DNA, including any viral DNA that may be present. Second, after it has stimulated its target cell to commit suicide, the activated T cell disengages and is stimulated to reproduce. Thus, additional T cells that express the same T-cell receptor are generated to continue the. battle against the invader after these T cells have been identified as a suitable weapon.

## Class II MHC Proteins

Not all T cells are cytotoxic. *Helper T cells, a different class, stimulate the pro-liferation of specific B lymphocytes and cytotoxic T cells and thereby serve as partners in determining the immune responses that are produced.* The importance of helper T cells is graphically revealed by the devastation wrought by AIDS, a condition that destroys these cells. Helper T cells, like cytotoxic T cells, detect foreign peptides that are presented on cell surfaces by MHC proteins. However, the source of the peptides, the MHC proteins that bind them, and the transport pathway are different. Helper T cells recognize peptides bound to MHC molecules referred to as class II. Their helping action is focused on B cells, macrophages, and dendritic cells. *Class II MHC proteins* are expressed only by these *antigen-presenting cells,* unlike class I MHC proteins, which are expressed on nearly all cells.

The peptides presented by class II MHC proteins do not come from the cytosol. Rather, *they arise from the degradation of proteins~that have been internalized by endocytosis.* Consider, for example, a virus particle that is captured by membrane-bound immunoglobulins on the surface of a B cell. This complex is delivered to an endosome, a membrane enclosed acidic compartment, where it is digested. The resulting peptides become associated with class II MHC proteins, which move to the cell surface. Peptides from the cytosol cannot reach class II proteins, whereas peptides from endosomal compartments cannot reach class I proteins. This segregation of displayed peptides is biologically critical. The association of a foreign peptide with a class II MHC protein signals that a cell has *encountered* a pathogen and serves as a call for *help.* In contrast, association with a class I MHC protein signals that a cell has *succumbed to* a pathogen and is a call for *destruction.*

## Foreign Peptides on Antigen-Presenting Cells

The overall structure of a class II MHC molecule is remarkably similar to that of a class I molecule. Class II molecules consist of a 33-kd *Z* chain and a noncovalently bound 30-kd δ chain. Each contains two extracellular domains, a transmembrane segment, and a short cytosolic tail. The peptide-binding site is formed by the $\chi_1$ and $\delta_1$ domains, each of which contributes a long helix and part of a δ sheet. Thus, the same structural elements are present in class 1 and class II MHC molecules, but they are combined into polypeptide chains in different ways. Class II MHC molecules appear to form stable dimers, unlike class I molecules, which are monomeric. The peptide-binding site of a class II molecule is open at both ends, and so this groove can accommodate longer peptides than can be bound by class I molecules; typically, peptides between 13 and 18 residues long are bound. The peptide-binding specificity of each class II molecule depends on binding pockets that recognize particular amino acids in specific positions along the sequence. Helper T cells express T-cell receptors that are produced from the same genes as those on cytotoxic T cells. These T-cell receptors interact with class II MHC molecules in a manner that is analogous to T-cell-receptor interaction with class I MHC molecules. Nonetheless, helper T cells and cytotoxic T cells are distinguished by other proteins that they express on their surfaces. In particular, helper T cells express a protein called CD4 instead of expressing CD8. *CD4* consists of four immunoglobulin domains that extend from the T-cell surface, as well as a small cytoplasmic region.

The amino-terminal immunoglobulin domains of CD4 interact with the base on the class II MHC molecule. Thus, helper T cells bind cells expressing class II MHC specifically because of the interactions with CD4. When a helper T cell binds to an antigen-presenting cell expressing an appropriate class II MHC-peptide complex, signaling pathways analogous to those in cytotoxic T cells are initiated by the action of the kinase Lck on ITAMs in the CD3 molecules associated with the T-cell receptor. However, rather than triggering events leading to the death of the attached cell, *these signaling pathways result in the secretion of cytokines from the helper cell.* Cytokines are a family of molecules that include, among others, interleukin-2 and interferon.

Cytokines bind to specific receptors on the antigenpresenting cell and stimulate growth, differentiation, and in regard to plasma cells, which are derived from B cells, antibody secretion. Thus, the internalization and presentation of parts of a foreign pathogen help to generate a local environment in which cells taking part in the defense against this pathogen can flourish through the action of helper T cells.

## Diverse MHC Proteins

MHC class I and II proteins, the presentears of peptides to T cells, were discovered because of their role in *transplantation rejection.* A tissue transplanted from one person to another or from one mouse to another is usually rejected by the immune system. In contrast, tissues transplanted from one identical twin to another or between mice of an inbred strain are acceptead. Genetic analyses revealed that rejection occurs'when tissues are transplanted between individuals having different genes in the major histocompatibility complex, a cluster of more than 75 genes playing key roles in immunity.

The 3500-kb span of the MHC is nearly the length of the entire *E. coli* chromosome. The MHC encodes class I proteins (presenters to cytotoxic *T* cells) and class II proteins (presenters to helper T cells), as well as class III proteins (components of the complement cascade) and many other proteins that play key roles in immunity. Human beings express six different class I genes (three from each parent) and six different class II genes. The three loci for class I genes are called HLA–A, –B, and –C; those for class II genes are called HLA–DP, –DQ, and –DR. These loci are *highly polymorphic:* many alleles of each are present in the population. For example, more than 50 each of HLA–A, –B, and –C alleles are known; the numbers discovered increase each year.

Hence, the likelihood that two unrelated persons have identical class I and II proteins is very small ($<10^{-4}$), accounting for transplantation rejection unless the genotypes of donor and acceptor are closely matched in advance, Differences between class I proteins are located mainly in the $\chi_1$ and $\chi_2$ domains, which form the peptide-binding site. The $\chi_3$ domain, which interacts with a constant $\delta_2$-microglobulin is largely conserved. Similarly, the differences between class II proteins cluster near the peptide-binding groove. Why are MHC proteins so highly variable? *Their diversity makes possible the presentation of a very wide range of peptides to T'cells.* A *particular* class I or class II molecule may not be able to bind any of the peptide fragments of a viral protein.

The likelihood of a fit is markedly increased by having several kinds (usually six) of each class of presenters in eacn individual. If all members of a species had identical class I or class II molecules, the population would be much more vulnerable to devastation by a pathogen that had evolved to evade presentation. The evolution of the diverse human MHC repertoire has been driven by the selection for individual members of the species who resist infections to which otler members of the population may be susceptible.

## Human Immunodeficiency Viruses

In 1981, the first cases of a new disease now called *acquired immune deficiency syndrome (AIDS)* were recognized. The victims died of rare infections because their immune systems were crippled. The cause was identified two years later by Luc Montagnier and coworkers. AIDS is produced by *human immunodeficiency virus (HIV),* of which two major classes are known: HIV-1 and the much Less common HIV-2. Like other *retroviruses,* HIV contains a single-stranded RNA genome that is replicated through a double-stranded DNA intermediate. This viral DNA becomes integrated into the genome of the host cell. In fact, viral genes are transcribed only when they are integrated into the host DNA. The HIV virion is enveloped by a lipid bilayer membrane containing two glycoproteins:gp41 spans the membrane and is associated with gp120, which is located on the external face.

The core of the virus contains two copies of the RNA genome and associated transfer RNAs, and several molecules of reverse transcriptase. They are surrounded by many copies of two proteins called p18 and p24. *The host cell for HIV is the helper T cell.* The gp120 molecules on the membrane of HIV bind to CD4 molecules on the surface of the helper T cell. This interaction allows the associated viral gp41 to insert its amino-terminal head into the host-cell membrane. The viral membrane and the helper-cell membrane fuse, and the viral core is released directly into the cytoso. Infection by HIV leads to the destruction of helper T cells because the permeability of the host plasma membrane is markedly increased by the insertion of viral glycoproteins and the budding of virus particles.

The influx of ions and water disrupts the ionic balance, causing osmotic lysis. The development of an effective AIDS vaccine is difficult owing to the antigenic diversity of HIV strains. Because its mechanism for replication is quite error prone, a population of HIV presents an everchanging array of coat proteins. Indeed, the mutation rate of HIV is more than 65 times as high as that of influenza virus. A major aim now is to define relatively conserved sequences in these HIV proteins and use them as immunogens.

## RESPONSES AGAINST SELF-ANTIGENS

The primary function of the immune system is to protect the host from invasion by foreign organisms. But how does the immune system avoid mounting attacks against the host organism? In other words, how does the immune system distinguish between self and nonself? Clearly, proteins from the organism itself do not bear some special tag identifying them. Instead, selection processes early in the developmental pathways for immune cells kill or suppress those immune cells that react strongly with self-antigens. The evolutionary paradigm still applies; immune cells that recognize self-antigens are generated, but selective mechanisms eliminate such cells in the course of development.

### Positive and Negative Selection in the Thymus

T cells derive their name from the location of their production — the thymus, a small organ situated just above the heart. Examination of the developmental pathways leading to the production of mature cytotoxic and helper T cells reveals the selection mechanisms that are crucial for distinguishing self from nonself. These selection criteria are quite stringent; approximately 98% of the thymocytes, the precursors of T cells, die before the completion of the maturation process. Thymocytes produced in the bone marrow do not express the T-cellreceptor complex, CD4, or CD8. On relocation to the thymus and rearrangement of the T-cell-receptor genes, the immature thymocyte expresses all of these molecules.

These cells are first subjected *to positive selection*. Cells for which the T-cell receptor can bind with reasonable affinity to either class I or class II MHC molecules survive this selection; those for which the T-cell receptor does not participate in such an interaction undergo apoptosis and die. The affinities of interaction required to pass this selection are relatively modest, and so contacts between the T-cell receptor and the MHC molecules themselves are sufficient without any significant contribution from the bound peptides (which will be derived from proteins in the thymus). *The role of the positive selection step is to prevent the production of T cells that will not bind to any MHC complex present, regardless of the peptide bound.* The cell population that survives positive selection is subjected to a second step, *negative selection.*

Here, T cells that bind with high affinity to MHC complexes bound to self-peptides expressed on the surfaces of antigen-presenting cells in the thymus undergo apoptosis or are otherwise suppressed. Those that do not bind too avidly to any such MHC complex complete development and become mature cytotoxic T cells (which express only CD8) or helper T cells (which express only CD4). The negative selection step leads to *self tolerance;* cells that bind an MHC-self-peptide complex are removed from the T-cell population. Similar mechanisms apply to developing B cells, suppressing B cells that express antibodies that interact strongly with self-antigens.

## Autoimmune Diseases

Although thymic selection is remarkably efficient in suppressing the immune response to self-antigens, failures do occur. Such failures results in *autoimmune diseases.* These diseases include relatively common illnesses such as insulin-dependent diabetes mellitus, multiple sclerosis, and rheumatoid arthritis. In these illnesses, immune responses against self-antigens result in damage to selective tissues that express the antigen. In many cases, the cause of the generation of self-reactive antibodies or T cells in unclear.

However, in other cases, infectious organisms such as bacteria or viruses may play a role. Infection leads to the generation of antibodies and T cells that react with many different epitopes from the infectious organism. If one of these antigens closely resembles a self-antigen, an autoimmune response can result. For example, *Streptococcus* infections sometimes lead to rheumatic fever owing to the production of antibodies to streptococcal antigens that cross-react w/ith exposed molecules in heart muscle.

## The Immune System and Cancer

The development of immune responses against proteins encoded by our own genomes can be beneficial under some circumstances. Cancer cells have undergone significant changes that often result in the expression of proteins that are not normally expressed. For example, the mutation of genes can generate proteins that do not correspond in amino acid sequence to any normal protein. Such proteins may be recognized as foreign, and an immune response will be generated specifically against the cancer cell.

Alternatively, cancer cells often produce proteins that are expressed during embryonic development but are not expressed or are expressed at very low levels after birth. For example, a membrane glycoprotein protein called *carcinoembryonic antigen (CEA)* appears in the gasterointestinal cells of developing fetuses but is not normally expressed at significant levels after birth. More than 50% of patients with colorectal cancer have elevated serum levels of CEA. Immune cells recognizing epitopes from such proteins will not be subject to negative selection and, hence, will be present in the adult immune repertoire. These cells may play a cancer surveillance role, killing cells that overexpress antigens such as CEA and preventing genetically damaged cells from developing into tumors.

## MAPLE SYRUP URINE DISEASE AND JAMAICAN VOMITING SICKNESS

In a rare autosomal recessive condition (discovered in 1954) the urine and perspiration has a maple syrup odor. High concentrations of the branched-chain 2-oxoacids formed by transamination of valine, leucine, and isoleucine are present, and the odor arises from decomposition products of these acids. The branched-chain amino acids as well as the related alcohols also accumulate in the blood and are found in the urine. The biochemical defect lies in the enzyme catalyzing oxidative decarboxylation of the oxoacids. Insertions, deletions, and substitutions may be present in any of the subunits

The disease which may affect one person in ~ 200,000, is usually fatal in early childhood if untreated. Children suffer seizures, mental retardation, and coma. They may survive on a low-protein (gelatin) diet supplemented with essential amino acids, but treatment is difficult and a sudden relapse is apt to prove fatal. Some patients respond to administration of thiamin at 20 times the normal daily requirement. The branched-chain oxoacid dehydrogenase from some of these children shows a reduced affinity for the essential coenzyme thiamin diphosphate. Polled hereford calves in Australia develop maple syrup urine disease relatively of often. One cause was established as a mutation that introduces a stop codon that causes premature termination within the leader peptide during synthesis of the thiamin diphosphate-dependent El suburut.

A similar biochemical defect in Trnuta of *'Bacillus subtilis* causes difficulties for this bacterium, which requires branched-chain fatty acids in its membranes. Branched acyl-CoA derivatives are needed as starter pieces for their synthesis. With the oxidative decarboxylation of the necessary oxoacids blocked, the mutant is unable to grow unless supplemented with branched-chain fatty acids. Because persons may be born with defects in almost any gene, a variety of other problems leading to accumulation of organic acids are also known. *Methylmalonic aciduria* and propionic acidemia and propionic acidemia. *Lactic acidemia* often results from a defect in pyruvate dehydrogenase.

A rare defect of catabolism of leucine is *isovaleric acidemia,* a failure in oxidation of isovaleiy 1-CbA. The symptoms of this disease are also present in the Jamaican vomiting sickness, caused by eating unripe ackee fruit. Although the ripe fruit is safe to eat, unripe fruit contains a toxin *hypoglycin A* with the following structure. It is metabolized to an acyl-CoA derivative as shown. This is an enzyme-activated inhibitor of the medium-chain fatty acyl-CoA dehydrogenase required for β oxidation of fatty acids.

The compound also inhibits isovaleryl-CoA dehydrogenase, causing an accumulation of isovaleric acid in the blood. Depression of the central nervous system by isovaleric acid in the blood could be responsible for some symptoms. However, death from the highly fatal Jamaican vomiting sickness comes from the hypoglycemic effect. Blood glucose levels may fall as low as 0.5 mM, one-tenth the normal concentration and patients must be treated by infusion of glucose.

## SULFUR COMPOUNDS OF GARLIC, ONIONS, SKUNKS, ETC.

Many familiar odors and tastes come from sulfur-containing com-potinds. Crushing onions or garlic releases the pyridoxal phosphate-dependent enzyme *alliinase.* In garlic it acts upon the amino acid *alliin* (accompanying scheme) releasing, by β elimination, a sulfenic acid that dimerizes to form *allicin,* a chemically unstable molecule that accounts for the odor of garlic.

Among the breakdown products of allicin is the nonvolatile ajoene, a compound with anticoagulant activity and perhaps accounting for one aspect of the purported medical benefits of garlic. Another is an antibacterial activity.

Onions contain an amino acid that is a positional isomer of alliin. When acted upon by alliinase it produces 2-propene sulfenic acid, which isomerizes to the *lachrimatory factor* that brings tears to the eyes of onion cutters.c This, too, decomposes to form many other compounds.

The defensive secretion of the striped skunk has intrigued chemists for over 100 years. The components were shown to contain sulfur, and one was incorrectly identified and was long accepted as being butyl mercaptan. Modern capillary gas chromatography by Wood as revealed the presence of seven major components, with the indicated structures. Two are simple volatile mercaptans, but three are thioacetates which hydrolyze in water only slowly, releasing smell for days or weeks from sprayed animals. Washing with mildly basic soap hastens the hydrolysis.

Many readers (~ 40%) may be aware that after eating asparagus a strong odor appears in their urine. These genetic "stinkers" secrete S-methyl thioacrylate, and related compounds, derived from a plant constituent.

The sulfur compound sulfuraphane, extracted from fresh broccoli, has received attention in recent years because of its strong action in inducing synthesis of quinone reductase and glutathione *S*-transferases that help detoxify xenobiotics and may have significant anticancer activity.

# Table of Contents

# HOW TO START A BUSINESS FOR FREE

## The Ultimate Guide to Building Something Profitable from Nothing

SILVER LAKE PUBLISHING
LOS ANGELES, CALIFORNIA

# How to Start a Business for Free

**The Ultimate Guide to Building Something Profitable from Nothing**

First edition


Silver Lake Publishing
2025 Hyperion Avenue
Los Angeles, California 90027

For a list of other publications or for more information, please call 1.888.663.3091. In Alaska and Hawaii and outside of the United States, please call 1.323.663.3082.



Library of Congress Catalogue Number: pending

How to Start a Business for Free
The Ultimate Guide to Building Something from Nothing
Includes index.
Pages: 308

ISBN: 1-56343-770-8
Printed in the United States of America.

# Introduction: Getting Started

There are many reasons to go into business for yourself. Maybe you want to expand your earning potential or you want flexible work hours and the best parking spot. Or, maybe you're just tired of working for someone else and commuting two hours a day.

Some people want to start their own business because their current employer is downsizing and it seems easier than pounding the pavement looking for a new job. Others want to start a business to gain the freedom they've always dreamed of because being "self-employed" affords you the freedom to control your own destiny.

Let's face it, everyone—from newspapers, magazines and TV to radio, Learning Annex seminars and the Internet—is touting why you should go into business for yourself. You've probably seen the signs on freeway on-ramps that entice you to think about self-employment. *Have you ever wanted to be your own boss? Set your own hours? Work from home? Want the freedom to work when you want? Are you thinking about starting your own company but don't know where to start?* The list goes on and on.

Being in business for yourself, however, isn't always what it's cracked up to be. Being in business for yourself often means working long hours—starting early in the morning and working late into the evening most days and even putting time in on weekends. Being in business for yourself means postponing that vacation to Jamaica because there is no one else to run the business when you're not there. And, the little things, like movies, golf, hiking or catching an occasional ball game, are no longer recreation when you own your own business. They become luxuries.

Whatever the reason, starting your own business requires an independent spirit and a strong sense of self-motivation to stand out from others in your marketplace. And there are plenty of other entrepreneurs in the United States willing to take that risk and prove to the world that they have those qualities. Why not? According to the U.S. Department of Labor, firms with fewer than 500 employees employ 55 percent of the private, non-farm work force, contribute 48 percent of all sales in the country and are responsible for 51 percent of the private gross domestic product. In addition, small business-dominated industries produced an estimated 68 percent of the 2 million new jobs created during 2000, according to the U.S. Departments of Labor and Commerce.

If that's not enough motivation to get your business plan up and running, according to the Small Business Administration, small businesses:

- represent more than 99 percent of all employers;
- employ 52 percent of the private workers;
- employ 61 percent of the private workers on public assistance;
- employ 38 percent of the private workers in high-tech occupations;
- provide virtually all of the net new jobs;
- provide 51 percent of the private sector output;
- represent 96 percent of all exporters of goods;
- receive 35 percent of federal contract dollars; and
- are home-based 53 percent of the time and are franchises 3 percent of the time.

This means if you decide to start your own business, you will not be alone...not by a long shot! Moreover, as you start to plan what kind of business you will start and how you will finance it, it's a good idea to find other entrepreneurs in your community to help you keep your head above water. Other entrepreneurs can help you avoid some of the common pitfalls tied to owning your own business. They've been down the same road you are about to take, so find them and ask as many questions as you can. We'll go into greater detail on mentoring resources in a later chapter, but it's never too early to ask for direction.

As you begin to talk to others who have made the small-business leap and look at other businesses that are operating in the economy around you, you'll see that the opportunities for entrepreneurship are abound, waiting for the taking. You can turn anything you love into a money-making opportunity, and this book will show you *how*. Hundreds of businesses can be started for little or no capital at all. For example, a house-sitting or tax-preparation business. Or maybe you're more into the dog-grooming or diversity training business. Take the founders of Nantucket Nectars for example, otherwise known as the "Juice Guys" who took a sloppy boat business and turned it into a multimillion-dollar company.

Tom Scott and Tom First didn't want the corporate job with the corporate car or the morning commute. They had flunked accounting, the only business-related course under their belts, but knew when they graduated from Brown University in 1989 that they wanted to live on Nantucket year-round and make something work—on their own. Scott had already worked in the harbor the previous year as a taxi driver. He didn't want to work for someone else, so he started Allserve, a floating convenience store, on a 19-foot Boston Whaler that drove around the harbor servicing the Nantucket boating community. Scott sold muffins, delivered newspapers, disposed of trash and even did some people's laundry—and he loved it because he was working for himself, outside, on a boat and making money.

The following summer, First joined Scott and the two expanded the business to include boat towing, repairs and rescue. Getting through the cold and slow winter, however, was rough. One night, First made a juice blend for dinner and within five minutes, the two Toms were joking with one another, "Let's sell this off the boat next summer. We'll call it Nantucket Nectars."

The rest is history. But it's not history without a few low points and some major hits. In the start-up phase, the two Toms did everything they could to stay afloat. They sold their juice concoctions off their Whaler and eventually out of a little storefront on the Straight Wharf of Nantucket (which is still there). They never lost sight of their goal to survive on Nantucket year-round and maintain a business. And, knowing that their floating convenience store was too seasonal to work in the long-run, however, the Toms settled on making the juice company work. Without fancy funding, the two Toms resorted to making money elsewhere until the business got going. They performed oddball jobs here and there, including shucking scallops, painting houses, bartending and pumping waste. They also focused on the quality of their juice product (innovative bottle design and flavors).

The two Toms paid bills slowly, collected receivables as fast as possible and paid themselves nothing. At one point, Tom Scott lived in his car or in a group house with no heat to make ends meet. Eventually though, the business grew and an angel investor (coincidentally a client they had serviced in the past) kicked in some money to help move the company along.

In 1999, Nantucket Nectars reported $60 million in sales. One thing, however, did have to change with such rapid growth: the company grew too large for Nantucket and moved to Cambridge, Massachusetts. But by then, the Juice Guys had made their mark in the juice world. They could sail over to Nantucket in their own boat whenever they needed some island healing.[1]

The lessons from the two Toms are clear: Success in business doesn't necessarily start with an expensive MBA or a windfall of venture capital. It starts by asking yourself the most basic of questions: What do you want in life and where and how do you want to live? Focus on those essentials and other things will likely follow. And it's always key to make friends, even in bars and harbors, whom you can call when you need some cash.

[1] Ocean Spray Cranberries, Inc. bought a large piece of the company in early 1998—which was later sold to Cadberry Schweppps—but Scott and First retain a large piece of the company's ownership and remain active in its management.

Never underestimate your own financial power. Between boot-strapping your business and asking Aunt Betty for a loan to start your company, digging deep into your own pockets first is a step toward getting what you need.

Raising capital outside your close network of friends and neighbors is a hard thing to do these days. Capital is scarce. The late 1990s had people wrongly believe that venture capital is easy to snatch up—for even the rookie entrepreneur in need of start-up cash (to also finance a fancy car and lavish parties.) That may have been the case back during the heady dot.com days, but those days are over and the current venture capital market is closer to how it usually works. It takes a lot more of a sales pitch to get noticed.

In reality, the vast majority of successful new businesses are self financed. Among the honored businesses that made *Inc.* magazine's 500 list over the past 20 years are Microsoft, Domino's Pizza, Oracle, Jenny Craig, Pete's Brewing, Jamba Juice, The Sharper Image, Princeton Review and Patagonia. When those companies made the list, all were privately owned and at least five years old. Few, if any, started with institutional capital. And, in 2001 only 3 percent of the *Inc.* 500 companies received venture-capital funds at start-up.[2]

This book answers the questions you may have about starting and running your own business, and attempts to resolve those fuzzy lines between writing a business plan and actually putting the plan into action. Starting and running your own business takes hard work. You need the wisdom, discipline, courage and persistence to know how to best protect your investment whether you're running a carpentry business or doing consulting work as a motivational speaker.

This book explains in simple terms the types of business structures that exist, and provides a working knowledge of how to develop the financing, marketing, product development and operations of your own business. Whether your idea for a business is big or small, you'll want to

[2] "Brief Profiles of 2001 *Inc.* 500 Companies"; *Inc.* magazine, 11/01.

be informed about things like networking resources, grass roots advertising, meeting the competition and license and registration.

None of the material here will make you an expert on running a multimillion dollar business. However, we hope to take some of the mystery out of the nuts and bolts of start-ups so that you can make conscious decisions in your endeavors—business or otherwise—minimize costs and maximize profits.

And minimizing costs is probably the most important aspect of starting a business. Since so many new businesses fail within the first year or two that they exist, surviving the early going can be an end in itself. Your struggle is to stay in the game while your name or brand gets established…until your channels to market get set…until you can attract the right people to help you grow.

This is where what I call the Free Mentality comes in to play. Obviously, starting a business requires spending some money along the way—but, because survival is so tough, saving every dollar possible is essential during the first years. You need to get everything that you can as cheaply as possible. Free, if you can.

Pressing that point is what this book is about.

# CHAPTER 1: CHOOSING A BUSINESS

Many start-up businesses thrive on little or no capital. They are the result of executing at least one good idea, whether it's a new product or a new or better personal or professional service; an advertising or fundraising campaign; or a concept for an accurate translation, proposal writing or lawn mowing service.

Exactly what kind of business can you start up on your own? The decision often is a personal one. The decision is one that relates to you and your life…and the things you enjoy. Running a business is not always easy, and there's no reason to struggle and work at it unless you're doing something you truly enjoy. In other words, you don't want to run a business you hate. So, choose something that you're interested in doing and that you like. If you can't stand dogs, don't run out and open a dog grooming business. Unless, of course, you've invented a whole new and innovative concept for dog grooming that inspires you.

You also don't have to limit yourself to just one type of business. If one idea doesn't make enough money, you can supplement it by broadening your business offerings and expanding into a related area.

Let's say you love butter, but you can't stand the way it's sold. You don't like the sticks, you don't like the tubs…or the convenient squirt-top bottles. Try inventing a better way to dispense butter or margarine, something you might call the Butterfly and start-up a kitchen appliance company.

Choosing a business starts with asking yourself a few basic questions. The most important question: What do you enjoy doing? Can it be translated into a business? Does it revolve around a service or product? Compile a list of businesses that you would like to start. The business can be derived from a creative idea or a unique and new invention or service. Or, it can be derived from something you already know how to do. Try not to think so much about money when you make this list. For now, let your imagination wander and see where it takes you. There will be plenty of time to hone that idea down to a realistic endeavor.

## Service-Based vs. Product-Based Businesses

An important thing to consider when you're brainstorming for business ideas is the type of business you want to start. In other words, decide whether you want to start a service-based business or an item or product-based business. These are the two basic types of businesses you can start quickly and easily.

The first, service-oriented businesses, allow you to perform a service for your client, such as pet sitting, dog walking, catering, writing, consulting, etc. Service-oriented businesses are easy to start with little seed money. All you need to get started is an idea for a service that you want to provide, some recommendations and a lot of word-of-mouth advertising.

The other type of business you can start is an item- or product-oriented business. With this type of business you provide an item or product for your client. You can offer customized shirts for hunters, handmade birdhouses or back scratchers. Any of these products could be the route to starting a successful business. One thing to keep in mind: Product-oriented businesses take a little more start-up capital than service businesses because you need to come up with the resources to make the products and find outlets to sell them, but they can be profitable and rewarding, too.

Although service-based businesses are designed to sell expertise and assistance rather than concrete or tangible goods like product-based businesses, the goals of service-based and product-based businesses overlap in many respects.

As with all businesses, economics play a pivotal role in the decisions you make as an entrepreneur. Although there's a risk involved in starting your own business—service- or product-based—and filing the paperwork for a business license and obtaining a legal trademark or applying for a grant seem like a hassle and a huge burden on your time, if you play your cards right, you'll be successfully rewarded for your time, energy and research.

When choosing a business, you must first determine the value of your business idea before you can do something useful with it. Weigh the potential value of your business idea against: 1) the probability of others seeing that value; and 2) the costs of securing and maintaining that idea.

Remember: There's nothing more subjective than *value*; what you may think is valuable is probably worthless to someone else. If you're not sure, ask around. Input from others can save you a lot of time and money in the long-run.

Another factor to consider when choosing a business: the Internet. The boom of the Internet Age adds another factor, and possibly more value, to your business idea. The Internet offers a new delivery system for transmitting business ideas—books, software, data, customer lists, advertising campaigns, manufactured products, professional services, newsletters, etc. We'll go into more detail on using the Internet to your benefit later in the book.

## Providing a Service

The most commonly recognized types of businesses that can be started for free are service-based businesses, which fall into the following categories:

- personal services;
- consulting services;

- training services;
- creative services; and
- business services.

Out of all the businesses you thought you would like to start, are any of them service-based? On the following pages we'll consider each type of service-based business in turn.

## Taking Care of People and Their Possessions

Most people have at least one chore they simply don't have time—or want—to do. By starting a personal service-oriented business, you can take care of that need. Ask around your neighborhood. Ask your old co-workers. Does anyone need a cook? Is someone looking for a house-sitter next month? If you love to cook, offer to cater a party or be an in-home chef. If you love pets, start your own pet-sitting or dog-walking business. Just think of what you love and what other people need. This type of business is easy to start and relatively cheap. You can post fliers in your neighborhood in an attempt to get some initial clients. The following are a few more ideas to help get you started.

❖ ***Dog Walker***. If you love animals and live in a neighborhood of busy professionals, a professional dog walking service may be a good business to start. Even if dog owners know their dogs can stay inside from morning to night without having any mishaps, with a little savvy marketing, you can make it clear that it is better for their dog to be out and about during the day. Would you want to be cooped up inside all day? And because your customers provide the dog and the leash, there's little start-up cost on your end. In fact, all it takes to start a dog-walking business are scheduling skills and good sales ability. Start by posting fliers or placing ads all around town. In most cases, the people who need your services are in your own neighborhood.

❖ ***Pet Sitter***. This is another area where a love for animals proves beneficial to the start-up process. Catering to the needs of pet owners needing help during the work week, vacations or illness can be a lucrative business. Although some pet owners unload their pets on a dog or cat kennel while they go on vacation, animals are always more comfortable in familiar surroundings. If you start a professional pet sitting business, you can offer several services, including overnight service or day service where you visit the client's house once or twice a day to feed the pet, walk it if necessary or play with it for a short time. You can also offer to care for the pet in your own home. Other responsibilities may include: trips to the vet or ensuring that the pet gets its medication. As with the dog walking business, the start-up cost for this business is minimal. Start by posting fliers or placing ads all around town. Ask neighbors or friends and family if they need anyone or know of someone who does.

- ***House Sitter***. An empty home is an easy and tempting target for burglars. Even if a homeowner remembers to put a hold on his newspaper subscription, sets automatic timers for the lights, radios and televisions or lowers the sound on the telephone ringers and answering machine, he doesn't fool burglars. Burglars often know when no one is home. One solution: a house sitter. A professional house sitter can provide a valuable service when a client takes a trip or an extended stay away from home.

  As a house sitter, you can provide the following services: pick up newspapers, mail and other things delivered to the house; mow the lawn or shovel the snow out of the driveway, depending on the season; check to ensure that the automatic timers are working on lights, stereos, televisions, etc., and that no light bulbs have burned out; park a car in the driveway; or check the house for any other problems that arise such as a freezer breaking down, storm damages, broken pipes, etc. You can also offer a plant-watering service, look after the owner's pool or offer to take care of their pets for another small fee. Like pet sitting and dog walking, there is little start-up cost for this business and, in addition to the fee for your services, this job can provide you with a roof over your head if you offer live-in services. You may even find yourself spending a long period of time bouncing from house sitting job to house sitting job without having to fork over money for a monthly rent check. Post fliers at the grocery store or ask businesses if you can post them on their bulletin boards. Friends and neighbors are also a good starting point. Ask around at work, your co-workers may know someone who needs your services.

A note on résumés: With most of these businesses, it's also important to prepare a résumé that outlines your work skills and experience. Your résumé must "sell you" to a prospective employer and show him that you have all the requirements for the position you are applying for. Your résumé should include information about the jobs you've held as well as your accomplishments, skills and experience. References are also an important part of a résumé. For a summary of the résumé resources available to you on the Internet, go to **http://jobstar.org/tools/resume/res-web.cfm**.

❖ ***Professional Organizer***. There are people in this world who do not possess a single organizational skill. And, there are others who work full-time while raising a family of five and simply don't have time to keep everything they own in order. As a professional organizer, you can step in, provide assistance and help control the situation. Do you enjoy designing filing systems? Do you look at a messy closet and consider it a challenge? If so, this business may be right up your alley. As a professional organizer you can provide clients with ideas, information, structure, solutions and systems to help their businesses function better. Professional organizers assist businesses with everything from time management, clutter control and space planning to event planning, personal shopping, financial management and public speaking. (We'll go into more detail on a few of these later in this section.) But, before starting this type of business, ask yourself the following: What types of organizing services will you offer? Will you specialize in any areas? How will you charge for your services?

While there are professional certifications for this profession, this business basically requires no capital investment to get started and can be started as a part-time business until you are ready to run it on a full-time basis. However, as in any service-based business, there are many variables involved in being a professional organizer. Things like marketing, skill and the amount of time you are willing to devote, as well as your geographic location, play a greater role in your success as a professional organizer. If your business is successful, you may even want to start a line of organizing products, including books, planners, newsletters, etc. If you need experience, find a friend or two who needs your services and offer to organize an area of their home for free in exchange for their word-of-mouth referrals. Or, contact your local Chamber of Commerce for a list of organizations dedicated to promoting and supporting the field of professional organizing.

If you want to find out more about getting certified in this field, visit the National Association for Professional Organizers' (NAPO) Web site at **www.napo.net** or call or write to them at P.O. Box 140647, Austin, TX 78714; phone: (512) 454-8626; fax: (512) 454-3036. NAPO is currently developing its own certification program (the completion date is unknown).

- ***Personal Move Coordinator***. If you live in a particularly transient area, such as an area near a military base or a large city, a move coordinator might be a good start-up opportunity for you. Even though there are professional moving companies that handle most of the packing and shipping involved with a move, there are plenty of other things that must be taken care of before the moving truck arrives. Often it is difficult for the person moving to take care of everything while working a full-time job. This is where you come in. Move coordinators take care of many tasks, including meeting and supervising contractors and movers, coordinating phone, gas, cable, DSL and electric hook-ups, arranging for appliance installation, organizing the house before the movers arrive or taking any unwanted belongings to a consignment store or charity.

  If a client is moving to a new area, you may also want to provide other services such as supplying school and financing services information or providing them with information on temporary housing needs.

- ***Personal Shopper/Errand Runner***. If you love to shop you might want to look into being a professional personal shopper. Believe it or not, there are people who would rather have a tooth pulled than go anywhere near a mall or fight midday traffic. In addition, people who are homebound, either because they are elderly or because they are ill, need someone to take care of errands like going to the dry cleaners, picking up groceries and shopping for cards and gifts. You can solve that problem by offering your services. This business requires great listening skills—you have to know the person's taste and budget before you use his or her money to find the right item for the best price—whether it's a gift idea, a business item, furniture, accessories or clothing. And, if you plan to travel overseas or anywhere in the United States, you can expand your services to include a greater number of items. There is little start-up cost for this business...all you need is an e-mail account, phone number or address where clients can contact you and, in most cases, a mode of transportation.

- ***Babysitter/Childcare Provider***. Taking care of someone else's children can be the easiest job in the world if you like kids and have a lot of patience. If you have a small child or small children, your local jurisdiction will probably let you keep approximately five children during the day as an in-home childcare provider. (Check with local laws before you bring all the kids in though.) This allows you to earn some money while you care for your own children and provide your services to parents who have to work but are uncomfortable taking their kids to a large daycare facility. You can also provide babysitting services before and after school, in the evenings and on weekends. You won't have a 9-to-5 work schedule, but parents are more comfortable hiring a mature and reliable adult than the 13-year-old Britney Spears look-alike from down the street.

  Some states, including Virginia, also provide respite pay support to parents with special needs children, so if you are interested in working with children with mental or physical disabilities, you can bring in a higher hourly rate than you could with watching non-disabled children.

  Depending on whether you have your own kids, you may be able to start this business for very little money. You'll probably want to provide a variety of toys, books and games to keep the children occupied, but if you already have kids, you won't need to go overboard. You may want to provide snacks for the kids, too. So factor in the cost of a few small meals. You don't have to get certified as a teacher but you may want to look into a basic first aid class or CPR training. It wouldn't hurt to look into liability and medical insurance either. Ads, fliers and word-of-mouth are your best marketing tools in this business. If clients are happy, they'll pass your name on to friends, neighbors and relatives looking for childcare.

❖ ***Catering/Personal Chef Service***. In this busy, fast-paced age, there are plenty of people who want healthy, wholesome, home-cooked meals that do not have the time or the energy to prepare them. If you are a talented chef, you can provide meals for your clients in the comfort of their own homes on a full-time basis or offer to purchase a client's groceries, go to their home one day a week and prepare and freeze enough meals to last the rest of the week. For most people, the latter is a more affordable option.

As a personal chef, you can also offer packages that allow the consumer to buy three days' worth of cooking or other amounts—estimate the fees for these services based on your time and effort. You can provide your own ideas for a menu or prepare a client's favorite recipes or other dishes that are requested. Other options include offering special weekly packages that meet various dietary restrictions, such as low-carb, low-fat, vegetarian or kosher meals. But best of all, there is little to no start-up cost for this business because your clients pay for the groceries and let you use their kitchen. And, as far as pots and pans and utensils go, you can either use your client's or your own. You may have better luck starting this type of business if you've graduated from a reputable cooking school, but this is another business where word-of-mouth referrals are key. If you offer your services to friends or neighbors for free in exchange for their referrals, you may be able to get this business started without formal training...provided you are an excellent cook.

## Using Your Creative Energies

If you have any artistic talent, be it acting, painting, photography or poetry, you may want to start a business using your artistic talent. These businesses depend on your creativity, but if you love to create, this could be the perfect type of business for you to start. In today's marketplace, there is a great demand for talented artists, writers, photographers and designers. If you have any of the following talents or any other artistic talent, see if you can't turn it in to a full-time, lucrative business.

- ***Artist/Craftsperson***. While it is difficult to make a living working as an artist, there are ways to use your artistic talents, such as painting or sculpting, to make enough money to support the fine art you really want to create. Get to know local interior decorators—they may know high-end clients who want specially painted details in their homes or some hand-painted tile work done in the kitchen and around the fireplace. If you are a painter or a photographer, you can provide portrait services for clients who want formal paintings or pictures of their children, home, pets or themselves.

  You can also team up with a local interior decorator and provide faux finishes for walls and other surfaces in homes. This is a lucrative business, because even though the faux finishes are cheaper than the real thing, people are willing to pay good sums of money to make sure their walls look *faux*, not *fake*!

  Another lucrative start-up for the fine arts major: the craftsperson. People of all income levels want to personalize their homes with decorative objects. Whether you are a talented woodworker, jeweler, metalsmith, quilt-maker or fabric painter, you can make a comfortable living by selling your wares at fine arts and crafts shows in your area...or offer an ordering service. Although you may have to spend some start-up funds to create some sample pieces, you can use those pieces to acquire new orders (and up-front payment) for additional work.

- ***Singer***. Turning your singing talent into a business isn't easy, but it can be extremely rewarding if you like to perform...and you're good. In addition to providing singing services for weddings and parties, you can land a job with a local Italian restaurant that has Opera Night and serenade diners.

Of course, you can also seek out more traditional singing gigs, including paid soloist or section leader work at a local church or synagogue, a job with a local professional singing group or the lead or backup singer for a local professional band, orchestra or studio gig. If you can emulate the sound and tone of Patsy Cline, Linda Ronstadt or Alannis Morissette, you may be able to land a few commercial gigs. But, be careful. Oftentimes, in today's litigious society, using the images or qualities of a celebrity abusively or without permission for commercial purposes could land you, but more realistically the company that you work for, in court facing a right of publicity lawsuit. In fact, Bette Midler sued Ford for using a voice like hers in its commercials for this very reason, among others. If you are going to take this route, it might be safer to stick to parties and other entertainment events. If you've had any professional classical training or extensive voice training or studies, another option is to offer your services as a vocal coach. You wouldn't need an office for this type of business. You could conduct business out of your home or at a client's home.

❖ ***Writer/Editor***. Freelance writers can find work in everything from Web content to print material. And, once the words are written and set to the page or computer screen, someone has to edit them to make them legible, engaging and clear. Writers and editors are always in demand at places like publishing companies, big corporations and technology firms. It costs very little to start a writing or editing business, particularly if most transactions are via e-mail.

By surfing electronic job boards on sites like **www.sunoasis.com**, **www.about.freelance.com** or **www.mediabistro.com**, a savvy writer or editor can bring in a number of freelance contracts for everything from editing a screenplay or movie poster and writing a story for a travel magazine to collaborating on a manuscript, ghostwriting a book or writing book proposals or grants. Of course, you might want to be literate in proofreaders' marks or have an English and/or journalism degree. Some businesses are fairly strict about this, but it's not always necessary. If you can prove that you're worth it, you may not need the degrees.

If there is a particular type of subject matter you enjoy, such as scientific writing or sociology, but you don't want to study it yourself, one way to get involved in that field is to offer your services as a freelance editor. The advantage to you is that you get to work with many different organizations and with a variety of subject matter. The advantage to the company you work for is that it can hire you by the project or by the hour, rather than paying for a full-time staff member to review and edit in-house work. Running a freelance editing business also means you have the opportunity to choose between doing substantive editing, where you work with the author to make the copy clearer and more readable, or copyediting where you proof the finished copy for grammar, spelling and formatting errors.

Another way to drum up business is to gather print materials, such as a menu from your favorite restaurant, a newsletter at your local community center or the brochure for a nearby garden center, write a bid letter suggesting ways you could improve the material and send it to the business. In many cases, these materials are written by freelance writers and if you're a better copywriter than they are, you might round up some clients.

Other business ideas that use your artistic talent:

- **Graphic Designer**. Graphic designers and writers often work together on projects. A good graphic designer with an eye for color, shape and function can bid out his or her services to many different businesses and work on everything from annual reports, T-shirts and book covers to posters, calendars and stationary. A graphic designer adds a more polished and professional look to these items. Typically, this start-up would require computer equipment and software. If you are not equipped to do this kind of work at home already, the start-up costs may be prohibitive. However, if you already own a good computer, high-quality printer and graphic design software such as Adobe

Illustrator, Quark Xpress, PageMaker and PhotoShop, you may be in a great position to make your graphic mark on the world.

- **Web Site Designer**. Of course you'll need a computer, but if you already have a computer and a knack for design, there are a lot of companies looking for talented Web site designers on a contract basis. Whether you know HTML (the language used to build Web sites) or have more advanced programming skills, like ASP or Cold Fusion, you may be able to parlay your skills into lucrative contracts to design and program.
- **Motivational Speaker**. If you think you have a story to tell that can inspire others and you are dynamic in front of large crowds, you might want to consider starting a motivational speaking business. You don't have to have any special education or experience to break into this career and succeed. Start by speaking for free to local schools and clubs to get your name out there, and then use that experience to garner bigger and bigger contracts, such as corporate retreats. Get on the roster of a speaker's bureau such as SpeakersQuest or Leading Authorities Speakers Bureau to drum up more business. Don't be discouraged if it takes some time to get these gigs going; it can take a great deal of experience and a number of references to get on those lists.
- **Photographer**. Start-up photography businesses are a great way to put your artistic talents to work for you. Clearly, the biggest challenge to starting such a business is the cost of the equipment, not to mention the development costs. But it is possible to start such a business with nothing more than a good 35mm camera and a supply of well-priced film. Rather than starting out by working in the wedding industry or portrait photography—which require larger lights, more equipment and studio space—you can start your business by clicking photos to sell to magazines and newspapers, either in your local area or around the world. Find a local freelance writer who needs photographs for their articles, and team up with them to provide art for their stories. Or hire yourself out to smaller, more casual family events, such as a family

reunion or graduation party, where the group wants good photographs but doesn't want to pay for a high-end photographer. In no time, you'll have earned the money to buy more equipment and to go after even more lucrative assignments. If you like to take scenic photos, frame some of the images and try selling them at craft shows or fairs. If word gets out, it won't be long before you'll be taking new orders.

## Consultants Make the World Go Round

If you're tired of the 9-to-5 work week and the office politics involved with working for a large corporation or are at the end of your career, but wish to stay active in your field, you may want to consider branching out on your own as a consultant either part- or full-time. A stable job no longer holds the same definition it once did. Today's businesses are more apt to churn employees through buyouts than carry them on through to retirement. A consulting business can buy you more time with your family, a better than average income and the freedom to work out of your own home. But, you'll need a marketable skillset and an area of expertise that others are willing to pay for.

If you're not ready to make the big jump to being out on your own, you may want to negotiate with your current employer to provide services on a contract basis. If you're on good terms with your current employer, this is the easiest way to transition into a new business...and obtain additional clients. The following are a few examples of consulting services that have little to no start-up business costs.

- ***Management Consultant***. Although every organization would like to say it runs just as it should, on occasion, someone from the outside with a different perspective is sought to evaluate a problem. Whether a company has a management/staff relations issue, a nonprofit or corporate board that must learn how to more effectively serve the organization or needs assistance in a search for a new executive director, you can help solve these problems. As a management consultant, you can dedicate yourself to add value through increased profits, greater sales, improved cash flow, enhanced return on investment, greater team productivity, etc. You can start a management consulting business for almost no initial costs, but you must have a working knowledge of the organizations that you plan to consult and a good network of contacts to establish a revenue stream and credibility.

- ***Political Consultant***. Political campaigns are cyclical and people who work in them generally act as temporary employees for the life of the campaign. However, a political consulting business can prove lucrative for someone with political savvy and good instincts. As a political consultant, you can work as a generalist—a jack-of-all-trades on a campaign—or a specialist in political fundraising, direct mail campaigns, polling, phone banks, focus groups and research, performance coaching for political speakers, radio, produce media commercials or grassroots organizing. Today, some political consultants even specialize in setting up political Web sites and online fundraising. This can be an exciting and lucrative business particularly if you align with winning candidates. The political world is a small one, and campaigns snap up talented consultants quickly when the election season starts. Other services you can offer as a political consultant: evaluate a client's local campaign structure; write, design and produce print and media materials; or provide strategic advice on all aspects of a client's campaign. You can also help candidates with event planning, financial filings and field organizing.

- ***Communications Consultant***. From public relations to customer service communications, social marketing programs for new organizations and resolving conflict, communications consultants work with corporations, associations, nonprofits and small businesses on all aspects of their communications strategy. Often, public relations agencies hire consultants to assist on large campaigns or to fill in for another staff member that's on vacation or leave from the office. Communications consultants can help to promote small businesses in the local media. This is a valuable service for small businesses that do not have the money or staff to take on their own marketing or public relations campaign, nor enough money to hire a larger agency with its attendant cost.

As a communications consultant, you can offer counseling or coaching for the senior management of an organization or train individuals or groups of employees on everything from communication skills to public speaking and advanced strategic communication. Additional services you can provide as a communications consultant: news lead monitoring; marketing, including database development, geo-demographic targeting and marketing plan development; and developing, writing and editing everything from annual reports, business plans and press releases to business correspondence, bios, speeches and brochures. And, if you have any expertise in telecommunications, information technology or computer networking, you may want to offer your services in these areas, too.

- ***Fundraising***. If you like working with the nonprofit sector and are good at raising money, it might be lucrative to start a fundraising business. You can work with organizations to help them develop their fundraising mechanisms—product fundraising or holding events—or actually raise the money yourself. Starting this type of business does not cost much, but can be risky. You have to be confident that you can deliver on your funding promises.

  Whether you work with scout groups, church or school groups, civic groups, daycare centers or fraternal organizations, if you have a good sense of what foundations, corporations and other resources are out there and how to target an organization's fundraising pitch, you have a good shot at bringing in the funds for your client. You can also consult clients on the following for an additional fee: how to choose a fundraising product and fundraising company, seasonal or new fundraising ideas and how to develop a fundraising campaign. Most professional fundraisers conduct the fundraising campaign for a fee—typically a percentage of the money the campaign raised. If you've done business with college organizations, hospitals and cultural organizations, use this to your advantage, saving financial data from these campaigns. Organizations in search of a reputable fundraiser often ask these types of organizations for referrals.

This business is relatively low in cost to start, but find out whether you need to be licensed or bonded. This is a requirement in many states. In North Dakota, for example, you must file an application with the Secretary of State for "Professional Fundraiser/Solicitor License" and it'll cost you $100. In California, commercial fundraisers (for charitable purposes) must file an annual registration form with the office of the Attorney General, certify a check for $200 and send a $25,000 bond. If you're interested in fundraising as a career, contact The Association of Fundraising Professionals (AFP), which represents 26,000 members throughout the United States, Canada and Mexico. The AFP advances philanthropy through advocacy, research, education and certification programs for fundraising professionals.

A few business more ideas:

- **Diversity Trainer**. As more and more companies and organizations recognize the need to embrace the diversity of their employees, they are hiring diversity or cross-cultural trainers to help facilitate training and discussions both on- and off-site of everything from conflict resolution and safeguarding against harassment to preparing employees to work in the global marketplace and racial, ethnic, cultural and gender diversity. For an extra fee, you can even offer training for employees relocating to a foreign country.

  Large companies, educational institutions and city governments often develop their own diversity departments, but there are plenty of other businesses, community organizations and educational settings hiring diversity trainers on a full- and part-time basis. Some diversity trainers choose to work with a group of other diversity trainers. This way, they can work together in marketing their business and partnering in workshop development and delivery. The start-up fees for the business vary. You can have a brochure made up and buy a mailing list or network with friends and family. People often hire someone they know or someone an acquaintance knows, but in most cases, credentials will win a client over. Fees may be determined by credentials (i.e., inexperienced trainers receive low fees compared to those with credentials). For more information on certification programs and schools for diversity trainers, search the Internet. (One school's site that also posts a complete discussion on becoming a diversity trainer: **www.diveristyuintl.com/courses.htm**.) You can charge by the hour or provide a total workshop package price. Pick a specialty such as conflict resolution or managing sexual attraction in the workplace and go from there. Market yourself. Give free talks and demonstrations to attract clients. Another way to get clients: Attend conferences and conduct workshops or presentations. Collect business cards.

- **Proposal and Grant Writer**. Nonprofits, corporations and individuals looking to drum up funding and business hire proposal and grant writers to conduct grant research and write, review and evaluate proposals. College professors who need grant proposals written for projects, other consulting firms that need to submit sales proposals to companies for consideration and nonprofits that are looking for a new source of funding often obtain the services of proposal and grant writers. To start this kind of business, you must have extensive experience writing different types of proposals and grants, and a knack for matching your skills to local organizations and individuals looking for your services. Publications like *The Chronicle of Philanthropy* and articles in your local newspaper can point you toward funding sources and the organizations who want those funds.
- **Literary or Talent Agent**. There are many writers and performers out there, and many of them want to spend most of their time being creative rather than selling themselves. That's where agents come in. If you have contacts in the publishing or performing fields, and you enjoy working with people, you might enjoy being a literary or talent agent. An agent is someone who markets books or talent and negotiates contracts, in exchange for a commission on a book or client's advance (the money a publisher or employer pays up front) and royalties (the money a book or client earns through sales). Usually agents get between 10 and 20 percent of the fee that their clients receive for any given job or publication. You might get away with charging for reasonable office expenses, such as photocopying manuscripts and postage, but make sure you inform your clients of this charge in advance.

Get a listing in annual agent listings such as *Literary Marketplace* or *Literary Agents of North America.* Both are annual publications that can be found at the library, and list literary agencies and agents, as well as book publishers and editors. You don't have to be in New York, the center of the book publishing industry, but it is advantageous particularly when meeting and talking with editors and publishers. Become a member of the Association of Author's Representatives or the Independent Literary Agents Association. Although you must meet certain criteria before joining, it's a good idea to know in advance what member qualifications are necessary. Visit **www.aar.online.org** and **http://literaryagents.org**. These organizations represent professional, reputable agents and encourage professional standards.

## Telling the World About Something

Are you a great teacher? Do your colleagues come to you when they need help? Are you positive, motivated and willing to share your knowledge with others? If so, you should get paid for it.

Corporations, nonprofits, community groups and government organizations need trainers for various purposes. Perhaps you are a computer whiz who can pass on knowledge to others. Or, you know a lot about media relations and want to train corporate officers to be more savvy in on-camera and print interviews. Maybe you can train nonprofit organizations to do better grassroots organizing. Whatever the specialty, be it vocal accompaniment, mathematics, business communications skills or customer service, if you can find a place to hold your classes at no cost—often the client organization will provide the location. You can start this business without laying out much cash at all. The following are only a few ideas.

- ***Tutor***. If you have the ability to communicate effectively and have expertise in one or more subject areas, you can offer instruction and support as a tutor in those areas, provide help with homework, run weekend, summer and vacation sessions or teach students test preparation skills. Interpersonal skills are a definite must for this type of business. If you can't work collaboratively with others, this might not be the business to start. You must have a strong working knowledge of a subject area to assist the tutee in learning the curriculum. Tutors often specialize in basic study skills, foreign languages, learning disabilities, math, writing, college entrance preparation and English as a second language, just to name a few. While you can certainly travel to a central location, such as a library or school or to the home of the students you tutor, tutoring can be performed face-to-face out of your own home or electronically via e-mail, chat rooms, fax or phone. If you decide to tutor (or run any business for that matter) in your home, you might want to look into small business insurance that covers you for liability. For roughly $200 per year for coverage, you'll be protected if an accident occurs in your home.

Tutoring can be a rewarding and enjoyable experience with flexible hours. As a tutor you can arrange the meeting times and days with the tutee and negotiate charges for your services. If there are a number of students requesting assistance in your area of expertise, offer to lead a weekly study group at the local library. To drum up more business, offer your services as a notetaker, scriber, scanner or reader for students with disabilities. Place ads for your service in school papers or post fliers on bulletin boards at the YMCA or other community meeting places. If you're tutoring in a specific area, your knowledge and experience in that area are typically enough to attract business. But, if you want to be recognized or recommended by a school district, look into getting your teaching certification or certification for the SAT and other test preparation tutoring. Remember: Tutors must be organized and able to manage tasks, maintain schedules and guide tutoring sessions appropriately so the tutee will learn the skills efficiently.

- ***Performing Arts Teacher***. If you are an accomplished pianist, worked as a music teacher at the local elementary for several years or are proficient on the guitar, flute or any other instrument, you might want to consider starting a performing arts teaching business. Similar to tutors, music and vocal teachers have worked out of their own homes for years, saving the cost of renting out a studio. If you have experience, it is likely that you already have the instrument or other equipment in your home to use for demonstrations. Offer advanced or beginners skills classes on everything from the trombone and flute to voice training and vocal and piano accompaniment. For an extra fee, teach your clients public performance skills and preparation for competitions. This business requires several clients to be lucrative and demands a high level of patience and rapport with students. If you are successful with a handful of clients, they'll pass your name on to others in need of assistance. Many will invite you into their home, or they will have the instrument for you to use and instruct on. The only cost to you will be driving to clients' homes.

- ***Arts and Crafts Instructor***. If you can knit, make jewelry or sculpt, teaching arts and crafts out of your home or at a central location can be a lucrative business. Often local adult education or enrichment services, community centers and community colleges are looking for instructors for classes. This business could easily be combined with a craftsperson business. Depending on your area of expertise, as an arts and crafts instructor, you can offer courses on everything from architecture, ceramics, needlework and illustration to industrial design, printmaking, textiles and sewing or quilting. Teach students to develop skills in creating unusual decoration and functional home accents. Offer tips on form, color and composition. Establish a course to create special holiday decorations at various times in the year. Other classes taught by arts and crafts instructors include beading, weaving, silk flowers, macramé, mosaics, bread-dough art, découpage, needlepoint, Hopi Indian pottery and wooden keychains.

  Costs to start this business are relatively low, but marketing yourself is the key to its survival. Post fliers or place ads in the local paper. Offer a discount to people who bring a friend. Offer a three-session drawing class that provides basic instructions and introduction to landscape, still-life and facial features. Anything to bring business in and get your name around. In most cases, it is normal for students to buy their own materials so the cost of supplies would be relatively low. And, if you set up shop at home or work for an adult education or enrichment center you alleviate the need to pay rent.

## Social Graces

This is another area where your expertise and enjoyment come into play. If you love to plan parties, write calligraphy or teach manners, there are people who will pay you well for your services. Put your social skills to work for you in these services.

- ***Etiquette Consultant***. Should the salad fork sit to the left or the right of the dinner fork? How long can you wait to write a thank-you note after receiving a gift? These are just some of the questions an image or etiquette consultant can answer for his or her clients. Other topics: how to network at a conference or how to give a killer handshake. Everyone from tearoom owners and operators to Fortune 500 businesses is interested in increasing profits and what better way to increase profits than by educating employees in etiquette? Today's businesses are concerned with the lack of professional polish within their organizations, having observed their employees dressing inappropriately or demonstrating poor manners while in the presence of clients or industry peers.

  Another suggestion: Offer classes for children on proper table manners, and the dos and don'ts of good manners. Many parents do not have time to teach extensive etiquette lessons to their kids, but want to make sure their kids mind their manners and behave properly in public. Private schools are good places to market your services if you choose this as a business, as well as community centers. If you decide that you'd rather work with businesses, offer courses on everything from business etiquette, international protocol, introductions and dining skills to materials and marketing, public relations, professional image development, appropriate wardrobe and professional speaking. For an extra fee, you can offer a one-day session teaching businesses e-mail etiquette.

This business is relatively cheap to start because there are no development costs, no inventory for you to purchase up front and presentations can take place at numerous locations and to various groups, requiring no office location. Venues where classes can occur include: resorts, hotels, country inns, restaurants, country clubs, museums, tearooms, community centers, etc. Remember the interests of clients. This type of business generates a lot of business by word of mouth. And, if your clients are happy, they'll tell their friends and co-workers who will call or e-mail you to find out more about your services.

- ***Wedding/Party Planner***. If you have strong organizational skills and good contacts in the hospitality industry, you can start a wedding or party planning business. Anxious brides need someone to provide them with advice, contacts and solutions for major and minor wedding crises, including everything from announcing the engagement, wedding or event to friends and family to determining the budget, securing the location and hiring a caterer. As a wedding planner you can assist with the gift registry, tuxedo rentals or entertainment, offer tips to make the planning process easier and offer assistance in the choice of photographer. As a party planner you can plan parties and events ranging from a four-year-old's birthday party to an office Christmas party. You can choose a specialty or do a little of everything, including: sending out invitations, hanging decorations, finding entertainment, etc. This job requires you to be well-organized and have impeccable communication and people skills. You'll have a long list of tasks to complete on time and within a client's budget. Be prepared to find creative solutions to fill clients' requests and to resolve a crisis. (After all, you never know what a jilted bride will grab and throw or when a DJ will get stuck in traffic.)

  Although this business can be started with almost no start-up funds, it is important to develop good relationships and compile an extensive database of vendors before you start, including everything from photographers, caterers, florists and car/limo rentals to bridal shops, venues, nail/hair salons and party rentals suppliers. Start by informing potential referrals about your business and ask them to send you materials about their services to share with your clients. In no time, you'll have an excellent collection of resources to employ when your clients ask you for help and you'll be able to negotiate the best deals for your clients. This type of business also generates business by word of mouth. And, if your clients are happy, they'll tell their friends and co-workers who will call or e-mail you to find out more about your services. Start by posting fliers in supermarkets, stores, libraries and community centers. Place an ad in the local paper or local newsletters. If you're still having a hard time getting the ball rolling, try volunteering or bartering your services with the people you know—your accountant, your veterinarian or the local florist. This is a great way to get exposure around town. Or offer to work for free for another wedding or party planner to gain the experience you need to drum up your own clients. Most wedding and party planners price their services one of two ways: on a percentage of the total budget (usually 10 to 15 percent); or on package prices, based on the services a client desires.

- ***Calligrapher***. From handwritten envelopes to individually designed placecards, calligraphers are in high demand for events such as weddings and gala dinners. You can also expand your business by hand lettering gift items, such as a framed poem from one sweetheart to another, certificates of appreciation, decorated wedding certificates or baby announcements. Other services a calligrapher can offer: invitations and announcements, business cards and logos, commemorative awards or decorations, ads, signs, calligraphy lessons and family trees. Strike up good relationships with wedding and event planners who can refer your services to potential customers. If your clients are happy, they'll tell their friends and co-workers who will call or e-mail you to find out more about your services. Depending on the task at hand, you can either charge by the line or by the hour. Offer a service for rush jobs, at extra charge, if you can work under pressure and put jobs together quickly.

## Providing Business Services

More businesses are turning to outside help as they downsize and look for ways to cut costs. If you can provide services for local businesses that can save them money or time, you can create a financially successful business with your talents and skills. The following examples are only a few of the business services you may want to consider.

- ***Tax Preparer***. Although it can be expensive to become a licensed tax preparer, if you are already working for a company that prepares taxes that has paid for your training and licensing, you can put that knowledge and certification to work for yourself in the evenings and on weekends during the tax season (from mid-January to mid-April). Some employers may require you to work for a certain period of time if they foot the bill to train you, but there is no reason why you can't start something for yourself on the side. A caveat: Your employer may have a non-compete clause in your contract that does not allow you to take clients with you. So, it might be worth starting your own business on the side before you quit your full-time job. This will allow you to build up a client base. Once you start working for yourself, hold informational seminars at your local library or community college. Offer a discount for your services to those who attend, and they'll be more willing to try your services.

  The qualifications that a business desires in a tax preparer usually revolve around a CPA license. Most business executives view the CPA credential to be an important criterion in their hiring process. So even if you won't be doing any auditing, you might want to consider getting a CPA license. And, if you do, advertise that fact. It's also helpful to join a professional organization like the American Institute of Certified Public Accountants (AICPA) found at **www.aicpa.org** or the National Association of Tax Practitioners (NATP) at **www.natptax.com**. Membership in a professional organization typically means that the accountant is current in the various areas of accounting and tax law changes. If you belong to a professional organization, advertise that fact, too. Even if you don't intend to get your CPA license, you should look into a reputable tax preparation class. H&R Block, a leading tax preparation company, offers a 12-week tax course for about $80 to anyone wanting to gain an understanding of how to prepare a tax return. The course is a write-off so don't worry about the cost. Besides, you'll earn that money back in no time from your first few clients. In addition to learning about the intricacies of preparing an individual tax return, the course also covers recent changes to the tax code, provides you with a better understanding of tax law and with strategies on how to save money now and in the future. For more information on the H&R Block Income Tax Course, call (888) 271-6343 or visit the company's Web site at **http://hrblock.com/taxcourses**.

If you do plan on getting your license, start studying now. The process can be long (for some it can take as many as five or more years!) and arduous...but it will pay off in the end. When you pass the exam, throw yourself a party. This way, your friends, family associates and acquaintances will know that you're a CPA. And, if you belong to any organizations or community groups (e.g., church groups), see if you can offer classes on money management and taxes (e.g., through the Sunday School). If you have to offer your services free at first to build a reputation, don't worry. It will pay off in the long run. Look into homeschooling organizations. They usually allow free ads in their newsletters.

You'll need a computer and printer, tax prep software, tax forms on CD and some file cabinets for organization. You may even want to invest in software, such as TurboTax Home and Office (usually around $75 including the state software), or TaxAct (costs around $95 including state software). Register with the IRS as a tax preparer and you'll receive their newsletter and Package X (a package of tax forms).

As a tax preparer for individuals, you can earn anywhere from $25 to $50 or more an hour working out of your own home. You can also prepare or assist in preparing tax returns for business clients year round to increase your earnings. Most people ask for referrals from friends and colleagues when tax time comes around. Before starting a job, set up an interview where you discuss the cost of services and estimate the cost for the job at hand. You should base fees on the complexity of the return. Never base fees on the size of the tax saving or refund and never guarantee a refund before completing a return.

It may help to have expertise in a few business or professional areas. If, for example, you are preparing a physician or child care worker's taxes, you would want to have experience in that particular area.

- ***Accountant/Bookkeeper***. If you are already working as an accountant for a company and enjoy what you do, you can head out on your own and open your own general accounting business. If you want to specialize in an area such as taxation or financial planning, you'll need more training. If you want to be a CPA, you'll need even more training (a four or five-year upper level education program). You must be good with numbers and have great organizational skills in order to be an adequate bookkeeper. The best market to target in this type of position is other small business owners who may not be able to afford a full-time accountant to handle their financial recordkeeping. You can work with businesses and individuals on everything from tax planning and preparation, to monthly financial reports, audits and billing and payroll services.

  The equipment you'll need for this business varies depending on who you're working for. If you're working for a small business, you can probably use their fax machine, copier and computer but you may want to invest in some of your own equipment as well, including software, reference books and the yearly tax guide (this alone will set you back about $500). As with other small start-ups, print fliers and place ads. Once you get a few clients you'll also be able to market yourself via word of mouth. Join business and community organizations to get things going.

- ***Financial Planner***. This is another business that is easy and cheap if you're already working for someone else in this capacity. However, if you have a knack for investing and a good way with people, you might consider starting a small financial planning business. As the promise of Social Security dims, there are many people who need help to ensure that they are able to retire comfortably. As a financial planner, you can work with individuals, families and small businesses to analyze, set and achieve financial goals. If you already have expertise in accounting, taxation, finance or business law, you may not need certification.

This type of business thrives on networking. It's a business primarily built on personal relationships that translate into word-of-mouth referrals to family, friends and co-workers. It wouldn't hurt to invest in a few fliers, advertisements or direct-mail campaigns to bring in more clients.

This business usually requires you to have your own computer with spreadsheet, word processing and database software. And, depending on the services you intend to offer, you may want to invest in financial and investment analysis software or useful trade publications.

- ***Medical or Legal Transcriptionist***. If you are a quick and accurate typist, you can begin a home transcription business with an investment in a tape player outfitted with transcription foot pedals. Doctors office staff members are often too busy to transcribe notes from meetings with patients and dictated letters, and this service can be extremely valuable. If you can find psychiatrists or psychologists who trust your adherence to confidentiality, you will discover that they often have more of this type of work than general practice doctors. The same principle holds true for law offices. Often there is more transcription work than the regular staff can handle, particularly in the area of depositions and dictated legal memos or other correspondence. Because your clients will not have to pay for the overhead of having a full-time transcriptionist on staff, you will be providing them a needed service at a cheaper price. As you add clients, you can expand your business and potentially turn the transcribing over to your employees.

Additional business services ideas:

- **Translator**. As the business economy continues to expand globally, more companies are looking for qualified, accurate translation services. If you are fluent in a language other than English, you can target companies that serve markets in a specific country or area of the world. You can also offer your services via a Web site or through e-mail, allowing you to pull in more clients from other areas.

- **Computer Consultant**. When it comes to computers, there are still people out there who have no idea how to do anything beyond turning the power on (some people even have a hard time with that). As a computer consultant, you can offer repair services, training and setup, as well as needs assessments and recommendation services for businesses or individuals looking to overhaul or replace their current computer systems. For an additional fee you can offer to design and install networks. If you already have a computer, once you have the necessary training or certification, you can teach others on client premises or out of your own home and charge anywhere from $25 to $150 an hour, depending on the services requested. If you want to make a little more cash, offer to teach on-premises workshops for local businesses. Or offer to teach an Internet family course and promote it through local schools, libraries and the YMCA.

  Place ads in local papers, online classifieds and the Yellow Pages. Print up fliers and post them on bulletin boards all around town, including supermarkets, community centers and gyms.

If you're still not sure about the type of business you want to start, think more about what you like to do as well as what service business is needed in your area. We've touched on a few, but other ideas include: online researcher, personal coach, personal trainer, program coordinator, nutritional consultant, dietician, seamstress or tailor, costume designer, make-up artist, event choreography, woodwork refinishing service, publicist, elder care, cleaning service, nanny service, etc. For more ideas, browse the miscellaneous job ads on sites such as **Craigslist.com** or **Yahoo.com**. Just think of what you love, and what other people need, and you're on your way to your own service-oriented business!

## Item or Product-Based Businesses

If services aren't your sort of thing, you might want to look into starting an item or product-based business. With this type of business, you offer an item or product for sale to businesses, stores or customers. Perhaps it's your own invention or something all your friends rave about, like your secret brownie recipe, mouth-watering salsa or homemade leash for your dog. Whatever it is, if people find value in it, you can provide it to them. These businesses are a little harder to start than service businesses, but once you get going, the sky's the limit!

People who start product-based businesses are usually crafty, inventive and good with executing ideas. They don't necessarily have to be good with working with people (at least not as much as someone who starts a service-based business) and those who'd rather spend their time alone can do so by being behind the scenes of creating a product—and hiring out the distribution of that product. When you picture Mrs. Fields, you think of her baking in the kitchen and not out in front of people, serving up the cookies. Of course, she may have had to do this in the beginning.

The hard part about product-based businesses is finding an economical way to make your product and a profitable way to get it to your market. Most start-up businesses of this kind require more capital because of the fact you are producing something that will—in most cases—require materials or ingredients. This shouldn't hinder you, however. Product-based businesses can be as rewarding and lucrative as any of the service-based businesses. Remember the story of the Juice Guys from the Introduction. Tom Scott and Tom First started with their juice and eventually found crafty ways of marketing their special juice to the masses. Now, their juice can be found in every corner store throughout the U.S. What the two Toms started with was a good product. And, anyone who has a good product can probably do something with it.

If you know that you'd rather start a product-based business, but don't have an idea for a product, there are many things you can do to jump-start your imagination. Try flipping through your favorite catalogs and think of things people can't do without but would pay money to purchase. Examples of catalogs or magazines that might get you started include Crate & Barrel, The Pottery Barn, Martha Stewart's *Living*, Sharper Image, Restoration Hardware, Pier 1 Imports and Ikea. You can also browse the Internet for ideas and see what's selling on Amazon.com and other large retail sites like Wal-Mart.com and Target.com or SurlaTable.com and Williams-Sonoma.com.

Other ideas to consider are:

- **Custom Clothing**. If you love to sew and have a knack for clothing design, try selling some of your products at craft fairs or flea markets. Whether you make shirts for hunting or motorcycle riding, horse jumping or baseball, or any type of specialty clothing or accessories, like hats, aprons or scarves, you can sell your items to boutiques, gift shops and specialty stores or at craft fairs or flea markets. Make your product appealing, unusual or unique and you'll have a market for your special items. If you have a tough time getting stores to respond, ask if they'll put a few items on display to sell on consignment. This is the next best way to get your product to the masses. If they won't help you out, ask around; there are stores out there selling items on consignment.
- **Greeting Cards/Stationary**. Do you have a flair for poetry, saying something heartwarming and sincere or are you hilarious? Do you have good drawing, painting or other artistic skills? If you do, you might try starting your own line of greeting cards. You can design and create a few different cards or styles of stationary for special occasions and sell them at local card shops, gift shops and specialty stores. If you have a talent for photography or graphic design, you could team up with a writer. You might want to concentrate on local landmarks or make cards for special days like anniversaries, weddings and birthdays.
- **Specialty Food Products**. Do your friends rave about your cakes, cookies or meatloaf? If you love to cook and make high-quality products, you can start your own business providing stores and restau-

rants with your homemade goodies. Many restaurants are too busy to make their own decadent desserts or homemade soups and stocks. You can provide them with products each day or every few days. Many people begin these types of businesses at home and then expand and move to a commercial kitchen space. You'll need to check your local health laws about making food in your own home. You can start out by simply making your prized dish and getting a local restaurant to try it out on customers. Ask the restaurant to add your dish to the menu for a week or weekend—and see what happens. Maybe the restaurant will want to buy a certain amount from you on a weekly basis. You can test your dish out for free the first time around, then negotiate the price per batch later on—as well as worry about the health-related laws you must follow.

- **Growing Herbs and Vegetables**. If you have some extra room in your yard and love to garden, put that talent to work growing herbs and vegetables to sell. Many restaurants are always looking for fresh herbs for their recipes, and more and more farmers' markets are springing up all the time. If you grow unusual herbs and vegetables that aren't easily found in your area, you can provide a needed product and exercise that green thumb. Consider Lucia McMillan Cleveland of San Luis Obispo, California, for example. She's otherwise known as the founder and spokesperson of The Spice Hunter, a flourishing enterprise famous for its unique line of over 150 spices and seasoning blends, that can be found in grocery, specialty and natural food stores nationwide.

  In 1980, Cleveland, who turned a hate for a job as a state water inspector and a love of homemade soup into a $20 million-dollar business that employs 100 people, took $10,000 out of her savings and created The Spice Hunter. One night, after a disappointing day on the job, Cleveland attempted to drown her sorrows in homemade soup...but was frustrated by the lack of packaged bouquet garni at the grocery store. So, she decided to make her own bouquets—and sell them. At the time, the company offered only a dozen spices and seasoning blends. But, Cleveland peddled her goods at booths at local farmers' markets and food fairs—and eventually coaxed local San

Luis Obispo grocery stores into carrying her blends. Today, the Spice Hunter's products are distributed to over 17,000 stores nationwide in the Gourmet, Natural Food and Grocery marketplace. Cleveland recently sold the company to C.F. Sauer, a privately held $500 million company from Richmond, Virginia. Cleveland is now retired except for international spice trips for The Spice Hunter.

Again, to drum up more ideas on the type of item- or product-related businesses you can start, browse the miscellaneous job ads on Web sites such as **Craigslist.com** and **Yahoo.com**. Other ideas include:

- apparel
- candles/soaps
- pottery
- make-up/nail polish
- perfume
- photography
- journals
- picture frames
- labels
- bath/spa products
- inventions
- kitchen appliances
- selling fruit
- jewelry
- handbags
- calendars
- seat covers
- placemats
- wrapping paper
- pastries
- arts & crats
- linens
- fruit grafting
- hair accessories
- watch cuffs
- organizers
- home furnishings
- maple syrup
- neighborhood newsletter
- sewing patterns

## Additional Resources For Business Ideas

If you're still not sure what type of business you'd like to start or you'd like to do some additional research on the businesses I offered as examples, check out the following business idea resources on the Internet.

- **www.ideaexplore.net**. This is the best Web site for new business ideas. IdeaExplore is a free Web site that has a large collection of new business and invention ideas, as well as new ways of thinking of ideas for improving things.
- **www.businesstown.com/businessopps/newbiz.asp.** This site includes lists of businesses you can start for under $1,000; $1,000 to

$5,000; $5,000 to $10,000 and so on. It also offers advice on finding business opportunities all around you, advice on what business ideas to follow up on and links to other Internet resources.

- **www.bizymoms.com/startup.html** is targeted specifically at stay-at-home, work-at-home mothers. If you fall into this category, this site might have the perfect idea for you.
- A list of popular businesses that can be started from home for very little money is located at **http://independentwife.hypermart.net/business.htm.**
- A Canadian guide to developing business ideas is located at: **http://bsa.cbsc.org/gol/bsa/interface.nsf/engdoc/0.html.**
- Reports from the Small Business Resource Center, including instructions on how to start everything from a catering service to a mobile locksmithing business, are located at **www.webcom.com/seaquest/sbrc/specrept.html.**
- Business information that you can understand without having to hire an attorney, including instructions on how to start everything from a poetry-on-demand business to a wake-up and reminder service, are located at **www.ideacafe.com.** Also includes a Q&A section with Idea Cafe experts where you can bounce your ideas off others.

## Other Sources for Business Information

- *Small Time Operator: How to Start Your Own Business, Keep Your Books, Pay Your Taxes, and Stay Out of Trouble* (Small Time Operator, 25th Edition)—by Bernard B. Kamoroff; Paperback, $17.95. ISBN: 0917510186.
- *Jump Start Your Business Brain: Win More, Lose Less, and Make More Money with Your New Products, Services, Sales & Advertising*—by Doug Hall, Tom Peters; Paperback, $16.99. ISBN: 1558706429.

- *If You're Clueless About Starting Your Own Business and Want to Know More*—by Seth Godin; Paperback, $15.95. ISBN: 1574100939.
- *The McGraw-Hill Guide to Starting Your Own Business: A Step-by-Step Blueprint for the First Time Entrepreneur*—by Stephen C. Harper; Paperback, $12.95. ISBN: 0070266875.
- *Start Your Own Business: The Only Start-Up Book You'll Ever Need*—by Rieva Lesonsky (Editor); Paperback, $24.95. ISBN: 1891984217.

## Conclusion

The key to choosing your business is choosing it well. The business needs to be in a field you like because there will be tough times—days when you have to call potential customers but don't feel like it...days when money is tight. Enjoying what you do helps you through these tough times.

Also, you need to choose carefully, so that the business you do generates revenue. Revenue—also called cash flow—is the lifeblood of business. If you don't have money coming in...there's not much else to your business. That's why you need to think more carefully in the early days of the start-up.

Although many of the businesses we examined in this chapter can be started for free as a stand-alone business, they can all be expanded as your revenue and client list grows and as you take on more work. You may need to move to the next level, hiring on one or more employees to help you fulfill your contracts. You'll have enough cash flow to finance the business, rather than having to find ways to start the business with little or no outside financing—and you'll be in a better position to expand in any direction you wish to take the business, be it online or out of state.

Success is around the corner—you may have to do some hard work to get there, but with help from this book and a dedication to work hard, you will reap the rewards of small business ownership.

# Chapter 2: Meeting the Customers ...and the Competition

Once you've decided on a type of business you want to start, it's important to do some market research. After all, you don't want to open a beauty parlor if there are already two such shops that are open on your block. Or, you don't want to have two book stores in the same strip mall.

No business factor is more perplexing or challenging than marketing—perhaps because it covers a wide array of activities and disciplines. In many businesses it includes sales, customer research and elements of product development. In others it includes assessing the competition, customer communication and market research.

This confusion leads some people away from giving sales and marketing the focus they deserve. Successful companies throw around terms like "market responsive" or "market driven" to explain their success, but in truth, in a world that depends on information, everyone must be market responsive. Customers don't buy your products or those of your competitors for mysterious reasons, and whether you're offering to prepare someone's taxes, make and sell your own greeting cards, walk someone else's dogs or tutor mathematics to high schoolers at the local community center, you must know who your competition is, who your customers are and what they want—even if your first client is your former employer. And, you have to be able to acknowledge if and when their wants and needs change—and be willing to meet those demands.

## Figuring Out What Works...and What Doesn't

Delivering a successful product or service faster and more efficiently to your customers requires lots of information. Delivering a successful product or service faster and more efficiently than your competitors requires a lot *more* information. Marketing—in its broadest definition—is that information. It's a means of figuring out what works and doesn't work for your business.

There are dozens of firms that will charge you high prices to gather this information. But you don't have to go far or pay big bucks to find out what you need to know. On the following pages, we'll outline some useful tools and walk you through this process considering everything from your industry sector's potential and your competition to your customers and compiling customer data that's meaningful to your business.

Marketing varies greatly depending on the business you choose to start. In an event or party planning business, for example, word of mouth is the best marketing tool. But marketing in this business also entails making up fliers and posting them on bulletin boards in supermarkets, pet stores and other local merchants or placing an ad or blurb in the local paper. It also entails networking with vendors, because event planning relies on your contacts and your ability to give customers what they want—even if it means hiring an 86-year-old stripper for a groom's bachelor party. Marketing in the doula/birthcoach business, on the other hand, entails networking with childbirth instructors, birthing centers and the maternity departments of local hospitals. It also entails huge expenses for advertising because being a successful labor coach relies on your ability to get your name around town, including printing up fliers and brochures to send to ob/gyn and pediatricians' offices, local La Leche League branches, lactation consultants, motherhood preparation classes and twins' groups or to post on bulletin boards at child care centers, preschools, mothers clubs and fitness centers.

No matter who they are or where they are, your customers can get just about anything they want at any time they want it. And, they can get the things or services they want from you or they can choose to go to the guy down the street, and usually on terms pretty close to their own. Let's face it, today's customers are accustomed to getting products and services faster and better and with high quality service. In fact, products and services are generally sold on one or more of three criteria:

- quality;
- value; and
- service.

Your challenge is to determine what combination of these criteria works most effectively for your business.

You'll learn quickly that there is as much to learn from the people who don't like your products or services as those who do. This means taking a hard look at your successes and your failures. So, before jumping in head first, you need to do a little market research...and where better to start than with the industry and the competition.

## Researching the Industry...and Competition

Knowing the industry and who your competitors are and being able to anticipate how a competitor will act or react can provide a significant advantage in planning your business strategy.

First things first. Are you entering an "established" or start-up industry? Chances are you're venturing down an already beaten path. Unless, of course, you invented some new and revolutionary product that breaks into a new industry or changes an old one. Some would argue that Starbucks established a new industry—that of the designer coffee—while others would say it merely revolutionized an old industry (i.e., coffee drinking). But, for the purposes of this book, we'll assume you're entering an industry that is already established. This means you'll have competitors, but before you begin to compile data about your competitors you might want to start your market research with your industry sector whether it be in manufacturing, transportation, hospitality, retail or entertainment.

For example, what do you know about your industry? Be specific.

You can start by compiling the following:

- A description of your industry niche;
- Financial and market objectives for the company;
- A description of your operations;
- The company's product line;
- A quick analysis of the competition;
- A description of likely customers; and
- An outline of the company's marketing plan.

### Industry Sector Market Research Questions

After gathering the above, ask yourself the following:

- What are the barriers, if any, to enter the industry?
- What are the sales and profit trends in your sector?
- What are the product trends? What's hot? What's cold?
- How do you expect your industry to change in the next year? In the next five years?
- Are there particular demographic factors at work in this industry?
- Are there seasonal buying patterns inherent to this industry?
- Can you find projections of growth trends for the industry (projections done by trade groups, security analysts, government)?
- What are the factors that will affect demand for your product or service (general business conditions, technological innovation, governmental factors, customer growth)?
- Are there any potentially adverse political, social, economic or environmental conditions?
- Are there significant barriers to growth in this industry?

Think hard about your answers to these questions. Different sectors will require different sorts of effort. If your sector is crowded, you need to investigate who's there and your position coming into the market. Knowing your marketplace has a direct impact on what you do and how your business performs. The answers to these questions should shape how you do business.

Don't let the competition sway you from trying to enter the same market. There are lots of success stories from people who entered crowded markets and managed to make names for themselves. Think of those Juice Guys again who got into the beverage business amid enormous companies like Pepsi and Coca-Cola. Although they later sold portions of their company to larger entities, they made a few bucks in those key transactions. Ben & Jerry still pump out ice cream from Vermont amid Haagen-Dazs, Dreyer's and Baskin-Robins 31 Flavors. If you're an amazing and innovative event planner, perhaps you'll become the next Martha Stewart and found an enterprise. It doesn't cost much to offer your service to people hungry for help and good advice.

Your goal is to do better what others do well. If you can find a weakness in the competition or a way to improve upon a great idea, you can muscle your way in. Service-based businesses are easier to start because they require little or no capital—other than the money you need for market research and initial marketing—but can be worth a lot at a later date. The largest caveat to starting a product-based business is the simple fact that you'll most likely have to spend some money in materials to produce your product or prototype. So, you'll have to bear some greater expense in the beginning. But you can minimize these costs and focus on the quality so that you can later entice others—perhaps some venture capitalists—to help your company grow. The Juice Guys were blending drinks in their bare bones kitchen for a while until they could find the means to grow their bootlegged start-up. You can do the same.

Keeping track of your competition is important as a sales tool and as a means of setting goals for product development or service. It is important to know who they are and what they do well, because your customers will know—even if you don't.

List all the major competitors you know about. If you don't know any or only know a few, you'll have to conduct some research. Use industry magazines, trade association contacts, financial reports like Standard & Poor's Industry Report, Dun & Bradstreet (many of these can be found in local libraries), annual reports of publicly held companies, online data and news services such as Nexis and Dow Jones or talk to industry experts. Some entrepreneurs even telephone competitors directly to find out about their sales volume, products and pricing policies.

Obtain catalogs or other marketing materials. Buy products or use services. Ask friends and acquaintances what they like and don't like about the competitors' product.

## Use the U.S. Census Bureau

One of the best resources for free business information is the U.S. Census Bureau (**www.census.gov**). Many people think the Census Bureau only counts people in cities, towns and rural areas across the nation every 10 years. But the Census Bureau tracks other numbers, too—including the number of businesses in your area.

Every five years, the Census Bureau takes an economic census that provides a detailed portrait of the economy from the national to the local level. An economic census took place in 2002. (An advance report comes out in early 2004, with subsequent, more detailed reports following. Some information doesn't come out until 2006.)

Currently, the 1997 economic census data is available. The Census Bureau released its initial report from that census in 1999 but continued to produce more and more detailed reports on aspects of that census through 2001. Results from the 1997 census provide information on businesses operating at more than 21 million locations, providing key information on everything from the number of businesses and employees and the value of shipments to sales, receipts, revenue and payroll.

If, for example, you wanted to open a real estate business in Wilmington, Delaware, and after conducting research through the Census

Bureau discover that there are already 428 real estate businesses in the region, you may opt to start another business. Perhaps a business that specializes in real estate appraising in the same geographic area (your research tells you that there are only 30 such businesses in the area). On the other hand, if you're thinking of opening a driving school in Oklahoma, the Census Bureau's data informs you that there are only eight such businesses in the entire state.

All the census data is available free on the Internet and some of it is available in print for a fee. To order census products or to obtain information on Census Bureau data, contact the Census Bureau Customer Services Center at (301) 763-INFO (4636).

## Zapdata.com

Another site that offers instant access to free business data is **www.zapdata.com**. This site allows subscribers to register for free, and then offers a wide variety of business information, including the following:

- instant access to top-quality, targeted prospect lists;
- valuable demographic and geographic business data;
- analysis reports that unlock key customer or prospect characteristics;
- real-time, single company profile information, including company SIC codes, key contacts and an online business directory;
- free facts on target markets; and
- industry reports.

While some of the services are offered for a fee, others are not. It's a worthwhile site to browse for valuable market information. The site also offers live customer support.

## Assessment of Competition

Competitor's Name: Location:

Parent Company: ☐ Subsidiary ☐ Division ☐ Branch

Product Lines(s):

| Year | Sales | Net Income | Total Assets | Equity |
|---|---|---|---|---|
| 200___ | | | | |
| 200___ | | | | |
| 200___ | | | | |

Estimated Market Share

Rate the following areas in order to determine major strengths and weaknesses of the competitor (suggest using "+"; "N"; "–"):

1. Name recognition ______________________
2. Product line ______________________
3. Quality ______________________
4. New products ______________________
5. Pricing ______________________
6. Marketing share ______________________
7. Financial condition ______________________

"+" Better than your company
"N" Neutral/About the same as your company
"–" Worse than your company

Briefly describe the competitor's reputation, competitive advantages and disadvantages and overall marketing strategy:

## Your Local Chamber of Commerce

Browsing the Web site or directory of your local chamber of commerce can also provide you with information about the businesses in your neighborhood, city or region. A directory of local and state chambers is located at **www.uschamber.com/Chambers/Chamber+Directory/default.htm**, and many chambers offer their information to potential business owners at little or no cost.

After reviewing publicly available data on competitors, prepare a detailed assessment of the competition. This assessment will help identify competitors' strengths and weaknesses in products, quality, service, price, etc., as well as the outlook of your entire sector.

Among the most important—and difficult—questions you want to answer about your market and your competitors include the following:

- What are the barriers, if any, to enter the industry?
- What is your position in your industry? Do you have competitors? If so, how many and is the number increasing or declining?
- Who are your competitors?
- What are the strengths and weaknesses of your competitors' products? Of their management teams?
- Who are the industry leaders and why? How do they mix quality, value and service?
- Will you be able to bring new products to the market more quickly than your competitors?

Begin by writing down everything you know about your competition and its products or services. Ask a couple friends or anyone who will be working with you to do the same. This way you'll have a variety of impressions; your friends or co-workers will probably have a different perspective than you. Pulling all these ideas together, you can build a universe within which you can place and define your product and how your customers will perceive you. You want to understand your competition so as to differentiate your product or service.

The worksheet on the previous page provides a means to keep track of competitors' progress in the market and to learn from their successes and failures.

After you have completed the worksheet and compiled a list of data on your sector and competition, you can begin to put the information to work for you.

Some important questions to consider:

- Can you counter each competitor's weakness with a strength in your product or service? How else can you turn their weaknesses to your advantage?
- Could they counter your weaknesses with their strengths?
- What parts of the market are your competition moving into? Moving away from? Do these trends have any bearing on your business?
- Are your competitors more stable and better-capitalized than you are? Does this matter much in your market?
- How does your company's market position compare to those of your competitors? Are you a market leader or a market follower? Is there an advantage to being one or the other (which makes more money)?

Other questions to consider in this process:

- Can your company expand the market in this industry?
- Does your market reward broad-based efforts or niche operations?
- What can you do to position the company for growth that takes advantage of its market and industry sectors? If growth prospects are limited, should you consider moving out of that market?

If you want to have a successful business, it is important that you do your homework. This means you must know as much as possible about both the business field you have chosen and your competitors. This may seem like a lot of work, but this knowledge is important to your business's success.

## Getting to Know Your Customers

You know what other businesses are out there that might draw clients and customers away from yours—but who does business with your competitors and why? Once you've analyzed your business from the perspective of the competition and the marketplace, you must move on to the most important factor—the customer.

As we mentioned earlier, delivering a successful product or service faster and more efficiently to your customers requires lots of information. But, you don't have to hire a consultant to do this research. Begin by defining your audience and product/service purpose, focusing on specific values you will provide to the customer.

Customer research seeks to learn what motivates people to buy your product or service. This is why anything that tracks sales in a detailed way helps you—a major reason why big retailers offer their own credit cards, encourage catalog orders and code their register receipts. It's also why direct mail marketers offer free gifts and why home shopping and computer online services make marketers' mouths water.

Since you probably don't have any customers yet, ask people whose opinions you respect (family, friends, co-workers, business acquaintances) to review your business idea with an objective eye. These people don't have to be experts in your industry. (After your business is established, you can ask your customers these same questions.) Describe the product or service—or, better yet, show them a sample—and ask for feedback. Find out how they perceive your product or service and your competition. What's the evidence of that need? How do they get information about competitors and their products—advertising, direct mail, word of mouth? Find out how they view quality, value and service in relation to your product—and in relation to your competitors' products. Ask them to describe their responses in some detail. Look for information on price ranges and barriers, format or style preferences, service requirements, probable buying patterns.

After reviewing the results you should have a basic idea about the problems the product or service will face—and have the chance to respond. Analyze the responses with a focus on what features your product should have and what would make it more attractive.

Also consider the factors beyond your control that affect your customers: geographic boundaries, demographic limits and cultural influences. The following Web sites offer data on some of these factors.

## PRIZM Data

PRIZM Cluster Data, found at **http://cluster2.claritas.com/YAWYL/Default.wjsp?System=WL**, provides a snapshot look at the biggest demographic groups in each region of the U.S. The data is divided by zip code—if you enter the zip code of the location of your business, you'll get a picture of the five largest groups of residents in that zip code.

Claritas, Inc., which collects and distributes the PRIZM data, separates residential areas into one of 62 clusters, such as Winner's Circle, made up by executive suburban families; Suburban Sprawl, young townhouse couples; and Inner Cities, inner-city single parent families.

The PRIZM data is not complete—the free data available on the Internet only shows the largest five clusters in each zip code. Also, if a zip code is primarily businesses, it might not be included in the data because it only documents residential clusters. Claritas, Inc. offers the full data set for a fee. However, in the early stages of your business, it might be more important just to get a snapshot glimpse of the largest groups that live around you.

## ESRI Business Information Solutions

ESRI at **www.infods.com** or **www.esribis.com** also offers free searchable data samples. ESRI uses the ACORN segmentation system, which categorizes consumers into groups similar to the PRIZM system clusters. The ACORN system divides all U.S. residential areas into 43 clusters and nine summary groups based on demographic characteristics such as income, age, household type, home value, occupation, education and other consumer behavior characteristics. ESRI offers a free ACORN segmentation system CD filled with demographic information about the 43 clusters. The CD also includes interactive Potential Purchase Index (PPI) charts that show the types of products and services that might be purchased by each ACORN segment.

This site also allows visitors to enter a zip code and obtain information about the top ACORN cluster in that area. However, the data it provides about that cluster is more in-depth than what is provided at the PRIZM site. It compares the total population and number of households to national information, breaks the population down by race and gender and provides valuable data about income, median housing values and median rent.

## The Right Site®

The Right Site®, developed by Easy Analytic Software Inc. and located at **www.easidemographics.com**, offers one million pages of free, easy-to-use demographic reports for the avid market researcher. This site allows the user to select an area from 12 different geography breakdowns—examples include TV markets, counties and metropolitan areas—and then select from a list of 12 different kinds of reports, including an Owner and Renter Occupied Houses Report and Analysis, a Quality of Life Report and Analysis and an Education Report and Analysis. This information is based on the 1990 census. Updated data is available on CD-ROM for a fee.

In exchange for some basic information about you, your company and how much demographic data you buy annually, you can access the EASI Free Special Reports, which include ring reports (demographic data in a one-mile, three-mile and five-mile ring around your business location) and other valuable reports.

## The Hispanic Market

The Hispanic Market Web site for Marketers offers a wealth of free information. Located at **www.hispanic-market.com**, the site includes a primer designed to help marketers better understand the Hispanic community, which makes up approximately 14 percent of the United States population. Other site features include a breakdown of the top 25 U.S. Hispanic markets, tips on advertising to target this growing community and information on the debate between whether the correct label for people of this community is Hispanic or Latino.

## Other Sources

You can't control marketplace forces, but you can minimize your marketing risks. When you own a reliable base of data about your customers and their tendencies, you can sell selectively to people most likely to buy a particular product. If you know that middle aged Philadelphia men buy red ties in February, you know all you need to know. Some additional resources include the following:

- ***American Demographics***. *American Demographics* is available by subscription, but it is available for free at **www.inside.com/default.asp?entity=AmericanDemo**. This magazine has a wealth of useful information—check out the Top Lines articles for information about demographic studies happening around the country, and click on the Indicators articles for quick pieces on trend forecasts. Back issues are also available through the Web site.
- ***Bureau of Labor Statistics Consumer Expenditure Surveys***. If you want to know what the average American consumer is spending, go to **www.bls.gov/cex** and peruse the Bureau of Labor Statistics (BLS) Consumer Expenditure Surveys. The BLS collects this data by using two types of surveys: 1) a diary survey, which is designed to track consumers' small expenditures, including food and beverages, both at home and in eating places, housekeeping supplies, tobacco, nonprescription drugs and personal care products and services; and

2) an interview survey, which tracks larger expenditures, including those for property, automobiles and major durable goods and those that occur on a regular basis, such as rent or utilities. If you are going into a consumer-oriented business, this data can give you a good idea of whether or not the consumers in your area have the available cash to spend on your services.

- ***Demographics of Web Users***. If you are going to market your business primarily to Internet users, check out **www.wilsonweb.com/webmarket/demograf.htm** for links to helpful articles on demographic information about those consumers. Links include results of a study of how much time Internet users spend doing various online activities, information about the gender gap on the Internet, and how Internet users spend twice as much time online at the workplace as they do at home.
- ***Nielsen Media Research***. Although the reports available on the Nielsen Media Research page are not free, they are not outrageously expensive and would be valuable if you plan to advertise your business on television, or if you plan to try to get free media coverage by marketing a story about your business to local television or radio stations. More information is available at **www.nielsenmedia.com/reports_available/reports.html**.

## The Power of Observation

Whether you are going to make and sell a product or provide a service, some of the best market research you can do is to simply observe consumers' habits. Try surveying potential customers or clients (use the same questions you asked your family and friends). If you offer an incentive and keep the survey short, they'll be more likely to fill it out.

Go to businesses that offer similar or competitive services and see what customers are buying. Talk to customers in and around the business—don't make it obvious that you're doing research, but don't be afraid to be friendly and curious. Find out what they like about the service or product line, what they don't like and why. Everyone has an opinion. All you have to do is tap into that valuable information.

## Customers and Product Development

Product development is the incremental process by which you make an idea into a product or service—and thereafter increase its quality and usefulness to your customer as time passes. Thus, product development has to do with existing products as well as with new; indeed, most companies develop their best new products from existing ones.

Routine customer comments offer sources of innovation. The company that translates the wishes of its customers into products appealing to a broad market succeeds almost automatically.

Successful products also come from looking at what's hot in your industry or at what your competitors do better than you do. This is another reason to keep tabs on your competitors. And, as technology develops, it becomes more and more difficult to stay on the cutting edge in any given industry. Stay on the mailing lists for competitors' products and read trade journals with an eye toward developments that signal new needs in your marketplace.

Product development is not only the most creative but perhaps the riskiest function of a business. This is why it is important to conduct research before you begin to know how well your product or service will do and know when to make changes. Ask yourself:

- Does the product come from a need you know your customers have? Or do you merely think it's a need? What's the evidence of that need?
- Are you counting on the quality of the product—not just value or level of service—to sell the product? If one of these three factors weighs more heavily than the others, which does?
- How many competing new products are in the market already? If interest centers on a competitor's product, is there room for you?

## Networking For Free

Ask any successful business owner, and he or she will tell you that your business will go nowhere if you don't network. Networking is nothing more than spreading the word about your business and making connections with other business owners and potential clients.

You can do some of this networking by going to local events and talking to other business owners in your area, but you can also spread your networking potential to a much wider audience using the Internet. There's a world out there waiting to hear about your business, so why pay for what you can get for free?

## Networking for Women

If you're a woman trying to get started in business, there are plenty of networking opportunities available for you. Women seem determined to help other women. Here is a brief list of the Web resources that can get you started networking with women who are also in business—and who may be able to open the door to new customers and clients to help your business grow.

- **Canadian Women's Business Network**, 3995 MacIsaac Drive, Nanaimo, BC, Canada V9T 3V5, (250) 518-0567, **www.cdnbizwomen.com**. This site offers message boards, articles and other resources for networking. The Member's Showcase section highlights new members and their business profiles.
- **DC Web Women, www.dcwebwomen.org/join/index.html**. Although this networking group is geared toward women who live in the Washington, DC area, the free mailing list is a great place to find out about freelance jobs. It's also good way to learn more about using the Web, although much of the Web-related discussion is relatively technical. The group's Web site includes bulletin board discussion forums on topics including a Women in Business column and a Legalese column that focuses on legal issues surrounding business on the Internet.

- **Home-Based Working Moms**, P.O. Box 500164, Austin, TX 78750, (512) 266-0900, **www.hbwm.com**. From the HBWM Discussion Room to the Member Directory, there are plenty of opportunities to find other mothers who work from home. And when it's time to put work away for the day, the site offers ideas for activities to do with your children and plans for quick, nutritious dinners.
- **National Association of Female Executives**, (800) 634-NAFE, **www.nafe.com**. This organization has more than 200 local chapters around the country with plenty of networking opportunities. Membership is only $29 and benefits include access to discounted services, including everything from insurance companies, office supply stores and magazine subscriptions to education services and shopping services, such as NAFE's Online Mall. Beyond networking, this organization could help you save a great deal of money as your business gets underway. And the membership is tax-deductible!
- **National Association of Women Business Owners**, 1411 K Street, N.W., Washington, DC 20005, (202) 347-8686, **www.nawbo.org**. With 82 chapters around the country—including one in Puerto Rico—this organization provides plenty of networking opportunities. Members have access to a discussion board section where they can learn about what women all around the U.S. are doing to keep their businesses successful and tell others about what they are doing themselves.
- **Women's Chamber of Commerce of Texas**, P.O. Box 2605, Austin, TX 78755-0051, (512) 476-4140 or (713) 665-1637 (Houston phone number), **www.womenschambertexas.com**. From quarterly Power Lunches to monthly meetings, this organization provides many opportunities for meeting other Texas female entrepreneurs.
- **Women's Enterprise, www.womens-enterprise.com**. This Web site offers highlights from the print version of this magazine. It provides news about women in business and includes a link to a chat section where you can talk online with other businesswomen. The site also includes a comprehensive list of women's business organizations with contact phone numbers.

- **Women Incorporated**, 333 South Grand Ave., Suite 2450, Los Angeles, CA 90071, (800) 930-3993, **www.womeninc.com**. Women Incorporated sponsors an annual networking conference, Uncommon Women on Common Ground, but also sponsors regional training sessions and other networking events. Their Web site's Biz to Biz section and the message boards offer online networking opportunities, as well.

In addition to these resources, check out the resources for women looking for mentors listed in Chapter 3. Many of those resources also offer networking opportunities for women who are members, and some even offer those same opportunities for non-members.

## Conclusion

Research goes a long way toward identifying who does business with you and why. In turn, this information prepares you to make the best and most effective use of your marketing and sales efforts like determining whether you should sell your product or services through a catalog.

By collecting information and adapting this knowledge to your markets, you can:

- Identify and define customer expectations regarding service;
- Translate expectations into clear, deliverable service features;
- Arrange efficient, responsive and integrated service delivery systems and structures;
- Monitor and control service quality and performance; and
- Provide quick, cost-effective response to customers' needs.

And, all of these factors put together create sales opportunities. If, for example, you enhance your service or product's usefulness to your customer by adding valuable features or options that they need, you create an interested listener ready to hear more about your products. Each time you propose an innovative approach to a real need, you create a sales opportunity. In order to do these things, you have to know how your customers will use your products.

Remove the barriers blocking communication between you and your customers. This usually takes the form of research and information-gathering. But whatever form it takes, it needs to focus on:

- Increasing the number of potential customers who come into contact with you;
- Increasing your conversion rate, so that more of these potential customers actually buy from you; and
- Make sure they buy again.

Research is a central part of starting a business in an information-based economy like ours. As intellectual property issues like copyrights, trademarks and branding grow in importance, having a good idea of where your business fits in its marketplace becomes more essential. In the following chapters, I'll touch on some examples of how important intellectual property can be to even a boot-strapped start-up.[1]

[1] For a more complete discussion of intellectual property issues, see Silver Lake Publishing's book *The Value of a Good Idea* (ISBN: 1-56343-745-7, 2002).

# Chapter 3: The Nuts and Bolts of Business Ownership

No matter what type of business you plan to start, it's important to understand the legal and financial requirements before you make your first moves. You may be able to start on a very small basis without filing paperwork for licenses or permits. Some people start this way—even though it's not, strictly speaking, legal. But, if you have any kind of success, you'll need to get legal at some point. So, you need to have a firm grasp of what "legal" means.

One caveat: You can apply for licenses and permits on your own. But, when it comes to other matters, be honest with yourself. Even if it's cheaper to do some legal work yourself, hiring a lawyer or an accountant to help with the structure or financial management of your business is smart if you've never been good at balancing your checkbook or if the mention of contracts gives you a migraine. After all, this is your business, so you want to do it right.

## Choosing a Business Structure

Most state governments provide a number of ways in which your business can be organized. Each has certain advantages and disadvantages for you as an owner. In some cases, the advantages have to do with how much tax you pay...and how you pay it. In others, they have to do with who controls the business, who's accountable to whom...and how. In others, specific issues like succession, liability or capital structure are the key concerns.

However you start your business, one of the first major decisions you must make is how to structure it. In most situations, there are six basic types of business structures:

- sole proprietorship;
- general partnership;
- limited partnership;
- closed (Subchapter S) corporation;
- open corporation; and
- limited liability company.

## Sole Proprietorship

The simplest form of business, a **sole proprietorship**, actually lacks much structure. It is a true business monarchy; many start-up enterprises begin this way. If you're starting a business for free, you'll almost always start this way.

A large percentage of sole proprietorships belong to those who are involved in the so-called "cottage industries"—businesses operated wholly or in part out of their owners' homes. These firms participate in many fields of business, but they usually are small and often a part-time pursuit, rather than the owner's sole source of income. Many freelancers and consultants who operate out of their home form this kind of business because it does not require any legal structure or special agreements to set up.

Another reason many people choose to structure their business this way: Sole proprietorships are inexpensive and easy to organize. There is minimal record-keeping involved and the owner pays no business taxes. However, sole proprietorships present a number of serious drawbacks. Sole proprietorships protect you the least from liability and bankruptcy because nothing separates your business from your personal assets. Other drawbacks of sole proprietorships include the following:

- **They are taxed at a higher rate**. All income from the business must be reported as personal income on your income tax forms. Therefore, it is taxed at a much higher rate than business income is usually. There also are no tax breaks for fringe benefits and insurance.
- **They have a harder time getting financing**. To grow, most businesses borrow money at one time or another, and it generally is much harder for a sole proprietorship to get a loan. If the company is incorporated, it can sell stock (give up equity in the business) to raise capital, if necessary.
- **The owner is exposed to unlimited personal liability**. A lawsuit can attach everything you own—not just the assets of the business. And although liability insurance is available, it can be very expensive and may not cover all potential losses.
- **They have a hard time building equity**. From a succession standpoint, it is hard to build equity for the next generation of a family-owned sole proprietorship. If you die, the business is automatically dissolved. What is left to the heirs is a **personal estate**, not equity in a business, stock in a company or a similar investment.

Generally, the problems posed by a sole proprietorship begin to outweigh the advantages when the income from your business reaches about $100,000 a year (that number can vary dramatically). Above that level, the protections offered by a more formal structure are probably worth their higher costs.

Many businesses start as a sole proprietorship but eventually incorporate.

## General Partnership

If you are going into business with someone else, you might want to consider forming a partnership. In fact, many modern companies shun sole proprietorships and start out as partnerships between people who have different sorts of skills. In a **general partnership**, two or more people join together to conduct a business, and each is jointly and severally liable for its operations.

Partnerships are almost as easy to form as sole proprietorships. Similar to a sole proprietorship, a partnership does not require any special registration or structure to set up. A written agreement isn't even required, although it's a good idea.

Assets, including cash, business-related deeds and bills of sale—as well as anything else the business will need in order to function—must be transferred into a partnership.

A partnership also can borrow money, often benefiting from the creditworthiness of several members.

Partnerships become a separate legal entity, but do not pay income taxes. Instead, they compute annual taxable income and file a partnership tax return. This return allocates the income (or loss) to each partner, who then must report it on his or her individual income tax return.

Like a sole proprietorship, a partnership makes all partners **jointly liable** for the debts and obligations of the business. A creditor can seek the assets of any or all partners. And any agreement among the partners to share that responsibility, although binding on them, is not binding on the creditor.

An important note: You can purchase liability insurance if you are in a sole proprietorship or partnership to limit the scope of your financial responsibility. However, it is always advisable to discuss these issues with an attorney. If you want to get around the high cost of retaining an attorney, see if your local law school offers a business advisory program, or ask your lawyer friend to look over your start-up information and offer any pointers to assist you with the process.

Partnerships often are formed by professionals, such as doctors, dentists, lawyers and accountants. They have a limited life, usually specified in the partnership agreement. If a partner dies, becomes incapacitated, goes bankrupt or simply withdraws, the partnership automatically terminates unless otherwise specified in the agreement.

A partner can transfer his or her interest to someone else or pass it along to an heir at the time of his or her death, unless forbidden to do so by the partnership agreement. A sound partnership agreement should provide for a buy-out in the event of a death or in the event the partnership should break up. If the agreement permits the remaining partner(s) the right of first refusal and specifies that the interest in the company cannot be sold to anyone who offers a lower amount, a fair price and a satisfactory buy-out generally can be arranged.

Remember: A written contract between the two or more partners is important. If you are forming a partnership, sit down with your partner(s) *before* you start the business and put everything in writing. Go over everything from how you will start the business (who is putting forward funds to get it started, any ideas that might be proprietary, etc.), and how you will run the business (who is responsible for handling the accounting, who is responsible for dealing directly with clients, etc.), to how you will end the business (how will profits be split or debts be handled).

Putting everything in writing from the start protects you from problems later. Even if you go into business with the person who has been your best friend since 4th grade, disagreements can occur down the road.

Estimates show that partnerships break up as frequently as marriages—about 50 percent of the time—and perhaps slightly more often. As with a marriage, a partnership breakup can be amicable or devastating, depending on the temperaments of the parties involved. Therefore, the careful selection of a partner is extremely critical.

If you are forming a partnership, look for a partner who:

- stimulates enthusiasm;
- stimulates new ideas;
- is easy to work with—not egocentric, autocratic or stubborn;
- can offer a different perspective and has complementary talents or experience;
- uses logic rather than emotion;

- shares the same goals; and
- is a good team worker.

## Limited Partnership

In a **limited partnership**, one or more general partners manage the business and are personally responsible for its debts, while the limited partners have no role in day-to-day business operations and are liable only to the extent of their investments for the company's financial obligations.

As in a proprietorship, this structure avoids double taxation—that is, the taxation of both the business income and the individual income.

Obviously, the limited partnership offers more benefits to a passive investor than to someone who wants to be actively involved in the operation of the business.

## Corporation

Corporations are the safest, generally the most versatile and, therefore, the most common form of business structure apart from the sole proprietorship.

Of course, not all newly formed corporations are new businesses. Many are proprietorships and partnerships that have moved up to a more sophisticated structure. (As I mentioned earlier, a lot of people incorporate solely for legal or tax reasons.)

Legally, a **corporation** is an entity totally separate from its investors. It is responsible for its own bills, files its own income tax returns and pays its own taxes. It can sue and be sued. It lives on indefinitely, regardless of who its stockholders may be at a given time.

The main advantage of a corporate form of business structure comes from the fact that the owners (stockholders) are fully sheltered from the liabilities of the company. This can be particularly valuable to those who are attempting to build a fast-growth company in which considerable risk may be involved.

This is not to say that the people running a corporation are completely absolved of responsibility. Management can be sued by stockholders for malfeasance or nonperformance.

The dramatic increase in the number of firms incorporating each year is due, in part, to the huge settlements being awarded in court cases these days. The high—and increasing—cost of malpractice insurance for professionals, such as doctors and lawyers, has convinced increasing numbers to form personal-service or professional corporations, even though this structure means a higher...and in some cases double...tax burden.

Advantages of a corporate business structure are:

- The issue of **stock**—actually, shares in the business. A company can sell off some of its shares when the price is high and buy back some of its shares when the price is low, thereby using its own stock as a medium for investment. Stock also can be used as security for loans, like cash in arranging a merger and as inducements in the hiring and retention of key personnel.
- Some **tax advantages**. In some cases, corporations pay lower income taxes than individuals. On the other hand, corporate income is taxed twice—once at the corporate level and again when the profits are distributed as dividends to stockholders, who pay taxes on them as personal income.
- **Professional detachment**. The people who run a corporation may not be the people who own it. The head of the firm actually may be an employee, hired by the stockholders to manage the company, in which case the manager may be free of personal and political issues.

## Subchapter S Corporation

The **closed, or Subchapter S, corporation** can be a useful vehicle for getting a new business started. That is precisely why the provision was put into law in 1958.

Under Subchapter S regulations, the company passes all of its gains and losses to the stockholders, enabling the stockholders to use any initial losses the business may incur during start-up to offset earnings from other sources—up to the amount that they have invested in the company.

Subchapter S corporations may have subsidiaries, providing they do not own more than 80 percent of the stock of a subsidiary. But they cannot have more than 35 stockholders.

Subchapter S corporations can be converted to open corporations; but, once that is done, owners are restricted as to when and how they can revert back to Subchapter S status.

## Limited Liability Company (LLC)

In the 1990s, the limited liability company (LLC), a new organizational form, gained legal status in many states. LLCs are similar to partnerships, except that the liability of partners is limited to their equity investment.

If you are planning on having employees and a larger business, you might consider the more formal structure of the LLC. In fact, many advisers now advise that any new business should be established as an LLC unless the peculiar facts and circumstances of that business indicate otherwise.

Typically, LLC members set forth a business purpose in their articles of organization or operating agreement. This agreement is ideal for incorporating a mission statement as part of the company purpose, as well. But its real purpose is to identify partners, the roles each plays and how any disputes or buy-outs will be resolved.

An LLC's operating agreement can provide for a dispute to be resolved by an appropriate mechanism. An agreement might even be drafted to require the buying and selling of ownership interests under specified terms and circumstances.

An LLC operating agreement can easily be prepared to establish a management decision-making process. Quorum requirements for members' or managers' meetings can be set uniformly for all decisions or specifically for one or more types of decisions. In addition, voting requirements can be made to fit the interests of the members. For example, certain issues may require a simple majority vote, others a super-majority or unanimous vote to approve.

## Resources for Choosing a Business Structure

If you're still not sure how you want to structure your business, there are many great resources on the Internet that can help you determine the type of business structure that is best for you and your business, including:

- MyCorporation.com, a commercial site that provides incorporation services for a fee, has a good list of answers to commonly asked questions about incorporating and other business structures at **www.mycorporation.com/inclearn.htm**.
- You can order a free handbook called *How to Incorporate Your Business Now* by visiting **www.corporate.com** or by calling (800) 877-4224.
- Nolo Self-Help Law has great resources—many of them downloadable for free or a small fee—on their Web site. For example, a free encyclopedia article at **www.nolo.com/encyclopedia/ sb_ ency. html# Subtopic16** includes extensive descriptions of all the types of business structures and what it takes to form your company under one of those structures.
- Download free e-books on incorporating in various states at **www.free2incorporate.com/corporation_states.htm**. This site also offers a number of articles on incorporating for the lowest possible cost.
- iVillage's state-by-state start-up guide is located at **www.ivillage.com/ topics/work/homebus**. This collection of links includes contact information for each state, as well as other helpful forms and informative items for when you are ready to get up and running.

- The Business Start Page offers a free mini-course at **www.bspage.com/1start/start.html** that includes planning tools for deciding whether to start your business as a sole proprietorship, partnership or corporation.
- Attorney Lee Madere Jr. has put together an extremely comprehensive and clear article on the different business structure options. The article can be found at **www.madere.com/bizstrct.html**.
- Another comprehensive overview appears on the American Express Small Business Services site at **http://home3.americanexpress.com/smallbusiness//resources/starting/structuring**.
- The Small Business Association (SBA) has a good collection of links on its site at **www.sba.gov/hotlist/businessnames.html** that will lead you to the sites for each state, and provide a starting point for finding information about not only the laws for incorporating in each state, but also information on other legal requirements for business, such as registering your business name.

## License and Registration

Once you have decided on and established your business structure, you need to make sure you are completely legal and registered at the local and state level. If you are a sole proprietor or in a partnership, the local level will probably be the only place you need to be concerned. However, check with your local business licensing office to make sure you do not need to register at any larger level than that office. With few exceptions, most cities and counties require any business—from home day cares and administrative/retail businesses to liquor stores and security companies—to obtain a valid business license, even if that business is already licensed in another city or county. In other words, you can't engage in business without first having procured a license from the city or county in which you operate. In most cases, you can't open a business checking account until you obtain a license because most banks require a business license in order to do this. So, you'll probably want to invest in one anyway.

What is a business license? **Business licenses** are annual non-regulatory licensing fees levied by individual cities and counties for the privilege of conducting business within a particular city or county jurisdiction. The primary purpose of a business licence is to ensure that the municipal land use regulations, building and fire codes and other community safety requirements are followed by businesses. In some cases, business licences are also used to collect important statistical information about local business activities and trends that impact city or county policy. Thus, if you have a business, you're probably required to have a valid business licence.

## Applying for a Business License

The first thing you'll have to do is fill out an application for the business license. The application will ask for basic information about your business, such as name, address, contact information, type of organization, etc. Depending on the type of organization, you may also be required to fill out a separate form relating only to Sole Ownership (Sole Proprietorship), Partnership, Corporation, Non-Profit, etc.

You must have a separate license for each separate type of business conducted at the same location and for each separate branch or business property location.

If you're not sure how to find a business license application in your area, search the Internet. In some cases, business licence application forms are available from a city's Business Licence or Business Services Division. In others, they're available from the Department or Secretary of State. Many states and counties now post the application on their Web sites. All you need to do is print the form and fill it out. At the very least, you'll be able to find a phone number to call for information about obtaining the application. (See Appendix A for a list of state specific information on where to get a business license.)

If you still can't find what you're looking for, use the old standby, the phone book. Look under the government listings or county listings. There's usually a listing for business licenses in the Frequently Called Numbers section. If it's not listed there, search under Licenses or Finance Department.

The application for the license will provide you with the basic instructions on how to apply (i.e., whether you need to present the application and payment in person, or mail it in).

Business license fees vary greatly from city to city, or depending on what type of business you have, your business square footage, etc. Some businesses will have to pay more than others. (The fee for a childcare operation is only $50.) The application fee typically costs between $50 to $125, and the license is usually valid for one year. (This fee is prorated if you open a new business part way through a year.)

Remember: The fee is not refundable once a licence has been approved and issued. So, if you change your mind about starting your business or your partner backs out of the enterprise, don't bother trying to get your money back.

After you have completed an application form and paid your licensing fee, the licence staff then reviews your application and forwards it to the appropriate agencies for approvals. Once you're approved, a business licence will be mailed to you. The time required to process a licence application varies but usually ranges within seven to 10 working days.

If you have a business licence, you'll automatically receive a renewal notice in the mail each year. Licence renewals can be paid by mail, at some financial institutions or in person at the Business Licensing Office.

You can obtain a business license for certain types of businesses that operate from home, but there are some restrictions. City zoning bylaws typically restrict the type of business activities that are allowed to operate in a residential neighborhood (to minimize disturbances to surrounding properties), limit the size of the area used for business purposes and regulates parking and storage impacts. In most cases, retail stores are specifically excluded as a home based business.

Remember: Business licenses are not transferable. It is your responsibility to advise the Business Tax Office of ownership changes, relocation or termination of business. And, if, at any time, you stop operating your business, notify the Business Licence Division immediately. Do this in writing and keep a copy. If you fail to inform them, you'll continue to receive renewal notices and could be liable for fees.

## The Temptation to Go Without

If you think you can get away with operating your business without a license, think again. In some cases, you can be fined up to $100 per day for operating without a valid business licence. In others, businesses that operate without first obtaining a license may be liable for a penalty of 50 percent of the current year's license fee, plus any prior year's unpaid license fee for the preceding 36 months.

When applying for a licence, bring your business licence fee, along with any relevant certificates or licences required by the provincial or federal governments to operate the business. Different permits and inspections are required depending on the type of business and its location. Ask about any permits, licenses and/or identification numbers that must be obtained prior to submitting an application for a business license.

The details of your business operation will determine the type of federal, state, county and city licenses, permits, certificates and approvals you must have to open your business. The important ones to ask about:

- **Alcoholic Beverage Control Licenses**;
- **Building Permits**;
- **Certificates of Appropriateness** (required for any new construction);
- **Certificates of Occupancy**;
- **Commercial Vehicle Permit**;
- **Fictitious Name or Doing Business As (DBA)**;
- **Firearm License**;

- **Fire Marshal Inspection**;
- **Hazardous Use Permits**;
- **Health Department Inspections**;
- **Home Occupation Permits**;
- **Seller's Permit**;
- **Sign Permits**; and
- **Zoning Inspection Permits**.

In some cities or counties, applications received for some business categories, such as pawn shops, weapons related businesses, and massage parlors/technicians, are referred to the police department for regulatory review and approval. Other businesses that might need approval: adult book stores, palm readers, dating or escort services, fortune-tellers, pool/billiard rooms, tattoo parlors, arcades, modeling services, check cashing services, adult theaters and acupressure/acupuncture services.

Most small and home-based businesses will only require a local business license or permit. (There are times when you will not need a license to conduct business. For example, if you are a consultant you probably won't need a license. However, most states have licensing requirements for certain types of businesses so check with your local government offices for information about licensing requirements.)

Certain businesses, such as attorneys, barbers, contractors, dentists, businesses serving food and social workers, will also require a state license. To find out if your business requires a state license, contact your local government offices. They should be able to give you information as to whether your business will require state licensing. Another good source of state specific information is your local library.

Few businesses require **federal licensing**. But, if you intend to provide investment advice or deal with firearms, you'll have to obtain a federal license.

## Employer Identification Numbers

An Employer Identification Number (EIN) is also known as a federal tax identification number, and is used to identify a business entity. Most businesses need an EIN. The EIN is your account number for the collection and reporting of taxes withheld and wages paid to the employees (if you have them). Most partnerships, corporations and trusts need an EIN. Sole proprietors generally need an EIN only if they have employees, have a Keogh pension plan or must pay certain federal excise taxes. For more details on federal and state EINs, see below.

If you are organized as a corporation or partnership, or if you hire or plan to hire employees, you will need to obtain a **Federal Employer Identification Number (FEIN)** for the purpose of withholding FICA taxes and Social Security taxes. To obtain an FEIN, contact the Internal Revenue Service (IRS) toll-free at (800) 829-1040 or download an application (SS-4) for an Employer Identification Number at **FTP.fedworld.gov/pub/irs-pdf/fss4.pdf**.

In some states, if you're organized as a corporation or partnership, or if you hire or plan to hire employees, you may also need to secure a **State Employer Identification Number (SEIN)** for the purpose of withholding State Income Tax, Disability and Unemployment Insurance. (Many states also use the federal EIN for state income tax reporting purposes.) Contact the local State Employment Development Department Division of Audits or your state department of revenue to find out how to get a number.

## Patents and Trademarks

Maybe your million-dollar business idea came to you while you were driving to that job you would quit in a heartbeat. Maybe it came to you while you were washing the dishes after dinner or while you were sitting around the kitchen table with your closest friends. Or maybe you had a dream and woke up in the middle of the night scrambling for a piece of paper to write down the plans for the product or service that no one else has executed yet. Regardless of how that idea came to you, if you don't trademark or patent it, you may lose out on the big bucks you hoped to make, or worse, you could even lose out on the idea itself, if someone else thinks of it and registers it first. To make sure that doesn't happen, it is important to figure out whether you have an idea that needs to be—or can be—registered and protected.

After you've secured a business license, necessary permits and a federal identification number, you should focus on protecting any ideas or symbols that represent your business's identity and image. This means registering any trademarks or copyrights…even if your trademark isn't a billion-dollar trademark like Coca-Cola, Home Depot or Ford.

## Who Grants Trademarks...and How?

Getting a trademark takes **time and money**. It's wise to first search the trademark database for pending and existing trademarks so you don't file for someone else's trademark. This can easily be done by going to **http://tess.uspto.gov**, by visiting the Trademark Public Search Library in Virginia or by visiting other such libraries throughout the country that house these databases (refer to Appendix B for help). Private search firms will also conduct a search for a fee. Then, you must fill out the application and pay a filing fee, which generally is $325 per class of goods or services. (Again, refer to Appendix B for information and contacts regarding the trademark application process.) It can take months before you get your federally registered mark, or "notice of allowance," and you might encounter a few bumps in the road that call for a trademark trial and appeal board. The **U.S. Patent and Trademark Office (www.uspto.gov)** does not decide whether you have the right to use a given mark, however, so even without a registration you can still use any adopted mark to identify the source of your goods and services. Once you obtain a federally registered mark, it's up to you to enforce your rights in the mark.

In the U.S., the Commissioner of Patents and Trademarks grants trademarks. When the Commissioner's office is considering an application for a trademark, it lists the mark on its Principal Register and issues a certificate of registration. This certificate provides the registrant with *prima facie* evidence of:

- the validity of the mark and its registration;
- the registrant's ownership; and
- the registrant's "exclusive right" to use the mark on or in connection with the goods and services specified in the certificate of registration.

The Commissioner does not register a mark unless it meets the requirements established by statute. With a certificate of registration, therefore, the registrant obtains evidence that its mark is not generic in the eyes of the relevant public and that its mark is not merely descriptive, but at a minimum is descriptive and has secondary meaning.

Through the certificate of registration, the Commissioner introduces his opinion that the application of the registrant was sufficient to demonstrate a valid mark. The Commissioner need not require evidence of secondary meaning if the applied-for mark is "inherently distinctive by being suggestive, arbitrary or fanciful."

However, the Commissioner and the Patent and Trademark Office (PTO) are not the final arbiters of what can be trademarked. Federal law vests ultimate adjudication of trademark disputes in the federal courts. (So, trademark holders who allege infringement may sue infringers in federal court and obtain monetary damages, equitable relief or both.)

More revealing, Congress expressly vested in federal courts the power to "determine the right to registration, order the cancelation of registrations, in whole or in part, restore canceled registrations and otherwise rectify the register with respect to the registrations of any party to the action."

When a **certificate of registration** is entered into evidence, it serves only as "*prima facie* evidence of the validity of the registered mark."

## How Patents Work

Legally, patents are the most complex form of intellectual property protection in the United States and most developed economies. Simply said, a patent is a form of monopoly that the government allows for a limited period of time in order to reward and encourage the development of new devices, products and technologies.

Again, the key concept here is **restriction**. A patent is designed to allow a patent holder to restrict the use of his or her product or process.

Of course, patent law has supported a blossoming technology-based economy and competitive markets full of entrepreneurial business. The law works well in maintaining a balance of economic power among inventors, investors, competitors and—ultimately, consumers.

A discussion of patents can take many forms. Most information published on patents is related to the complicated and involved process of applying for one. That's not my focus here; there are plenty of hefty patent books around already. For the forms, you can go directly to the Patent Office and get most of the paperwork you need.[1]

Instead, I'll consider how patents can be used to protect the value of an idea or invention that a person or company develops.

Patents are substantively different than copyrights or trademarks. For one thing, the patent application process is long and complex; it's neither as quick, or as easy as filing a copyright or trademark. For another, the law doesn't rule so completely in patent issues; lawyers and judges have to rely on nonlegal experts in most patent disputes.

Patents usually deal with sophisticated technology, requiring an in-depth understanding of complex technology that often exceeds the average patent attorney's technical savvy.

---

[1]For more information about patents in general or how to file an application, visit the Patent and Trademark Office at *www.uspto.gov*. Or, refer to Appendix B.

## Types of Patents

When you file a patent, you must explain your invention in detail, declare that you are the original and first inventor of the subject matter and pay a fee. Completing this application isn't simply a process of filling in the blanks; to the contrary, **claims drafting**, or writing about your invention in a manner that makes a patent enforceable, is an acquired skill. The application must explain how the invention differs from **prior art**, or existing technology, and it must describe how the invention can be used. The one who decides whether to grant or deny a patent is called the **examiner**, and he bases his decision on the **claims**, or the parts that define the invention. The three types of patents that patent law protects are:

1) **Utility patents**: Any new process, method, machine, manufacture or composition of matter, or any new and useful improvement thereof;
2) **Design patents**: New, original and ornamentation design for an article of manufacture, including the article's appearance; and
3) **Plant patents**: Distinct and new varieties of plants that have been invented or discovered and asexually reproduced.

## Patent and Trademark Registration

The place to check this is at the homepage of the United States Patent and Trademark Office (USPTO), which is located at **www.uspto.gov**. This site is loaded with information about applying for a patent or a trademark. There are registration fees involved in applying for either, and unfortunately, there is no way around that. However, you should be able to download everything you need from the site so you can file it without using a lawyer or other professional to help you with this. However, consider having someone who is familiar with current patent and trademark law review your application before you submit it just to make sure you complete everything correctly. A small mistake can be enough to hold up your application or cause it to be rejected, and that will cost you valuable time and money while you revise or restart the application process.

## Patent and Trademark Databases

There are several useful searchable databases located at: **www.uspto.gov/web/menu/search.html**. For example, the U.S. Trademark Electronic Search System (TESS) contains more than 2.9 million pending, registered and dead federal trademarks, and it is updated Tuesday through Saturday with the latest data. For example, if you want to open The Sunshine Bar and Grill, and you think your concept might be franchisable down the road, but wanted to make sure no one else has trademarked that name, you could search on that in the database. As of this writing, that name was not included in the database and was likely available for use.

It is important to note, however, that just because a business name, slogan or graphic does not appear in the database when you look it up, that does not mean you are guaranteed to get to use it. Someone else could be putting together his very own trademark application for The Sunshine Bar and Gill somewhere else in the country, and if he registers the name first, it's no longer available to you. The USPTO staff will do its own search once your application is filed to ensure that there is no duplication of trademark.

The Patent database also offers a great deal of information. It offers "the full text of all U.S. patents issued since January 1, 1976, and full-page images of each page of every U.S. patent issued since 1790 through the most recent weekly issue date (usually each Tuesday)," according to the Web site. If you have developed a great device that you think will revolutionize the world as we know it, it's best to check this database prior to filling out any forms. Make sure someone else didn't have the same great idea before you did.

Once you have determined that you are ready to submit a patent and/or trademark application, the easiest spot to visit is the USPTO's Electronic Business Center, which is located at **www.uspto.gov/ebc/index.html**. This section offers everything from downloadable forms to an online status system that allows you to see where your application is in the USPTO pipeline. Everything you need to file an application or do follow ups is clearly explained for new users.

## Employment Laws

While you may have supervised people in a previous job, your own business is probably the first time you've really had to manage hiring and retention of and separation from employees. You'll also have to manage a work force diverse in terms of race, disability and gender, as well as meet the needs of workers with diverse lifestyles (i.e., single parents, unmarried employees with spousal equivalents, gay couples, job-sharers and two-income families). Without a Human Resources or Personnel department to handle these things, you will face some challenges.

Businesses have to obey a complex body of federal and local laws that control how people are hired, managed and—if necessary—fired. Various federal and some state and local laws set out to regulate the workplace. Anti-discrimination law, such as Title VII of the Civil Rights Act of 1964, for example, dictates that you hire the best people without consideration for outside prejudices. Occupational safety law dictates that you furnish employees with a workplace that is free from recognized hazards that are causing or are likely to cause death or serious physical harm. The body of employment law has created a level of regulatory compliance that leaves most people confused. Employment law isn't always a clear-cut issue. So, common sense doesn't always apply. An employer with no bias in his or her mind can be found guilty of discrimination.

Some employers are afraid to even address the issue of color or gender at work. With charges of racism or sexual harassment so commonplace...the topics seem best left alone. But you can avoid all this mess by staying informed.[2]

Chances are that you're going to have to get the business off the ground by yourself, or, at best, with the help of free labor. Translation: You

[2] For a detailed discussion of hiring and firing issues and other workplace law, see Silver Lake Publishing's book *Rightful Termination* (ISBN: 1-56343-067-3, 1996).

need help from friends and family. Most start-up businesses don't have enough steady money to hire their own employees. But if you have enough start-up funds or contracts to get employees on board, you must be prepared, which, again, means you must be informed.

If you're large enough that you need employees, you need a good handle on the legal requirements involved in the hiring and firing process, including information on both the employer's and employee's rights and responsibilities.

## Hiring an Independent Contractor

You may not want to bring on anyone full-time when you're getting started. The solution: bring in independent contractors. If you follow the rules for designating someone as an independent contractor, it can be a good way to get the help you need without providing benefits or paying taxes on the employees. However, the IRS has very strict definitions about what constitutes an independent contractor and when that person crosses the line and becomes an employee, so be sure to familiarize yourself with these rules before hiring someone. An independent contractor can save your business time and money, but if the IRS or any other government agency decides that your independent contractor is actually an employee, you could face extremely unfortunate financial consequences (i.e., be required to pay back withholding taxes and interest if they decide that you've misclassified employees). For information on how to determine whether an individual providing services is an independent contractor or employee see the IRS Web site at **www.irs.gov** or go to **www.irs.gov/businesses/small/industries/article/0,,id=98873,00.html**.

## The EEOC—What It Means to You

There are plenty of people who will rant and rave about how the Equal Employment Opportunity Commission (EEOC) is an enemy of business. They scream that the Americans with Disabilities Act (ADA) was a result of a conspiracy against the National Federation of Independent Businesses.

Ranting doesn't accomplish much, though. The ADA, Title VII of the Civil Rights Act of 1964 and class action lawsuits are all part of running a business in the real world. The EEOC is a fact of life in the real world.

Employer-employee relations now form a large body of law that has become a highly competitive and lucrative legal specialty. Hundreds of thousands of charges brought against employers and thousands of precedent-setting court decisions have almost completely altered the traditional principles of employment law. Each time the courts define—in detail—what constitutes unfair, unethical or discriminatory behavior, the risk of a company being sued rises.

Even now, nearly 40 years after the height of the civil rights movement, employment law is still evolving. In recent years, there's been a shift away from discrimination cases brought on the basis of rigid categories like race, gender and age to cases dealing with less certain categories like disabilities and sexual harassment.

In addition, suits about discrimination in hiring used to outnumber suits about firing. Today, the reverse is true—by a factor of three or more. Most observers attribute this shift to a politicized workplace, since it's unlikely that an employer who *would not* discriminate in hiring workers *would* discriminate in firing them.

The reasons for the increases in employee litigation are more complex than just an increase in the number of laws and regulations governing employer-employee relationships. There is a greater awareness of these laws by workers, and a seemingly greater willingness among people to turn to litigation as a way to resolve disputes with an employer. Worst of all, many employers invite these lawsuits by nervously avoiding politically incorrect topics.

But employees know their rights. A whole generation of conditioning by popular media and scatter shot governmental regulation has made employees very certain about their rights—even when they are mistakenly certain. And they are certainly willing to exercise these rights.

Contrary to a popular myth, it's not just executives who sue their employers. EEOC studies suggest that an hourly or minimum wage employee will initiate legal action as often as a highly paid executive.

To the degree that employment law has protected the innocent worker, it has been a positive change in the work force. But it also represents a real and measurable threat to even the fairest and most generous employer. Employee litigation is epidemic, and employers who are not armed with a working knowledge of the law and how it applies to them are vulnerable to potentially ruinous lawsuits.

No matter how fairly and equitably your treat your employees, no matter how well you follow legal advice, no matter how consistent and fair in your treatment of employees, discrimination remains a business risk.

Meanwhile, insurance coverage for these disputes is sometimes unreliable. Discrimination claims are usually contested by insurers. And policies are typically written with specific exclusions for so-called "employment practices liability."

You can purchase separate employment practices liability insurance. But this is also an unpredictable prospect. There is no standard EPL policy form. Thus, not only do policy terms and conditions vary from insurer to insurer, but even the name given to the various policies is not uniform.

Although policy language varies, the most significant provisions are common to virtually all EPL policies, at least in some form.

Most EPL policies provide coverage for the three basic employment-related actions that can result in liability—wrongful termination, discrimination and sexual harassment. A caveat: the definitions of key terms can vary significantly from policy to policy.

Many employers try to avoid workplace liability by means of evasive efforts like heavy use of temporary workers or employee leasing. But none of these tactics work—you can face a sexual harassment or ADA claim from a temp or a leased employee just as easily as from a traditional one.

Smaller companies can keep employment levels low enough so that laws like the ADA or Family and Medical Leave Act (FMLA) won't affect them. As long as you have fewer than 50 employees, these laws usually don't apply. But other employment laws apply, even if you have just one employee.

Laws the EEOC enforces that probably apply to you and your business include the following:

- **Title VII of the Civil Rights Act of 1964**, which prohibits employment discrimination based on race, color, religion, sex (including sexual harassment and pregnancy) or national origin and protects employees who complain about such offenses from retaliation;
- **Age Discrimination in Employment Act (ADEA)**, which protects workers age 40 and older from discrimination based on age. This law makes job descriptions important. The ADA prohibits employers from discriminating against people with disabilities who are qualified for a position and able to perform its essential functions;
- **Rehabilitation Act of 1973**, which prohibits employment discrimination against people with disabilities in the federal sector;
- **Title I of the Americans with Disabilities Act (ADA)**, which prohibits employment discrimination against people with disabilities in the private sector and state and local governments; and
- **Sections of the Civil Rights Act of 1991**.

All employers with more than 15 staff, public, private or nonprofit, come under the EEOC's Uniform Guidelines on Employee Selection Procedures.[3]

All of these employers can be sued by the EEOC for "discrimination" if the racial, ethnic and gender mix of new hires diverges sufficiently from that of all other qualified applicants—for example, if the percentage of blacks hired is lower than the percentage of blacks applying.

---

[3] For a more detailed discussion on EEOC guidelines and compliance issues, see Silver Lake Publishing's book *Mastering Diversity* (ISBN: 1-56343-102-5, 1995).

The EEOC's goals are, at their core, admirable. Its enforcement of those goals is sometimes problematic. But, as a business owner it's your job to meet this challenge and keep ahead. The EEOC can investigate your management policies and practices if it suspects discrimination. Or if someone has complained. So, if you have employees you should have a working knowledge of how these laws are enforced and how you can comply. For more information on the EEOC and tools to help you handle any questions about employment law and to help you take control of these issues, visit the Commission's Web site at **www.eeoc.gov**.

## Department of Labor Laws

Regardless of the number of employees in your employ, you must be familiar with your federal and state labor laws. The U.S. Department of Labor (DOL) makes it easy to find out what the federal laws are through the eLaws Advisor, a service located on the DOL Web site at **www.dol.gov/elaws**. This service uses an expert artificial intelligence system that will interact with you to help answer any questions that you may have about the labor laws that apply to you and your business.

Even if you these laws don't apply to you because you don't have 50 or more employees, there are several Department of Labor laws to pay attention to, including the:

- **Family and Medical Leave Act (FMLA)**. This act states that workers at companies with 50 or more employees are entitled to as much as 12 weeks of leave for their own serious illnesses, the serious illness of a family member or for parental leave. The FMLA does not require employers to pay a worker during leave, but it does require employers to protect the worker's job (the same job or an equivalent one). If your business is in California, you could be subject to the state's new paid leave law, which allows up to six weeks of partly paid leave for workers to care for a newborn or a seriously ill family member. Eligible employees are paid fifty percent of their regular wages for up to six weeks while absent under the new law. To date, no other states have followed suit, but the trend hasn't stopped short. In fact, 27 states currently have pending legislation dealing with paid family and medical leave, including Georgia and Massachusetts. Visit the DOL's Web site (**www.dol.gov**) for more information on a particular state.

- **Fair Labor Standards Act (FLSA)**. The FLSA is the country's basic wage and hour law. It has many components, including those that set the minimum wage and overtime provisions. This law does apply to you; it does not depend directly upon the number of employees you employ. The FLSA covers an individual employee whose work affects interstate commerce, or it can apply to all employees working for an employer that is covered as an enterprise that is involved in interstate commerce. And, the DOL and the courts have attached broad meaning to the term "interstate commerce." For more information on this broad definition and the FLSA, visit the Wage and Hour Division's Web site at **www.wagehour.dol.gov**.
- **Uniformed Services Employment and Re-employment Rights Act (USERRA)**. This act provides protections for initial hiring and adverse employment actions by an employer if the actions relate, even in part, to the employee's military service (National Guard or Reserve). The protection also extends to potential witnesses of a discriminatory action on the part of the employer. For more on the USERRA, contact Employer Support of Guard and Reserve (ESGR) Ombudsmen Services toll-free at the national ESGR Headquarters at (800) 336-4590 (ask for Ombudsmen Services).
- **Occupational Safety and Health (OSH) Act**. In an effort to protect the health and safety of every worker, the Congress passed the OSH Act on December 29, 1970. The act requires every employer engaged in business affecting commerce to comply with OSHA (Occupational Safety and Health Administration) regulations and to provide a safe and healthful workplace for his or her employees. For more information on the health and safety standards promulgated under this act, visit the OSHA Web site at **www.osha.gov** or **www.osha-slc.gov/OCIS/standards_related.html** for more on OSHA standards and related documents.

- **Consolidated Omnibus Budget Reconciliation Act (COBRA).** This act requires employer-sponsored group health plans to offer beneficiaries, who would otherwise lose coverage when leaving a job–whether they were fired or quit—the opportunity to elect continuation of their health plan coverage. Qualification also extends to those affected by the death or divorce of the employed spouse or parent. The federal law extends the coverage for 18 months. COBRA only applies, however, to employers with 20 or more employees, at least half of the time during the preceding year. However, some states, such as California, have enacted state law (Cal-COBRA) that provide for continuation of coverage to employees of small employers who were not governed by COBRA. Coincidentally, the Cal-COBRA also extends the maximum period of COBRA coverage under insured health care plans to 36 months for qualifying events occurring on or after January 1, 2003. Visit the DOL's Web site (**www.dol.gov**) for more information on COBRA and health insurance.

While the majority of the laws that control employment issues are enforced at the federal level, most states have developed their own laws. In many cases, these state laws create new protected groups or extend protections that federal laws have already made. Municipal ordinances can be even more aggressive.

The majority of these state laws mirror the content and structure of the federal laws. In some cases, though, the state laws stand apart. The difference is usually that more groups are protected by state laws than by federal laws.

To find out more about these laws or any others laws that may apply to you and businesses in your state, check with your local business licensing office when you register for your license to find out how to contact your state's labor department. Alternatively, visit your state's Web page—most will have a link to the labor department, if not other contact information such as an address or phone number.

## Other Employment Resources

One place to obtain a good primer on hiring your first employee on the Internet: **www.bankrate.com/brm/news/biz/biz_ops/19991206.asp**. **Bankrate.com** also offers a useful, step-by-step plan detailing how to decide whether or not you're ready to bring that person on board, and if so, how to do it correctly.

A collection of helpful articles on the topic appears on the Inc. Web site at **www.inc.com/leadership_and_strategy/index.html**. This guide touches on topics ranging from recruiting and screening to interviewing and dealing with employment tax issues.

Trying to figure out what to pay that employee? The U.S. Department of Labor offers a helpful guide to minimum wage and overtime pay at **www.dol.gov/asp/programs/handbook/minwage.htm**.

A great selection of human resource forms, manuals and other helpful documents is available at AllBusiness.com. The site includes resources on everything from the proper handling of American with Disabilities Act requirements and a checklist of questions you are not allowed to ask your prospective employees to information on drug testing and downloadable I-9 forms. For more information on these resources, go to **www.allbusiness.com/cmt/information/top_level.jhtml?fname=1610**.

Once you have your employees on board, you want to keep them there. After all, the time you've invested in training them is worth a lot of money to you and you want these employees to remain loyal to you. Some resources that include information and help with retention of employees include the following:

- BusinessKnowhow.com offers a list of tips about how to retain employees at **www.businessknowhow.com/manage/retention.htm**.
- Read about the challenges of retaining—and recruiting—employees at **www.entrepreneurs.about.com/smallbusiness/entrepreneurs/library/weekly/aa071900a.htm**.

- Non-monetary rewards are always a good way to retain employees. Get some ideas for these types of rewards by reading **www.business-survival.com/reports/NonMoneyRewards.html**.
- In the second of a two-part series that appeared in *USA Today*, business columnist Rhonda Abrams focused on retention issues. Read the column at **www.usatoday.com/small/strategies/rhonda19.htm**.
- You'll be competing for the perfect employee against many other businesses, including very large ones. ZDNet offers an article on competing against big companies at **www.zdnet.com/smallbusiness/stories/general/0,5821,2571104,00.html**.
- MoreBusiness.com offers five tips on retaining employees at **www.morebusiness.com/running_your_business/management/d906062469.brc**.

If the worst-case scenario happens and you have to let an employee go for one reason or another, turn to these resources for ideas and help with the task:

- The Business Owners Toolkit has a firing checklist at **www.toolkit.cch.com/tools/trmntp%5Fm.asp**.
- BusinessTown.com emphasizes getting to the point of the conversation quickly in the article that appears at **www.businesstown.com/people/firing-howto.asp**.
- A tutorial on the subject is available at **www.learnthat.com/courses/business/fireemployee**.

As an employer, you have to take control of your workplace environment. The laws don't tell you how to do it, employees can't and regulators don't have the commitment to try.

If you suspect—or worse yet, if you have no idea—whether your employment practices could be called into question, it's a good time to review your methods and the laws that apply to you.

Another tool that many employers overlook or misjudge: the employee handbook. Used well, this can set the ground rules for employment without creating contractual obligations. Used badly, it sets nothing and creates unwanted contracts.

A few points that can help keep things going well:

- include a prominent disclaimer, near the front of the handbook, stating that the book is not an employment contract;
- spell out clearly that employment is "at-will" and that employees can be terminated at any time and for any reason. Any modification of the "at-will" status is only binding if it's in writing and signed by a designated person;
- state that policies in the handbook can be amended at any time;
- list certain kinds of behavior—including discrimination, bias and sexual harassment—that you do not tolerate. Commit to taking disciplinary action (though it's best to keep the action non-specific) immediately upon discovering the behavior;
- outline a procedure that allows an employee to report prohibited or illegal behavior either through his or her supervisor or through a designated supervisor outside of his or her area;
- to the extent feasible, offer flexibility in matters like work schedules and sharing responsibilities. Managed carefully, these benefits cost you little and create a great sense of value among employees;
- state what you are for as well as what you forbid. Commit to a non-discriminatory workplace that's consistent with existing diversity laws.

Another helpful step: Include a form for the employee to sign that indicates he or she has received and read it and understands its terms.

Last, write job descriptions for each kind of job you have. These descriptions don't have to be long—but they should describe the basic requirements and responsibilities of a job.

If the job demands a certain kind of physical capability, describe it. If it includes particular pressures, name them. You have latitude to use your business judgment in writing these descriptions, as long as you do so in good faith and in advance of any kind of legal challenge.

The bottom line? When it's all said and done, your business will have reduced exposure to discrimination claims, lower absenteeism and turnover, full use of human resources, fewer conflicts and possible market growth—to name a few.

Remember: Don't shy away from these topics. Take control—just as you can any other part of your business. If you do, you don't have to worry about firing a problem person because she is a racial minority, or hesitate to hire a decent candidate because he's in a wheelchair. And, use this information as a primer. If you're already facing an diversity problem—or a discrimination lawsuit—don't try to use this book in place of a lawyer.

## Conclusion

In this chapter, I've discussed permits and licenses in some detail. And I've described the different legal forms a business can take. But I've only outlined the various legal and regulatory issues that you face when you start hiring employees. When you're dealing with these issues, it's very hard to do anything for free. The best way to control expenses with your legal structure is to be sole proprietor—avoid partners and complex ownership at the start. The best way to control license and permit costs is to keep your activities focused and run out of your home, if at all possible. And the best way to control personnel costs is to avoid hiring employees for as long as you can.

# Chapter 4: Getting Money for Free

Now that you've decided what kind of business you want to start, you need to find the money to start it. Many entrepreneurs loathe the money-raising aspect of starting a business. Lenders can be tough—even insulting—to people trying to borrow money for a small business. That's why it's a good idea to run your business for free—without borrowing—for as long as you can.

Regardless of what kind of business you plan to go into, it will cost some money to get it underway. That amount may be limited—some business resources say most small businesses can be started for under $5,000—but it still may be more than you want to put on your Visa or than you can afford to take out of your savings account.

At some point, a growing business will need to raise money. So you need to know a little about how business finance works.

How much money do you actually need? It's important to think through all the finances of your business before you get started so there are no surprises down the road. Also, no matter what kind of money you go after, be it loans, venture capital, angel funding or other types of financial assistance, you will need to approach your funder with a solid plan about why you need the money, where it is going to go and how you think you will make the money back as your business prospers.

## Raising Capital

Every business needs money in order to grow.

The hardest part of the start-up process for many people is coming up with the idea for the business. And, once you've started, it's easier to raise money...but you can't be shy. Business owners can no longer limit their activities to being solely manufacturers, retailers or providers of services; they must also become seekers of capital. They must go out and seek capital.

Although there are numerous sources of capital out there, to many individuals the means of tapping them are generally unknown. To most people, finance is a mystery—and it is easy to understand why.

Banks and other financial institutions often don't provide clear explanations of the basis on which they make capital available. To those seeking funds, the operations conducted in these institutions seem mysterious and clouded by double-talk and insider jargon.

The need for proper financing has become increasingly heightened in recent years. As sales dollar volume figures have increased, markets have broadened, and it has been necessary to obtain more working capital to sustain these higher levels. And that's not the half of it. Taxes take a much greater portion of earnings, decreasing the availability of funds arising from company profits, which could otherwise be used to finance growth. Capital planning has therefore inevitably become—far more than it used to be—a factor that requires almost continuous attention during the establishment of any business.

Fortunately, the availability of capital has kept pace with today's greater and more varied needs. In this chapter, I'll attempt to bridge this gap between need and availability and attempt to explain simply the details of financing and identify the procedures to follow in order to obtain such capital, including some of the following:

- How to calculate your working capital and your need for the same;
- How to make sure that your lending officer sees the incentives in sponsoring your loan;

- How to make sure that your financial statement reflects your business in the best possible light;
- How to analyze your financial statement from the financier's point of view and how to enhance the picture;
- How to determine cash flow and use it to best the advantage in obtaining certain types of financing;
- How to obtain financing in excess of the net worth of your company;
- How to tap into the venture capital (VC) market;
- How to take advantage of Small Business Administration (SBA) and other Loan Programs;
- How to obtain financing through Small Business Investment Companies (SBICs);
- How to determine whether a government grant program is best for you and your growing business; and
- How to take advantage of angel funding.

## Types of Financing: A Little Background

In a dynamic economy, change is so rapid and continuous that new sources of business finance are created today that didn't exist a decade ago.

The field of business finance has also changed—every year.

Financial institutions are in constant competition, devising new funding mechanisms to put more of their money to work and meet the changing needs of their clients. One result of this competition has been the creation of a broad spectrum of business finance whose analysis is a must for every seeker of capital.

There are three generic sources of capital: public, institutional and private financing.

**Public financing** involves the issuance of securities to more than a very small group of investors and, with the exception of intrastate issues, requires registration with the Securities and Exchange Commission (SEC). This type of financing almost always involves the services of stock brokerage firms, which refer to this end of their business as **investment banking**. As a start-up entrepreneur you probably won't need to concern yourself with public financing because it usually applies to established businesses that have already traveled the other routes of financing.

**Institutional financing**, financing more appropriate to your needs, multiplies your options—so it comes as no surprise that start-up businesses make heavy use of institutional financing. In doing so they tap into:

- commercial banks;
- insurance companies;
- commercial finance companies;
- pension funds;
- the Small Business Administration;
- factors;
- venture capitalists;
- industrial loan banks and real estate investment trusts (REITs); and
- investment bankers who can arrange private placements.

This list gives you an idea of the wide variety of sources for institutional investing.

In one respect, however, most of these institutions stay on one side of a very definite line; they stop short of what you might call 100 percent financing. With the exception of venture capitalists, they may advance as much as 90 percent of the capital required to get an enterprise up and running, but many people erroneously believe that they may occasionally lend all the necessary money. Instead, they want to see some of the owner's capital at risk, too.

**Private financing** is the third—and oft most important—source of capital. The groups from which this capital can come range from relatives and friends or a small group of hopeful investors, to substantial backers of new ideas and operations known as venture capitalists. If you don't have any friends or family (or they can't afford to help you out), professionals in the larger cities—attorneys, accountants and particularly financial consultants—are often helpful in providing access to private investors.

Venture capitalists don't operate under strict rules governing security and return. They invest their money with a frank view of the risks and with meager or no security, in a gamble for a substantial return. Unlike institutions, which primarily seek interest income, venture capitalists seek an ownership interest. They want to cash in on the increase in the value of the business operation once it establishes itself in its market.

In short, they want capital gains, not interest income. This makes for subtle differences in the way they approach a business venture—and for subtle differences in the way you approach them with your idea. I'll go into more detail on venture capitalists later, but remember that what you need to keep in mind, in presenting your idea to a venture capitalist, is the fact that the venture capitalist is primarily interested in substantial capital gains in the future.

## Three Types of Business Capital

For now, let's focus on the three types of funding that apply to most start-up businesses:

- **Equity capital**;
- **Working capital**; and
- **Growth capital**.

An understanding of each is crucial for one practical reason: In order to obtain financing, you must know the nature of what you seek. Bankers won't make these distinctions for you. They'll consider exactly what you apply for—and reject your application if you don't qualify for the particular type of financing you have specified. Yet you might qualify for an entirely different form of financing. But you should know this *before* you meet with a banker.

It's important to understand that you have options. Again, look for creative ways to get what you need to grow your business *before* you meet with the bankers.

**Equity capital** commonly represents the original investment in the business plus retained earnings. Technically, on a balance sheet, it reflects ownership of the enterprise as held by principals and other investors. It also represents the total value of the business, since all other financing amounts to some form of borrowing that the business must repay.

So, a banker who asks you *What do you have in the business?* wants to know about your equity capital. Banks generally don't loan equity capital. That's why you need to start your business for free. But banks will consider lending the other two kinds of funding that small companies need.

The need for **working capital** arises from the ongoing activities of business. As sales increase, so do accounts receivable—that is, money owed to you and the business by customers but not yet received. The enterprise needs money to carry accounts receivable because no business can precisely match income and outgo. Your business also needs money to satisfy the costs of increasing inventory (if you have any) to meet rising demand.

Fortunately, you can usually obtain working capital on a steady, revolving basis. And, although your needs for working capital fluctuate over time, the need always exists. Although borrowed funds may be used for working capital for fairly long periods of time, the amount fluctuates, depending on the cyclical aspects of a particular business.

For example: Gift and toy makers build up inventory during the summer and fall in order to meet the demands of pre-Christmas shipping costs. Their need for working capital follows the pattern of any industry whose peak sales coincide with the Christmas shopping season. The manufacturer needs money to produce, market and ship well before the holiday season, and then more money to carry accounts receivable until retailers pay for their shipments. The need for working capital in these industries peaks in November and December and dips in the middle of January, when checks from retailers flow in and create a high liquidity. At that time, the loans are reduced as the need for working capital decreases.

Whatever the timing or industry, working capital plays the same role. It provides you with funds to carry your business through its period of greatest cash need. It provides a ready source of outside money when you most need it.

**Growth capital** differs from working capital because the need for it does not spring from the cyclical nature of the business. Some lenders lump the two together, but the need for growth capital comes from the desire to expand the business, improve production facilities, develop and market new products or even cut costs. You justify the financing when you can project greater profits to result from the capital. In making growth capital available, institutions don't look toward seasonal liquidity but rather to increased profits from which the business may repay the loan.

Every business needs all three types of capital if it succeeds—equity capital for permanent needs, working capital for seasonal needs and growth capital for expansion. You can't expect any single financing program, maintained for a short period, to meet every future need.

If you ask for a working capital loan, your bank will want to know whether you can reduce or eliminate the loan during your annual period of greatest liquidity. If you ask for growth capital, your bank will want you to demonstrate how the money will buy fixed assets or perhaps additional marketing to yield a profit substantial enough to repay the loan over time.

If you don't make it clear that you need either working or growth capital, the bank may explain that the loans it makes are temporary and that it can't lock its money into a business.

Lending institutions don't invest in businesses, stockholders do. Lending institutions simply advance businesses money on which they earn interest.

## Looking for Equity Capital

If you are starting from scratch, with nothing but an idea in mind, you must decide what equity capital structure to use. Ask yourself the following:

- How much capital does the business need?
- What should you give in return for that capital?
- What form should the capital take?

Remember: When you draw up a financing plan, it must appeal to potential investors' interests as well as to your own. The plan must include two basic considerations:

- Potential investors must have enthusiasm for your plans and confidence in the management of the enterprise.
- They must find the offer of an equity interest, whether as stockholders or partners, attractive when compared to other investment possibilities.

In satisfying these requirements, you pave the way to obtain equity capital from partners or backers. Your most important tool, however, is an accurate presentation.

A good presentation can get you far. If you want to start your own line of chocolate chip cookies, devise a clever promotional campaign, too. Be prepared to offer the financier information on everything from who your product targets and the type of packaging you want to use to any ideas you have lined up for endorsements. Bring in samples of the products you will offer. Also be prepared to discuss how much money you think you'll need and what you're willing to give up to investors in return for it.

Remember: Your success—and the success of your business—depends on a carefully prepared presentation and business plan, including the following elements:

- A description of the product or service. (Remember, pictures and graphics are worth a thousand words.)
- An estimate of the size of the market, along with projections for sales volume over several years. (Where all your research finally comes in handy.)
- Cost breakdowns, pricing policies, a break-even analysis and profit projections.
- Past sales volume and earnings (in the case of an existing business).
- The nature of the competition.
- Information about the management of the enterprise, including general background and details of experience bearing directly on the venture.
- References, including trade, banking and personal.
- Sources of supply.
- The specific financing necessary to get the venture going.

Of course, depending on the type of business you plan to start, you may want to tailor this list—adding or losing elements—to fit your needs.

## How Much Do You Need?

So how much do you really need? You must come up with a specific figure; ball park numbers don't work. Investors want to know exactly how you arrived at the numbers you've come up with. Show them.

If you ask for too little or for too much, it shows that you haven't thought things through. And too little capital can be as bad as none at all. All of the funding may disappear if you ask your investors for an insufficient capital sum. On the other hand, you don't want to ask for a ridiculously high sum. Any investor will be good with evaluating the practicality of your numbers—so don't try to fool him or her. Your goal is to come off as trustworthy and realistic as possible.

Stay away from big, round numbers. If you ask for "half a million," this tells investors that you've made little effort to delineate the real need. To the serious investor, such unrealistic thinking immediately proclaims a lack of experience with substantial sums of money and raises serious doubt as to your ability to handle such sums. In the face of such requests, investor confidence is lost almost immediately.

Another common mistake: adding any standby or reserve cash to the total capital you're asking for far in excess of any reasonable amount for this purpose. This type of thinking appears frequently. If you need \$200,000—but ask for \$350,000 because you want to keep about \$150,000 in the bank, over and above the company's actual needs just in case you should need it, a sophisticated investor will turn you down.

Whatever the amount you come up with, it must be just right—neither too little nor too much. Conduct a cash-need flow analysis that covers the costs of the following:

1) **Physical plant assets**.
2) **Office equipment**.
3) **Supplies**.
4) **General and Administrative Costs**.
5) **Capital needed to carry you to break even**.

This should give you a good idea of the amount of cash you will have to come up with to start your business. You may not need all of these expenses when you first get started, but they provide you with an idea of what you will need in the future in order for your business to take off.

In order to make a meaningful projection, you predict the growth of income on a monthly basis, usually over one or two years.

Even if you are fortunate enough to possess purchase orders for your service or product before start-up, these sales do not create income until you complete each order—that is, until you provide the service or make the product, ship it to your customer and transfer title (which occurs, incidentally, even in over-the-counter sales). It usually takes several months after inception before a business can deliver goods or services.

Gradually, sales approach a plateau through a series of monthly increases. Only you can project this trend, based on your own estimate of pricing and volume.

After you compute your beginning capital requirement, you can begin to add other expenses to the list, such as money to carry **inventory** and money to carry **accounts receivable**. You can calculate these only after you compute your beginning capital requirement.

As your sales begin and continue to climb you have to find a way to carry the increasingly larger inventories and accounts receivable.

When this happens, you have reached the point at which you need working capital. Your equity capital has carried you to break-even, and you now need working capital.

## Small Business Loans

As, I've said before, banks and other institutional leaders don't usually loan money to cover the equity capital of a new business. And venture capitalists aren't as easy as they used to be. But there are some sources that will loan money—or help arrange a loan—to a small business.

Loans are another method of obtaining money to start and run your business. Loans differ from venture capital in that the lender will usually require you to be personally liable for the funds you receive, while venture capital only requires you to give a percentage of the business revenues or profits to the funder.

## SBA Loan Programs

The Small Business Administration (**www.sba.gov**) is one of the best sources of start-up and ongoing financing for new businesses. The SBA does not give money out directly, but works with local banks to guarantee new loans. Generally, the SBA will offer a guaranty for 80 percent of loans up to a certain amount and 75 percent of higher amounts. Borrowers generally pay approximately 3 percent of the loan to the SBA in guaranty fees.

When a lender offers an SBA-guaranteed loan, it agrees to certain interest rate levels. Although the rates are negotiable and may be fixed or variable, the rates cannot go above SBA maximums. Those maximums are 2.25 percent over the prime rate for loans of less than seven years and 2.75 percent over the prime rate for loans of seven years and longer. Loans under $50,000 may be subject to slightly higher rates.

To qualify for many SBA loans, businesses must be a certain size. While these size requirements may be unrealistic for a start-up, they are definitely within reach for a successful business experiencing growing pains soon after start-up. For example, a wholesale business is defined as a small business if it has 100 employees or fewer, and an agriculture business is defined as a small business if it brings in $500,000 to $19 million in revenue each year. Check with your local SBA office or local lender to determine if your new business qualifies for these loan programs.

All necessary forms for SBA programs can be found at: **www.sba.gov/library/forms.html**. Check it out for loan applications, personal financial statement documents and other useful and necessary forms. Some forms can be filled out and submitted online, but others require Adobe Acrobat Reader (available for free at **www.adobe.com**) to open and print the forms from your computer.

- ***The MicroLoan Program***. The MicroLoan Program, the smallest of the SBA loan programs, was developed to increase the availability of very small loans to prospective small business borrowers. Under this program, the SBA makes funds available to nonprofit intermediaries who in turn make loans to eligible borrowers in amounts that range from under $100 to a maximum of $25,000. The average loan size is $10,000. The maximum term allowed for a loan is six years. However, loan terms vary according to the size of the loan, the planned use of funds, the requirements of the intermediary lender and your needs. Interest rates vary, depending upon the intermediary lender. Rates are generally competitive. Each non-profit lending organization has its own loan requirements, but must take as collateral any assets bought with the MicroLoan. In most cases, the personal guarantees of the business owners are also required.

A list of MicroLoan intermediary lenders is located at: **www.sba.gov/financing/microparticipants.html**.

❖ ***7(A) Loan Guaranty***. The 7(A) Loan Guaranty is one of the SBA's major small business loan programs. The maximum amount the SBA can provide security for is generally $750,000, so it is unlikely that a borrower could use this program to borrow more than $1 million.

In order to be considered for the 7(A) Loan Guaranty program, applicant businesses must:

- operate for profit;
- operate in the United States or its possessions;
- have reasonable owner equity to invest; and
- use personal assets and other financing sources first.

Businesses that are ineligible for this program include the following:

- real estate investment and other speculative activities;
- lending activities;
- pyramid sales plans;
- illegal activities;
- gambling activities; and
- charitable, religious or certain other nonprofit institutions.

For more information on this program, go to **www.sba.gov/financing/fr7aloan.html**.

❖ ***SBALowDoc***. Small businesses can get up to $150,000 under this program. This program also provides faster turnaround on loan approval—100 percent of loan applications are approved or denied by the SBA within 36 hours or less. Your chances of obtaining this loan are pretty good; the SBA guarantees 75 to 80 percent of loans through this program.

To be eligible for the SBALowDoc Program, you must use the money to start or grow a business and you cannot have more than 100 employees. If this is an existing business, annual sales for the preceding three years cannot exceed $5 million. Also, the application requires you to have good credit and be of good character. For more information on this program, go to **www.sba.gov/financing/frlowdoc.html**.

- *SBAExpress*. This program is very similar to the SBALowDoc Program, but under its regulations, the SBA only guarantees 50 percent of the loan amount. However, the maximum loan amount is only $150,000 and the turnaround for approval is 36 hours. For more information on this program, visit **www.sba.gov/financing/frfastrak.html**.
- *CommunityExpress*. This pilot program targets businesses in low- to moderate-income areas. Although it provides the same 36-hour turnaround as the SBALowDoc and SBAExpress programs, only selected National Community Reinvestment Coalition (NCRC) lenders offer this program. The NCRC, which was formed in 1990 by 16 national, regional and local organizations, was set up to develop and harness the collective energies of community reinvestment organizations from across the country so as to increase the flow of private capital into traditionally underserved communities. You can find a list of approved vendors at: **www.sba.gov/financing/frcomexp.html#participation**.

  Business owners can receive up to $250,000 under this program. For more information, visit **www.sba.gov/financing/frcomexp.html**.
- ***Community Adjustment and Investment Program***. The United States Community Adjustment and Investment Program (CAIP) was developed to help communities that saw job losses resulting from trade changes after the North American Free Trade Agreement (NAFTA)

was implemented. The terms of this loan program are similar to the SBA 7(A) Guaranty terms, but unlike other programs, this one doesn't require you to pay a guaranty fee. Information about this program and a list of communities eligible for this program are available at **www.sba.gov/financing/frcaip.html**.

❖ ***Certified Development Company (504) Loan***. The CDC-504 loan program funds capital improvements for certain small businesses. To qualify, a business cannot have more than $6 million in fixed assets (i.e., property, equipment, etc.) and more than $2 million in revenue. Most businesses just starting out fit that qualification. For more information, go to **www.sba.gov/financing/frcdc504.html**.

❖ ***CAPLines***. There are several loan opportunities that fall under this umbrella program. The CAPLines loans are intended to help small businesses keep up with their working capital needs. Most of the categories of loans can be revolving, which means that same amount could be borrowed again once the initial loan is paid off, or non-revolving.

The categories of loans are:

- **Seasonal Line**. Designed to help cash flow during peak seasons. You can use this to buy inventory when you have many orders coming in but have not been paid for those orders.
- **Contract Line**. For labor and materials used in a specific contract.
- **Builders Line**. Similar to the contract line, but designed specifically for small general contractors or builders working on commercial or residential building projects. For this line of credit, the project serves as the collateral.
- **Standard Asset-Based Line**. You can draw from this line of credit based on your existing assets. You simply repay the loan based on your cash cycle. The lender can charge additional fees for this type of credit line because the lender has to do more work to determine the amount of collateral you have available.

- **Small Asset-Based Line**. This line is similar to the standard asset-based line, but it has a maximum of $200,000 in credit, and the lender can waive the stricter requirements to monitor your collateral once you have demonstrated you can repay the credit line consistently.

For more information, go to **www.sba.gov/financing/frcaplines.html**.

❖ ***International Trade Loans***. International Trade Loans of up to $1,250,000 are designated for businesses that plan to trade internationally and/or those businesses that are affected by import competition. The loan must be used to fund improved or expanded facilities for exporting to another part of the world, and the business plan submitted as part of the loan process must show a reasonable plan to cover the loan with profits from the business. For more information, visit **www.sba.gov/financing/frinternational.html**.

## Small Business Investment Companies

The Small Business Association (SBA) also licenses and regulates **Small Business Investment Companies (SBICs)** that invest venture capital in small businesses. SBICs are all for-profit businesses that hope to make money on the investment as a start-up business becomes successful.

As I mentioned earlier, venture capital is funds received by a business owner for the start-up or maintenance of a business.

There are two types of SBICs: **regular SBICs** and **specialized SBICs**, which invest in small businesses owned by minority or socially and/or economically disadvantaged entrepreneurs. There are SBICs in every state, as well as in Guam and Puerto Rico. A directory of all the licensed SBICs is located on the Internet at **www.sba.gov/gopher/Local-Information/Small-Business-Investment-Companies**.

SBICs offer the following types of financing:

- **Seed Financing**. A small amount of funding designed to help an entrepreneur put together a plan and qualify for further start-up capital.

- **Start-Up**. This funding helps businesses who have not sold a product yet, but that have a management team and concept in place and have a product developed.
- **Early Stage**. This funding helps businesses move from product testing and early marketing to manufacturing and selling the product.
- **Expansion Financing**. This funding helps businesses who have begun selling their product on a wider scale, but need more support to get the product to a wide enough market so they can make a profit.
- **Later Stage Financing**. This funding allows profitable or break-even companies to expand further.
- **Management Buy-Out/Leverage Buy-Out/Acquisition Financing**. This funding allows companies to purchase other companies or product lines.

Between 1990 and 1999, the overwhelming majority of funding through this program went to businesses that had been in existence for less than three years, according to the SBA. In 1999, SBIC program licensees provided $2,243,200,000 in financing to start-up businesses. Note: Of all the financing given out to businesses by SBIC program licensees during 1999, 30.6 percent of it went to manufacturing businesses and 32.8 percent went to service businesses.

A guide to seeking SBIC financing is available online at **www.sba.gov/INV/howtoseek.html**. A primer for acquiring venture capital is located at **www.sba.gov/library/pubs.html#fm-5**. The primer is downloadable in text, Adobe PDF and Microsoft Word97 formats.

## Getting a Grant

The federal government supports innovative research through **Small Business Innovative Research (SBIR)** and **Small Business Technology Transfer (STTR)** programs, which are offered through several government departments. It is worth noting that these grants are sometimes not offered to businesses that are just getting started, but they could be applied for and used to fund a business once it is underway.

The SBIR Program is a highly competitive three-phase award system that provides qualified small business concerns with opportunities to propose innovative ideas that meet the specific research and development needs of the federal government.

According to the SBA, the three-phase system works this way:

- **Phase I** is a feasibility study to evaluate the scientific and technical merit of an idea. Awards are for periods of up to six months in amounts up to $100,000.
- **Phase II** is to expand on the results of and further pursue the development of Phase I. Awards are for periods of up to two years in amounts up to $750,000.
- **Phase III** is for the commercialization of the results of Phase II and requires the use of private sector or non-SBIR Federal funding.

Phase One grants are usually less than Phase Two grants, which are awarded for further and deeper research. In order to be considered for a Phase II project you must have been a Phase I awardee, and likewise for Phase III. These grants often favor women- and minority-owned businesses, and when reviewers look at the grant proposals, they look for innovative ideas that have potential for commercialization. In other words, the end result of the research must be something that you can sell.

In order to be considered for SBIR funding, your business must have less than 500 employees and you must be the recipient of a competitively awarded SBIR funding agreement (contract or grant) entered into between an SBIR participating federal agency and a small business concern for the performance of experimental, developmental or research work funded by the federal government.

The SBA itself does not make any awards under this program. It has the authority and responsibility for monitoring and coordinating the government-wide activities of the SBIR Program and reporting its results to Congress. But the federal agencies actually responsible for selecting SBIR topics, releasing SBIR solicitations, evaluating SBIR proposals and awarding SBIR funding agreements on a competitive basis include:

- Department of Agriculture;
- Department of Commerce;
- Department of Defense;
- Department of Education;
- Department of Energy;
- Department of Health and Human Services;
- Department of Transportation;
- Environmental Protection Agency;
- National Aeronautics and Space Administration; and
- National Science Foundation.

To be eligible for an award of SBIR funding agreements, a small business must:

- be independently owned and operated;
- principal place of business is located in the United States; and
- at least 51 percent owned or in the case of a publicly owned business, at least 51 percent of its voting stock is owned by United States citizens or lawfully admitted permanent resident aliens.

Like the SBIR program, the STTR program is a highly competitive three-phase program that reserves federal research and development funding for small businesses in partnership with nonprofit research institutions to move ideas from the laboratory to the marketplace, foster high-tech economic development and address the technological needs of the federal government.

According to the SBA, the three phases of the STTR Program are:

- Phase I is the start-up phase for the exploration of the scientific, technical and commercial feasibility of an idea or technology. (Awards are for periods of up to one year in amounts up to $100,000.)

- Phase II is to expand Phase I results. Research and development work is performed and commercialization potential is considered. (Awards are for periods of up to two years in amounts up to $500,000.)
- Phase III takes Phase II innovation from the lab into the marketplace. (There is no STTR funding in this phase.)

The federal departments and agencies responsible for proposals and awarding STTR funding agreements include the following:

- Department of Defense;
- Department of Energy;
- National Aeronautics and Space Administration;
- Department of Health and Human Services; and
- National Science Foundation.

For more general information about the SBA's SBIR and STTR programs, go to **www.sba.gov/sbir/indexfaqs.html**.

The **National Institutes of Health (NIH)** also offers an SBIR Program that offers grants from $100,000 to $750,000. For information and application forms, go to **http://grants.nih.gov/grants/funding/sbir.htm**.

The **Department of Defense (DoD)** also has an SBIR program and an STTR program to support **defense-related technology research**. These programs fund more than a half billion dollars each year in early-stage research and development projects at small technology companies. These projects must serve a DoD need and have commercial applications. This program offers a "Fast Track" option for small businesses that can get outside investor funding. In this scenario case, the DoD will match each dollar the investor puts in with $1 to $4. For information and application forms, go to **www.acq.osd.mil/sadbu/sbir**.

For information about specific departmental forms, visit the following:

- Department of the Air Force: **www.afrl.af.mil/sbir/index.htm**
- Department of the Army: **www.aro.army.mil/arowash/rt**

- Department of the Navy: **www.navysbir.brtrc.com**
- Ballistic Missile Defense Organization: **www.winbmdo.com/scripts/MAIN.asp**
- Defense Advanced Research Projects Agency: **www.darpa.mil/sbir**
- Special Operations Acquisition and Logistics Center: **http://soal.socom.mil/index.cfm?page=SADBU**

The **Department of Agriculture** offers SBIR grants to small businesses proposing research that will improve agriculture technology and yield. Phase one grants are for a maximum of $70,000. For information, visit: **www.reeusda.gov/sbir**.

The **Department of Education** offers SBIR grants of up to $50,000 in phase one and up to $300,000 in phase two. For information, visit **www.ed.gov/offices/OERI/SBIR**.

The **Department of Commerce** offers SBIR grants through the National Oceanic and Aeronautics Administration (program information available at **www.rdc.noaa.gov/~amd/sbir.html**) and the National Institute of Standards and Technology (program information available at **http://patapsco.nist.gov/ts_sbir**).

To find out more about other programs, go to the following:

- Department of Energy: **http://sbir.er.doe.gov/sbir**.
- Department of Transportation: **www.volpe.dot.gov/sbir**.
- National Aeronautics and Space Administration: **http://sbir.nasa.gov**.
- National Science Foundation: **www.eng.nsf.gov/sbir**.
- Environmental Protection Agency: **http://es.epa.gov/ncerqa/sbir**.

## Additional Government Resources

The federal government makes grant funds available for many types of businesses. Here are some examples of programs that offer funding for specific reasons, as well as resources for finding more of these programs when you're ready to get started. Keep in mind that the Congressional appropriations process—which determines whether or not these programs, or new ones, get funded—happens every year, so it is important to make sure the program in which you are interested still exists!

- **Adult Education National Leadership Activities**. A wide range of grants are available to assist businesses dedicated to improving adult basic education. Funds from the U.S. Department of Education's Office of Vocational and Adult Education are available for applied research, development, demonstration, dissemination, evaluation and related activities that contribute to the improvement and expansion of adult basic education nationally. Funds for a project are usually granted for a 12- to 18-month period. The agency funded close to $101,000,000 in projects during FY 2000.

  For more information about this program, write to the Division of Adult Education and Literacy, Office of Vocational and Adult Education, U.S. Department of Education, 600 Independence Avenue, SW., Washington, DC 20202-7242.

- **Advanced Technology Program**. If you plan to start a business that deals with highly technical products, this might be a program worth closer examination. Since its first competition in 1990, the ATP has made 285 awards to single companies and 146 to joint ventures—two companies that have joined together to work on the same project. Proposals are selected on the basis of their scientific merit and their potential for broad-based economic benefits.

  The program holds general competitions, where any technology project is eligible, and highly-focused competitions, such as Digital Video in Information Networks and Adaptive Learning Systems. According to the program administrators, ATP funds are used to develop a wide range of technologies in areas such as x-ray lithography, data storage, machine tool control, electro-optics, superconductivity, printed wiring boards, flat panel displays, handwriting recognition, semiconductors, biotechnology, ceramics, composites, computer-aided design and manufacturing and DNA diagnostics.

This is a competitive program, but if you win a grant, it can be extremely lucrative. Awards have ranged from $482,000 to $31,500,000, with the average award being $3,200,000.

To obtain a copy of The ATP Proposal Preparation Kit, call the ATP toll-free "hotline" at (800) ATP-FUND or (800) 287-3863. The Kit is also available under the heading Publications on the ATP's Web site at **http://atp.nist.gov**. Or, write to the Advanced Technology Program, National Institute of Standards and Technology, 100 Bureau Drive Stop 4701 Gaithersburg, MD 20899-4701.

- **Bureau of Indian Affairs (BIA)**. Part of the U.S. Department of the Interior, the BIA offers a loan guaranty program to support entrepreneurship and economic development on Indian reservations and in primarily Native American communities around the country. The maximum loan the BIA will guaranty is $500,000 for an individual, partnership or a corporation, and the borrower must be a member of a federally recognized Native American or Alaska Native group or tribe.

  Information about the program is available at Bureau of Indian Affairs Public Affairs Office, 1849 C Street, NW Washington, DC 20240, phone: (202) 208-3711, fax: (202) 501-1516, **www.doi.gov/bia/ecodev/loanpgm.html**.

- **Computer and Information Science and Engineering Grant Program**. If you can prove that your business provides research and access that helps people understand or use computers better, you can apply for the Computer and Information Science and Engineering Grant Program (CISE) through the National Science Foundation. This program made more than 1,500 awards during 2000 and the number continues to grow every year. The average grant given is nearly $75,000 for use over a six- to three-year period.

  For more information on this program, contact the Assistant Director, Computer and Information Science and Engineering, National Science Foundation, 4201 Wilson Blvd., Arlington, VA 22230, phone: (703) 306-1900. Or, visit their Web site at **www.cise.nsf.gov**.

- **Economic Adjustment Assistance**. The Economic Adjustment Division of the U.S. Department of Commerce's Economic Development Administration provides grant funds for businesses in economic development districts; states, cities or other political subdivisions of a state or a consortium of political subdivisions; Indian tribes or a consortium of Indian tribes; institutions of higher learning or a consortium of such institutions; or public or nonprofit organizations or associations acting in cooperation with officials of a political subdivision of a state. The program gave out close to $110,000,000 during 2001.

  This program generally provides 50 percent of the funds needed for a given business or project. However, in some cases, the program administrators will grant more funds to businesses that cannot provide matching funds or that are located in extremely disadvantaged areas. For more information about this program, contact the Economic Adjustment Division, Economic Development Administration, Room H7327, Herbert C. Hoover Bldg., Department of Commerce, Washington, DC 20230, phone: (202) 482-2659. You can also find the Economic Development Administration online at **www.doc.gov/eda**.

- **The U.S. Department of Housing and Urban Development (HUD)**. HUD makes a serious effort to award contracts to women-owned and economically disadvantaged businesses. Current contract solicitations are posted on HUD's Web site at **www.hud.gov/cts/ctsoprty.html**. You can also sign up for e-mail updates to find out when any new contract solicitations or amendments to current solicitations are posted to the site by visiting **www.hud.gov/grants/index.cfm**.

- **National Endowments for the Arts and Humanities**. If you are a visual or performing artist, a writer or starting a business that is somehow involved in the humanities, you may be able to get grant funding from among the wide variety of offerings at the **National Endowment for the Arts (www.nea.gov)** and the **National Endowment for the Humanities (www.neh.gov)**. Visit either program's Web site to determine what grant funding is available this fiscal year and whether you are eligible for it. For example, NEA grants for creative writers are available for poetry writers and prose writers in alternate years.

Beyond the basic grant and loan programs I've just listed, there is a whole sub-economy based on contracts, grants and other assistance that the government offers to small business—especially those started or owned by women, members of certain racial or ethnic groups and other traditionally under-represented entrepreneurs. Whole books are dedicated to scouting these programs. I'll take just a quick look at a few.

## Catalog of Federal Domestic Assistance

The **Catalog of Federal Domestic Assistance (CFDA)** is great resource for government grants and programs that can provide your business with start-up and continuation funds. The CFDA is available online at **www.cfda.gov**, but can also be ordered in hard copy, CD-ROM or diskette versions. The CFDA Web site includes helpful—and free—information on writing grant proposals.

The hard copy version is available for $87 per year. When you order this, you'll receive the catalog in June and a supplement in December. The CD-ROM is issued twice a year—in June and December. A single copy of the CD-ROM is $50 and an annual subscription that includes both versions is $85. The CD-ROM also includes browsing software that allows you to search the catalog by keyword and category, and the Federal Assistance Award Data System (FAADS), which allows you to see who received funding, where they live and what programs were funded. The diskette version is also issued twice a year, and $85 buys you both the June and December issues of the CFDA. However, the diskette version does not include search software. Order forms for any of the three versions are available at **www.cfda.gov/public/cat-order.htm**, or you can place a credit card order or get more information by calling (202) 708-5126. You can also fax orders to (202) 512-2250 or mail order forms to Superintendent of Documents, P.O. Box 371954, Pittsburgh, PA 15250-7954.

## FedBizOpps

FedBizOpps at **www.FedBizOpps.gov** is the single government point-of-entry (GPE) for federal government procurement opportunities over $25,000. Government buyers are able to publicize their business opportunities by posting information directly to FedBizOpps via the Internet. Through one portal—FedBizOpps (FBO)—commercial vendors seeking federal markets for their products and services can search, monitor and retrieve opportunities solicited by the entire federal contracting community.

## Commerce Business Daily

Another great resource: Commerce Business Daily, the government's daily listing of U.S. government procurement invitations, contract awards, subcontracting leads, sales of surplus property and foreign business opportunities. Approximately 500 to 1,000 notices appear in the Commerce Business Daily each day (notices appear only once). You can order a print subscription to the publication by writing to Superintendent of Documents, P.O. Box 371954, Pittsburgh, PA 15250-7954. You can also fax orders to (202) 512-2250 or call (202) 512-1800 between 7:30 A.M. and 4:30 P.M. eastern time. The cost for first class postage service for one year is $324 (six months is $162). The cost for regular postage service for one year is $275 (six months is $137.50). However, it is also available for free on the Internet at **http://cbdnet.access.gpo.gov**.

Remember: You must meet the government or private granting agencies guidelines and reasons for giving grants. Research is the key to finding this information, because the amounts of money available and qualifications are always changing. You have to be willing to work diligently! Use your initial research to check the guidelines of granting agencies to determine the types of grants they are making that you might qualify for.

The following are some good general grant Web sites to check out:

- At-a-Glance Guide to Grants. **http://www.adjunctnation.com/money/links**.

- The Foundation Center. This Web site has excellent links to foundation homepages and other useful sites for grant seekers. **www.fdncenter.org**.
- The Complete Grants Database. **www.grantselect.com**.

## Foundations

Government funded grant or loan programs are not your only option, there are thousands of private and corporate foundations with money available to give as grants or loans. While much of that money goes to non-profit organizations, there are some foundations that will provide grants to businesses and individuals. Tailor your grant proposal to describe how you will use the money to implement something that will benefit your community at large and the residents of your area. However, if your business is providing a valuable service in line with the goals of the foundation, you will have a better chance of being a successful applicant.

The following is a list of private foundations, including information on the types of grants that they offer to businesses and individuals.

- **The Abell Foundation**. This foundation offers the Abell Venture Fund, a $25 million venture capital fund devoted to investing in companies located in Baltimore, Maryland, or willing to relocate there. According to the Foundation's Web site, they "prefer to invest in businesses led by a strong management team, capable of reaching $30 to $50 million in sales within five years, and with either a significant competitive advantage or the benefit of being the first in a large industry." For more information, contact: Abell Venture Fund, 111 S. Calvert Street, Suite 2300, Baltimore, MD 21202. The fund can be found online at **www.abell.org/venturef.htm**.
- **Ewing Marion Kauffman Foundation**. Although most 7th grade girls don't want to have much to do with their mothers, if you are a mother looking to start a business and have a pre-teen daughter, this foundation has a program that may be right for you. The **Mother And Daughter Entrepreneurs In Teams (MADE-IT)** program

offers mothers and their 7th grade daughters the opportunity to develop and start a business. The goal of the program is to help the family earn money to send the daughter to college, but the mentoring process is worth all the effort. The program includes a week-long summer institute that allows the teams to develop valuable negotiation and other business skills that will help them succeed. For information about the program, visit the Web site at **www.emkf.org/Entrepreneurship/programs/made_it.cfm**, or contact the program at MADE-IT, Kauffman Center for Entrepreneurial Leadership, Ewing Marion Kauffman Foundation, 4801 Rockhill Road, Kansas City, MO 64110; phone: (816) 932-402; fax: (816) 932-1100; e-mail: made-it@emkf.org.

- **Sobrato Family Foundation**. One of the areas on which this foundation focuses is community and economic development in Santa Clara, California, or southern San Mateo and southern Alameda Counties, also in California. Although they do not provide start-up funds, if you are an entrepreneur in that region and have an idea that will help your business grow and also benefits the community, you may be able to develop a successful grant proposal for this foundation. The contact person for questions or applications is Margaret Wiley, Grants Manager, Sobrato Family Foundation, 10600 N. De Anza Blvd., Suite 200, Cupertino, CA 95014; phone: (408) 446-0700, Ext. 146; fax: (408) 446-2896; e-mail: margaret@sobrato.com. You can find the foundation online at **www.sobrato.com/foundation**.

## Is VC Funding For You?

At the surface, venture capital or VC, funding sounds like a terrific opportunity. Go to a group of investors or an investment management company, make your pitch and take home millions of dollars to start your business. However, the go-go years of the dot.com bubble have passed. And the market for venture capital financing has become a lot more difficult.

Only five out of 1,000 businesses that take their ideas before venture capitalists receive funding, according to the Profit Dynamics Inc. *2000 Venture Capital Survey*.

It is important to consider whether you really need VC funding or not. Generally, the criteria is whether your idea is a "multi-million dollar idea" or not. Is this something that will become a national corporation? Are you willing to open your idea up to additional large investments and the additional scrutiny that will bring? Are you looking to grow your business quickly and make substantial capital investments in office or manufacturing space, inventory and computer equipment? If the answer to all these questions—not just one or two—is yes, then perhaps VC funding is what you're looking for.

Several years ago, Ian Morrison, former president of the Institute for the Future and author of *The Second Curve* (a great study of the growth plateaus that bedevil businesses) said: "By the year 2000, 99 percent of American business will be on the Internet—but only 3 percent will know why."

Too often business decisions are made according to a crowd mentality. Like lemmings, we follow the masses, and are driven forward before we know our destination. Nothing speaks more clearly to this than the wave of bad retail dot.com companies funded with frenzied venture capital in late 1999 and early 2000. These foolish VC bets were proof of Morrison's prophesy. The bankers wanted to be in what they called the Internet's retail space. But they didn't know *why*. All they knew was that everyone else was pumping money into that marketplace.

Greedy start-ups and their VC moneymen poured into the marketplace, presuming to know what their customers wanted. And, in late 2000 and early 2001, many of them went belly-up. Why? Because consumers weren't interested in hundreds of niche retail variations of Amazon.com—as the collapse of Pets.com, eToys.com and scores of other ventures attest.

These followers fixated on Internet commerce—the technology tool—instead of the underlying businesses. The bottom line for new venture seekers: proof of concept (that your business can work) is absolutely necessary.

## Venture Capital: How to Get It

One of the most important things you can do when looking for venture capital of any kind is your homework. This is where all your research comes back into the picture. Ask yourself, *Does the company you're targeting fund the type of business you're trying to start?* If not, you need to look elsewhere. Similarly, if the company has funded many similar businesses, check to see if any of those businesses have already begun working with a nearly identical business idea as yours—it's unlikely a VC firm will fund the same idea twice.

According to the Profit Dynamics Inc. survey, 42 percent of venture capitalists say they prefer to be contacted initially through e-mail. The best thing to send, according to the survey data, is a two- to three-page executive summary—56 percent of VCs like to see that summary first. Twenty-two percent prefer to see a 10- to 15-page mini-business plan. (For more on how to write an executive summary and business plan, go to Chapter 5.)

It is important to note, however, that you should have a business plan ready before you make that initial contact with the VC firm. (I'll go into more detail on fine tuning this in the next chapter.) There is nothing worse than contacting such a firm and then having to scramble to pull together a complex plan without any preparation time. Also, that lack of preparation will reflect poorly on you as a business owner. Have everything prepared ahead of time and ready to go before you even send that first e-mail. If you're prepared it shows and you'll be much more likely to get your foot in the door and one step closer to landing the funding.

## Finding a VC Firm

A large number of venture capital firms used to focus on technology start-ups—that is where the hot economic growth was taking place, and these firms wanted to make their money back plus a healthy profit. But, the tides have changed and venture capital firms are no longer throwing money at the dot.coms. They're investing their money elsewhere—and this could be good for you.

A list of firms to examine and consider can be found in Appendix C to this book, including their area of interest. Be sure to look at each firm closely—each one has a different area of interest and many fund businesses at different stages. Some fund start-ups, but others do not provide funding until a business has become more established.

## Other Sources of Funding

With all the game shows on television, it should come as no surprise that there is a game show that caters to entrepreneurs. Guests of **The Money Hunt**, based in Connecticut, have raised $250,000,000 in capital since appearing on their show, according to the show's producers. Contestants who make it through the show's selection process pitch their business idea and face eight minutes of questioning from a panel of experts. The winner, selected by the panel, is the person who outpitches the other contestant on that day's show. That successful salesperson takes home $100,000 in start-up money and additional prizes, including business software and services to help them get their company underway. To register for an online audition, visit their Web site at **www.moneyhunt.com**.

If you're not interested in going on TV, but would be interested in a lower-key way to pitch your idea to the public, there are a number of bulletin boards available on the Internet that offer entrepreneurs and potential funders a place to exchange ideas and information. One good one is **The Angel Capital Electronic Network** (**https://ace-net.sr.unh.edu/pub**). Other good sources of funding opportunities are your local community redevelopment organizations and local economic development boards. While many may not have regular grant opportunities, there may be time when one or the other will offer grant funding to deserving start-ups—that type of aid provides an economic boost to the area and directly benefits the community those groups try to serve.

## Non-Profit Entrepreneurial Funds

Although there are not many non-profit organizations that simply provide start-up funds with no strings attached, some non-profits have developed programs that can provide new business owners with low-interest loans and other sources of funding as well, including the following:

- ***Ewing Marion Kauffman Foundation***, 4801 Rockhill Road, Kansas City, MO 64110-2046, phone: (816) 932-1000. **www.emkf.org**.
- ***The Coleman Foundation, Inc.***, 575 W. Madison Street, Suite 4605, Chicago, IL 60661, phone: (312) 902-7120. **www.colemanfoundation.org**.
- ***The Edward Lowe Foundation***, P.O. Box 8, Cassopolis, MI 49031-0008, phone: (800) 232-LOWE (5693). **www.lowe.org**.
- ***Foundation for Enterprise Development***, 7911 Herschel Ave., Ste. 402, La Jolla, CA 92037, Phone: (858) 459-4662; or 2020 "K" Street, NW, Ste. 400, Washington, DC 20036, Phone: (202) 530-8920; or 8 East Fourth St., Jamestown, NY, 14702-3050, phone: (716) 488-1911. **www.fed.org**.
- ***Appalachian Regional Commission***, 1666 Connecticut Avenue, NW, Suite 700, Washington, DC 20009-1068, phone: (202) 884-7799. **www.arc.gov**.
- ***Women's Business Development Center***, 8 South Michigan Avenue, Suite 400, Chicago, IL 60603, phone: (312) 853-3477.
- ***The Jim Rouse Entrepreneurial Fund***, 6751 Columbia Gateway Drive, Columbia, MD 21046, phone: (410) 313-6530. **www.jref.org**.
- ***The National Minority Supplier Development Council, Inc.***, 15 West 39th Street, New York, New York 10018, phone: (212) 944-2430. **www.nmsdcus.org**.
- ***US Pan Asian American Chamber of Commerce***, 1329 18th Street. NW, Washington, DC 20036, phone: (202) 296-5221. **www.uspaacc.org**.

- ***National Black Chamber of Commerce***, 1350 Connecticut Ave. NW, Suite 825, Washington, DC 20036, phone: (202) 466-6888. **www.nationalbcc.org**.
- ***National Association of Women Business Owners***, 1595 Spring Hill Rd., Ste. 330 Vienna, VA 22182, Phone: (703) 506-3268. **www.nawbo.org**.
- ***National Women's Business Council***, 409 3rd Street, S.W. Suite 210, Washington, DC 20024, phone: (202) 205-3850. **www.nwbc.gov**. (Springboard competitions for venture capital funding sponsored by the National Women's Business Council. **www.springboard2000.org**.)
- ***Trickle Up***, 121 West 27th Street, Suite 504, New York, New York 10001, phone: (212) 362-7958. **www.trickleup.org**.
- ***The National Financial Network on WomenInc.com***, **www.womeninc.com/business_center/national_financial_network.html**.
- ***Montana Women's Capital Fund***, Lisa Gentri, Director, 54 North Last Chance Gulch, P.O. Box 271, Helena, MT 59624, phone: (406) 443-3144.
- ***Nevada Self Employment Trust***, Virginia M. Hardman, Project Director, 1600 E. Desert Inn Road, Suite 209E, Las Vegas, NV 89121, phone: (702) 734-3555.

## Angel Funding

There are between 1.5 and 2.5 million people in the United States who qualify as **angel funders**, which the Small Business Association defines as those who have from $250,000 to $5 million to invest in businesses. These funders differ from venture capital firms because they are individuals and generally want to invest less than the average VC firm shells out. The problem is that because these funders are just regular people, they are often more difficult to locate than VC firms.

However, there are tools available and places you can post information to get your business and business ideas noticed. If someone with funding believes you have a business worth financing, you may be able to find that money that will take you from struggling to running smoothly. The following are just a few places to look for your business's guardian angel.

- ***The New York Angels Investor Program***. New media industry businesses in the New York tri-state area are eligible to present their business plans to the New York Angels Investor Program. The program serves both entrepreneurs and investors, and works as a facilitator and clearinghouse to help businesses that fit that criteria find angel funding more easily.

  Investors who participate in the program must commit to investing at least $25,000 to a qualified business each year. Qualified businesses should submit no more than four pages of an executive summary to angels@nynma.org. Once a month, the program invites seven to 10 companies to present their plans to a roomful of investors. For more information visit their Web site at **www.nynma.org/business_services**.

- ***The Capitol Investor's Club***. Some of the most powerful technology names in the Washington, D.C., area make up the 18-member Capitol Investor's Club. Steve Case, CEO of AOL; Mario Morino, Chairman of Potomac Knowledgeway and the Morino Institute; and Michael Saylor, Chief Executive of MicroStrategy Corporation are just some of the faces seen around the table at this group, which provides angel funding for regional start-ups in need of a boost. To date, they have not funded many start-ups, but they are committed to funneling investment dollars back into Washington, DC's high-tech market, and they hold monthly dinners during which one entrepreneur is invited to give a 15-minute pitch. For more information about the club, reference the following article that appeared in the *Washington Post* at **www.washingtonpost.com/wp-srv/business/longterm/techboom/boom3.htm**.

- ***The Mid-Atlantic Venture Association***. According to the Mid-Atlantic Venture Association (MAVA), there are more than 1,000 venture-backed companies within 50 miles of Washington, DC. This association helps to ensure that that number continues growing, both through intimate meetings between entrepreneurs and groups of investors and larger activities such as the Mid-Atlantic Venture Fair held each fall. For more information, visit **www.mava.org** or contact MAVA at 2345 York Road, Timonium, MD 21093, phone: (410) 560-5855.

❖ ***The Environmental Capital Network***. For companies that want to develop environment-friendly products and services, the Environmental Capital Network may be a place to check out. Located at **www.ecn-capital.org**, the network offers funding forums and other services to "green" businesses and investors. For more information, contact The Environmental Capital Network, 416 Longshore Drive, Ann Arbor, MI 48105, phone: (734) 996-8387.

❖ **Angel Money**. The Angel Money team, at **www.AngelMoney.com**, is seeking projects in which to invest. Although they are only interested in technology projects and entrepreneurs who are ready to make the leap to work on their business full-time, they are a source of small, private start-up funding.

❖ *ACE-Net*. The SBA's Office of Advocacy developed ACE-Net, the **Access to Capital Electronic Network**, as a low-cost way for accredited investors to find businesses to fund. Although it costs $450 per year to list your business on the site, it could be a worthwhile investment if you have run out of other funding options and need wider exposure.

There are restrictions as to who can be approved for listing on the site, however. Not just anyone can pay the $450 to get listed. Qualifications for site members include the following:

- Only entrepreneurs who can sell security interest in their companies can enroll with ACE-Net. This means that if you have a corporation or a limited liability company (LLC), you may be listed on the ACE-Net Company Database.

- You cannot be listed if you have a sole proprietorship, general or limited partnership, joint venture, "blank check" or development stage company, or are in an oil or gas business.

The ACE-Net Entrepreneur Application requires you to certify that you meet this criteria.

❖ ***The Capital Network***. A more expensive angel funding network is The Capital Network, located at **www.thecapitalnetwork.com**. This network charges entrepreneurs who want to participate $750 for six months of time working with the organization, but a simple application form allows the group to screen potential businesses to see whether or not they have a chance to be funded successfully.

## Funding That's Closer to Home

Last, but certainly not least, never forget that there are probably smaller investors right in your own backyard who can help you find the funding you need to start your business. If you only need $5,000 to get underway, for example, you may be able to find five people who can afford to invest $1,000 apiece to support your business.

The advantage to finding these small funders is that it can be an easier process. First, if someone can afford to invest a small amount of money in your company or business, they can also afford to risk that small amount. Therefore, you will be more comfortable and feel less pressure to turn the investment around rapidly. Also, it is much easier to ask someone for a relatively small amount of money than it is to go through the process of raising large amounts of money through grants and loans. There is much less paperwork, and your chances of success may be greater.

Who should you ask? Start with family, friends and coworkers and consider other business owners whose businesses may complement yours. Perhaps you want to open a printing shop, and you know someone who owns a small publishing house. Perhaps he would be able to invest a small amount of money to help you get started in return for a potential return on his investment and discounts on his printing jobs.

However, be cautious about who you ask. If you know someone cannot afford to invest in your company, do not ask him—you don't want to make him feel pressured to give you money that he doesn't have. And make sure you have a solid business plan and have completely thought through each aspect of your business before you make your pitch, as you would if you were going after funding from any other source. You will not be able to sell anyone else on your idea if you're not completely sold yourself—after all, to get them to invest, you want to excite your prospects. Be ready to show them projected numbers that demonstrate that you will be able to turn their investment around and make them some money. You want this to be a win-win situation.

Some good books to read for ideas on raising capital include the following:

- ***The Corporate Finance Sourcebook*** published by National Register Publishing Company (ISBN: 0872179230, September 1999). This annual book includes sources of financing arranged by industry, geographic area and method of financing. This is a particularly expensive resource, so this is one worth finding in your local public or university library.
- ***Entrepreneur* magazine's *Guide to Raising Money*** by John Wiley (ISBN: 0471179957, January 1998). This book includes a list of microbusiness-friendly banks, which is defined as banks that have 25 percent of total assets in loans under $100,000, arranged by state.
- ***Financing Your Business Dreams With Other People's Money: How and Where to Find Money for Start-up and Growing Businesses*** by Harold R. Lacy (ISBN: 1890394114, 1998).
- ***Financing Your Business With Venture Capital: Strategies to Grow Your Enterprise With Outside Investors*** by Frederick D. Lipman (ISBN: 0761514600, November 1998).
- ***Finding Money: The Small Business Guide to Financing*** by Kate Lister and Tom Harnish (ISBN: 0471109835, 1995).
- ***Free Money for Small Businesses and Entrepreneurs*** by Laurie Blum (ISBN: 047110387X, January 1995).

- ***Free Money From the Federal Government for Small Businesses and Entrepreneurs*** by Laurie Blum (ISBN: 0471130095, January 1996).
- ***Galante's Complete Venture Capital And Private Equity Directory*** (ISBN: 1893648001, 1996). This is another expensive resource that is worth finding in a library.
- ***SBA Loans: A Step-By-Step Guide*** by Patrick D. O'Hara (ISBN: 0471207527, January 2002).

See the next chapter for some good start-up and operating expense budget tools available on the Internet.

## Conclusion

Don't think you're alone if you feel uneasy about raising money for your start-up business. Even experienced entrepreneurs hate this part. Bankers aren't usually visionaries—and they don't consider scrappy start-ups a choice use of their precious funds. But, at some point, you're probably going to need some form of financing.

In most cases, the best financing you'll find will be some form of a loan guarantee from the Small Business Administration or a similar organization. Thank the gods of commerce for these groups—they do give small businesses with imperfect balance sheets a chance.

Even if you work with the SBA, though, you're going to need to know the basic concepts and jargon of business finance. In this chapter, I've outlined the basics—enough so you can sound informed when you make your presentation.

In the next chapter, I'll drill down into the details of *running* your business.

# Chapter 5: Moving from the Plan to Operations

This book is about starting a business for free, but it wouldn't be complete without some information on forming a business plan. I'll outline the basic necessities of a business plan here, but if you want your business to be successful—and eventually operate from some place other than your backyard—you'll probably want to form a more formal plan sometime down the road. And, in order to that, you'll need some idea of how to plan for that success.

Every entrepreneur knows that you don't succeed in the market by drawing up great business plans. You succeed by *implementing* those plans—if you know what it takes to succeed and then execute. In other words, you have to out score your opponent. This means many things—making money, dominating markets, providing products and services of value, making a contribution to society, giving others the opportunity to achieve success and security.

Whatever the goal, you must first set it out in a business plan. But that's not the only thing you must do. Even though your business plan looks great when you put it together on paper, when you actually get started, the real world intrudes. You discover that things don't play out as you imagined. You discover, in other words, the difference between the business plan and the game plan—the difference between ends and means. You discover the necessity of adjusting the means by which you keep your operation headed toward its goals.

Again, to do this, you need good information.

Can you make a good marketing decision without knowing how a similar product or service in the marketplace sells now? *No.*

You must know where and how the actual experience of your company deviates from the projections in your business plan—financially, in your marketing and sales efforts, operationally, and in your product development efforts. You must know where you are in order to get where you want to be—before time runs out in the game.

To do this, you need a business plan and some idea of how to implement that plan.

A lot of small businesses fail to do this. In fact, many start operations without having a formal business plan in place. And, even those that do often find their projections so optimistic as to be useless once they commenced operations. Others discover their plans so incomplete as to be ineffective.

Some companies find that their plans are lacking because they didn't originally consider the factors most important to running a new company—finance, marketing, product development and operations. In fact, very few use balance sheets or profit analyses in their plans. Fewer still list the products and services they offer, or pricing information. Benchmarks like cash flow, advertising and promotional spending and capital expenses show up rarely, if at all.

Some small business owners attribute their success to things other than their business plan, including investments in technology, employee training or marketing. But among those small businesses that succeed, most base operations on a formal plan.

## Preparing a Plan

No matter what kind of business you plan to start, you must spend some time planning your strategy. By spending some time outlining the various steps you will take to lay the groundwork for your business and get it moving, you will also prepare yourself for a much stronger, healthier business. A little planning will help you a lot in the long run.

A benefit of planning is the increased accuracy of financial projections. Accurately predicting income allows you to create realistic goals for growth and expansion.

As an entrepreneur, you probably have a vision for your company, but without a plan, *you* may falter, because it's hard to make good decisions without a good framework. A successful plan should walk the reader through each step of the planning process, from executive summary to industry analysis to exit strategy.

Business plans formulate broad goals, but they don't tell you how to set your business up and run it day to day so as to reach your goals. They don't tell you how to implement your business plan and measure your performance. If your business plan isn't on target, you won't get financing. That's an unfortunate loss of opportunity. But if your plan doesn't work, your company might fail after or even before it's started. You'll end up losing a lot of money—your own and that of your investors.

## Getting Started

From a tightly constructed executive summary that pinpoints what makes your business unique, to the more detailed financial statements, marketing plan and an analysis of the competition, your business plan should clearly define what makes your business different from the competition. What do you offer that no one else does, and why should someone fund that?

A business plan, by necessity, must set a broad target for the success of a company without knowing what day to day work will be like or what practical challenges will emerge. You must find tools that apply management theory and your goals to the practical challenges of starting and running a company.

The primary elements of a most business plans cover everything from introductory elements, business description and the market to development and production, sales and marketing, management and financials. But, when creating a business plan, you have to start with three broad goals:

- To create vision and mission statements that define your purpose;
- To communicate these statements clearly and effectively; and
- To measure and encourage progress.

If you accomplish these three goals, you've gone a long way toward realizing a fourth: To build a company with a strong sense of purpose and improve your prospects for success.

Achieve this and you'll be on your way to leading a successful business.

## Creating a Vision and a Mission

The first thing to accomplish when writing your business plan is to write your company's executive summary. If this doesn't succeed, your business idea will never get financing. Investors won't invest in something that doesn't capture their attention. You can use the summary as a template for the entire business plan, so it helps to write it before you jump into anything else. The main purpose of the **executive summary** is to help focus your ideas, so keep this in mind when writing the summary. Keep it short, too. An executive summary should be no longer than two pages. Use the executive summary on the following pages as a guide when creating your own executive summary.

Next, using the executive summary, create your company's vision statement and mission statement. The two are different things—but closely related.

A **vision statement** expresses what you want your company to be in the business world. A **mission statement** expresses what your company does to achieve its vision. The vision statement expresses the end;

## BigFreebies.com ♦ Executive Summary

BigFreebies' objective is to create a community of interested, involved consumers by providing them with free access to the largest compilation of valuable free goods and services on the internet. Although other Web sites provide freebie information, they offer mostly small value items and coupons. BigFreebies is the only site that has a comprehensive directory of free goods and services of substantial value (such as - Free Grants, Free Expert Advice, and How to Live Rent-Free).

BigFreeebies uses this online library of free products and services to attract a large audience. Free is the most powerful word in marketing, providing the best response in direct mail or print ads. "The response rate to a free sample...might be five or 10 times... (other methods)," Permission Marketing by Seth Godin. BigFreebies leverages this marketing technique by combining it with the unlimited access of the Internet. By providing this unique, valuable information, BigFreebies develops a community of interested consumers, which is very attractive to marketers.

We intend to develop the BigFreebies "brand" to make it a leading provider of important and valuable free goods and services on the Internet. By using strategic alliances and proprietary marketing techniques, BigFreebies intends to attract a large number of visitors. As usage of the internet continues to increase, this will provide an attractive forum for advertisers to market their products.

The unique content of BigFreebies attracts visitors, who while using the Web site, are offered free goods and services of our advertisers and affiliates through promotional and free trial offers. Our clients' offers will be targeted to consumers that are most likely to use them. This is important for companies introducing new products or services, and attract new customers. A network of consumers interested in receiving targeted messages has proven to be the most effective means of marketing.

BigFreebies provides marketers advantages over other forms of advertising, since their performance-based marketing solution is cost-efficient. Marketers pay only for actual leads or interest generated, providing advertisers an effective and targeted method of marketing products to consumers, who have already shown an interest in their products. Additionally advertising start-up costs are lower, and campaigns can be initiated quicker than traditional methods. BigFreebies also. BigFreebies' also benefits marketers. Marketers are receptive to use the BigFreebies marketing model since it provides these significant benefits with little risk. This is particularly important to clients who want to test market a new product or marketing technique quickly without incurring substantial costs. BigFreebies content will also complement the products of the companies marketing on our site, allowing a company to both enforce their offer and enhance their brand identification.

BigFreebies will receive revenue from our clients for lead generation; the number of visitors ("click-throughs") delivered to affiliates' Web sites; through banner advertisements; and from commissions from sales occurring on affiliates' Web sites.

BigFreebies provides substantial advantages for both consumers and marketers. Consumers are drawn by the uniqueness of the site's content, and the substantial benefits from free goods and services they receive, and marketers have access to a powerful performance-driven direct marketing tool. BigFreebies community of users will remain interested and loyal because of benefits they receive from valuable, high quality free offers; while advertisers receive a targeted, receptive network of consumers.

### BUSINESS OBJECTIVES OF BIGFREEBIES

- Provide substantial value to our visitors through its unique and attractive content;
- Provide visitors with a user-friendly Web site, so they can find information quickly & easily;

- Increase Web site traffic and promote the BigFreebies brand;
- Increase client base by offering custom marketing, good customer service and an efficient sales team
- Increase our revenues by providing our clients with the best responsiveness to their offers; and
- Maintain our "low expense" advantage over competitors.

**COMPETITIVE ADVANTAGES OF BIGFREEBIES**

- The uniqueness and desirability of its Web site;
- Its attractiveness and availability of content;
- The low cost of operation and maintenance; and
- Clean design and easy navigation.

**MARKETING PLANS**

**Target Audience.** Since valuable free goods and services is the a topic that is universally popular with the largest possible of consumers and internet users, our target audience is virtually unlimited. Our technology allows our clients to target users that may have a specific interest in their offer.

**Obtaining Site Visitors and Traffic.** BigFreebies intends to obtain site visitors and direct traffic to the Web site by: Branding of BigFreebies; Search engine/Directory Submission; Syndicating Content; Hosting Multiple Domain Names; Banner Exchanges and Links; E-Mail Newsletter; Buy/Rent Opt-In Addresses; Public Relations; Articles for Magazines, E-Zines.

**Return Visits to the BigFreebies Web Site ("Stickiness of the Site").** BigFreeebies has initiated programs to ensure that visitors return often to the Web site. These programs include contests and prizes that will be awarded for frequent use of the site, and the constant addition of new content and BigFreebies on a regular basis. We promote repeat visits by continually improving usability of our Web site and regular e-mail communications. We also encourage our advertisers to offer freebies that will be beneficial to our visitors, and monitor our advertisers offers to make certain that they provide the goods or services they have promised.

**Competition.** There are numerous sites that offer small value coupons and newsletters. However, BigFreebies is the only site that specializes in finding the latest in valuable free products and services, and none of the sites has the online research that BigFreebies supplies to its visitors. This specialty produces information of great value to our visitors, as well as give us a clear, distinctive, competitive edge over all other sites. Our ability to operate in an more efficient and cost-effective manner also provides us with a substantial benefit over our competitors who have engaged many times more employees and expenses than BigFreebies. We have also designed many features to both attract visitors to our site and encourage repeat visits that are unavailable through our competitors.

**SOURCES OF REVENUE FOR BIGFREEBIES**

Performance based marketing - "click through" revenue; Lead generation and dynamic marketing; On-Site Advertising; Commissions on sales of products through affiliates sites; Merger with a strategic partner.

BigFreebies provides substantial advantages for both consumers and marketers. Consumers are drawn by the uniqueness of the site's content, and the substantial benefits from free goods and services they receive, and marketers have access to a powerful performance-driven direct marketing tool. The competitive advantages of BigFreebies give us a clear, distinctive edge over all others.

the mission statement, the means. The vision statement sets the goal. It should articulate your company's long-term goals. The mission statement involves more explanation and tells you how to reach that goal. It summarizes the company's standards and goals and then describes the most likely means by which the company can realize its vision.

Examples of vision statements:

*Our vision is to be the first of a network of high quality, low cost sites that will provide various types of free information, goods and services. We will create sites that will be attractive, sought after and desired by Internet users and consumers. We intend to pioneer a new breed of profitable Internet businesses that can be launched with reasonable start-up costs using ability, ingenuity, hard work, strategic alliances and virtual marketing techniques in the place of substantial capital.*

Or:

*Our vision is to be world class in the eyes of our customers at creating well-designed, effective and exciting work environments.*

You must get your ideas across clearly and concisely. If your vision statement is clear and concise, it will be easier to convey to investors. Research. Do your homework.

If you have a vision statement that's 65 pages long, it's too long. A classic vision statement has several characteristics: it is clear, understandable and aligned with the company's values; it involves people throughout the organization; it is memorable; it is linked to customer needs; and it requires the organization to work hard to obtain its goals.

Fifty words or less should suffice. But if you need more than fifty words, try to keep it simple.

A complete mission statement, on the other hand, clearly and fully describes which factors—and, if necessary, which resources—are most critical to supporting the business strategy.

An example of a mission statement:

*Company ABC competes enthusiastically in a free enterprise system. Our style is aggressive and our practices honest; our conduct is legal and ethical. We are motivated by a fair return on investment and are committed to strengthening the company through reinvestment. We believe that customers will be attracted by what we offer and this will ensure our continued success.*

*To completely satisfy our customers is our primary mission. We listen to our customers and understand their changing needs. We achieve their satisfaction by quickly translating their needs into products and services that are world class and that emphasize quality, design, innovation, and value. We are convinced that the success of our business depends on satisfied customers... .*

*Company ABC members [employees] are the most important resource of our company. The diversity of their races, genders, talents and personalities enables us to be more innovative, dynamic and flexible. Our members endorse the practice of continuous improvement, believing it offers the best path to pride in their work, greater job security, customer satisfaction and success for our company. Our corporate culture offers a participative environment that supports teams and individuals. Company ABC encourages member development and achievement through recognition, rewards and opportunities for career growth.*

*In order to achieve total customer satisfaction, Company ABC methods of operation are shaped by our dedication to quality. Corporate-wide quality initiatives result in superior products and services for our customers. At Company ABC we combine smart thinking with hard work to eliminate wasted time, effort and materials... . Our philosophy of quality includes the preservation of our environment and the protection of resources. Our pursuit of quality extends to our communities, where we build for the future by investing in the quality of life.*

A mission keeps a company on track in good times and bad. This is the statement of your decision to act, and a definition for what direction that action will take. You cannot succeed in business unless you set a direction.

A company needs a mission statement that is easy to understand. It formulates what an organization wants to be and stimulates specific goals for the organization. It needs to be something useful and applicable to daily operations. You—and, more importantly, your co-workers—should feel comfortable using your vision statement in everyday conversation.

Why did you start your business? What did you want to accomplish? What did you want to leave behind?

Try several drafts. Ask people, friends, family, co-workers, to look at each draft. Does your mission explain the following:

- Who you are as a company?
- Where you want to make your mark?
- How high you want to shoot?
- What you believe in?
- Does it embody the spirit of where you want your company to head?

The results are positive when you take the time to plan out your strategy. If you're too close to a situation to look objectively at what you ought to be striving for, you can't see the unknown hurdles ahead. You can have a great idea, but if you're not focused—if you don't have a vision or a mission—the obstacles will get in your way.

A mission statement is the best tool for direction. Some people get so busy managing today's business that tomorrow's gets pushed aside. Goals are most useful when they help you decide what you should do today to help you achieve what you want for your company tomorrow.

## Business Description

A successful business plan should include a company overview that includes a brief statement about your company, establishing the following:

- company name;
- date the company was founded;
- a description of what the business does (e.g., provider of professional organizing services, manufacturer of specialty food products, etc.);
- business structure; and
- location.

If your business operates in an industry that must abide by certain regulations, such as the toxic waste, weapons and armaments, generic engineering or explosives industries, this statement should also include any information on the government agencies regulating your business. Remember to provide information on how the agencies regulate your business (e.g., your business must document and account for uses and disposal of toxic materials, conduct background checks of employees with access to toxic materials, etc.). And, if any permits or inspections are necessary to operate your business, list these too.

Strategic alliances with larger, more established businesses are another factor that investors take into account when determining whether to give you funds for your business. Any leverage from relationships you have with others is appealing to investors, particularly if it improves your work performance. Describe each company and the details of the alliance. Is it a joint venture, a distribution agreement, original equipment manufacturer relationship, etc.? What is their position in the marketplace? How does your company benefit from the relationship? Are there any risks involved in the alliance?

## Development and Production

Following this brief statement, describe, in more detail the product or service that you provide and make reference to whether you're in the seed, start-up or growth stage of business. Investors like to know this from the get-go. Also document whether you've just developed your product, took on your first client or booked your first international order.

Even if you haven't booked any business, state your goals in as much detail as possible. For example, you expect to achieve X dollars in sales and X dollars in pre-tax profits in 2004 and achieve X dollars in sales and X dollars in pre-tax profits in 2005. Explain how you will achieve your goals. Describe how you will use the funds and what you will need to do to make the business succeed in a general sense, such as: marketing your new product, developing a new product that captures the needs of your customers, determining the right Internet strategy, building or expanding the business's facilities to meet increased demand, etc.

List the details about your product or service. List the products or services your company provides. How does your product or service work? What consumers needs are addressed by your product or service? What value do you add to the product or service that none of your competitors has? If there are more than one, list them in order of highest sales or significance in product line. Get detailed. If you offer a house sitting service, list all the services you provide and your fees for each (e.g., plant watering $60, lawn mowing $75 price, pet care $50, etc.).

If your company plans to expand its services or product line in the future, make note of that. Indicate whether your products are in the introductory, growth or mature stage or whether you plan to follow your product or service up with another. List any critical factors involved with the production or delivery of your service or product. Explain what your product has to offer and its advantage in the marketplace. Do you have a patent, a well-recognized brand name, etc.? These are important factors that should be included in your plan.

## The Market

You should also define your market(s). What markets do you plan to compete in? Will you sell your products in other markets? Go back to the research you performed in Chapter 1. The information you gathered about the market's performance is key here. What is your position in the marketplace? How will the market perform in the future? What is your pricing strategy? What distribution channels do you intend to use (e.g., wholesalers, catalogs, mass merchant retailers, consolidators, etc.)? Use figures and site the sources. Sadly, this is the most crucial but ill-prepared section on most business plans. Be sure your information is timely. You don't want to use figures from 1982 if you're starting a business in 2004.

Marketing is about information. Provide investors with useful information about your industry sector, your competition, services you need to provide relevant customer demographics and anything else that relates to what you make and how you make it.

Basically, you need to provide information on the levels of capacity, capability and performance (and what—if any—changes in policy, procedures and practices) you need to get your business started. Ask yourself the following:

- How much equipment, process capacity and facility space is needed?
- Where should this design, equipment, process capacity and facility space be located?
- How many and what kind of people are needed in each function?
- What kinds of design, management information and control, materials and resources and distribution systems are needed to support your projections?
- Do your operating policies, procedures and practices support the plan?

Focus on what marketing experts call the six P's: product, price, packaging, purchasing trends, public response and profitability.

Don't make the mistake (that many small businesses do) of thinking of marketing as a process that stops when the actual selling of a product begins. While marketing may play a bigger role in the early part of a product or service's life cycle, it should continue throughout the life of your business or that particular product or service.

Provide information on your customers (e.g., who they are, where they are, how your reach them, if they buy your product or service from someone else, how you market to them, etc.). Will you advertise, attend trade shows, conduct on-site product sampling or launch a promotional campaign? How does the way you promote your product or service translate into a competitive advantage for your company?

Think about the future. How will you market and sell the product? How much money will it cost—and generate? What will you have to change internally to make the product?

Offer up any new ideas and evaluate their immediate and long-term value. Can you get higher-quality products to market faster and at a lower cost than your competitors can? Factor in your own historical performance as a competitor.

Other information to include in your business plan:

- **Competition**. List your competitors and explain why your product is unique in the marketplace. Investors want to see something unique, proprietary or protected about your product or service. Do you have a secret ingredient or hold a patent on an invention? If you have no direct competition but there are alternatives to your product or service in the marketplace, list those, too. Provide a brief description of each competitor, including: product, price, location, promotion, management and financial position.
- **Risks vs. Opportunity**. List any market risk, pricing risk, product risk, management risk, etc. Explain how you will overcome these risks and why it will work. Having a strategy for dealing with every risk from limited operating history, limited resources and market uncertainties to production uncertainties and limited management experience is a good way to attract investors. Likewise, if you have any opportunities to become a major force in the market, detail this in your plan as well. Will you partner with a larger company who knows the market? Dominate a niche in the marketplace and become a major force in your industry? Improve performance in the field?

- **Business Management**. Who will help you achieve your plan? Do you have any employees? Do you have someone in charge of research and development? If so, provide a brief statement of each person's background and/or experience and what he or she brings to the table. What contacts do they contribute to the business? Explain how your employees will work together to achieve your business plan.
- **Financials**. How much do you need and why exactly do you need it? Investors want to know exactly what they're getting into before they throw a fistful of cash your way. Provide a breakdown of how the funds will be spent. For example, X dollars for complete development, X dollars to purchase equipment, X dollars to market our new product or service, etc. Describe why you need the funds and what you are willing to give up in return. Focus on the traditional performance parameters that most companies have to meet—unit cost, return on investment, cash flow and profit margin. The challenge in setting financial objectives often has less to do with numbers you target than with translating those numbers to investors. What is your exit strategy? How much time will be required to pay back a loan or provide a return for investors? You will also want to include a snapshot of your current financial position, no matter how weak. Be clear. If the information the investor needs is hard to find, you may not get your money. Remember: Include everything from sales, gross profit and pre-tax dollars to assets, accounts payable, liabilities, research and development costs and book value.

Provide illustrative material to give an investor a better feel for your company, including clips from industry publications, information on relevant patents and market research data.

Other tips for successful business plans:

- Attempt to keep the document to a 30-page maximum.
- Make the executive summary easy to read and fun—this is the part that gets read.
- Make your financial forecasts realistic, but optimistic.

- The quality of your management is crucial.
- Product/services must be unique, but you must thoroughly describe your competition. If there is no perceived competition, there may not be a market.
- List any potential problems and the risks of business failure. If you can't think of any, you haven't thought this through thoroughly.
- Tailor the business plan to your audience. A business plan for investors is different than a business plan to be read by employees.

Remember: There's no need to reinvent the wheel—there are plenty of sample business plans available. You can use those samples to develop your own plan easily. Simply find a plan that is somewhat similar to what you envision and rewrite it to match the details of your business idea. When you are finished, you'll have something you can give to investors, but you will also have something to reference once your business gets going. Your plan can serve as a road map to keep you on the right track as you move through your first year in business and beyond.

Once you have looked over some sample plans and gotten an idea of what style plan you want to write, start compiling your data. Experts say it can take between two and eight weeks to write a truly comprehensive, quality business plan—there is much to be done, between gathering supporting documents, writing down all the inspiration that's been floating around in your head and asking others to review it to make sure it makes sense to someone other than you.

That may seem like a great deal of time to spend on a single plan, but remember, you will be able to not only use this plan to get funding and get underway, but also to remind yourself, as you progress in your business, of where you have been and where you are trying to go. There are many tools available to help you take that plan from the idea stage to a finished product. See Appendix D for a list of some of the tools available to you.

## Free Business Plans on the Internet

Your best starting place for examples of business plans (many of which are free) is the Internet. These sample business plans can give you a good start, and help brew many ideas of your own. But, be careful not to copy them word for word, and add your own facts, figures and passion to convey your business ideas to potential investors.

- Located at **www.bplans.com**, **The Planning Resource Center**, powered by Palo Alto Software, offers a comprehensive list of sample plans, including plans for the Take Five Sports Bar, Salvador's Sauces, Trend Setter's Hair Studio and Computer Consulting.
- A sample plan for a fictitious manufacturing company is located at **www.sb.gov.bc.ca/smallbus/workshop/sample.html**, part of the **Canada/British Columbia Business Service Centre's** Web site.
- The **Small Business Institute** lists plans for an ice cream parlor, Mexican restaurant, painting contractor, counseling service, dentist and manufacturing business at **www.busplan.cc/sample/sample.htm**.
- A set of business plans, including **how to open a bookstore**, **how to open a confectionary store** and **how to open a youth center**, are located at **http://bedandbreakfast.hypermart.net**.
- Case study **sample plans for a manufacturing company, a service provider and a retailer** are located at **www.toolkit.cch.com/text/P02_9000.asp**.
- A sample business plan for the **Cute Cookies Corporation** is available under the MBA Tools section of **www.ventureline.com**.
- **Alpha Services** offers sample business plans in word processing format for $89.95 plus shipping. These plans are formatted to meet many of the requirements of financial institutions for submitting plans for loan applications, including for Small Business Administration (SBA) loans. (See Chapter 4 for more about the loans offered through the SBA.) They are located on the Internet at **www.alphaservices.com**. However, sample plan formats are available at **www.alphaservices.com/**

**plan_ex1.html**. Those formats are very basic outlines, but they can be used to see if you have completed all the necessary elements of your plan.

- A complete plan for **Rainbow Kites, Inc.**, a fictitious company, is located at **www.businesstown.com/planning/creating-rainbow.asp**.

## Software Programs

There are also a number of software companies that make programs that help you write and assemble the ideal business plan. And, even if you cannot afford or do not wish to purchase specific software for this purpose, you can usually browse through several sample plans at their sites. The following are only a few of the places you can find these plans:

- **Jian Software** has a complete business plan workshop at **www.jian.com/workshop/index.htm**.
- At **www.bplans.com/sp**, Palo Alto Software lists all the sample plans located at The Planning Resource Center.
- At **www.planware.org**, you'll find software to help you write a business plan, as well as software that will help you prepare financial projections and compile cash-flow forecasts.
- Business Resource Software, Inc., at **www.brs-inc.com**, offers three different types of business planning software, including Plan Write for Business, Plan Write Expert Edition and Business Insight. The sight also offers several other types of software geared towards a business's marketing, pricing and selling strategy.
- At **www.businessplans.org**, you'll find everything you need to know about writing a business plan, including software, resources and sample plans.

## Books to Help You Put Your Plan Together

The following is a list of books that might be helpful as you put your plan together:

- ***The Business Planning Guide: Creating a Plan for Success in Your Own Business*** by David H. Bangs, Jr. (ISBN: 1574100998, June 1998).
- ***Business Plans Handbook: A Compilation of Actual Business Plans Developed by Small Businesses Throughout North America*** by William H. Harmer, ed., and Terrance W. Peck, ed. (ISBN: 0787620777, August 1999) . This book has become an annual serial publication. It is expensive, but is a good resource for examples of plans for businesses similar to or just like the business you are trying to get underway.
- ***How to Write a Business Plan*** by Mike P. McKeever (ISBN: 0873378636, November 2002).
- ***The Instant Business Plan Book: 12 Quick-And-Easy Steps to a Profitable Business*** by Gustav Berle and Paul Kirschner (ISBN: 0940673886, March 1997).
- ***The Total Business Plan: How to Write, Rewrite and Revise*** by Patrick D. O'Hara (ISBN: 0471078298, December 1994).

## Mentoring: How Free Advice Can Help

There are many different organizations that support small businesses and business in general. These organizations can be good resources for networking and mentoring by other business owners who are further along in the process than you are. Although some have membership fees, in most cases the fees are tax-deductible. Check with your accountant to be sure. Also, many organizations that have membership fees offer member and non-member rates to attend conferences and events—if you do not want to be a member and don't expect to attend many events, it might be more cost-effective to just pay the non-member rate to attend a few events that are of use to you.

The **Small Business Administration's Small Business Development Center (SBDC) program** provides free and low-cost mentoring services to entrepreneurs around the country. There are nearly 1,000 SBDC service centers around the country and in Guam, Puerto Rico, Samoa and the U.S. Virgin Islands, so there is bound to be one near you and your new business. Each state has a headquarter center, which is often located at a university or college, and most states have service centers strategically located around the state to provide most business owners with the best opportunity for assistance.

The SBDCs help start-up businesses with an array of issues, including financing, planning, organization, production and other technical assistance problems. Counselors and staff at the SBDCs will also help business owners locate venture capital funding, learn about international trade opportunities and available grants.

To find the SBDC closest to you, visit ***www.sba.gov/gopher/Local-Information/Small-Business-Development-Centers*** and click on the link for your state or territory. You will download a text document that includes contact information for each service center within that state. The information includes e-mail addresses, mailing addresses, phone and fax numbers and the names of each center's director. If you do not have Internet access, you can get the same information by calling the main number for each state or territory's SBDC. For a list of numbers to call for more information in each state, see Appendix E.

Many of the SBDCs have Web sites that include a range of resources including handouts from seminars, guides to entrepreneurship, links to helpful services on the Internet and online registration for free consulting services. That list of Internet links is located at **www.sba.gov/hotlist/sbdc.html**. Although each link points to sites geared toward entrepreneurs in different states, some of the links and other materials offered at each of the sites apply to business owners who live anywhere in the country, so they're worth checking out.

The following is a list of organizations that have national, regional or local chapters and that offer mentoring and many other resources for small business owners:

Coastal Enterprises, Inc. provides technical assistance and mentoring services for Maine residents, particularly low-income residents, who want to start businesses. The organization fosters self-employment through many initiatives. You can contact them at Coastal Enterprises, Inc. 36 Water Street, P.O. Box 268, Wiscasset, ME 04578. Phone: (207) 882-7552. Internet: **www.ceimaine.org**.

Entrepreneurs from Washington, DC, Northern Virginia or Maryland can register to become part of the Dingman Center's Mentor Program. To be eligible for the program, however, you have to have at least a first draft of a business plan. There is a $25 fee to apply to the program, and mentoring services cost $45 per hour. You can contact them at the Dingman Center's Mentor Program 4321 Hartwick Road, Suite 300, College Park, MD 20740. Phone: (301) 403-4290.

Although E-Mentoring on AskTheEmployer.com (located at **www.askanemployer.com/mentor.asp**) primarily provides career-mentoring services, it can also be helpful for self-employed entrepreneurs. Simply select the area of business in which you are in and the state in which you live, and the database will provide a list of volunteers who have agreed to mentor by e-mail.

Residents in and around London, Ontario, who wish to start a business can gain access to the London Entrepreneurial Education Association Small Business Mentoring Program, which matches local business owners with new entrepreneurs. For $25 (Canadian), entrepreneurs register with the program, and then are only responsible for any costs incurred when they meet with their members, such as travel to and from the meeting. You can contact them at the London Entrepreneurial Education Association Small Business Mentoring Program 1764 Oxford Street East, London, Ontario, Canada, N5V 3R6. Phone: (519) 659-2882. Internet: **www3.sympatico.ca/mentoring/index.html**.

The Louisiana Business Incubation Association offers a directory of LBIA members at **www.ecol.org/Lbiamem.htm**. Business incubators often provide mentoring services for local start-ups, as well as training and funding opportunities.

At Mentors Peer Resources (**www.mentors.ca/findamentor.html**), after filling out a form its database will attempt to match you with an appropriate mentor. This site also has good information on how to find a mentor in your own community.

Reach4It.com at **www.reach4it.com** offers online mentoring and charges no fees for its services.

One of the largest free mentoring resources is the Service Corps of Retired Executives (SCORE), which works in partnership with the Small Business Administration to provide training and mentoring to entrepreneurs. There are 389 SCORE chapters in the United States and its territories, with more than 11,000 volunteers working to help prospective business owners get their dreams off the ground. In 1998, SCORE volunteers gave more then one million hours of time to free or reduced cost programs sponsored by chapters. If you do not live near a chapter, mentoring and counseling is available by e-mail. SCORE's Web site also offers how-to workshops, articles and tips on various aspects of starting and running a business. Contact them at Service Corps of Retired Executives (SCORE) 409 3rd Street, S.W., 6th Floor, Washington, DC 20024. Phone: (800) 634-0245. Internet: **www.score.org**.

The Southern Oregon Women's Access to Credit (SOWAC) organization, which serves both men and women, provides guidance and support to those who want to own their own business. SOWAC provides financial and educational assistance and has an active mentoring program. Contact them at Southern Oregon Women's Access to Credit, 33 N. Central Avenue, Suite 209, Medford, OR 97501. Phone: (541) 779-3992. Internet: **www.sowac.org**.

Despite the name, the Women's Economic Development Agency provides mentoring and educational services to all individuals—not just women. Contact them at Women's Economic Development Agency 675 Ponce de Leon Avenue, Atlanta, GA 30308. Phone: (404) 853-7680.

## Organizations for Minorities

There are many organizations that cater primarily to minority business owners. If you qualify to be a member of one of these organizations, joining can put you in touch with other successful business people that want to see you succeed. Those potential mentors can help you find funding and avoid the pitfalls that they may have discovered along the way.

Asian Women in Business was founded in 1995, and supports Asian women entrepreneurs through "information, education and networking opportunities." Contact them at Asian Women in Business 1 West 34th Street, Suite 200, New York, NY 10001. Phone: (212) 868-1368. Internet: **www.awib.org**.

Founded as an organization to support low-income Hispanic women as they develop the skills they need to become self-sufficient, Mi Casa has expanded its mission to include entrepreneurial support for this group of women. Its Business Development Training Program is a comprehensive educational program that helps women get the tools they need to succeed with their start-up business. The program includes one-on-one mentoring services. You can contact this organization at Mi Casa 571 Galapago Street, Denver, CO 80204. Phone: (303) 573-1302. Internet: **http://micasadenver.org**.

The National Black Chamber of Commerce promotes and helps sustain African-American small businesses throughout the United States. Programs offered include funding workshops, computer training and other helpful resources. Write or call them at the National Black Chamber of Commerce 2000 L Street, Suite 200, Washington, DC 20036. Phone: (202) 466-6888. Internet: **www.nationalbcc.org**.

The Organization of Native American Business & Entrepreneurial Network (ONABEN) serves business owners in Oregon and Washington. Contact them at the Organization of Native American Business & Entrepreneurial Network 520 Southwest 6th Avenue, Suite 914, Portland, OR 97204. Phone: (503) 243-5015.

The U.S. Hispanic Chamber of Commerce promotes Hispanic businesses and supports them through technical assistance and other mentoring services. Contact them at U.S. Hispanic Chamber of Commerce 1019 19th Street NW, Suite 200, Washington, DC 20036. Phone: (202) 842-1212. Internet: **www.ushcc.com**.

The U.S. Pan Asian American Chamber of Commerce was founded in 1984 to represent all Asian ethnic groups involved or relating to trade. In addition to providing advocacy, education, information and networking opportunities to its members, the organization also promotes activities that will further the business and professional interests of members, collects, evaluates and disseminates information of interest to members and represents, expresses and gives effect to the opinions of members with respect to trade, finance, commerce, industry and related issues. It also conducts charitable, educational and similar programs for the benefit of members as well as the Asian American community. Contact them at U.S. Pan Asian American Chamber of Commerce 1329 18th Street NW, Washington, DC 20036. Phone: (202) 296-5221. Internet: **www.uspaacc.org**.

Minorities and women who live in Mendocino County, California, can get one-on-one mentoring and assistance through West Company, a microenterprise incubator that helps get small businesses on their feet. This is also a good place to go for funding if you are part of the community this organization serves. Contact them at West Company 367 N. State St., Suite 201, Ukiah, CA 95482. Phone: (707) 468-3553; or 306 E. Redwood Ave., Suite 2, Fort Bragg, CA 95437. Phone: (707) 964-7571. Internet: **www.westcompany.org**.

## Organizations for Women

In addition to organizations geared toward offering assistance to minorities, there are many formal—and informal—organizations for women seeking mentors. However, even if you are not a woman, call or write to the organization for information about other groups in the same area. It wouldn't hurt to ask if they also offer their services to men either—you'd be surprised how many do.

Why are there so many of these organizations? Although there are many small businesses operating in the U.S. today, only a small percentage of them are run by women. That percentage is growing quickly, however, thanks to what may be a new "old girl's network." Women business owners seem very interested in helping other women get their business started. Women often refer business to each other and offer free services to each other to support this fast-growing economic power group. Discriminatory? Some would say so, but others say it's just a way of balancing out a long-standing inequality between women and men in the business world.

If you're a woman entrepreneur looking for a mentor, see Appendix F for a list of organizations that offer mentoring resources to women.

By networking with members of those organizations, you may be able to find someone who would want to be a mentor to you as you get your business started.

## Creating a Budget

Most people hate budgets. If you do, you probably run your business using your daily bank balance as your only financial tool. Some people do well for years in this manner. But sooner or later or when your payroll grows beyond a handful of workers, you'll need to have more knowledge about how much it costs to make your products, what profit margins you need to pay the bills, what your company is worth and you'll need to put things on paper.

Budgeting takes some of the risk out of running your own business by minimizing the guesswork that comes with winging it. Budgeting gives you a blueprint for action. It tells you what to expect and alerts you to trouble when the unexpected happens. It also measures your success.

If you're starting a business you should know at least the basics of budgeting. Profits aren't everything in business, but without them, no business survives long enough to reach its goals. Fittingly, therefore, almost everything in budgeting stems from the simple formula for determining profit: Sales minus expenses.

A budget helps to get you from the generalities of your business plan to the specifics of day to day operations. By setting priorities, the budget makes clear what your finances permit you to do to reach your goals. It translates your vision and mission statements into action, by identifying the tasks necessary to reach your objectives. It allows you to explore the costs of reaching those goals and budget for them, outlining the time and resources you must commit in order to reach the goals. The budget quantifies your plan in dollars.

A budget also sends clear messages to lenders and investors, who don't do business with people whose position they can't understand. They want to see budgets before throwing any money in your direction.

But remember that you don't win when bankers or investors give you money. You win when you pay them off. For new businesses the first order of business is to pay off lenders. The surest way to do this is to identify basic financial factors and measure them, so that you can make mid-course corrections.

## Using a Budget Notebook

Create a Budget Index that documents every expense item, including which vendors you use and general ledger numbers to correctly categorize each expense.

The chart on page 164 and 165 (closest to the left margin) represents an index for a budget notebook. The main categories are section topics and have their own dividers.

Each line item listed will have its own page in the notebook, detailing that particular expense. Each company will have its own additional items, but these are the main categories that are almost universal.

Analyze your sales projections and expenses in detail. Don't pass over any item. Categorize items for reporting purposes. Will the cost of renting your office be listed under equipment rents or repairs and maintenance?

Review your expenses and project your sales from the bottom up—and put it in writing. The budget notebook will be the one place to find all the answers. You can keep copies of contracts in this notebook to show expense commitments you have made for periods of time that may be beyond the current year.

Set up your notebook in a three-ring binder first—don't worry about what you'll put in it. Prepare the dividers and have one blank sheet of paper for each item with its title on the top. To begin writing, start with the easiest items first, usually some recurring, consistent expenses. For example, how much do you pay in rent each month? Add to this any information you need regarding your lease, such as the starting and ending dates, and when increases occur. This is a good time to review your lease and look for any hidden costs that will need to be a part of your budget.

The budget notebook starts with sales projections, but you might want to include other sections, such as:

- *Cost of Goods Sold*—the direct costs and any other expenses incurred in making your product;
- *Sales and Marketing Expenses*—what it costs you to market and sell your product;
- *Overhead Expenses*—most of the other expenses incurred in operating an office, such as personnel not in other categories, facilities costs and administrative items such as office supplies, etc.; and
- *Income Taxes*—I don't discuss estimating taxes in this book, but you should have a section in your notebook for tax planning, and discuss this with your CPA.

List all the products/services you plan to offer. How much are you going to sell each unit of product/service for? How much do your competitor's charge? Do you discount prices for volume or other criteria? How much do you have to make per unit to be profitable? What are your customers willing to pay for your products? Are some willing to buy at a number higher than others? What projections did you make in your business plan? Project the number of units you expect to sell this year. Project as conservatively as possible. Base your projections on how some of your competitors have performed.

Multiply the number of units you project will be sold by the average price you plan to sell your product/service for to get an idea of what sales dollars the product would be expected to bring.

Are sales projections particularly aggressive for some products and not for others? Do these differences accurately reflect the positions of various products in their sales cycles? Are there other significant sources of income not taken into account by product sales (such as shipping and handling)?

Even if sales of your products generally don't appear to have any dramatic fluctuations from month to month, certain products may. This may signal a particular buying segment that orders at a particular time of year. Knowing this may help you spend your marketing dollars for this buying group at the right time.

## Expense Budgeting

This process gives you a format for writing down and tracking expected expense items that will appear on each page in the Budget Notebook, as well as a way to compare these numbers each year.

Our example on the next page projects budget expenses for "rents," including office building leases, parking, warehousing and equipment rentals. Project what you expect to pay for these items.

On a blank sheet of paper first list the type of item, and next to it list the vendor of that item. Lastly, make an estimate of how much you think you will spend on that item this year. In the future, this estimate can be made by looking at how much you spent last year and making an educated guess as to whether this will go up or down. This will also give you an idea of what the expense figure for this year might be and whether there is a tendency to over- or under-budget for this item.

Whenever possible, list the important financial points of the lease (i.e., automatic price increase clauses and other hidden costs) to make this process easier next year, and to let the accounting department know if

## Budget Index

| | General Ledger Account Numbers |
|---|---|
| Sales Projections | |
| Cost of Goods Sold | |
| Materials purchased | |
| Salaries & wages | |
| Production supplies | |
| Temporary help | |
| Shipping supplies | |
| Mailing & shipping | |
| Sales & Marketing Expenses | |
| Salaries | |
| Sales commissions | |
| Direct mail | |
| Advertising | |
| Publicity | |
| Consulting | |
| Other sales & marketing expense | |
| Overhead Expenses | |
| *Personnel* | |
| Salaries | |
| Bonuses | |
| Payroll taxes | |
| Group life & health insurance | |
| Workers' comp. insurance | |
| Employee benefit plans | |
| Officers' salaries | |
| Employment expense | |
| Training | |
| Temporary help | |
| *Facilities* | |
| Rents | |
| Property tax | |
| Repairs & maintenance | |
| Utilities | |
| Property & liability insurance | |
| *Administration* | |
| Accounting services | |
| Automobiles | |
| Bank charges | |

| | General Ledger Account Numbers |
|---|---|
| | |
| Computer supplies | |
| Contributions | |
| Depreciation & amortization | |
| Dues & subscriptions | |
| Legal services | |
| Licenses | |
| Miscellaneous | |
| Office supplies | |
| Other professional services | |
| Retirement plans | |
| Telephone | |
| Travel | |
| Income Taxes | |

these payments are expected to increase, stop or decrease at any point during the year.

Have all important expense items been captured in the worksheet? Do any numbers seem out of proportion to the value of the service you are paying for? Do you plan to use less of a given product or service over the course of the year? Do you plan to pay a specific amount for a given product or service each month, or does it vary by the quantity used?

Prepare a page for each item listed in your budget notebook in this manner (i.e., insurance, payroll, bonuses, legal services, expenses, etc.).

Categorizing your expenses this way is important to begin to look at your overall profit picture as a number you can control. From the information gathered, you can begin to analyze your profit picture by putting your numbers into the major categories I mentioned earlier:

- Sales
- Cost of goods sold
- Sales and marketing expenses
- Overhead expenses (including administration, personnel, facilities, etc.)

You may also want to include a include a section for income before taxes and net income.

Remember: When you put your budgets and projections together, build some tolerance for variance into your numbers. Your budget stops being accurate the moment you finish it. Other tips include:

- Remain flexible and keep an attentive eye on your market.
- Control expenses and reshape your finances to fit your needs.
- Learn the value of understanding the impact of finances on your business.
- Evaluate your projections regularly, particularly when adding capital (yours or somebody else's) to your business.

There are many tools available on the Web and many books on the market to help you determine how much money you need. One of the best Web tools I've come across is the *CCH Business Owner's Toolkit Guidebook—Develop a Cash Flow Statement*, which is located at **www.toolkit.cch.com/text/P06_6400.asp**. This worksheet provides blanks for all the possible expenses your business will face as it gets going, including big ticket items like office space rent and utilities, and small items like postage and office supplies. It also allows you to incorporate your personal expenses, which is important if you are to be a sole proprietorship. When you first get started, the money for your business and personal expenses are likely to be coming out of the same pot—you may not be able to afford to pay yourself a salary immediately—so it is important to consider your personal expenses in this amount. However, do not take those into account when you are applying for loans or grants—those financial statements should strictly include information about your business expenses and projected income.

The following are some good start-up and operating expense budget tools available on the Internet:

- A start-up expense calculator is available on the Oxygen.com Web site located at **www.oxygen.com/tools/calculators/startupexp_calc.html**.

## Sample Expense Budget Page

YEAR: 20___

**RENTS**

| | | |
|---|---|---|
| Main Building | ABC Properties | $ 196,806 |
| ($16,198/mo. increases 3% on 8/1/04) | | |
| Parking | Parking Lots Inc. | $ 480 |
| (12 spaces at $40/mo.) | | |
| Warehouse | SuperStorage | $ 2,400 |
| ($200/mo.) | | |

**Equipment Rentals**

| | | |
|---|---|---|
| Postage meter rental | Pitney/Bowes | $ 835 |
| ($412.50 twice a year, due 1/1 and 7/1) | | |
| Copy machines | Xerox | $ 4,272 |
| ($356/mo. expires 4/4/05) | | |
| Telephone | Bell Communications | $ 14,628 |
| ($1,219/mo. expires 7/1/07) | | |

TOTAL: $ 219,421

Budget last year: $ 195,008

Actual last year: $ 201,962

- A cost-estimation shareware program is just one of the many helpful financial programs available for download from the Small Business Administration Web site at **www.sba.gov/library/sharewareroom-financing.html**.

One important thing to keep in mind: You pursue two objects in controling expenses—to eliminate "default spending," or spending by habit, and to set priorities for the spending that you must do.

## Handling the Money (Accounting)

Whether you have employees or are flying solo, you must set up a system to track where your money is going, whether its on its way in or out.

There are several services on the Internet that allow you to track your financial data from anywhere in the world. There are a couple of advantages to doing your finances online. First, it means you are not responsible for backing up the data. These companies are in business to store that data for you and they are responsible for keeping it in tact. Also, it means that you can access your information while traveling or working offsite, which can be helpful if you do a lot of traveling for business or otherwise. However, while security is probably not a problem, it is still a concern. Your financial data is important, and you do not want everyone to access it. So, remember to inquire about a company's security system and ask for customer references or testimonials before you agree to do business with them.

Also, keep in mind that the Internet is an unstable world. Always keep printouts or other backups of your data in case the company you select goes out of business and try to find a company that has been in business for a while. The following is a list of online organizations that offer financial data tacking:

- NetLedger, located at **www.netledger.com**, provides their service for $9.95 per month. They offer a free 30-day trial for new users, so it's possible to check them out before you start spending your money.

This service has been named the best by ZDNet and other prominent reviewers.

- QuickBooks, located at **www.quickbooks.com**, offers either desktop software or a Web-based financial management system. Prices for the products vary.
- Peachtree is another popular software package that started out as a desktop program and has now also added a Web component at **www.epeachtree.com**.
- Jungle **www.vjungle.com**, includes accounting capabilities for free in their complete suite of business services.

If you want to handle your finances right on your desktop, check out one of the three major small business accounting packages:

- MYOB. **www.myob.com**;
- QuickBooks. **www.quickbooks.com**; or
- Peachtree. **www.peachtree.com**.

## Conclusion

When you've finished brainstorming your ideas for each section of your business plan, you should have a good idea of the best practical goals you can set for your company. You should be able to look toward the horizon without tripping over any obstacles at your feet. Now you're ready to put it in writing.

Once you have your business plan in place, you endeavor to set performance standards and accomplish your business's corporate objectives and action plans. Performance standards can accomplish several goals—the most important among these: Monitoring on a regular basis the progress your company makes. These standards serve as yardsticks allowing you to measure how well you do the things that win business.

What is also important is to show that you have the ability to deliver on your plan (proof of concept), usually best shown through the founder's past experience and your marketing plan. Also you must demonstrate that the business plan and profit objectives have a reasonable chance of success.

The most successful new ventures that investors are particularly interested in, are ideas, businesses or technology that is distinctive, fills a new or existing need better than others or has a competitive edge, so make sure the business plan addresses this.

It also wouldn't hurt to have an experienced person to help you develop your concept. My view of the best qualities for this person is a combination of business experience, street smarts, and being able to effectively evaluate a firms products, management and marketing ability to determine if they have a good opportunity to be successful.

Don't give up quickly; you could be subject to the normal VC "wild goose chase" (There are many billion-dollar companies now, that were turned down by the first VC firms they approached); there could be a problem with your concept, or just the way you're presenting it.

Always carefully consider the burdens that come along with venture capital such as proving your business to others, giving away a large percentage, and having to spend time on requests of VC partners for information, financials, etc. instead of spending 100 percent of your time on developing your business.

I'd also recommend that new entrepreneurs get the book: *Getting Everything You Can Out of What You Got* by Jay Abraham (ISBN: 0312284543, October 2001) . I found it to be one of the best books I've ever read for entrepreneurs.

Remember: Practicality is important, too, if not the essential aspect of setting performance standards. Set realistic goals...don't go overboard. Realistic goals distinguish a workable business plan from an unworkable one.

In the next chapter, I'll look at the mechanics of moving business from the earliest stages to the next step—guided by your well-formed business plan.

# Chapter 6: Office Space, Equipment & Other supplies

As an entrepreneur, you've probably focused solely on how to get your business off the ground. But no matter what kind of business you start, you will need an office of some kind. Whether that office is in your home or out of your home, in your basement or in a spare room, in a leased commercial office space or in a business incubator, an important piece of planning and starting your business is to decide where you will work.

## There's No Place Like Home

With the advent of better technology, more and more people are setting up shop in their houses, eliminating long commutes to work, office politics to bear and best of all, nitpicking bosses.

If you have space to create an office in your own home, and if you—and your family—are comfortable with that idea, you may be able to save yourself a great deal of overhead in the early years of your business. But where do you begin? Have you given any thought to where you would work in your home? Do you know how to turn a room into a workspace? Where will you locate the office? Do you have small children at home?

There aren't really any guidelines to putting an office together in your home, but there are a several things you'll want to consider before setting up shop. Let's face it, shuffling back and forth from the den to the dining room won't work forever. And, there are other things to consider such as the equipment needed, the size of the equipment and even where the equipment will be located, etc. When setting up a home office, use the following as a guide to help you decide what room works best and how to set it up.

## Office Location

First: Where will you locate your office? For many entrepreneurs working out of their homes, a home office goes where it happens to fit—crammed into a corner, dumped on the kitchen table, occupying half of the den or a guest room or on a card table set up next to the washer and dryer in the basement. This decision is crucial to the success of your business. Many people overlook this factor and assume that a little corner of a room is all they need. They're wrong. In most cases, people end up moving from room to room...to room to room.

But before you map out any prospective home office, first make certain that the home office you have in mind is, in fact, legal where you live. In some communities customers are prohibited from visiting a home office. In others, you can't hire outside help. Ensure that there are no legal constraints on the business you plan to start out of your home.

Remember: If you decide to run a business out of your home on the sly, you could find yourself face to face with a code enforcement officer eager to shut you down. Although many home-based business owners (particularly those with low or no traffic businesses) don't bother with the technicalities, the prudent thing to do is to call or visit your local city hall, and talk to someone familiar with the codes regarding conducting local business. Sometimes, to be absolutely legal, you must obtain a zoning exception or variance. Of course, similar to any other legal area, if questions or problems arise, you should contact a knowledgeable attorney to help you.

No matter where your home office is in your house, know this: its location will play a pivotal role in the success of your business. A central location to work will allow you to conduct your business efficiently and function more productively. Remember: It's always better to have a dedicated room for your office—a room specifically designed to be an office that offers an Internet connection, fax and phone lines, built-in bookshelves and filing systems. But, if you have to set up an office in just part of a room, for example, if the only extra room you have is your spare bedroom, consider putting a convertible sofa in one part of the room and a desk and bookshelves and filing cabinets on the other side of the room.

By arranging them carefully, you can even set up a border for your office so it's easier for guests to stay in the room when they visit.

Other factors to consider when deciding on a location for the office: Are you comfortable working in the basement? Or do you need lots of light and air to keep your energy flowing? Do you have small children at home? If so, you might want to pick a room that isn't near their play area. Is there a phone jack nearby? If you decide to set up a separate business line, you don't want to incur the expense of installing new phone jacks and running the phone line to your office. It's much cheaper if the phone jack is already there. (I'll go into more detail about setting up additional phone lines later in this chapter.)

If you live in a balmy, summer climate, you might want to take this into consideration when deciding where to locate your home office. An office with direct southern exposure may be better for baking rather than word processing in the late afternoon. Likewise, living in a rainy area may make a basement office unsuitable, particularly if you have a great deal of electronic equipment.

## Office Size and Set-Up

Most architects recommend that the ideal home office be at least 10 by 10 feet. The space should contain adequate space for a desk, chair and computer, as well as areas for storage and shelf space for files or books. If you have enough room, you may want to consider a filing cabinet, too. The actual size of the office will depend on the nature of your business and the number of people who will use the office. If you'll be assembling or manufacturing products, you may need more furniture, such as a long table or countertop. And if you intend to have clients visit, you may need to purchase a couch or set up a meeting area.

Another factor to consider: Does the area have a door? If it doesn't you may want to set up shop in a room that does. Even if the perfect spot is the alcove off the living room, if you don't have a door, the other residents of your home may distract you. In addition, it's highly likely that you have to put in some long hours, including working nights and on weekends. The rest of your family will carry on as usual, and if you can't shut a door to signal to them that you are working, you may find it hard to separate your work life from your home life. You may want to consider a separate entrance to your office for this reason, too. Even if you do not plan to meet clients at your home office, there's always a chance that you may have to do so. If a client comes to your home, you will need to walk them to your office and if the path to your office goes through your children's playroom and all the toys are spilling from their toy chest, your level of professionalism will diminish. Think about where your customers or clients will enter and where you'll execute business. If your customers have to walk through parts of your home that don't exactly exude professionalism, your may want to rethink the office's location. Remember: Consider safety and what your customers see along the way. You may not think it's that important, but think about the impression you're offering by what your customers or clients see.

Think about the future. Is your office scalable and ready for expansion? Don't limit your businesses future to the here and now. If your business succeeds, eventually you'll outgrow the barest of equipment and facilities. As your business expands, you'll need more equipment and furniture to accommodate its growth. Eventually, you'll need more outlets for faxes, copiers and scanners.

Estimate where you think you'll be in a few years and how that space could handle the growth of your business. If your guest bedroom has a good-sized walk-in closet, plan how you will use that if it's needed in the future. Will you install built-in shelves to accommodate books or add a filing system for storing all your business files?

## Budgeting for an Office

What is your budget for the office? The amount of money you can spend will determine the type of furniture and storage facilities you can purchase for your office. Today, some people are incorporating home offices into the construction plans of their home. Signing a lease or laying out cash for a new home isn't high on anyone's list. But if you're looking for a new home, it wouldn't hurt to look for a place with office space if you're planning on running a business out of your home. Look for spare bedrooms, roomy basements or extra attic space. Scope out power outlets, particularly in older homes that may not be outfitted as well as newer structures. But you don't need a wad of cash to set up a good office. If built-ins are out of your limit, modular office systems, including everything from desks to computer tables and bookshelves to cabinets, are a more affordable solution. Look around your current home. For about $40 a new slipcover will put life into that old couch that you threw in the basement or attic. Other places to find good bargains on furniture include rummage sales, the swap meet, local Salvation Army or thrift stores.

## Office Lighting

Are you comfortable working in the basement? Or do you need lots of light and air to keep your energy flowing? Lighting is a key issue when setting up any office. Corporate offices put great emphasis on the quality of lighting...and you should, too. Poor lighting causes eyestrain and low productivity. If your basement doesn't have good lighting, you may want to rethink the location of your office or install the appropriate fluorescent, incandescent or recessed lighting. And, if you plan to use a computer, avoid setting your monitor up in areas that produce glare such as the back of a window or skylight.

## We're a Happy Family

Think about your family ties. Ask yourself: Will your family be impacted by your office? If so, how? How will your office be impacted by family? Choose a location for your office so as to minimize noise and intrusion on business matters. The best solution: a door that closes. If you want to be professional, eliminate any proximity to yelling kids or barking dogs.

Are you starting a home business to be close to your family? If you are, you may want to rethink your strategy. A home-based business can create conflicts within families...and for the business.

Yes, a home-based business allows for family togetherness. Home employment provides a working parent the opportunity to spend more time with the family. And, in many instances, children have the opportunity to see what their parents do for a living, and to learn about the business firsthand. But, those who seek the benefit of being close to family often find themselves in a dismaying dilemma. As they are starting their business, they must work harder; and by focusing more on work, they discover themselves devoting less time to the family. Even while working at home, they still work hard, maybe even harder than they did in the corporate world. And, since their hard work is presumably for the gain of the entire family, they expect their family members to understand and allow them the space and freedom to work; an expectation that isn't always met.

As a result, a home business can create conflicts within your family. If your only activity is your business, it is likely that your family and social life will suffer. Some people even get divorced because of problems resulting directly from the pressures of starting and running a business. Some are able to successfully juggle their responsibilities, while others let the stress of the business take over their lives.

Problems with family and friends can spell disaster for your business. It's important to generate family support for your home office because this support will make your adjustment easier and your work arrangement more pleasant. To avoid potential conflicts, ask yourself the following: *Would a home-based business allow you to balance work and family obligations? Can you expect your family to be supportive?*

You will need the support of your friends and family and you may not get that support if you alienate everyone by not being sociable or not making time for them. But if you're not careful, the very entrepreneurial nature that got you into the business may drive you to lose those individuals who support you and your business endeavors. You need to balance each role in your life—business person, mom or dad, husband or wife.

## Tax Advantages of a Home-Based Business

Often you can expect to pay as much as half of your home business's billing rate just to cover overhead costs. But the good news is that operating a home-based business can also yield tax advantages. Because you do business out of your home, you may be entitled to deduct a portion of the operating expenses and the depreciation of your home. You may also be entitled to deduct expenses from using a vehicle for your business, including gas, insurance, depreciation, etc. And, the IRS recently lowered its requirements for a home office deduction. Now, your home office doesn't have to be the only place that you have an office, it just has to be a place where you do a substantial amount of work.

If you meet specific criteria, you may be able to deduct a percentage of your regular expenses, such as rent, interest, taxes, insurance, repair and maintenance, etc. To qualify for deductions from the business use of your home, the IRS requires that part of your home must be set aside *regularly* and *exclusively* for the business. In other words, your workspace must be used as either :

- Your principal place of business for any trade or business in which you engage;
- As a place to meet and deal with clients or customers in the normal course of your business; and

- In connection with your trade or business if you are using a separate structure that is not attached to your home or residence (e.g. a studio, garage or barn).

If you hire any employees to work out of their own homes, they may be entitled to deduct expenses for the business use of their homes, too. There is one catch, though: they must work at home *for your convenience* and not because it's appropriate or helpful for them.

To determine what percentage of your home operating expenses and depreciation is deductible, use one of the following:

1) Divide the area used for your business by the total area of your home. So, if your home measures 7,000 square feet and you are using 700 square feet for your home office, you will be able to deduct 10 percent of expenses that go towards rent, mortgage interest, depreciation, taxes, insurance, utilities, repairs, etc.; or
2) Divide the number of rooms used for your business by the total number of rooms in your home. While this method may seem easier, it is key that all of the rooms in your home are approximately the same size for this method to work.

Once you determine the percentage of your home expenses that is deductible, multiply this figure by each of your expenses to obtain the dollar amounts of your deductions. (For example, 20 percent times a $1,000 home utilities expense equals a $200 business utilities deduction.) Expenses that benefit only your business (i.e., painting or remodeling the office) are 100 percent deductible. Expenses that benefit your home and but aren't related to the business (i.e., lawn care or the installation of a new whirlpool tub) are *not* deductible.

If you need help or more information, consult your accountant, bookkeeper or local IRS office to get information on full compliance with IRS regulations, particularly if you own your home. If you decide to sell your home, the home expense deductions you've taken for the business will have a bearing on how and when capital gains on the sale are to be recognized.

Remember: Have a plan in place for your office. Where do you see yourself and your business in two months, two years or 10? Will you eventually move to a larger space? Will you need new furniture? Do you plan to add on a room to the office? A proper plan for a home-based business can also minimize any conflicts between business and family. Do you have rules set up for your family to abide by when it comes to the business? Will your children run screaming into your office when you're on the phone with a client or supplier? Do you expect them to tiptoe around?

Once you have answered these questions, you will have a better idea of whether or not to move forward with an office at home. But what if you decide that working at home is not for you? There are times when a home office won't be right for you. Not every house or apartment is suited for productive workspace. If you're intent on a home office, but there's simply inadequate room or your family or roommates aren't exactly up for the idea, don't try to make the impossible fit. If it doesn't work, it doesn't work. Make alternative arrangements. Look into subletting or renting another space. Later, in the chapter, I'll discuss business incubator programs that are available to help you find a cheaper way to set up an office without resorting to the home office setup.

## Leasing Office Space

No matter what kind of business you start, there may come a time when you can no longer work in your home. Perhaps it's too distracting or too isolating. Or, maybe you need to have meetings with clients and you want a space larger than the left corner in your basement . Or maybe you want a more professional atmosphere. Whatever the reason, you will want to find a place to locate your business for as little as possible.

It's not easy to find these places, but this list of ideas should give you a place to start. Some of them are free, but others charge a monthly fee for the privilege of working in their space. However, that fee does not just buy you space—often that fee includes mentoring services, equipment and other bonuses that save you on costly overhead. Also, many of these organizations are flexible about the timing of your payments—after all, they want to see you succeed in business. That's why they were founded!

For most start-up entrepreneurs, leasing office space is a major commitment. The cost of leasing office space can significantly affect your business's bottom line, particularly if you have little start-up capital to begin with. However, the more prepared you are before starting the process, the more successful the outcome.

It's important to know what you want or don't want before you begin looking for office space or begin the negotiating process. The key to getting what you need is to define what you need and want at the outset of your search for office space. What features are you looking for? Site accessibility? Freeway access? A particular layout? A specific move-in date? What is your time frame for the process? Are you looking for offices or open work areas? Do you know the maximum and minimum amount of square footage that fits your needs? Do you need the groundfloor for your business? It's also a good idea to comparison shop before you begin to familiarize yourself with market trends and prices. Use the following as a guide for what should be considered in your search.

List the three best features of your current workspace. Then list its three worst features. Use these as a guide when you're looking for space.

Other things to take into consideration:

- Location;
- Employees;
- Size of offices and work areas;
- Security;
- Equipment;
- Restrooms;
- Parking;

- Seating capacity;
- Conference rooms; and
- Storage.

First, find a knowledgeable real estate agent. Before hiring an agent, ask your prospective broker questions to ensure he or she can perform for you. Ask the broker how much office leasing experience the person has and how much experience he or she has with your particular type of business. Does he or she have any legal and financial business knowledge? If so, can the person review the lease contract for legal and business issues, or will you need further counsel? Find out if he or she is experienced in lease negotiations. If not, you may want to ask for another agent. Is the agent familiar with the area in which you want to locate your business? Is he or she involved in scheduling property inspections? How does the agent go about finding properties? Remember to discuss his expected commission. You don't want to walk away not knowing how much the agent charges, particularly if you have a budget.

After you've chosen an agent, be sure to give him the specific information you outlined above, including details on everything from basic space requirements, size and layout, budget/cost, lease and extension options and expansion to termination, geographical area, price and term, image and quality and growth projections

If you're just starting out, you probably won't need a space planner. But, if your business has outgrown its last space and you're looking for something much larger, a space planner or interior designer might be able to help you find a space that's right for you and your business.

When you've narrowed the list down to two or three spaces, review and redraw space plans, listing all efficiencies or inefficiencies of each property. Ask yourself the following:

- Is the location near potential customers? Near competitors?
- Are there any zoning regulations or signage restrictions?

- Are there adequate support services nearby (e.g., suppliers, printers, distribution centers, etc.)?
- Is insurance affordable at this location?
- Are there any building or health codes that would affect your business?
- Does the location have the amenities that you need?

When you've made your final selection, establish terms required with your broker. Verify all terms and conditions with everyone involved in the process and don't sign anything that you haven't reviewed with your broker. Your broker should be able to help you prepare and present a counteroffer, and approve, renegotiate or select another alternative.

Determine the length of the lease and have a monetary range in mind. Most leases are three years. If you expect considerable growth, ask to add a clause in your lease that addresses this.

How much rent can you afford to pay each month? Ask your realtor for the current market rental costs in the area in which you wish to lease, or to renegotiate at your current location. Most businesses typically pay between four and five percent of their total operating costs on rent.

One good way to lower your costs is to inquire about any square footage available where a lease has fallen through. In many cases, large corporations rent entire floors and then realize that they do not need as much space or decide to rent elsewhere. In this scenario, the building usually sublets the space for less than the going rate in that building.

Ask about any other costs associated with the lease. For example, are you responsible for heating and air conditioning?

Write down everything that you would like to see in the lease. Have a clear cut idea of what you want. Do you know which things are negotiable and which things are not? Remember: the more you need, the less negotiating room you will have.

Be ready to compromise, you're probably not going to get everything you want. Know what you can easily give up.

At the very least, be familiar with the current market trends for the location, the particulars of the real estate you view and what you want. These are crucial to obtain the lease you want. And, remember, don't be afraid to walk if the terms are not acceptable to you. This is why it's important to be prepared before you begin the leasing process.

## Sharing Office Space...and Overhead Costs

Of course, you don't have to purchase or lease any office on your own. If you cannot qualify for one on your own or do not want to participate in an incubator program, there are other options, such as sharing office space. (A business incubator—see page 187—provides entrepreneurial, business development and financial experience to a business. Generally their goal is to accelerate the formation, growth and success rates of scalable, technology-based businesses that have proprietary or patentable intellectual property that is being commercialized through the sale of new products or services.) While networking, see if you can identify other entrepreneurs with complementary businesses and/or skills who might want to share office space.

If you're reluctant to take on a roommate, you may want to think again. Sharing an office can give a small business the same benefits of a larger and more diverse one and offers many advantages, particularly if your businesses are related in some manner. For example, if you've opened a food supplement store catering to the fitness-minded, you might want to think about partnering up with a physical therapist, yoga instructor, acupressurist or dietitian. Or, if you're starting a Web development business, find someone who owns a Web content writing and editing business and/or someone who specializes in Web marketing. With some creative arranging, you could share the cost to rent a space and share overhead such as the cost of an assistant, phone service and utilities.

Even if your businesses don't complement each other, sharing an office lends itself to an atmosphere of sharing, networking and supporting one another. These elements will come in extremely handy during the early stages of your business.

If you're looking at a space that is too big and the rent too high, approach a friend and ask him if he would like to share the space and related expenses. You may even bring each other work and collaborate on projects, and their family and friends may also become your regular customers and vice versa. And, if you keep separate office hours, you can even cover for one another during absences, answering questions and taking names of prospective clients.

Sharing office space with another business can help your business—but make sure you choose wisely. There are some disadvantages with sharing an office. Like having a college roommate, you have to tolerate differences of opinion and habits. If your friend Joe wants to go in on a space with you and he wants wood floors, but you want tile...you'll have some things to work out. But the problems aren't always small, so it's important to choose your roomie wisely. And, if you don't want your roomie to get up and leave in the middle of the lease because it could drain your funds, agree to give one another three months notice before leaving. This way, you'll have plenty of time to find someone else to take his place.

Things to look for when you're looking for someone to share an office:

- **Synergy**. Look for a business that complements your own (refer to the examples on the last page).
- **Respect**. Any parties involved should be able to tolerate differences and communicate. (If you do partner up with a friend, remember to keep business and personal relationships separate. If you don't, you could wind up with no roommate and a soured friendship.)
- **Compatibility**. Personal and ethical compatibility are important for the success of both businesses. Treat your partner's clients and vendors in the same manner that you would treat your own.

- **Responsibility**. Stick to all agreements and responsibilities, even the most simple and obvious ones such as a rule against interrupting or talking loud when your office mate is on the phone.

As in any leasing situation, develop a checklist of elements vital to selecting the "right" office space for both businesses. Brainstorm to determine all of the elements the location must have, inside and out, before beginning your search. This saves time and energy and also helps ensure success in finding a place that is appropriate to the needs of both businesses.

Assess what both businesses need, would like to have if possible and what you can afford. Not all your criteria may be available so prioritize your list so that the space you choose consists of the most important factors.

Take into consideration the types of space that you will need for both businesses. Private offices? Workstations? File areas? Reception areas? Conference rooms? Storage? Bathroom? Come up with a rough square footage estimate and a list detailing the kinds of space you need now and in the future.

Discuss cost. When deciding if you can afford to lease a space, determine the total amount it will cost you to move in and to maintain your lease plus the amount of money it will take to renovate the space so that it will meet your needs. Is it affordable? If not, look for another space.

Remember: Figure out who is going to pay for what, and what each person's priorities are. For instance, if you work only eight hours a day, while he works 15 hours a day, the electric bill will need to be divided up fairly, based on usage.

## Subletting

Many landlords charge a hefty security deposit so it might be more cost-effective for you to sublet space. You can sublet in one of two ways: 1) You can take over part of an existing office (one or more private offices) and share facilities (i.e., reception area, kitchen, conference room, copy room, bathroom, etc.).; or 2) You can sublet an entire space from a business that has moved to another floor or building. By doing this, you won't have to pay a security deposit and you may see a substantial decrease in your rent. If the lessor is still under contract you may be able to haggle to cut cost of rent. Remember: subletting is rarely long-term, so be prepared to find suitable replacement space, if you're forced to relocate.

When subletting you can sometimes acquire office furnishings and architectural elements from the existing tenant, including: existing network wiring, an alarm, phone system, furniture or more.

## Executive Suites

Another, less expensive option for smaller companies: the executive suite. Executive suites are essentially shared offices with services provided by a management firm. They are a great way for small companies to get off to a fast start and involve little risk. They also allow for flexibility and rapid growth.

Executive suites often feature the following:

- Private offices with utilities and janitorial services;
- Reception area/receptionist;
- Conference rooms;
- Personalized telephone answering;
- Mail handling; and
- Coffee/food services.

For an extra cost, some even offer:

- Furniture/equipment rental;
- Secretarial/word processing services;
- Computer related services and equipment;

- Administrative and accounting services;
- Postal meters/scales, UPS/Federal Express;
- Fax/Telex/Photocopy equipment; and
- High speed Internet access.

An executive suite allows you to enjoy professional reception, secretarial and administrative support personnel without having to hire, train or provide employee benefits, purchase office equipment and budget for maintenance problems.

## Business Incubators

As more entrepreneurs try to get underway in businesses of their own, more communities are setting up business incubators—affordable office and industrial space, business and management assistance that also includes shared support services such as mentoring and training seminars. They often have tenant boards that approve applications for space in the incubator. Although there is no guarantee of success in business, they can make the difference between a successful and unsuccessful start-up. According to the National Business Incubation Association's (NBIA) *1998 State of the Business Incubation Industry*, North American incubators have created nearly 19,000 companies still in business. The NBIA also says that 87 percent of graduates from these programs stayed in business at least five years.

Business incubators are an alternative to setting up shop at home or leasing office space. They typically provide the following:

- A network of relationships (i.e., other entrepreneurs and customers or suppliers);
- Financing assistance (i.e., obtaining loans or gaining access to federal and state research and development funds);
- Business and technical assistance (i.e., in-house expertise and network support);

- Shared business services (i.e., telephone answering, bookkeeping, word processing, secretarial/receptionist services, fax and copy machines, computers, business libraries, etc.); and
- Flexible space/leases at below market rates.

Business incubators are generally operated by universities, colleges and community colleges, for-profit businesses and economic development agencies, local governments or consortia of all of these organizations. They typically serve the needs of light manufacturing and service firms and those developing new products or engaged in research and development, but may also include construction-related, sales and marketing or wholesale and distribution firms.

Incubators vary in the services they offer and in the charges to their tenants, so before you choose an incubator, you should know what to look for. Use the following as a guideline for prospective incubators:

- How is the facility managed? Is it managed well? Does it have support from sponsoring organizations? If so, who are these sponsors and why do they support the incubator?
- Inquire about the policies and procedures of the incubator. If your application is accepted, how long can you remain a tenant? Is there a graduated structure as your business matures or does the incubator take royalties or ownership rights in return for reduced charges? How simple is the exit process if your business fails?
- Talk to others who have used the incubator in the past. How long did they use the incubator before moving to their own space? Talk to current tenants of the facility to find out what they think of the incubator.
- What services does the incubator provide? Does it offer on-site assistance and access to contacts and community business services? Does it provide seminar or training programs? Are there costs for these services?

- What are the charges for space and services? How do they compare to market rates in the area? What are the lease requirements? Is there room for business growth?

For a list of business incubators around the country, see Appendix G. For more information on business incubators, contact the National Business Incubation Association at: 20 E. Circle Drive, Suite 190, Athens, OH 45701; (614) 593-4331; or the Small Business Administration at: Business Initiative, Education, and Training, 409 Third Street, SW, Sixth Floor, Washington, DC 20416, (202) 205-6665.

## Arts Incubators

Across the United States, more and more arts organizations are setting up incubator space for local arts groups, too. This space is often given—rent-free—to arts organizations or individual artists who are getting on their feet but need space to operate and create their art.

Different incubators have different application processes and selection criteria. Be sure to investigate those closely before applying. But this could be a way to have a place to do your art without having to pay for space.

The following is a brief list of arts incubators around the country, including information on what you have to do to become part of their incubator programs:

- **The Arts Incubator**. This organization, located in Arlington County, Virginia, centralizes resources to get the most facilities for the most artists. Much of the work they do is to find free rehearsal and studio space. This program concentrates on what the artists and arts groups can give back to the community. As their statement of principles says:

    > Applicants' proposed means of giving back to the community play a role in decisions about which artists and groups receive assistance. For this reason, the Arts Incubator does not offer so-called "buying time" fellowships.

> Instead, artists are required to exhibit or perform within the community and to engage it through outreach activities. Individuals and groups must include public activities as part of proposed projects in order to be considered for assistance. Likewise, inclusion of community service in proposed projects is significant in evaluating whether an organization will receive Arts Incubator support.

For information about the program, contact the Gunston Arts Center, 2700 South Lang Street, Arlington, VA 22206-3106, (703) 228-6960. More information is available at: **www.arlingtonarts.org/incubator/story.html**.

- **Arts Bridge**. This organization, located in Chicago, Illinois, has provided support for program participants since 1987. Current participants include a documentary film production company, a professional a cappella group and two dance companies. For information about the program, contact Arts Bridge at 2936 North Southport, Suite 210, Chicago, Illinois 60657-4210, (773) 296-0948. More information is available at: **www.artsbridge.org/programs.htm**.
- **Cultural Arts Council of Houston/Harris County**. This organization sponsors the MODE program, which provides space in an arts incubator that includes office space and equipment, a shared conference room and mentoring support. There is a monthly fee for the service, but the Cultural Arts Council subsidizes that fee. MODE has an application process, and a panel decides who gets to join the program. Not every member of the incubator program has to be a 501(c)(3), or non-profit, organization. For information about the program, visit: **www.cachh.org/cachhfaq.html** or contact the Cultural Arts Council at 3201 Allen Parkway, Houston, TX 77019, (713) 527-9330.
- **The Arts Council of New Orleans**. This arts council operates the Entergy Arts Business Center, which offers a range of business services and mentoring support to arts organizations and individuals. There is a membership fee to receive the services offered, but that fee can

be waived for low- to moderate-income individuals. Tenant members have access to mentoring services, low-cost office space and additional benefits. Contact the ACNO at 225 Baronne Street, Suite 1712, New Orleans, LA 70112, (504) 523-1465. More information is available at: **www.louisiana-arts.com**.

- **The San Jose Arts Incubator**. For a low monthly fee, arts organizations can become tenants in the Arts Incubator Office (AIO) program. However, this program requires that participants be non-profit organizations and there is a residency requirement for the program. Contact the Office of Cultural Affairs, City of San Jose, 4 North Second Street, San Jose, CA 95113, (408) 277-5144. More information is available at: **www.sanjoseculture.org/art_inc/aiogline.html**.
- **MetroArts of the Capital Region**. This organization has an extensive arts incubator program. Contact MetroArts at 123 Forster Street, Harrisburg, PA 17102, (717) 238-1887. More information is available at **www.metroarts-pa.org**.

## Office Equipment and Supplies

Once you've decided where your office is going to be located, you need to figure out how to stock it with supplies and equipment and how to pay for overhead items like utilities.

Sure, a shiny new desk or a top-of-the-line scanner would be wonderful, but not every business owner can afford all new furniture and equipment when he or she is just starting out. However, there are plenty of ways to get good office furniture without spending a lot of money.

The equipment you require is directly linked to the home-based business you choose. Telephones, fax machines, photocopiers and computers will be the main equipment in your office. However, depending on your business, you may want a more sophisticated computer system equipped with huge memory and multimedia capabilities, a scanner, a laser printer or publishing and graphics software. But you don't need these things from day one.

Separate your needs and wants. Remember, you're trying to start a business, not run one into the ground, so initially purchase only what you need. Make a list of everything you need to start the business. Don't go overboard with high-end computer systems and the latest high-tech gizmos.

Only buy equipment that you will use in the early stages of your start-up. And, if you have to buy some equipment, look around for low-cost used models or purchase new equipment on an installment plan. But before you buy anything, look around the house for things you already own that are usable. Or check the classified ads and garage sales. Both are good, inexpensive sources for typewriters, computers, answering machines, etc. Also consider what you need on-site and what is available through an outside service. Buy only what is absolutely necessary for start-up, and wait until the business is off the ground to get the extras. In your start-up phase, it is better to keep your capital expenditures as low as possible.

If you can't communicate (via phone, fax, e-mail or voicemail) effectively, it will be hard to get anything done, let alone market yourself as a professional. So think about installing at least one extra phone line (and maybe two, if you get a lot of faxes or use the Internet frequently and you don't have a DSL line) so that you don't tie up the home line with business calls. You won't want to miss important calls just because you're trying to send an e-mail.

Go with a high-quality, noise-free cordless phone, it will allow you to roam while talking and receive calls while you're out front talking with the UPS man or in the kitchen fixing your lunch.

Having an extra phone line is a standard business deduction so don't worry about the additional cost. Besides, at the end of the day, you can switch your business line to ring only in your office, or you can turn off the ringer so as not be disturbed in the rest of your home.

If you do use the Internet a lot, you may want to pay for a service that will give you a speedier Internet connection, like DSL, an ISDN or a cable modem. Call an electrician who specializes in installing high-tech systems or your local cable company. They can help you figure out what you need and usually offer package deals.

Office supplies are always cheaper when you buy in bulk, so consider setting up a buying club among your business colleagues for the smaller things. If you know other home-based or small-business entrepreneurs, see if they would like to pool resources with you and go in on larger orders of products you all need, including copier paper, legal pads, pens and staples.

## Office Furniture

Remember: You want to make your office comfortable and cozy, but useful. You want your home office to reflect you both personally and in a professional way. So, before you run out and purchase that antique wood desk you've had your eye on, run to your nearest office supply shop for a standard issue desk and chair, think about how you intend to use the space. Take some time to look at the space and analyze what you will need. Collect all the relevant data you will need to make intelligent decisions. Carefully measure the space of your work area, particularly if you're planning on installing a wall-to-wall system because every inch will be critical. Also measure windows, including the height from the floor to the sill, doors and closets. Will you use the closet to provide filing space? Or, do you need shelving to house your samples or inventory? Does your business require a lot of paperwork? Do you need bookshelves? Or do you need to install other kinds of equipment? Will you have frequent visitors? Do you need conference space? Can you solve this with a couch and table or do you need a desk that can be turned into a conference table to seat five or more people?

Note where your electrical outlets are located. If your workspace has prominent baseboards, crown molding or heating ducts, you should measure these, too. Other things you might want to consider: How big is your monitor? Do you have a horizontal or vertical CPU? What are the sizes of your key electronic equipment such as your printer, scanner, fax machine, speakers or telephone? You will need this information to plan a workspace that truly meets your needs.

Other questions to ask yourself: Is it part of a room that has other functions, like a guest bedroom, a den or a playroom? How much of your work will you want to "hide" when you're done for the day/evening?

Even if your home office has to serve multiple functions—an office during the day, a computer/homework center at night or a guest room when someone is visiting—there are ways to make it work. For example, buy furniture that works double duty. If you plan to have guests stay in your office over the weekend, consider a futon couch that turns into a double bed or consider installing a Murphy bed that opens up into a double or queen size bed. Or look for a coffee table that can serve as a conference table or desk in the day and fold back down to service guests at night. Be careful, though—you could lose some or all of your home office deduction, if you are planning on taking it, but consult a tax advisor for more information before jumping to conclusions. The rules of tax deductibility of a home office changed again at the beginning of 2003 to become more favorable and provide more deductions for home offices, so be certain to discuss this with your accountant to see how it will impact (and hopefully save money) for you.

Carefully plan your home office layout and remove any furniture that is not serving a purpose aside from taking up valuable office space. That includes a dining room chair you may be using instead of an ergonomically correct office chair (or at least one that is adjustable and offers more support than your present chair).

Once you figure out how you're going to work in your space, you can search for furniture that's appropriate. Do you have a space problem? Is your home office space in an unusual shape? If you do have the cash, hire a carpenter to build a desk or bookshelves that solves the space issues. If you think things through ahead of time, you're more likely to end up with a working environment that works for both the business and your family.

Define your furniture requirements by what you need to work effectively. Ask yourself the following: What kind of furniture do we need? Tables for meetings? Cabinets for files? Shelves for books? Carts for fax machines or printers? Generate a list of the minimum furniture you need. Avoid buying things that you do not need. Buy furniture for its functionality, not for image. Space, for one, is a problem for most home offices. Try to maximize every square inch of your office.

Most home-based entrepreneurs make do with whatever furniture and equipment is readily available. Others make one piece of furniture serve more than one purpose. For example, the top of a lateral file cabinet can double as an extra surface for a paper cutter, fax machine, copier, etc.

The home office market has grown enough to make the furniture solutions much more interesting than they were even five years ago, so you may even want to consider leasing or renting furniture, particularly if you expect to upgrade your furniture quickly as your business expands. However, if you can pay cash, use cash. You'll want to accrue as little financing and leasing expenses as possible before you get the business on its feet. Paying cash for the furniture will also keep your balance sheet clean of liabilities and keep your credit line open for other more important purchases.

Office furniture is available at several price levels. Manufacturers such as Ethan Allen produce high ticket desks, file cabinets, bookshelves, computer workstations and office chairs. Ikea also offers some moderately-priced office furniture. If you're budget conscious, however, I suggest checking the local classifieds for any close out sale or used furniture sale.

Whether your office looks brand-new and spiffy, or old and well worn, the important thing is that it fits your purposes. Consider function over aesthetics. Your office may look like a million bucks but it may be designed to hinder rather than help you make a fortune.

## Furniture Leasing Companies

If you opt for leasing your furniture, first, check with local business furniture leasing companies. Often, they will have used furniture for sale at prices well below original retail, even though the furniture may not have been used for more than a year or two and is in great shape.

Some national furniture leasing companies to check with include:

- Aaron Rents Office Furniture (**www.aaronrentsoffice.com**);
- Brook Furniture Rental (**www.bfr.com**);
- Cort Furniture Rental (**www.cort1.com**);
- Globe Business Resources (**www.glbe.com**);
- Office Furniture Rental Alliance (**www.ofralliance.com**); and
- Swingles Furniture Rental, Inc. (**www.swingles.comindex.html**).

## Discount Office Supplies

Another great resource for gently used office furniture is thrift shops. Check your local Goodwill or Salvation Army outlet for furniture that might need just a slight overhaul—maybe a well-placed nail or new fabric for a seat cushion. These stores are also great sources of old filing cabinets.

And don't rule out using what you have around your house. Once you have your business successfully up and running, you can invest in great furniture, but at the beginning, there's nothing wrong with using a card table and a kitchen chair as your desk set-up. If you have extra cardboard boxes around, you can convert them into file boxes by using hanging file frames to keep everything neat. Don't like the look of the plain boxes? Buy some fabric or wrapping paper on sale and wrap the boxes in something more eye-catching.

So, you know what your need, but what are the most important equipment you can purchase for your office? Before getting caught up in choosing software, a computer and other gadgets, invest some time in the purchase of a good chair, a good desk, a door that closes and storage space. These are the most important equipment you can purchase for your office. Remember: It's okay to start simple, but plan for the future. You will want to grow your office as you grow your business. Use the following as a guide when hunting for these items:

- **Chair**. Don't spend a mere $25 for a chair you'll sit on for most of the day. Look for a chair that is fully adjustable to accommodate the changes in the ways you sit, type or wheel to and away from your desk. If your chair isn't comfortable, your chair will become burdensome as time goes on. Look for a chair that has good, adjustable, lower back (lumbar) support. Test drive the chair before making the actual purchase. Most good and reliable furniture stores offer a three-day trial period.
- **Desk**. The desk you use is almost as important as the chair you settle on. The desk should provide enough room for your computer keyboard to rest at or below your elbow level. You should not have to reach upwards to type and should have room to stretch your legs. Try to find a desk with as much surface area as you can afford and fit into your office.
- **Door**. A door that closes is probably the most important piece of home office equipment. A door will keep out unruly children or other family members and friends and defines your workspace.
- **Storage**. Ensure that you have adequate filing or inventory capacity. If you have an organized system, you'll be more apt to file papers or inventory right away so you can easily retrieve items that you need right away. Use sturdy shelves to store books, reference materials and supplies nearby. Store similar items together and place these items strategically. Books you refer to often should be within reach while supplies and other materials you use less frequently should be placed on higher or lower shelves.

Fortunately, as the number of home offices continues to rise, furniture manufacturers have introduced a wide array of furniture to meet various work styles. For those who like to move around as they work, there are some interesting new options.

Think about your health and safety, too. Always consider comfort and ergonomics. The wrong desk and chair can lead to ergonomic disorders such as backache, headaches, eyestrain and other irritations and inconveniences. Fatigue, loss of concentration and irritability can also be attributed to the use of the wrong furniture.

If you are going to spend a lot of time at your desk or in front of the computer, invest in a really comfortable chair and set up appropriate lighting. If money is no object, splurge on ergonomically designed furniture and office systems with everything efficiently and conveniently at your fingertips. Modular furniture makes it easy to move your office around as your needs change.

## Office Supplies on the Internet

While there are certainly plenty of discount office supply stores out there, one of the most convenient and cheapest ways to get office supplies is on the Internet. Many Web-based supply stores offer benefits like free shipping and discounts when you spend a certain dollar amount. Here is a list of sites where you can get these types of benefits—and more:

- **www.onvia.com**—Free shipping on all products.
- **www.officemax.com**—Free shipping on all orders over $50.
- **www.works.com**—Cooperative buying program that can cut up to 30 percent off your monthly supply bill. Free shipping on all products for a limited time.
- **www.benjaminofficesupply.com/NewDefault.asp**—Free nationwide delivery.
- **www.serviceshere.com**—This is a business supply auction site, so with some digging and careful bidding, you can get good deals on supplies.
- **www.contractprice.com**—Name your price for office supplies at this site.
- **www.onlineofficesupplies.com**—Free shipping.

## Free Computer Equipment

Your ability to compete as a home business comes from your ability to match the technical equipment of large corporations. If you need a computer for the graphic design business you're starting, you'll want to use the same software used by the largest publishing houses and graphic design shops. Take this into consideration before you go out and buy a computer—even if you can get a huge savings by purchasing refurbished computer equipment direct from the top manufacturers online (**www.dell.com** is one example). Fortunately most of this software is more affordable today than it was even a few years ago.

Before you buy a new computer or replace your current one, it's safe to choose your software first. If you don't, you could wind up with a computer that can't run the actual software you need; can't run it fast enough or that has a screen that is too small and won't allow you to move around in the spreadsheet program or graphics package that you need for your business.

Look over the instruction manual before buying anything. And, if you're making your purchase on E-Bay, another top source for purchasing computer equipment at a huge discount, request an owner's manual in order for the purchase to take place. (It wouldn't hurt to check out the seller's rating either. If you're doling out the money for a computer, this could prove important to the legitimacy of the transaction. Contact other people who have conducted transactions with this buyer, and ask them how it went.) The manual should include a quick start section and a tutorial, or state that such help is available online. The instruction manual should also be written in plain language and well organized and indexed. If it isn't, you'll waste a lot of time looking for what you need or trying to figure out what you're doing.

If your business requires the use of computers, you'll want to look into having at least one, if not two, back-up systems. Without one, one measly power outage could lead to the loss of a lot of data.

## Free Laser Printers

Xerox is currently trying to promote its line of networkable color printers, and that means a great benefit for small business owners. Visit **www.freecolorprinter.com** to apply for one of three printers worth $3,000 to $6,000.

If you are approved, you will receive the printer to use for free for three years and you can return it at any time. At the end of the three years, the printer is yours to keep. The program also includes a free technical support program and free black ink.

The application requires you to estimate how many pages you will print each month, and once you have received the printer, you will be required to use the printer at that level. If your printing use falls below that level in any given month, you have to pay a $75 fee. The program administrators monitor this by requiring you to print out and submit a usage report generated by the printer each month. You are also required to buy your color ink and any maintenance materials through the freecolorprinter.com Web site, but those costs are at market rates and are very reasonable.

## Free Software

The following is a list of sites where you can download software for free. You can find free Web browsers, free accounting programs, free spreadsheet programs, free word processing programs, free utilities—everything you'll need to get your business up and running.

Be wary: Only some of the software is freeware, which means you can download and use it without ever paying a fee. Some of the software is shareware, which means that if you try it out and like it, you are on your honor to pay a fee of anywhere from $5 to $50 for the software license.

Oftentimes if you opt to buy, the license will give you access to a wider range of features than you if you did not pay for and register the software. However, this system allows you to use the software for free until you decide whether you want to keep it or not, and that means you will not waste money on software you do not need or will not use. Check out the following sites for plenty of programs:

- **http://shareware.cnet.com**;
- **www.jumbo.com**;
- **www.zdnet.com/downloads**;
- **www.freewarefiles.com**;
- **www.galttech.com**;
- **www.ptf.com**;
- **http://shareware.about.com/compute/shareware/cs/businesssoftware/index.htm**;
- **www.freewarehome.com**;
- **www.freewareweb.com**;
- **www.freeware-downloads.com**;
- **http://freeware.intrastar.net**;
- **www.freewarefilez.com**;
- **www.freehound.com**;
- **www.freewarenet.com**;
- **www.executive.com/freeware/freeware.asp**;
- **www.pcwin.com/freeware**;
- **www.5star-shareware.com**; and
- **http://home.netscape.com/computing/download/index.html**.

## Additional Resources

Additional lists of free computer equipment and software are located at the following Web sites:

- **www.geocities.com/RodeoDrive/1084/computer.html**;
- **http://members.fortunecity.com/franknarf/pc.html**; and
- **http://4freestuff.4anything.com/4/0,1001,3565,00.html**.

## Utility Bills

Utility bills are going to be a big part of your overhead costs. After all, you can't run a business with no gas, electricity or lighting. But with a little time and effort, you'll find that there are cost-cutting procedures that are effective and could save you a lot of money in the long run.

Want to slash your utility bills? Review your bills quarterly and comparison shop. You may even want to ask around, other companies might be getting a better deal than you. Chart your expenses and act on your findings.

Want to keep your phone costs down? Switch your long-distance telephone service if you can find one with lower rates. Long-distance companies spend millions of dollars to retain customers and attract new ones. Sometimes you won't even have to switch. Call your phone company and ask for a better plan, or see if they offer a cash bonus or some other great deal to stay with them. If they say no and you threaten to switch providers, they may offer you that better deal. Why? They have no incentive to reduce costs unless you threaten to leave.

Browse the Internet for companies offering free services. One company, **Tel3.com**, offers a free 50-minute phone card—no strings attached. And, because it knows how confusing rates and fees can be, they offer two simple and cost-effective choices: 1) a card that gives you a free week of calling every month (during the rest of the month you pay 4.9 cents a minute on domestic calls); and 2) a free 50-minute phone card (you'll pay $1 for shipping and handling). Another company, **DialPad.com**, lets you make free phone calls from your Web browser to anywhere in the U.S. Although the sound quality isn't perfect, it's comparable to using a cell phone.

Another option: Use your e-mail instead of picking up the phone every time you want to reach out and touch someone. In some instances, this can be more efficient and cost-effective.

If you still can't find a good deal, visit **Go2Call.com**. This site allows you to quickly compare rates and features of the many providers who can place an Internet call. They can probably provide you with good deals on products and accessories like headsets and microphones, too. In the end, you'll realize tremendous savings over traditional phone rates.

You also can save money on your local phone bill in only a few minutes by asking your provider to block "900" calls that charge several dollars per minute. This is particularly important if you're running a home-based business and your kids have access to your work phone.

Comb your monthly phone bill for any items you don't want to pay for. You'd be surprised at the number of people still paying $4 or more per month to rent their phones from the phone company, even though they could buy a phone for as little as $10 from their local Wal-Mart or Radio Shack stores.

There are a number ways you can cut your electricity bills. Replace bulbs in lamps that are on for more than two hours a day with compact fluorescent bulbs. Fluorescent lights have come down in cost recently, are much more energy-efficient and last a lot longer than normal bulbs, so you won't have to change them as frequently. If you don't like the harsh white light they emit, try looking for a fluorescent bulb that gives off a very warm yellowish light (yes, they make them now). And, for those of you who are into energy conservation and the environment, the Rocky Mountain Institute reports that a fluorescent bulb will prevent the emission of 1000 pounds of carbon dioxide from electrical power plants.

If you work out of your home, try using a 20-watt desk lamp instead of 60-watt light bulbs that light an entire room. Something as small as this can save you roughly $5 on electricity for every 500 hours you spend at your desk.

Another tip: Unplug equipment or devices that you haven't used in the past month. Even if they aren't turned on, they're probably using juice just to stay warm.

If you can get away with using laptops, by all means do so. A typical computer system uses roughly $35 to $140 worth of electricity per year. By using a laptop computer, you can reduce this cost by as much as 85 percent.

If you do work from home, and you've established a reliable payment record, you might be able to get your deposit back from the power company (you probably paid it when you moved in). Usually you can get it back after you've lived in your home for one or two years. And, they'll typically pay you 6 percent yearly interest on the deposit.

Ask about any promotional packages offered by the company. You might be able to enroll in a *Time of Use* program. Some power companies offer programs that charge more for electricity during prime times and less during off hours. If you switch to this program, the power company will install a new meter. In some cases, homeowners can save as much as $500 a year with this idea. Ask about their programs for businesses, too.

Try fans, too. They can cut your air conditioning bills significantly. But when the air conditioning is on, reduce the amount of heat that enters your home. The same goes for heaters. In seldom-used rooms or storage areas, keep the door closed (if the room has a thermostat, turn it down or off). Shut down air conditioners at night.

Replace inefficient heating and air-conditioning systems. If your furnace is only operating at 65 percent efficiency and you replace it with a 90 percent-efficient furnace, you could save as much as $27 for every $100 you spend on heat. The local utility company might even be willing to help you pay for the new unit or give you a low-cost loan.

And, once you're up and running, devise a cost-control blueprint to bolster your bottom line. Examine your utility bills: water costs, gas and electricity, lighting, etc. If you see that they've risen significantly from when you started the business, and not what you expected due to increased business, your costs could be easier to cut than you think. Check for leaky pipes and inefficient lights. With a few basic repairs and adjustments, you can probably lower your water and gas bills immediately. Look for new lighting. And, don't worry about shelling out a few more bucks. It'll probably end up paying for itself in six months or less.

Remember: You don't have to hire an outside consultant such as a utility or telecommunications auditor to help control costs. This is something you can easily do yourself. Besides, the time you spend searching for a specialist could be better spent on other areas of the business.

Some utilities will give you a do-it-yourself audit kit to help you find more ways to save on electricity. In some places, such as areas where electrical capacity is in short supply (e.g., the East Coast and California, power companies may help pay to make your home (and your home-based business) more energy-efficient, perhaps even 50 percent of your costs.

If you have kids, ask them to help you with your utility-slashing crusade. Offer them a portion of the savings to help control the costs. You'll be surprised at what they'll find if there's something in it for them.

## Energy Star Small Business Financing Options

The Department of Energy's Energy Star Small Business Financing options encourage small business owners to use energy more wisely, which in turn lowers utility bills—at home or at your office—and thereby increases profits. After all, the easiest way to earn money as a business owner is not to have to spend it.

A list of helpful no- and low-cost ways to use energy more wisely is located on the Internet at **www.eren.doe.gov/energytips/handidea.html**, and a list of six areas where you can make major savings is located at **www.eren.doe.gov/energytips/six.html**. Call your local electric or gas company. Oftenimes, they'll send a representative in for free to see where you can make no- or low-cost changes that will reduce your energy bills.

Don't be afraid to call and ask for this service! It costs you nothing and can save you hundreds of dollars each year if you implement changes.

There is also information on **www.eren.doe.gov/energytips/financing.html** about how to get low- or no-cost financing on energy equipment through your local utility company. From Guaranteed Savings to leasing options, there is a range of ideas for lowering your energy bills.

## Conclusion

Moving into office space and equipping that space with tools are often the definitive steps in a new business's growth. Unfortunately, some small companies squander months or years of hard-won profit in a few bad decisions about where to move or what to buy. The critical challenge is to keep the lean, efficient perspective of a broke start-up, even as your business grows. This chapter has offered some tactics for keeping a lean focus.

# CHAPTER 7: BUILDING A WEB SITE FOR YOUR BUSINESS

## Why You Need a Web Site

As growing numbers of consumers and potential clients or customers access the Internet through their home or work computers, the opportunity for expansion via the Web continues to grow. In fact, Internet analysis firm Forrester Research, Inc., predicted that by 2001 more than half of all U.S. households will be online, more than one-third will have made purchases online and one in 10 will have banked or invested online.

But what does a Web site mean for you and your business? Web presence is important to any business—even if most of your business is not conducted online. A Web site can also add value to your product or service. It can give you critical marketing exposure and will put you in touch with customers who might not know about you otherwise. And, even though your product/service can be acquired through other than online channels, the Internet provides your customers with a more attractive (faster, more convenient, easier) way to shop and buy.

But how can you tap into this market without spending thousands of dollars? It's possible, though you have to move carefully. In this chapter, I'll show you how.

There are great opportunities available for small businesses on the Internet. Web sites are a great place to advertise basic information about your company, including the services you offer, samples of your work and contact information. But even more appealing is the ability to serve current customers more efficiently and effectively.

If you don't offer a retail product, you may think that you don't want to join in the e-commerce revolution—perhaps you don't want to be the next Amazon.com or even the next small online retailer. But even if you don't sell any products online, your Web site can bring in customers from all over the region in which you live and maybe around the world, depending on your area of expertise and what services you offer. You owe it to yourself to see if those customers are out there.

## Getting Started

Ask yourself: What should your Web site do? What are your objectives for the site? Plan your Web site accordingly. In other words, write a "business plan" for your Web site that adheres to these objectives. Do you want to drive traffic from the Web site to the physical store? Do you want to expand the community of your customers? Do you want to attract 100,000 visitors in the first year your site is up? Bring in 50,000 hits a month by the year's end? Sell an average of $150,000 worth of merchandise or services per month by year's end?

Also ask yourself: If you're selling kitchen gadgets, do you want to offer a content/commerce site that provides content to the culinary community and then sell your merchandise online? Do you want to provide culinary products that aren't for sale anywhere else online, at bargain prices? Do you want to partner up with other culinary sites such as a recipe site or cooking magazine's site? Can you partner with Sur La Table or Crate & Barrel to give your online store a physical presence in these cooking stores? Do you plan to sell targeted advertising to culinary magazines, kitchen and culinary retail sites and restaurants or culinary schools?

These are important things to know before building a Web site. A Web site can offer many advantages. It can extend the reach of the store to others outside your area and add to the revenue base. But, if you're going to set up a site, give it attention. A site offering basic functions is better than none, but a good site will produce revenue and enhance the image of the business.

Before you begin to build a Web site, you should have a clear understanding of the ways a Web presence might help your business. This will help you evaluate your need for a Web site. Ask yourself the following:

- Is your competition online?
- Are your customers Web-savvy?
- Would your business benefit from communicating with customers on a 24-hour basis?
- Would your marketing efforts be more successful if you had a Web site?
- Would the business benefit from direct sales and distribution?
- Would your product or service benefit from online promotion and/or sales?

If you've answered *yes* to one or more of the above questions, a Web site would probably be beneficial to your business. You don't need to be a large corporation to have a Web site. In fact, the smaller you are, the more important it is to have a Web presence. Why? For one, customers can view your products, price list, promotional information, etc., anytime they want. And, you can make immediate changes and additions to your product, pricing and sales information. Some customers know exactly what they want and when they need it and instead of requiring them to visit your business or wait until you show up to take their order, they could simply visit your Web site to reorder a product or contact you.

## Registering a Domain Name

The first step in setting up a Web site is registering a domain name. A **domain name** is the unique name that identifies a Web site. Most domain names are meaningful, easy to remember and identify a person, business, service or product. Every Web site you've ever been to and every e-mail you've ever sent or received has a domain name in its address that is part of the Domain Name System (DNS), a global network of servers that helps users find their way around the Internet by translating Uniform Resource Locators (URLs) to numerical Internet Protocol (IP) addresses, which computers on the Internet use to communicate with each other. Without DNS, we'd have to memorize a complicated string of numbers instead of URLs or e-mail addresses.

There are currently almost 17,800,000 Web domain names registered, according to DomainStats.com. Of those, nearly 9,500,000 have a dot.com domain name.

Domain names aren't free. But, if you are a serious about your business, a domain name is the best investment you can make. Even if you don't plan to set up your Web site right away. Domain name registrars set their own registration and renewal fees, but it doesn't cost much to register a domain name (typically between $60 and $80) and **registration is good for two years**. Some registrars even offer deals if you register a name for three or more years. And others bundle domain name registrations with other services, such as e-mail forwarding, Web site pointing and URL redirection and Under Construction pages. If you have a professional design your site, they'll probably have domain registration companies to recommend, but you can also search for registrars online. Up until 1999, the domain name registration business was a government-approved monopoly. But now that the market is nonexclusive more there are more than 100 companies offering to register domain names. My favorite domain name registrar is **www.registerfly.com**, but there are many others out there, so you might want to shop around. Amazon.com just added domain names to the long list of services it offers to consumers.

When registering a domain name there are a few guidelines to follow, including:

- You can use the characters *a* to *z* in upper or lower case and *0* to *9* in any combination;
- You can also use hyphens as long as they're not at the start or end of your domain name;

- You can register an all-numeric or all-character domain name; and
- You cannot exceed 63 characters, excluding the characters used in the domain name extension (also known as the Top Level Domain), such as .com, .biz, .info, .net, .org.

The biggest challenge in registering your domain name will be selecting an available domain name that you want. When choosing a domain name, think about how the name can affect your business. Choose a name that will represent your business and stay on the minds of customers.

Many businesses choose a dot.com name that is the logical extension of their established brand, for example, BananaRepublic.com or CrateandBarrel.com. Others choose domain names that tell visitors something about what they do. TruckerLawyers.com, for example, is a site aimed at helping truckers who have suffered a work-related injury or sickness find a qualified lawyer to handle their work injury or illness claims. And, WorldKnit.com is a site about knitting—offering fiber artists knitting, weaving, spinning and felting supplies. It also offers interviews, articles and a glossary on knitting. If none of these strategies floats your boat, another strategy is to select a domain name made up of totally unique terms or use an existing word in a new context. Amazon.com, the online retailer of books, CDs, videos, DVDs, toys and games, electronics, kitchenware, computers and more, is a prime example of the branded word approach.

What about domain extensions? Well, there are generic domain name extensions (.com, .org and the like), country-level domains (.co.uk, .fr, .jp etc.) and alternative domains (.biz, .info, .name etc.) , just to name a few. Traditionally, .net has been used by Internet companies, .org has been used by noncommercial organizations. The domain extension .biz is reserved for businesses. For more information on .biz restrictions, contact your registrar or visit **www.nic.biz**, the .biz registry operator. But you'll probably want to go with one of the most common extensions, a .com, .info, .net or .org. Generally anyone can use these domains because they have the least amount of restrictions. You won't be able to use a .gov or a .uk because these are restricted to government agencies and to the United Kingdom.

If you're thinking of registering the ".org" of a well-known .com domain name or vice-versa in an attempt to get some free traffic, you may want to think again. In some cases, this will land you in court. Stay away from a ".com" that is trademarked. But, if you feel you must use this tactic, shoot for a generic ".com" domain name or a dictionary word that is not trademarked. The same goes for registering a misspelling of a popular site's domain name (i.e., ammazon.com, ebayy.com). "Typosquatting" has landed many people in court. Most of the cases were decided in favor of the site owner, but it's probably smart to steer clear of these anyway to avoid any legal fees.

The best rule to follow: Choose a domain name associated with your industry or line of business. A domain that uses your company's name is easy to remember.

After you've come up with several domain names that you would consider registering (and, I say several because one won't do—chances are, someone out there has already snatched up your top choice), your next step is choosing a **registrar** to register the domain name. If you already have a Web host, you can go to a registrar and apply for a domain name. If you do not have a Web host, look for a registrar that allows you to park your domain name at a temporary Web site. This way, for a reasonable fee, you can quickly secure your domain name before someone else snatches the name you want.

My favorite Web host is also **www.registerfly.com**, but just as domain name registrars, there are hundreds of others. To compare them, browse the Internet using a search engine such as **www.dogpile.com** or **www.Google.com.** Enter "Web hosts" into the search engine and hit enter.

While each registrar may have slightly different requirements in terms of the information you will need to provide when registering a domain name, a registrar generally requires you to provide various contact and technical information, including the following:

1) Registrant—the company or individual to whom the domain name actually belongs.

2) Administrative Contact—a person authorized to make changes to the domain name, such as altering the address associated that domain name. In most cases, this will be you.
3) Technical Contact—a person authorized to make certain changes to the domain name, such as altering the DNS servers associated with that domain name. This also will be you.
4) Billing Contact—the person to whom all bills and other correspondence will be sent. Again, this will be you.
5) DNS Server Settings. This is where you specify the primary and secondary DNS server settings you would like to associate with the domain name.

Some registrars may streamline this process, not requiring billing contact information, for example, but most generally require the information listed above. A word of caution: If a registrar does not ask for registrant information, reread their terms and conditions carefully...some unscrupulous registrars take money from people and set them up as the admin, technical and billing contacts, but put *themselves* as the registrant (which effectively means that the registrar owns—and has exclusive control of—the domain name).

You will also be asked for payment information and additional information, such as a user name and password (to log in and make changes to your domain name information). The registrar will then submit the information you provided to a central directory known as the registry, which provides Internet users with the information necessary to send you e-mail or to find your Web site.

While you may be tempted to use a false name or address when registering a domain name (to keep your personal information out of the public domain) there are several disadvantages to this approach. First, it's usually against the registration rules of most registrars. There have been several cases in which domain owners have been forced to provide accurate contact information at short notice or risk losing their domains. Another reason: You may not find out about important changes that could affect your domain name (e.g., if you don't provide an accurate mailing address, you may miss the deadline to renew your domain name registration). In addition, if you don't provide accurate information, you cannot be contacted for legal challenges. While this may seem like a positive, you could lose your domain name by default for failure to respond to a complaint.

If privacy is your main concern, register the domain name in the name of the domain name (this isn't exactly kosher, but most registrars seem to let this one slide) instead of your own (e.g., if you register "blueberry-pancake.com" then enter "blueberry-pancake.com" as the name of the domain owner). Or, consider renting a mailbox with a company that offers such services so that the address is not traceable back to you.

File all information related to your domain name in a safe place; an e-mailed or mailed confirmation is easy to lose. If this happens, you'll have no record of the domain name you registered or the registrar you used when you want to make changes later.

If your registrar offers any other safeguards, take advantage of those. Every registrar offers different services, so it's smart to read up on all the safeguards your registrar offers. Ask your registrar if they offer a service that allows you to lock a domain name registration so that any change requests not authorized are automatically refused.

Carefully read all e-mail messages relating to your domain name. Some registrars automatically authorize actions such as a domain name transfer if an e-mail sent to confirm the transaction is not acted upon within a specified number of hours or days. So, if you're going to the Bahamas for three weeks, don't blab about it to your favorite discussion group...no matter how tempting. Or, be prepared to have your laptop in tote because any determined domain name hijacker could be on the prowl to take control of your domain names if you fail to respond to your registrar's e-mail requesting confirmation to deny transfers or other modifications to your domain name. If you do suspect any suspicious activity, report it immediately. If you receive an odd e-mail relating to changes on your domain name, or you notice a change on your domain's record (an unfamiliar registered e-mail address) that was not authorized, alert your registrar. Tell them what happened and forward any documentation you have relating to the matter.

Other tips when it comes to domain names:

- **Use a secure e-mail address when registering a domain name.** Don't use a free e-mail account, such as Yahoo! or Hotmail. Some free e-mail services suspend user accounts after 30 days of inactivity. This means that someone else could wind up with the e-mail address you used to own. Another reason not to use these accounts: these services also tend to shut down without notice when they run out of funds. You would be unable to block changes requested on your domain that you didn't authorize if the company shuts down. Don't use your work e-mail either. You may think that your job is secure, but in this day and age, you might not be working at the same company when you domain name comes up for renewal.
- **Make use of domain name monitoring services** that provide you with information about any potential hijacking attempt on your domain names. Some of these companies will monitor up to 10 domain names, alerting you to any attempt to hijack or otherwise change your domain name information.
- **Stay on top of renewals.** Most people lose control of their domain names through neglect or carelessness, rather than through a third party. Write your domain name's expiration date on your calendar and highlight it with a big red pen. Renew your domain name several days or weeks in advance. Payments sent at the last minute may be delayed, causing your domain name to be released to someone else. Remember: The domain name that you register is only yours if you keep paying the renewal fee.
- **Carefully read everything during the registration process** and proofread the domain name you want to register. If you make an error and register the wrong name or change your mind about registering it after paying, you can't cancel your registration. Under the current Domain Name Service Agreement that all accredited registrars oper-

ate under, all sales are final. If you want to register a different domain name, you have to pay again.

## How Can I Build My Own Web Site?

A critical question to give thought to is who will develop your site—will you do it yourself, have it done by a partner or employee or hire someone else to do it?

Many people think that the best way to build to a Web site is to go out and hire a professional. If you can pay for a professional, it might be easier to hire someone to set up your Web site than trying to build your own site.

Web sites can be set up fairly simply and cost-efficiently. But they are also complex to design, time-consuming and expensive. When deciding who will develop your site, estimate your budget for the Web site. You can do this by evaluating your site purpose and operation needs. If, for example, you want a basic site with low volume buying and selling functions, little content and navigation functions, you can purchase or rent Commercial Server Provider space for about $1,000 to $5,000. But, if you want a site that is capable of a greater volume of transactions, such as catalogs and databases that are integrated to other systems in your business, you would need your own server that can cost anywhere from $5,000 to $20,000. High volume selling sites with complex design, navigation and integration requirements, intranet/extranet functionality and high levels of security often run between $20,000 to $100,000, so you might want to steer clear of those—especially if you're just starting out.

After evaluating your needs—and your budget—you'll have a better understanding of the kind of site you can afford to set up and whether you will need to hire a professional to do so.

Many people who try to build their own sites soon realize that learning site design and construction is more difficult than they anticipated. But if you're low on funds and are familiar with Hypertext Markup Language or HTML (the computer language used to create documents that can be read by a Web browser), there's no reason why you can't set up your own site.

The key point here being: Learn HTML first. Sure there are some very good programs for novices out there; but sooner or later you'll want to learn HTML so if your program errs, you'll know what is wrong or why it is doing, or not doing, what you want it to do. If you know HTML and how it works, you'll be able to look at the source code for your page and understand what you are looking at. You don't need to be an expert at HTML, you just need to have a working knowledge of the language so you can find your way around a page full of source code.

Don't get discouraged if things don't go as you expect. Find yourself a few good HTML references, such as *HTML 4 Dummies* by IDG Books, *HTML: A Beginner's Guide* by Wendy Willard, *Html 4 How-To: The Definitive Html 4 Problem-Solver* by John Zakour, et al. or *How to Do Everything with HTML* by James H. Pence.

Remember: Building your own site not only helps to control costs, it also ensures that you know and understand the features on your site. And, this means that you'll know more about what you are doing—or not doing—to attract customers.

## Prefabricated Site Templates

If you've thought about building a Web site for your business and don't have the time to invest in learning HTML, you've probably thought about purchasing a software product that promises to get your site up and running quickly. In most cases, these products offer prefabricated Web page layouts (also called templates or wizards) that can be tailored for your particular business.

There are many advantages to using these templates. First, they're relatively easy to use. The most obvious benefit is that even the technological novice can launch a site quickly using these tools. In addition, using template-driven tools to build a site yourself is usually less expensive than hiring someone to do it for you. A disadvantage: pre-fabricated Web sites limit the level of customization you can achieve. This means other businesses using the same product may wind up with the same layouts and color schemes and a site similar to your own.

If you decide to use templates, keep the following in mind: Carefully examine and review your layout choices. Look for a design and layout that matches your business. An example: If you run a party planning business, go for something fun and colorful. If, on the other hand, you're a diversity trainer for large corporations, such a site might not work for you. Many template packages offer a selection of backgrounds, font treatments and colors to choose from, so review your choices before making a decision.

You may even want to research competitors' Web sites to see what benefits and features they offer before choosing a particular template. If there are any sites that contain benefits and features that you would like to incorporate and apply to your site, use them as a guideline when building your site. If you plan to offer a shopping cart, visit the sites of successful retailers like (amazon.com, rei.com and walmart.com) and take notes on the content, design aspects of the site and the functions they include.

A word of caution: Don't borrow someone else's stuff and use it in your Web site. Ask for permission from the owner first. Using someone else's stuff without permission is against the law. This is why it's better to learn HTML. This way, you learn how to create graphics, javascripts, etc., without having to steal them from someone else. If you do see something that you want to use, ask. Designers or programmers may want to see their work being used by others on the Internet so they may be willing to give or sell you the right to use their work.

Once you find a template that you like, ask the template manufacturer for a list of sites that use the same template. If they want to sell their product, they should be able to provide you with a list of customers. Evaluate how the sites perform. Do they load quickly? How is the navigation structure set up? Will it work for your business?

Ensure that the template you select supports the features you would like for your Web site. If you want to sell your products or services via shopping cart, include animation or allow visitors to download programs or audio files, look for a template that supports these features. Ask yourself the following: Do you want to create a search function within your site? Do you want to link to other sites? Do you want to include an index for the entire site? Do you want to include printable order forms in PDF-format? or video graphics and complex auction or transaction/payment pages?

Look into the steps involved in updating your site, too. What's the process for modifying the content of your Web site? Can you convert word processing documents into Web format? Will it be necessary to hire someone to do this? The answers to these questions may affect your decision.

Check for extras. Most packages offer additional features such as buttons and arrows or automatically create e-mail order forms. These extras may be enough for you to choose one package over another.

## Hiring a Professional

If you decide to outsource your Web site production, ask other small e-biz owners for referrals for designers and programmers they have hired. Look at their Web sites, too. This should give you a clear idea of the quality of the Web developer's work.

Surf the Internet for professional Web site developers and e-commerce business solution providers. You can hire a small firm that will work with you to design exactly what you want or choose a big software company that provides standardized design packages. It's always better to have a few options to choose from, so try to find three to five companies that you would consider hiring to design your Web site.

A number of companies specialize in developing Web sites for a particular industry. These companies may be worth the extra cost because they provide Web design expertise and are familiar with your industry, which will enable them to create a successful Web site for your company and even assist with an Internet advertising campaign.

There are several advantages to using external resources, including: you can get your site up and running sooner; the designer can work with you to incorporate more complex features, such as easy checkout and pay pages; and the designer has a lot more experience with planning and implementing a site.

But even with these positives, there are negatives, too: increased costs for implementing a Web site; the designer may not be knowledgable about your products or services; and the designer may not be familiar about the promotional techniques used by your business.

## Building a Site Map

Once you've decided whether you are building the site on your own or hiring someone to help you with your site development, it's best to design a site map and navigation scheme that best fits your e-business objectives. The navigation scheme should be easy-to-follow and intuitive. Think of it as a step-by-step outline of how your site will be organized. It outlines what visitors see on your home page and when moving through the various pages of your site.

Keep in mind: You want a fast-loading, extremely user-friendly interface to encourage users to browse and click through as many pages on your site as possible. This is key to your Web strategy. No one will want to visit or use your site if it's too difficult to navigate. Make it simple and appealing to visitors and encourage online purchases with features such as pop-up coupon offers or rebates.

Arrange the information in such a way as to make sure that each user can browse through products and find relevant product content easily.

What are the goals of your Web site? Obviously, you'll want to provide customers with as much information about your product or services as you can. (Be careful, a text-heavy Web site is a turnoff to many visitors.) But what other things do you want to include on your Web site? Do you want a page devoted to a question and answer section? Ask the cook section? Buyers guide? Links to information on product or recipe reviews? Amazing cooking tips, etc.?

Think about your customers. Who are they and how will they benefit from your Web site? Professional youngsters because of their relatively high Internet use? People outside your area with Internet access that have come in contact with the physical store or learned of it by word-of-mouth or advertising? Online shoppers using search engines or browsing multiple sites for the best price or access to discontinued or hard-to-find products? Internet learners or newbies who will increase their use of the Internet over time? How will your Web site need to be adjusted for your consumers? If you're offering online tutoring for elementary schoolers, will your site use graphics and fonts that are appealing to youngsters?

What site benefits and features do you plan to incorporate into your Web site? List three to five content features and benefits that you want to incorporate into the site. In addition to providing content that is related to your products, think about providing other content, such as weather reports, daily news bulletins, etc. These added benefits create repeat visitors because they can access everything they need from your site. The following is a list of content ideas you might want to incorporate into your Web site:

- An extensive and thorough question and answer section that will encourage customers to return to the site whenever they have an inquiry.
- A section that provides additional advice for items not covered by the question and answer section.
- Web-based promotions, such as pop-up coupons, that encourage visitors to purchase your products or services through the Web site.
- Contact information, including company name, its mission/vision, products and service descriptions, an interactive product demo, benefits, prices, customer case studies and testimonials, etc.
- A quick assessment of the contents of your site (i.e., an index of table of contents).
- Articles, forms, lists or glossaries.
- A letter from the president of the company.

- Privacy statement, terms and conditions.
- Help section.
- Search function.
- Jobs/Employment section.
- Press releases.
- Investor relations.
- Shopping cart.
- Library of training and education materials.
- Community bulletin board.
- Meet Our Friends page that hyperlinks to the Web sites of other clients, vendors and friends.

After you determine what you want to incorporate into the site, specify exactly what content is required and whether that content already exists or must be created. Creating content for your site often demands more than just converting print material into HTML. If, however, you already own the content, your must convert this content to online content. Translation: make it short and sweet. Try to keep any text on the main page of the site short by using few words, active verbs, bullet points and short phrases. Often, visitors scan information for the main points that they're looking for, so avoid long paragraphs. If you still feel you need to make lengthy text available to visitors, offer the expanded version of the text in print or via a downloadable PDF format. Whatever you do, don't clutter your home page with the text.

If you don't own the content you need and it's necessary, you can acquire the content through an exchange, lease or outright purchase. To cut down on costs, offer to post it on your site in exchange for a button link or advertising for the content owner.

The next step is to identify the site functions (e.g., buttons, click-throughs, databases, security, shopping carts, input boxes, etc.) you need or plan to use on the site.

Ask yourself, "What do I want this site to do for my business?" Then ask, "How can I accomplish this goal through specific site features?" An example: If you're a retail site, you will want your Web site to be a selling tool, so your goals for the site might include:

- to create interest in the product or service by showing how it fulfills a need for the customer.
- to provide information needed to make a purchase decision, including product or service features and benefits, customer testimonials, pricing and shipping information, frequently asked questions, etc.
- to make it easy to buy and obtain the product or service, including security and step-by-step purchase and delivery instructions.
- to provide the information needed for customers to contact you after the sale.

The color schemes, graphics, type style and overall appearance of your site are as important as the language content of your site. You want the site to appeal to your target audience—not bore them. A site intended for teens would not be appealing to a more mature audience of senior citizens, so plan accordingly. Again, who is your target audience? Think about the site features that you feel cater to that audience.

List the top three to five design features and benefits that you want to incorporate into the site. The following is a list of features you might want to incorporate into your site:

- Form fields, such as buttons, drop-down menus, input boxes, check boxes, on-off "radio" buttons, etc.
- Navigation tools, such as hypertext links, clickable buttons, icons and image maps.
- Online purchasing or shopping cart functions, including secure server technology, digital authentication, encryption, merchant software and electronic software that makes it possible to securely transmit data.
- Software that allows data entered by visitors to go to your server and be processed by the applications designed into your site (i.e., Cold Fusion, Visual Studio, etc.).

- Databases that store information for retrieval, like an online registration feature (i.e., MSAccess, FoxPro, Oracle, Sybase).
- Cookies (tiny files generated by a Web server and placed by a Web site operator or Internet vendor onto a visitor's computer ready for future access) and other Web site-tracking devices used to collect personal information gathered about Internet users are a preferred mechanism for online advertising. The cookies allow companies to monitor user activities on their Web sites and provide monthly analyses of Web site traffic, including in which country and region users are located, the length of time spent at the sites and the various pages they visited. Some cookies also enable companies to ascertain the URL of the Web site's users visited just prior to visiting the pharmaceutical sites, as well as the "query string" of any search users conducted to get to the site. Cookies can also capture personally identifiable information, such as names, addresses, telephone numbers, dates of birth, sex, insurance status, medical conditions and occupations, despite its agreement with other companies not to collect such information. (In recent years, privacy complaints related to cookies have raised legal issues as Internet use increases. But, if a company doesn't intentionally access a computer without authorization to obtain information via an interstate or foreign communication, the company's hasn't done any harm. In most cases, users can prevent a company from obtaining this information by requesting an opt-out cookie, or by configuring their browsers so as to prevent any cookies from being placed on their computers. However, companies often hide this information in the fine print of a privacy policy posted on the site. Businesses using cookies may have to follow stricter guidelines regarding the collection of personal data from online users in the future. In fact, a federal bill is currently pending that, if enacted, would ban the collection of personal information considered "sensitive" absent prior consent from users. Your best bet is to offer consumers more information about their privacy policies and an opt-out from the profiling.)

Identify at least three to five functionality features that you would like to use on your site. Most of these functions can be leased from vendors set up to handle e-commerce storefronts (what the visitor sees) and backroom operations (what makes the site function). For larger volumes or unique business situations, you may wish to develop the functions you need from a combination of available resources.

Create a prototype site. Make a simple drawing of a sample home page that illustrates the key features of your site, including any elements mentioned above. Review the proposed page. Does it fit with your overall goals?

Visualize customers moving through your site. List what they will see on each page of the site, including: views, features and functions. Complete a list for each scenario a customer may encounter on your site. Is the site easy to navigate? The site navigation scheme makes it possible for a visitor to navigate through your site. A home page typically presents a border or frame of buttons or icons that take the visitor directly to various areas of your site. The navigation scheme should stay consistent throughout the site. Your site navigation helps visitors know what's the most important parts of your site and allows them to get there with ease, no matter where they are in the site.

Define clear starting and ending points. Set a timeline. Obviously the project will end when your site is up and running online and functioning according to your plans. But have a specific date in mind and commit to make something happen by that date.

Keep in mind that you will also have to plan for any upgrades or enhancements to your site in the future. In fact, a site maintenance plan is as important as your original Web site plan. You'll want to keep site content up-to-date, interesting and in Web-suitable style. A maintenance plan anticipates the need to modify and update content and improve site performance as necessary. If you track and monitor how visitors use your site, a site maintenance plan will also allow you to drop sections that visitors don't find appealing in favor of new content or features. This plan helps you make editorial and technical changes to the site, such as making sure that all links are active and accurately described. Make sure that new content fits with the goals of the business and maintains the design standards established when the site was created. This plan can also address how to respond to emergencies, such as a server problem, etc.

Keep the following in mind when creating a Web site: First impressions count. Your home page is the thing visitors will see, so make it count. Devote the time necessary to "get it right." A poorly designed Web site won't draw customers, but an engaging, attractive site encourages visitors to explore your site and what you have to offer.

If your Web site is fast and easy to navigate, visitors will more than likely return to your site. Visitors prefer clean and simple sites that get them to what they want quickly. Elaborate graphics that take eons to load or pages of text are a surefire way to lose or turn away visitors.

Remember: Don't forget to include a quick assessment of the contents of your site. This will serve as a table of contents or index for your site, so make sure that it's thorough and allows a visitor to effortlessly link to other areas of the site. There's nothing worse than landing on a Web site that has no index or search function. Keep in mind: Web sites with built-in barriers, such as membership sign-in requirements, also lose visitors who don't want to sign up or fear being spammed in the future.

## Free Tools To Build a Web Site

Still stumped? Don't worry, if you don't know how to build a Web site, there are several tools out there to help you get started. The following is list of sites that can help you go from Web novice to Web master in no time. If you've never put a Web page together before, the templates on these pages will get you up and running in no time.

- **http://geocities.yahoo.com/home**—This site uses the step-by-step Yahoo! PageWizards to get your site up and running quickly. When you're ready for more advanced HTML work, the FileManager option allows you to code your page by hand.
- **www.ivillage.com**—Although you have to sign up for membership to build a Web site on this site, the template program makes it simple to set up a Web site on this site.

- **www.megspace.com/webmasters.html**—This site offers Web site templates to help out newbie Web site builders.
- **www.00server.com**—This site offers a Quick Site Builder that gets you up and running quickly.
- **www.aboveworld.com**—Above World's site offers a template-based Web page builder.
- **www.angelfire.com**—Angelfire is one of the big free Web page sites. They also offer a free template-based Web site builder.
- **www.homestead.com**—Homestead is another of the large free Web page sites. Its templates are geared toward sites that will use sophisticated e-commerce and promotion tools.
- **www.50megs.com**—50Megs.com offers the largest amount of free Web space of any of the free providers. There is also a template system available to build your site.
- **www.BizLand.com/b1b034b8ea903420/index.html**—BizLand is focused on providing free small business Web sites. The site includes Web-building tools and also provides site statistics at no charge that tell us how many people have viewed your Web site or specific Web pages on your site, which is something that many of the other free Web providers do not do.

Also, check with the free Internet access providers listed on page 234-235 of this chapter. Many offer free Web space and templates in case you need help putting your Web site together.

The following is a list of sites you can turn to for information about more advanced HTML techniques. This is just a sampling—if you know people who design Web sites, ask them to recommend sites and books you can use to learn more advanced techniques.

- **www.boutell.com/faq**—FAQs about browsers, CGI-bin programming and source code.
- **www.ncsa.uiuc.edu/General/Internet/WWW/HTMLPrimer.html**—A beginner's guide to HTML.

- **http://werbach.com/barebones**—The Bare Bones Guide to HTML includes all the names of HTML tags in existence.
- **www.davesite.com/webstation/html**—This site offers an interactive guide to HTML so you can learn how to program your page by actually working with code.
- **http://members.aol.com/teachemath/class.htm**—This site teaches Web page developers at the beginner and intermediate levels how to program in HTML.
- **www.doghause.com/top15.html**—Looking for some ideas on how to design a really good Web page? This site lists the top 15 design errors Web designers make.
- **http://hotwired.lycos.com/webmonkey/teachingtool**—This teaching tool provides information on HTML basics.
- **http://snowwhite.it.brighton.ac.uk/~mas/mas/courses/html/html.html**—This site offers information on more advanced techniques, including frames and CBI-bin scripts.

## Free-Commerce: The Virtual Incubator

Similar to a standard business incubator (page 187) in all respects, a virtual incubator also provides entrepreneurial, business development and financial experience to a businesses, but it is all done via the Web. As we mentioned earlier, all businesses and entrepreneurs require support, advice, mentoring and information, particularly in the start-up phase of the business. Virtual incubators provide support services, such as online mentoring, access to resource information and resource providers, etc., to participants online anywhere. The following are just a few; browse the Internet for other options.

- **The Public Webmarket**, which is located at **www.civicnet.org/webmarket**, offers Web space for small, rural retail businesses to offer their wares on the Internet. This virtual incubator was started as a small experiment in 1995. You can reach the Center for Civic Networking, which sponsors the Webmarket, at 650 Mullis St. #203, Friday Harbor, WA 98250, (360) 378-1925, or through its Web contact form at **www.civicnet.org/ccnfeedback.htm**.

- **The Blue Ridge Web Market**, which is located at **www.brwm.org/main.html**, is modeled after The Public Webmarket. Retailers and business owners who operate their businesses in the Western North Carolina counties of Madison, Buncombe, Henderson and Transylvania are eligible to post their items in booths as part of the Web Market. The site is hosted and maintained by the Mountain Area Information Network, which can be reached at (828) 255-0182.
- **Vegenet**, located at **www.vegenet.com**, is a Web marketplace for small agricultural producers in Hawaii. For information on becoming a seller, call (808) 929-8322. This service was also modeled after the Public Webmarket.

## Internet Store Hosting

Now that we've delved into the confusing world of the Internet, and simplified that concept a bit, it's time to talk about Internet Store Hosting.

An Internet "host" is a company that leases server hard drive space and provides you with everything you need to set up an online storefront. When you choose a host, the computer files that make up your Internet Store reside on their server.

Remember, as with any other type of service, different hosts provide different services, and it's important to know what each one offers before making a decision.

There are two methods you can use to start an Internet Store: 1) Internet Malls; and 2) ECommerce Hosting Providers.

If you use an Internet Mall, your store appears among others. It's in a mall that offers nearly everything an Internet consumer could ever want to purchase. If you're just starting a business and trying to stick to a budget, an Internet mall is a smart choice; by using an Internet Mall Provider (typically costs about $50 or less a month), your store will be right in the middle of things...and may get more traffic this way, which, in turn, means more money.

Internet malls are also great for people who have virtually no Web site experience. You can create a store for an Internet mall fairly easily if you use an Internet Browser, such as Microsoft's Internet Explorer. In most cases, you can get this for free with the purchase of a home computer. In addition, most mall providers have templates for easy-to-build storefronts. The browser allows you to choose the color and style of the text and background that will appear on your store's site, add links to other pages within the store and create product pages with images and descriptions of your products.

An Internet mall also allows you to access and edit your store from any location with an Internet connection. But, the best part of opening a store in an Internet mall: Your store is up and running in front of millions of shoppers within 24 hours. Internet malls already have shoppers visiting that mall, who will see your store the second you open it. This means you can bring in customers right away, while you continue to work on or promote your site.

One disadvantage to an Internet mall: most stores in the mall look the same. Even though you can change the background and text and arrange the pages how you want them, if you're using the same template, they all end up looking similar.

One good example: The Yahoo! Store (**http://store.yahoo.com**), is probably the largest Internet Mall Provider out there. Yahoo! even offers a free "30 day test drive" store. They do ask for a credit card, but if you e-mail them after about 25 days and let them know that you don't want the store, they won't charge your card. This is great way to get used to Internet store hosting and test-drive your store ideas without having to take the full leap. If you do decide that the store is just the way you want it, you can open the store with confidence. The Yahoo! store offers 50 item stores for $50/month, and you can add as many extra products as you like for just 10 cents per product per month.

If you're more of a "do-it-yourselfer" and want more control over the way your site looks and functions, you might want to look into using an **ECommerce Hosting Provider** instead of an Internet Mall. With an ECommerce Hosting Provider, you can create a unique store without being limited by templates. However, ECommerce Hosting won't provide you with built-in traffic and it's a bit pricier. But, if you can afford it, this service will set you up with all the basic Internet hosting needs (rented space on a server, a shopping cart and, in most cases, some kind of Merchant Account software that allows you to collect money from your customers).

With this service, you'll probably have to create your own HTML pages, too, but some companies offer to supply pre-created pages (that are more flexible than templates offered by Internet Malls) that can be modified. A program like Microsoft FrontPage is perfect for creating pages for an ECommerce Hosting Provider because the software is easy to learn and use.

Remember: Going this route provides you with less traffic than in an Internet Mall, so be prepared to find customers. Think of the ways that you can draw customers to your store. How will your customers find your site? It'll take some work, but you will need to promote your site. The following is a list of the most common ways to promote your site:

- **Banner Exchange Advertising**. Look into banner exchange programs. Most are free and operate on one basic premise: If you scratch my back, I'll scratch yours. In other words, if you put an ad for another business's site on yours, that business will do the same for you. Be careful. If you partner up with too many sites that aren't getting any traffic at all, it won't help you. Ask the company how many hits (traffic) it gets a month, a week or even a day. And, try to look for businesses that are relevant to your own product or service lines.
- **E-Mail Campaigns**. Look for an e-mail campaign to buy into. This is a marketing method that works very well (when it's done right). But be careful. A cheap e-mail campaign may lead your mass e-mails to servers that reject them because the addresses are no longer valid, or

because they are viewed as unsolicited junk mail. Before you buy in, ask if the e-mail campaign sends e-mail to "opt-in" members, or people that have requested to receive e-mail from the e-mail campaign provider (these are typically better because you avoid getting into trouble for spamming).

- **Search Engine Positioning**. Choose your meta tags or keywords used to position your company in search engine listings wisely. Thousands upon thousands of companies are jockeying for position with Web masters all over the Internet who are trying for the top listings. Your goal is to be on the first three pages (or the Top 30 listings) of Internet search results on any given search engine. You can pay search engines for top listings, but it's going to cost you...a lot. If you don't have a ton of money, look for companies that provide search engine positioning for a reasonable rate.

There are only a handful of search engines that really matter. Try to choose some that you've actually used yourself. Some good ones: Yahoo!, Google, Altavista, AOL, Netscape, etc. If a search engine is too small, it's not going to do you any good. Also, find out if they will guarantee your positioning.

One example of a good ECommerce Hosting Provider: Verio ECommerce Hosting (**http://hosting.verio.com/index.php/ecommerce.html**). They are fairly large and offer several great hosting packages.

Just like a physical store, your online store allows you to make sales directly to consumers. And just like a landlord provides you with space to display your goods, an Internet host will provide your Internet store with this same "virtual" space online. So, basically it comes down to whether you want more control over the appearance of your site or whether you want a large amount of traffic from day one. If you do have the funds and want to build your own site, and are looking for some pointers on things such as database and automation programming for a directory to your site, etc., you might want to look into hiring a professional Internet programmer.

If you're interested in running an e-commerce business but aren't sure what to sell, check out Internet host **www.vstore.com**. This service allows you to set up an online store using products available through Vstore's inventory. The site has a Web development feature that makes it easy to set up the online store, even if you don't know how to use HTML or other programming languages. Most of the success stories on the site sound like they come from people who are using Vstore.com's services as a side business rather than their primary income, though. This might be a way to get your feet wet in e-tailing before jumping full force into online businesses.

If you're ready to set up your own full-service e-commerce business, try **www.freemerchant.com**. This site offers everything you need to get up and running as an e-commerce sales person, including site development tools, but for absolutely no cost. Freemerchant.com even claims to have no billing department. The site makes its money by setting up partnerships and referral deals with other e-businesses. Services offered for free through the site include:

- Internet Store Hosting;
- Secure Shopping Cart;
- Internet Store Builder;
- Unlimited Store Catalog Size;
- Traffic Logs (these tell you how many people have viewed your Web site or specific Web pages on your site);
- E-mail Account;
- Merchant to Merchant Banner Exchange;
- Shipping Calculator;
- Tax Calculator;
- Coupon Creator;
- Discounted Corporate Services & Offers; and
- Technical Support.

This site has garnered great reviews from on- and off-line publications including the *Los Angeles Times*, *USA Today*, *The New York Times* and *PC Magazine*.

## Free Internet Service Providers

Internet Service Providers (ISPs) do what their name describes; they enable you to connect to the Internet. Without an ISP, your computer would function similar to a telephone without having a carrier for access to other telephone lines. An ISP also allows you to send and receive e-mail and set up your own Web site for business or personal reasons.

Similar to telephone companies, there are many ISPs, all charging different rates and providing varying kinds (and quality) of service. You might want to visit **www.cnet.com**, which does a good job rating ISPs and their services.

Some ISPs offer free services, but the number offering these services for free has diminished over the years. And because ISPs constantly change their services—by discontinuing, curtailing or adding annoying ads to their services—it's best to read the fine print of their contracts and their service (if possible) before you sign up.

Internet access doesn't cost much, but with the proliferation of free Internet access and e-mail services, it's hardly worth paying good money to get on the Web. The following is a list of Internet Service Providers (ISPs) and some of the services they offer for free.

A word of caution: Because these services are free, they are sometimes unreliable. For this reason, it's a good idea to sign up for accounts with more than one service so you are guaranteed to be able to get on the Internet whenever you need to...for free.

- **www.address.com**—Provides free Internet access, free e-mail, free weather updates and free URL forwarding. Plans to provide free Web site hosting, instant messaging and free chat.

- **www.netzero.net**—NetZero uses banner ads and a toolbar to track users' Web browsing habits. In return, you get free Internet access and free e-mail. It currently supports PC users.
- **www.juno.com**—Provides free e-mail and Internet access. Premium Internet access is available through Juno for $9.95 per month. Currently only supports PC computers and they have no plans to develop a Macintosh version. Also includes 12 MB of Web space (this is typically enough for a personal Web site or a small business site).
- **www.senior.com**—Free Internet access for Windows-based systems.
- **www.spiegel.com**—This well-known catalog company has expanded its services to include free Web access. The service includes access to Yahoo!'s free e-mail service, but requires that the user have a PC system. One catch: An advertising bar must remain on screen at all times.

If you already have Internet access through another service, but are simply looking for free e-mail services, any of the access services listed above can probably help you. However, for a list of additional free, Web-based e-mail services, go to **www.emailaddresses.com**. This directory also includes links to other free services, including free online calendars, free online storage, which is great for backing up your home files, and free Web hosting.

## E-Mail Lists

There are plenty of free e-mail lists on everything from religious icons and sports memorabilia and software to hardwood out there—if you aren't careful, you'll find them clogging your e-mail inbox. However, there are some that can provide good information without overwhelming you. You can always unsubscribe to these lists as quickly as you can subscribe to them. Try out various lists to see which ones are right for you and then continue to subscribe to the ones that are most helpful.

Some e-mail lists are discussion lists and some are e-zines, which do not allow you to interact with other list members. Both types of lists can provide a great benefit, however. E-zines can provide you with a large amount of information in an easily digestible form. Sometimes the lists include links to articles that appear online—you can decide whether you read the full articles or not.

Discussion lists are the way to go if you're looking to network. Many of these lists have like-minded members who will share information about contracts for which you can apply, ways to solve business challenges and ideas for strengthening your business.

There are a few large e-mail list sites. It's worth browsing through these sites to see if any of the lists described fit your needs. These are the places to look for discussion lists on a topic specific to your business or on a more general topic. Many of these lists are available in digest format, which means you will receive one e-mail with all the messages from the day compiled, rather than receiving an e-mail every time someone sends one to the list.

- **CataList**. Search for the right list from among more than 32,000 public lists at: **www.lsoft.com/lists/listref.html**.
- **Yahoo! Groups**. Many existing lists and the opportunity to build new lists at: **http://groups.yahoo.com**.
- **List Universe**. Search for lists at: **www.list-universe.com**.
- **Topica**. Nearly 91,000 lists available at: **www.liszt.com**.
- **Publicly Accessible Mailing Lists**. Search for mailing lists at: **www.neosoft.com/cgi-bin/paml_search**.
- **Reference.com**. More than 150,000 discussion groups are listed at: **www.reference.com**.
- **SparkLIST**. More than 400 lists available at: **http://sparklist.net**.
- **Subscription-Service.com**. **http://subscription-service.com**.

## E-Zines and Newsletters

E-zines are simply newsletters that you receive online rather than via e-mail. They include everything from fan sites dedicated to favorite celebrities to Web logs (or "blogs") that focus on politics or current events. In their early years, these e-zines and blogs were amateur efforts; but in the early 2000s, smart companies started using them as alternative marketing tools.

Here are some e-zines and newsletters that are worth subscribing to. Most of these do not allow you to post messages to the list, but some allow you to ask questions or have other interactive features.

- AimDIRECT Entrepreneur Ezine, **www.alacarim.com/aimdirect-ezine**. This list, which mails to subscribers each Friday, includes tips on marketing your business on the Internet. When you subscribe, the newsletter offers you the opportunity to include a free three-line ad in an upcoming newsletter.
- A Marketing Tip A Day, **www.ideasiteforbusiness.com/survey.htm.** Receive one marketing tip each day in your Inbox, or receive all seven for each week on Sunday.
- American Express Small Business Insider, **http://home3.americanexpress.com/smallbusiness/newsletter/?aexp_nav=hp_ads**. This informative monthly e-newsletter provides information and insight about doing business better.
- BandRadio.com's e-mail List, **www.bandradio.com**. Subscribe to a list that provides information for independent artists and musicians.
- Web Marketing Today, **www.wilsonweb.com/wmt**. More than 97,000 people subscribe to this bi-monthly e-zine. Articles focus on site promotion, developing a marketing plan, and fulfilling requests for business when your marketing plan kicks in, among other topics.

## Message Boards

Message boards are sections of the Internet where people with similar interests leave information and notes about a particular subject for each other. They can serve various purposes for a small business. They can be a resource of information—and they can be a guerrilla marketing tool for services...and some products.

Some useful message boards include:

- **Becoming A Virtual Assistant, www.delphi.com/virtualassist/start**. Although these message boards focus on the virtual assistant business, the topics discussed—how to switch from being an employee to being a business owner, unique marketing tips and what products to offer, for example—include discussion topics and information that are of use to entrepreneurs of all kinds. Virtual assistants handle day-to-day administrative tasks for other entrepreneurs and companies. Because they work as independent contractors, they can save businesses money while still taking care of the tasks that the business owner does not have time to take care of.
- Message board on **www.herwebbiz.com**. This site provides a community portal where you'll search topics, submit news and enter a forum for joining discussions of your interest. There are general formus, as well as ones focused on business, health and family and administration. The site also provides a place for you to store an e-mail account.
- Business Strategies, **www.delphi.com/busstrat/start**. From financing and planning to sales and marketing, message posters on this forum tackle all stages of starting and managing a business.
- Business N@tion's Business Forums, **www.businessnation.com/bbs/index.sht**. This moderated board has a wide variety of topic areas, from women in business to accounting, credit and taxes. If your question or discussion issue does not match an existing topic on the boards, you can set up a new topic that fits your needs.
- Consulting Forum, **www.delphi.com/n/main.asp?webtag=wf-consulting&nav=start**. This message board forum, sponsored by

WetFeet.com, focuses on the challenges of operating a consulting business. Some of the discussion that goes on in this message board forum is related to working with particular consulting firms, but the majority of the messages are about what it takes to get into the consulting business and what it takes to be successful.

- Entrepreneur Forum, **www.delphi.com/ab-entrepreneur/start**. This forum, sponsored by the Entrepreneur section of About.com, includes lots of free resources for new entrepreneurs. From free classes to publicity opportunities, there are great resources available through these message boards. Plus, there are plenty of discussions of the nuts and bolts of operating a business, whether that business is operated out of your home or elsewhere.
- IdeaCafe's Cyber Schmooz Forums, **www.ideacafe.com/html/CS.html**. This site offers wide range of message boards on topics from Start-Up Stew (Focusing on start-up issues and challenges) to Tech Talk (Focusing on business technology issues of all kinds). The boards are good places to browse for ideas or to discuss any problems you might be facing in your business.
- InternetBusiness4U Message Boards, **http://internetbusiness4u.community.everyone.net/commun _v3/scripts/directory.pl**. From marketing on the Internet to business opportunities, this message board is a good place to discuss the challenges of doing business online.
- Small Business Information, **www.delphi.com/ab-sbinfo/start**. The Small Business area of About.com sponsors this forum. It includes an Introductions area for forum visitors to tell others about the type of business they're running, an area focused on start-up issues, and lots of other general information.
- The StartUp Network, **www.delphi.com/startupnetwork/start**. Much of the discussion in these message boards revolves around Internet start-ups, but it does have interesting resource boards and offers general discussion on business issues.
- Those Who Can, Consult, **http://boards2.ivillage.com/messages/get/wdcanconsult1.html**. This message board, which requires that

posters be iVillage.com members, is a great place to discuss getting into the consulting business. This board is a facet of the Those Who Can, Consult free online course, offered through iVillage.com's Leadership School. The course can help you identify the skills you have and start planning a consulting business.

- Women.com's Small Biz Boards, **http://messages.women.com/messages/MCat2/MBoard86///wwwthreads.pl?Level=Board&ID=86**. Women.com has a comprehensive small business board with many topics for discussion. Whether you're just deciding what business you want to run or need to flesh out you business plan, this is where to look for ideas and suggestions.
- Workaholics4Hire, **www.delphi.com/workaholics/start**. If you want to work from home and are interested in picking up telecommuting contracts in different areas, this forum may be helpful. It is also a good place to simply discuss the challenges of working from a home office and solutions for those problems.
- Work At Home Forum, **www.delphi.com/homebasedwork/start**. From Success Stories to Grips, this forum deals with all issues about working from home. Users report that it is an active forum that is well-managed and maintained, which means there is not a lot of junk on these boards.

## Marketing Your Web Site

So your site is up and running...now what? Do you know how to take full advantage of the power of the Internet? After your site is up and running, you need to promote your business on the Internet. Promotion is critical to the success of your online business. If you don't take the proper steps to create...and maintain...your Web site, no one will know that it's there. Promoting a Web site combines traditional promotion methods and promotion methods specific to the Internet. Again, set goals for yourself. What do you want to accomplish in the first 30 days? Use the following list to jump-start some ideas for promoting your own Web site.

- Will you be able to communicate your company name, its strong points, products and services, benefits, prices, customer list, etc. to new customers and prospects?
- Do you want to increase non-local sales?
- Do you plan to generate revenue from your Web site? Through direct sales (e.g., revenue from selling products or services)? Indirect sales (e.g., revenue from sales of advertising space or associate buttons)? Licensing/selling content (e.g., revenue from licensing or selling your content to other sites or businesses, including feature articles, weekly columns, surveys, studies, etc.)? Other?
- Do you want to sell products on a 24-hour basis?
- Do you plan to reduce cost/dependence on printed product information?
- Do you plan to improve customer service? Respond to visitor/customer feedback?
- Will you provide product descriptions, frequently asked questions, price information and order forms directly from your site?
- Do you plan to provide an interactive product demo or customer case studies and testimonials so that prospective customers can experience your product or service online?
- Do you want to submit your site to several major search engines? Submit your site to the sites of national organizations in your industry?

It does you no good to put a new business in place if you don't advertise, and unless you are savvy about cheap advertising these costs may drive your business into the ground faster than you would have ever imagined. So, how do you market your business without breaking the bank?

## Customer Feedback

It's important to follow-up on and carefully consider customer input. If you have an "e-mail us" function that encourages site visitors to ask questions, report problems or suggest new features, service enhancements and quality improvements, follow up on their responses. Don't ask for input if you're not going to respond promptly and effectively. Internet-savvy purchasers have high expectations when it comes to turn-around time and customer service. If your business is still relatively small or you can't afford to outsource this activity, be willing to add an 800 number for phone-in orders and questions. And, be prepared to deal with "customer relations" and customer service. If you have very angry customers, you'll want someone working for you that can put out those fires and keep people happy.

A lot of ISPs will provide you with site-tracking tools, but you can also purchase site-tracking software programs for your own server. Review any trends in site activity. Ask yourself the following:

- Where are people coming from to get to your site? Is one search engine or a particular link providing most of your traffic?
- What do visitors do once they get to your site? Look for patterns of use within the site. Some parts of your site may get heavy traffic while others have very little, suggesting a content or layout change to streamline navigation.
- What do visitors look for on your site? If you have a search function, you'll be able to see what topics get the most inquiries. This will also provide you with a list of key words when you register your Web site with search engines. Add polls or mini-surveys on the site to learn about visitor preferences.
- Who's visiting the site? Collect demographic information/purchaser statistics. These often provide you with information on things such as where your customers live, when they buy, what products they buy, etc.

The low-down on cookies: As I mentioned earlier, some people feel that it's important to respect the privacy of visitors by avoiding the invasive placement of cookies or other means of identifying users. But, cookies are a key way to track a user's every move…and increase your business. For more on cookies and how they work, go back to page 224.

## Tackling the Search Engines

If you have a business Web site up and running, you need to make sure you register it with search engines—directly or through www.registerit.com. Before you start this process, you need to come up with 10 keywords that people might use to search for your site and a 20-word description of what your site has to offer. The following are several links to services that will submit your Web site to major search engines for free:

- **http://websitepromotion-12600.hypermart.net**;
- **http://usa-www.hypermart.net/free.htm**;
- **http://submitplus.bc.ca**;
- **www.addme.com**;
- **www.submitexpress.com**;
- **www.siteowner.com**;
- **www.submitawebsite.com**;
- **http://submit.superb.net**; and
- **http://online-biz.com/promote** (this site includes links to a number of other sites that provide free submission services).

Other ways to make the Internet and its technology work for you:

- Get a link to another Web site in exchange for providing a link from your site to their's.
- Get banner ads on other Web sites that fit your target market profile.
- Send customers, prospects and vendors an e-mail that informs them about your site.
- Send press releases via e-mail.
- Contribute content to an electronic magazine or trade journal that specializes in your target market.

Another great way to market your Web site is to offer a newsletter of your own, but make sure that it doesn't go out too often—twice a month should suffice. It also shouldn't give away too much information—you want to use the newsletter as an opportunity to tell people about news and information that helps them understand why they need your services, and then use it to drive traffic to your Web site.

There are several free services available that allow you to run a free listserv without having all the software and server space yourself to run one in-house. (A listserv allows a group of individuals with a common interest who have provided you with their e-mail address to receive information on e-zines. One of the best services I've come across is Yahoo! Groups **http://groups.yahoo.com**. Also, try **www.ListServe.com and www.topica.com**.

## Grassroots Advertising

The best way to get new business is to spread your name around by word of mouth. If people know who you are, they are more likely to utilize your services.

Make a list of everyone you know. Go through all your address books and your Rolodex and dig up their phone numbers. Then make notes for each person—why would they want to know about your business? How could they help? Who do they know who might be interested in your services?

Armed with your list and notes, make phone calls to each one. Keep the calls brief, and remember—these are not sales calls, they're just information calls. You're letting people know that your Web site is open for business. Tell them what you're doing, ask them if they have any questions and if they would keep you in mind and refer you to anyone else they know who might need your services. You'll be surprised how quickly this can turn into some lucrative business opportunities.

Another great way to get some grassroots advertising is to volunteer your services to a local community organization. In return, ask that they acknowledge your business's Web site and its services publicly. Clearly, you do not want to seem like you're only giving your time and services for the name recognition, but if you are tactful and straightforward about increasing your business, fellow volunteers or the organization itself may begin to give you business.

## Other Ways to Get Your Name Out

There are so many ways to get your name out in the community and to your potential customers it's hard to fit them all in this book. Here are some ideas for free or cheap ways:

- Any time you or your business do anything newsworthy—your first official day in business, when you hire your first employee, when you donate time to a charity, send a press release to the business section of your local papers and include your Web site address. Keep a list of all the papers where your customers live and update your contacts from time to time. Don't forget local weekly newspapers—often those readers will become your best customers!
- Send postcards or letters that announce your Web site to customers, prospects and vendors, encouraging them to visit your Web site. You may even want to offer them a coupon or discount for any purchases they make via the Web site. Offer 10 or 15 percent off their next order if they refer a new customer. If they like your services, you may find that you end up with a number of referrals to boost your business base.
- Advertise in community newspapers, magazines, industry newsletters, etc. Always include your Web site address in the ad copy. Another way to promote your site: write articles or columns about some aspect of your business for newspapers or trade journals and include your Web presence as part of the content. Do you have something to say about a current event? If there is a news story about which you can write a letter to the editor or a column that will provide a newsworthy opinion and still mention or highlight your business without

appearing to be too self-serving, you might be able to get your piece published in a local newspaper at no cost to you.

- Make a one-page flyer that you can copy onto colored paper. Include information about your business and services, a discount coupon and contact information, including your Web site address. Pass that flyer out to people who might be interested in your services. For example, if you're starting a pet-sitting or dog-walking business, pass out the flyer to people in the parking lot of your local pet store. Or, if you plan to go into the computer training business, pass out your flyer outside a computer or business store. Be careful, though—you want to make sure you are not breaking any laws by doing this. Check with your local police department, the shopping center management company (often they own the parking lot and sidewalks outside the stores), and code enforcement agency to make sure you're not doing something illegal.
- Add your Web site address and features to your business's current marketing materials, including brochures, business cards, letterhead, newsletters, press releases, purchase orders, renewals, product spec sheets, etc. If you're worried about the cost of doing this, print up stickers with the relevant information and use those on the materials until you use up the existing inventory. Don't be stingy with your business cards. After all, they are relatively inexpensive and quick to print. Give people two or more cards and ask them to give the extra card to someone else who might need your services. You never know how far that chain of referrals might take you.
- Have an inexpensive magnetized sign made for the side of your car that displays your logo and Web site address. Even if your business is on the Internet, most of your customers will probably be local. You can get your name out this way while driving around and doing your errands, and by using a magnetized sign, you'll do no permanent damage to your car. Or, if you have an interesting logo for your business, print up inexpensive bumper stickers or small window stickers and include them in mailings to current and prospective clients.

- Trade services with your dentist or doctor. Offer to design a brochure or flyer for them, or offer to perform some other service in exchange for them to keep your cards or brochures displayed at their offices. Always include your Web site address. A word of caution: Check with your tax advisor or on the IRS Web site (**www.irs.gov**) to make sure you understand the tax ramifications of such barter deals. They won't cost you any money, but you might have to report them.

- If your local school or other charitable organization offers a silent auction, donate a gift certificate for your business or a certain amount of service. You will receive free publicity in the program and possibly a free mention about your business and its Web site at the auction itself. Even those who don't bid on your services might be interested in becoming a client or customer of your business down the road. Be sure to provide plenty of business cards that organizers can spread out on the table next to the bidding sheet.

- Don't be stingy with your business cards. After all, they are relatively inexpensive and quick to print. Each time you meet someone to whom you plan to give your card, give them two or more cards and ask them to give the extra card to someone else who might need your services. You never know how far that chain of referrals might take you.

- Ask your vendors to recommend you. If you usually work with a particular copy shop, office supply salesperson or computer supplier, ask them to recommend you to other companies with which they do business. It won't cost them anything, and if you're giving them business, they will be more willing to send business back your way.

- Put your Web site address is in the signature of every e-mail you send, whether that e-mail goes to your friends, family or business associates. You never know where that e-mail will end up or who'll click through that link to your Web site. You could be one click away from a new customer!

The following is a list of Web sites with unique marketing ideas for your business.

- **www.ideasiteforbusiness.com/unusual.htm.**
- **http://members.aol.com/WandR10395/tips.htm#mktg.**
- **www.frugalfun.com/22waysmarketing.html.**
- **http://autumnwinds.com/bbbltd/bbc/market.html.**
- **www.drnunley.com/MARKET.htm.**
- **www.smallbiz2000.com/articles/article0005.html.**
- **http://loska.com/successconnection.**
- **www.moytura.com/internetmarketing/neat-ideas.htm.**
- **http://realm.net/net/5/Win00e-5-busimark.htm.**
- **www.wilsonweb.com/webmarket.**
- **www.ideasformarketing.com.**

## Conclusion

The Internet revolution is still new enough that it's hard to say what role it will end up playing for most businesses. Certainly, some entrepreneurs and VC lenders have over-estimated the effects. But it's true the net is a leveling factor. It can even the field of competition between small companies and big ones. So small companies should use it aggressively.

An important note: Creating a Web site for your business potentially impacts all other activities of your business model, so make sure you modify your ongoing operations to take advantage of your Web site and avoid potential problems. Integrate existing operations and your Web site. Ask yourself the following: Do you have a way to collect customer feedback/ suggestions? Can you track orders from the Internet? Can you accept online payments? Do you have someone assigned to provide sales support? Are product descriptions and prices listed on your site? Can you process and fulfill online orders using your existing order entry system? Can you fulfill orders in a timely manner? Does your existing return and cancellation policy work for online sales? Can you assist customers with online shopping and purchase procedures? Are you prepared to answer inquiries from customers who visit your site?

If you can answer these questions and answer yes to most, you're well on your way to maintaining an effective and lucrative Web site.

# Conclusion: The Free Mentality

We live in a time of great uncertainty for most businesses. That's bads news for some large companies—it's a great opportunity for scrappy start-ups.

This book has been about that word *scrappy*. It's about stretching every dollar you invest in a business...or that your business generates...to improve your new business's chances of surviving its first year. And surviving long after that.

Of course, the strict truth is that you can't start a business—completely—for free. At some point, every business requires some investment. But the Free Mentality is essential to any new business.

What's the Free Mentality? It's the boot-strapping mindset that saves money wherever it can. It shakes the printer cartridge, when it's running low on toner, to last for one more memo. It uses Internet chat rooms and message boards as advertising platforms. It finds a good sublet for half the price of standard office space.

It does all of this to make sure the start-up business survives. Lasts another month...gets through a tough quarter...makes it past a rotten week. It saves a nickel here and a dime there because somewhere out in the future...maybe a few months away, maybe a couple of years away, every start-up business has a break-even point. Once it reaches that point, the business generates enough cash-flow to support itself and its owner.

The tough part is reaching that point.

As I've said throughout this book, my goal has been to offer some strategies for using the Free Mentality to help get your start-up business to its break-even point...and even to keep that mentality once it gets there. In addition to that, I've offered some specific references for finding free information, services and business support. Use these, use these. And e-mail me (dc@davidcaplan.com) any other good ones you find. I'll include them in future editions of this book.

You never know when the $40 you save on Internet access this month buys you the extra time you need next year to land a six-figure account.

It can and you can.

# Appendix A: Where to Get a Business License

- **Alabama**—Corporations Division, Office of Secretary of State; **www.sos.state.al.us/business/corporations.cfm**.
- **Alaska**—Division of Banking Securites and Corporations, Department of Community and Economic Development; **www.dced. state. ak.us/bsc/corps.htm**.
- **Arizona**—Corporation Commission Name Search; **www.cc. state.az.us/corp/filings/namesrch.htm**.
- **Arkansas**—Online Filing System for Arkansas; **www.sosweb. state. ar.us/ofs/index.html**.
- **California**—Secretary of State Business Service Center; **www.ss. ca. gov/business/business.htm**.
- **Colorado**—Secretary of State Business Center; **www.sos.state.co. us/pubs/business**.
- **Connecticut**—Tax Connecticut Department of Revenue Services, Starting a Business; **www.drs.state.ct.us/taxassistance/start.html**.
- **Delaware**—Division of Corporations; **www.state.de.us/corp/corp. htm**.
- **District of Columbia**—**www.ci.washington.dc.us**.
- **Florida**—Business & Industry, Vendor Search; **http://fcn.state.fl.us/ owa_spurs/owa/spurs_www.vendor_search.criteria**.

- **Georgia**—Secretary of Stat, Business Entity Information; **www.sos.state.ga.us/corporations/corpsearch.htm**.
- **Hawaii**—Business Name Search; **www.ehawaiigov.org/DCCA/biz-name/html**.
- **Illinois**—Secretary of State, Corporate/LLC Information Search; **www.sos.state.il.us/depts/bus_serv/corpnames.html**.
- **Idaho**—Business Entity Search; **www.state.id.us/apps/sos/corp/search.html**.
- **Indiana**—Secretary of State, Corporations; **www.in.gov/sos/business/corps/searches.html**.
- **Iowa**—Corporation Search; **www.sos.state.ia.us/corp/corp_search.asp**.
- **Kansas**—Kansas Department of Commerce & Housing; **www.kansascommerce.com/0301majbus.html**.
- **Kentucky**—Business Filing Forms; **www.sos.state.ky.us/busser/busfil/forms.htm**.
- **Louisiana**—Commercial Divison, Corporations Section; **www.sec.state.la.us/comm/fss-index.htm**.
- **Maine**—Doing Business in Maine; **www.state.me.us/business/corp.html**.
- **Maryland**—Business Information Network; **www.state.md.us**.
- **Massachusetts**—Corporations Division; **www.state.ma.us/sec/cor**.
- **Michigan**—**www.michigan.gov/emi/1,1303,7-102-117_399.00.html**.
- **Minnesota**—Type of Filings; **www.sos.state.mn.us/business/typeoffile.html**.
- **Mississippi**—Secretary of State, Business Services Division; **www.sos.state.ms.us/busserv/busserv.asp**.
- **Missouri**—Business Registration Information; **http://mosl.sos.state.mo.us/bus-ser/corpdata.html**.

- **Montana**—Secretary of State, Business Services; **www.state.mt.us/sos/Business_Services/business_services.html**.
- **Nebraska**—Secretary of State; **www.nol.org/home/SOS/corps/corpform.htm**.
- **New Hampshire**—Businesses and Organizations; **http://webster.state.nh.us/revenue**.
- **New Jersey**—**http://soswy.state.wy.us/corporat/corporat.htm**.
- **New Mexico**—Corporations Inquiry; **www.nmprc.state.nm.us/ftq.htm**.
- **New York**—Department of State; **www.dos.state.ny.us**.
- **Nevada**—**www.state.nv.us**.
- **North Carolina**—**www.state.nc.us**.
- **North Dakota**—New Business Registration; **www.state.nd.us/businessreg**.
- **Ohio**—Secretary of State; **www.state.oh.us/sos/info.html**
- **Oklahoma**—**www.sos.state.ok.us/**
- **Oregon**—Secretary of State; **www.filinginoregon.com**.
- **Pennsylvania**—Entrepreneur's Guide; **www.paopen4business.state.pa.us**.
- **Pennsylvania**—Corporations, Department of State; **www.dos.state.pa.us/corps/corp.html**.
- **Pennsylvania**—Business Forms; **www.paopen4business.state.pa.us**.
- **Rhode Island**—First Stop Business Center; **www.state.ri.us/bus/frststp.htm**.
- **South Carolina**—Secretary of State; **www.scsos.com/corporations.htm**.

- **South Dakota**—Business Startup Package; **www.sdgreatprofits.com/start-up/startup.htm**.
- **Tennesee—www.state.tn.us**.
- **Texas—http://soswy.state.wy.us/corporat/corporat.htm**.
- **Utah**—Department of Commerce, Business Entity List; **www.commerce.state.ut.us**.
- **Virginia**—Office of the Clerk, Virginia State Corporation Commission; **www.state.va.us/scc/division/clk/index.htm**.
- **Washington—www.wa.gov**.
- **West Virginia—www.state.wv.us**.
- **Wisconsin**—Department of Financial Institutions; **www.wdfi.org**.
- **Wyoming**—Corporations; **http://soswy.state.wy.us/corporat/corporat.htm**.

For more state specific data, visit **www.isquare.com/states/states.cfm**. The site offers state specific contact information on everything from your local SBA office, incorporation info and local chamber of commerce, to your local department of commerce office, postal business center and taxing authority.

# Appendix B: List of Patent & Trademark Resources

In the U.S., intellectual property is managed by the Library of Congress in Washington, DC and the Patent and Trademark Office (PTO) in Arlington, Virginia. You can contact these offices in a variety of ways, as the following lists show. Although the online system is well-equipped to deliver forms, manuals and information to you for every type of intellectual property, there are other avenues to use. If ever you feel lost in the mess, consult an attorney who specializes in intellectual property and who can help you weave your way through the paperwork.

## Trademarks

General inquiries about trademarks, as well as the products and services of the U.S. Patent and Trademark Office, should be mailed to General Information Services Division, U.S. Patent and Trademark Office, Crystal Plaza 3, Room 2C02, Washington, DC 20231.

- **Online**: Go to *www.uspto.gov* and go to Trademarks. Applicants are encouraged to use e-mail for trademarks.
- **Filing**: To file an application online using the Trademark Electronic Application System (**TEAS**): *www.uspto.gov/teas/index.html*. This is the preferred method.
- **Patent and Trademark Depository Library**: If you do not have Internet access, you can access TEAS at any PTDL throughout the

United States. These are libraries that provide many PTO services, located in regional areas. Information about the Patent and Trademark Depository Library Program, as well as a list of these libraries, are available online at *www.uspto.gov/web/offices/ac/ido/ptdl/index.html*

- **Mail or Hand-Delivery**: Send or deliver correspondence to the Commissioner for Trademarks, Box-New App-Fee, 2900 Crystal Drive, Arlington, VA 22202-3513.
- **Automated Telephone Line**: To obtain a printed form, call (703) 308-9000 or (800) 786-9199. You may NOT submit an application by fax.
- **Trademark Trial and Appeal Board**: To contact this division, call (703) 308-9300 or write to 2900 Crystal Drive, Arlington, VA 22202.
- **Trademark Applications and Registrations Retrieval (TARR)**: To retrieve information about pending and registered trademarks, go to *http://tarr.uspto.gov.*

## Useful Phone Numbers

- **Assignment Division**, for recording assignments: phone (703) 308-9723; fax (703) 308-7124.
- **Certification Division**, for certified copies of registrations: phone (703) 308-9726; fax (703) 308-7048.
- **Copy Sales Department**, for copies of files and registrations: phone (703) 305-8716; fax (703) 308-8759.
- **Government Printing Office**, for copies of the Official Gazette and other USPTO publications: phone (202) 512-1800; fax (202) 512-2250.
- **Intent to Use/Divisional Unit**, for filing Statements of Use, Extension Requests and Requests to Divide Applications: phone (703) 308-9550; fax (703) 308-7196.

- **Office of the Commissioner for Trademarks**, for filing petitions to the Commissioner: phone (703) 308-8900; fax (703) 308-7220.
- **Post Registration Division**, for filing post registration documents: phone (703) 308-9500; fax (703) 308-7178.
- **Publication and Issue Division**, for original certification of registration: phone (703) 308-9401; fax (703) 305-4100.
- **Trademark Assistance Center**, for general trademark information and printed application forms: phone (703) 308-9000; fax (703) 308-7016.
- **Trademark Trial and Appeal Board**, for filing notices of opposition and petitions to cancel registrations: phone (703) 308-9300; fax (703) 308-9333.

## Patents

Like trademarks, general inquiries should be mailed to General Information Services Division, U.S. Patent and Trademark Office, Crystal Plaza 3, Room 2C02, Washington, DC 20231.

- **Online**: Go to *www.uspto.gov* and go to Patents.
- The **Patent Assistance Center (PAC)** at the U.S. Patent and Trademark Office provides information services to the public concerning any general questions regarding patent examining policies and procedures. You can reach this office at 800-PTO-9199 (800- 86-9199) or 703-308-HELP (4357), Monday - Friday 8:30 A.M. to 5:00 P.M. (Eastern Time Zone). You can fax this center at (703) 305-7786.

  Because the patent process is particularly challenging, it's best to visit the PAC first and be directed to further resources.
- **The Office of Independent Inventor Programs (OIIP)** was established in March 1999 in order to meet the special needs of independent inventors. The OIIP establishes new mechanisms to better disseminate information about the patent and trademark process and

to foster regular communication between the USPTO and independent inventors.

- **Mailing Address**: You can write the OIIP at the Director—United States Patent and Trademark Office, Office of Independent Inventor Programs, Box 24, Washington, DC 20231.
- **Telephone**: (703) 306-5568
- **Fax**: (703) 306-5570
- **E-mail**: independentinventor@uspto.gov

*Remember*: The USPTO's home page is *www.uspto.gov.* And you can send e-mail directly at *uspto@uspto.gov*, and indicating "Patent" or "Trademark" in the Subject box.

# APPENDIX C: LIST OF RESOURCES

## Technology

- **21st Century Venture Partners**, www.21st-century.com, Two South Park, Second Floor, San Francisco, CA 94107, Phone: (415) 512-1221.
- **Accel Partners**, www.accel.com, 428 University Avenue, Palo Alto, CA 94301, Phone: (650) 614-4800.
- **Advanced Technology Ventures**, www.atv-ventures.com, 485 Ramona Street, Palo Alto, CA 94301, Phone: (650) 321-8601; or 281 Winter Street, Suite 350, Waltham, MA 02451, Phone: (781) 290-0707.
- **Alta Partners, Inc.**, www.altapartners.com, One Embarcadero Center, Suite 450, San Francisco, CA 94111, Phone: (415) 362-4022.
- **Altos Ventures**, www.altosvc.com, 2882 Sand Hill Road, Suite 100, Menlo Park, CA 94025, Phone: (650) 234-9771.
- **The AM Fund**, www.amfund.com, 1716 Briarcrest, Suite 507, Bryan, TX 77802, Phone: (409) 846-6072; or 4600 Post Oak Place, Suite 100, Houston, TX 77027, Phone: (713) 627-9111; or 8911 Capital of Texas Highway, Westech 360, Suite 2310, Austin, TX 78759, Phone: (512) 342-2024.
- **ARCH Venture Partners**,www.archventure.com, 8725 W. Higgins Road, Suite 290, Chicago, IL 60631, Phone: (773) 380-6600; or 45

Rockefeller Plaza, Suite 2071, New York, NY 10111, Phone: (212) 332-5053; or 1000 Second Avenue, Suite 3700, Seattle, WA 98104, Phone: (206) 674-3028; or 1155 University, S.E., Albuquerque, NM 87106, Phone: (505) 843-4293; or 6801 N. Capital of Texas Highway, Suite 225, Austin, TX 78731, Phone (512) 795-5830.

- **Aspen Ventures**, www.aspenventures.com, 1000 Fremont Avenue, Suite 200, Los Altos, CA 94024, Phone: (650) 917-5670.
- **Atlas Venture**, www.atlasventure.com, 222 Berkeley Street, Boston, MA 02116, Phone: (617) 859-9290; or 1600 El Camino Real, Suite 290, Menlo Park, CA 94025, Phone: (650) 614-1444.
- **The Aurora Funds, Inc.**, www.aurorafunds.com, 2525 Meridian Parkway, Suite 220, Durham, NC 27713, Phone: (919) 484-0400.
- **Austin Ventures**, www.austinventures.com, 114 West 7th Street, Suite 1300, Austin, TX 78701, Phone: (512) 485-1900.
- **Associated Venture Investors**, www.avicapital.com, One First Street, Suite 2, Los Altos, CA 94022, Phone: (650) 949-9862.
- **BancBoston Ventures**, www.bancboscap.com/bbv/index.asp, 100 Federal Street, P.O. Box 2016, Boston, MA, 02110.
- **Battery Ventures**, www.battery.com, 20 William Street, Suite 200, Wellesley, MA 02481, Phone: (781) 577-1000; or 901 Mariner's Island Boulevard, Suite 475, San Mateo, CA 94404, Phone: (650) 372-3939.
- **Bay Partners**, www.baypartners.com, 10600 North De Anza Boulevard, Suite 100, Cupertino, CA 95014-2031, Phone (408) 725-2444.
- **Benchmark Capital**, www.benchmark.com/home.html, 2480 Sand Hill Road, Suite 200, Menlo Park, CA 94025, Phone: (650) 854-8180.
- **Bessemer Venture Partners**, www.bessemervp.com, 535 Middlefield Road, Suite 245, Menlo Park, CA 94025, Phone: (650) 853-7000; or 83 Walnut Street, Wellesley Hills, MA 02481, Phone: (781) 237-6050.

- **Brentwood Venture Capital**, www.brentwoodvc.com, 3000 Sand Hill Road, Bldg. 1, Suite 260, Menlo Park, CA 94025, Phone: (650) 854-7691; or 11150 Santa Monica Blvd., Suite 1200, Los Angeles, CA 90025, Phone (310) 477-7678; or 1920 Main Street, Suite 820, Irvine, CA 92614, Phone: (949) 251-1010.
- **The Canaan Team**, www.canaan.com/canaanteam/index.html, 105 Rowayton Avenue, Rowayton, CT 06853, Phone: (203) 855-0400; or 2884 Sand Hill Road, Suite 115, Menlo Park, CA 94025, Phone: (650) 854-8092.
- **Columbia Capital**, www.colcap.com, 201 North Union Street, Suite 300, Alexandria, VA 22314, Phone: (703) 519-2000.
- **Charles River Ventures**, www.crv.com, Bay Colony Corporate Center, 1000 Winter Street, Suite 3300, Waltham, MA 02451, Phone: (781) 487-7060; or 3180 Porter Drive, Palo Alto, CA 94304.
- **CID Equity Partners**, www.cidequity.com, One American Square, Suite 2850, Box 82074, Indianapolis, IN 46282, Phone: (317) 269-2350; or 41 South High Street, Suite 3650, Columbus, OH 43215, Phone: (614) 222-8185; or 2 North LaSalle Street, Suite 1705, Chicago, IL 60602, Phone: (312) 578-5350; or 312 Elm Street, Suite 2600, Cincinnati, OH 45202, Phone: (513) 381-4748.
- **Commonwealth Capital**, www.ccvlp.com, 20 William Street, Suite 225, Wellesley, MA 02481, Phone: (781) 237-7373.
- **Crosspoint Venture Partners**, www.crosspointvc.com/index_ns.html, The Pioneer Hotel Building, 2925 Woodside Road, Woodside, California 94062, Phone: (650) 851-7600; or 18552 MacArthur Boulevard, Suite 400, Irvine, CA 92612, Phone: (949) 852-1611.
- **Divine Interventures**, www.divineinterventures.com, 4225 Naperville Road, Lisle, IL 60532, Phone: (630) 799-7500.
- **Draper Fisher Jurvetson**, www.drapervc.com, 400 Seaport Court, Suite 250 Redwood City, CA 94063, Phone: (650) 599-9000,

- **Edison Venture Fund**, www.edisonventure.com, 1420 Springhill Road, Suite 420, McLean VA 22102, Phone: (703) 903-9546; or 1009 Lenox Drive #4; Lawrenceville NJ 08648, Phone: (609) 896-1900; or 1700 Race Street, Philadelphia PA 19103, Phone: (215) 299-7400 x239.
- **El Dorado Ventures**, www.eldoradoventures.com, 2884 Sand Hill Road, Suite 121, Menlo Park, CA 94025, Phone: (650) 854-1200.
- **Enterprise Partners**, www.ent.com, 7979 Ivanhoe Avenue, Suite 550, La Jolla, CA 92037-4543, Phone: (858) 454-8833.
- **Euclid Partners**, www.euclidpartners.com, 45 Rockefeller Plaza, Suite 907, New York, NY 10111, Phone: (212) 218-6880.
- **Geocapital Partners**, www.geocapital.com, 2 Executive Drive, Suite 820, Fort Lee, NJ 07024, Phone: (201) 461-9292.
- **Highland Capital Partners**, www.hcp.com, Two International Place, Boston, MA 02110, Phone: 617-531-1500; or 555 California Street, Suite 3100, San Francisco, CA 94104, Phone: (415) 981-1230.
- **Hummer Winblad Venture Partners**, www.humwin.com, 2 South Park, 2nd Floor, San Francisco, CA 94107, Phone: (415) 979-9600.
- **IDG Ventures**, www.idgventures.com, 650 California Stree, 24th Floor, San Francisco, CA 94108, Phone: (415) 439-4420.
- **InnoCal**, www.innocal.com, 600 Anton Blvd., Suite 1270, Costa Mesa, CA 92626, Phone: (714) 850-6784.
- **Institutional Venture Partners**, www.ivp.com/intro1_snd1.html, 3000 Sand Hill Road, Building 2, Suite 290, Menlo Park, CA 94025, Phone: (650) 854-0132.
- **Intersouth Partners**, www.intersouth.com, 3211 Shannon Road, Suite 611, Durham, NC 27707, Phone: (919) 493-6640.
- **InterWest Partners**, www.interwest.com, 3000 Sand Hill Road, Building 3, Suite 255, Menlo Park, CA 94025, Phone: (650) 854-8585; or Two Galleria Tower, 13455 Noel Road, Suite 1670, Dallas, TX 75240, Phone: (972) 392-7279.

- **Kleiner Perkins Caufield & Byers**, www.kpcb.com, 2750 Sand Hill Road, Menlo Park, CA 94025, Phone: (650) 233-2750; or Four Embarcadero Center, Suite 1880, San Francisco, CA 94111, Phone: (415) 421-3110.
- **Menlo Ventures**, www.menloventures.com, 3000 Sand Hill Road M, Building 4, Suite 100, Menlo Park, CA 94025, Phone: (650) 854-8540.
- **Mohr, Davidow Ventures**, www.mdv.com, 2775 Sand Hill Road, Suite 240, Menlo Park, CA 94025, Phone: (650) 854-7236.
- **Mobius Venture Capital**, www.sbvc.com, 200 West Evelyn Ave., Suite 200, Mountain View, CA 94041, Phone: (650) 962-2000 or 100 Superior Plaza Way, Suite 200, Superior, CO 80027, Phone: (303) 642-4000.
- **Morgan Stanley Venture Partners**, www.msvp.com, 1221 Avenue of the Americas, 33rd Floor, New York, NY 10020, Phone: (212) 762-7900; or 3000 Sand Hill Road, Building 4, Suite 250, Menlo Park, CA 94025, Phone: (650) 233-2600.
- **Murphree Venture Partners**, www.murphco.com, 1100 Louisiana, Suite 5225, Houston, TX 77002, Phone: (713) 655-8500.
- **New Enterprise Associates**, www.nea.com, 2490 Sand Hill Road, Menlo Park, CA 94025, Phone: (650) 854-9499; or One Freedom Square, 11951 Freedom Drive, Suite 1240, Reston, VA 20190, Phone: (703) 709-9499; or 1119 St. Paul Street, Baltimore, MD 21202, Phone: (410) 244-0115.
- **Oak Investment Partners**, www.oakinv.com, 525 University Avenue, Suite 1300, Palo Alto, CA 94301, Phone: (650) 614-3700; or One Gorham Island, Westport, CT 06880, Phone: (203) 226-8346; or 4550 Norwest Center, 90 South Seventh Street, Minneapolis, MN 55402, Phone: (612) 339-9322.
- **Onset Ventures**, www.onset.com, 2400 Sand Hill Road, Suite 150, Menlo Park, CA 94025, Phone: (650) 529-0700.

- **Olympic Venture Partners**; www.ovp.com, accepts inquiries by e-mail at info@ovp.com.
- **Piper Jaffrey Companies**, www.piperjaffray.com, 222 South Ninth Street, Minneapolis, MN 55402, (612) 342-6000.
- **Redleaf Group**, www.redleaf.com, 14395 Saratoga Avenue, Suite 130, Saratoga, CA 95070, Phone: (408) 868-0800; or 100 First Avenue, Suite 950, Pittsburgh, PA 15222, Phone: (412) 201-5600.
- **Sequel Venture Partners**, www.sequelvc.com, 4430 Arapahoe Avenue, Suite 220, Boulder, CO 80303, Phone: (303) 546-0400.
- **Sequoia Capital**, www.sequoiacap.com, 3000 Sand Hill Road, Bldg. 4, Suite 280, Menlo Park, CA 94025, Phone: (650) 854-3927.
- **Sevin Rosen Funds**, www.srfunds.com, Two Galleria Tower, 13455 Noel Road, Suite 1670, Dallas, TX 75240, Phone: (972) 702-1100; or 169 University Avenue, Palo Alto, CA 94301, Phone: (650) 326-0550.
- **Sierra Ventures**, www.sierraven.com, 3000 Sand Hill Road, Building Four, Suite 210, Menlo Park, CA, Phone: (650) 854-1000.
- **TA Associates**, www.ta.com, High Street Tower, Suite 2500, 125 High Street, Boston, MA 02110, Phone: (617) 574-6700; or 70 Willow Road, Suite 100, Menlo Park, CA 94025, Phone: (650) 328-1210; or One Oxford Center, Suite 4260, Pittsburgh, PA 15219-1407, Phone: (412) 441-4949.
- **Technology Crossover Ventures**, www.tcv.com, 575 High Street, Suite 400, Palo Alto, CA 94301, Phone: (650) 614-8200; or 160 West 86th Street, Suite 12B, New York, NY 10024, Phone: (212) 277-3900; or 56 Main Street, Suite 210, Millburn, NJ 07041, Phone: (973) 467-5320.
- **Technology Funding**, www.techfunding.com, 2000 Alameda de las Pulgas, San Mateo, CA 94403, Phone: (800) 821-5323.
- **Telos Venture Partners**, www.telosvp.com, 2350 Mission College Blvd., Suite 1070, Santa Clara, CA 95054, Phone: (408) 982-5800.

- **Trident Capital**, www.tridentcap.com, 505 Hamilton Avenue, Suite 200, Palo Alto, CA 94301, Phone: (650) 289-4400; or 11150 Santa Monica Blvd., Suite 320, Los Angeles, CA 90025, Phone: (310) 444-3840; or 272 East Deerpath, Suite 304, Lake Forest, IL 60045, Phone: (847) 283-9890; or 200 Nyala Farms, Westport, CT 06880, Phone: (203) 222-4594.
- **Trinity Ventures**, www.trinityventures.com, 3000 Sand Hill Road Building 4, Suite 160, Menlo Park, CA 94025, Phone: (650) 854-9500.
- **US Venture Partners**, www.usvp.com, 2180 Sand Hill Road, Suite 300, Menlo Park, CA 94025, Phone: (650) 854-9080.
- **VenGlobal Capital**, www.venglobal.com, 20195 Stevens Creek Boulevard, Suite 110, Cupertino, CA 95014, Phone: (408) 861-1035.
- **Weiss, Peck & Greer**, www.wpgvp.com, Suite 3130, 555 California Street, San Francisco, CA 94104, Phone: (415) 622-6864.
- **Western States Investment Group**, www.wsig.com, 919 Towne Center Drive, Suite 310, San Diego, CA 92122, Phone: (858) 678-0800.
- **Weston Presidio Capital**, www.westonpresidio.com; or Pier 1, Bay 2 San Francisco, CA 94111 Phone: (415) 398-0770; or 200 Clarendon Street, 50th Floor Boston, MA 02116, Phone: (617) 988-2500.

## Communications

- **Advanced Technology Ventures**, www.atv-ventures.com, 485 Ramona Street, Palo Alto, CA 94301, Phone: (650) 321-8601; or 281 Winter Street, Suite 350, Waltham, MA 02451, Phone (781) 290-0707
- **Atlas Venture**, www.atlasventure.com, 222 Berkeley Street, Boston, MA 02116, Phone: (617) 859-9290; or 1600 El Camino Real, Suite 290, Menlo Park, CA 94025, Phone: (650) 614-1444.

- **Austin Ventures**, www.austinventures.com, 114 West 7th Street, Suite 1300, Austin, TX 78701, Phone: (512) 485-1900.
- **Battery Ventures**, www.battery.com, 20 William Street, Suite 200, Wellesley, MA 02481, Phone: (781) 577-1000; or 901 Mariner's Island Boulevard, Suite 475, San Mateo, CA 94404, Phone: (650) 372-3939.
- **Columbia Capital**, www.colcap.com, 201 North Union Street, Suite 300, Alexandria, VA 22314, Phone: (703) 519-2000.
- **Commonwealth Capital**, www.ccvlp.com, 20 William Street, Suite 225, Wellesley, MA 02481, Phone: (781) 237-7373.
- **Edison Venture Fund**, www.edisonventure.com, 1420 Springhill Road, Suite 420, McLean VA 22102, Phone: (703) 903-9546; or 1009 Lenox Drive #4; Lawrenceville NJ 08648, Phone: (609) 896-1900; or 1700 Race Street, Philadelphia PA 19103, Phone: (215) 299-7400 x239.
- **Euclid Partners**, www.euclidpartners.com, 45 Rockefeller Plaza, Suite 907, New York, NY 10111, Phone: (212) 218-6880.
- **Highland Capital Partners**, www.hcp.com, Two International Place, Boston, MA 02110, Phone: 617-531-1500; or 555 California Street, Suite 3100, San Francisco, CA 94104, Phone: (415) 981-1230.
- **Kleiner Perkins Caufield & Byers**, www.kpcb.com, 2750 Sand Hill Road, Menlo Park, CA 94025, Phone: (650) 233-2750; or Four Embarcadero Center, Suite 1880, San Francisco, CA 94111, Phone: (415) 421-3110.
- **Menlo Ventures**, www.menloventures.com, 3000 Sand Hill Road M, Building 4, Suite 100, Menlo Park, CA 94025, Phone: (650) 854-8540.
- **Oak Investment Partners**, www.oakinv.com, 525 University Avenue, Suite 1300, Palo Alto, CA 94301, Phone: (650) 614-3700; or One Gorham Island, Westport, CT 06880, Phone: (203) 226-8346; or 4550 Norwest Center, 90 South Seventh Street, Minneapolis, MN 55402, Phone: (612) 339-9322.

- **Onset Ventures**,www.onset.com, 2400 Sand Hill Road, Suite 150, Menlo Park, CA 94025, Phone: (650) 529-0700.
- **Olympic Venture Partners**, www.ovp.com, accepts inquiries by e-mail at: info@ovp.com.
- **Sevin Rosen Funds**, www.srfunds.com, Two Galleria Tower, 13455 Noel Road, Suite 1670, Dallas, TX 75240, Phone: (972) 702-1100; or 169 University Avenue, Palo Alto, CA 94301, Phone: (650) 326-0550.
- **Sierra Ventures**, www.sierraven.com, 3000 Sand Hill Road, Building Four, Suite 210, Menlo Park, CA, Phone: (650) 854-1000.
- **TA Associates**, www.ta.com, High Street Tower, Suite 2500, 125 High Street, Boston, MA 02110, Phone: (617) 574-6700; or 70 Willow Road, Suite 100, Menlo Park, CA 94025, Phone: (650) 328-1210; or One Oxford Center, Suite 4260, Pittsburgh, PA 15219-1407, Phone: (412) 441-4949.
- **Telos Venture Partners**, www.telosvp.com, 2350 Mission College Blvd., Suite 1070, Santa Clara, CA 95054, Phone: (408) 982-5800.
- **VenGlobal Capital**, www.venglobal.com, 20195 Stevens Creek Boulevard, Suite 110, Cupertino, CA 95014, Phone: (408) 861-1035.
- **Western States Investment Group**, www.wsig.com, 919 Towne Center Drive, Suite 310, San Diego, CA 92122, Phone: (858) 678-0800.
- **Weston Presidio Capital**, www.westonpresidio.com, Pier 1, Bay 2 San Francisco, CA 94111 Phone: (415) 398-0770; or 200 Clarendon Street, 50th Floor Boston, MA 02116, Phone: (617) 988-2500.

## Financial and Business Services

- **The Canaan Team**, www.canaan.com/canaanteam/index.html, 105 Rowayton Avenue, Rowayton, CT 06853, Phone: (203) 855-0400; or 2884 Sand Hill Road, Suite 115, Menlo Park, CA 94025, Phone: (650) 854-8092.

- **CID Equity Partners**, www.cidequity.com, One American Square, Suite 2850, Box 82074, Indianapolis, IN 46282, Phone: (317) 269-2350; or 41 South High Street, Suite 3650, Columbus, OH 43215, Phone: (614) 222-8185; or 2 North LaSalle Street, Suite 1705, Chicago, IL 60602, Phone: (312) 578-5350; or 312 Elm Street, Suite 2600, Cincinnati, OH 45202, Phone: (513) 381-4748.
- **TA Associates**, www.ta.com, High Street Tower, Suite 2500, 125 High Street, Boston, MA 02110, Phone: (617) 574-6700; or 70 Willow Road, Suite 100, Menlo Park, CA 94025, Phone: (650) 328-1210; or One Oxford Center, Suite 4260, Pittsburgh, PA 15219-1407, Phone: (412) 441-4949.

## Health Care Management

- **Advanced Technology Ventures**, www.atv-ventures.com, 485 Ramona Street, Palo Alto, CA 94301, Phone: (650) 321-8601; or 281 Winter Street, Suite 350, Waltham, MA 02451, Phone (781) 290-0707.
- **Bessemer Venture Partners**, www.bessemervp.com, 535 Middlefield Road, Suite 245, Menlo Park, CA 94025, Phone: (650) 853-7000; or 83 Walnut Street, Wellesley Hills, MA 02481, Phone: (781) 237-6050.
- **The Canaan Team**, www.canaan.com/canaanteam/index.html, 105 Rowayton Avenue, Rowayton, CT 06853, Phone: (203) 855-0400; or 2884 Sand Hill Road, Suite 115, Menlo Park, CA 94025, Phone: (650) 854-8092.
- **CID Equity Partners**, www.cidequity.com, One American Square, Suite 2850, Box 82074, Indianapolis, IN 46282, Phone: (317) 269-2350; or 41 South High Street, Suite 3650, Columbus, OH 43215, Phone: (614) 222-8185; or 2 North LaSalle Street, Suite 1705, Chicago, IL 60602, Phone: (312) 578-5350; or 312 Elm Street, Suite 2600, Cincinnati, OH 45202, Phone: (513) 381-4748.
- **Euclid Partners**, www.euclidpartners.com, 45 Rockefeller Plaza, Suite 907, New York, NY 10111, Phone: (212) 218-6880.

- **Kleiner Perkins Caufield & Byers**, http://www.kpcb.com, 2750 Sand Hill Road, Menlo Park, CA 94025, Phone: (650) 233-2750; or Four Embarcadero Center, Suite 1880, San Francisco, CA 94111, Phone: (415) 421-3110.
- **New Enterprise Associates**, www.nea.com, 2490 Sand Hill Road, Menlo Park, CA 94025, Phone: (650) 854-9499; or One Freedom Square, 11951 Freedom Drive, Suite 1240, Reston, VA 20190, Phone: (703) 709-9499; or 1119 St. Paul Street, Baltimore, MD 21202, Phone: (410) 244-0115.
- **Sequel Venture Partners**, www.sequelvc.com, 4430 Arapahoe Avenue, Suite 220, Boulder, CO 80303, Phone: (303) 546-0400.
- **Weston Presidio Capital**, www.westonpresidio.com, Pier 1, Bay 2 San Francisco, CA 94111 Phone: (415) 398-0770; or 200 Clarendon Street, 50th Floor Boston, MA 02116, Phone: (617) 988-2500.

## Life Sciences/Biomedical/Pharmaceuticals/ Medical Devices

- **Alta Partners, Inc.**, www.altapartners.com, One Embarcadero Center, Suite 450, San Francisco, CA 94111, Phone: (415) 362-4022.
- **ARCH Venture Partners**, www.archventure.com, 8725 W. Higgins Road, Suite 290, Chicago, IL 60631, Phone: (773) 380-6600; or 45 Rockefeller Plaza, Suite 2071, New York, NY 10111, Phone: (212) 332-5053; or 1000 Second Avenue, Suite 3700, Seattle, WA 98104, Phone: (206) 674-3028; or 1155 University, S.E., Albuquerque, NM 87106, Phone: (505) 843-4293; or 6801 N. Capital of Texas Highway, Suite 225, Austin, TX 78731, Phone (512) 795-5830.
- **Atlas Venture**, www.atlasventure.com, 222 Berkeley Street, Boston, MA 02116, Phone: (617) 859-9290; or 1600 El Camino Real, Suite 290, Menlo Park, CA 94025, Phone: (650) 614-1444.
- **The Aurora Funds, Inc.**, www.aurorafunds.com, 2525 Meridian Parkway, Suite 220, Durham, NC 27713, Phone: (919) 484-0400.

- **BancBoston Ventures**, www.bancboscap.com/bbv/index.asp, 100 Federal Street, P.O. Box 2016, Boston, MA, 02110.
- **Bessemer Venture Partners**, www.bessemervp.com, 535 Middlefield Road, Suite 245, Menlo Park, CA 94025, Phone: (650) 853-7000; or 83 Walnut Street, Wellesley Hills, MA 02481, Phone: (781) 237-6050.
- **Brentwood Venture Capital**, www.brentwoodvc.com, 3000 Sand Hill Road, Bldg. 1, Suite 260, Menlo Park, CA 94025, Phone: (650) 854-7691; or 11150 Santa Monica Blvd., Suite 1200, Los Angeles, CA 90025, Phone (310) 477-7678; or 1920 Main Street, Suite 820, Irvine, CA 92614, Phone: (949) 251-1010.
- **The Canaan Team**, www.canaan.com/canaanteam/index.html, 105 Rowayton Avenue, Rowayton, CT 06853, Phone: (203) 855-0400; or 2884 Sand Hill Road, Suite 115, Menlo Park, CA 94025, Phone: (650) 854-8092.
- **CID Equity Partners**, www.cidequity.com, One American Square, Suite 2850, Box 82074, Indianapolis, IN 46282, Phone: (317) 269-2350; or 41 South High Street, Suite 3650, Columbus, OH 43215, Phone: (614) 222-8185; or 2 North LaSalle Street, Suite 1705, Chicago, IL 60602, Phone: (312) 578-5350; or 312 Elm Street, Suite 2600, Cincinnati, OH 45202, Phone: (513) 381-4748.
- **Enterprise Partners**, www.ent.com, 7979 Ivanhoe Avenue, Suite 550, La Jolla, CA 92037-4543, Phone: (858) 454-8833.
- **Euclid Partners**, www.euclidpartners.com, 45 Rockefeller Plaza, Suite 907, New York, NY 10111, Phone: (212) 218-6880.
- **Highland Capital Partners**, www.hcp.com, Two International Place, Boston, MA 02110, Phone: 617-531-1500; or 555 California Street, Suite 3100, San Francisco, CA 94104, Phone: (415) 981-1230.
- **Institutional Venture Partners**, www.ivp.com/intro1_snd1.html, 3000 Sand Hill Road, Building 2, Suite 290, Menlo Park, CA 94025, Phone: (650) 854-0132.

- **Intersouth Partners**, www.intersouth.com, 3211 Shannon Road, Suite 611, Durham, NC 27707, Phone: (919) 493-6640.
- **InterWest Partners**, www.interwest.com, 3000 Sand Hill Road, Building 3, Suite 255, Menlo Park, CA 94025, Phone: (650) 854-8585; or Two Galleria Tower, 13455 Noel Road, Suite 1670, Dallas, TX 75240, Phone: (972) 392-7279.
- **Kleiner Perkins Caufield & Byers**, www.kpcb.com, 2750 Sand Hill Road, Menlo Park, CA 94025, Phone: (650) 233-2750; or Four Embarcadero Center, Suite 1880, San Francisco, CA 94111, Phone: (415) 421-3110.
- **Menlo Ventures**, www.menloventures.com, 3000 Sand Hill Road M, Building 4, Suite 100, Menlo Park, CA 94025, Phone: (650) 854-8540.
- **Morgan Stanley Venture Partners**, www.msvp.com, 1221 Avenue of the Americas, 33rd Floor, New York, NY 10020, Phone: (212) 762-7900; or 3000 Sand Hill Road, Building 4, Suite 250, Menlo Park, CA 94025, Phone: (650) 233-2600.
- **New Enterprise Associates**, www.nea.com, 2490 Sand Hill Road, Menlo Park, CA 94025, Phone: (650) 854-9499; or One Freedom Square, 11951 Freedom Drive, Suite 1240, Reston, VA 20190, Phone: (703) 709-9499; or 1119 St. Paul Street, Baltimore, MD 21202, Phone: (410) 244-0115.
- **Oak Investment Partners**, www.oakinv.com, 525 University Avenue, Suite 1300, Palo Alto, CA 94301, Phone: (650) 614-3700; or One Gorham Island, Westport, CT 06880, Phone: (203) 226-8346; or 4550 Norwest Center, 90 South Seventh Street, Minneapolis, MN 55402, Phone: (612) 339-9322.
- **Onset Ventures**, www.onset.com, 2400 Sand Hill Road, Suite 150, Menlo Park, CA 94025, Phone: (650) 529-0700.
- **Olympic Venture Partners**, www.ovp.com, accepts inquiries by e-mail at info@ovp.com.

- **Sequel Venture Partners**, www.sequelvc.com, 4430 Arapahoe Avenue, Suite 220, Boulder, CO 80303, Phone: (303) 546-0400.
- **Sequoia Capital**, www.sequoiacap.com, 3000 Sand Hill Road, Bldg. 4, Suite 280, Menlo Park, CA 94025, Phone: (650) 854-3927.
- **TA Associates**,www.ta.com, High Street Tower, Suite 2500, 125 High Street, Boston, MA 02110, Phone: (617) 574-6700; or 70 Willow Road, Suite 100, Menlo Park, CA 94025, Phone: (650) 328-1210; or One Oxford Center, Suite 4260, Pittsburgh, PA 15219-1407, Phone: (412) 441-4949.
- **US Venture Partners**, www.usvp.com, 2180 Sand Hill Road, Suite 300, Menlo Park, CA 94025, Phone: (650) 854-9080.
- **Weiss, Peck & Greer**, www.wpgvp.com, Suite 3130, 555 California Street, San Francisco, CA 94104, Phone: (415) 622-6864.
- **Western States Investment Group**, www.wsig.com, 919 Towne Center Drive, Suite 310, San Diego, CA 92122, Phone: (858) 678-0800.
- **Weston Presidio Capital**, www.westonpresidio.com.

## Miscellaneous

- **Venture Investors**, ww.ventureinvestors.com, University Research Park, 505 South Rosa Road, Madison, WI 53719, Phone: (608) 441-2700.
- **Weston Presidio Capital**, www.westonpresidio.com, Pier 1, Bay 2 San Francisco, CA 94111 Phone: (415) 398-0770; or 200 Clarendon Street, 50th Floor Boston, MA 02116, Phone: (617) 988-2500.

## Physical Sciences

- **ARCH Venture Partners**, www.archventure.com, 8725 W. Higgins Road, Suite 290, Chicago, IL 60631, Phone: (773) 380-6600; or 45 Rockefeller Plaza, Suite 2071, New York, NY 10111, Phone: (212) 332-5053; or 1000 Second Avenue, Suite 3700, Seattle, WA 98104,

Phone: (206) 674-3028; or 1155 University, S.E., Albuquerque, NM 87106, Phone: (505) 843-4293; or 6801 N. Capital of Texas Highway, Suite 225, Austin, TX 78731, Phone (512) 795-5830.

## Retail and Other Consumer-Oriented Businesses

- **Oak Investment Partners**, www.oakinv.com, 525 University Avenue, Suite 1300, Palo Alto, CA 94301, Phone: (650) 614-3700; or One Gorham Island, Westport, CT 06880, Phone: (203) 226-8346; or 4550 Norwest Center, 90 South Seventh Street, Minneapolis, MN 55402, Phone: (612) 339-9322.
- **TA Associates**, www.ta.com, High Street Tower, Suite 2500, 125 High Street, Boston, MA 02110, Phone: (617) 574-6700; or 70 Willow Road, Suite 100, Menlo Park, CA 94025, Phone: (650) 328-1210; or One Oxford Center, Suite 4260, Pittsburgh, PA 15219-1407, Phone: (412) 441-4949.
- **US Venture Partners**, www.usvp.com, 2180 Sand Hill Road, Suite 300, Menlo Park, CA 94025, Phone: (650) 854-9080.
- **Weston Presidio Capital**, www.westonpresidio.com, Pier 1, Bay 2 San Francisco, CA 94111 Phone: (415) 398-0770; or 200 Clarendon Street, 50th Floor Boston, MA 02116, Phone: (617) 988-2500.

## Additional Venture Capital Resources

- **www.vfinance.com** includes a database of VC firms organized by business sector.
- **Yahoo.com** maintains a comprehensive list of VC firms at http://dir.yahoo.com/Business_and_Economy/Business_to_Business/Financial_Services/Finance_and_Receivables/Financing/Corporate_Finance/Venture_Capital.

- An extensive list of funds in alphabetical order is located at **http://advocacy-net.com/venturemks.htm.**
- Search for VC and other types of funding at **www.businessfinance.com**.

# Appendix D: List of Business Plan Tools

- Download free demo versions of **Business Plan Pro**, **Marketing Plan Pro** and **Cash Plan Pro** from **www.bplans.com**. Developed by Palo Alto Software, the demos will help you determine if you want to take it to the next level and purchase software to develop your plans. This site also offers a **MiniPlan** function that allows you to put together a test plan for your idea. It includes a test of your objectives, helps you define a mission and other helpful steps.

  Another helpful tool available on this site is the **Plan Wizard**. This tool takes five minutes to go through and asks questions like "How new is your company?" and "At what stage in the development of the plan are you at?" At the end of the Plan Wizard process, you receive a list of sample plans that most closely match your business stage and criteria. All plans are downloadable in a PDF format, viewable online or available through the Business Plan Pro software.

- The **Small Business Administration** has an extremely helpful business plan tutorial located at **www.sba.gov/starting/indexbusplans.html**. This tutorial, which can be downloaded or viewed as a text file, walks you through each step of the business plan writing process. It goes beyond the basics of business plan writing to in-depth coverage of strategies for marketing and reaching out to consumers and offers a comprehensive discussion of financial planning for your business.

- Links to several helpful articles, including "**Creating Your Business Plan**," "**Structuring Your Business Plan**" and "**Planning for Profits**" are located at **www.businesstown.com/planning/creating.asp**.
- **Invest-Tech** offers a white paper on business planning located at **www.planware.org/bizplan.htm**. It includes a downloadable (in Microsoft Word format) 36-page business plan guide and template and offers good advice on why you need a plan and each stage of the plan. The site includes a free financial planner at www.planware.org/online.htm and shareware versions of some of the software offered by this company, including **Exl-Plan**, a Microsoft Excel-based program that helps you project revenue and expenses for your business's first five years.
- **American Express's Small Business Services** Web site offers helpful and important information for any stage of your business. One of its best tools: the **Small Business Exchange Business Plan Workshop** at **http://home3.americanexpress.com/smallbusiness/resources/starting/biz_plan/index.shtml?aexp_nav=hp_ads**. This workshop walks you through the primary elements of a business plan: introductory elements; business description; the market, development and production, sales and marketing, management and financials. It also allows you to test your planning skills on a fictional business plan located at **http://home3.americanexpress.com/smallbusiness/resources/starting/biz_plan/try**. It examines the plan for Bella's Biscotti, a start-up that hopes to supply biscotti to gourmet shops in a fictional metropolitan city and its suburbs. During the exercise, you are presented with a step in the planning process and a description of three options for what Bella's Biscotti's owner could do next. You select what you think is the best option and the workshop explains why that option is or is not the best of the three.
- **BizPlanIt.com's Virtual Business Plan** at **www.bizplanit.com/vplan.htm** walks you through each step of the planning process, from executive summary to industry analysis to exit strategy. Each section includes a synopsis of the basic principles of developing that plan sec-

tion, a list of common mistakes to avoid and additional advice. BizPlanIt.com also offers a monthly e-mail newsletter that includes tips on improving your business plan. A sign-up form for the newsletter is located at **www.bizplanit.com/free/newsletter1.htm**.

- The **Canada/British Columbia Business Service Centre** has a business planning outline on the Internet at **www.sb.gov.bc.ca/smallbus/workshop/busplan.html**. It walks you through the planning process, outlining each piece of the business plan.
- The **Planning Your Business Module at the Business Owner's Toolkit** includes information on what a plan can do for you, what documents and information to gather before putting the plan together, tips on writing the plan and how to make it more appealing for the reader, and how to use the completed plan most effectively. This tool is located at **www.toolkit.cch.com/text/P02_0001.asp**.
- The **Entrepreneurial Edge** has a list of answers to "frequently asked questions" about writing a business plan posted at **http://edge.lowe.org/fmpro?-db=library.fp5&-format=library/docs/index.htm&record=7573&-find**.
- In conjunction with Bloomberg, **Learn2.com** offers a 12-step tutorial to writing your own business plan. The plan is located at **www.learn2.com/bloom/0603.html** and also includes links to other helpful plans, such as how to obtain a business license, how to rent commercial space and how to hire employees.
- **Deloitte & Touche**, a major management consulting firm, offers a primer on writing a business plan at **www.us.deloitte.com/growth/guidebooks/busplan.htm**. This article outlines the features of a plan and what each feature's most important elements are.
- A **free short course on writing business plans**—put together by a professional marketing communications expert—is located at **www.users.cloud9.net/~kvivian/html/business_plans_.html**. Check it out for tips on writing either a technology-oriented or traditional business plan.

## College Programs

College and university marketing and business programs often develop sample plans. And other college programs do more than simply providing a sample business plan—some also post tools that can help you as you develop your plan. Here are a few of the most helpful:

- The **Howard University Small Business Development Center** has put together a comprehensive outline of what needs to be in a business plan at **www.ntia.doc.gov/opadhome/mtdpweb/busplano.htm**. This outline includes lists of questions—if you can answer them and put the answers in narrative form, you'll be one step closer to developing a thorough business plan.
- The **University of Colorado at Boulder's Center for Entrepreneurship** has posted three business plan templates in Microsoft Word 95 format. They are easy to use and modify with your own business information. They are located at **http://bus.colorado.edu/faculty/lawrence/documents/templates.htm**.
- Examples of the best business plans entered in the **MOOT CORP® Competition** are located at **www.businessplans.org/MootCorp.html**. The competition invites MBA candidates from the best business schools in the United States to submit their business plans to panels of investors. This site also includes a selection called the Best of the Best. These four plans offer the best examples of an executive summary, business concept, financial tables and proposed offer. Plans from the list of 1999 winners include a textile company, a customized advertising service for the Internet and a recreational tourism company.

# APPENDIX E: LIST OF SBDCs

- University of Alabama, Birmingham, AL, (205) 943-6750
- University of Alaska/Anchorage, Anchorage, AK, (907) 274-7232
- Maricopa County Community College, Tempe, AZ, (480) 731-8720
- University of Arkansas, Little Rock, AR, (501) 324-9043
- California Trade and Commerce Agency, Sacramento, CA, (916) 324-9538
- Office of Business Development, Denver, CO, (303) 892-3794
- University of Connecticut, Storrs, CT, (860) 486-4135
- University of Delaware, Newark, DE, (302) 831-2747
- Howard University, Washington, DC, (202) 806-1550
- University of West Florida, Pensacola, FL, (850) 595-6060
- University of Georgia, Athens, GA, (706) 542-6762
- University of Guam, Mangialo, Guam, (671) 735-3590
- University of Hawaii at Hilo, Hilo, HI, (808) 974-7683
- Boise State University, Boise, ID, (208) 462-3877
- Dept. of Commerce & Community Affairs, Springfield, IL, (217) 524-5856

- Economic Development Council, Indianapolis, IN, (317) 264-2820
- Iowa State University, Ames, IA, (515) 292-6351
- Fort Hays State University, Hays, KS, (785) 296-6514
- University of Kentucky, Lexington, KY, (606) 257-7668
- Northeast Louisiana University, Monroe, LA, (318) 342-5506
- University of Southern Maine, Portland, ME, (207) 780-4420
- University of Maryland, College Park, MD, (301) 403-8303
- University of Massachusetts, Amherst, MA, (413) 545-6301
- Wayne State University, Detroit, MI, (313) 964-1798
- Dept. of Trade and Economic Development St. Paul, MN, (651) 297-5770
- University of Mississippi, University, MS, (601) 232-5001
- University of Missouri, Columbia, MO, (573) 882-0344
- Department of Commerce, Helena, MT, (406) 444-4780
- University of Nebraska at Omaha, Omaha, NE, (402) 554-2521
- University of Nevada,Reno, Reno, NV, (775) 784-1717
- University of New Hampshire, Durham, NH, (603) 862-6975
- Rutgers University, Newark, NJ , (973) 353-1927
- Santa Fe Community College, Santa Fe, NM, (505) 428-1362
- State University of New York, Albany, NY, (518) 443-5398
- University of North Carolina, Raleigh, NC, (919) 715-7272
- University of North Dakota, Grand Forks, ND, (701) 777-3700
- Department of Development, Columbus, OH, (614) 466-2711
- S.E. Oklahoma State University, Durant, OK, (405) 924-0277
- Lane Community College, Eugene, OR, (541) 726-2250

- University of Pennsylvania, Philadelphia, PA, (215) 898-1219
- Inter American University, Hato Rey, PR, (787) 763-6811
- Bryant College, Smithfield, RI, (401) 232-6111
- American Samoa Community, Pago Pago, Samoa, 001 684-699-9155
- University of South Carolina, Columbia, SC, (803) 777-4907
- University of South Dakota, Vermillion, SD, (605) 677-5287
- University of Memphis, Memphis, TN, (901) 678-2500
- Tennessee Board of Regents, Nashville, TN, (615) 366-3931
- Dallas Community College, Dallas, TX, (214) 860-5835
- University of Houston, Houston, TX, (713) 752-8425
- Texas Tech University, Lubbock, TX, (806) 745-3973
- University of Texas at San Antonio, San Antonio, TX, (210) 458-2450
- Salt Lake City Community College, Salt Lake City, UT, (801) 957-3489
- Vermont Technical College, Randolph Center, VT, (802) 728-9101
- University of the Virgin Islands, St. Thomas, US VI, (340) 776-3206
- Dept. of Economic Development, Richmond, VA, (804) 371-8251
- Washington State University, Pullman, WA, (509) 338-7765
- Governor's Office of Community and Industrial Development, Charleston, WV, (304) 558-2960
- University of Wisconsin, Madison, WI, (608) 253-7794
- University of Wyoming, Laramie, WY, (307) 766-3505

# Appendix F: List of Women's Mentoring Organizations

- The **American Business Women's Association**, founded in 1949, has spent more than 50 years providing workplace skills and development opportunities for women around the country. ABWA has more than 545,000 members. Contact them at American Business Women's Association 9100 Ward Parkway, P.O. Box 8728, Kansas City, MO 64114-0728. Phone: (800) 228-0007. Internet: **www.abwahq.org**.
- **American Business Women International** was founded in 1995 and has offices in California, New Mexico and Illinois. Contact them at American Business Women International 4829 Corrales Road, Corrales, NM 87048. Phone: (800) 606-ABWI. Internet: **www.abwi.org**. ABWI supports women at all levels of business who are interested in international trade issues.
- From one-on-one mentoring to training, the **American Woman's Economic Development Corporation** offers myriad services to women developing their business. Although the organization is based in New York, it has offices in Southern California, Washington, DC and Connecticut. American Woman's Economic Development Corporation 71 Vanderbilt Avenue, Suite 320, New York, NY 10169. Phone: (212) 692-9100.
- **Ann Arbor Community Development Corporation Women's Initiative for Self Employment (WISE)** 2008 Hogback Road, Suite 2A, Ann Arbor, MI 48105. Phone: (313) 677-1400.

- **Asian Women in Business (AWIB)** was founded in 1995 and supports Asian women entrepreneurs through "information, education and networking opportunities."Asian Women in Business 1 West 34th Street, Suite 200, New York, NY 10001. Phone: (212) 868-1368. Internet: **www.awib.org**.
- Along with the mentoring services **Coastal Enterprises, Inc.** provides to all Maine residents, the organization has a Women's Business Development Program targeted specifically at low-income women and women business owners.Contact them at Coastal Enterrpises, Inc. 36 Water Street, P.O. Box 268, Wiscasset, ME 04578. Phone: (207) 882-7552. Internet: **www.ceimaine.org**.
- The **Denver Business Women's Network** supports businesswomen in the Denver, Colorado, area. The site includes a list of member businesses, that is helpful if you're trying to find someone in a similar business to talk to or ask for mentoring. Contact them at: Denver Business Women's Network P.O. Box 211284, Denver, CO 80221. Phone: (303) 448-4948. Internet: **www.dbwn.org**.
- The **EMPOWER Pyramid Career Services** organization works one-on-one with start-up women business owners. Contact them at EMPOWER Pyramid Career Services 2400 Cleveland Ave., NW, Canton, OH 44709. Phone: (216) 453-3767.
- The **Florida Women's Business Development Center** offers a mentoring program and additional educational and networking programs for women entrepreneurs.Florida Women's Business Development Center EAS 2611, University Park, Miami, FL 33199. Phone: (305) 348-3951, ext.3903.
- The **Forum for Women Entrepreneurs** (FWE) is one of the best mentoring organizations for women who want to break into the high-tech world. The FWE, which also has offices in Los Angeles and Seattle, has more than 850 members, most of who are referred to the organization by other members or women involved in technology. Contact them at Forum for Women Entrepreneurs 2600 Campus Drive, Suite 200, San Mateo, CA 94403. Phone: (650) 357-0333. Internet: **www.fwe.org**.

- **Greater Columbus Women's Business Initiative** 37 North High Street, Columbus, OH 43215-3065. Phone: (614) 225-6082.
- The **Massachusetts Center for Women and Enterprise, Inc.** is a non-profit educational organization that targets women business owners. Scholarships are available for low-income women who want to attend the training provided by the center, and there are also mentoring services available.Contact them at Massachusetts Center for Women and Enterprise, Inc. 45 Bromfield Street, 6th Floor, Boston, MA 02108. Phonc: (617) 423-3001, ext. 222.
- Founded as an organization to support low-income Hispanic women as they develop the skills they need to become self-sufficient, **Mi Casa** has expanded its mission to include entrepreneurial support for this group of women. The Business Development Training Program is a comprehensive educational program that helps women get the tools they need to succeed with their start-up business. The program includes one-on-one mentoring services.Contact them at Mi Casa 571 Galapago Street, Denver, CO 80204. Phone: (303) 573-1302. Internet: **http://micasadenver.org**.
- The **Mississippi Women's Economic Entrepreneurial Project**, established by the National Council of Negro Women, offers mentorship and other services to women in this designated Rural Enterprise Zone. Contact them at Mississippi Women's Economic Entrepreneurial Project 106 West Green Street, Mound Bayou, MS 38762. Phone: (601) 741-3342.
- Local women entrepreneurs from the Washington, DC-area offer programs and workshops and other mentoring opportunities for up-and-coming women business owners at the **National Women's Business Center**. The Crestar Coaching Series, designed for women in the early stages of their businesses, is available by application to women who want more intensive, personal mentoring. Contact them at the National Women's Business Center 1001 Connecticut Avenue, NW, Suite 312, Washington, DC 20036. Phone: (202) 785-4WBC. Internet: **www.womensbusinesscenter.org**.

- The **Northwest Ohio Women's Entrepreneurial Network (NOWEH)** is developing a program that would allow an expert panel of business owners to review new business plans. Contact them at Northwest Ohio Women's Entrepreneurial Network 5555 Airport Highway, Suite 210, Toledo, OH 43615. Phone: (419) 381-7555.
- Although there is a membership fee to belong to the **Ohio Business Women's Resource Network**, that fee entitles business owners to mentoring and other educational services.Contact them at Ohio Business Women's Resource Network 77 South High Street, 28th Floor, Columbus, OH 43266-0101. Phone: (614) 466-2682.
- With chapters from California to Massachusetts and from Bermuda to Korea, the **Organization of Women in International Trade** promotes women doing business in international trade has something to offer women—and men—in many areas of the world. However, if there isn't a chapter near you, the organization offers a Virtual Chapter through their Web site at **www.owit.org.**, complete with message boards and other resources for locating mentors and networking. This organization promotes mentorship among women around the world.
- The **Pennsylvania Women's Business Development Center (WBDC)** uses local successful woman entrepreneurs to mentor and work with start-up business owners. Contact them at Pennsylvania Women's Business Development Center 1315 Walnut Street, Suite 1116, Philadelphia, PA 19107-4711. Phone: (215) 790-9232.
- Women and minorities who live in Mendocino County, California, can get one-on-one mentoring and assistance through **West Company**, a microenterprise incubator that helps get small businesses on their feet. This is also a good place to go for funding if you are part of the community this organization serves. Contact them at West Company 367 N. State St., Suite 201, Ukiah, CA 95482. Phone: (707) 468-3553; or 306 E. Redwood Ave., Suite 2, Fort Bragg, CA 95437. Phone: (707) 964-7571. Internet: **www.westcompany.org**.
- The **Women's Business Assistance Center, Inc.** offices provide counseling and mentoring services to women business owners, along

with myriad additional training and other assistance programs. Contact them at Women's Business Assistance Center Inc. Alabama Office: 1301 Azalea Rd., Mobile, AL 36660. Phone: (334) 660-2725. Internet: **http://ceebic.org/~wbac/WBC%20So%20AL2.html**; or Florida Office: 6235 N. Davis Hwy, Pensacola, FL 32504. Phone: (850) 484-2765. Internet: **http://ceebic.org/~wbac/WBC%20FL2.html**.

- The **Women's Business Center White Earth Reservation Tribal Council** is located on the White Earth Reservation in Minnesota, and offers mentoring services and several training seminars to Native American women who live on the reservation. Contact them at Women's Business Center White Earth Reservation Tribal Council North Main Street, P.O. Box 478, Mahnomen, MN 56557. Phone: (218) 935-2827.
- **Women's Business Development Center** 8 South Michigan Avenue, Suite 400, Chicago, IL 60603. Phone: (312) 853-3477.
- The **Women's Business Institute**, which serves North Dakota, Minnesota and the surrounding areas, offers mentoring services at a reduced fee for institute members through its Business Success Teams. Contact them at Women's Business Institute 320 N 5th St. Suite 203, P.O. Box 2043, Fargo, ND 58107-2043. Phone: (701) 235-6488.
- **Women's Business Owners Corporation** 18 Encanto Drive, Palos Verdes, CA 90274-4215. Phone: (310) 530-7500.
- The **Women's Enterprise Development Corporation** offers assistance, training and mentoring to women business owners in California. Contact them at Women's Enterprise Development Corporation 100 West Broadway, Suite 500, Long Beach, CA 90802. Phone: (562) 983-3747.
- **Women Entrepreneurs for Economic Development** 1683 N. Claiborne Ave., Suite 100, New Orleans, LA 70016. Phone: (504) 947-8522.
- **Women Entrepreneurs, Inc.** Bartlett Building, 36 East 4th Street, Cincinnati, OH 45202. Phone: (513) 684-0700.

- Young women in the Greater Washington, DC-area can join the **Women's Information Network (WIN)**, which offers one of the best mentoring activities available in the region. The annual Women Opening Doors for Women dinner offers women entrepreneurs an opportunity to spread the word about their business and ask prominent women in the Washington, DC business community to serve as mentors to them. The evening begins with a large reception for all attendees and speakers, and then the women break off into small groups for dinner at a private home to hear additional speakers in their focus area. The June 2000 dinner focus areas included: Women Entrepreneurs: The Color of Money (B. Smith, chef and cookbook author, was one of the invited speakers); Women in Management Consulting; SHE-Commerce: Netting Success Online; Dot-Com Queens: Women Making Waves in the Internet Start-up Craze; and One P.R. Life To Live, Women in Communications. WIN also sponsors events all year long that allow women to mix and mingle with others who could serve as their mentors. Contact them at Women's Information Network 1800 R Street NW, Unit C-4, Washington, DC 20009. Phone: (202) 347-2827. Internet: **www.winonline.org**.

- **Women's Initiative for Self-Employment** offers one-on-one counseling and mentoring services and other support to women business owners in English and Spanish. Courses offered include a business assessment workshop, a business skills workshop and a workshop on writing a business plan. Contact them at Women's Initiative for Self-Employment 450 Mission Street, Suite 402, San Francisco, CA 94105. Phone: (415) 247-9473.

- The **Women in New Development Bi-County Community Action Programs, Inc.** provides technical assistance and mentoring to new and young businesses in certain counties in Minnesota. Contact them at Women in New Development Bi-County Community Action Programs, Inc. P.O. Box 579, Bemidji, MN 56601. Phone: (218) 751-4631.

- **Women's Opportunity & Resource Development, Inc.** provides development opportunities for small businesses through group trainings.

The trainings are provided by local business experts who mentor the participants. Contact them at Women's Opportunity & Resource Development, Inc. 127 N. Higgins, Missoula, MT 59802. Phone: (406) 543-3550.

- The **Working Women's Money University (WWMU)** calls itself an "entrepreneurial training camp" for women. Although the focus is on raising money to get businesses going, the trainings conducted by WWMU will expose business owners to mentors through the professional entrepreneurs who run the programs. Contact them at Working Women's Money University 234 Quadrum Drive, Oklahoma City, OK 73108. Phone: (405) 232-8257.

In addition to these resources, check out the resources on networking for women listed in Chapter 2. Networking is an excellent way to find a mentor.

# Appendix G: List of Business Incubators

The following is a list of business incubators around the country. Contact the ones near you for more information about the services that they offer. If one has a particular focus, that is listed with the contact information.

- Northeast Alabama Entreprenuerial System, 1400 Commerce Blvd., Suite 1, Anniston, AL 36207. Phone: (256) 831-5215. Internet: **www.neaes.org/index.html**. Focuses on service and light manufacturing businesses.
- Office for the Advancement of Developing Industries (OADI), 2800 Milan Court, Birmingham, AL 35211. Phone: (205) 943-6560. Internet: **www.uab.edu/oadi**. Focuses on high technology companies.
- Entrepreneurial Center, 110 12th Street North, Birmingham, AL 35203. Phone: (205) 250-8000. Internet: **www.entrepreneurialctr.com**. Focuses on service and light manufacturing businesses.
- Center for Entrepreneurial Excellence, Mobile, AL. Phone: (334) 660-7002. Internet: **www.ceebic.org**.
- Montgomery Area Small Business Incubator, 600 South Court Street, P.O. Box 79, Montgomery, AL, 36101. Phone: (334) 240-6863. Internet: **www.montgomeryincubator.org**. Focuses on service and light manufacturing businesses.
- Ozark Technology Center, 1807 U.S Hwy 231 S., Ozark, Al 36360. Phone: (334) 774-4952. Internet: **www.bizincubator.org**. Multi-use small business incubator.

- Sitka Business Incubator, 303 Lincoln Street, Suite 3, Sitka, AK 99835. Phone: (907) 966-3301. Internet: **www.incubationworks.com**.
- Arizona Technology Incubator, 1435 N. Hayden Road, Scottsdale, AZ 85257. Phone: (480) 990-0400. Internet: **www.accessarizona.com/community/groups/ati/index.html**. Focuses on high technology businesses.
- Tucson Technology Incubator, The University of Arizona, Science and Technology Park, 9040 South Rita Road, Suite 1100, Tucson, AZ 85747. Phone: (520) 663-3597. Internet: **www.tucsonincubator.org/home.htm.**
- Genesis Technology Incubator, (Mailing Address) 1 University of Arkansas, Fayetteville, AR 72701-1201, (Physical Address) 700 Research Center Boulevard, Fayetteville, AR 72701. Phone: (501) 575-7227.
- Arkansas Biotechnology Incubator, University of Arkansas for Medical Sciences, Biomedical Biotechnology Center, 4301 West Markham Street, Slot 718, Little Rock, AR 72205. Phone: (501) 686-6696. Internet: **http://biotech.uams.edu**.
- CALSTART Project Hatchery Business Incubator, 3360 E. Foothill Boulevard, Pasadena, CA 91107. Phone: (626) 744-5600. Internet: **www.calstart.org/calindex3.html**.
- Business Technology Center of Los Angeles County, 2400 Lincoln Avenue, Altadena, CA 91001. Phone: (626) 296-6300. Internet: **www.labtc.org**.
- EC2, 746 West Adams Boulevard, Los Angeles, CA 90089-7727. Phone: (213) 743-2344. Internet: **www.ec2.edu**. Focuses on companies developing new media content or communications technology.
- Communications Technology Cluster, 2201 Broadway, 2nd Floor, Oakland, CA 94612-1932. Phone: (510) 836-8985. Internet: **www.ctcluster.com**. Focuses on communications technology firms.
- Center for Applied Competitive Technologies at San Diego City College, 1313 Twelfth Avenue, San Diego, CA 92101. Phone: (619)

230-2080. Internet: **www.cact-sd.org**. Focuses on technology businesses.

- San José Software Business Cluster, 2 North First Street, 4th Floor, San Jose, CA 95113. Phone: (408) 535-2701. Internet: **www.sjsbc.org**. Focuses on software start-ups.
- Colorado Venture Centers, 1610 Pierce Street, Texas Building, Lakewood, CO 80214. Phone: (303) 237-3998. Internet: **www.coloradoventure.org**. Focuses on businesses involved in biotech, software, environment, advanced materials and alternative energy.
- Boulder Technology Incubator, Marine Street Science Building, 3215 Marine Street, Boulder, CO, 80303. Phone: (303) 492-8585. Also located at 1821 Lefthand Circle, Suite B, Longmont, CO, 80501. Phone: (303) 678-8000.
- Enterprise North Florida Corporation, 7400 Baymeadows Way, Suite 201, Jacksonville, FL 32256, (904) 730-4700. Focuses on technology-based companies.
- Bay County Small Business Incubator, 2500 Minnesota Avenue, Lynn Haven, FL 32444. Phone: (850) 271-1107. Internet: **http://enfc.org**. Focuses on newly formed or expanding service industry, light manufacturer, assembler, or research & development firms.
- Enterprise Development Corporation of Florida, 3998 FAU Boulevard, Suite 200, Boca Raton, Florida 33431. Phone: (561) 620-8494. Internet: **www.edc-tech.org**. Focuses on technology-based companies.
- Seminole Technology Business Incubation Center, 1445 Dolgner Place, Sanford, FL 32771. Phone: (407) 321-3495. Internet: **www.seminoleinc.com**. Focuses on technology-based companies.
- Florida/NASA Business Incubation Center, 5195 South Washington Avenue, Titusville, FL 32980. Phone: (321) 269-6330. Internet: **www.trda.org/fnbic**. Focuses on technology-based companies.

- The Fulton County Business Incubator, 5534 Old National Highway, Building H, Suite 300, Atlanta, GA 30349. Phone: (404) 559-9466. Internet: **www.fcbi.org**. Focuses on the telecommunication, information technology and related industries.
- Augusta-Richmond County Small Business Incubator, 3140 Augusta Tech Drive, Augusta, GA 30906-3381. Phone: (706) 792-9044. Internet: **www.arcsbi.com**. Focuses on service, manufacturing, and research & development businesses.
- The South DeKalb Business Incubator, 2632 Rainbow Way, Decatur, GA 30034. Phone: (404) 241-3522. Internet: **www.sdbusinc.org/index.htm**. Includes the Incubator Without Walls program for businesses that are not quite ready to take the plunge for their own office space.
- Manoa Innovation Center, 2800 Woodlawn Drive, Suite 100, Honolulu, HI 96822. Phone: (808) 539-3600. Internet: **www.htdc.org/mic/mic.html**. Focuses on high technology businesses.
- Idaho Innovation Center, 2300 North Yellowstone, Idaho Falls ID 83401. Phone: (208) 523-1026. Internet: **www.iictr.com**.
- The Bonner Business Center, 804 Airport Way, Sandpoint, ID 83864. Phone: (208) 263-4073. Internet: **www.sandpoint.org/bbc**. Focuses on light manufacturing, assembly, wholesale distribution, research and development, manufacturer's representatives, and service companies. Also has special food preparation areas for businesses involved in specialty food product development.
- CSI Business Incubator, College of Southern Idaho, 315 Falls Avenue, P.O. Box 1238, Twin Falls, ID 83303-1238. Phone: (208) 733-9554, ext. 2450. Internet: **www.csi.cc.id.us/support/isbdc/incubator/businc.htm**.
- North Central Idaho Business Technology Incubator, 121 Sweet Avenue Moscow, ID 83843. Phone: (208) 885-3801. Focuses on high technology and biotechnology.

- Idaho State University Business and Technology Center, Campus Box 8044, 1651 Alvin Ricken Drive, Pocatello, ID 83201. Phone: (208) 236-2430.
- Upper Snake River Valley Incubator, 310 N. 2nd East, Rexburg ID 83440. Phone: (208) 356-4524, ext. 322.
- North Idaho Business Center for Innovation & Development, 11100 Airport Drive, Hayden, ID 83835. Phone: (208) 772-0584.
- Dunn Richmond Economic Development Program Small Business Incubator, Office of Economic and Regional Development, Southern Illinois University, Carbondale, IL 62901. Phone: (618) 536-4047. Internet: **www.siu.edu/~econdev/bip/bipindex.htm**.
- Chicago Southland Development Inc., 1655 Union, Chicago Heights, IL 60411. Phone: (708) 754-6960. Focuses on a range of businesses, from service to light manufacturing.
- Business Center of Decatur, 2121 S. Imboden Ct., Decatur, IL 62521. Phone: (217) 423-2832. Focuses on office, manufacturing and warehouse businesses.
- Technology Innovation Center, 1840 Oak Avenue, Evanston, IL 60201. Phone: (847) 864-0800. Internet: **www.theincubator.org**.
- Technology Commercialization Laboratory, Urbana, IL. Internet: **www.aces.uiuc.edu/~tcl**.
- Evansville's Small Business Incubators, Evansville, IN.
- Venture Out Business Center, 975 Industrial Drive, Madison, IN 47250. Phone: (812) 273-6510. Internet: **www.vobc.com/page7.html**. Focuses on light manufacturing and service-oriented companies.
- Iowa State Innovation System, 2501 North Loop Drive, Suite 600, Ames, IA 50010. Phone: (515) 296-PARK. Internet: **www.isupark.org**.
- Enterprise Center of Johnson County, 9875 Widmer Road, Lenexa, KS 66215. Phone: (913) 438-2282. Internet: **www.ecjc.com**.

- Louisiana Business and Technology Center, South Stadium Drive, Baton Rouge, LA 70803-6100. Phone: (225) 334-5555. Internet: **www.bus.lsu.edu/lbtc**.
- The Enterprise Center of Louisiana. Phone: (337) 896-9115. Internet: **www.ecol.org**.
- The NOBID Enterprise Center, 13801 Old Gentilly Road, New Orleans, LA 70129. Phone: (504) 254-2600. Internet: **www.nobid.org**.
- UMBC Research Park and Technology Center, 1000 Hilltop Circle, Baltimore, MD 21250. Phone: (410) 455-1000. Internet: **www.umbc.edu/Business/Research**.
- Technology Advancement Program, 387 Technology Drive, University of Maryland, College Park, MD 20742. Phone: (301) 314-7803. Internet: **www.erc.umd.edu/TAP**.
- The Enterprise Center at Salem State College, 121 Loring Ave., Salem, MA 01970. Phone: (978) 542-7528. Internet: **www.enterprisectr.org**. Focuses on technology, service and light manufacturing businesses, but also works with micro-enterprises. Space is available at market rates, but that includes educational programs and mentoring.
- MBI International, P.O. Box 27609, Lansing, MI 48909-0609. Phone: (517) 337-3181. Internet: **www.mbi.org**. Focuses on biotechnology research and development.
- Genesis Business Centers, Ltd., 3989 Central Ave. N.E., Suite 530, Columbia Heights, MN 55421. Phone: (612) 782-8576. Internet: **www.genesiscenters.com**. Focuses on high technology start-ups.
- University Technology Enterprise Center, Ltd., 1313 Fifth Street SE, Minneapolis, MN 55414. Phone: (612) 379-3800. Internet: **www2.pro-ns.net/~utec**.
- The Golden Triangle Enterprise Center, One Research Boulevard, Suite 201, Starkville, MS 39759. Phone: (662) 320-3990. Internet: **www.gtec.org**. Focuses on high technology.

- Mississippi Enterprise for Technology, Building 1103, Suite 140A, John C. Stennis Space Center, MS 39529-6000. Phone: (228) 688-2083. Internet: **www.mset.org/incubator.html**.
- The Technology Development Center, University of Nebraska Technology Park, 4701 Innovation Drive, Lincoln, NE 68521-5330. Phone: (402) 472-4200. Internet: **www.unebtechpark.com/TDC.asp**.
- Stevens Technology Ventures Incubator, 610 River Street, Hoboken, New Jersey 07030-5053. Phone: (201) 216-5366. Internet: **http://attila.stevens-tech.edu/tvi.**
- Ground Floor Ventures, 720 Monroe Street, Suite E-209, Hoboken, NJ 07030. Phone: (201) 420-4446. Focuses on women owned and/or operated software and Internet technology businesses.
- Center for Environmental Sciences & Technology, University at Albany, CESTM B203, 251 Fuller Road, Albany, NY 12203. Phone: (518) 437-8600. Internet: **www.albany.edu/pr/CESTMINCU.html**.
- Western New York Technology Development Center, Inc., Baird Research Park, 1576 Sweet Home Road, Amherst, NY 14228. Phone: (716) 636-3626. Internet: **http://wings.buffalo.edu/wnytdc**.
- The Case Center, 2-212 Center for Science and Technology, 111 College Place, Syracuse University, Syracuse, NY 13244. Phone: (315) 443-1060. Internet: **www.cat.syr.edu**. Focuses on high technology start-ups.
- Incubator Program at Rennselaer Polytechnic Institute, 1223 Peoples Ave., Troy, NY 12180. Phone: (518) 276-6658. Internet: **www.rpi.edu/dept/incubator/homepage/index.html**. Focuses on technology businesses.
- Cincinnati Business Incubator, Inc., 1634 Central Parkway, Cincinnati, OH 45210. Phone: (513) 362-2703. Internet: **http://cbincubator.org**. Focuses on women- and minority-owned businesses.

- Lewis Incubator for Technology, 4440 Warrensville Center Road, Cleveland, OH 44128, (216) 586-3888 or 21000 Brookpark Road, MS 501-6, Cleveland, OH 44135. Phone: (216) 433-5300. Internet: **www.liftinc.org**.
- Dayton/Miami Valley Entrepreneurs Center, Suite #106, 3155 Research Blvd., Kettering, OH 45420. Phone: (937) 258-5400. Internet: **www.techincubator.org**. Focuses on technology enterprises.
- Allen Economic Development Group, 147 N. Main Street, Lima, OH, 45801. Phone: (419) 222-7706. Internet: **www.aedg.org**. This organization is developing a business incubator program.
- Youngstown Business Incubator, 241 Federal Plaza West, Youngstown, OH 44503. Phone: (330) 746-5003. Internet: **www.ybi.org**. Focuses on technology-based enterprises.
- Pontotoc Area Vo-Tec School, 601 West 33rd, Ada, OK 74820. Phone: (405) 436-0180, ext. 2244. Internet: **www.pontotoc.com**. Focuses on manufacturing products.
- Meridian Technology Center, 1312 South Sangre Road, Stillwater, OK 74074-1899. Phone: (405) 377-3333. Internet: **www.meridian-technology.com./Home_Nav3.asp**. Focuses on entrepreneurs, early-stage technology companies, service companies and companies seeking to commercialize new products.
- Oregon Innovation Center, P.O. Box 1510, Redmond, OR 97756. Phone: (541) 504-2929. Internet: **www.innovationcenter.org**. Focuses on technology-based start-ups.
- Ben Franklin Technology Partners, 200 N. Third Street, Suite 400, Harrisburg, PA 17101. Phone: (717) 234-1748. Internet: **www.net.bfp.org**. This organization has four centers around the state.
- University City Science Center, 3624 Market Street, Philadelphia, PA 19104. Phone: (215)387-2255. Internet: **www.ucsc.org**.
- West Philadelphia Enterprise Center, 4548 Market Street, Philadelphia, PA 19139. Phone: (215) 895-4000. Focuses on small businesses and urban entrepreneurs.

- Centre County Business Incubator Program, 200 Innovation Blvd., Suite 201, State College, PA 16803. Phone: (814) 234-1829. Internet: **www.cbicc.org**.
- William C. Goodridge Business Resource Center, 140 Roosevelt Avenue, York, PA 17404. Phone: (717) 852-0408. Internet: **www.starinc.com/goodridge**.
- Technology 2020 Business Incubation Program, 1020 Commerce Park Dr., Oak Ridge, TN 37830. Phone: (865) 220-2020. Internet: **www.tech2020.org**.
- Austin Technology Incubator, 3925 West Braker Lane, Austin, TX 78759. Phone: (512) 305-0000. Internet: **www.ic2-ati.org**.
- Software Commercialization and Innovation Center, Bryan, TX. Phone: (409) 775-9173. Internet: **www.scic.org**.
- Fort Worth MedTech Center, 1606 Mistletoe Boulevard, Fort Worth, TX 76104. Phone: (817) 921-2205. Internet: **www.medtech.org**. Focuses on medical and high technology start-ups.
- NetStrategy. Focuses on Internet start-ups. Internet: **www.netstrategycorp.com/default.asp**.
- Waco Business Resource Center, 401 Franklin, Waco, TX 76707. Phone: (254) 754-8898. Internet: **www.brc-waco.com**.
- Hampton Roads Technology Incubator, 24 Research Drive, Hampton, VA 23666. Phone: (757) 865-2141. Internet: **www.hr-incubator.org**.
- Virginia Biotechnology Research Park, 800 East Leigh Street, Richmond, VA 23219. Phone: (804) 828-5390. Internet: **www.vabiotech.com**.
- New Century Venture Center, 1354 Eighth Street, SW, Roanoke, VA 24015. Phone: (540) 344-6402. Internet: **www.ncvc.com**.
- Applied Process Engineering Laboratory, 350 Hills Street, Suite #101, Richland, WA 99352. Phone: (509) 372-5146. Internet: **www.apel.org**.

- Tri-Cities Enterprise Association, 2000 Logston Boulevard, Richland, WA 99352. Phone: (509) 375-3268. Internet: **www.owt.com/tea**.
- Unlimited Future, Inc., 1650 Eighth Avenue, Huntington, WV 25713. Phone: (304) 697-3007. Internet: **www.unlimitedfuture.org**.
- Madison Enterprise Center, 100 South Baldwin St., Madison, WI 53703. Phone: (608) 256-6565. Internet: **www.cwd.org/mec/index.html**. Focuses on new and expanding light industrial businesses.
- Superior Business Center, 1423 North 8th Street, Superior, WI 54880. Internet: **www.superbus.com**.
- The Laramie County Enterprise Center, 1400 East College Drive, Cheyenne, WY 82007-3298. Phone: (307) 778-1299. Internet: **www.lccc.cc.wy.us/cebd/index.htm**.

# Index

NO POSTAGE
NECESSARY
IF MAILED
IN THE
UNITED STATES

**BUSINESS REPLY MAIL**

FIRST-CLASS MAIL PERMIT NO. 73996 LOS ANGELES CA

POSTAGE WILL BE PAID BY ADDRESSEE

SILVER LAKE PUBLISHING
2025 HYPERION AVE
LOS ANGELES CA 90027-9849

NO POSTAGE
NECESSARY
IF MAILED
IN THE
UNITED STATES

**BUSINESS REPLY MAIL**

FIRST-CLASS MAIL PERMIT NO. 73996 LOS ANGELES CA

POSTAGE WILL BE PAID BY ADDRESSEE

SILVER LAKE PUBLISHING
2025 HYPERION AVE
LOS ANGELES CA 90027-9849